D1768543

Scattering Theory in Quantum Mechanics

LECTURE NOTES AND SUPPLEMENTS IN PHYSICS
John David Jackson and David Pines, *Editors* (Nos. 1–13)

Volumes of the Series published from 1962–1974 are not officially numbered. The parenthetical numbers shown are designed to aid librarians and bibliographers to check the completeness of their holdings.

(1)	John David Jackson	Mathematics for Quantum Mechanics: An Introductory Survey of Operators, Eigenvalues, and Linear Vector Spaces, 1962
(2)	Willem Brouwer	Matrix Methods in Optical Instrument Design, 1964
(3)	R. Hagedorn	Relativistic Kinematics: A Guide to the Kinematic Problems of High-Energy Physics, 1964 (3rd printing, with corrections, 1973)
(4)	Robert S. Knox and Albert Gold	Symmetry in the Solid State, 1964
(5)	David Pines	Elementary Excitations in Solids: Lectures on Phonons, Electrons, and Plasmons, 1964 (3rd printing, 1977)
(6)	Gabriel Barton	Introduction to Dispersion Techniques in Field Theory, 1965
(7)	David Bohm	The Special Theory of Relativity, 1965
(8)	David Park	Introduction to Strong Interactions: A Lecture-Note Volume, 1966
(9)	Hans A. Bethe and Roman W. Jackiw	Intermediate Quantum Mechanics, 1968 (3rd printing, with corrections, 1974)
(10)	Gordon Baym	Lectures on Quantum Mechanics, 1969 (5th printing, with corrections, 1977)
(11)	K. Nishijima	Fields and Particles: Field Theory and Dispersion Relations, 1969 (3rd printing, with corrections, 1974)
(12)	B. H. Bransden	Atomic Collision Theory, 1970
(13)	R. D. Sard	Relativistic Mechanics: Special Relativity and Classical Particle Dynamics, 1970

LECTURE NOTES AND SUPPLEMENTS IN PHYSICS
David Pines, *Editor*

Volumes published from 1975 onward are being numbered as an integral part of the bibliography:

Number		
14	Hans Frauenfelder and Ernest M. Henley	Nuclear and Particle Physics, A: Background and Symmetries, 1975
15	H. W. Wyld	Mathematical Methods for Physics, 1976
16	Werner O. Amrein, Josef M. Jauch, and Kalyan B. Sinha	Scattering Theory in Quantum Mechanics: Physical Principles and Mathematical Methods, 1977

Other volumes in preparation

Scattering Theory in Quantum Mechanics

Physical Principles and Mathematical Methods

WERNER O. AMREIN
University of Geneva

JOSEF M. JAUCH

KALYAN B. SINHA
University of Geneva

1977
W. A. Benjamin, Inc.
ADVANCED BOOK PROGRAM
Reading, Massachusetts
London · Amsterdam · Don Mills, Ontario · Sydney · Tokyo

WILLIAM MADISON RANDALL LIBRARY UNC AT WILMINGTON

CODEN:LNSPB

Library of Congress Cataloging in Publication Data

Amrein, Werner O
Scattering theory in quantum mechanics.

(Lecture notes and supplements in physics ; 16)
Bibliography: p.
Includes index.
1. Scattering (Physics) 2. Scattering (Mathematics)
3. Quantum theory. I. Jauch, Josef Maria, 1914-1974, joint author. II. Sinha, Kalyan B., joint author
III. Title.
QC20.7.S3A47 539.7'54 77-20853
ISBN 0-8053-0202-6
ISBN 0-8053-0203-4 pbk.

American Mathematical Society (MOS) Subject Classification Scheme (1970): 47-01, 47A40, 47A55, 47A70, 47B25, 81A09, 81A10, 81A45, 81A81

Reproduced by W. A. Benjamin, Inc., Advanced Book Program, Reading, Massachusetts, from camera-ready copy prepared by the office of the authors.

Copyright © 1977 by W. A. Benjamin, Inc.
Published simultaneously in Canada.

All rights reserved. No part of this publication may be reproduced, stored in a retrieval system, or transmitted, in any form or by any means, electronic, mechanical, photocopying, recording, or otherwise, without the prior written permission of the publisher, Addison-Wesley Publishing Company, Inc., Advanced Book Program, Reading, Massachusetts 01867, U.S.A.

Manufactured in the United States of America

ABCDEFGHIJ-HA-7987

QC20
.7
.S3
.A47

One of the authors (K.B.S) dedicates this book to Akhila, Raju and Jaya.

179188

CONTENTS

FOREWORD

Everyone concerned with the teaching of physics at the advanced undergraduate or graduate level is aware of the continuing need for a modernization and reorganization of the basic course material. Despite the existence today of many good textbooks in these areas, there is always an appreciable time-lag in the incorporation of new viewpoints and techniques which result from the most recent developments in physics research. Typically these changes in concepts and material take place first in the personal lecture notes of some of those who teach graduate courses. Eventually, printed notes may appear, and some fraction of such notes evolve into textbooks or monographs. But much of this fresh material remains available only to a very limited audience, to the detriment of all. Our series aims to fill this gap in the literature of physics by presenting occasional volumes with a contemporary approach to the classical topics of physics at the advanced undergraduate and graduate level. Clarity and soundness of treatment will, we hope, mark these volumes, as well as the freshness of the approach.

Another area in which the series hopes to make a contribution is by presenting useful supplementing material of well-defined scope. This may take the form of a survey of relevant mathematical principles, or a collection of reprints of basic papers in a field. Here the aim is to provide the instructor with added flexibility through the use of supplements at relatively low cost.

The scope of both the Lecture Notes and Supplements is somewhat different from the FRONTIERS IN PHYSICS Series. In spite of wide variations from institution to institution as to what comprises the basic graduate course program, there is a widely accepted group of "bread and butter" courses that deal with the classic topics in physics. These include : mathematical methods of physics, electromagnetic theory, advanced dynamics, quantum mechanics, statistical mechanics, and frequently nuclear physics and/or solid-state physics. It is chiefly these areas that will be covered by the present series. The listing is perhaps best described as including all advanced undergraduate and graduate courses which are at a level below seminar courses dealing entirely with current research topics.

The above words were written in 1962 in collaboration with David Jackson who served as co-editor of this Series during its first decade. They serve equally well as a Foreword for the present volume which seeks to provide for both physicists and mathematicians an account of a number of aspects of quantum scattering theory with emphasis on the use of mathematically rigorous methods to derive quantities of physical interests. It was conceived by the late Josef Jauch, a distinguished theoretical physicist, who was especially interested in scattering theory as a sequel to his book, "Foundations of Quantum Mechanics."

Work on the book with Werner Amrein had reached the preparatory stage at the time of Professor Jauch's untimely death in 1974; Amrein then asked Kalyan Sinha to join him in undertaking the challenging task of completing the text-monograph, according to Jauch's "grand design." The authors have made every effort to make the material accessible to the graduate student, both by providing complete proofs of the basic propositions found herein and by including problems, which are intended to assist the reader in becoming familiar with the concepts and methods of mathematical scattering theory. It gives me pleasure to welcome them as contributors to this Series.

DAVID PINES

PREFACE

Much of our knowledge about the physical world is derived from scattering experiments. Such experiments have led, for example, to the discoveries of the atomic nucleus, of nuclear fission, of new particles in modern accelerators and to the determination of the structure of crystals. The theoretical description of many of these phenomena is best given in the framework of quantum theory and is traditionally approached in one of the following three ways. The first one consists of studying solutions of the time-independent Schrödinger equation and of relating both the interaction and the scattering data to the asymptotic behaviour of such solutions at large distances. The second approach, the so-called stationary scattering theory, is an abstraction of the first one. This method had for a long time stayed on a formal level because of the difficulties in treating the continuous spectrum of the Hamiltonian. The last, and in our thinking the most satisfactory method, is the time-dependent one in which the temporal development of states and their behaviour in the remote past and the distant future plays the central role.

These three methods of course describe the same physics and, from a purely mathematical point of view, they also lead to a determination of the spectral structure of certain self-adjoint operators, in particular the so-called Schrödinger operators.

The purpose of this book is to explain the basic physical concepts of quantum scattering theory, to develop the necessary mathematical tools for their description, to display the interrelation between the three methods mentioned above and to derive the properties of various quantities of physical interest with mathematically rigorous methods. The book is designed for both physicists and mathematicians and can be used as a textbook in the physics or mathematics curriculum on the graduate level. It can also serve as an initiation to research in mathematical scattering theory.

The presentation is mathematical in nature, and the principal results are given in the form of propositions. With few exceptions, complete proofs are provided. We use almost exclusively Hilbert space methods, sometimes at the expense of not obtaining the strongest possible result. Further results and some of the proofs omitted in the text are given in a section entitled "Notes and Supplementary Material", which is appended to many chapters and may be skipped in the first reading. The book also contains about 180 problems (many of them with a hint indicating a method of solution). They serve two purposes : to familiarize the reader with concepts and methods, and to supplement certain parts of the proofs in the text. Relatively difficult problems are marked with a dagger (†).

It has not been possible to include all branches of scattering theory, which has been a very active field of research in recent years. Notable omissions are the inverse scattering problem and classical scattering. As applications we consider only non-relativistic scattering theory in three dimensions. Also omitted are the theory of resonance scattering, bounds on the number of eigenvalues of Schrödinger operators and properties of the corresponding eigenfunctions.

The prerequisites for this book are some knowledge of real and complex analysis and of the general principles of quantum mechanics. In particular the theory of the Lebesgue integral on $\mathbb{R}^n$ is basic to the development of the material. The other necessary mathematical tools from functional analysis, namely Hilbert space, linear operators and spectral theory for self-adjoint operators, are exposed in some detail.

The chapters have been organized in increasing order of complexity, and the book may be read in this order from the beginning to the end. However this does not quite reflect the logical structure which is as follows :

- Mathematical tools (Chapters 2, 3-1 and 5) and basic quantum mechanics (Section 3-2),
- General principles and abstract formulation of scattering theory for simple and multichannel scattering systems (Chapters 4, 6, 7, 13-2 and 14),
- Applications to non-relativistic potential scattering, containing more technical points (Chapters 8-13, 15, 16).

The reader who is already familiar with Hilbert space methods and quantum mechanics may, after getting acquainted with Chapter 1 and the notations from Chapters 2 and 3, go on directly to Chapter 4. Chapters 8-12 should be read in that order, though Chapter 13 may be inserted after Chapter 8.

The material of Chapters 1, 2-1, 2-2, 3-1, 3-3, 4, 5, 6, 7-3, 7-4, 8-1, 8-2, 10-1, 10-2, 11-1, 11-4, 12-1 with only the simplest proofs was taught by one of us (W.O.A.) in a one-semester course (35 hours) to physics students. We think that a complete coverage of the book will take three semesters. For a two-semester course, the instructor should choose between treating more topics and giving all the proofs involved.

Due to lack of space, certain equations have not been displayed and long expressions had to be broken at the end of a line. We apologize for any inconvenience this may cause. The symbol # denotes the end of a proof. Books occuring in the bibliography are referred to by capital letters in a square bracket (e.g. [R] for Royden's book), whereas research papers are referred to by the author's name and an appropriate number. The bibliography is far from complete. We have often included only a very recent or representative paper.

It is a pleasure to remember here the people who in one way or another have contributed to the preparation of this book. In particular we thank Professor Marcel Guenin for help and encouragement, Professor Constantin Piron and Professor Peter Rejto for discussions, and our colleagues for assistance in the proof-reading. Above all, our thanks go to Mrs. Claudine Brügger for her infinite patience and great skill in typing the difficult manuscript and to

Mr. Vaughan Jones for his careful reading and criticism of the book. We are grateful to the Physics Department of the University of Geneva for providing the necessary facilities, and to the Swiss National Science Foundation for financial support to one of the authors (K.B.S.).

Finally we must express our deep regrets that the late Professor Josef Jauch has not lived to see the completion of this volume. Professor Jauch devoted a considerable part of his scientific activities to scattering theory and stimulated a research group in this field in Geneva for many years. He had conceived a book on quantum scattering theory as a sequel to his "Foundations of Quantum Mechanics" and invited the first author to be his collaborator. At the time of his unexpected death in August 1974, the work was still in a preparatory stage. We, the undersigned, then joined to elaborate the details of the contents and of the presentation. The responsibility for any omissions and shortcomings of the text is therefore entirely ours, and we can only hope that our approach also represents Professor Jauch's conception of the subject.

WERNER O. AMREIN
KALYAN B. SINHA

PART I

INTRODUCTORY MATERIAL

CHAPTER 1 : PHYSICAL HEURISTICS

This chapter begins with a description of the physical situations encountered in scattering experiments (Section 1-1). Section 1-2 outlines the different types of scattering such as elastic, inelastic, multichannel, and rearrangement scattering. The results of a scattering experiment are expressed in terms of certain observable quantities such as cross sections, lifetimes (or level widths) and branching ratios. These quantities are defined for the simplest systems and examples are given (Section 1-3). In preparation for the transcription of the physical situation into a mathematical language we outline briefly in Section 1-4 the physical characteristics of a scattering system.

1-1 DESCRIPTION OF SCATTERING EXPERIMENTS

Long before the modern scientific era the Ionian physiologists formulated the basic question of natural philosophy : How can the multitude of natural phenomena be explained in terms of the primeval substance and its properties. This basic question in its various forms has more or less determined the orientation of western science until the present.

In the modern version this question appears in the form : what are the primary constituents of matter and how do they interact with each other ? The answer to this question is obtained almost exclusively through scattering experiments. The theory which relates the results of such experiments to the interaction between primary constituents is scattering theory.

Though much has been learned, especially during the last fifty years, we are far from a complete understanding of the primary constituents of matter. In fact as time progresses and as the large accelerators continue to increase our factual knowledge of the subatomic world, we get the impression that we have entered into a vast and largely unknown territory of which we see as yet only dim outlines. We are no longer sure today whether those units of matter which we have confidently called elementary particles are truly what we thought them to be, whether indeed this concept makes any sense at all.

We know very little yet about the laws of interaction between these elementary particles. There is no direct access to these laws. We can only infer them from the world of phenomena as they manifest themselves in the laboratories of the experimental physicists. It is found empirically that these phenomena can be expressed most conveniently and concisely in terms of certain parameters such as charge, mass, spin, isotopic spin, strangeness, etc. The object of the theory is to establish the connection between the world of phenomena thus described and the laws of interactions between elementary particles.

We are thus in search of a relation between two worlds, the world of phenomena on the one hand and the world of elementary particles on the other. This connection is fruitful because it enables us to describe and thereby understand the complex and varied world of phenomena in terms of a much simpler and permanent world of elementary laws. A well-known example is the extremely simple law of interaction between charged particles, which describes the whole range of atomic spectroscopy and of atomic and radiative scattering.

The connection between the two worlds is established by scattering theory which is the object of this book. The theory is essentially mathematical in nature. It rests on mathematical abstractions, which are the ultimate building blocks of the theory, for instance the basic postulates of quantum mechanics and the asymptotic condition which will be described in Section 1-4. It is related to the world of phenomena by its rules of interpretation. The connection between the postulates of the theory and the observable quantities is established, as far as possible, by mathematically rigorous deductions which are the essence of the mathematical scattering theory.

Inasmuch as most scattering experiments which are of interest in fundamental physics are strongly influenced by quantum effects, scattering theory must be regarded as a branch, and in fact one of the most important ones, of quantum mechanics. It is concerned with systems in special states, the scattering states, which occur in typical scattering experiments.

In this chapter we shall give a brief description of the principal scattering experiments in order to obtain a survey of the quantities which ultimately link the observations with the formal part of the theory.

There are many different varieties of scattering experiments. All of them have certain elements in common which we shall exhibit in a simple case illustrated schematically in Figure 1.1. This representation exhibits the four

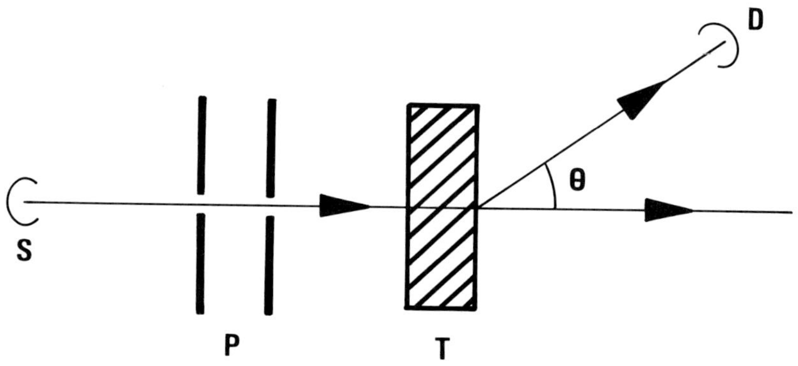

Figure 1.1 : Schematic representation of a typical scattering experiment.

essential parts which can be identified in nearly every scattering experiment. These four ingredients are the source S, the preparing apparatus P, the target T and the detector D.

The source S produces the particles which are to interact with the particles in the target T. It is important that the source can repeatedly produce particles under practically identical and well defined conditions, because all scattering experiments involve repeated measurements on identically prepared systems. The preparing apparatus P (e.g. a collimator, a beam spectrometer or a polarizer) serves to define the initial conditions of the incident particles, in particular their momenta, with as large a precision as is compatible with an intensity sufficient to give a reasonable counting rate.

The target T contains the particles which are supposed to interact with the incident particles. The conditions of the target can have very important effects on the counting rate and they must be known and considered in the attempt to relate the observations to the individual interactions. For instance if the target is thick then multiple scattering can occur, which affects the angular distribution of the detected particles in a very significant way. If the target has a crystalline structure, then interference effects will become important and they lead to characteristic diffraction patterns which depend on the structure of the target just as much as on the interaction itself. If the target particles are themselves in motion, as for instance in a gas target, then important effects due to this motion must be accounted and corrected for before the result of the measurement can be used for the study of individual interactions. The easiest interpretation is possible with a thin target having a random distribution of scattering centers at rest.

If the target itself is in a well defined state of motion we have two colliding beams of particles. A typical arrangement would be as indicated in Figure 1.2.

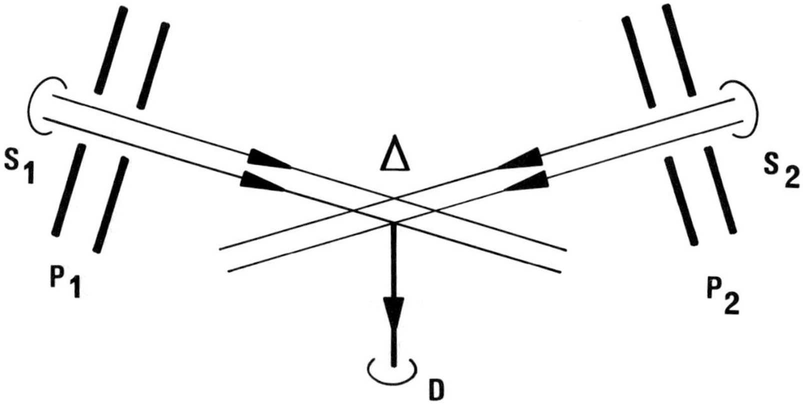

Figure 1.2 : Schema of a colliding beam experiment.

Here there are two sources S_1 and S_2 and two preparing apparatus P_1 and P_2 arranged in such a way that the two beams collide in a space region Δ. The detector D is then placed so as to detect particles which are scattered from either beam into a new direction. In this situation it is clearly not natural to distinguish one of the beams from the other and a symmetrical treatment for both beams is more appropriate for the description of the experiment.

However if the particles have non-zero rest mass it is possible to describe the experiment of Figure 1.2 in accordance with the general schema of Figure 1.1 by transforming into a new reference system for which one of the beams is at rest. This transformation will lead to a scattering experiment such as sketched in Figure 1.1 with the important difference that the detector D is then itself in motion with respect to the target. Only in an experiment with particles of vanishing rest mass (e.g. photons) is this transformation not possible and the symmetrical treatment of the two beams is imperative.

The detector D is usually placed in such a way as to detect only particles which are scattered by the target. That is, the detector should not respond if the target is removed. This is not always possible in practice, either because the incident beam is not sufficiently well collimated or because there is residual scattering in the surrounding material. Such effects give rise to corrections which can be determined by careful calibration of the detector.

It is also important that the detector be placed sufficiently far from the target so that the interaction between the scattered particles and the target is negligible when the particles are detected. The detector itself will in all cases have a finite angle of resolution. For a detailed analysis of the experiment it is desirable to have this angle as small as possible. There are practical limits to this since too small a resolution angle may be incompatible with a sufficient statistical accuracy. The same considerations limit the precision of the preparing apparatus.

1-2 DIFFERENT TYPES OF SCATTERING

So far we have made no distinction between different types of scattering. This we shall do now. The simplest scattering process is elastic scattering between different particles. This process may be symbolically represented by an expression such as

$$a + b \to a + b \ . \tag{1.1}$$

This expression indicates that the particles a and b in some initial state are scattered into particles a and b in some final state. The term elastic refers to the fact that the total kinetic energy of the particles before and after the collision is the same. A special case of elastic scattering is the scattering between identical particles

$$a + a \to a + a \ . \tag{1.2}$$

In quantum mechanics this process is more complicated to treat since identical systems can only occur in symmetrical or anti-symmetrical states depending on whether they satisfy Bose-Einstein or Fermi-Dirac statistics. Furthermore the detector will not distinguish between particles, which must be considered when the experiment is compared with the theory.

When the incident particles or the target have internal degrees of freedom it is possible that during the scattering process one of the particles undergoes a change of its internal state. If this takes place then the kinetic energy of the particles before and after the scattering is no longer the same, and one speaks of inelastic scattering. The more frequent case is the excitation of an internal state with an energy above that of the initial state of the system. In that case the final kinetic energy is less by the amount of the excitation energy. This is hypoelastic or endoergic scattering. It is, however, also possible that some metastable excited state of one of the scattering particles is deexcited during the scattering process. In this case the final kinetic energy is larger than the initial one, and we may speak of hyperelastic or exoergic scattering.

Another type of scattering associated with internal degrees of freedom is observed when these degrees are degenerate in energy. This is for instance the case for particles with spin. In that case the scattering process would still be described by (1.1) but account must be taken of the effect of scattering on the internal degrees of freedom. This gives rise to a wide class of interesting polarization phenomena which can give significant information on the spin-dependence of the interaction.

An entirely different type of scattering is observed if the outgoing particles differ in number and kind from the incoming ones. For example the constituents of the incident beam and the scattering centers in the target may not be elementary particles but composite structures such as α-particles, hydrogen atoms, etc., and during the scattering process the interaction may break up such composite systems into some of their constituent parts or arrange their constituent parts into new composite systems. One then speaks of rearrangement scattering and may write schematically

$$a + b \rightarrow c + d + e + \dots \tag{1.3}$$

We shall see later that scattering theory can be developed in such a way that elastic scattering, inelastic scattering, hyperelastic scattering and rearrangement scattering can be treated in a unified way. This unified theory is called multichannel scattering theory.

A typical example of multichannel scattering is the scattering of a deuteron d by a fixed center of force. A deuteron can be either scattered elastically or it can be broken up into its constituent parts (a proton p and a neutron n), so that we have the two possibilities

$$
d \begin{cases} \nearrow \; d \\ \searrow \; p + n. \end{cases}
$$

Each different possible collection of particles and composite systems after the collision determines a so-called scattering channel. In the preceding example there are only two scattering channels.

Another reason for a change in type and number of particles is the creation of new kinds of particles during the scattering process. These creation processes are especially frequently observed in high energy scattering. At sufficiently high energy they are in fact always present. A collision between two nucleons N_1 and N_2, for instance, may proceed according to the schemes

$N_1 + N_2$	$\nearrow\; N_1 + N_2$	elastic channel
	$\nearrow\; N_1 + N_2 + \pi$	various inelastic channels
	$\searrow\; N_1 + N_2 + K + \bar{K}$	
	$\searrow\; \ldots\ldots\ldots$	

Here new particles such as π and K mesons may be created, and certain final channels clearly represent more than a simple rearrangement of the constituents of N_1 and N_2. The usual multichannel scattering theory does not include such collisions, and their theoretical description is beyond the scope of this book.

In all cases that we have considered so far the initial number of particles or composite systems which participate in an individual scattering process is always two. It is logically possible to consider scattering processes with more than two initially participating particles. Such situations are rarely encountered in the laboratory on account of the great experimental difficulties for preparing such initial states. They play a certain role in the theory of dense gases. But in elementary particle physics they are not important.

Another case of importance, however, is that of one incident free particle only. If the particle is stable, the proton for instance, nothing of physical interest will happen. The evolution of the state of a single stable free particle does not give any information on any interaction. If it is unstable, however, then it will decay into some unstable or stable fragments, and it is perfectly possible to regard such a decay as some sort of degenerate scattering process according to the schema

$$a \to b + d + \dots \tag{1.4}$$

In this case no target is needed to observe the process. It is possible to consider such a decay of an unstable particle as part of an ordinary scattering process, since the unstable particle must have been produced sometime in the past in an ordinary scattering process. This way of looking at unstable systems is especially useful if the lifetime of the unstable particle is very short. We then have a case of resonance scattering (formation and decay of a resonance). This is often observed and well known in nuclear physics where resonances are associated with some short lived excited state of an intermediate nucleus.

If the unstable particle is sufficiently long lived it is possible to produce ordinary scattering effects with it and develop a theory where the particle is treated with a sufficiently good approximation as a stable particle.

1-3 THE OBSERVABLE QUANTITIES

Every physical theory must have an interpretation in terms of certain physically observable effects. These effects are usually expressed by giving the numerical values of certain quantities. The principal quantities which occur in scattering theory are the scattering cross section, the lifetime of an unstable particle or the resonance width and the branching ratio for multichannel processes. For the reasons outlined in the previous section the lifetime is in a sense a secondary quantity which can be related to the behavior of the cross section in some appropriate scattering process.

The branching ratio (i.e. the probability of scattering into a particular channel) is not really an independent quantity, since it is known if the individual cross sections for the different channels are known. Often it is possible on the basis of symmetry considerations, without a detailed theory, to obtain expressions for the branching ratios without the knowledge of the value of the cross section. Here we shall restrict our discussion to the notion of the cross section.

In order to have a definite case before us, we return for a moment to the simplest but typical case depicted in Figure 1.1, where we assume in addition that the target consists of only one scatterer. The intensity of the scattered particles which is observed by the detector will be proportional to the intensity of the incident beam. If we denote by $N(\Delta\omega)$ the number of particles scattered per unit time into the solid angle $\Delta\omega$ subtended by the detector D, and by N_o the number of incident particles per unit time and unit surface of the target, then we have the fundamental relation

$$N(\Delta\omega) = \sigma N_o \Delta\omega. \tag{1.5}$$

The factor of proportionality σ defined by (1.5), which will be a function of the scattering angle θ and the azimuthal angle ϕ associated with the direction of the incident beam, is called the scattering cross section for the solid angle $\Delta\omega$. If this angle becomes infinitesimal, we get the differential scattering cross section

$$\frac{d\sigma}{d\omega}(\theta,\phi) = \lim \frac{N(\Delta\omega)}{N_o \Delta\omega} \text{ as } \Delta\omega \text{ shrinks to the point } (\theta,\phi). \tag{1.6}$$

If the target consists of n scatterers and no coherence effects are present, one will have instead of (1.5) that $N(\Delta\omega) = \sigma n N_0 \Delta\omega$. In the case of inelastic and multichannel scattering or for colliding beam experiments the above definition will have to be appropriately generalized, which we shall do later without difficulty.

The <u>total scattering cross section</u> σ_{tot} is obtained from the differential one by integrating the latter over the unit sphere, i.e.

$$\sigma_{tot} = \int_0^{\pi} \sin\theta \, d\theta \int_0^{2\pi} d\phi \, \frac{d\sigma}{d\omega}(\theta,\phi). \tag{1.7}$$

It should be remarked that the differential cross section and the total cross section also depend on the other parameters which serve to characterize the experimental setup. One such parameter is the kinetic energy of the particles in the prepared beam. If these particles (or those forming the target) have spin, one might also observe their polarizations before and after the interaction takes place, and the cross section would then also be a function of the polarizations.

The name "cross section" has been taken over from classical mechanics into quantum mechanics. It can be seen from (1.6) and (1.7) that the dimension of $[d\sigma/d\omega](\theta,\phi)$ and of σ_{tot} is the square of length. If one considers the scattering of a particle in classical mechanics by a spherically symmetric potential $V(r)$, one finds that $\sigma_{tot} = \pi R^2$, where R is the smallest number such that $V(r) = 0$ for all $r > R$ [N]. Thus in this case σ_{tot} is just the area of the obstacle (characterized by $V(r)$) that presents itself to the incident particle. It should be noted that here σ_{tot} is independent of the kinetic energy of the incoming particle. This disagrees with results from scattering experiments in microphysics. In quantum scattering theory the total cross section is in general energy dependent.

One of the main purposes of scattering theory is the calculation of the differential scattering cross section in terms of the interaction between the particles of the incident beam and those of the target. This interaction is assumed to exist, but it may not be known explicitly at the outset. A measurement of $d\sigma/d\omega$ will then lead to a determination of the properties of the interaction, e.g. by checking for which form of the interaction one obtains the closest agreement between the theoretical and the experimental values.

There is another theoretical approach which is known as the <u>inverse scattering problem</u>. In this theory one assumes that certain experimental data (in fact the so-called scattering amplitude rather than directly observable quantities) are known, and one proves the existence of an interaction (usually a potential) that leads to precisely the given data when inserted as the interaction in the direct scattering theory indicated above. This book does not deal with the inverse scattering problem. The reader is referred to [CS].

1-4 THE PHYSICAL CHARACTERISTICS OF SCATTERING SYSTEMS

The preceding description of typical scattering experiments shows that the essential characteristics of a scattering system may be expressed as follows :

In a scattering process we may distinguish three stages in the temporal evolution of the system. In the first stage the state of the system is prepared in the remote past. During this stage the incident particle and the target particle are supposed to be so far removed from each other that their mutual interaction is negligible and therefore has practically no effect on the evolution of either particle. This assumption is always made in the interpretation of typical scattering experiments where the initial conditions are the values of a set of physical quantities that are supposed to characterize free particles (e.g. momenta, spins, masses). Therefore one expects that in the remote past the state of the system evolves according to the laws pertaining to free particles.

During the second stage the particles interact with each other and the evolution is governed by an equation of motion for which the interaction term plays an essential role. It is in fact this interaction which produces the scattering.

During the third stage one encounters the same situation as in the first one. In fact after the scattering has occurred the particles separate from each other so that again the interaction has practically no effect on the future evolution of the states. In the distant future the detector observes the new state of the particles which was produced by the scattering.

The requirement that the states describing scattering events must be characterizable at large negative and large positive times by quantities pertaining to free particles or to scattering fragments is called the asymptotic condition. In order to express this condition in a mathematical language, we need to study the description of the time evolution of quantum-mechanical systems and to introduce a topology (or a notion of convergence) which will serve to express the difference between the actual system and a free system in the remote past and the distant future. This will be done in the course of the next two chapters.

CHAPTER 2 : HILBERT SPACE AND LINEAR OPERATORS

In the preceding chapter we have given a description of the scattering process in physical terms. In this chapter we shall begin to transcribe this description into the mathematical language of quantum mechanics. We have seen that in the distant past and the remote future the time evolution of a scattering system is physically indistinguishable from that of a system of free particles. Our first task is then to specify the mathematical concepts and theorems needed for the description of particles, especially of free particles, and of their temporal evolution. This is the topic of the present and the following chapter.

The basic mathematical object which is needed for the description of particles in quantum mechanics is the Hilbert space. In Section 2-1 we introduce the abstract Hilbert space, discuss some of its elementary properties and illustrate these notions in L^2-spaces. In Section 2-2 we present various simple results concerning linear operators in Hilbert space. The emphasis is on getting acquainted with bounded and with self-adjoint operators. Section 2-3 is devoted to some particular classes of operators, namely compact, Hilbert-Schmidt and trace class operators. Finally in Section 2-4 we introduce the tensor product and the direct sum of Hilbert spaces.

The results of Sections 2-3 and 2-4 are not used before Chapter 7, and the reader may skip these sections here and familiarize himself with their material when arriving at Chapter 7.

2-1 THE ABSTRACT HILBERT SPACE AND ITS CONCRETE REALIZATIONS

The abstract Hilbert space $\mathcal{H}$ is a collection of objects called vectors, denoted by f, g,..., which satisfy the following three axioms.

I. $\mathcal{H}$ is a linear vector space with complex coefficients.

This means that to every pair of vectors $f, g \in \mathcal{H}$ there is associated a third vector $(f + g) \in \mathcal{H}$. Furthermore to every vector f and every complex number α there corresponds another vector $\alpha f \in \mathcal{H}$. The following rules are postulated :

$$f + g = g + f, \quad (f + g) + h = f + (g + h), \tag{2.1}$$

$$\alpha(f + g) = \alpha f + \alpha g, \quad (\alpha + \beta)f = \alpha f + \beta f, \quad \alpha(\beta f) = (\alpha\beta)f, \tag{2.2}$$

$$1 \cdot f = f. \tag{2.3}$$

There exists a unique vector θ in $\mathcal{H}$ such that for all $f \in \mathcal{H}$

$$\theta + f = f, \quad 0 \cdot f = \theta. \tag{2.4}$$

II. <u>There exists a strictly positive scalar product in H.</u>

The scalar product (f,g) is a function of pairs of vectors $f,g \varepsilon H$ with values in the set C of complex numbers and satisfies the following conditions*) :

$$(f,g) = \overline{(g,f)}, \tag{2.5}$$

$$(f,g+\alpha h) = (f,g) + \alpha(f,h) \text{ for all complex } \alpha, \tag{2.6}$$

$$||f|| \equiv (f,f)^{\frac{1}{2}} > 0 \quad \text{unless } f = \theta. \tag{2.7}$$

(The bar in (2.5) denotes complex-conjugation.)

III. <u>The space H is complete</u> in the norm defined by (2.7) : Whenever $\{f_n\}$, n = 1, 2,..., is a Cauchy sequence in the sense that $||f_n - f_m|| \to 0$ as $n,m \to \infty$, there exists a vector $f \varepsilon H$ such that $||f_n - f|| \to 0$ as $n \to \infty$.

In this book we shall deal only with separable Hilbert spaces, i.e. we shall postulate also the following axiom :

IV. <u>The space H is separable</u>. This means that there exists a sequence $\{f_n\} \varepsilon H$ (n = 1, 2,...) with the property that it is dense in H. We recall that a subset D of H is <u>dense</u> in H if, for any $f \varepsilon H$ and any $\eta > 0$, there exists at least one element f_η in D such that $||f - f_\eta|| < \eta$.

The last two requirements are topological in nature. They limit the size of the space in opposite directions. The first one can always be satisfied by a standard technique of

---

*) In conformity with the practice of most physicists we shall consider the scalar product (f,g) to be linear in g and anti-linear in f, whereas the convention in the mathematical literature is usually the opposite.

adjunction of suitable limit elements; i.e. if a collection of vectors verifies Axioms I and II, one can convert it into a Hilbert space by adding suitable limit elements (this will be used in Section 2-4). The second one is a genuine restriction, requiring that there be a countable dense set $\mathcal{D}$ in H. So far non-separable spaces have not been needed in quantum mechanics. We may add that the property of denseness in a metric space may be visualized by considering the example of the real line R with the usual Euclidean metric where the set of rational numbers is dense.

Before we introduce the L^2-spaces as concrete examples of Hilbert spaces, we collect some additional definitions and a few elementary relations that follow from the above axioms. These will be useful at many later stages of this book. The reader may also look at Problem 2.1 for other examples of Hilbert spaces.

We begin with two simple inequalities. The first one is the Schwarz inequality :

$$|(f,g)| \leq \|f\| \|g\| . \tag{2.8}$$

For its proof we distinguish two cases. (a) If $f = g$, (2.8) holds with the equality sign by the definition (2.7). (b) If $f \neq g$, we may assume for instance that $g \neq \theta$. For any complex number α one has (cf. also Problem 2.2) $0 \leq \|f + \alpha g\|^2 = (f + \alpha g, f + \alpha g) = \|f\|^2 + |\alpha|^2 \|g\|^2 + \alpha(f,g) + \bar{\alpha}(g,f)$. (2.8) is a particular case of this obtained by setting $\alpha = -(g,f)/\|g\|^2$ and multiplying the resulting inequality by $\|g\|^2$.

As a consequence of (2.8) one obtains the Minkowski

inequality which is also called the triangle inequality :

$$\|f + g\| \leq \|f\| + \|g\|. \tag{2.9}$$

Its proof is simple : $\|f + g\|^2 = \|f\|^2 + \|g\|^2 + (f,g) + (g,f)$
$\leq \|f\|^2 + \|g\|^2 + 2|(f,g)| \leq \|f\|^2 + \|g\|^2 + 2\|f\|\,\|g\|$
$= (\|f\| + \|g\|)^2$.

The triangle inequality implies together with axioms I and II that the Hilbert space is a normed linear space. The norm $\|f\|$ of a vector f is a measure of the distance between f and the zero vector θ, or in other words $\|f - g\|$ is a mesure of the distance between f and g. Since the vectors of $\mathcal{H}$ will be interpreted as the pure states of some physical system, two states f and g are practically indistinguishable if $\|f - g\|$ is very small. This suggests that one could use this norm in order to express the asymptotic properties of a scattering system, viz. the practical indistinguishability of the real system from a free system when $|t|$ is very large. This will be done in Chapter 4. The precise mathematical statement will be a limit relation : the larger $|t|$ the less the real state should differ from a freely evolving state. For this reason we shall now look at convergence properties of sequences $\{f_n\}$ of elements of $\mathcal{H}$.

In order to define the convergence of a sequence of vectors of $\mathcal{H}$, one resorts to the simpler notion of convergence of a sequence of (real or complex) numbers. So far we have introduced two kinds of numbers constructed from elements of $\mathcal{H}$, the norm of a vector and the scalar product between two vectors. Each of these can be used to define a topology on $\mathcal{H}$.

The convergence of a sequence of vectors in the norm $\|\cdot\|$ has already been used in formulating Axiom III. In Hilbert space theory this is called strong convergence. A sequence of vectors $\{f_n\}$ converges strongly to a limit vector f if $\|f - f_n\| \to 0$ for $n \to \infty$. We then write $f_n \to f$ or $\text{s-lim } f_n = f$ as $n \to \infty$. A necessary and sufficient condition for strong convergence is that the sequence be Cauchy in the sense defined in Axiom III. The sufficiency is nothing but Axiom III, and the necessity is an easy consequence of the triangle inequality : If $f_n \to f$, then $\|f_n - f_m\| = \|f_n - f + f - f_m\| \leq \|f_n - f\| + \|f - f_m\| \to 0$ as $n,m \to \infty$. This inequality can also be applied to verify the uniqueness of the limit vector f (cf. Problem 2.3).

The convergence in H obtained by means of the scalar product is called weak convergence. A sequence $\{f_n\}$ converges weakly to a limit f if for every $g \in H$ the sequence of scalar products $\{(f_n,g)\}$ converges to (f,g). If this is the case we write $\text{w-lim } f_n = f$ as $n \to \infty$. The Cauchy criterion is also valid for weak convergence, i.e. $\{f_n\}$ converges weakly if and only if for every $g \in H$ the sequence $\{(f_n,g)\}$ is a Cauchy sequence of complex numbers. If $\{f_n\}$ is a Cauchy sequence in this sense, there exists a unique vector $f \in H$ such that $\text{w-lim } f_n = f$. We shall not use this criterion, but we may add that the uniqueness of the weak limit is an immediate consequence of Proposition 2.2 below.

Strong convergence implies weak convergence, but the converse is not true. In fact one has the following relation which is often very useful :

PROPOSITION 2.1 : $\text{s-}\lim_{n\to\infty} f_n = f$ if and only if $\text{w-}\lim_{n\to\infty} f_n = f$ and $\lim_{n\to\infty} \|f_n\| = \|f\|$.

Proof : It involves the definition of strong and weak convergence and the inequalities of Schwarz and Minkowski.

(i) Suppose $f_n \to f$. We get with (2.8)

$$|(f_n,g) - (f,g)| = |(f_n - f,g)| \leq \|f_n - f\| \|g\| \to 0$$

for every $g \in H$, i.e. $\text{w-}\lim f_n = f$. By using (2.9) we deduce

$$\|f_n\| \leq \|f_n - f\| + \|f\| \quad \text{and} \quad \|f\| \leq \|f - f_n\| + \|f_n\|,$$

and hence $\|f\| - \|f - f_n\| \leq \|f_n\| \leq \|f\| + \|f - f_n\|$. Since $\|f - f_n\| \to 0$ for $n \to \infty$, it follows that the limit of $\|f_n\|$ as $n \to \infty$ exists and is equal to $\|f\|$.

(ii) $\|f - f_n\|^2 = \|f\|^2 + \|f_n\|^2 - (f,f_n) - (f_n,f)$.

If $\text{w-}\lim f_n = f$ and $\|f_n\| \to \|f\|$, the right-hand side converges to $\|f\|^2 + \|f\|^2 - (f,f) - (f,f) = 0$, i.e. $f_n \to f$. #

An example of a weakly convergent sequence which does not converge strongly is an infinite orthonormal sequence. Before we can verify this statement, we have to introduce the notion of orthogonality. Two vectors f and g are said to be orthogonal to each other if $(f,g) = 0$. Similarly two subsets M_1 and M_2 of H are mutually orthogonal if $(f_1,f_2) = 0$ for all $f_1 \in M_1$ and all $f_2 \in M_2$. An important relation concerning mutually orthogonal vectors is the following :

$$\|\sum_{i=1}^{n} f_i\|^2 = \sum_{i=1}^{n} \|f_i\|^2 \quad \text{if} \quad (f_i,f_j) = 0 \quad \text{for all } i \neq j. \tag{2.10}$$

This is easily verified by writing the left-hand side as a scalar product and using the linearity (2.6) of the scalar product.

An <u>orthonormal sequence</u> of vectors $\{h_i\}$ is characterized by the property that $(h_i, h_j) = \delta_{ij}$, where $\delta_{ij} = 1$ if $i = j$ and $\delta_{ij} = 0$ if $i \neq j$. If $f \varepsilon H$, we then have from (2.6) and (2.10)

$$0 \leq \|f - \sum_{i=1}^{n} (h_i, f) h_i\|^2 = \|f\|^2 + \sum_{i=1}^{n} |(h_i, f)|^2 - 2 \sum_{i=1}^{n} |(h_i, f)|^2,$$

hence $\sum_{i=1}^{n} |(h_i, f)|^2 \leq \|f\|^2$. Since this holds for each n, we obtain <u>Bessel's inequality</u>

$$\sum_{i=1}^{\infty} |(h_i, f)|^2 \leq \|f\|^2. \tag{2.11}$$

(2.11) implies that $(h_i, f) \to 0$ as $i \to \infty$ for every $f \varepsilon H$, i.e. $\{h_i\}$ converges weakly to zero. But $\{h_i\}$ cannot converge strongly, since $\|h_n - h_m\|^2 = \|h_n\|^2 + \|h_m\|^2 = 2$ if $n \neq m$.

An orthonormal set of vectors $\{e_i\}$ is called an <u>orthonormal basis</u> of H if the set of finite linear combinations of vectors belonging to $\{e_i\}$ is dense in H. In a separable Hilbert space an orthonormal basis is always a countable set. This can be seen as follows : Let μ belong to some index set, let $\{e_\mu\}$ be any orthonormal set in H, i.e. $(e_\mu, e_\nu) = 1$ if $\mu = \nu$ and $(e_\mu, e_\nu) = 0$ if $\mu \neq \nu$. Let $D = \{f_i\}$ be a countable dense set in H. For each μ, there exists $i = i(\mu)$ such that $\|e_\mu - f_{i(\mu)}\| < \frac{1}{2}$. Since $\sqrt{2} = \|e_\mu - e_\nu\| \leq \|e_\mu - f_{i(\mu)}\| + \|e_\nu - f_{i(\mu)}\| < \frac{1}{2} + \|e_\nu - f_{i(\mu)}\|$, we must have $i(\nu) \neq i(\mu)$ if $\nu \neq \mu$. Thus there is a one-to-one correspondence between $\{e_\mu\}$ and a subset of the set of positive integers, i.e. $\{e_\mu\}$ is countable.

The existence of an orthonormal basis can be established by choosing a subset of linearly independent vectors of a countable dense set D and applying to it the Schmidt orthogonalization process [K, Ch. I.6.3.], [RS, Thm. II. 7]. The dimension of a Hilbert space is equal to the number N of vectors of an orthonormal basis (N does not depend on the choice of a particular basis). Our axioms are equally valid for finite and infinite-dimensional spaces. However if the space if finite-dimensional the last two axioms are a consequence of the others. Furthermore in the finite-dimensional case the strong and the weak topology coincide (Problem 2.4).

An arbitrary vector in H can always be expanded in a given basis. A little care is needed though if H is infinite-dimensional, since only finite linear combinations of vectors are admitted by axiom I. Infinite linear combinations are understood as strong limits of finite linear combinations. Thus the expansion of $f \in H$ in an orthonormal basis $\{e_i\}$ means that the sequence of vectors $\{f_n\}$, where $f_n = \sum_{i=1}^{n} (e_i, f) e_i$, converges strongly to f as $n \to \infty$. (f_n is seen to be a linear combination of the first n vectors of the basis $\{e_i\}$, and the coefficient of e_i is nothing but the "component" (e_i, f) of f along e_i.) The fact that the above is a complete expansion is expressed by the Parseval relation which states that the equality sign holds in (2.11) if $\{h_i\}$ is an orthonormal basis (Problem 2.5) :

$$\|f\|^2 = \sum_{i=1}^{\infty} |(e_i, f)|^2. \tag{2.12}$$

(2.12) says that the square of the length of the vector f is equal to the sum of the squares of the absolute values of its components along the vectors e_i of an arbitrary orthonormal basis.

Another simple result in relation with the concept of

orthogonality which will frequently be used is the following :

PROPOSITION 2.2 : Let $\mathcal{D}$ be a dense set in $\mathcal{H}$ and $f \varepsilon \mathcal{H}$. If $(f,g) = 0$ for all $g \varepsilon \mathcal{D}$, then $f = \theta$.

Proof : Suppose $f \neq \theta$. Given $\eta > 0$, there exists $f_\eta \varepsilon \mathcal{D}$ such that $\|f - f_\eta\| < \eta \|f\|^{-1}$. Then, using also (2.8), we have $(f,f) = |(f,f - f_\eta) + (f,f_\eta)| = |(f,f - f_\eta)| \leq \|f\| \, \|f - f_\eta\| < \eta$. Since η is arbitrary, this implies $\|f\|^2 = 0$, hence $f = \theta$ by (2.7). #

A concept that we shall need in the next section is that of a linear manifold. This is a subset $\mathcal{M}$ of $\mathcal{H}$ that satisfies Axiom I but not necessarily Axiom III ($\mathcal{M}$ will always verify Axioms II and IV, since it is a subset of $\mathcal{H}$, cf. Problem 2.6). A subset of $\mathcal{H}$ that satisfies all four axioms will be called a subspace[*)].

In a finite-dimensional Hilbert space a linear manifold is also a subspace, since Axiom III is not an independent postulate. An example of a linear manifold that is not a subspace in an infinite-dimensional Hilbert space is the set $\mathcal{M}$ of all finite linear combinations of a countably infinite number of linearly independent vectors $\{f_i\}$. It is seen that the sum of two elements of $\mathcal{M}$ and the product of an element of $\mathcal{M}$ and a complex number are again finite linear combinations of the vectors $\{f_i\}$, and the other postulates of Axiom I hold because $\mathcal{M}$ is a linear subset of $\mathcal{H}$.

---

*) In some books the designation "subspace" is used also for a manifold, and a subspace in the sense of our definition is then called a closed subspace.

The closure of a linear manifold (i.e. the manifold obtained by adding to M all the limit points, in the sense of Axiom III, of strong Cauchy sequences of vectors belonging to M) is a subspace of H. In the above example the closure of M is strictly bigger than M, since it contains also certain infinite linear combinations of the vectors $\{f_i\}$. (For a specific example, suppose that the f_i form an orthonormal basis $\{e_i\}$, and let $f = \sum_{k=1}^{\infty} k^{-1} e_k$. Then f belongs to H, since by (2.12) $\|f\|^2 = \sum_{k=1}^{\infty} k^{-2} < \infty$, but $f \notin M$.)

An important example of a closed linear manifold (i.e. of a subspace) is the orthogonal complement $N^{\perp}$ of a subset N of H, i.e. the set of all vectors $f \in H$ such that $(f,g) = 0$ for all $g \in N$. The proof that such a set is a subspace is simple and is left to the reader. In this connection it is also worth noticing the following fact known as the projection theorem ([AG],[K],[RS]) : If M is a subspace and $M^{\perp}$ its orthogonal complement, then every vector f in H has a unique decomposition $f = f_1 + f_2$ with $f_1 \in M$ and $f_2 \in M^{\perp}$. A simple consequence is the following : If M is a linear manifold such that the only vector of H that is orthogonal to M is the vector θ, then M is dense in H.

The case where a linear manifold M is dense in H (i.e. where the closure of M is equal to H) will be of particular importance in the next section for the definition of unbounded linear operators in H. An example of a dense linear manifold is the set of all finite linear combinations of vectors of a basis $\{e_i\}$ of H (Problem 2.5).

As a last general remark we mention a theorem which

we shall have occasion to use in later chapters. It concerns bounded linear functionals on a Hilbert space H. By definition such a functional is a linear map Φ from H into C (i.e. $\Phi(f)\varepsilon C$ for each $f\varepsilon H$, and $\Phi(\alpha f + g) = \alpha\Phi(f) + \Phi(g)$ for all $\alpha\varepsilon C$ and $f,g\varepsilon H$) which is bounded with respect to the norm in H, i.e.

$$|||\Phi||| \equiv \sup_{f\neq\theta} \frac{|\Phi(f)|}{||f||} < \infty.$$

If g is a fixed vector in H, one may associate with it a bounded linear functional Φ_g on H by $\Phi_g(f) = (g,f)$. The boundedness of Φ_g follows from (2.8) :

$$|||\Phi_g||| = \sup_{f\neq\theta} \frac{|(g,f)|}{||f||} \leq \sup_{f\neq\theta} \frac{||g||\,||f||}{||f||} = ||g||.$$

In fact one has $|||\Phi_g||| = ||g||$, since equality holds in (2.8) for $f = g$. It is an interesting property of Hilbert space that the converse is also true :

PROPOSITION 2.3 : Let $\Phi : H \to C$ be a bounded linear map. Then there exists a uniquely determined vector $g\varepsilon H$ such that $\Phi(f) = (g,f)$ for all $f\varepsilon H$, and $|||\Phi||| = ||g||$.

This result is known as the Riesz representation theorem. We shall omit its proof, since it is given in numerous books on functional analysis (e.g. [K],[RS]). The idea of the proof is indicated in Problem 2.7.

So far we have considered the abstract Hilbert space. While many of the formal developments of scattering theory can be given in the abstract, the interpretation of the theory in terms of observable quantities usually requires the choice of some concrete realization of the space. The most important such realization for scattering theory is the space $L^2(\mathcal{R}^n)$. It consists of all Lebesgue measurable*) complex-valued functions defined on n-dimensional Euclidean space $\mathcal{R}^n$ which are absolutely square-integrable, i.e. such that**)

$$\int_{\mathcal{R}^n} |f(\underline{x})|^2 \, d^n x < \infty. \tag{2.13}$$

The set of all such functions is a linear vector space if one defines addition and multiplication by scalars as follows :

$$(f_1 + f_2)(\underline{x}) = f_1(\underline{x}) + f_2(\underline{x}), \quad (\alpha f)(\underline{x}) = \alpha f(\underline{x}).$$

The scalar product between two such functions is defined by

$$(f,g) = \int_{\mathcal{R}^n} \bar{f}(\underline{x}) g(\underline{x}) d^n x. \tag{2.14}$$

This integral is finite, since $|\bar{f}(\underline{x})g(\underline{x})| \le \frac{1}{2}|f(\underline{x})|^2 + \frac{1}{2}|g(\underline{x})|^2$.

This scalar product verifies (2.5) and (2.6). Concerning its strict positivity there is a certain complication to which one must pay attention in many of the mathematical developments of the theory. This stems from the fact that

$$\|f\|^2 \equiv \int_{\mathcal{R}^n} |f(\underline{x})|^2 \, d^n x - 0 \tag{2.15}$$

does not imply that $f(\underline{x}) = 0$ for all $\underline{x}$. It only implies that $f(\underline{x}) = 0$ almost everywhere (a.e.) with respect to the Lebesgue measure on $\mathcal{R}^n$. This means that $f(\underline{x})$ may differ from zero and may in fact assume arbitrary values on a set of Lebesgue measure zero.

---

*) For measure and integration theory the reader may consult e.g. [L],[R].

**) Vectors in $\mathcal{R}^n$ will be denoted by $\underline{x}$.

The Hilbert space $L^2(\mathcal{R}^n)$ does therefore not consist of the individual functions themselves but rather of classes of equivalent functions. Two functions are defined to be <u>equivalent</u> if they differ only on a set of measure zero. We may formally introduce the following notation : Let V be the set of individual functions satisfying (2.13) and V_o the subset of V satisfying (2.15). Then the Hilbert space we wish to define is the quotient space $V/V_o = L^2(\mathcal{R}^n)$. The element θ of $L^2(\mathcal{R}^n)$ is given by the class V_o which contains in particular the function $f \equiv 0$. It is quite elementary to verify that the operations of addition, multiplication by scalars, and scalar product can be transferred to the classes since they are independent of the representative elements inside the classes with which the operations are carried out.

These remarks concerning the quotient space V/V_o may strike a physicist as pedantic, since it is in most cases possible to transfer all operations in the Hilbert space $L^2(\mathcal{R}^n)$ to individual functions (a practice which we shall frequently follow in the traditional manner). There are occasional situations where the above remark is essential and must be borne in mind.

The completeness of $L^2(\mathcal{R}^n)$ is a classical result of analysis known as the Riesz-Fischer Theorem. We shall not give a proof of it in this book (cf. for instance [RN],[R], [RS] or [L]). The separability of $L^2(\mathcal{R}^n)$ can be established in different ways (cf. e.g. [RN],[SO]). One possible method is to show that a general function in $L^2(\mathcal{R}^n)$ can be approximated arbitrarily well by a finite linear combination of characteristic functions*) of n-dimensional rectangles whose end points have rational coordinates (the details may be read in [RN, Section 32]). Separability then follows since the set of all such characteristic functions is countable.

The characteristic function of an n-dimensional rectangle can be approximated in $L^2(R^n)$-norm arbitrarily well by an infinitely differentiable function vanishing outside the rectangle; this is done by changing the characteristic functions near the edges of the rectangle into a smooth function (Problem 2.8). It follows that the set $C_0^\infty(R^n)$ **) of all infinitely differentiable functions of compact support is dense in $L^2(R^n)$. This result is often useful in scattering theory.

Apart from the linear manifold $C_0^\infty(R^n)$ we shall also need another linear manifold denoted by $S(R^n)$. A function f belongs to $S(R^n)$ if it is infinitely differentiable and if f and its partial derivatives of all orders decrease faster than any negative power of x as $x \to \infty$ ***). More precisely, $f \varepsilon S(R^n)$ if f is infinitely differentiable (i.e. $f \varepsilon C^\infty(R^n)$) and if for every 2n-tuple of non-negative integers $\{j_1,\ldots,j_n,m_1,\ldots,m_n\}$ one has

$$\sup_{\underline{x}\varepsilon R^n} \left| (x_1)^{j_1} \cdots (x_n)^{j_n} \frac{\partial^{|m_1+\cdots+m_n|}}{\partial x_1^{m_1} \cdots \partial x_n^{m_n}} f(x_1,\ldots,x_n) \right| < \infty. \quad (2.16)$$

---

*) The characteristic function of a set S in R^n is defined by $\chi_S(\underline{x}) = 1$ if $\underline{x}\varepsilon S$ and $\chi_S(\underline{x}) = 0$ if $\underline{x} \notin S$.

**) If $S \subseteq R^n$ is an open set, then f is said to belong to $C_0^\infty(S)$ if it has partial derivatives to all orders and there exists a compact set in S outside which f is identically zero.

***) Here we have defined $x = |\underline{x}| \equiv (\sum_{i=1}^n x_i^2)^{\frac{1}{2}}$.

Such functions are also called functions of rapid decrease. An example is the function $\exp(-x^2)$. $S(R^n)$ has the following interesting properties :

PROPOSITION 2.4 : (a) $S(R^n)$ is dense in $L^2(R^n)$.
(b) $S(R^n)$ is invariant under Fourier transformation.

Proof : The denseness follows because $S(R^n)$ contains $C_0^\infty(R^n)$ and the latter is dense in $L^2(R^n)$. The Fourier transformation will be defined here for functions belonging to $S(R^n)$ and extended to a larger class of functions in the next section. If $\underline{k}$ and $\underline{x}$ are two vectors in R^n, we define $\underline{k}\cdot\underline{x} \equiv k_1x_1 + k_2x_2 + \cdots + k_nx_n$. If $f\varepsilon S(R^n)$, we may define a new function $\tilde{f} : R^n \to C$ by the formula

$$\tilde{f}(\underline{k}) = (2\pi)^{-n/2} \int d^n x e^{-i\underline{k}\cdot\underline{x}} f(\underline{x}). \qquad (\underline{k}\varepsilon R^n) \qquad (2.17)$$

Proposition 2.4(b) asserts that $\tilde{f}$ belongs again to $S(R^n)$. In order to simplify the notation, we indicate the proof of this for the case $n = 1$ and remark that the proof for $n > 1$ is essentially the same (Problem 2.9).

Let $f\varepsilon S(R^1)$. The proof will rely on the following relation which is an immediate consequence of (2.16) :

$$|(1 + x^2) \frac{d^r}{dx^r} x^m f(x)| \leq c \text{ for all } x\varepsilon R, \qquad (2.18)$$

where c may depend on r and m. From (2.17) one obtains

$$\frac{d^m}{dk^m} \tilde{f}(k) = (-i)^m (2\pi)^{-\frac{1}{2}} \int dx e^{-ikx} x^m f(x). \qquad (2.19)$$

Here we have interchanged the derivatives and the integral on the right-hand side. This is justified provided that for each m the improper integral in (2.19) converges (as a limit of integrals over an increasing sequence of finite intervals) uniformly in k [L, Vol. I, p. 252]. This is the case since (2.18) implies that

$$\left| \int_{|x|\geq R} dx e^{-ikx} x^m f(x) \right| \leq \int_{|x|\geq R} dx \frac{c}{1+x^2} \leq \int_{|x|\geq R} dx \frac{c}{x^2} = \frac{2c}{R} ,$$

which converges to zero as $R \to \infty$ uniformly in k. It follows that $\tilde{f}$ is infinitely differentiable. One also obtains from (2.19) by integrating by parts

$$k^r \frac{d^m}{dk^m} \tilde{f}(k) = (-i)^{m+r} (2\pi)^{-\frac{1}{2}} \int dx e^{-ikx} \frac{d^r}{dx^r} (x^m f(x)).$$

Hence

$$\sup_{k\in\mathbb{R}} \left| k^r \frac{d^m}{dk^m} \tilde{f}(k) \right| \leq (2\pi)^{-\frac{1}{2}} \int dx \left| \frac{d^r}{dx^r} x^m f(x) \right| \leq (2\pi)^{-\frac{1}{2}} \int dx \frac{c}{1+x^2} .$$

The last integral is seen to be finite. Hence $\tilde{f}$ belongs to $S(\mathbb{R}^1)$. #

The result of Proposition 2.4(b) can still be strengthened. In fact the Fourier transformation is a mapping from $S(\mathbb{R}^n)$ onto $S(\mathbb{R}^n)$. This can be seen by defining first the inverse Fourier transformation on $S(\mathbb{R}^n)$ by

$$\hat{f}(\underline{x}) = (2\pi)^{-n/2} \int d^n k e^{i\underline{k}\cdot\underline{x}} f(\underline{k}), \quad f\in S(\mathbb{R}^n). \tag{2.20}$$

The only difference between (2.17) and (2.20) is the sign of the exponent. The same reasoning as in the preceding proof implies that $\hat{f}\in S(\mathbb{R}^n)$. If we can show that (2.20) is in fact the inverse transformation of (2.17), i.e. that $\hat{\tilde{f}} = f$, we shall have proved that the Fourier transformation is a bijection of $S(R^n)$ onto itself.

For this, let $f,h\in S(R^n)$ and $\alpha > 0$. By using (2.17) and making appropriate substitutions for the integration variables, one gets

$$\int h(\alpha\underline{k})\tilde{f}(\underline{k})e^{i\underline{k}\cdot\underline{y}}d^n k = (2\pi)^{-n/2}\int d^n k h(\alpha\underline{k})\int d^n x e^{-i\underline{k}\cdot(\underline{x}-\underline{y})}f(\underline{x})$$

$$= \alpha^{-n}\int d^n x\tilde{h}(\alpha^{-1}(\underline{x}-\underline{y}))f(\underline{x}) = \int d^n u\tilde{h}(\underline{u})f(\alpha\underline{u} + \underline{y}). \qquad (2.21)$$

As $\alpha \to 0$, the integrand of the left-hand side converges pointwise to $h(0)\tilde{f}(\underline{k})\exp(i\underline{k}\cdot\underline{y})$. It is also absolutely majorized, uniformly in $\alpha \geq 0$, by the supremum over $\underline{y}\in R$ of $|h(\underline{y})||\tilde{f}(\underline{k})|$. Since the latter function is an integrable function of $\underline{k}$, the Lebesgue dominated convergence theorem (Proposition 2.35) implies that

$$\lim_{\alpha\to 0}\int h(\alpha\underline{k})\tilde{f}(\underline{k})e^{i\underline{k}\cdot\underline{y}}d^n k = h(0)\int\tilde{f}(\underline{k})e^{i\underline{k}\cdot\underline{y}}d^n k = h(0)(2\pi)^{n/2}\hat{\tilde{f}}(\underline{y}).$$

Similarly one gets for the last member of (2.21)

$$\lim_{\alpha\to 0}\int\tilde{h}(\underline{u})f(\alpha\underline{u} + \underline{y})d^n u = f(\underline{y})\int\tilde{h}(\underline{u})d^n u.$$

The desired result $\hat{\tilde{f}}(\underline{y}) = f(\underline{y})$ is obtained by verifying for a particular function $h\in S(R^n)$ that $(2\pi)^{n/2}h(0) = \int\tilde{h}(\underline{u})d^n u$. This is possible for instance for $h_o(\underline{x}) = \exp(-\frac{1}{2}x^2)$. Then (Problem 2.10) $\tilde{h}_o(\underline{k}) = \exp(-\frac{1}{2}k^2) = h_o(\underline{k})$, hence

$$\int\tilde{h}_o(\underline{u})d^n u = \int h_o(\underline{u})d^n u = (2\pi)^{n/2}\tilde{h}_o(0) = (2\pi)^{n/2}h_o(0).$$

Thus we have shown that (2.20) defines the inverse of (2.17). We shall henceforth also denote the mapping $f\mapsto\tilde{f}$ by F, i.e. we shall write $\tilde{f} = Ff$. We then have $f = F^{-1}\tilde{f} = \hat{\tilde{f}}$, where F^{-1} is given by (2.20). In the next section we shall extend F and F^{-1} to operators on the entire space $L^2(R^n)$.

A last important result is the remark that F and F^{-1} are isometric on $S(R^n)$, i.e.

$$\|\tilde{f}\| = \|f\| = \|\hat{f}\| \qquad \text{for } f\varepsilon S(R^n), \tag{2.22}$$

$$(\tilde{g},\tilde{f}) = (g,f) = (\hat{g},\hat{f}) \qquad \text{for } f,g\varepsilon S(R^n). \tag{2.23}$$

(2.22) is a special case of (2.23). The first equality in (2.23) is a particular case of (2.21) : wet set $\alpha = 1$, $\underline{y} = 0$ and $h = \bar{\tilde{g}}$; the left-hand side then equals $(\tilde{g},\tilde{f})$. By noticing that $\bar{\tilde{f}} = \hat{\bar{f}}$, one obtains $\tilde{h} = \hat{\bar{g}} = \bar{g}$, i.e. the right-hand side of (2.21) equals (g,f) for the particular choice of α, $\underline{y}$ and h made above. This establishes the first part of (2.23). The proof of its second part is left as an exercise (Problem 2.11).

2-2 LINEAR OPERATORS IN HILBERT SPACE

A linear operator in a Hilbert space H is a linear mapping between vectors of H. As an example we have already seen the Fourier transformation F in $L^2(R^n)$ which was defined on all vectors belonging to the dense set $S(R^n)$ and which is obviously linear. We have seen that this operator does not change the L^2-norm of a vector f. Many operators that are important in physical applications do not have such a simple property, and indeed they may be such that the norm of certain image vectors may exceed that of the corresponding initial vector by an arbitrarily large amount (such operators will be called unbounded). A well-known example from elementary quantum mechanics is the position operator Q for a particle in one-dimensional space. Here $H = L^2(R)$, and Q is the operator of multiplication by the variable x in $L^2(R)$:

$$(Qf)(x) = xf(x).$$

If we take for instance $f(x) = (1+|x|)^{-1}$, then $f \in L^2(R)$ but $\|Qf\| = \infty$, i.e. $Qf \notin L^2(R)$. Thus one expects that in general a linear operator will be defined only on some subset of the Hilbert space. In the above example this domain of definition D(Q) of the operator Q can be taken as the set of those f in $L^2(R)$ such that xf(x) is also an element of $L^2(R)$. It is easily seen that this subset D(Q) is a manifold, and it is also dense in $L^2(R)$ (it contains all functions belonging to $S(R)$ which is dense in $L^2(R)$).

A linear operator*) is defined by giving its domain, i.e. a linear manifold D(A) in H, and a linear mapping A from D(A) into H. Linearity means that, if $f, g \in D(A)$ and $\alpha \in C$, then $A(\alpha f + g) = \alpha Af + Ag$. If A is a linear operator, then D(A) contains the vector θ, since $\theta = 0 \cdot f$ for arbitrary $f \in D(A)$. One then has $A\theta = 0 \cdot (Af) = \theta$. The following notation will also be used : If M is a subset of D(A), then AM is the set of vectors f in H such that f = Ag for some g in M. The set AD(A) will be called the range of the operator A.

Two linear operators A and B are equal if and only if D(A) = D(B) and Af = Bf for all $f \in D(A)$. A linear operator A' is called an extension of A if $D(A) \subseteq D(A')$ and A'f = Af for

*) A linear operator will usually be simply called an operator.

all $f \in D(A)$. Here A' coincides with A on D(A), but it may be defined on a larger domain than A. In this case we shall write $A \subseteq A'$. One may also call A the restriction of A' to D(A).

For some operators A there is a natural way of defining an extension $\bar{A}$. One takes a strong Cauchy sequence $\{f_n\}$ in D(A). If the sequence $\{Af_n\}$ is also Cauchy, and if one denotes by f and g the strong limits of $\{f_n\}$ and $\{Af_n\}$ respectively, it is natural to define $\bar{A}f = g$. Since f is not necessarily in D(A), one may define an extension $\bar{A}$ of A by applying the above definition to all Cauchy sequences $\{f_n\}$ in D(A) which are such that $\{Af_n\}$ is also Cauchy. However, this construction makes sense only if the element g is independent of the choice of a particular Cauchy sequence $\{f_n\}$ converging to f, i.e. if the following condition is satisfied : Whenever $\{f_n\}$ and $\{f'_n\}$ are two Cauchy sequences in D(A) converging strongly to the same limit f and $\{Af_n\}$ and $\{Af'_n\}$ are also Cauchy, then $\text{s-lim } Af_n = \text{s-lim } Af'_n$. Since A is linear, this condition is easily seen to be equivalent to the following one (Problem 2.12) : Whenever $\{f_n\} \in D(A)$, $f_n \to \theta$ and $\{Af_n\}$ is strongly Cauchy, then $Af_n \to \theta$.

An operator verifying one of these two equivalent conditions is said to be closable, and then the above extension $\bar{A}$ is called the closure of A. A is said to be closed if it is identical with its closure, i.e. if $A = \bar{A}$.

The following is a different criterion for an operator to be closed :

LEMMA 2.5 : Let A be a linear operator in $\mathcal{H}$. The following two statements are equivalent :

(a) A is closed.

(b) Whenever a sequence $\{f_n\}$ verifies (i) $f_n \varepsilon D(A)$, (ii) $f_n \to f$, and (iii) $Af_n \to g$, then $f \varepsilon D(A)$ and $Af = g$.

Proof : The result is a simple consequence of the definition of the closure.

(i) Suppose $A = \bar{A}$, and let $\{f_n\}$ verify (i)-(iii). Since A is closable, one has $f \varepsilon D(\bar{A})$ and $\bar{A}f = g$. Since $A = \bar{A}$, this means that (b) is verified. Hence (a) implies (b).

(ii) Suppose (b) holds. Then A is closable, since the hypotheses $\{f_n\} \varepsilon D(A)$, $f_n \to \theta$ and $Af_n \to g$ then imply $g = A\theta = \theta$. From the construction of $\bar{A}$ and (b) one sees that $D(\bar{A}) \subseteq D(A)$, i.e. $\bar{A} = A$. #

The operators that one encounters in applications are usually closed or closable. An operator A may have more than one closed extension (an example of such an operator will be given further on in this section). If an operator A has a closed extension B, then it is closable. To see this, suppose $\{f_n\} \varepsilon D(A)$, $f_n \to f$ and $Af_n \to g$. Since $A \subseteq B$ and B is closed, Lemma 2.5 implies that $f \varepsilon D(B)$ and $g = Bf$. Since the vector Bf is independent of the sequence $\{f_n\}$ converging to f, this shows that A is closable.

If A is closable, then its closure $\bar{A}$ is its smallest closed extension, i.e. if A' is an arbitrary closed extension of A, then $\bar{A} \subseteq A'$ (Problem 2.13). An example of a non-closable operator is given in Problem 2.14.

We shall make additional comments on closed extensions

further on in this section. Here we shall indicate an important class of closable operators, namely the bounded operators. A linear operator A is said to be <u>bounded</u> if there exists a number $M < \infty$ such that $\|Af\| \leq M\|f\|$ for all $f \varepsilon D(A)$. If there exists no such M, A is said to be <u>unbounded</u>. For bounded A one defines its <u>norm</u> $\|A\|$ as*)

$$\|A\| = \sup_{\substack{f \varepsilon D(A) \\ f \neq \theta}} \frac{\|Af\|}{\|f\|} . \tag{2.24}$$

One then has for every $f \varepsilon D(A)$:

$$\|Af\| \leq \|A\| \, \|f\|. \tag{2.25}$$

As a consequence of this inequality, one has the following result : Let A be bounded and $\{f_n\}$ a strong Cauchy sequence in D(A). Then $\{Af_n\}$ is also strongly Cauchy. In fact

$$\|Af_n - Af_m\| \leq \|A\| \, \|f_n - f_m\| \to 0 \text{ as } n, m \to \infty.$$

This can be used to prove that a bounded operator is always closable. This result is contained in the following proposition which will prove to be useful in scattering theory.

<u>PROPOSITION 2.6</u> : If A is a bounded linear operator on a Hilbert space $\mathcal{H}$, it has a unique bounded extension $\bar{A}$ to the subspace spanned by D(A) (i.e. to the closure $\overline{D(A)}$ of D(A)). $\bar{A}$ is closed, and $\|\bar{A}\| = \|A\|$. In particular, if D(A) is dense in $\mathcal{H}$, then $D(\bar{A}) = \mathcal{H}$.

<u>Proof</u> : We verify that the closure $\bar{A}$ of A has the required

---

*) It is customary to use the same symbol $\|\cdot\|$ for the norm of an operator and for the norm of vectors in $\mathcal{H}$. It is seen from (2.24) that the former is defined in terms of the latter.

properties.

(i) Let $\{f_n\}$ and $\{f'_n\}$ be two strong Cauchy sequences in D(A) converging to the same limit f. Then $\{Af_n\}$ and $\{Af'_n\}$ are also Cauchy, and by (2.25) and (2.9)

$$\|Af_n - Af'_n\| \leq \|A\| \|f_n - f'_n\| \leq \|A\| (\|f_n - f\| + \|f - f'_n\|),$$

which converges to zero as $n \to \infty$. Hence $\{Af'_n\}$ converges to the same vector as $\{Af_n\}$, i.e. A is closable. It also follows that every f in $\overline{D(A)}$ belongs to the domain of $\bar{A}$. Hence $D(\bar{A}) = \overline{D(A)}$.

(ii) Suppose A' is another bounded extension of A defined on $D(A') = \overline{D(A)}$. As in (i), A' is closable and $D(\overline{A'}) = \overline{D(A')}$. Since D(A') is closed, this implies $D(\overline{A'}) = D(A')$, i.e. A' is a closed operator. It follows that A' is an extension of $\bar{A}$. But $D(A') = D(\bar{A})$, so that $A' = \bar{A}$. This proves the uniqueness of the extension.

(iii) We have from (2.24) and the fact that $A \subseteq \bar{A}$

$$\|\bar{A}\| = \sup_{f\varepsilon D(\bar{A}),f\neq\theta} \frac{\|\bar{A}f\|}{\|f\|} \geq \sup_{f\varepsilon D(A),f\neq\theta} \frac{\|Af\|}{\|f\|} = \|A\|.$$

The opposite inequality, i.e. $\|\bar{A}\| \leq \|A\|$, is obtained by letting f and $\{f_n\}$ be as in (i) and by applying twice Proposition 2.1 :

$$\|\bar{A}f\| = \lim_{n\to\infty} \|Af_n\| \leq \|A\| \lim_{n\to\infty} \|f_n\| = \|A\| \|f\|. \ \#$$

As a consequence of Proposition 2.6, one may always consider a bounded operator to be defined on a subspace. If D(A) in Proposition 2.6 is not dense in H, this subspace is strictly smaller than H. One may then extend A to a bounded operator $\tilde{A}$ defined on all of H by setting $\tilde{A} = \bar{A}$ on $\overline{D(A)}$ and by identifying $\tilde{A}$ with an arbitrary bounded operator B on the orthogonal complement $D(A)^{\perp}$ of $D(A)$. ($\tilde{A}$ is then defined everywhere : An arbitrary f in H can be decomposed uniquely into a sum of an element f_1 in $\overline{D(A)}$ and of an element f_2 in $D(A)^{\perp}$, and by linearity one has $\tilde{A}f = \bar{A}f_1 + Bf_2$.) The simplest possibility is to take B = 0 on $D(A)^{\perp}$, which we shall do for some operators used in scattering theory.

As an example where Proposition 2.6 can be applied we mention the Fourier transformation F defined in Section 2-1. We there had $D(F) = S(R^n)$, and $\|F\| = 1$ from (2.22). Hence F may be extended to a bounded operator of norm 1 defined on all of $L^2(R^n)$. We shall henceforth denote this extension also by F. This extension is still an isometric operator, i.e. $\|Ff\| = \|f\|$ for all $f \varepsilon L^2(R^n)$ (Problem 2.15). Also it is still given by (2.17) for vectors f in $L^2(R^n)$ which are also in $L^1(R^n)$, i.e. which satisfy in addition to (2.13) the condition

$$\|f\|_1 \equiv \int_{R^n} |f(\underline{x})| d^n x < \infty.$$

Indeed, for such a vector f the integral in (2.17) exists for each $k \varepsilon R^n$, and one can show that the vector f defined in this way is identical with Ff (the reader may look up details in Section 2-5). For a general f in $L^2(R^n)$ the integral in (2.17) need not make sense ($f \varepsilon L^2(R^n)$ does not imply that f is integrable); then one has to define Ff as $Ff = \text{s-lim}\, Ff_m$ for $m \to \infty$, where $f_m \varepsilon L^2(R^n) \cap L^1(R^n)$ and $f_m \to f$. A particular way of choosing f_m is to set $f_m(\underline{x}) = f(\underline{x})$ for $x \leq m$ and $f_m(\underline{x}) = 0$ for $x > m$. The strong convergence as $m \to \infty$ of this particular sequence $\{Ff_m\}$ to $Ff \equiv \tilde{f}$ is also called <u>convergence in the mean</u>, and one uses the following notation for this method of defining the Fourier transformation in $L^2(R^n)$:

$$\tilde{f}(\underline{k}) = (2\pi)^{-n/2} \ell.\text{i.m.} \int d^n x \exp(-i\underline{k}\cdot\underline{x}) f(\underline{x}) \tag{2.26}$$

(some other details about this are given in Section 2-5).

Similarly F^{-1} can be extended to $L^2(R^n)$, and again we shall denote this extension by F^{-1}. It is isometric and given by (2.20) for $f \in L^2(R^n) \cap L^1(R^n)$. Also the extended operator F^{-1} is still the inverse of the extended operator F : FF^{-1} and $F^{-1}F$ are bounded extensions of the identity operator I_o defined on $D(I_o) = S(R^n)$; by Proposition 2.6 the only bounded extension to all of $L^2(R^n)$ of I_o is the identity operator I given by If = f for all f in H. Hence $FF^{-1} = I$ and $F^{-1}F = I$.

The sum of two operators A and B is defined in a natural way as follows : $D(A + B) = D(A) \cap D(B)$, and $(A + B)f = Af + Bf$ for $f \in D(A + B)$. In particular, if $D(B) = H$, then $D(A + B) = D(A)$. From the triangle inequality (2.9) one easily deduces that for bounded operators (Problem 2.16)

$$\|A + B\| \leq \|A\| + \|B\|. \tag{2.27}$$

One may similarly define the product of A and B : $f \in D(AB)$ if and only if $f \in D(B)$ and $Bf \in D(A)$, and $(AB)f = A(Bf)$ for $f \in D(AB)$. (2.25) implies that for bounded operators

$$\|AB\| \leq \|A\| \, \|B\|. \tag{2.28}$$

The operator multiplication is not commutative, i.e. AB may be different from BA (a well-known example is that of the operators $P = -id/dx$ and Q = multiplication by x in $L^2(R)$; another example is the unilateral shift operator and its adjoint which will be discussed later in this section).

It should be remarked that, even if both A and B are

densely defined, D(A + B) or D(AB) may consist only of the vector θ. For this reason one has to be very cautious when adding or multiplying unbounded operators. We shall discuss in later chapters some of the difficulties that arise in scattering theory due to the unboundedness of some important operators. When both A and B are bounded and defined everywhere, it is not necessary to worry about domain problems.

A linear operator A is <u>invertible</u> if Af = Ag, f and g in D(A), implies f = g, or equivalently if Af = θ, fεD(A), implies f = θ. The <u>inverse</u> A^{-1} is then well defined and given as follows : $D(A^{-1})$ = AD(A) (i.e. the range of A) and $A^{-1}(Af) = f$. It is easy to see that A^{-1} is also linear. Also $A^{-1}A$ is then the restriction of the identity operator I to D(A) and AA^{-1} is the restriction of I to $D(A^{-1})$. If A is bounded and D(A) = H, then $A^{-1}A = I$. The following result is a direct consequence of the definitions : If A is closed and invertible, then A^{-1} is also closed (Problem 2.17).

We shall now introduce the concept of the <u>adjoint</u> operator A* of a linear operator A. For this we assume that D(A) is dense in H. We first define the domain D(A*) : a vector gεH belongs to D(A*) if there exists a vector g*εH such that

$$(g,Af) = (g^*,f) \qquad \text{for all } f\varepsilon D(A). \tag{2.29}$$

The mapping A* is then defined as A*g = g*. Thus (2.29) may be rewritten as

$$(g,Af) = (A^*g,f) \text{ for all } f\varepsilon D(A) \text{ and all } g\varepsilon D(A^*). \tag{2.30}$$

A* is well defined, i.e. the vector g* in (2.29) is unique.

In fact, if g_1^* has the same property as g^*, then $(g_1^* - g^*, f) = 0$ for all $f \varepsilon D(A)$, and since $D(A)$ is dense, Proposition 2.2 implies that $g_1^* = g^*$. Clearly A^* is also linear.

The adjoint of a linear operator A is always a closed operator. In fact, if $g_n \varepsilon D(A^*)$, $g_n \to g$ and $A^* g_n \to h$, then for any f in D(A)

$$(g , Af) = \lim_{n\to\infty} (g_n , Af) = \lim_{n\to\infty} (A^* g_n , f) = (h , f),$$

which shows that $g \varepsilon D(A^*)$ and that $A^* g = h$.

Another simple property of the adjoint is the following (Problem 2.18) : If A is closable and D(A) dense, then

$$A^* = (\bar{A})^* \equiv \bar{A}^*. \tag{2.31}$$

If $D(A^*)$ is also dense in $\mathcal{H}$, then $A^{**} \equiv (A^*)^*$ exists. One then has the following result :

<u>PROPOSITION 2.7</u> : Let A be a linear operator such that D(A) and $D(A^*)$ are dense in $\mathcal{H}$. Then A is closable and $\bar{A} = A^{**}$.

We shall not give a complete proof of this proposition here. Notice that it follows immediately from the definition (2.30) that A^{**} is an extension of A (Problem 2.19). Since A^{**} is the adjoint of an operator, it is closed. Hence A has a closed extension and is therefore closable.

The proof that A^{**} coincides with the closure of A requires the notion of the graph of A. This method is indicated in Section 2-5. The converse of Proposition 2.7 is also true : if A is closable and D(A) dense, then $D(A^*)$ is also dense. We shall not use this result and therefore omit its proof (cf. e.g. [RS],[RN]).

In the next proposition we specify the properties of the adjoint of a bounded operator.

PROPOSITION 2.8 : Let A be a bounded operator with $D(A) = \mathcal{H}$. Then A^* is bounded, $D(A^*) = \mathcal{H}$ and $\|A^*\| = \|A\|$. In addition $A^{**} = A$.

Proof : Let g be a fixed vector in $\mathcal{H}$. From (2.8) and (2.25) it follows that $|(g , Af)| \leq \|A\| \|f\| \|g\|$. Hence the correspondence $f \mapsto (g , Af)$ defines a bounded linear functional Φ on $\mathcal{H}$ with $|||\Phi||| \leq \|A\| \|g\|$. It then follows from Proposition 2.3 that there exists $g^* \varepsilon \mathcal{H}$ such that (2.29) holds, and that $\|g^*\| \leq \|A\| \|g\|$.

Since g was arbitrary, we have shown that $D(A^*) = \mathcal{H}$, and that $\|A^*\| \leq \|A\|$. To prove the converse inequality, we apply the same reasoning to A^* in the place of A to deduce that $\|A^{**}\| \leq \|A^*\|$. Since A^{**} is an extension of A and $D(A) = \mathcal{H}$, one has $A^{**} = A$. Hence the last inequality becomes $\|A\| \leq \|A^*\|$, which proves that $\|A\| = \|A^*\|$. #

If A is bounded with $D(A) = \mathcal{H}$ and D(B) is dense, it follows from (2.30) that

$$(\alpha A)^* = \bar{\alpha} A^*, \quad (A + B)^* = A^* + B^* \text{ and } (AB)^* = B^*A^*. \qquad (2.32)$$

If both A and B are unbounded, the last two equalities need not hold, since the domains of the respective left-hand and right-hand member may be different. As a good exercise for becoming familiar with the notion of the adjoint the reader may show that in the general case (Problem 2.20) :

$$B^*A^* \subseteq (AB)^*. \tag{2.33}$$

The following is also easy to verify :

$$\text{if } A \subseteq B, \qquad \text{then } B^* \subseteq A^*. \tag{2.34}$$

We shall now discuss some special types of operators that we shall encounter throughout this book. We begin with orthogonal projections*), denoted here by F. They are defined by the requirements

$$D(F) = \mathcal{H} \text{ and } F^2 = F = F^*. \tag{2.35}$$

Their most interesting property is that the set of all orthogonal projections is in one-to-one correspondence with the family of all subspaces of $\mathcal{H}$.

To prove the above assertion, let F be a projection and define $M_F = F\mathcal{H}$. Thus, if $f \varepsilon M_F$, there exists $g \varepsilon \mathcal{H}$ such that $f = Fg$. Hence $Ff = F^2 g = Fg = f$. On the other hand, if f is such that $Ff = f$, then obviously $f \varepsilon M_F$. Thus $M_F = \{f \varepsilon \mathcal{H} | Ff = f\}$. M_F is clearly a linear manifold. To show that it is a subspace, i.e. that it is strongly closed, let $\{f_n\} \varepsilon M_F$ be a strong Cauchy sequence, $f_n \to f$. Let $g \varepsilon \mathcal{H}$. Then $(Ff - f, g) = (f, F^*g) - (f,g) = \lim[(f_n, F^*g) - (f_n, g)] = \lim[(Ff_n, g) - (f_n, g)] = (\theta, g) = 0$ as $n \to \infty$. It follows from Proposition 2.2 that $Ff - f = \theta$. Thus $f \varepsilon M_F$, which proves that M_F is strongly closed.

If $h \varepsilon M_F^{\perp}$ and $g \varepsilon \mathcal{H}$, we have from (2.30) $(Fh,g) = (F^*h,g) = (h,Fg) = 0$. Thus, by Proposition 2.2, $Fh = \theta$. This shows

*) We shall not use non-orthogonal projections in this book. Therefore an orthogonal projection will simply be called a projection.

that F is the orthogonal projection onto M_F. Conversely, given a subspace M, one may define a linear operator F as follows : Let $\{e_i\}$ be an orthonormal basis of M and

$$Fg = \sum_i (e_i, g)e_i. \tag{2.36}$$

It is easy to verify that this operator is an orthogonal projection (Problem 2.21) with range M.

A particular case is that of a one-dimensional subspace $M = \{\alpha f \mid \alpha \in C\}$, where f is a fixed unit vector*) in H. In this case we shall denote by F_f the corresponding orthogonal projection defined by (2.36). The action of F_f may be written as

$$F_f g = (f,g)f.$$

We mention two other simple properties of projections :

<u>LEMMA 2.9</u> : Let F be an orthogonal projection. Then

(a) $\|F\| = 1$ unless $F = 0$, (2.37)

(b) $(f,Fg) = (Ff,Fg)$ for all f,g in H. (2.38)

<u>Proof</u> : Let $f \in H$. One may write $f = Ff + (I - F)f$. Here $Ff \in M_F$ and $(I - F)f \in M_F^{\perp}$. Thus one has with (2.10) $\|f\|^2 = \|Ff\|^2 + \|(I - F)f\|^2 \geq \|Ff\|^2$, which shows that $\|F\| \leq 1$. The fact that $\|F\| = 1$ follows by taking $\theta \neq f \in M_F$, which is possible unless F is the zero operator on H.

To prove (b), one uses (2.35) :

$$(f,Fg) = (f,F^2 g) = (f,F^*Fg) = (Ff,Fg). \; \#$$

---

*) A <u>unit vector</u> is a vector f verifying $\|f\| = 1$.

For an example, let $H = L^2(R^n)$ and let Δ be a measurable subset of R^n. Then $L^2(\Delta)$ is defined to be the set of those equivalence classes of functions in $L^2(R^n)$ which have support in Δ, i.e. whose representatives are zero almost everywhere (with respect to Lebesgue measure) on the complement of Δ. It is easy to see that $L^2(\Delta)$ is a subspace of $L^2(R^n)$, i.e. $L^2(\Delta)$ is itself a Hilbert space. The orthogonal projection F_Δ of $L^2(R^n)$ onto $L^2(\Delta)$ may be written as follows

$$(F_\Delta f)(\underline{x}) \equiv (\chi_\Delta f)(\underline{x}) = \chi_\Delta(\underline{x})f(\underline{x}), \tag{2.39}$$

where χ_Δ is the characteristic function of the set Δ. Clearly $D(F_\Delta) = H$ and $F_\Delta^2 = F_\Delta$. The fact that $F_\Delta^* = F_\Delta$ is obtained by writing (2.30) in the form of integrals.

Next we consider partial isometries, denoted here by Ω. These are important in scattering theory, as we shall see in Chapter 4. They are defined by the requirements

$$D(\Omega) = H \text{ and } \Omega^*\Omega = E, \ E \text{ a projection.} \tag{2.40}$$

For such an operator one has with (2.30) and (2.38)

$$(\Omega f,\Omega g) = (f,\Omega^*\Omega g) = (f,Eg) = (Ef,Eg). \tag{2.41}$$

Hence, if $f,g \in EH$, then $(\Omega f,\Omega g) = (f,g)$. This shows that such an operator is isometric on a part of H, namely on EH, i.e. it preserves the length of vectors of EH and the angles between vectors of EH. This explains the designation "partial isometry". It also follows from (2.41) that Ω is zero on the orthogonal complement of EH : if $f \in (EH)^\perp$, then $\|\Omega f\|^2 = \|Ef\|^2 = 0$, whence $\Omega f = \theta$. This may also be written as $\Omega(I - E)f = \theta$ for all $f \in H$, i.e. $\Omega = \Omega E$.

A partial isometry may also be defined by the property (2.41) :

PROPOSITION 2.10 : If Ω is a linear operator with $D(\Omega) = H$ and E a projection such that $\|\Omega f\| = \|Ef\|$ for all $f \varepsilon H$, then $\Omega^*\Omega = E$.

Proof : Since $\|\Omega f\| = \|Ef\|$ for all $f\varepsilon H$, one has $\|\Omega\| = \|E\|$, i.e. $\|\Omega\| = 1$ unless $E = 0$. By Proposition 2.8 $\|\Omega^*\| = \|E\|$, and Ω^* is defined everywhere. Now one obtains from the polarization identity (Problem 2.22) and (2.38) that for any $f,g\varepsilon H$

$$(\Omega^*\Omega f,g) = (\Omega f,\Omega g) =$$

$$\tfrac{1}{4}\{\|\Omega(f+g)\|^2 - \|\Omega(f-g)\|^2 - i\|\Omega(f+ig)\|^2 + i\|\Omega(f-ig)\|^2\} =$$

$$\tfrac{1}{4}\{\|E(f+g)\|^2 - \|E(f-g)\|^2 - i\|E(f+ig)\|^2 + i\|E(f-ig)\|^2 =$$

$$(Ef,Eg) = (Ef,g).$$

Hence $\Omega^*\Omega f - Ef$ is orthogonal to every $g\varepsilon H$, i.e. $\Omega^*\Omega f = Ef$ by Proposition 2.2. #

A particular case of a partial isometry is obtained by setting $E = I$. Ω is then isometric on all of H and is called an isometry.

The adjoint Ω^* of a partial isometry is also a partial isometry. In fact $F \equiv \Omega^{**}\Omega^* = \Omega\Omega^*$ is a projection, since $F^2 = \Omega\Omega^*\Omega\Omega^* = \Omega E\Omega^* = \Omega\Omega^* = F$ and $F^* = (\Omega\Omega^*)^* = \Omega^{**}\Omega^* = F$. Some additional properties of partial isometries that are essential for later chapters are collected in the following proposition.

PROPOSITION 2.11 : Let Ω be a partial isometry, $\Omega^*\Omega = E$, and define $F = \Omega\Omega^*$. Then

(a) $||\Omega|| = ||\Omega^*|| = 1$ unless $E = 0$.

(b) $\Omega E = \Omega$, $E\Omega^* = \Omega^*$, (2.42)

$F\Omega = \Omega$, $\Omega^* F = \Omega^*$. (2.43)

(c) The range of Ω is a subspace, and F is the orthogonal projection onto this subspace.

(d) The restriction of Ω to the subspace EH is invertible, and its inverse is given by Ω^* (more precisely by the restriction of Ω^* to FH).

Proof : (a) has been shown in the proof of Proposition 2.10. (b) $\Omega E = \Omega$ has already been proved. By using (2.35) and (2.32), one obtains from this $E\Omega^* = E^*\Omega^* = (\Omega E)^* = \Omega^*$. This proves (2.42). Next we have $F\Omega = \Omega\Omega^*\Omega = \Omega E = \Omega$. Also, from (2.42) : $\Omega^* F = \Omega^*\Omega\Omega^* = E\Omega^* = \Omega^*$, which proves (2.43).
(c) The definition $F = \Omega\Omega^*$ shows that the range of F is contained in that of Ω : If $f \in H$ and $f = Fg$, then $f = \Omega(\Omega^* g) \in \Omega H$. Thus $FH \subseteq \Omega H$. Similarly it follows from $F\Omega = \Omega$ that $\Omega H \subseteq FH$. Thus $\Omega H = FH$, and (c) is proved.
(d) Suppose $f \in EH$ and $\Omega f = \theta$. From (2.41) one then has $||f||^2 = ||\Omega f||^2 = 0$, i.e. $f = \theta$. Thus the restriction of Ω to EH is invertible. Since $\Omega^*\Omega f = f$ for f in EH, it is clear that its inverse is given by the restriction of Ω^* to FH. #

For a general partial isometry Ω, the projections E onto the so-called initial set EH of Ω and F onto the final set or range FH of Ω are both different from the identity operator. For an isometry, one has $E = I$, but F may still be different from I. In a finite-dimensional Hilbert space, the range of an isometry is the entire space (i.e. $F = I$), since it is a subspace that has the same dimension as the entire space. In an infinite-dimensional Hilbert space an isometry Ω may map the entire space onto a proper subspace of infinite dimension. An example of such an operator is the unilateral shift operator : Let $\{e_i\}_{i=1}^{\infty}$ be an orthonormal basis of H, define $\Omega e_i = e_{i+1}$ and extend this definition to finite linear combinations of the vectors $\{e_i\}$ by linearity. It follows from (2.10) that, if $f = \sum_{i=1}^{N} \alpha_i e_i$, then $\|\Omega f\|^2 = \|f\|^2$. By Problem 2.15 and Proposition 2.10, the closure of the above operator defines an isometric operator Ω with $D(\Omega) = H$. One sees that ΩH is the proper subspace spanned by $\{e_2, e_3, \ldots\}$, and e_1 is orthogonal to ΩH. Thus $F = I - F_{e_1} \neq I$, where F_{e_1} is the orthogonal projection onto the subspace determined by e_1. (One may also explicitly calculate Ω^* : $\Omega^* e_1 = \theta$ since $e_1 \varepsilon (\Omega H)^{\perp}$, and $\Omega^* e_i = e_{i-1}$ if $i \geq 2$ by Proposition 2.11(d).)

An isometry for which F is also the identity operator is called a unitary operator. Thus U is unitary if

$$D(U) = H \text{ and } U^*U = UU^* = I. \tag{2.44}$$

PROPOSITION 2.12 : Let U be unitary. Then

(a) The range of U is equal to H.

(b) U is invertible and $U^{-1} = U^*$. (2.45)

(c) $\|Uf - g\| = \|f - U^*g\|$ for all $f, g \varepsilon H$. (2.46)

Proof : (a) and (b) follow immediately from Proposition 2.11. (c) is easily verified by writing the norms as scalar products and using (2.44). #

We next consider symmetric and self-adjoint operators. A is called symmetric if D(A) is dense in H and $A \subseteq A^*$, i.e. if

$$(Af,g) = (g,Af) \text{ for all } f,g\in D(A). \tag{2.47}$$

A special case is that of a self-adjoint operator which is characterized by $A = A^*$. The condition $A = A^*$ is a severe restriction for the case of unbounded operators, since in addition to (2.47) it requires the domain of A^* to be exactly the same as that of A. For bounded operators this difficulty does not arise, since every bounded symmetric operator with $D(A) = H$ is self-adjoint (this follows immediately from the definitions). If A is bounded and symmetric and D(A) is only dense in H, then the closure $\bar{A}$ of A is self-adjoint. In fact one can easily check that in this case $\bar{A} = A^*$, which then implies together with (2.31) that $\bar{A} = \bar{A}^*$.

Since A^* is closed, the requirement $A \subseteq A^*$ implies that A has a closed extension. This means that a symmetric operator A is always closable. In addition $D(A^*)$ is dense, so that by Proposition 2.7 $\bar{A} = A^{**}$. By applying (2.34) to $A \subseteq A^*$ and using (2.31), one obtains $A^{**} \subseteq A^* = \bar{A}^*$. It follows that the closure $\bar{A}$ of a symmetric operator A is also symmetric. A self-adjoint operator $(A = A^*)$ is always closed. If one extends a symmetric operator A to a larger domain, the domain of the adjoint of the extension A' will be smaller than (or possibly equal to) $D(A^*)$, since a vector belonging to $D(A'^*)$ has to satisfy more conditions than a vector belonging to $D(A^*)$, as can be seen from (2.30). Thus one will have $A \subseteq A'$ and $A'^* \subseteq A^*$. Since $A \subseteq A^*$, it may happen that for certain extensions A' one will have $A' = A'^*$, i.e. that A' will be self-adjoint. For this reason the study of closed extensions of a symmetric operator is an important problem in functional analysis. We shall return to it in Chapter 8.

We wish to point out here that the number of self-adjoint extensions of a symmetric operator may be zero, finite, countably or even uncountably infinite.

The preceding remark may be illustrated by means of differential operators. We wish to associate various symmetric operators with the formal operator id/dx acting on functions defined on various subsets of $\mathcal{R}$. Let $\mathcal{D}$ be the set of all functions in $L^2(\mathcal{R})$ which are absolutely continuous on each finite interval [a,b] and such that $f\varepsilon L^2(\mathcal{R})$. Now consider first the Hilbert space $L^2(0,1)$. We define A_o to be the operator $(A_o f)(x) = if'(x)$, where $f\varepsilon D(A_o)$ if $f\varepsilon L^2(0,1)\cap\mathcal{D}$ and $f(0) = f(1) = 0$. By integrating by parts one may verify that A_o is symmetric; it is also closed, and $D(A_o^*) = L^2(0,1) \cap \mathcal{D}$ (no boundary condition is involved !). Hence $D(A_o^*)$ is strictly larger than $D(A_o)$. This operator A_o has an uncountable number of self-adjoint extensions, which are obtained by replacing the boundary condition $f(0) = f(1) = 0$ by the less restrictive condition $f(0) = \exp(i\phi)f(1)$ with $\phi\varepsilon[0,2\pi)$. Every ϕ determines a different self-adjoint extension of A_o [RN, no. 119],[AG, no. 49].

Similarly one may associate with the formal operator id/dx a symmetric operator A_1 in $L^2(0,\infty)$: f belongs to $D(A_1)$ if $f\varepsilon L^2(0,\infty)\cap\mathcal{D}$ and $f(0) = 0$. This operator has no self-adjoint extension at all [AG, no. 49]. Thirdly one may define A_2 in $L^2(\mathcal{R})$ by $(A_2 f)(x) = if'(x)$ and $D(A_2) = \mathcal{D}$. This symmetric operator is self-adjoint, i.e. it has exactly one self-adjoint extension.

Clearly the restriction of a self-adjoint operator B

to a dense subset of its domain is a symmetric operator A. If the dense subset is sufficiently large, B will just be the closure of this restriction. In that case A is said to be essentially self-adjoint. Generally a symmetric operator A is said to be <u>essentially self-adjoint</u> if $\bar{A}$ is self-adjoint. An equivalent definition of essential self-adjointness is easily seen to be $A^* = A^{**}$ (Problem 2.23). An essentially self-adjoint operator has one and only one self-adjoint extension. In fact, assume A' to be a self-adjoint extension of an essentially self-adjoint operator A. Then $A' = A'^* \subseteq A^* = A^{**} = \bar{A}$. Since $\bar{A}$ is the smallest closed extension of A, A' must be equal to the closure $\bar{A}$ of A.

The notion of essential self-adjointness is important since in applications one is often given a non-closed symmetric operator. If such an operator can then be shown to be essentially self-adjoint, it follows that it determines a unique self-adjoint operator.

It is useful to have a criterion for a symmetric operator to be self-adjoint or essentially self-adjoint. Such a criterion can be formulated in terms of the ranges $(A \pm i)D(A)$ of the operators $(A \pm i)$ *). We first prove an auxiliary result :

<u>LEMMA 2.13</u> : Let A be a symmetric operator, $\bar{A}$ its closure. Then the ranges of $\bar{A} \pm i$ are the closures of the ranges of $A \pm i$ respectively.

---

*) We sometimes use the notation $A + \alpha$ for the operator $A + \alpha I$ $(\alpha \in C)$.

Proof : (i) If $f \in D(\bar{A})$, there exists a sequence $\{f_n\} \in D(A)$ with $f_n \to f$ and $Af_n \to \bar{A}f$. Thus $(A \pm i)f_n \to (\bar{A} \pm i)f$, which shows that $(\bar{A} \pm i)D(\bar{A})$ is contained in the closure of the range of $(A \pm i)$ respectively.

(ii) Since A is symmetric, we have for $f \in D(A)$

$$\|(A \pm i)f\|^2 = \|Af\|^2 + \|f\|^2 \pm i(Af,f) \mp i(f,Af)$$

$$= \|Af\|^2 + \|f\|^2. \qquad (2.48)$$

Now suppose for instance that g belongs to the closure of $(A + i)D(A)$. Then there exists a sequence $\{f_n\} \in D(A)$ such that $(A + i)f_n \to g$. It follows from (2.48) that both $\{f_n\}$ and $\{Af_n\}$ are strong Cauchy sequences. Therefore, by Lemma 2.5, $h \equiv \text{s-lim}\, f_n \in D(\bar{A})$, and $g = \bar{A}h + ih$. Thus g belongs to the range of $\bar{A} + i$. This shows that the range of $(\bar{A} + i)$ is closed. The lemma follows by combining this with the result of (i). #

PROPOSITION 2.14 : A symmetric operator A in $\mathcal{H}$ is self-adjoint if and only if the range of both of the operators $A \pm i$ is $\mathcal{H}$.

Proof : (i) Suppose $(A \pm i)D(A) = \mathcal{H}$. Let $g \in D(A^*)$. Since $(A - i)D(A) = \mathcal{H}$, there exists $h \in D(A)$ such that $(A^* - i)g = (A - i)h$. Since A is symmetric, one has $(A - i)h = (A^* - i)h$, which leads to $(A^* - i)(g - h) = \theta$. Now for any $f \in D(A)$

$$0 = ((A^* - i)(g - h),f) = (g - h,(A + i)f).$$

Since $(A + i)D(A) = \mathcal{H}$, it follows from Proposition 2.2 that $g - h = \theta$, i.e. $g \in D(A)$. Thus $D(A^*) \subseteq D(A)$. Since $A \subseteq A^*$, we must have $A = A^*$.

(ii) Suppose $A = A^*$. By Lemma 2.13, $(A \pm i)D(A)$ are closed subspaces of $\mathcal{H}$. Thus, if for instance $(A + i)D(A) \neq \mathcal{H}$, there exists $g \in \{(A + i)D(A)\}^{\perp}$. It follows that $g \in D((A + i)^*) =$

$D(A^* - i) = D(A^*)$, and $(A^* - i)g = \theta$. Thus $(g,g) = (g,-iA^*g) = (iAg,g) = (iA^*g,g) = (-g,g) = -(g,g)$, whence $(g,g) = 0$, i.e. $g = \theta$. This proves that $(A + i)D(A) = H$. #

COROLLARY 2.15 : A symmetric operator A in H is essentially self-adjoint if and only if the range of both of the operators $A \pm i$ is dense in H.

Proof : By Lemma 2.13, range $(A \pm i)$ is dense in H if and only if range $(\bar{A} \pm i) = H$, i.e. according to Proposition 2.14 if and only if $\bar{A}$ is self-adjoint. #

In the next proposition we specify a class of operators that are always self-adjoint, namely the maximal multiplication operators by real functions in L^2-spaces. A certain converse of this will be given in Chapter 5 where we shall see that every self-adjoint operator is unitarily equivalent to a multiplication operator in some (more general) L^2-space.

The statement of the proposition involves the notion of the essential supremum of a measurable function $\psi : \Delta \to R$, where Δ is a measurable subset of R^n. ψ is said to be essentially bounded if there exists a number $M < \infty$ such that $|\psi(\underline{x})| \leq M$ for almost all $\underline{x}$ in Δ (with respect to Lebesgue measure). The essential supremum of ψ is the infimum of all numbers M verifying the above condition and will be denoted by $\|\psi\|_\infty$. The set of all essentially bounded measurable func- defined on Δ is denoted by $L^\infty(\Delta)$.

PROPOSITION 2.16 : Let Δ be a measurable set in R^n and ψ a real-valued measurable function defined on Δ which is finite almost everywhere (with respect to Lebesgue measure). Define an operator A in $L^2(\Delta)$ by

$$D(A) = \{f\varepsilon L^2(\Delta) | \psi(\underline{x})f(\underline{x})\varepsilon L^2(\Delta)\}$$

and $$(Af)(\underline{x}) = \psi(\underline{x})f(\underline{x}) \text{ for } f\varepsilon D(A).$$

Then (a) A is self-adjoint.

(b) A is bounded if and only if ψ is essentially bounded, and in that case $||A|| = \operatorname*{ess\,sup}_{\underline{x}\varepsilon\Delta} |\psi(\underline{x})| = ||\psi||_\infty$.

Remarks : (i) A is called maximal since D(A) is the maximal subset in $L^2(\Delta)$ on which multiplication by ψ makes sense in $L^2(\Delta)$.

(ii) The condition that ψ be a measurable function ensures that $\psi(\underline{x})f(\underline{x})$ is measurable, which is necessary if the latter function is to belong to $L^2(\Delta)$.

Proof : (i) Clearly D(A) is a linear manifold. We show that it is also dense. For this, we define the sets $\Delta_m \equiv \{\underline{x}\varepsilon\Delta | |\psi(\underline{x})| \leq m\}$, m = 1,2,... . Let $f\varepsilon L^2(\Delta_m)$. Then $f\varepsilon D(A)$, since

$$\int_{\Delta_m} |\psi(\underline{x})|^2 |f(\underline{x})|^2 d^n x \leq m^2 \int_{\Delta_m} |f(\underline{x})|^2 d^n x \leq m^2 ||f||^2. \quad (2.49)$$

Thus each of the subspaces $L^2(\Delta_m)$ of $L^2(\Delta)$ belongs to D(A).

Now if $h\varepsilon L^2(\Delta)$, then $\chi_{\Delta_m} h\varepsilon L^2(\Delta_m)$, and if $h\varepsilon[L^2(\Delta_m)]^\perp$, then $\chi_{\Delta_m} h\varepsilon[L^2(\Delta_m)]^\perp$. Thus if h is in $L^2(\Delta)$ as well as in $[L^2(\Delta_m)]^\perp$ for each m, $\chi_{\Delta_m} h$ is the zero vector in $L^2(\Delta_m)$ for each m by Proposition 2.2, i.e. $h(\underline{x}) = 0$ a.e. in $\bigcup_m \Delta_m$. Since the complement of $\bigcup_m \Delta_m$ in Δ is a set of measure zero (the set of points where $\psi(\underline{x})$ is not finite), we have $h(\underline{x}) = 0$ a.e. in Δ, i.e. $h = \theta$. Thus the only vector orthogonal to D(A) in $L^2(\Delta)$ is the vector θ, which means that D(A) is dense in $L^2(\Delta)$ (see the statement following the projection theorem on page 28).

(ii) We now prove (b). If ψ is essentially bounded and M its essential supremum, then one deduces as in (2.49) that $\|A\| \leq M$. If $M_o < M$, let $\Delta(M_o) \equiv \{\underline{x}\varepsilon\Delta \mid |(\psi(x)| \geq M_o\}$. The measure of $\Delta(M_o)$ is positive. If f is a function which is zero outside $\Delta(M_o)$, then $\|Af\| \geq M_o \|f\|$, i.e. $\|A\| \geq M_o$. Hence $M_o \leq \|A\| \leq M$ for each $M_o < M$, i.e. $\|A\| = M$.

If ψ is not essentially bounded, let $m > 0$ and $\Delta(m) = \{\underline{x}\varepsilon\Delta \mid |\psi(x)| \geq m\}$. $\Delta(m)$ has positive measure; hence as above there exists f in $L^2(\Delta)$ such that $\|Af\| \geq m \|f\|$. Since this is true for each $m > 0$, A is unbounded.

(iii) Finally we prove (a). Clearly A is symmetric, i.e. $(f,Ah) = (Af,h)$ for all $f,h\varepsilon D(A)$. For $f\varepsilon L^2(\Delta)$, define $f_\pm(\underline{x}) = f(\underline{x})[\psi(\underline{x}) \pm i]^{-1}$. Since ψ is real, one has $|\psi(\underline{x}) \pm i|^{-1} \leq 1$ and $|\psi(\underline{x})||\psi(\underline{x}) \pm i|^{-1} \leq 1$. These inequalities imply as in (2.49) that $f_\pm\varepsilon L^2(\Delta)$ and $f_\pm\varepsilon D(A)$ respectively. Clearly $(A \pm i)f_\pm = f$. This means that range $(A \pm i) = \mathcal{H}$, and by Proposition 2.14 A is self-adjoint. #

To conclude this section we consider convergence properties of sequences of linear operators. We shall mainly be concerned with sequences of bounded operators and therefore restrict the general definitions to this case.

In order to define the notion of convergence of a sequence of operators, one has recourse to the notion of convergence of a sequence of vectors. Thus if $\{A_n\}$ is a sequence of bounded operators with $D(A_n) = \mathcal{H}$ for all n, we say that A is the <u>strong limit</u> of A_n if for each $f\varepsilon\mathcal{H}$ the sequence of vectors $\{A_n f\}$ converges strongly to the vector Af, i.e. if

$$\lim_{n\to\infty} \|Af - A_n f\| = 0 \quad \text{for all } f \in H.$$

We then write $A_n \to A$ or $A = \text{s-}\lim_{n\to\infty} A_n$. Similarly A is the <u>weak limit</u> of $\{A_n\}$ if

$$\lim_{n\to\infty} (f, A_n g) = (f, Ag) \quad \text{for all } f, g \in H.$$

In this case we write $A = \text{w-}\lim_{n\to\infty} A_n$.

A third type of convergence is the <u>convergence in operator norm</u>. A sequence of bounded operators $\{A_n\}$ converges to A in this sense if $\|A_n - A\|$ converges to zero as $n \to \infty$. This is also called <u>uniform convergence</u> since it is equivalent to the requirement that $\text{s-}\lim A_n f = Af$ as $n \to \infty$ uniformly on the set $\{f \in H \mid \|f\| = 1\}$. We shall write $\text{u-}\lim A_n = A$ as $n \to \infty$. It is clear that the uniform convergence of a sequence $\{A_n\}$ to A implies its strong convergence to A. Also, by Proposition 2.1, the strong convergence of $\{A_n\}$ to A implies weak convergence of $\{A_n\}$ to A.

The Cauchy criterion is valid for each of the three kinds of convergence introduced above; e.g. if $\{A_n\}$ is a sequence of bounded operators with $D(A_n) = H$ and if for each $f \in H$ the sequence of vectors $\{A_n f\}$ is strongly Cauchy, then there exists a bounded linear operator A such that $A_n \to A$. The proof of this is based on the following interesting fact : if $\{A_n\}$ is a sequence of weakly convergent bounded operators, i.e. such that the sequence of scalar products $\{(f, A_n g)\}$ is a Cauchy sequence of complex numbers for all $f, g \in H$, then the sequence of norms $\{\|A_n\|\}$ is bounded, i.e. there exists $M < \infty$ such that $\|A_n\| \le M$ for all $n = 1, 2, \ldots$. A fortiori the above property is true for a sequence of strongly convergent and for a sequence of uniformly convergent bounded operators. Similarly a weakly convergent sequence of vectors $\{f_n\}$ is uniformly bounded. This result is known as the uniform boundedness principle and will not be proved here (cf. [AG, no. 29],[RN, no. 84],[RS, Thm. 6.1]). We shall occasionally use it to prove abstract theorems, but in most applications of these theorems the uniform boundedness of the occuring sequences can easily be verified directly.

The formulation of the asymptotic properties of scattering systems is based on strong convergence. For this reason we shall prove here the following two propositions concerning strong convergence. They will be used on various occasions. The first one asserts that, in order to establish the strong convergence of a uniformly bounded sequence of operators $\{A_n\}$, it is sufficient to verify the strong convergence of $\{A_n f\}$ for a fundamental set of vectors f in H (a subset N of H is called fundamental if the linear manifold consisting of all finite linear combinations of vectors belonging to N is dense in H. In particular an orthonormal basis of H and a dense subset of H are fundamental in H).

PROPOSITION 2.17 : Let $\{A_n\}$ be a sequence of linear operators with $D(A_n) = H$ and $\|A_n\| \leq M < \infty$ for all n = 1, 2, ... Let N be fundamental in H and suppose that $\{A_n f\}$ converges strongly for every $f \in N$. Then there exists a bounded linear operator A with $D(A) = H$, $\|A\| \leq M$ and $\text{s-lim}\, A_n = A$ as $n \to \infty$.

Proof : Let M be the dense linear manifold consisting of all finite linear combinations of vectors belonging to N. Then $\{A_n f\}$ converges strongly for every $f \in M$. The idea is to prove strong convergence of $\{A_n\}$ on H by approximating an arbitrary $g \in H$ by an element of M and then using the strong convergence of $\{A_n\}$ on M.

Let $\hat{A}$ be the linear operator defined by $D(\hat{A}) = M$ and $\hat{A}f = \text{s-lim}\, A_n f$ $(n \to \infty)$ for $f \in M$. We notice that $\|\hat{A}\| \le M$, since by Proposition 2.1 $\|\hat{A}f\| = \lim \|A_n f\| \le M\|f\|$ for all $f \in M$. Let A be the unique extension of $\hat{A}$ to all of H (Proposition 2.6). Then $\|A\| \le M$.

If $g \in H$, there exists a sequence $\{f_m\} \in M$ such that $f_m \to g$. From the triangle inequality (2.9) we find for any m and n

$$\|Ag - A_n g\| \le \|Ag - Af_m\| + \|\hat{A}f_m - A_n f_m\| + \|A_n f_m - A_n g\|$$
$$\le M\|g - f_m\| + \|\hat{A}f_m - A_n f_m\| + M\|g - f_m\|.$$

Given $\delta > 0$, one may choose first m so large that $\|g - f_m\| < \delta/4M$. Since $f_m \in M$, there exists $N < \infty$ such that for all $n \ge N$: $\|\hat{A}f_m - A_n f_m\| < \delta/2$. Thus $\|Ag - A_n g\| < \delta$ for all $n \ge N$, i.e. $A_n g \to Ag$. #

<u>PROPOSITION 2.18</u> : Let $\{A_n\}$, $\{B_n\}$, A and B be bounded operators defined on all of H. (a) If $\text{s-lim}\, A_n = A$ and $\text{s-lim}\, B_n = B$, then $\text{s-lim}\, A_n B_n = AB$ as $n \to \infty$. (b) If $\text{u-lim}\, A_n = A$ and $\text{u-lim}\, B_n = B$, then $\text{u-lim}\, A_n B_n = AB$ as $n \to \infty$.

<u>Proof</u> : It is based on the triangle inequality. If $f \in H$, then

$$\|ABf - A_n B_n f\| = \|(A - A_n)Bf + A_n(B - B_n)f\| \le \|(A - A_n)Bf\| + \|A_n\|\,\|(B - B_n)f\| \le \|(A_n - A)Bf\| + M\|(B-B_n)f\|. \qquad (2.50)$$

Here we have used the fact mentioned before Proposition 2.17 that there exists $M < \infty$ such that $\|A_n\| \le M$ for all n. Each term on the right-hand side of (2.50) converges to zero as $n \to \infty$ by the hypotheses, which proves that $A_n B_n f \to ABf$, establishing (a). The proof of (b) is left as an exercise (Problem 2.24). #

As an immediate consequence of this proposition, one sees that the action of a bounded operator can always be interchanged with strong limits :

<u>COROLLARY 2.19</u> : Let A be a bounded operator with $D(A) = \mathcal{H}$. Suppose $D(B_n) = \mathcal{H}$ and $\text{s-lim } B_n = B$ as $n \to \infty$. Then $\text{s-lim } AB_n = AB$ and $\text{s-lim } B_n A = BA$ as $n \to \infty$.

The following result is a direct consequence of the definitions (Problem 2.24) : If a sequence of bounded and everywhere defined operators $\{A_n\}$ converges strongly to A, then the adjoint sequence $\{A_n^*\}$ converges weakly to A^*. The adjoint sequence need not be strongly convergent though. We shall encounter this question again in Chapter 4 and discuss some special cases there.

<u>PROPOSITION 2.20</u> : Let A be everywhere defined and $\|A\| < 1$. Then $I - A$ is invertible. Its inverse is bounded, defined everywhere, given by the uniformly convergent series (called the <u>Neumann series</u>)

$$(I - A)^{-1} = \sum_{n=0}^{\infty} A^n = I + A + A^2 + \dots \qquad (2.51)$$

and verifies $\|(I-A)^{-1}\| \leq (1 - \|A\|)^{-1}$.

The proof of this result is left as an exercise (Problem 2.25).

Finally, we add a few comments concerning strong convergence when unbounded operators are involved. There are two cases that can be considered here. Firstly, a sequence $\{A_n\}$ of bounded operators may converge strongly to an unbounded operator, i.e. one may have $A_n f \to Af$ for every $f \in D(A)$, where A may be an unbounded operator. In this case the sequence of norms $\{\|A_n\|\}$ cannot be bounded. An example of this will be seen in Section 2-4. Secondly the operators A_n may themselves be unbounded and converge strongly on a subset of $\mathcal{H}$ which is common to their domains, i.e. the sequence of vectors $\{A_n f\}$ may be strongly Cauchy for each f in $\bigcap_{n=N}^{\infty} D(A_n)$ for some N (which may depend on f). The limit may define a bounded or an unbounded operator. Whenever unbounded operators are involved in statements using strong convergence, we shall explicitly specify the set of vectors on which convergence takes place. When we speak simply of strong convergence of operators, it is always understood that only bounded operators occur.

2-3 COMPACT OPERATORS

In this section we define compact operators, Hilbert-Schmidt operators and trace class operators and derive some consequences of these definitions. The aim is not to give a complete theory of these classes of operators but rather to assemble those of their properties that will be needed in later chapters. A few of the lengthier proofs will be given only in Section 2-5.

Consider an operator of the form

$$Tf = \sum_{i=1}^{N}(e_i,f)h_i, \text{ with } D(T) = H, \tag{2.52}$$

where $\{e_i,h_i\}$ are 2N vectors in H and $N < \infty$. The range of T is the finite-dimensional subspace spanned by $h_1,\ldots,h_N$, and $Tf = \theta$ if f is orthogonal to the subspace spanned by $e_1,\ldots, e_N$. Thus T may be viewed as an operator that acts on the finite-dimensional subspace M spanned by $\{e_i,h_i\}$ in the sense that it is zero on $M^{\perp}$ and its range also lies in M. T may be described by a matrix acting on vectors in M. An operator of the form (2.52) with $N < \infty$ is called a <u>finite rank operator</u>. (An equivalent definition is obtained by requiring that the range of T be finite-dimensional.)

We first give some simple properties of finite rank operators. Henceforth $\mathcal{B}(H)$ will denote the set of all bounded and everywhere defined linear operators in H.

<u>LEMMA 2.21</u> : Let T be a finite rank operator. Then
(a) T^* is a finite rank operator.
(b) If $B \in \mathcal{B}(H)$, then BT and TB are finite rank operators.
(c) If T_1 is of finite rank, then so is $T + \alpha T_1 (\alpha \in C)$.

<u>Proof</u> : (b) and (c) are immediate from the definition. To prove (a), let $f,g \in H$. It follows from (2.30) and (2.52) that

$$(f,T^*g) = (Tf,g) = \sum_{i=1}^{N}(f,e_i)(h_i,g) = (f,\sum_{i=1}^{N}(h_i,g)e_i).$$

Hence by Proposition 2.2

$$T^*g = \sum_{i=1}^{N}(h_i,g)e_i, \tag{2.53}$$

which proves that T^* is of finite rank. #

An operator A in $\mathcal{B}(\mathcal{H})$ is said to be <u>compact</u>[*)] if there exists a sequence $\{T_N\}$ of finite rank operators such that $\|A - T_N\| \to 0$ as $N \to \infty$. It follows from this definition that each finite rank operator is compact and that in a finite-dimensional Hilbert space every operator is compact. We shall denote by $\mathcal{B}_0$ the set of all compact operators. Let us prove some consequences of the preceding definition.

<u>PROPOSITION 2.22</u> :

(a) A is compact if and only if A^* is compact.

(b) If $A\varepsilon\mathcal{B}_0$ and $B\varepsilon\mathcal{B}(\mathcal{H})$, then $AB\varepsilon\mathcal{B}_0$ and $BA\varepsilon\mathcal{B}_0$.

(c) If A_1 and A_2 are compact, so is $A_1 + \alpha A_2$ ($\alpha\varepsilon\mathbb{C}$).

(d) If $\{A_n\}\varepsilon\mathcal{B}_0$, $A\varepsilon\mathcal{B}(\mathcal{H})$ and $\|A-A_n\| \to 0$ as $n\to\infty$, then $A\varepsilon\mathcal{B}_0$.

<u>Proof</u> : Let $\{T_N\}$ be a sequence of finite rank operators converging uniformly to A.

(a) This follows from the fact that $\|A-T_N\| = \|A^*-T_N^*\|$ (cf. Proposition 2.8) and Lemma 2.21(a).

(b) By Lemma 2.21(b) BT_N and T_NB are finite rank operators. Now by (2.28) $\|BA-BT_N\| \leq \|B\|\,\|A-T_N\| \to 0$, which shows that BA is compact; similarly one proves $AB\varepsilon\mathcal{B}_0$.

(c) This follows from the triangle inequality (2.27) and Lemma 2.21 (c).

---

*) This definition is equivalent to the customary definition of compactness : A is compact iff for every bounded sequence $\{f_n\}$ the sequence $\{Af_n\}$ has a strongly convergent subsequence [RS, Section VI.5]. This latter characterization of compactness is valid in more general metric spaces than Hilbert space, but the one given above is more convenient for our purposes.

(d) Let $\eta > 0$. For each n we choose a finite rank operator $T_{N(n)}$ such that $||A_n - T_{N(n)}|| < \eta/2$. We next choose n so large that $||A-A_n|| < \eta/2$. Then by (2.27) $||A-T_{N(n)}|| \leq ||A-A_n|| + ||A_n - T_{N(n)}|| < \eta$. Hence $||A-T_{N(n)}|| \to 0$ as $n\to\infty$, which proves that A is compact. #

PROPOSITION 2.23 : Let $\{f_n\}$ be a sequence of vectors which converges weakly to θ, and let A be a compact operator. Then $\{Af_n\}$ converges strongly to θ.

Proof : Since $\{f_n\}$ is uniformly bounded by the uniform boundedness principle, there exists $M < \infty$ such that $||f_n|| \leq M$ for all n. Given $\eta > 0$, choose a finite rank operator T of the form (2.52) such that $||A-T|| < \eta/2M$. It follows from (2.9) and (2.52) that

$$||Af_n|| \leq ||(A-T)f_n|| + ||Tf_n|| < \eta/2 + \sum_{i=1}^{N} |(e_i,f_n)|\,||h_i||.$$

Since $\text{w-lim}\, f_n = \theta$, there exists n_o such that $|(e_i,f_n)|\,||h_i|| < \eta/2N$ for all $i = 1,\ldots, N$ and all $n > n_o$. Hence $||Af_n|| < \eta$ for $n > n_o$, which proves that $||Af_n|| \to 0$ as $n \to \infty$. #

Compact operators share the following property with operators acting in a finite-dimensional space : Consider the equation $f-Af = g$ where g is a given vector, A is compact and f is to be determined. Then either the homogeneous equation $f-Af = \theta$ has a non-trivial solution or else the equation $f-Af = g$ has, for any given $g\in H$, a unique solution $f\in H$, namely $f = (I-A)^{-1}g$ ($(I-A)^{-1}$ is then bounded and defined everywhere). The preceding result is known as the Fredholm alternative and follows immediately from Proposition 2.24

below, the proof of which can be found in Section 2-5. If A is for instance a compact integral operator in an L^2-space (cf. (2.65) for its definition), the above is a result about the existence and the uniqueness of the solution of an integral equation. Of course the operator $(I-A)^{-1}$ cannot in general be written down explicitly. However, if for instance $\|A\| < 1$, the solution may be written as a power series in A by using Proposition 2.20.

For a non-compact operator the above alternative need not hold, since $(I-A)^{-1}$ may exist but be an unbounded operator, so that $f-Af = g$ is not solvable in $\mathcal{H}$ for every g. This happens for instance if A is self-adjoint and the point $\lambda = 1$ belongs to its continuous spectrum (Problem 2.26).

<u>PROPOSITION 2.24</u> : Let A be a compact operator and $z \varepsilon \mathcal{C}$, $z \neq 0$. Then either the equation $Af = zf$ has a solution $f \neq \theta$ in $\mathcal{H}$ or $(zI-A)^{-1}$ exists and belongs to $\mathcal{B}(\mathcal{H})$.

We shall now discuss some spectral properties of compact operators. (A general definition of the spectrum of an operator is given in Section 5-6). In an infinite-dimensional Hilbert space the spectrum of a compact operator consists of isolated*) non-zero eigenvalues**) and of the point $z = 0$. The latter may itself be an eigenvalue or an accumulation point of eigenvalues or both. This result is the content of

---

*) z_o is an <u>isolated</u> eigenvalue if there exists a neighborhood of z_o containing no eigenvalue other than z_o.

**) The number $z \varepsilon \mathcal{C}$ is an <u>eigenvalue</u> of a linear operator A if there exists a vector $f \neq \theta$ in D(A) such that $Af = zf$. f is called an <u>eigenvector</u> of A.

Proposition 2.25 the proof of which is also deferred to Section 2-5.

PROPOSITION 2.25 : If $A \in \mathcal{B}_o$, then each non-zero eigenvalue of A is of finite multiplicity (i.e. the corresponding space of eigenvectors is finite-dimensional). Furthermore the only possible accumulation point of the eigenvalues of A is the point $z = 0$.

If $A \in \mathcal{B}_o$ is self-adjoint, it is possible to choose an orthonormal basis $\{e_i\}$ of $\mathcal{H}$ such that each e_i is an eigenvector of A, i.e. verifying $Ae_i = \lambda_i e_i$ for some (real) λ_i. This is nothing but the spectral theorem for a self-adjoint operator A for the case where A is compact and will be deduced below. It should be said here that for a non-compact self-adjoint operator B such a basis need not exist, since B may also have continuous spectrum (the relevant details will be explained in Chapter 5). We first give a characterization of the eigenvalues and the eigenvectors of a general self-adjoint operator and some lemmas needed to prove the spectral theorem.

PROPOSITION 2.26 : Let A be self-adjoint. Then

(a) All eigenvalues of A are real.

(b) If $Af_1 = \lambda_1 f_1$ and $Af_2 = \lambda_2 f_2$ with $\lambda_1 \neq \lambda_2$, then f_1 is orthogonal to f_2.

Proof : (a) Suppose $Af = \lambda f$ with $f \neq \theta$. By using (2.30) we then obtain $\lambda(f,f) = (f,\lambda f) = (f,Af) = (Af,f) = (\lambda f,f) = \bar{\lambda}(f,f)$, which implies that $\bar{\lambda} = \lambda$.

(b) One obtains as in (a) that

$$(\lambda_1 - \lambda_2)(f_1,f_2) = (Af_1,f_2) - (f_1,Af_2) = 0.$$

Since $\lambda_1 \neq \lambda_2$, this implies that $(f_1,f_2) = 0$. #

LEMMA 2.27 : Let E be a subset of H and denote by M the subspace spanned by E (i.e. the closure of the set $\mathcal{D}$ of all finite linear combinations of vectors belonging to E). Let $A,B\varepsilon\mathcal{B}(H)$ and suppose that $Af = Bf$ for all $f\varepsilon E$. Then $A = B$ on M.

Proof : The hypotheses imply that $Af = Bf$ for all $f\varepsilon\mathcal{D}$. Hence $Af = Bf$ for all f in M by the uniqueness of the closure (Proposition 2.6). #

LEMMA 2.28 : Let $\{F_k\}$, $k = 1,\ldots,n$ be a set of projections with mutually orthogonal ranges, i.e. verifying $F_jF_k = \delta_{jk}F_k$, and let $\alpha_k\varepsilon C$. Then for all $f\varepsilon H$

$$\|\sum_k \alpha_k F_k f\|^2 \leq \sup_k |\alpha_k|^2 \, \|f\|^2. \tag{2.54}$$

Proof : This follows by applying (2.10) and Bessel's inequality (2.11) :

$$\|\sum_k \alpha_k F_k f\|^2 = \sum_k |\alpha_k|^2 \|F_k f\|^2 \leq \sup_k |\alpha_k|^2 \sum_j \|F_j f\|^2 \leq \sup_k |\alpha_k|^2 \|f\|^2. \ \#$$

LEMMA 2.29 : Let A be a compact self-adjoint operator, and suppose that A has no non-zero eigenvalue. Then $A = 0$.

The proof of this lemma will be indicated in Section 2-5. We shall use it to establish the spectral theorem for compact self-adjoint operators. Suppose $A = A^*$ and $A\varepsilon\mathcal{B}_o$. Let $\{\lambda_k\}$ be an enumeration of all non-zero eigenvalues of A such that $|\lambda_{k+1}| \leq |\lambda_k|$ for all $k = 1,2,\ldots$. We denote by M_k the subspace spanned by all eigenvectors corresponding to the eigenvalue λ_k, by M_o the subspace $M_o = \{f|Af = \theta\}$ and by

$E\{\lambda_k\}$ the orthogonal projection whose range is M_k. One has dim $M_k < \infty$ for $k \neq 0$ by Proposition 2.25 and $E\{\lambda_j\}E\{\lambda_k\} = \delta_{jk}E\{\lambda_k\}$ by Proposition 2.26 (i.e. M_k is orthogonal to M_j if $k \neq j$). The spectral theorem may now be stated as follows.

PROPOSITION 2.30 : Suppose A is self-adjoint and compact. For each $k = 0,1,\ldots$, let $\{e_k^{(i)}\}$ be an orthonormal basis of M_k. Then the set $\{e_k^{(i)}\}_{i,k}$ is an orthonormal basis of H. Furthermore

$$A = \sum_{k\neq 0} \lambda_k E\{\lambda_k\}, \tag{2.55}$$

the sum being convergent in the uniform operator topology.

Proof : (i) Let M be a subspace which is invariant under A, i.e. such that $Af\in M$ for each $f\in M$. Let $g\in M^{\perp}$, $f\in M$. Then $(Ag,f) = (g,Af) = 0$, which shows that $Ag\in M^{\perp}$, or that $M^{\perp}$ is also invariant under A.

(ii) Let M_+ be the subspace spanned by $\{e_k^{(i)}\}_{i,k}$ with $k \neq 0$, F the orthogonal projection with range M_+ and $F' = I-F$. Clearly A leaves M_+ invariant, hence $AM_+^{\perp} \subseteq M_+^{\perp}$ by (i). This means that

$$F'AF' = AF'. \tag{2.56}$$

Two consequences of (2.56) are : (a) The operator AF' is self-adjoint. (b) If $AF'f = \lambda f$ for some $\lambda \neq 0$, then $F'f = f$ and hence $Af = \lambda f$. Therefore, since all eigenvectors of A corresponding to a non-zero eigenvalue lie in M_+, AF' is a self-adjoint compact operator having no non-zero eigenvalue. Thus $AF' = 0$ by Lemma 2.29, or in other words $M_+^{\perp} = M_0$. This shows that the eigenvectors of A (including those in M_o) span H.

(iii) It is easily seen that the series in (2.55) converges uniformly by using Lemma 2.28 and the fact that $|\lambda_k| \to 0$ as $k \to \infty$ if A has an infinite number of non-zero eigenvalues. Its limit is some operator B in $B(H)$. If e is any eigenvector of A, then clearly $Ae = Be$. The fact that $A = B$ on H now follows from the result of (ii) and Lemma 2.27. #

We next derive a canonical form for an arbitrary compact operator. If $A \varepsilon B_o$, then $A^*A \varepsilon B_o$ by Proposition 2.22. Furthermore A^*A is self-adjoint and positive, i.e. $(f, A^*Af) = ||Af||^2 \geq 0$ for all $f \varepsilon H$. The preceding identity also shows that $A^*Af = \theta$ implies $Af = \theta$.

It follows that all eigenvalues of A^*A are real and non-negative, and each non-zero eigenvalue has finite multiplicity. Let $\mu_1 \geq \mu_2 \geq \cdots$ be an enumeration of the non-zero eigenvalues of A^*A such that each of them appears as many times as its multiplicity. Let $\{e_{j,o}\}$ be an orthonormal basis of the subspace $M_o = \{f | A^*Af = \theta\}$. By Proposition 2.30 there exists an orthonormal set $\{e_k\}$ such that $\{e_k, e_{j,o}\}$ is a basis of H and such that

$$A^*Ae_k = \mu_k e_k \,, \quad A^*Ae_{k,o} = \theta. \tag{2.57}$$

Let $\lambda_k = \mu_k^{1/2}$. The numbers $\{\lambda_k\}$ are called the <u>singular values</u> of the compact operator A. The following characterization of compact operators is a generalization of (2.52).

PROPOSITION 2.31 (Canonical expansion of compact operators) : Let $A \epsilon \mathcal{B}_o$. Then there exist two orthonormal sets $\{e_k\}$ and $\{h_k\}$ such that for all $f \epsilon \mathcal{H}$

$$Af = \sum_k \lambda_k (e_k, f) h_k, \tag{2.58}$$

where $\{\lambda_k\}$ are the singular values of A and the infinite sum (viewed as the limit of a sequence of operators) converges in operator norm.

Proof : Let $\{e_k, e_{j,o}\}$ be the orthonormal basis used in (2.57), and define $h_k = \lambda_k^{-1} Ae_k$. One has

$$(h_j, h_k) = \lambda_j^{-1} \lambda_k^{-1} (e_j, A^*Ae_k) = \lambda_j^{-1} \lambda_k (e_j, e_k) = \delta_{jk}.$$

Thus $\{h_k\}$ is an orthonormal set.

Define A_N by (2.58) with the sum running from $k = 1$ to $k = N$. Let $M > N$. By using (2.10) and (2.11) one obtains

$$\|(A_M - A_N)f\|^2 = \sum_{k=N+1}^{M} \lambda_k^2 |(e_k, f)|^2 \leq \lambda_{N+1}^2 \sum_{k=N+1}^{M} |(e_k, f)|^2$$

$$\leq \lambda_{N+1}^2 \|f\|^2.$$

Since $\lambda_N \to 0$ as $N \to \infty$, $\{A_N\}$ is a Cauchy sequence in the uniform operator topology. Denote by B its limit. Then

$$Be_j = \lim_{N \to \infty} \sum_{k=1}^{N} \lambda_k (e_k, e_j) h_k = \lambda_j h_j = Ae_j.$$

Clearly $Be_{j,o} = \theta$, and we have already seen that $Ae_{j,o} = \theta$. Hence B and A coincide on an orthonormal basis, i.e. $B = A$ by Lemma 2.27. #

In order to prove that a given operator is compact,

one may try to approximate it uniformly by a sequence of finite rank operators or by a sequence of operators that are already known to be compact (Proposition 2.22(d)). For this the Hilbert-Schmidt operators are very useful. Firstly they form a subset of the class of compact operators, and secondly in an L^2-space they have a simple characterization as integral operators (cf. Proposition 2.33), so that in many instances it is easy to decide whether a given operator belongs to the Hilbert-Schmidt class or not.

To define this class of operators, we introduce the Hilbert-Schmidt norm $\|A\|_{HS}$ of an operator A in $\mathcal{B}(\mathcal{H})$:

$$\|A\|_{HS}^2 = \sum_k \|Ae_k\|^2, \tag{2.59}$$

where $\{e_k\}$ is an orthonormal basis of $\mathcal{H}$. A is said to be a Hilbert-Schmidt operator if $\|A\|_{HS} < \infty$, and the set of all Hilbert-Schmidt operators will be denoted by $\mathcal{B}_2$.

In the above definition the quantity $\|A\|_{HS}$ appears to depend on the choice of an orthonormal basis $\{e_k\}$. We shall now show that the sum in (2.59) is the same for each orthonormal basis of $\mathcal{H}$. For this, let $\{g_k\}$ be an arbitrary orthonormal basis. By using the Parseval relation (2.12), (2.30) and again (2.12) one obtains

$$\begin{aligned}\sum_k \|Ag_k\|^2 &= \sum_k \sum_i |(e_i,Ag_k)|^2 = \sum_i \sum_k |(A^*e_i,g_k)|^2 \\ &= \sum_i \|A^*e_i\|^2,\end{aligned} \tag{2.60}$$

where the change of the order of summation is permitted because only non-negative terms are involved. Since $\{e_i\}$ may be thought of as being fixed, (2.60) shows that the value of $\sum_k \|Ag_k\|^2$ is the same for each orthonormal basis $\{g_k\}$. (2.60) also implies that

$$\|A\|_{HS} = \|A^*\|_{HS}. \tag{2.61}$$

PROPOSITION 2.32 :

(a) $A \in \mathcal{B}_2$ if and only if $A^* \in \mathcal{B}_2$.

(b) $$\|A\| \leq \|A\|_{HS}. \tag{2.62}$$

(c) Every Hilbert-Schmidt operator is compact, i.e. $\mathcal{B}_2 \subseteq \mathcal{B}_0$.

(d) If $A \in \mathcal{B}_2$ and $B \in \mathcal{B}(\mathcal{H})$, then $AB \in \mathcal{B}_2$ and $BA \in \mathcal{B}_2$.

(e) If $A_1 \in \mathcal{B}_2$ and $A_2 \in \mathcal{B}_2$, then $(A_1 + \alpha A_2) \in \mathcal{B}_2$ $(\alpha \in \mathcal{C})$.

Proof : (a) follows from (2.61). To prove (b), fix $f \neq \theta$ and choose an orthonormal basis $\{e_k\}$ such that $e_1 = f/\|f\|$. Then

$$\frac{\|Af\|^2}{\|f\|^2} = \|Ae_1\|^2 \leq \sum_k \|Ae_k\|^2 = \|A\|^2_{HS}.$$

Since this inequality holds for each $f \neq \theta$, (b) follows from (2.24).

(c) Let $A \in \mathcal{B}_2$ and fix an orthonormal basis $\{e_k\}$. For each $N < \infty$ we define a finite rank operator T_N by $T_N f = \sum_{i=1}^N (e_i, f) Ae_i$. Then $T_N e_k = 0$ if $k > N$ and $T_N e_k = Ae_k$ if $k \leq N$. Thus $\|A - T_N\|^2_{HS} = \sum_{k=N+1}^\infty \|Ae_k\|^2$. It follows from (2.62) and the hypothesis $A \in \mathcal{B}_2$ that

$$\lim_{N\to\infty} \|A - T_N\|^2 \leq \lim_{N\to\infty} \sum_{k=N+1}^{\infty} \|Ae_k\|^2 = 0,$$

which proves that A is compact.

(d) We have

$$\|BA\|_{HS}^2 = \sum_k \|BAe_k\|^2 \leq \|B\|^2 \sum_k \|Ae_k\|^2 = \|B\|^2 \|A\|_{HS}^2. \quad (2.63)$$

Hence $BA\varepsilon\mathcal{B}_2$. Since $A^*\varepsilon\mathcal{B}_2$ by part (a) and $B^*\varepsilon\mathcal{B}(\mathcal{H})$ by Proposition 2.8, this also implies that $B^*A^*\varepsilon\mathcal{B}_2$. By applying part (a) again, we get $(B^*A^*)^* = AB\varepsilon\mathcal{B}_2$.

(e) One has for $f,g\varepsilon\mathcal{H}$

$$\|f + g\|^2 \leq \|f + g\|^2 + \|f - g\|^2 = 2\|f\|^2 + 2\|g\|^2. \quad (2.64)$$

It follows that $\|A_1 + \alpha A_2\|_{HS}^2 \leq 2\|A_1\|_{HS}^2 + 2|\alpha|^2 \|A_2\|_{HS}^2$. #

We now consider integral operators in $\mathcal{H} = L^2(\mathcal{R}^n)$. Let $K_A : \mathcal{R}^{2n} \to \mathcal{C}$ be a measurable function. Then the operator A given by

$$(Af)(\underline{x}) = \int d^n y K_A(\underline{x},\underline{y}) f(\underline{y}) \quad (2.65)$$

is called an <u>integral operator</u> and K_A the <u>kernel</u> of A. The domain of A consists of those functions f in $L^2(\mathcal{R}^n)$ for which the integral in (2.65) exists for almost every $\underline{x}$ and for which the function $(Af)(\underline{x})$ defined by (2.65) belongs to $L^2(\mathcal{R}^n)$.

The function K_A is said to be a <u>Hilbert-Schmidt kernel</u> if

$$M_A \equiv \int d^n x d^n y |K_A(\underline{x},\underline{y})|^2 < \infty. \quad (2.66)$$

One has the following interesting result :

<u>PROPOSITION 2.33</u> : Let $\mathcal{H} = L^2(\mathcal{R}^n)$.

(a) If A is an integral operator with Hilbert-Schmidt kernel K_A, then A is a Hilbert-Schmidt operator and

$$||A||^2_{HS} = \int d^n x d^n y |K_A(\underline{x},\underline{y})|^2. \tag{2.67}$$

(b) If A is a Hilbert-Schmidt operator, then A is an integral operator with Hilbert-Schmidt kernel.

The proof of this result is given in Section 2-5. One can see from it that an analogous theorem is valid in a general L^2-space. If one evaluates $||A||_{HS}$ in the basis used in (2.57), one obtains

$$||A||^2_{HS} = \sum_k |(e_k, A^*Ae_k)| = \sum_k \lambda_k^2. \tag{2.68}$$

Thus $||A||^2_{HS}$ is the sum of the squares of the singular values of A.

We now pass on to the trace class operators. For their definition we first introduce the absolute value $|A|$ of a compact operator A. This is done by starting from the orthonormal basis $\{e_k, e_{j,o}\}$ used in (2.57) and defining $|A|e_k = \lambda_k e_k$, $|A|e_{j,o} = \theta$, where $\{\lambda_k\}$ are the singular values of A. This definition is extended by linearity to the set $\mathcal{D}$ of finite linear combinations of these basis vectors, and $|A|$ is the closure of the operator defined in this manner on $\mathcal{D}$ (Proposition 2.6).

We now define the <u>trace norm</u> of $A \varepsilon \mathcal{B}_o$ by

$$|||A|||_1 = \sum_k \lambda_k = \sum_k (e_k, |A|e_k) \equiv \mathrm{Tr}|A|. \tag{2.69}$$

The <u>trace</u> of a positive operator B is defined as $\mathrm{Tr}B = \sum_k (g_k, Bg_k)$, where $\{g_k\}$ is an orthonormal basis of $\mathcal{H}$. Since $\mathrm{Tr}B = ||B^{1/2}||^2_{HS}$, it is independent of the basis $\{g_k\}$[*)]. An

---

*) The square-root of a bounded positive operator will be defined in Lemma 5.5.

operator A in $\mathcal{B}_0$ is said to be of trace class if $|||A|||_1$ is finite. The set of all trace class operators will be denoted by $\mathcal{B}_1$.

The derivation of a fair number of the deeper results in scattering theory involves trace class operators. Since we shall not reproduce such arguments in this book, we refrain from proving here the basic properties of trace class operators and simply mention a few of them. A notable one is that an operator belongs to the trace class if and only if it can be written as the product of two Hilbert-Schmidt operators. Such a factorization is the usual method of showing that an operator is trace class. We collect this factorization property in the following proposition. Its proof is based on the canonical expansion of compact operators and can be found in Section 2-5.

PROPOSITION 2.34 :

(a) A belongs to $\mathcal{B}_1$ if and only if $A = BC$ with $B,C\varepsilon\mathcal{B}_2$.

(b) If $A\varepsilon\mathcal{B}_1$ and $T\varepsilon\mathcal{B}(\mathcal{H})$, then $AT\varepsilon\mathcal{B}_1$ and $TA\varepsilon\mathcal{B}_1$.

We may now define the trace of an arbitrary trace class operator by

$$\operatorname{Tr} A = \sum_k (g_k, Ag_k) \tag{2.70}$$

where $\{g_k\}$ is an arbitrary orthonormal basis of $\mathcal{H}$. We have to show that Tr A is finite and independent of the basis $\{g_k\}$. For this we write $A = BC$ with $B,C\varepsilon\mathcal{B}_2$. By using the polarization identity (2.97) one easily gets that

$$4 \sum_k (g_k, Ag_k) = 4 \sum_k (B^* g_k, Cg_k)$$
$$= \|B^*+C\|^2_{HS} - \|B^*-C\|^2_{HS} - i\|B^*+iC\|^2_{HS} + i\|B^*-iC\|^2_{HS}.$$

One sees that the last member of this equation is independent of the basis $\{g_k\}$ and finite by Proposition 2.32. Together with (2.61) the above identity also implies that, if $B, C \in \mathcal{B}_2$, then Tr BC = Tr CB.

For an operator A in $\mathcal{B}(\mathcal{H})$ to be of trace class it is not sufficient that the sum in (2.70) be convergent (or even absolutely convergent) for some orthonormal basis $\{g_k\}$. A counter-example is the unilateral shift operator in an infinite-dimensional Hilbert space defined by $\Omega g_k = g_{k+1}$, where $\{g_k\}$ is a fixed orthonormal basis. One has $(g_k, \Omega g_k) = 0$ for all k, so that the sum in (2.70) is absolutely convergent. However $\Omega \notin \mathcal{B}_1$ ($\Omega \in \mathcal{B}_1$ would imply $\Omega^*\Omega = I \in \mathcal{B}_1$). In order to conclude that $A \in \mathcal{B}_1$ the sum in (2.73) must be absolutely convergent for every orthonormal basis of $\mathcal{H}$.

We end this section with the following definitions : A $\mathcal{B}(\mathcal{H})$-valued functions $s \mapsto A(s)$ is called <u>strongly continuous</u>, <u>weakly continuous</u>, <u>norm continuous</u> or <u>continuous in Hilbert-Schmidt</u> norm if s-lim A(s) = A(t), w-lim A(s) = A(t), u-lim A(s) = A(t) or $\|A(s)-A(t)\|_{HS} \to 0$ respectively as $s \to t$.

2-4 DIRECT SUMS AND TENSOR PRODUCTS OF HILBERT SPACES

In this section we indicate two methods of constructing from a given family $\mathcal{H}_1, \ldots, \mathcal{H}_n$ of Hilbert spaces a new

Hilbert space H. The idea is essentially to take as elements of H the set $H_1 \times H_2 \times \cdots \times H_n$ of n-tuples $\{f_1, \ldots, f_n\}$ formed of elements f_i of H_i and to define a scalar product between such n-tuples by taking either the sum or the product of the respective scalar products in H_i. This leads to the direct sum and the tensor product of $H_1, \ldots, H_n$. The former is important in multichannel scattering theory, and the latter is involved whenever one deals with a quantum mechanical system which is composed of several subsystems.

Let us begin with two Hilbert spaces H_1 and H_2. The <u>direct sum</u> of H_1 and H_2, denoted by $H_1 \oplus H_2$, is defined as follows : The elements of $H_1 \oplus H_2$ are pairs of vectors $\{f_1, f_2\}$ with $f_i \varepsilon H_i$, and the scalar product between two such pairs is

$$(\{f_1,f_2\},\{g_1,g_2\}) = (f_1,g_1)_{H_1} + (f_2,g_2)_{H_2}. \tag{2.71}$$

Addition and multiplication by scalars are defined by

$$\{f_1,f_2\} + \{g_1,g_2\} = \{f_1 + g_1, f_2 + g_2\}, \tag{2.72}$$

$$\alpha\{f_1,f_2\} = \{\alpha f_1,\alpha f_2\}. \tag{2.73}$$

We leave it to the reader to check that $H_1 \oplus H_2$ is a Hilbert space, i.e. that Axioms I-IV of Section 2-1 are verified (Problem 2.30). One may remark that both H_1 and H_2 may be considered as subspaces of $H_1 \oplus H_2$: H_1 may be identified with the set of pairs of the form $\{f_1,\theta_2\}$ with $f_1 \varepsilon H_1$ and θ_2 the zero vector of H_2, and similarly for H_2. $H_1 \oplus H_2$ is simply the orthogonal sum of H_1 and H_2. In particular, if $\{e_i\}$ is an orthonormal basis of H_1 and $\{h_j\}$ an orthonormal basis of H_2, then the set of pairs $\{\{e_i,\theta_2\}, \{\theta_1,h_j\}\}$ is an orthonormal basis of $H_1 \oplus H_2$. Thus dim $(H_1 \oplus H_2)$ = dim H_1 + dim H_2. We also remark that a Hilbert space H may always be viewed as the direct sum of a subspace M of H and of its orthogonal complement : $H = M \oplus M^{\perp}$.

One may similarly define the direct sum of a finite or countable number of Hilbert spaces H_k. We shall denote this direct sum by $H = \oplus H_k$ $(k = 1,2,\ldots)$. Its elements are sequences $\{f_1,f_2,\ldots\}$ with $f_k \varepsilon H_k$ such that

$$\sum_k \|f_k\|^2_{H_k} < \infty . \tag{2.74}$$

Addition and multiplication by scalars is defined componentwise as in (2.72) and (2.73), and each H_k may again be considered to be a subspace of H. The proof of the completeness of a countably infinite direct sum of Hilbert spaces is similar to that for ℓ^2 (Problem 2.1). The space ℓ^2 gives an example for a countably infinite direct sum, since it may be viewed as a direct sum of one-dimensional Hilbert spaces.

For each k, let A_k be a bounded linear operator in H_k with $\|A_k\| \leq M < \infty$ for all k. We may define an operator A in $H = \oplus H_k$ by

$$A\{f_1,f_2,\ldots\} = \{A_1f_1,A_2f_2,\ldots\}. \tag{2.75}$$

We shall use the notation $A = A_1 \oplus A_2 \oplus \cdots = \oplus A_k$ for such an operator. Of course the operators of this form do not exhaust the set of bounded operators on $\oplus H_k$. They are characterized by the property that they leave each H_k invariant. The following rules follow immediately from the above definitions :

$$[\oplus A_k] + [\oplus B_k] = \oplus(A_k + B_k), \tag{2.76}$$

$$[\oplus A_k][\oplus B_k] = \oplus A_k B_k \text{ and } [\oplus A_k]^* = \oplus A_k^*. \qquad (2.77)$$

If the sequence $\{A_k\}$ is not uniformly bounded with respect to k or, more generally, if the operators A_k are unbounded, one may similarly define an operator $A = \oplus A_k$ by (2.75) with $D(A) = \{\{f_1,f_2,\ldots\} | f_k \varepsilon D(A_k)$ for each k and $\sum_k ||A_k f_k||^2_{H_k} < \infty\}$.

Let us now turn to the tensor product $G = H_1 \otimes H_2$ of two Hilbert spaces H_1 and H_2. For this we again consider pairs $\{f_1,f_2\}$ with $f_i \varepsilon H_i$ and try to define a scalar product between two such pairs by

$$(\{f_1,f_2\},\{g_1,g_2\})_G = (f_1,g_1)_{H_1}(f_2,g_2)_{H_2}. \qquad (2.78)$$

Here one is faced with two difficulties. The first one has to do with the linear structure of a Hilbert space. In fact, since $\{g_1,g_2\} + \alpha\{h_1,h_2\}$ has to belong to G, (2.6) requires that

$$(\{f_1,f_2\},\{g_1,g_2\} + \alpha\{h_1,h_2\}) =$$
$$(f_1,g_1)(f_2,g_2) + \alpha(f_1,h_1)(f_2,h_2).$$

Now in general the right hand side cannot be written in the form (2.78), i.e. as the scalar product between two elements of $H_1 \times H_2$. Thus, in order to obtain the linear structure of the tensor product, one has to introduce new elements not contained in $H_1 \times H_2$. This is done by first adding to $H_1 \times H_2$ all finite linear combinations of elements of $H_1 \times H_2$ and extending the scalar product (2.78) by linearity to these new elements.

The second difficulty turns up when one considers (2.7). It is seen that for instance $||\{f_1,\theta_2\}||_G = ||\{\theta_1,f_2\}||_G =$

0 and $\|\alpha\{f_1,f_2\} + \beta\{g_1,f_2\} - \{\alpha f_1 + \beta g_1,f_2\}\|_G = 0$ for any $f_1,g_1\varepsilon H_1$, $f_2\varepsilon H_2$ and $\alpha,\beta\varepsilon C$. In order to ensure that Axiom II is verified, one will therefore consider as vectors of $H_1 \otimes H_2$ the equivalence classes of the elements already introduced, two elements being equivalent if their difference has norm zero (that this does define an equivalence relation follows from (2.9) which is valid for the norm $\|\cdot\|_G$ since the Schwarz inequality (2.8) used in its proof can be established wihout using (2.7) [RN, no. 83]). The equivalence class of the pair $\{f_1,f_2\}$ will be denoted by $f_1 \otimes f_2$ and the set of all equivalence classes by $H_1 \hat{\otimes} H_2$. Finally one completes $H_1 \hat{\otimes} H_2$ by the standard method of completing a metric space [RS, Theorem I.1.3] to obtain the Hilbert space $H_1 \otimes H_2$.

We shall now indicate a more explicit way of constructing $H_1 \otimes H_2$. Let $\{e_i\}$ and $\{h_j\}$ be orthonormal bases of H_1 and H_2 respectively. Consider the set of pairs $\{e_i,h_j\}_{i,j}$. They clearly form an orthonormal set with respect to the scalar product (2.78). The <u>tensor product</u> $G = H_1 \otimes H_2$ may then be defined as a Hilbert space in which the above orthonormal set forms an orthonormal basis.

Up to an isomorphism, this latter definition coincides with the first one and is also independent of the choice of a particular basis in H_1 or H_2. To see this, it suffices to verify that each vector of the form $f_1 \otimes f_2$ with $f_i\varepsilon H_i$ can be completely expanded with respect to the orthonormal set $\{e_i \otimes h_j\}$. Now one has $f_1 = \sum\alpha_i e_i$, $f_2 = \sum\beta_j h_j$ with $\sum_{i,j}|\alpha_i\beta_j|^2 = \|f_1\|^2\, \|f_2\|^2$, and (2.78) implies that $(e_i \otimes h_j, f_1 \otimes f_2)_G = \alpha_i\beta_j$. Hence

$$\sum_{i,j} |(e_i \otimes h_j, f_1 \otimes f_2)_G|^2 = \|f_1\|^2 \|f_2\|^2 = \|f_1 \otimes f_2\|_G^2,$$

which proves the Parseval relation. It follows that we must have $f_1 \otimes f_2 = \sum_{i,j} \alpha_i \beta_j e_i \otimes h_j$.

There is no canonical way of identifying H_1 or H_2 with subspaces of $H_1 \otimes H_2$. However, if $f_2 \varepsilon H_2$ is a fixed vector such that $\|f_2\| = 1$, the set of vectors $f_1 \otimes f_2$ with f_1 ranging over H_1 is a subspace of $H_1 \otimes H_2$ that is isomorphic to H_1. It will be denoted by $H_1 \otimes f_2$. We also remark that $\dim H_1 \otimes H_2 = \dim H_1 \cdot \dim H_2$.

As an example we consider the tensor product $G_{mn} = L^2(R^m) \otimes L^2(R^n)$. Let $\{e_i(\underline{x})\}$ and $\{h_j(\underline{y})\}$ be orthonormal bases of $L^2(R^m)$ and $L^2(R^n)$ respectively ($\underline{x} \varepsilon R^m, \underline{y} \varepsilon R^n$). Then the set of functions $\{e_i(\underline{x}) h_j(\underline{y})\}$ forms an orthonormal set in $L^2(R^{m+n})$; indeed

$$\begin{aligned}(e_i h_j, e_r h_s)_{L^2(R^{m+n})} &= \int d^m x\, d^n y\, \bar{e}_i(\underline{x}) \bar{h}_j(\underline{y}) e_r(\underline{x}) h_s(\underline{y}) \\ &= (e_i \otimes h_j, e_r \otimes h_s)_{G_{mn}}.\end{aligned}$$

It is an interesting fact that the above set of functions is indeed an orthonormal basis of $L^2(R^{m+n})$, which means that $L^2(R^m) \otimes L^2(R^n)$ is naturally isomorphic to $L^2(R^{m+n})$ (identify $f_1 \otimes f_2 \varepsilon L^2(R^m) \otimes L^2(R^n)$ with $f_1 f_2 \varepsilon L^2(R^{m+n})$.) We shall use this result on various occasions, but we leave it to the reader to verify that $g = \theta$ is the only vector in $L^2(R^{m+n})$ which is orthogonal to the set $\{e_i(\underline{x}) h_j(\underline{y})\}$ (Problem 2.31).

Let A_k be linear operators in H_k ($k = 1,2$). One may define an operator denoted $A_1 \hat{\otimes} A_2$ in $G = H_1 \otimes H_2$ by

$$(A_1 \hat{\otimes} A_2)(f_1 \otimes f_2) = A_1 f_1 \otimes A_2 f_2 \quad \text{for } f_k \varepsilon D(A_k) \tag{2.79}$$

and extending this definition by linearity to the set of all finite linear combinations of vectors of the form $f_1 \otimes f_2$ with $f_k \varepsilon D(A_k)$. This set will be written as $D(A_1) \hat{\otimes} D(A_2)$. If each $D(A_k)$ is dense in the respective Hilbert space $\mathcal{H}_k$, $A_1 \hat{\otimes} A_2$ is densely defined. A particular case is an operator of the form $A_1 \hat{\otimes} I$ whose action differs from the identity only in the first component. In order that the above definition makes sense, one has to verify that, whenever $\sum_{i=1}^m \alpha_i\{f_1^i, f_2^i\}$ and $\sum_{k=1}^n \beta_k\{g_1^k, g_2^k\}$ define the same vector in $\mathcal{H}_1 \hat{\otimes} \mathcal{H}_2$, then so do $\sum_{i=1}^m \alpha_i\{A_1 f_1^i, A_2 f_2^i\}$ and $\sum_{k=1}^n \beta_k\{A_1 g_1^k, A_2 g_2^k\}$, where $f_j^i, g_j^k \varepsilon D(A_j)$. For this, let $u_k \varepsilon D(A_k)$. (2.30) and (2.78) then imply that

$$\sum_i \alpha_i(\{u_1,u_2\},\{A_1 f_1^i, A_2 f_2^i\}) = \sum_i \alpha_i(\{A_1^* u_1, A_2^* u_2\},\{f_1^i, f_2^i\})$$
$$= \sum_k \beta_k(\{A_1^* u_1, A_2^* u_2\},\{g_1^k, g_2^k\}) = \sum_k \beta_k(\{u_1,u_2\},\{A_1 g_1^k, A_2 g_2^k\}).$$

Since the set of vectors $\{u_1 \otimes u_2 | u_k \varepsilon D(A_k)\}$ is fundamental in $\mathcal{H}_1 \otimes \mathcal{H}_2$, $\sum_i \alpha_i\{A_1 f_1^i, A_2 f_2^i\} - \sum_k \beta_k\{A_1 g_1^k, A_2 g_2^k\}$ must be in the equivalence class of $\{\theta,\theta\}$ by virtue of Proposition 2.2.

If $A_1 \hat{\otimes} A_2$ is closable, we denote its closure by $A_1 \otimes A_2$. If $A_1 \varepsilon \mathcal{B}(\mathcal{H}_1)$ and $A_2 \varepsilon \mathcal{B}(\mathcal{H}_2)$, then $A_1 \hat{\otimes} A_2$ is bounded (see Problem 2.35), hence $A_1 \otimes A_2 \varepsilon \mathcal{B}(\mathcal{H}_1 \otimes \mathcal{H}_2)$. If only bounded operators are involved, one has the following rules :

$$\alpha(A_1 \otimes A_2) = (\alpha A_1) \otimes A_2 = A_1 \otimes (\alpha A_2), \tag{2.80}$$

$$(A_1 \otimes A_2)(B_1 \otimes B_2) = A_1 B_1 \otimes A_2 B_2, \tag{2.81}$$

$$(A_1 \otimes A_2)^* = A_1^* \otimes A_2^*. \tag{2.82}$$

If A_1 is self-adjoint, so is $A_1 \otimes I$ (Problem 2.39). For other properties of $A_1 \hat{\otimes} A_2$, see Problems 7.7, 11.3 and 14.8.

The construction of the tensor product $G = \otimes H_k$ (k = 1,...,n) of a finite number of Hilbert spaces is done in complete analogy with that given above for n = 2 by using instead of (2.78) the definition

$$(\{f_1,\ldots,f_n\},\{g_1,\ldots,g_n\})_G = (f_1,g_1)(f_2,g_2)\ldots(f_n,g_n). \quad (2.83)$$

Operators of the form $A_1 \hat{\otimes} \ldots \hat{\otimes} A_n$ can be defined by an obvious modification of (2.79). Infinite tensor products will not be used in this book.

2-5 NOTES AND SUPPLEMENTARY MATERIAL

A. We add some comments regarding the extension of the Fourier transformation from $S(R^n)$ to $L^2(R^n)$. To simplify the notation we set n = 1. Suppose $f \varepsilon L^1(R) \cap L^2(R)$, and let $\tilde{f}$ be defined by (2.17). Then if $g \varepsilon S \equiv S(R)$,

$$\begin{aligned} |\int dk \bar{\tilde{f}}(k)\tilde{g}(k)| &= (2\pi)^{-\frac{1}{2}}|\int dk \int dx \bar{f}(x) e^{ikx}\tilde{g}(k)| \\ &= |\int dx \bar{f}(x) g(x)| = |(f,g)| \leq \|f\|\,\|g\|, \end{aligned} \quad (2.84)$$

where the interchange of the order of integration is permitted since the integrand is absolutely integrable. Hence the first integral in (2.84) defines a bounded linear functional on S which can be extended by continuity to H. By Proposition 2.3 there exists $f_0 \varepsilon L^2(R)$ such that $\int dk\{\bar{\tilde{f}}(k) - \bar{\tilde{f}}_0(k)\}\tilde{g}(k) = 0$ for all $g \varepsilon S$.

Let [a,b] be a bounded interval. By taking a uniformly bounded sequence $\{\tilde{g}_n\} \varepsilon S$ with support in [a-1, b+1] converging pointwise to the characteristic function of [a,b], one gets from the dominated convergence theorem that

$\int_a^b dk\{\bar{\tilde{f}}(k) - \bar{\tilde{f}}_o(k)\} = 0$, which implies $\tilde{f}(k) = \tilde{f}_o(k)$ on $[a,b]$ (cf. [R, Lemma 5.7]). Hence $f = f_o \varepsilon L^2(\mathcal{R})$.

Let $\{f_m\} \varepsilon \mathcal{S}$ be such that $f_m \to f$. Then by (2.84) and (2.23), $(\tilde{f},\tilde{g}) = (f,g) = \lim(f_m,g) = \lim(\tilde{f}_m,\tilde{g}) = (\mathcal{F}f,\tilde{g})$ for all $g\varepsilon\mathcal{S}$, so that $\tilde{f}$ defined by (2.17) is identical with the vector $\mathcal{F}f$ by Proposition 2.2.

If one requires only $f\varepsilon L^2(\mathcal{R})$, then, as pointed out in Section 2-2, its Fourier transform has to be defined through convergence in the mean. One writes $f = \text{s-lim } f_m$ with $f_m(x) = f(x)$ for $|x| \le m$ and $f(x) = 0$ for $|x| > m$ and defines $\tilde{f} = \text{s-lim } \tilde{f}_m$ $(m \to \infty)$. $\tilde{f}_m$ is well defined, since $f_m \varepsilon L^1(\mathcal{R}) \cap L^2(\mathcal{R})$ as a consequence of (2.8) :

$$\|f_m\|_{L^1} = \int_{-m}^m dx|f(x)| \le \{\int_{-m}^m dx|f(x)|^2\}^{\frac{1}{2}}\{\int_{-m}^m ds\}^{\frac{1}{2}}$$
$$\le \|f\| (2m)^{\frac{1}{2}} < \infty. \qquad (2.85)$$

B. As the Lebesgue dominated convergence theorem is often invoked in this book, we shall state here a simple version of it for the convenience of the reader. For its proof one may consult e.g. [MS, page 169], [R, Thm. 4.15].

PROPOSITION 2.35 : Let Δ be a Lebesgue measurable set in $\mathcal{R}^n$. Let $\{f_t\}_{t\varepsilon\mathcal{R}}$ be a family of measurable functions from $\mathcal{R}^n$ to $\mathcal{C}$ and $g\varepsilon L^1(\Delta)$ such that $|f_t(\underline{x})| \le g(\underline{x})$ on Δ and $\lim f_t(\underline{x}) = f(\underline{x})$ for almost all $\underline{x}$ in Δ as $t \to \tau$, $\tau\varepsilon[-\infty,\infty]$. Then $f\varepsilon L^1(\Delta)$ and

$$\lim_{t\to\tau} \int_\Delta f_t(\underline{x})d^n x = \int_\Delta f(\underline{x})d^n x.$$

C. The graph of an operator. The definition of the closure of an operator A involves sequences $\{f_n\}\varepsilon D(A)$ such that both $\{f_n\}$ and $\{Af_n\}$ are strongly Cauchy. It is useful to combine these two Cauchy sequences into one mathematical object. For this one introduces the <u>graph</u> of A which is defined to be the set $\Gamma(A)$ of all pairs $\{f,Af\}$ with f ranging over D(A). It is natural to regard $\Gamma(A)$ as a subset of $\mathcal{G} \equiv \mathcal{H} \oplus \mathcal{H}$ (this is similar to plotting the graph of a function $f : \mathbb{R} \to \mathbb{R}$ in the plane $\mathbb{R}^2$). Since A is linear, $\Gamma(A)$ is a linear manifold in $\mathcal{G}$. In general a linear manifold $\mathcal{M}$ in $\mathcal{G}$ will be the graph of some operator if and only if the elements $\{f_1,f_2\}$ of $\mathcal{M}$ are uniquely determined by their first argument f_1. Since $\mathcal{M}$ is linear, this condition is equivalent to the requirement that $\mathcal{M}$ contain no element of the form $\{\theta,g\}$ with $g \neq \theta$.

The usefulness of regarding $\Gamma(A)$ as a subset of $\mathcal{G}$ resides in the following identity :

$$\|\{f,g\} - \{f_n,Af_n\}\|_{\mathcal{G}}^2 = \|f-f_n\|^2 + \|g-Af_n\|^2 .$$

One may deduce from this that, if A is closable, then $\Gamma(\bar{A}) = \overline{\Gamma(A)}$, and that A is a closed operator if and only if $\Gamma(A)$ is a closed subspace of $\mathcal{H} \oplus \mathcal{H}$.

The adjoint operator may also be specified in terms of graphs. To do this one introduces the following unitary operator U in $\mathcal{G}$: $U\{f,g\} = \{-g,f\}$. One then sees that the pair $\{g,g^*\}$ verifies (2.29) if and only if $\{g,g^*\}$ is orthogonal to $U\Gamma(A)$ in $\mathcal{G}$. Since $\Gamma(\bar{A})$ is just the closure of $\Gamma(A)$, this means that $\Gamma(A^*) \oplus U\Gamma(\bar{A}) = \mathcal{G}$ if A is closable.

Proposition 2.7 can now be proved by applying twice the preceding identity. (i) Upon replacing A by A* and using

(2.31), we obtain $\Gamma(A^{**}) \oplus U\Gamma(A^*) = G$. (ii) Upon applying U to both of its members, using the identity $U^2 = -I$ and the fact that $\Gamma(\bar{A}) = -\Gamma(\bar{A})$ as sets, we find that $U\Gamma(A^*) \oplus \Gamma(\bar{A}) = G$. By comparing the two equations thus obtained, we infer that $\Gamma(A^{**}) = \Gamma(\bar{A})$, whence $A^{**} = \bar{A}$.

D. We conclude by giving a number of proofs that were omitted in Section 2-3.

Proof of Proposition 2.24 : Since A/z is compact, it suffices to consider the case z = 1. So let us assume that the only solution of $Af = f$ is $f = \theta$. Then I-A is invertible, and it remains to show that $(I-A)^{-1} \varepsilon \mathcal{B}(\mathcal{H})$.

The idea of the proof is to reduce the problem to a finite-dimensional subspace. For this one chooses a finite rank operator T such that $\|A-T\| < 1$. Then $(I-A+T)^{-1} \varepsilon \mathcal{B}(\mathcal{H})$ by Proposition 2.20. Define $Y = T(I-A+T)^{-1}$ and denote by $\mathcal{M}$ its range. Clearly Y is a finite rank operator, so that $\mathcal{M}$ is a subspace (Problem 2.38) of dimension $n < \infty$. The following identity is also easily checked :

$$(I-A) = (I-Y)(I-A+T). \tag{2.86}$$

Denote by F the projection whose range is $\mathcal{M}$. Then $FY = Y$, implying that $F-YF = F(I-YF)$. Thus the operator F-YF maps $\mathcal{M}$ into itself. Suppose there exists $g \varepsilon \mathcal{M}$ such that $(I-Y)g = (F-YF)g = \theta$. Define $f = (I-A+T)^{-1}g$. It then follows from (2.86) that $(I-A)f = (I-Y)g = \theta$. Hence $f = \theta$ by the original assumption, which implies $g = (I-A+T)f = \theta$. This shows that the operator F-YF, considered as a map in $\mathcal{M}$, is invertible. Since this operator is given by a $n \times n$ matrix, the inverse matrix defines a bounded operator in M which will be denoted by $(F-YF)_M^{-1}$. It satisfies

$$(I-Y)(F-YF)_M^{-1} = F \text{ and } (F-YF)_M^{-1}(F-YF) = F. \qquad (2.87)$$

Now define $Z = (F-YF)_M^{-1}F(I+Y-YF) + (I-F)$. By what we have shown above, $Z\in\mathcal{B}(H)$. By using $Y = FY$, $Y(I-Y) = (I-Y)Y$ and (2.87), one obtains

$$Z(I-Y) = (F-YF)_M^{-1}(F-YF)(I-Y) + (F-YF)_M^{-1}(I-Y)FY + (I-F)(I-Y) =$$

$$= F(I-Y) + FY + (I-F)(I-Y) = I.$$

Thus $I-Y$ is invertible, and Z is an extension of $(I-Y)^{-1}$. Similarly one finds $(I-Y)Z = I$, implying that [range $(I-Y)$] = $H = D(Z)$. Hence $Z = (I-Y)^{-1}$. It now follows from (2.86) that $(I-A)^{-1} = (I-A+T)^{-1}(I-Y)^{-1}$, which is in $\mathcal{B}(H)$. #

Proof of Proposition 2.25 : Let $\{f_n\}$ be an infinite sequence of linearly independent eigenvectors of A, $\{z_n\}$ the corresponding sequence of eigenvalues. Choose an orthonormal set $\{e_i\}$ such that e_n is a linear combination of $f_1,\dots,f_n$, which can be done by the Schmidt orthogonalization procedure [K, Ch. I.6.3]. Thus

$$e_n = \sum_{i=1}^n \alpha_{ni}f_i,$$

and f_k is a linear combination of $e_1,\dots,e_k$. Now

$$Ae_n = \sum_{i=1}^n \alpha_{ni}z_if_i = \sum_{i=1}^{n-1} \alpha_{ni}(z_i-z_n)f_i + z_ne_n$$

$$= \sum_{k=1}^{n-1} \beta_{nk}e_k + z_ne_n$$

and $$\|Ae_n\|^2 = \sum_{k=1}^{n-1}|\beta_{nk}|^2 + |z_n|^2.$$

Since w-lim $e_n = \theta$ as $n\to\infty$, we have by Proposition 2.23 that $\|Ae_n\|^2 \to 0$ as $n\to\infty$, i.e. $\lim z_n = 0$ as $n\to\infty$.

This shows that an infinite sequence of different eigenvalues necessarily converges to $z = 0$. If z is an eigenvalue of A of infinite multiplicity, one takes for $\{f_n\}$ an infinite sequence of linearly independent vectors verifying $Af = zf$. Then $z_n = z$ for all n, and by the above we have $z = 0$. Thus the only possible eigenvalue of infinite multiplicity is $z = 0$. #

<u>Proof of Lemma 2.29</u> : (i) We first notice the following consequences of (2.9) and (2.27) respectively

$$\Big| \|f\| - \|g\| \Big| \leq \|f - g\| \quad \text{for} \quad \text{all } f,g\in \mathcal{H}, \tag{2.88}$$

$$\Big| \|A\| - \|B\| \Big| \leq \|A - B\| \quad \text{for} \quad A,B\in \mathcal{B}(\mathcal{H}), \tag{2.89}$$

which are obtained by arguments similar to those in part (i) of the proof of Proposition 2.1.

(ii) Let $\{T_N\}$ be a sequence of finite rank operators converging uniformly to A. It may be seen as in (2.50) that $\|A^*A - T_N^*T_N\| \to 0$ as $N \to \infty$. Hence $A^*A = A^2$ is the uniform limit of a sequence $Y_N \equiv T_N^*T_N$ of self-adjoint finite rank positive operators.

It follows from (2.30) that $Y_N^*f = \theta$ for all f in $(\text{range } Y_N)^{\perp}$. Hence Y_N is zero on $(\text{range } Y_N)^{\perp}$. By part (i) of the proof of Proposition 2.30, range Y_N is invariant under Y_N, i.e. Y_N can be decomposed into the sum of a self-adjoint operator on range Y_N and the zero operator on $(\text{range } Y_N)^{\perp}$.

(iii) Let $\{e_{iN}\}$, $i = 1,\ldots, M(N)$ be a set of mutually orthogonal eigenvectors of Y_N in range Y_N (i.e. $Y_N e_{iN} = \lambda_{iN} e_{iN}$ with $\lambda_{iN} \geq 0$) which span range Y_N (for a self-adjoint operator in a finite-dimensional space, such a set of eigenvectors always exists, cf. [H]). Then Y_N may be written as

$$Y_N f = \sum_{i=1}^{M(N)} \lambda_{iN}(e_{iN}, f)e_{iN} . \tag{2.90}$$

Let us denote by λ_N the largest eigenvalue of Y_N and by e_N one of the corresponding eigenvectors ($\|e_N\| = 1$). As in the proof of Lemma 2.28 one deduces from (2.90) that $\|Y_N\| \leq \lambda_N$. Since $\|Y_N e_N\| = \lambda_N$, we have in fact $\|Y_N\| = \lambda_N$. Since $\|A^2 - Y_N\| \to 0$ as $N \to \infty$, one obtains from (2.89) that $\lim \lambda_N = \lim \|Y_N\| = \|A^2\| \equiv \rho$ as $N \to \infty$.

(iv) We have by (2.88) and (2.9)

$$\begin{aligned} \big|\|A^2 e_N\| - \rho\big| &= \big|\|A^2 e_N\| - \|\rho e_N\|\big| \leq \|(A^2 - \rho)e_N\| \\ &\leq \|(A^2 - Y_N)e_N\| + \|(Y_N - \lambda_N)e_N\| + |\rho - \lambda_N|\,\|e_N\| \\ &\leq \|A^2 - Y_N\| + |\rho - \lambda_N| \to 0 \text{ as } N \to \infty. \end{aligned} \tag{2.91}$$

This implies that $\|A^2 e_N\| \to \rho$ as $N \to \infty$.

(v) Assume $\rho \neq 0$. Since A is assumed to have no non-zero eigenvalues, we have $(A \pm \rho^{\frac{1}{2}})^{-1} \varepsilon \mathcal{B}(\mathcal{H})$ by virtue of Proposition 2.24. Hence $(A^2 - \rho)^{-1} = (A - \rho^{\frac{1}{2}})^{-1} \cdot (A + \rho^{\frac{1}{2}})^{-1} \varepsilon \mathcal{B}(\mathcal{H})$, i.e. range $(A^2 - \rho) = \mathcal{H}$.

Let $f \varepsilon \mathcal{H}$. Then there exists $g \varepsilon \mathcal{H}$ such that $f = (A^2 - \rho)g$. As $\|(A^2 - \rho)e_N\| \to 0$ by one of the inequalities in (2.91), we get $(f, e_N) = ((A^2 - \rho)g, e_N) = (g, (A^2 - \rho)e_N) \to 0$ as $N \to \infty$. Thus e_N converges weakly to zero. By Proposition 2.23, $\|A^2 e_N\| \to 0$ as $N \to \infty$, a contradiction. Hence $\rho = 0$, i.e. $\|A^*A\| = 0$. Therefore $\|Ag\|^2 = (g, A^*Ag) = 0$ for every $g \varepsilon \mathcal{H}$, i.e. $A = 0$. #

<u>Proof of Proposition 2.33</u> : (a) (i) Let

$$\Delta = \{\underline{x} \varepsilon R^n \big| \int |K_A(\underline{x}, \underline{y})|^2 d^n y = \infty\}.$$

The Lebesgue measure of Δ is zero as a consequence of (2.66). Let $f\in L^2(\mathcal{R}^n)$ and $\underline{x}\notin\Delta$. The integral in (2.65) may be viewed as a scalar product in $L^2(\mathcal{R}^n)$, so that by applying to it the Schwarz inequality (2.8) one gets

$$|(Af)(\underline{x})|^2 \leq \int d^n y |K_A(\underline{x},\underline{y})|^2 \int d^n y |f(\underline{y})|^2.$$

This implies $\|Af\|^2 \leq M_A \|f\|^2$, i.e. $D(A) = \mathcal{H}$ and $A\in\mathcal{B}(\mathcal{H})$.

(ii) Let $A\in\mathcal{B}(\mathcal{H})$, let $\{e_k\}$ be an orthonormal basis of $\mathcal{H}$ and define $\alpha_{jk} = (e_j, Ae_k)$. Then $Ae_k = \sum_j \alpha_{jk} e_j$, and by (2.59) and (2.12)

$$\|A\|^2_{HS} = \sum_k \|Ae_k\|^2 = \sum_{k,j} |\alpha_{jk}|^2. \tag{2.92}$$

(iii) If $\{e_k\}$ is an orthonormal basis of $L^2(\mathcal{R}^n)$, the complex conjugate functions $\{\bar{e}_k\}$ also form an orthonormal basis. Hence the functions $\{e_j(\underline{x})\bar{e}_k(\underline{y})\}$ form an orthonormal basis of $L^2(\mathcal{R}^n) \otimes L^2(\mathcal{R}^n) = L^2(\mathcal{R}^{2n})$ (cf. Section 2-4). Since K_A is in $L^2(\mathcal{R}^{2n})$, it may be expanded with respect to the latter basis :

$$K_A = \sum_{j,k} \beta_{jk} e_j \bar{e}_k.$$

The Parseval relation in $L^2(\mathcal{R}^{2n})$ gives

$$M_A = \|K_A\|^2_{L^2(\mathcal{R}^{2n})} = \sum_{j,k} |\beta_{jk}|^2. \tag{2.93}$$

Now

$$\alpha_{jk} \equiv (e_j, Ae_k)_{L^2(\mathcal{R}^n)} = \int d^n x d^n y \bar{e}_j(\underline{x}) K_A(\underline{x},\underline{y}) e_k(\underline{y})$$

$$= (e_j \bar{e}_k, K_A)_{L^2(\mathcal{R}^{2n})} = \beta_{jk}. \tag{2.94}$$

By combining (2.92) - (2.94) we get (2.67), and since $M_A < \infty$ by hypothesis we have $A\in\mathcal{B}_2$.

(b) Define α_{jk} as in part (ii) above, and let

$$K_N(\underline{x},\underline{y}) = \sum_{j,k=1}^{N} \alpha_{jk} e_j(\underline{x})\bar{e}_k(\underline{y}).$$

K_N is a finite linear combination of functions in $L^2(\mathcal{R}^{2n})$, i.e. $K_N \varepsilon L^2(\mathcal{R}^{2n})$. Let $N' > N$. One has as in (2.93) that

$$\int d^n x d^n y |K_N(\underline{x},\underline{y}) - K_{N'}(\underline{x},\underline{y})|^2$$

$$= \sum_{j=1}^{N} \sum_{k=N+1}^{N'} |\alpha_{jk}|^2 + \sum_{j=N+1}^{N'} \sum_{k=1}^{N'} |\alpha_{jk}|^2$$

$$\leq \sum_{j=1}^{\infty} \sum_{k=N+1}^{\infty} |\alpha_{jk}|^2 + \sum_{j=N+1}^{\infty} \sum_{k=1}^{\infty} |\alpha_{jk}|^2.$$

Since $A\varepsilon\mathcal{B}_2$, the double sum in (2.92) is convergent. This implies that $\{K_N\}$ is a strong Cauchy sequence in $L^2(\mathcal{R}^{2n})$. Therefore it has a limit function $K(\underline{x},\underline{y})$ in $L^2(\mathcal{R}^{2n})$, and by part (a), $K(\underline{x},\underline{y})$ defines a Hilbert-Schmidt operator B in $L^2(\mathcal{R}^n)$.

It remains to show that B = A. For this one deduces as in (2.94) that

$$(e_j,Be_k)_{L^2(\mathcal{R}^n)} = (e_j\bar{e}_k,K)_{L^2(\mathcal{R}^{2n})} = \lim_{N\to\infty}(e_j\bar{e}_k,K_N)_{L^2(\mathcal{R}^{2n})} = \alpha_{jk}.$$

Hence $Be_k = \sum_j \alpha_{jk} e_j$, i.e. $Be_k = Ae_k$ for all k. The fact that B = A is now a consequence of Lemma 2.27. #

<u>Proof of Proposition 2.34</u> : (i) Suppose $A\varepsilon\mathcal{B}_1$. Let $\{e_k\},\{h_k\}$, $\{\lambda_k\}$ be as in Proposition 2.31. Define B and C by

$$Bf = \sum_j \lambda_j^{\frac{1}{2}}(e_j,f)h_j, \qquad Cf = \sum_k \lambda_k^{\frac{1}{2}}(e_k,f)e_k.$$

Then

$$BCf = \sum_j \lambda_j^{\frac{1}{2}}(e_j,Cf)h_j = \sum_{j,k} \lambda_j^{\frac{1}{2}} \lambda_k^{\frac{1}{2}}(e_j,e_k)(e_k,f)h_j$$

$$= \sum_k \lambda_k(e_k,f)h_k = Af.$$

Since $\sum\lambda_k < \infty$, $\lambda_k \to 0$ as $k \to \infty$. Thus one sees as in the proof of Proposition 2.31 that B as well as C is the uniform limit of a sequence of finite rank operators, i.e. $B, C \in \mathcal{B}_0$. By using also (2.68) we then get

$$\|B\|_{HS}^2 = \|C\|_{HS}^2 = \sum_k (\lambda_k^{\frac{1}{2}})^2 = |||A|||_1 < \infty,$$

which proves the "only if" part of (a).

(ii) Suppose A = BC with $B, C \in \mathcal{B}_2$. A is compact by Proposition 2.32. Consider its canonical expansion (2.58). Then one gets from (2.69), (2.58) and (2.8) that

$$|||A|||_1 = \sum_k \lambda_k = \sum_k (h_k, Ae_k) = \sum_k (B^* h_k, Ce_k)$$

$$\leq \sum_k \|B^* h_k\| \, \|Ce_k\| \leq (\sum_k \|B^* h_k\|^2)^{\frac{1}{2}} (\sum_j \|Ce_j\|^2)^{\frac{1}{2}}$$

$$\leq \|B^*\|_{HS} \|C\|_{HS} = \|B\|_{HS} \|C\|_{HS} < \infty, \qquad (2.95)$$

which proves (a). (b) follows from (a) and Proposition 2.32(d). #

PROBLEMS

2.1 : Verify that the following are Hilbert spaces :
(i) The set $\mathbb{C}^n$ of all n-tupels $\alpha = \{\alpha_1, \ldots, \alpha_n\}$ of complex numbers ($\alpha_i \in \mathbb{C}$) with the scalar product

$$(\alpha, \beta) = \sum_{i=1}^n \bar{\alpha}_i \beta_i . \quad (\alpha, \beta \in \mathbb{C}^n) \qquad (2.96)$$

(ii) The set ℓ^2 of all infinite sequences $\alpha = \{\alpha_1, \alpha_2, \ldots\}$ of complex numbers which satisfy $\sum_i |\alpha_i|^2 < \infty$, with the scalar product given by (2.96) for $n = \infty$ [AG, Section 4].

2.2 : Verify that (i) $\|\theta\| = 0$, $(f, \theta) = 0$ for all $f \in \mathcal{H}$, (ii) $(\alpha f, g) = \bar{\alpha}(f, g)$, $(f + \alpha h, g) = (f, g) + \bar{\alpha}(h, g)$.

2.3 : Prove that a sequence $\{f_n\}$ can converge strongly to at most one vector f.

2.4 : (a) If $f_n \to f$, $g_n \to g$, then $(f_n, g_n) \to (f, g)$ as $n \to \infty$.

(b) If w-lim $f_n = f$ and $\|f_n\| \le \|f\|$ for all n, then s-lim $f_n = f$ as $n \to \infty$. (c) Show that in a finite-dimensional Hilbert space weak convergence implies strong convergence.

<u>2.5</u> : Let $\{e_i\}$ be an orthonormal basis of $\mathcal{H}$ and $f \varepsilon \mathcal{H}$. Define $f_n = \sum_{k=1}^{n} (e_k, f) e_k$. (i) Show that $f_n \to f$ as $n \to \infty$. (ii) Prove Parseval's relation (2.12). (Hint : Use (2.11) to show that $\{f_n\}$ is a Cauchy sequence. Use Proposition 2.2 to prove that its limit is f. Use Proposition 2.1 for (ii).)

<u>2.6</u> : Verify that a linear manifold $\mathcal{M}$ satisfies axioms II and IV. (Hint : Let $\{f_n\}$ be a dense sequence in $\mathcal{M}$, $\bar{\mathcal{M}}$ the closure of $\mathcal{M}$ and $\eta_m = 1/m$. Define $f_{n,1}$ from the projection theorem and show that the sequence $\{f_{n,1}\}$ is dense in $\bar{\mathcal{M}}$. For each n choose a sequence $\{g_{nm}\} \varepsilon \mathcal{M}$ such that $\|g_{nm} - f_{n,1}\| < \eta_m$).

<u>2.7</u>† : Prove Proposition 2.3. (Hint : Show that the set $\mathcal{N} = \{f | \Phi(f) = 0\}$ is a subspace. If $\Phi \not\equiv 0$, there exists $h \varepsilon \mathcal{N}^{\perp}$ such that $\Phi(h) = 1$. Define $g = h / \|h\|^2$. If $f \varepsilon \mathcal{H}$, then $f - \Phi(f) h \varepsilon \mathcal{N}$, which implies $(g,f) = \Phi(f)$.)

<u>2.8</u>† : In $L^2(\mathbb{R})$, let $\chi_{[a,b]}$ be the characteristic function of the interval [a,b], and let $\eta > 0$. Show that there exists a function g in $C_0^\infty(\mathbb{R})$ such that $\|\chi_{[a,b]} - g\| < \eta$. (Hint : Replace $\chi_{[a,b]}$ near a and b by a function similar to $\exp[x^2(x^2 - M^2)^{-1}]$, $|x| \le M$.)

<u>2.9</u> : Prove Proposition 2.4(b) for n > 1.

<u>2.10</u> : Let α be a complex number with Re $\alpha > 0$. Make the appropriate choice of the branch of $\sqrt{\alpha}$ and show that the Fourier transform of the function $\exp(-\alpha|\underline{x}|^2/2)$ is $\alpha^{-n/2}\exp(-|\underline{k}|^2/2\alpha)$.

<u>2.11</u> : Prove the second equality (2.23).

<u>2.12</u> : Verify the equivalence of the two conditions for an operator to be closable given before Lemma 2.5.

<u>2.13</u> : Let A be closable, $\bar{A}$ its closure and A' a closed extension of A. Prove that $\bar{A} \subseteq A'$.

<u>2.14</u> : Let $\{e_i\}$ be an orthonormal basis of an infinite-dimensional Hilbert space $\mathcal{H}$, and define a linear operator A as follows : D(A) is the set of all finite linear combinations of vectors of $\{e_i\}$, and $Ae_k = ke_1$. Show that A is not closable. Verify also that D(A*) is not dense in $\mathcal{H}$. Find a sequence of bounded operators converging strongly to A on D(A). (Hint : Consider the sequence $\{f_n\}$ with $f_n = n^{-1} e_n$).

2.15 : Let Ω be a linear operator such that $||\Omega f|| = ||f||$ for all $f \in D(\Omega)$, and denote by $\bar{\Omega}$ the closure of Ω. Verify that $||\bar{\Omega}f|| = ||f||$ for all $f \in D(\bar{\Omega})$.

2.16 : Prove equations (2.27) and (2.28).

2.17 : Prove the following statements. (a) If A is closed and invertible, then A^{-1} is closed. (b) If in addition A^{-1} is bounded and defined everywhere and B is closed, then AB is closed. (c) If A is closed and C bounded with $D(C) = \mathcal{H}$, then A+C is closed. (d) If $\bar{A}$ and B are bounded with $D(\bar{A}) = D(B) = \mathcal{H}$ and AB is densely defined, then $\overline{AB} = \bar{A}B$.

2.18 : Verify the assertion (2.31).

2.19 : Show that, if A** exists, then $A \subseteq A^{**}$.

2.20 : Verify (2.32), (2.33) and (2.34).

2.21 : Show that (2.36) defines an orthogonal projection.

2.22 : Prove the polarization identity

$$4(f,g) = ||f+g||^2 - ||f-g||^2 - i||f+ig||^2 + i||f-ig||^2. \qquad (2.97)$$

Show that, for $f,g \in D(A)$, (f,Ag) can be similarly expressed as a sum of four terms of the form $\beta(h,Ah)$ with $\beta \in \mathbb{C}$.

2.23 : Suppose $A \subseteq A^*$. Show that $\bar{A}$ is self-adjoint if and only if $A^* = A^{**}$.

2.24 : (a) Suppose that $D(A_n) = D(A) = \mathcal{H}$, A_n and A are bounded and $A_n \to A$. Show that $\{A_n^*\}$ converges weakly to A* as $n \to \infty$. (b) Prove Proposition 2.18(b).

2.25 : Prove Proposition 2.20. (Hint : Use (2.28) and the identity $I-A^{n+1} = (I+A+\ldots+A^n)(I-A)$.)

2.26 : Let $\mathcal{H} = L^2(\mathbb{R})$ and let Q be the maximal multiplication operator defined by $(Qf)(x) = xf(x)$. Show that $(I-Q)^{-1}$ exists but is unbounded.

2.27 : Give an example of a compact operator which is not in the Hilbert-Schmidt class.

2.28 : Find necessary and sufficient conditions for a projection and for a partial isometry to be compact, Hilbert-Schmidt or trace class.

2.29 : (Polar decomposition) : Let A be closed. Then there exists a positive self-adjoint operator $|A|$ with $D(|A|) = D(A)$ and a partial isometry Ω with initial set $\overline{|A|\mathcal{H}}$ and final set $\overline{A\mathcal{H}}$ such that $A = \Omega|A|$. (This may be viewed as a generalization of the polar decomposition $z = \exp(i\phi)|z|$ of complex numbers). Verify the above statement for the case where A is compact by using Proposition 2.31.

2.30 : Verify that the space $\mathcal{H}_1 \oplus \mathcal{H}_2$ satisfies axioms I-IV.

2.31†: Prove that there is an isomorphism from $L^2(\mathcal{R}^m) \otimes L^2(\mathcal{R}^n)$ onto $L^2(\mathcal{R}^{m+n})$, see Section 2-4. (By an isomorphism between two Hilbert spaces $\mathcal{H}_1$ and $\mathcal{H}_2$ we mean a linear map j from $\mathcal{H}_1$ onto $\mathcal{H}_2$ such that $(jf,jg)_2 = (f,g)_1$ for all $f,g\in\mathcal{H}_1$.)

2.32†: Show that
(a) the direct sum $A = \oplus A_k$ of a sequence of self-adjoint operators $\{A_k\}_{k=1}^\infty$ is self-adjoint,
(b) the direct sum $F = \oplus F_k$ of a sequence of projections $\{F_k\}_{k=1}^\infty$ is a projection,
(c) $\otimes F_k$ is a projection in $\otimes \mathcal{H}_k$ (k=1,...,n).

2.33 : Let E and F be two projections. Show that the following three statements are equivalent : (i) $E\mathcal{H} \subseteq F\mathcal{H}$, (ii) $EF = E = FE$, (iii) $E \leq F$ (i.e. $F-E \geq 0$).

2.34 : Show that $A^*A\in\mathcal{B}_0$ if and only if $A\in\mathcal{B}_0$. (Hint : Use Proposition 2.31.)

2.35 : (a) Show that $\|A \hat{\otimes} I\| = \|A\|$. (Hint : Use an orthonormal basis of $\mathcal{H}_1 \otimes \mathcal{H}_2$). (b) If $A\in\mathcal{B}(\mathcal{H}_1)$, $B\in\mathcal{B}(\mathcal{H}_2)$, then $\|A \otimes B\| = \|A\|\,\|B\|$. [Hint : Use (a)]. (c) Suppose that $\|A_n\| \leq M$, $\|B_n\| \leq M$, $A_n \to A$ and $B_n \to B$. Show that $A_n \otimes B_n \to A \otimes B$. (Hint : Use Proposition 2.17.)

2.36 : (Cyclicity of the trace) : Let $A\in\mathcal{B}_1$, $B,C\in\mathcal{B}_2$, $D\in\mathcal{B}(\mathcal{H})$. Then $\mathrm{Tr}BCD = \mathrm{Tr}DBC = \mathrm{Tr}CDB$ and $\mathrm{Tr}AD = \mathrm{Tr}DA$.

2.37 : (a) Let A be Hilbert-Schmidt or trace class. Show that the infinite sum in (2.58), viewed as the limit of a sequence of operators, converges in Hilbert-Schmidt or trace norm respectively. (b) Let $B \in \mathcal{B}_1$. Then $\|B\| \leq \|B\|_{HS} \leq |||B|||_1$.

2.38 : Show that the range of a finite rank operator is a (closed) subspace.

2.39 : If A is self-adjoint in $\mathcal{H}_1$, then $A \hat{\otimes} I$ is essentially self-adjoint in $\mathcal{H}_1 \otimes \mathcal{H}_2$. If furthermore $\mathcal{H}_2$ is finite-dimensional, then $A \hat{\otimes} I = A \otimes I$.

CHAPTER 3 : ONE-PARAMETER UNITARY GROUPS AND FREE PARTICLES

The main purpose of this chapter is to introduce the quantum mechanics of free particles. For this, we first give a general framework to describe a quantum mechanical system, sometimes called the quantum kinematics, and then the dynamics of such a system. Section 3-1 is mathematical in nature and enunciates Stone's theorem for continuous one-parameter unitary groups. In Section 3-2 we briefly sketch the description of a quantum mechanical system in a complex Hilbert space along with some physically motivated examples. Finally Section 3-3 is devoted entirely to studying a particular one-parameter unitary group, viz. the so-called free evolution group and its various ramifications.

3-1 STONE'S THEOREM

Continuous one-parameter unitary groups play very important roles in quantum mechanics. In particular, they are useful to describe the time evolution of quantum mechanical systems, as we shall see in Sections 3-2 and 3-3. Unitary groups arise out of the need to represent symmetry transformations on states of a system, and readers interested in pursuing this question are referred to [P], Wigner [1]. A strongly continuous one-parameter unitary group is defined to be a mapping U from the real line into the set $\mathcal{B}(\mathcal{H})$ of all bounded operators on $\mathcal{H}$ having the following properties :

(a) strong continuity : $\operatorname{s-lim}_{\tau\to 0} (U_{t+\tau} - U_t) = 0$ for all $t\in\mathcal{R}$, (3.1)

(b) unitarity : $U_t^* = U_t^{-1}$ for all $t\in\mathcal{R}$, (3.2)

(c) group property : $U_tU_s = U_sU_t = U_{t+s}$, $t,s\in\mathcal{R}$
and $U_o = I$. (3.3)

We shall attempt to justify the continuity property in the next section. It is however important to state here that for a group, strong continuity at any $t\in\mathcal{R}$ is equivalent to strong continuity at $t = 0$, and that for unitary groups, strong continuity is equivalent to weak continuity (Problem 3.1). The main theorem in this section is Stone's theorem which associates a self-adjoint operator with every continuous one-parameter unitary group. The converse of this theorem is also true, viz. that every self-adjoint operator generates a continuous one-parameter unitary group.

PROPOSITION 3.1 : (Stone's Theorem) : Let $\{U_t\}$, $-\infty < t < \infty$, be a strongly continuous one-parameter unitary group in $\mathcal{H}$. Define a linear operator A (called the infinitesimal generator of $\{U_t\}$) as follows :

$$D(A) = \{f \mid \text{s-lim } i\tau^{-1}(U_\tau - I)f \text{ exists as } \tau \to 0\},$$

and $$Af = \text{s-lim } i\tau^{-1}(U_\tau - I)f \quad \text{for } f \in D(A). \tag{3.4}$$

Then D(A) is dense and A is self-adjoint.

We give the proof of Stone's theorem in Section 3-4 and that of the converse theorem in Chapter 5. An immediate and interesting consequence of Stone's theorem is the following corollary.

COROLLARY 3.2 : (a) Let W be a unitary operator and let U_t, A, D(A) be as in Proposition 3.1. Furthermore, suppose W commutes with U_t for all $t \in \mathcal{R}$. Then W leaves D(A) invariant and

$$WA = AW. \tag{3.5a}$$

(b) In particular, choosing W to be U_t itself, so that the commutation property is trivially satisfied, it follows that U_t leaves D(A) invariant and

$$U_t A = AU_t. \tag{3.5b}$$

Proof : Since W commutes with U_τ, one has

$$Wi\tau^{-1}(U_\tau - I)f = i\tau^{-1}(U_\tau - I)Wf.$$

For all $f \in D(A)$, the left hand side of the above equation has a strong limit as $\tau \to 0$, viz. WAf. Therefore, by Proposition 3.1, $Wf \in D(A)$ and the limit of the right hand side is AWf, implying $WD(A) \subseteq D(A)$ and $WA \subseteq AW$. Since $U_t W = WU_t$ implies $W^{-1}U_t = U_t W^{-1}$, identical conclusions are reached with W^{-1} replacing W, i.e. $W^{-1}D(A) \subseteq D(A)$ and $W^{-1}A \subseteq AW^{-1}$.

Upon multiplication by W, one obtains together with the first set of inclusions that $D(A) \subseteq WD(A) \subseteq D(A)$ and $A \subseteq WAW^{-1} \subseteq AWW^{-1} = A$, which shows that all these inclusions must in fact be equalities. #

3-2 DESCRIPTION OF QUANTUM MECHANICAL SYSTEMS

As in classical mechanics, the description of a quantum mechanical system needs two basic physical concepts, that of the states and that of the observables of the system. Such a description can be made conveniently in a Hilbert space, the properties of which we have studied in Chapter 2. Here we shall not give the experimental motivations leading to this Hilbert space structure. Nor shall we describe any other parallel or more basic formalisms of quantum mechanics. For some of these questions the reader is referred to [EM], [M] or [P].

A state of the system is represented by a positive*) trace class operator ρ in $\mathcal{H}$, called the density operator (or density matrix when a specific basis is referred to), which also satisfies $\mathrm{Tr}\,\rho = 1$. For the definition of trace class operators, the reader is referred to Section 2-3. We may deduce that the non-zero spectrum of ρ consists of positive eigenvalues λ_i, satisfying $\sum_i \lambda_i = 1$, where each λ_i is repeated as many times as the multiplicity of the eigenvalue. In the special case $\rho^2 = \rho$, ρ is a projection with one-dimension-

---

*) A linear operator A is said to be positive, written $A \geq 0$, if $(f,Af) \geq 0$ for all $f \in D(A)$.

al range. Such a state is called pure, in contrast with a general density operator ρ which is said to be a mixed state. In this book we shall be concerned almost exclusively with pure states.

If we denote by F_f the projection operator onto the one-dimensional subspace generated by f, then it is clear that to each pure state F_f one may associate a unit ray in H defined by $\{\alpha f \mid |\alpha| = 1\}$, where $\|f\| = 1$. This correspondence between pure states and unit rays is one-to-one. For the purposes of this book, however, we shall identify a pure state with a unit vector in its range, though this is not strictly accurate. For further discussion of this point, the reader should consult [M] or [P].

The observables of the system are represented by self-adjoint operators in H. In the following, we shall identify an observable with its representative self-adjoint operator A in H. In practice, to assign an operator to a given observable, one is guided by physical intuition derived from the corresponding classical problem. However, there are exceptions to this general practice.

The expectation value of an observable A in a pure state f is defined as

$$\mathrm{Exp}_f(A) = (f, Af) \tag{3.6}$$

whenever it exists. In most cases of physical interest, the operator A happens to be unbounded, making $\mathrm{Exp}_f(A)$ undefined whenever $f \notin D(A)$.

As we shall see in Chapter 5, every self-adjoint

operator A admits a spectral resolution, viz. there exists a unique family of projections $\{E_\lambda\}_{\lambda\varepsilon R}$ with the following properties :

(i) $E_\lambda = E_{\lambda+0} = \underset{\eta\to+0}{\text{s-lim}}\, E_{\lambda+\eta}$,*) (3.7)

(ii) $\underset{\lambda\to-\infty}{\text{s-lim}}\, E_\lambda = 0,\ \underset{\lambda\to+\infty}{\text{s-lim}}\, E_\lambda = I$, (3.8)

(iii) If $\lambda \leq \mu$, then $E_\lambda E_\mu = E_\mu E_\lambda = E_\lambda$. (3.9)

(iv) A vector f belongs to D(A) if and only if $\int_{-\infty}^{\infty} \lambda^2 d(f,E_\lambda f) < \infty$,

and for such an f and any $g\varepsilon H$, one has

$$(g,Af) = \int_{-\infty}^{\infty} \lambda\, d(g,E_\lambda f)\ . \tag{3.10}$$

This family of projections $\{E_\lambda\}$ is called the <u>spectral family</u> of the self-adjoint operator A. With the help of the spectral family $\{E_\lambda\}$, one can define a probability measure $p^A_{(a,b],f}$ on R as

$$p^A_{(a,b],f} = \int_a^b d(f,E_\lambda f), \tag{3.11}$$

where $-\infty < a < b < \infty$ and $f\varepsilon H$, $\|f\| = 1$. That it is a measure is obvious from the definition, and the fact that

$$p^A_{(-\infty,\infty),f} = \int_{-\infty}^{\infty} d(f,E_\lambda f) = \|f\|^2 = 1$$

makes it a probability measure. One interprets $p^A_{(a,b],f}$ as the probability that a measurement of the observable A on the system in the state f will yield a value in the interval (a,b]. This is sometimes known as "Born's probabilistic interpretation" in the quantum mechanics literature and forms

---

*) The notation $\eta\to+0$ means that η tends to zero through positive values.

the basis of all calculations and hence all predictions of the theory. For a state vector $f\in D(A)$, the expectation value of the observable A in the state f is given by (3.6) and (3.10) as

$$\mathrm{Exp}_f(A) \equiv (f,Af) = \int_{-\infty}^{\infty}\lambda d(f,E_\lambda f) = \int_{-\infty}^{\infty}\lambda\, P^A_{(d\lambda),f} \quad ,$$

which is in conformity with the usual definition of the expectation in probability theory.

Now we give three examples of quantum systems with corresponding Hilbert spaces, states and observables.

Example 3.3 : Spinless particle in a box in one dimension.

The relevant Hilbert space is $L^2[a,b]$, i.e. the vector space of the equivalence classes of all Lebesgue measurable complex-valued functions defined on the interval [a,b] which are absolutely square-integrable. The (pure) states of the particle are given by unit vectors $f\in L^2[a,b]$. One relevant observable is the position which is represented by the operator Q defined as follows :

$$D(Q) = \{f\in L^2[a,b] \mid \int_a^b |xf(x)|^2 dx < \infty\} \qquad (3.12)$$

and for $f\in D(Q)$: $(Qf)(x) = xf(x)$. It follows from Proposition 2.16 that Q is a bounded and everywhere defined self-adjoint operator. Its spectral family F_y is given by

$$(F_y f)(x) = \chi_{[a,y]}(x)f(x), \qquad (3.13)$$

where $\chi_I(x)$ is the characteristic function of the interval I. It is easy to verify the properties (3.7) - (3.10) for F_y (Problem 3.2). And the probability that a measurement of the position of the particle in the state f will yield a value in the interval I is given by

$$p^Q_{I,f} = \int_I d(f, F_x f) = \int_I |f(x)|^2 dx. \qquad (3.14)$$

This leads to the commonly found statement that a particle in the state f has position probability density $|f(x)|^2$.

Another observable of physical interest is the momentum P. To define it, we first introduce an operator $\hat{P}$ by

$$D(\hat{P}) = \{f \varepsilon L^2[a,b] \,|\, f \text{ absolutely continuous}, \\ f' \varepsilon L^2[a,b],\ f(a) = f(b) = 0\},$$
$$(\hat{P}f)(x) = -i(h/2\pi)f'(x), \text{ for all } f \varepsilon D(\hat{P}), \qquad (3.15)$$

where h is Planck's constant. Thus defined, $\hat{P}$ is closed and symmetric but not self-adjoint. In fact it has uncountably many self-adjoint extensions P_ϕ, given by the boundary conditions $f(a) = e^{i\phi}f(b)$, $\phi \varepsilon [0, 2\pi]$, as has been pointed out in Section 2-2. The box may be viewed as an apparatus confining the particle to the interval [a,b] and may be used to measure the momentum of the particle. The choice of such an apparatus will determine the boundary condition at the walls of the box and hence the observable P_ϕ (see [P]). For other physical aspects of these observables, see [RO].

<u>Example 3.4</u> : <u>Spinless particle in n-dimensional infinite space</u>.

The Hilbert space of interest here is $L^2(R^n)$ which was introduced in Section 2-1. The (pure) states of the particle are given by unit vectors $f \varepsilon L^2(R^n)$. The position observable is given by an n-component operator $\{Q_j\}_{j=1}^n$ as

$$D(Q_j) = \{f \varepsilon L^2(R^n) \Big| \int |x_j f(\underline{x})|^2 d^n x < \infty\}$$

and

$$(Q_j f)(\underline{x}) = x_j f(\underline{x}) \text{ for } f \varepsilon D(Q_j). \tag{3.16}$$

In contrast with the first example, the Q_j's are all unbounded self-adjoint operators, whose spectral families $\{F_{j,y}\}$ are given by

$$(F_{j,y} f)(\underline{x}) = \chi_{j,y}(\underline{x}) f(\underline{x}). \tag{3.17}$$

Here $\chi_{j,y}$ is the characteristic function of the region $\{\underline{x}\varepsilon R^n | -\infty < x_j \leq y_j$, other components of $\underline{x}$ unrestricted$\}$ in n-space. As before, we arrive at the interpretation of $|f(\underline{x})|^2$ as the position probability density of the particle in the state f.

The <u>momentum observable</u> is represented by an n-component operator P_j defined as follows.

$$D(P_j) = \{f\varepsilon L^2(R^n) \big| \int |k_j \tilde{f}(\underline{k})|^2 d^n k < \infty\}$$

and (3.18)

$$(FP_j f)(\underline{k}) = (h/2\pi) k_j \tilde{f}(\underline{k}), \text{ for all } f\varepsilon D(P_j),$$

where $\tilde{f}$ is the Fourier transform of f in $L^2(R^n)$ as defined in (2.26). The P_j's are all unbounded self-adjoint operators. In fact, each P_j is unitarily equivalent to the position operator Q_j. The unitary equivalence is implemented by the Fourier transformation F. This leads naturally to the interpretation of $|\tilde{f}(\underline{k})|^2$ as the momentum probability density of the particle in the state f (Problem 3.3).

It is simple to verify that $S(R^n) \subseteq D(P_j)$. In fact by Proposition 2.4(b), if $f\varepsilon S(R^n)$, then $\tilde{f}\varepsilon S(R^n)$, and since such an $\tilde{f}$ is continuous and rapidly decreasing at infinity, the integral in the definition of $D(P_j)$ converges. Also for $f\varepsilon S(R^n)$, it follows from the definition (3.18) and (2.20) that

$$(P_j f)(\underline{x}) = (2\pi)^{-n/2} \int e^{i\underline{k}\cdot\underline{x}} (FP_j f)(\underline{k}) d^n k$$

$$= (2\pi)^{-n/2} (h/2\pi) \int e^{i\underline{k}\cdot\underline{x}} k_j \tilde{f}(\underline{k}) d^n k = -i(h/2\pi) \partial f(\underline{x})/\partial x_j . \qquad (3.19)$$

Therefore on $S(R^n)$ the momentum operator is $-ih/2\pi$ times the operation of differentiation, which is the customary representation. However it should be emphasized that this restriction of the momentum operator to $S(R^n)$ is only essentially self-adjoint and not self-adjoint. Often (3.18) is said to define the generalized derivative in $L^2(R^n)$.

Example 3.5 : Particle with spin s (s = 0, 1/2, 1, 3/2, ⋯) in 3-dimensional infinite space.

The appropriate Hilbert space is $L^2(R^3, C^{2s+1})$, i.e. the space of (2s+1)-component functions $f = \{f_m \varepsilon L^2(R^3)$ for each $m = -s, -s+1, \cdots, s-1, s\}$ with inner product

$$(f,g) = \sum_m \int \overline{f_m}(\underline{x}) g_m(\underline{x}) d^3x . \qquad (3.20)$$

The position and momentum operators are defined as before, e.g.,

$$D(Q_j) = \{f \varepsilon L^2(R^3, C^{2s+1}) \,\big|\, \int |x_j f_m(\underline{x})|^2 d^3x < \infty, \text{ for each } m\}$$

and

$$(Q_j f)_m(\underline{x}) = x_j f_m(\underline{x}) . \qquad (3.21)$$

Now we introduce a new set of observables called spin, given by a triplet of operators S_1, S_2, S_3. This is done by defining first

$$(S_3 f)_m(\underline{x}) = (h/2\pi)\, m\, f_m(\underline{x});\ m = -s, -s+1, \ldots, s-1, s$$

$$(S_+ f)_m(\underline{x}) = (h/2\pi) \sqrt{(s+m)(s-m+1)}\, f_{m-1}(\underline{x}) \qquad (3.22)$$

and

$$(S_- f)_m(\underline{x}) = (h/2\pi) \sqrt{(s-m)(s+m+1)}\, f_{m+1}(\underline{x}) .$$

It is easy to see that S_3 and $S_\pm$ are bounded linear operators defined everywhere. S_3 is self-adjoint and $S_+^* = S_-$. One then defines bounded self-adjoint operators $S_1 = \frac{1}{2}(S_+ + S_-)$ and $S_2 = -\frac{1}{2}i(S_+ - S_-)$. It can be verified that S_1, S_2, S_3 satisfy the commutation relation $[S_1, S_2] = i(h/2\pi)S_3$ and its cyclic permutations. We also introduce the total spin operator $\underline{S}^2 \equiv S_1^2 + S_2^2 + S_3^2 = S_+S_- - (h/2\pi)S_3 + S_3^2$ which, being a sum of bounded self-adjoint operators, is self-adjoint. Its action in $L^2(R^3, C^{2s+1})$ is as follows (Problem 3.4) :

$$(\underline{S}^2 f)_m(\underline{x}) = s(s+1)(h/2\pi)^2 f_m(\underline{x}).$$

Remark : The introduction of these Hilbert spaces and some of the associated observables may seem a little arbitrary at this stage. Actually there are deeper reasons for the appearance of all the observables introduced so far, viz. the existence of related symmetry groups. In Example 3.4 for instance, one can define space translation symmetry and obtain P_j as the infinitesimal generator of the unitary representation of this symmetry group in $L^2(R^n)$ [P].

So far we have discussed the description of a quantum mechanical system and its relevant observables. This is the analogue of classical kinematics: a particle is described by giving its position and momentum. The next important step is to give the dynamics of the system or equivalently the equation of motion of the system. As already emphasized, here and in the sequel we shall be interested only in the dynamics of pure states. For a description of the dynamics of a general state, the reader is referred to [M, p. 81].

The time-evolution or the dynamics of a conservative

quantum mechanical system is given by a one-parameter unitary group as explained in the introduction to Section 3-1. Let the group be denoted by $\{U_t\}_{t\in R}$. If the unit vector f represents the state at time t = 0, then the state f_t at time t is given by

$$f_t = U_t f. \qquad (3.23)$$

It is clear from the unitarity of U_t that $\|f_t\| = \|f\| = 1$. It is a natural requirement that the expectation value of any observable in a state f at time t be a continuous function of t. This means that $\mathrm{Exp}_{f_t}(A) = (f_t, Af_t)$ is continuous in t. Making a special choice of A, namely a one-dimensional projection F_g, this leads to the conclusion that $|(U_t f,g)|^2$ is continuous in t. Without changing the expectation value or any predictions of the theory, we can choose a phase factor $\xi(t), |\xi(t)| = 1$, such that $(\xi(t)U_t f,g)$ is a continuous function of t and such that $U_t' \equiv \xi(t)U_t$ is still a group [SI]. As mentioned before (Problem 3.1), weak and strong continuity are equivalent for a unitary group. In view of all this we may assume that the time-evolution of a system is given by (3.23) where U_t is a strongly continuous one-parameter unitary group. We shall call equation (3.23) the <u>Schrödinger equation</u>.

Now by Stone's theorem (Proposition 3.1), it follows that there exists a self-adjoint (in general unbounded) operator H given as $Hg = \text{s-lim } i(h/2\pi)t^{-1}(U_t - I)g$ as $t \to 0$. This operator plays the same role in quantum mechanics as the Hamiltonian function plays in classical mechanics, viz. the infinitesimal generator of time evolution group, and we shall call this operator the <u>Hamiltonian operator</u>.

Let $f \in D(H)$. Then $f_t \equiv U_t f \in D(H)$ by Corollary 3.2(b) and f_t satisfies the differential equation

$$\frac{d}{dt} f_t = -i(h/2\pi)^{-1} H f_t, \tag{3.24}$$

where the derivative is understood as a strong limit. This equation is known in the physics literature as the Schrödinger equation. It has a role parallel to that of Hamilton's equations in classical mechanics. It is a first order differential equation (in a Hilbert space) whose solutions are the trajectories of the (pure) states. In most of this book, by Schrödinger equation we shall mean the integrated form (3.23) and not the differential form (3.24). From (3.24) and (3.23) one sees that, at least formally, the group U_t may be written in terms of its infinitesimal generator as $U_t = \exp(-2\pi i Ht/h)$. A precise definition of the exponential function of a self-adjoint operator will be given in Proposition 5.11.

Let f be any eigenvector of H, i.e. $Hf = (h\lambda/2\pi)f$ for some $\lambda \in \mathcal{R}$. Then $U_t f = \exp(-it\lambda)f$ (this follows from the functional calculus developed in Proposition 5.11). $U_t f$ and f differ by a phase factor and hence define the same ray and state. In other words, the expectation value of any (bounded) observable A in the state f_t is the same for all times, because $\mathrm{Exp}_{f_t}(A) = (U_t f, AU_t f) = (f, Af) = \mathrm{Exp}_f(A)$. Such a state is called a <u>stationary state</u>. Conversely, suppose $\mathrm{Exp}_{f_t}(A) = \mathrm{Exp}_f(A)$ for every (bounded) observable. By taking $A = F_g$, where g is any unit vector in $\mathcal{H}$, we have $|(U_t f, g)|^2 = |(f,g)|^2$. In particular, for every vector g such that $(f,g) = 0$, one has $(U_t f, g) = 0$. By the projection theorem, $U_t f$ belongs to the one-dimensional subspace generated by f, i.e. $U_t f = \alpha(t) f$, where $\alpha(t)$ is a complex-valued continuous one-parameter group. Such a group has a unique representation as an exponential function. Furthermore, since $|\alpha(t)| = 1$, this implies that $\alpha(t) = \exp(-it\lambda)$ for some real λ (Problem 3.6). In the equation $U_t f = \exp(-it\lambda)f$, the right-hand side clearly admits a strong derivative at $t = 0$, so that by Proposition 3.1 $f \epsilon D(H)$ and $Hf = (h/2\pi)\lambda f$, i.e. f is an eigenvector of H with eigenvalue $(h/2\pi)\lambda$.

We shall call an observable A a <u>constant of motion</u> if $\text{Exp}_{f_t}(F_\lambda) = \text{Exp}_f(F_\lambda)$ for all $f \epsilon \mathcal{H}$ and all real λ and t, where $\{F_\lambda\}$ is the spectral family of A. From the polarization identity (see Problem 2.22) it follows easily that (Problem 3.5)

$$U_t^{-1} F_\lambda U_t = F_\lambda \text{ for all } \lambda \text{ and } t \text{ in } \mathbb{R}. \tag{3.25}$$

By Problem 3.11 and Proposition 5.9, this is equivalent to

$$E_\mu F_\lambda = F_\lambda E_\mu \text{ for all } \lambda, \mu \epsilon \mathbb{R}, \tag{3.26}$$

where $\{E_\mu\}$ is the spectral family of the infinitesimal generator H of U_t. When (3.26) is satisfied, one says that the two (possibly unbounded) operators A and H commute. Since H certainly commutes with itself in this sense, the Hamiltonian is a constant of motion and is interpreted as the <u>energy observable</u> of the system.

Let A be any observable (not necessarily a constant of motion) and $\{F_\mu\}$ its spectral family. Then the probability measure $p^A_{I,f_t} \equiv \int_I d(f_t, F_\lambda f_t)$ of the observable A in the evolving state f_t at time t is identical with the probability measure of the evolved observable $A_t \equiv U_t^* A U_t$ at time t in the fixed initial state f. This is so because

$$\int_I d(f_t, F_\lambda f_t) = \int_I d(U_t f, F_\lambda U_t f) = \int_I d(f, U_t^* F_\lambda U_t f)$$

and the family $\{U_t^* F_\lambda U_t\}$ for any fixed t is the spectral family of the self-adjoint operator A_t (Problem 3.5). In the physics literature, these two points of view are distinguished and called the Schrödinger picture (evolving state vector) and the Heisenberg picture (evolving observables) respectively. The equations of evolution (both the differential and the integrated forms) that we have so far given are equations for the state vector and hence appropriate to the Schrödinger picture. One can give equivalent equations of evolution in the Heisenberg picture as follows : One differentiates formally the definition $A_t = U_t^* A U_t$ with respect to t and obtains

$$\frac{dA_t}{dt} = i(h/2\pi)^{-1}(HA_t - A_t H) \equiv i(h/2\pi)^{-1}[H, A_t]. \qquad (3.27)$$

This is called the Heisenberg equation of motion. (This equation is analogous to Hamilton's equation in classical mechanics written in terms of Poisson brackets, i.e. $dA/dt = -\{H,A\}$, where A is any classical observable and H is the Hamilton function.) When both A and H are bounded, the equation makes sense on any vector in $\mathcal{H}$. The general case when A and H are unbounded self-adjoint operators is more complicated, and it is precisely for this reason that we shall not use the differential version of either the Schrödinger or the Heisenberg equation.

We still have not reached a stage where we can actually compute the motion of the particle given the initial state f at time t = 0. What we have done is similar to writing down Newton's equation $\underline{F} = m\ddot{\underline{q}}$, but we cannot start computing until we know the force $\underline{F}$. In fact, we can distinguish between various classical theories by what we take for $\underline{F}$ or equivalently for the Hamilton function $H(\underline{p},\underline{q})$. Likewise, we can have various quantum mechanical theories by making appropriate choices for H. For every physical system, the choice of H is dictated by the underlying dynamical symmetry group and by the corresponding classical Hamilton function if one has any. For example, in non-relativistic physics the principle of Galilean relativity leads to the invariance of the theory under the Galilean symmetry group and this in turn restricts the Hamilton function for a single particle in non-relativistic classical mechanics to be of the form $H(\underline{p},\underline{q}) = \underline{p}^2/2m + V(\underline{q})$, where V is called the potential function and $\underline{p}^2/2m$ is the kinetic energy. Then various non-relativistic theories will be distinguished by the choice of the potential function V. For a more elaborate discussion of these points, the reader is referred to [P]. In quantum mechanics, one has the task of making H a self-adjoint operator after one has obtained a form for it from physical considerations. A study of this question is deferred until Chapter 8, and in this section we shall be content with giving a few simple examples.

Example 3.6 : Free non-relativistic spinless particle in $\mathcal{R}^n$.

Classically the Hamilton function is given by $H = \underline{p}^2/2m$, where m is identified with the inertial mass of the particle. Reminding ourselves that the appropriate Hilbert space for the description of such a quantum mechanical system is $L^2(\mathcal{R}^n)$, in which the momentum operators P_j are defined by (3.18), we write the Hamiltonian operator for a free particle as $K_o = (2m)^{-1} \sum_{j=1}^{n} P_j^2$, defined as follows :

$$D(K_o) = \{f \varepsilon L^2(\mathcal{R}^n) \mid \int |\underline{k}^2 \tilde{f}(\underline{k})|^2 d^n k < \infty\} \tag{3.28}$$

and $$(\mathcal{F}K_o f)(\underline{k}) = (2m)^{-1}(h/2\pi)^2 \underline{k}^2 \tilde{f}(\underline{k}), \text{ for } f \varepsilon D(K_o). \tag{3.29}$$

By Proposition 2.16, K_o is a self-adjoint operator. Furthermore, if $f \varepsilon \mathcal{S}(\mathcal{R}^n)$, then it is easy to show that

$$(K_o f)(\underline{x}) = -(2m)^{-1}(h/2\pi)^2 (\Delta f)(\underline{x}), \tag{3.30}$$

where Δ is the Laplacian $\sum_{j=1}^{n} \partial^2/\partial x_j^2$. Thus for $f \varepsilon \mathcal{S}(\mathcal{R}^n)$ one can write down the free Schrödinger equation in differential form as $df_t/dt = -(2m)^{-1}(h/2\pi)^2 \Delta f_t$. In the next section we shall return to this example and establish various properties of the above operator K_o.

Example 3.7 : Spinning electron in a constant magnetic field.

It is experimentally observed and also theoretically predicted that an electron carries with it a magnetic (dipole) moment given by $\underline{\mu} = g\mu_B \underline{S}$, where $\mu_B = eh(4\pi mc)^{-1}$ is the Bohr magneton. g is the gyromagnetic ratio of the electron (very close to 2) and $\underline{S}$ is the spin-triplet observable, with $s = \frac{1}{2}$. In this case, again by analogy with the classical theory, the Hamiltonian operator is formally given by the energy of interaction between the moment associated with the electron and the magnetic field, i.e. $H = -\underline{\mu}.\underline{\mathcal{B}}$, where $\underline{\mathcal{B}}$ is the magnetic field intensity vector. The relevant Hilbert space being $L^2(\mathcal{R}^3, \mathcal{C}^2)$, we conclude that $H = -g\mu_B|\underline{\mathcal{B}}|S_3$, where we have chosen the third direction in $\mathcal{R}^3$ to be the one parallel to the field $\underline{\mathcal{B}}$. Since S_3 is a bounded self-adjoint operator in

$L^2(\mathbb{R}^3, \mathbb{C}^2)$, so is H (see Example 3.5).

Example 3.8 : Spinless relativistic free particle in $\mathbb{R}^n$.

We know that the energy of a classical relativistic free particle is given by $K_0^R = (m^2c^4 + \underline{p}^2c^2)^{\frac{1}{2}}$, where $\underline{p}$ is its momentum, c is the speed of light and m is the rest mass. So the Hamiltonian operator is defined as :

$$D(K_0^R) = \{f\varepsilon L^2(\mathbb{R}^n) \mid \int \underline{k}^2|\tilde{f}(\underline{k})|^2 d^n k < \infty\} \tag{3.31}$$

and for $f\varepsilon D(K_0^R)$, $(\mathcal{F}K_0^R f)(\underline{k}) = [m^2c^4 + (h/2\pi)^2\underline{k}^2c^2]^{\frac{1}{2}}\tilde{f}(\underline{k})$.

As before, Proposition 2.16 shows that this is a self-adjoint operator.

3-3 FREE NON-RELATIVISTIC DYNAMICS

In this section we study in detail various properties of the operator K_0, the Hamiltonian of a spinless non-relativistic free particle, introduced in the last section. We also obtain a certain convenient description of the free time-evolution operator U_t associated with K_0 which will be very useful in scattering theory. We end the section with some results on asymptotic position probability measures for $t\to\pm\infty$.

For convenience of presentation, from now on we shall set $h/2\pi = 2m = 1$, keeping in mind that, whenever physical quantities are computed and a comparison with experimental results is attempted, $h/2\pi$ and m have to be reintroduced. This simplifies various expressions, e.g. for $f\varepsilon D(K_0)$,

$(\mathcal{F}K_o f)(\underline{k}) = \underline{k}^2\tilde{f}(\underline{k})$ instead of (3.29). We have seen that K_o is a self-adjoint operator and that its restriction to $\mathcal{S} \equiv \mathcal{S}(\mathcal{R}^n)$ is $-\Delta$. Now we prove

PROPOSITION 3.9 : K_o restricted to $\mathcal{S}(\mathcal{R}^n)$ is essentially self-adjoint.

Proof : We recall from Corollary 2.15 that a symmetric operator A is essentially self-adjoint if and only if the ranges of the operators $(A \pm i)$ are dense in $\mathcal{H}$. $K_o|_{\mathcal{S}}$ *) is symmetric since it is the restriction of the self-adjoint operator K_o to a dense subset $\mathcal{S}$ of its domain $D(K_o)$. Since for $g \in \mathcal{S}$ $(\mathcal{F}(K_o \pm i)g)(\underline{k}) = (\underline{k}^2 \pm i)\tilde{g}(\underline{k})$ and multiplication by $(\underline{k}^2 \pm i)$ maps $\mathcal{S}$ onto itself in a one-to-one fashion, it follows that the operators $(K_o|_{\mathcal{S}} \pm i)$ map $\mathcal{S}$ onto itself, once we remember from Proposition 2.4 (b) that the Fourier transformation leaves $\mathcal{S}$ invariant. Therefore, the ranges of $(K_o|_{\mathcal{S}} \pm i)$ are $\mathcal{S}$ and hence dense. This completes the proof. #

Remark : It is also true that K_o restricted to $C_o^\infty(\mathcal{R}^n)$ is essentially self-adjoint. For this it suffices to show that $K_o|_{\mathcal{S}} \subseteq \overline{K_o|_{C_o^\infty}}$, because we know that $K_o|_{C_o^\infty} \subseteq K_o|_{\mathcal{S}}$ and therefore $\overline{K_o|_{C_o^\infty}} \subseteq \overline{K_o|_{\mathcal{S}}} = K_o$. To this end we need to construct for every $g \in \mathcal{S}(\mathcal{R}^n)$ a sequence $g_n \in C_o^\infty(\mathcal{R}^n)$ such that $\text{s-lim}\, g_n = g$ and $\text{s-lim}\, K_o g_n = K_o g$, which can be shown to be possible (Problem 3.7).

The self-adjoint operator K_o is not a differential operator in the usual sense because a function $f(\underline{x})$ in $D(K_o)$

*) $A|_D$ denotes the restriction of the operator A to a subset $D \subseteq D(A)$.

need not be differentiable. Nevertheless, for $n = 3$ the functions in $D(K_0)$ have a certain regularity property as proven in

PROPOSITION 3.10 : Let $n = 3$. Then every $f \in D(K_0)$ is (equivalent to) a bounded, uniformly continuous function of $\underline{x}$.

Proof : Let $\alpha > 0$. Then

$$\int_{R^3} (\underline{k}^2 + \alpha^2)^{-2} d^3k = 4\pi \int_0^\infty (k^2 + \alpha^2)^{-2} k^2 dk = \pi^2/\alpha.$$

By applying the Schwarz inequality in $L^2(R^3)$ we obtain that

$$[\int |\tilde{f}(\underline{k})| d^3k]^2 = [\int |(\underline{k}^2 + \alpha^2)\tilde{f}(\underline{k})| (\underline{k}^2 + \alpha^2)^{-1} d^3k]^2$$

$$\leq \frac{\pi^2}{\alpha} \int |(\underline{k}^2 + \alpha^2)\tilde{f}(\underline{k})|^2 d^3k = \frac{\pi^2}{\alpha} \|(K_0 + \alpha^2)f\|^2 < \infty.$$

Therefore $\tilde{f} \in L^1(R^3)$. It now follows from (2.20) that $f(\underline{x})$ is bounded :

$$|f(\underline{x})| \leq (2\pi)^{-3/2} \int |\tilde{f}(\underline{k})| d^3k \leq c\, \alpha^{-1/2} \|(K_0 + \alpha^2)f\|$$

$$\leq c\, (\alpha^{-1/2} \|K_0 f\| + \alpha^{3/2} \|f\|). \tag{3.32}$$

Similarly one infers the continuity of $f(\underline{x})$ from (2.20) and the Lebesgue dominated convergence theorem (see also Problem 3.10). #

Remark : The above result is a special case of the Sobolev inequality [SO] and depends essentially on the dimension 3 of the space.

We now study properties of the one-parameter unitary group corresponding to K_0. These will be frequently used in later chapters. We define for $t \in R$ and $f \in L^2(R^n)$

$$(FU_t f)(\underline{k}) = \exp(-i\underline{k}^2 t)\tilde{f}(\underline{k}) \tag{3.33}$$

and first prove

PROPOSITION 3.11 : $\{U_t\}$ defined by (3.33) forms a continuous one-parameter unitary group in $L^2(R^n)$, and its infinitesimal generator is K_0.

Proof : (i) The group property of $\{U_t\}$ is an immediate consequence of the definition (3.33). Its continuity follows provided that we can show that U_t is weakly continuous at $t = 0$ (see Problem 3.1). This is done as follows :

$$(f,(U_t-I)g) = \int \bar{\tilde{f}}(\underline{k})[\exp(-i\underline{k}^2t)-1]\tilde{g}(\underline{k})d^nk.$$

The integrand converges pointwise to zero as $t\to 0$ and is uniformly (with respect to t) majorized by $2|\tilde{f}(\underline{k})\tilde{g}(\underline{k})|$, which is integrable since f and g are in $L^2(R^n)$. Then by the Lebesgue dominated convergence theorem, $(f,(U_t-I)g)$ converges to zero as $t\to 0$.

(ii) (3.33) implies that $\|U_tf\|^2 = \|\tilde{f}\|^2 = \|f\|^2$ and $(f,U_tg) = (U_{-t}f,g)$. The first of these two identities shows that U_t is defined everywhere, and the second one that $U_t^* = U_{-t}$. Hence $U_t^*U_t = U_{-t}U_t = I$ and $U_tU_t^* = I$, i.e. U_t is unitary.

(iii) One has for $\alpha\in R$

$$|it^{-1}[\exp(-it\alpha)-1] - \alpha| = |\int_0^{\alpha}[\exp(-its)-1]ds| \leq 2\alpha. \quad (3.34)$$

Now let $f\in D(K_0)$. Then $\|[it^{-1}(U_t-I) - K_0]f\|^2 =$

$$\int|\tilde{f}(\underline{k})|^2|it^{-1}[\exp(-it\underline{k}^2)-1] - \underline{k}^2|^2 \, d^nk. \quad (3.35)$$

The integrand in (3.35) converges pointwise to zero, and by (3.34) the integrand is majorized uniformly in t by $4|\underline{k}^2\tilde{f}(\underline{k})|^2$, which is integrable. Hence the Lebesgue dominated convergence theorem allows us to conclude that

$$\underset{t\to 0}{\text{s-lim}}\,[it^{-1}(U_t - I) - K_0]f = \theta \text{ for all } f\varepsilon D(K_0).$$

Therefore the infinitesimal generator of U_t is equal to K_0 on $D(K_0)$. Since K_0 is self-adjoint, the infinitesimal generator of U_t must be K_0. #

U_t defined by (3.33) gives the evolution operator for a free non-relativistic particle. Passing to the $\underline{x}$-representation we write

$$(U_t f)(\underline{x}) = \ell.\text{i.m.}(2\pi)^{-n/2}\int e^{i\underline{k}\cdot\underline{x}}e^{-ik^2 t}\tilde{f}(\underline{k})\, d^n k. \tag{3.36}$$

The following lemma and its corollary play a crucial role in scattering theory.

<u>LEMMA 3.12</u> : Let $f\varepsilon L^1(R^n)\cap L^2(R^n)$ and $t \neq 0$. Then

$$(U_t f)(\underline{x}) = (4\pi it)^{-n/2}\int \exp(i|\underline{x}-\underline{y}|^2/4t)f(\underline{y})\, d^n y, \tag{3.37}$$

where the branch of the square-root is chosen so that

$$(4\pi it)^{-n/2} = |4\pi t|^{-n/2}\cdot\begin{cases}\exp(-in\pi/4) & \text{if } t > 0\\ \exp(in\pi/4) & \text{if } t < 0.\end{cases}$$

<u>Proof</u> : (i) We shall prove the lemma for $n = 1$, since the extension to the general case is straightforward. The right-hand side of the relation (3.36) is an ordinary integral if $f\varepsilon S(R)$. In particular, for $f_a(x) \equiv \exp[-(x-a)^2]$, an elementary calculation shows that

$$\begin{aligned}(U_t f_a)(x) &= (2\pi)^{-\frac{1}{2}}\int e^{ikx}\, e^{-ik^2 t}\tilde{f}_a(k)\, dk\\ &= (4\pi)^{-\frac{1}{2}}\int e^{ikx}\, e^{-ik^2 t}e^{-k^2/4}\, e^{-ika}\, dk\\ &= (1 + 4it)^{-\frac{1}{2}}\exp\{-(x-a)^2/(1 + 4it)\},\end{aligned} \tag{3.38}$$

where for complex α we have chosen the branch of the square root so that $\mathrm{Re}\sqrt{\alpha} > 0$ and used the fact that the Fourier transform of $\exp(-\alpha x^2/2)$ is $\alpha^{-\frac{1}{2}}\exp(-k^2/2\alpha)$ (see Problem 2.10). On the other hand,

$$(4\pi it)^{-\frac{1}{2}}\int \exp(i|x-y|^2/4t)f_a(y)dy = (2it)^{-\frac{1}{2}}\exp(i|x-a|^2/4t)\cdot$$

$$(2\pi)^{-\frac{1}{2}}\int \exp[-i(y-a)(x-a)/2t]\exp[-(y-a)^2(1-i/4t)]dy$$

$$= (1+4it)^{-\frac{1}{2}}\exp[-(x-a)^2/(1+4it)]. \qquad (3.39)$$

Comparing (3.38) and (3.39) we arrive at (3.37) for each f_a and hence for the set $\mathcal{D}$ consisting of finite linear combinations of such functions, i.e.

$$(U_t f)(x) = (4\pi it)^{-\frac{1}{2}}\int \exp(i|x-y|^2/4t)f(y)dy \qquad (3.40)$$

for all $f\in\mathcal{D}$.

(ii) Next we claim that $\mathcal{D}$ is dense in $L^2(\mathbb{R})$. To see this let g be any vector orthogonal to f_a for all a. Then by (2.23) $\int \bar{\tilde{g}}(k)\exp(-ika-k^2/4)dk = 0$. This means that the Fourier transform of the function $\bar{\tilde{g}}(k)\exp(-k^2/4)$ at a equals zero. By the unitarity of F, $\tilde{g} = 0$ a.e. or $g = \theta$. Hence $\mathcal{D}$ is dense by the denseness criterion on page 28.

(iii) We know from Proposition 3.11 that $\|U_t f\| = \|f\|$. Therefore, the relation (3.40) can be extended to the whole of $L^2(\mathbb{R})$ and written

$$(U_t f)(x) = \ell.i.m.(4\pi it)^{-\frac{1}{2}}\int \exp[i(x-y)^2/4t]f(y)dy, \qquad (3.41)$$

just as was the case with the Fourier transformation. In particular, if $f\in L^1(\mathbb{R})\cap L^2(\mathbb{R})$, then the right-hand side makes sense as an ordinary integral and (3.37) is valid. #

COROLLARY 3.13 : Let $f \in L^1(R^n) \cap L^2(R^n)$, $t \neq 0$. Then $U_t f$ is a bounded function of $\underline{x}$ and

$$\sup_{\underline{x} \in R^n} |(U_t f)(\underline{x})| \leq |4\pi t|^{-n/2} \|f\|_1 \qquad (3.42)$$

The proof follows immediately from (3.37).

Remark 3.14 : We can rewrite (3.41) in n dimensions as

$$(U_t f)(\underline{x}) = \text{l.i.m.} \int K(\underline{x}-\underline{y};t)\, f(\underline{y})\, d^n y, \qquad (3.43)$$

where $$K(\underline{x}-\underline{y};t) = (4\pi it)^{-n/2} \exp[i(\underline{x}-\underline{y})^2/4t]. \qquad (3.44)$$

The integral kernel $K(\underline{x}-\underline{y};t)$ is called the free propagator.

Remark 3.15 : An interesting consequence of the inequality (3.42) is the following. Let Δ be a measurable set in R^n with Lebesgue measure $|\Delta| < \infty$ and F_Δ the projection operator in $L^2(R^n)$ associated with the set Δ by (2.39). Then the probability of finding the freely evolving particle at time t in the set Δ tends to zero as $t \to \pm\infty$ or equivalently,

$$\|F_\Delta U_t f\|^2 \to 0 \text{ as } t \to \pm\infty \text{ for all } f \in L^2(R^n). \qquad (3.45)$$

For a proof of this, one notes that by (3.42) one has for all $f \in S(R^n)$

$$\|F_\Delta U_t f\|^2 = \int_\Delta |(U_t f)(\underline{x})|^2\, d^n x \leq |4\pi t|^{-n} |\Delta|\, \|f\|_1^2,$$

which tends to zero as $t \to \pm\infty$. Since $S(R^n)$ is dense in $L^2(R^n)$ and since $\|F_\Delta U_t\| \leq 1$, we arrive at (3.45) by using Proposition 2.17.

For further discussion, it is useful to factorize U_t into a product of two simpler operators as defined in the following lemma.

LEMMA 3.16 : Suppose $t \neq 0$. Define two linear operators C_t and Q_t in $L^2(R^n)$ by

$$(C_t f)(\underline{x}) = (2it)^{-n/2}\exp(i\underline{x}^2/4t)\tilde{f}(\underline{x}/2t)$$
$$(Q_t f)(\underline{x}) = \exp(i\underline{x}^2/4t)f(\underline{x}) \qquad (3.46)$$

Then C_t and Q_t are unitary operators and

$$U_t = C_t Q_t \, . \qquad (3.47)$$

Proof : The unitarity of Q_t is obvious and that of C_t follows from the unitarity of the Fourier transformation. Since U_t and C_tQ_t are both unitary operators, we need to establish their equality on a dense domain and then extend the result to the whole of $L^2(R^n)$ by Lemma 2.27. Let $f \in S(R^n)$. Then by (3.37) we write

$$(U_t f)(\underline{x}) = (2it)^{-n/2}(2\pi)^{-n/2}\int \exp(i|\underline{x}-\underline{y}|^2/4t)f(\underline{y})d^n y$$

$$= (2it)^{-n/2}\exp(i\underline{x}^2/4t)(2\pi)^{-n/2}\int \exp(-i\underline{y}\cdot\underline{x}/2t)\exp(i\underline{y}^2/4t)f(\underline{y})d^n y$$

$$= (2it)^{-n/2}\exp(i\underline{x}^2/4t)(2\pi)^{-n/2}\int \exp(-i\underline{y}\cdot\underline{x}/2t)(Q_t f)(\underline{y})d^n y$$

$$= (2it)^{-n/2}\exp(i\underline{x}^2/4t)\ (FQ_t f)(\underline{x}/2t) = (C_tQ_t f)(\underline{x}).$$ #

The next proposition establishes the asymptotic behaviour of U_t for $t \to \pm\infty$.

PROPOSITION 3.17 : Let $f \in L^2(R^n)$. Then

$$\lim_{t\to\pm\infty} \|U_t f - C_t f\| = 0. \qquad (3.48)$$

Proof : Using Lemma 3.16 and the unitarity of C_t, we have

$$\|U_t f - C_t f\|^2 = \|C_tQ_t f - C_t f\|^2 = \|Q_t f - f\|^2$$
$$= \int |\exp(i\underline{x}^2/4t) - 1|^2 |f(\underline{x})|^2\, d^n x.$$

Since the integrand is bounded by the integrable function $4|f(\underline{x})|^2$ and converges pointwise to zero as $t\to\pm\infty$, an application of the dominated convergence theorem leads to the desired result. #

The Proposition 3.17 leads naturally to the following observation. Let Δ be a measurable subset of R^n and f be a vector in $L^2(R^n)$. Then

$$\left|\int_\Delta |(U_t f)(\underline{x})|^2 d^n x - \int_\Delta |(C_t f)(\underline{x})|^2 d^n x\right| = \left|\,\|F_\Delta U_t f\|^2 - \|F_\Delta C_t f\|^2\right|$$

$$= (\|F_\Delta U_t f\| + \|F_\Delta C_t f\|)\left|\,\|F_\Delta U_t f\| - \|F_\Delta C_t f\|\,\right|. \tag{3.49}$$

By applying the inequality (2.88) and observing that $\|F_\Delta\| \leq 1$, we see that the last member of (3.49) is majorized by $2\|f\|\,\|U_t f - C_t f\|$, which by virtue of Proposition 3.17 goes to zero as $t\to\pm\infty$. This means that given an initial state $f\in L^2(R^n)$, $\|f\| = 1$, the position probability measure over any measurable subset $\Delta \subseteq R^n$ for the freely evolving state $U_t f$ can be asymptotically replaced by that of the state $C_t f$, i.e.

$$\lim_{t\to\pm\infty} \int_\Delta |(U_t f)(\underline{x})|^2 d^n x = \lim_{t\to\pm\infty} \int_\Delta |(C_t f)(\underline{x})|^2 d^n x$$

in the sense that if the limit on one side exists, so does the other and they are equal.

Now

$$|(C_t f)(\underline{x})|^2 = |2t|^{-n} |\tilde{f}(\underline{x}/2t)|^2, \tag{3.50}$$

so that for large positive or negative times the position probability density of the freely evolving state f can be replaced by $|2t|^{-n}|\tilde{f}(\underline{x}/2t)|^2$, which is simply the probability that it has momentum $\underline{x}/2t$, the correct classical momentum to get from the origin to the point $\underline{x}$ in time t.

As an example, we compute the asymptotic probability $P^{\pm}_{free}(f,C)$ that the particle with initial state f, evolving freely, will be found as $t\to\pm\infty$ in a cone C with apex at the origin. From the definition it follows that

$$\begin{aligned} P^{\pm}_{free}(f,C) &= \lim_{t\to\pm\infty} \int_C |(U_t f)(\underline{x})|^2 d^n x \\ &= \lim_{t\to\pm\infty} \int_C |2t|^{-n} |\tilde{f}(\underline{x}/2t)|^2 d^n x = \int_{\pm C} |\tilde{f}(\underline{k})|^2 d^n k. \end{aligned} \tag{3.51}$$

The last step follows by making the change of variable $\underline{k} = \underline{x}/2t$ in the integral over C. This maps C onto itself or its reflection through the origin depending on whether $t > 0$ or $t < 0$. This explains the appearance of "$\pm C$" in the final result. This expression conforms to the intuition that for large positive (or negative) times the probability that the particle will be in C is the same as the probability that its momentum lies in the cone +C (respectively -C). A generalization of (3.51) for interacting particles will be given in Sections 7-3 and 15-3.

3-4 NOTES AND SUPPLEMENTARY MATERIAL

A. Proof of Theorem 3.1 :

(i) Define

$$R_{+i} f = i\int_0^{\infty} e^{-s} U_s f \, ds$$

and

$$R_{-i} f = -i\int_{-\infty}^{0} e^{s} U_s f \, ds, \text{ for } f\in\mathcal{H}. \tag{3.52}$$

At this point we shall not dwell on the definition of these integrals, except to say that they can be defined as strong limits of Riemann sums of vectors, similar to the ordinary Riemann integrals. More will be said on this in Section 4-4. From the definition it is clear that $\|R_{\pm i} f\| \leq \|f\|$ for all

$f \in H$, or in other words, $R_{\pm i}$ are bounded linear operators defined everywhere.

(ii) Now we show that the strong derivative of U_t at $t = 0$ exists on the ranges of $R_{\pm i}$. Let f be any vector in H. Then

$$i\tau^{-1}(U_\tau - I)R_{+i}f = -\tau^{-1}\int_0^\infty e^{-s}U_{s+\tau}f\,ds + \tau^{-1}\int_0^\infty e^{-s}U_s f\,ds$$

$$= -\tau^{-1}(e^\tau - 1)\int_\tau^\infty e^{-\sigma}U_\sigma f\,d\sigma + \tau^{-1}\int_0^\tau e^{-s}U_s f\,ds$$

$$= \tau^{-1}(e^\tau - 1)\{iR_{+i}f + \int_0^\tau e^{-\sigma}U_\sigma f\,d\sigma\} + \tau^{-1}\int_0^\tau e^{-s}U_s f\,ds.$$

By continuity of $e^{-s}U_s f$, the second term tends to f as $\tau \to 0$. Similarly, the first term tends to $iR_{+i}f$. This leads to the conclusion that the strong derivative of U_t exists on $R_{+i}H$. A similar conclusion follows for $R_{-i}H$, and we have

$$AR_{\pm i}f = \pm iR_{\pm i}f + f \text{ for all } f \in H. \tag{3.53}$$

(iii) Next we prove that $R_{+i}H$ is dense in H. Suppose g is orthogonal to $R_{+i}H$, i.e. $(g, R_{+i}f) = 0$ for all $f \in H$. Now, since $U_t U_s = U_s U_t = U_{t+s}$ it follows from the definition (3.52) that U_t commutes with R_{+i} and hence that U_t leaves $R_{+i}H$ invariant. Therefore

$$(g, U_t R_{+i} f) = 0 \text{ for all } f \in H. \tag{3.54}$$

Then using (3.52) and (3.54), one gets

$$\int_0^\infty e^{-s}(g, U_{s+t}f)ds = e^t \int_t^\infty e^{-\sigma}(g, U_\sigma f)d\sigma = 0,$$

or

$$\int_t^\infty e^{-\sigma}(g, U_\sigma f)d\sigma = 0\,. \tag{3.55}$$

Since (3.55) is true for all t, a differentiation with respect to t along with the observation that the integrand is a continuous function leads to the conclusion that $(g, U_t f) = 0$

for all t. In particular $(g,f) = 0$. Since f is an arbitrary vector in $\mathcal{H}$, it follows that $g = \theta$, so that by the statement following the projection theorem on page 28, $R_i\mathcal{H}$ is dense in $\mathcal{H}$. A similar conclusion follows for $R_{-i}\mathcal{H}$.

So far we have shown that $R_{\pm i}\mathcal{H} \subseteq D(A)$ and that $R_{\pm i}\mathcal{H}$ are dense in $\mathcal{H}$, so that A is a densely defined linear operator.

(iv) Here we prove that A is symmetric. Let $f,g \in D(A)$. Then $(Af,g) = \lim_{t\to 0} (it^{-1}(U_t - I)f,g) = \lim (f,-it^{-1}(U_{-t}-I)g) = (f,Ag)$. Now for all $f \in D(A)$, $\|(A\pm i)f\|^2 = \|Af\|^2 + \|f\|^2$ by (2.48). Therefore, $(A\pm iI)f = \theta$ implies $\|f\| = 0$, hence $f = \theta$. Thus $(A\pm i)$ are also invertible.

(v) Finally we claim that the operators $(A\pm i)$ map $D(A)$ onto $\mathcal{H}$. The equation (3.53) can be rewritten as

$$(A\mp i)R_{\pm i}f = f, \text{ for all } f \in \mathcal{H}. \tag{3.56}$$

It follows that $(A\mp i)$ map $R_{\pm i}\mathcal{H}$ onto $\mathcal{H}$. A fortiori they map $D(A)$ onto $\mathcal{H}$, since $R_{\pm i}\mathcal{H} \subseteq D(A)$. It now follows from Proposition 2.14 that A is self-adjoint, which completes the proof. #

We may add that (3.56) has the following interesting consequences (obtained upon multiplication by $(A\mp i)^{-1}$) :

$$R_{\pm i} = (A\mp i)^{-1} \text{ and hence } R_{\pm i}\mathcal{H} = D(A). \tag{3.57}$$

In fact, similar statements can be made for any complex number z (Im z ≠ 0) in place of ±i, viz. $(A-z)^{-1} \in \mathcal{B}(\mathcal{H})$ and

$$(A-z)^{-1} = \begin{cases} i\int_0^\infty e^{izs}U_s ds; & \text{Im } z > 0 \\ -i\int_{-\infty}^0 e^{izs}U_s ds; & \text{Im } z < 0. \end{cases} \tag{3.58}$$

The operator $(A-z)^{-1}$ is called the resolvent of the self-adjoint operator A and denoted by $R_z(A)$ or simply be R_z when only one operator A is involved. We shall return to resolvents and their properties in Chapter 5. The verification of the relation (3.58) is left as an exercise (Problem 3.8).

B. We end this section with the following remarks with reference to Corollary 3.13. For this purpose, we define L^p-spaces ($p \geq 1$) in a manner similar to that of Section 2-2. They consist of all Lebesgue measurable complex-valued functions on R^n which satisfy the following condition, viz.

$$\int_{R^n} |f(\underline{x})|^p d^n x < \infty . \tag{3.59}$$

It is well-known that the set of all such functions forms a vector space under the usual rules of addition and multiplication by complex scalars. As in the case of L^2, one says that two functions are equivalent if they differ only on a set of Lebesgue measure zero. Then $L^p(R^n)$ is defined to be the space of equivalence classes of such functions, satisfying (3.59). It can be shown that $L^p(R^n)$ is complete with respect to the norm defined as follows :

$$||f||_p = (\int |f(\underline{x})|^p d^n x)^{1/p}. \tag{3.60}$$

One important property of L^p-spaces is the Hölder inequality. It states that if $p^{-1} + q^{-1} = 1$ and if $f \varepsilon L^p(R^n)$ and $g \varepsilon L^q(R^n)$, then the product function $\bar{f} \cdot g$ (defined as $(\bar{f} \cdot g)(\underline{x}) = \overline{f(\underline{x})} g(\underline{x})$) belongs to $L^1(R^n)$ and furthermore,

$$|\int \bar{f}(\underline{x}) g(\underline{x}) d^n x| \leq ||f||_p ||g||_q . \tag{3.61}$$

It is clear that the Schwarz inequality for the Hilbert space $L^2(R^n)$ appears as a special case of (3.61) when $p = q = 2$.

One has similarly a generalisation of (3.42). Let $1 \leq p < 2$, $p^{-1} + q^{-1} = 1$ and $f \epsilon L^p(\mathcal{R}^n) \cap L^2(\mathcal{R}^n)$. Then it can be shown using (3.61) and Hausdorff-Young inequality, that $U_t f \epsilon L^q(\mathcal{R}^n) \cap L^2(\mathcal{R}^n)$ and

$$\|U_t f\|_q \leq c\ |t|^{-n(p^{-1}-\frac{1}{2})} \|f\|_p . \tag{3.62}$$

For a proof of (3.62), the reader may consult [RS II, Section IX.7].

The operators C_t and Q_t, used to prove (3.51), were introduced by Dollard [3].

PROBLEMS

3.1 : Show that for one-parameter unitary groups $\{U_t\}$, strong continuity at arbitrary t is equivalent to strong continuity at t = o and that strong continuity is equivalent to weak continuity. (Hint : Use Proposition 2.1.)

3.2 : Verify that the operators defined by the relation (3.13) satisfy the properties (3.7)-(3.10).

3.3 : Starting from the definition (3.18), construct the spectral family for the n-component momentum operator $\{P_j\}$ and show that $|\tilde{f}(\underline{k})|^2$ is the momentum probability density.

3.4 : Verify from the definition (3.22) (with $h = 2\pi$) that
(a) each S_j is self-adjoint and $S_+^* = S_-$,
(b) $[S_1, S_2] = iS_3$ and its cyclic permutations,
(c) $(\underline{S}^2 f)_m(\underline{x}) = s(s+1)f_m(\underline{x})$ and that $\underline{S}^2$ is a bounded self-adjoint operator.

3.5 : Let A be an observable and $\{F_\lambda\}$ its spectral family. (a) Show that under suitable domain conditions the family $\{U_t^* F_\lambda U_t\}$, for each fixed t, is the spectral family of the self-adjoint operator $A_t \equiv U_t^* A U_t$. (b) If furthermore, $\mathrm{Exp}_{f_t}(F_\lambda) = \mathrm{Exp}_f(F_\lambda)$ for all $f \epsilon \mathcal{H}$ and $\lambda, t \epsilon \mathcal{R}$, then $U_t^* F_\lambda U_t = F_\lambda$.

3.6 : Let $\alpha(t)$ be a complex-valued continuous one-parameter group. Prove that $\alpha(t)$ admits a unique representation as

$\alpha(t) = \exp(\beta t)$, where β is a complex number. (Hint : Show that continuity implies differentiability by using the identity $\int_t^{t+\tau}\alpha(s)ds = \alpha(t)\int_0^\tau \alpha(s)ds$.)

3.7 : Let g be a function in $\mathcal{S}(\mathbb{R}^n)$. Construct a sequence of functions g_n in $C_0^\infty(\mathbb{R}^n)$ such that s-lim $g_n = g$ and s-lim $K_o g_n = K_o g$. (Hint : Write $g_n(\underline{x}) = \phi(\underline{x}/n)g(\underline{x})$, where ϕ is a C_0^∞-function such that $0 \leq \phi(\underline{x}) \leq 1$ and $\phi(\underline{x}) = 1$ for $|\underline{x}| \leq 1$, and use the Lebesgue dominated convergence theorem.)

3.8† : Let U_t,A be as in Proposition 3.1. Show that for any complex number z (Im z $\neq$ 0), the operator $(A-z)^{-1}$ is bounded and defined everywhere and verifies (3.58).

3.9 : Show that for $z \notin [o,\infty)$, $(K_o-z)^{-1}$ is a bounded operator with domain $\mathcal{H}$. Verify that for any $f,g \in \mathcal{H}$ and Im z > 0, $(f,(K_o-z)^{-1}g) = i\int_0^\infty dt\, \exp(izt)(f,\exp(-iK_o t)g)$.

3.10 : Prove the uniform continuity in Proposition 3.10.
Hint : Verify and use the inequality

$$|e^{i\alpha}-e^{i\beta}| \leq 2^{1-\nu}|\alpha-\beta|^\nu \text{ for } 0 \leq \nu \leq 1,\ \alpha,\beta \in \mathbb{R}. \quad (3.63)$$

3.11 : Let U_t,A and D(A) be as in Proposition 3.1 and $B \in \mathcal{B}(\mathcal{H})$. Assume that $BU_t = U_t B$ for all t. Prove that B leaves D(A) invariant and that, for $f \in D(A)$, BAf = ABf, i.e. $BA \subseteq AB$. If furthermore $B^{-1} \in \mathcal{B}(\mathcal{H})$, then BA = AB. (See also page 187).

PART II

GENERAL FORMULATION OF SINGLE-CHANNEL SCATTERING SYSTEMS

CHAPTER 4 : TIME-DEPENDENT SCATTERING THEORY

In the first section of this chapter we discuss the standard formulation of the asymptotic condition of scattering theory for the case of single-channel scattering systems and deduce some notable consequences of this condition, in particular properties of the scattering operator. In Section 4-2 we assume that the Hamiltonian is invariant under a symmetry group and study some properties of the scattering operator under this additional hypothesis. In Section 4-3 we derive some consequences of the asymptotic condition for the time evolution of observables, which will give rise to a discussion of other possible ways of formulating that condition. Section 4-4 is mathematical in nature and deals with integration theory for vector-valued and operator-valued functions. This is used in Section 4-5 to transcribe the definition of the wave and scattering operators into integral representations, and these in turn will be the basis for deriving the equations of time-independent scattering theory in Chapter 6.

4-1 THE ASYMPTOTIC CONDITION

The aim of this section is to present a precise mathematical formulation of the asymptotic condition of scattering theory. We consider here the simplest possible case, namely that where a scattering system is completely described by two continuous unitary one-parameter groups $\{U_t\}$ and $\{V_t\}$ acting in the Hilbert space $\mathcal{H}$ formed by all possible (pure) states of the physical system under consideration. The group $\{U_t\}$ is interpreted as describing the time-evolution of the states of the system in the absence of a perturbation or of an interaction between its constituents and will be called the <u>unperturbed evolution group</u>. The group $\{V_t\}$ is assumed to represent the time-evolution with the interaction turned on and will be called the <u>total evolution group</u>.

The asymptotic condition will be mathematically formulated below in terms of an abstract pair of unitary one-parameter groups. Physically these two groups are of course not unrelated, and from the interpretation given above one sees that the relation between $\{U_t\}$ and $\{V_t\}$ will usually be given in terms of the infinitesimal generators of these two groups which are uniquely determined by Proposition 3.1. These two self-adjoint operators will henceforth be denoted by H_0 and H respectively, i.e. we write (see page 111)

$$U_t = \exp(-iH_0 t) \qquad \text{and} \qquad V_t = \exp(-iHt).$$

H_o is interpreted as the energy operator in the absence of interaction and will be called the unperturbed Hamiltonian. H (called the total Hamiltonian) is the sum of H_o and of the interaction Hamiltonian V, i.e. $H = H_o + V$. In many cases both H_o and V are unbounded self-adjoint operators, and then the sum $H_o + V$ defined on $\mathcal{D} \equiv D(H_o) \cap D(V)$ need not be a self-adjoint operator (it is symmetric if $\mathcal{D}$ is dense, cf. Problem 4.1). The total Hamiltonian H will then be a self-adjoint extension of $H_o + V$. One sees that, if $H_o + V$ is essentially self-adjoint, then the total evolution group is uniquely determined by giving H_o and V. On the other hand, if $H_o + V$ is not essentially self-adjoint, $H_o + V$ will have more than one self-adjoint extension, and in that case a further condition will be needed in order to define the total evolution. We shall return to the question of self-adjointness in Chapter 8.

As an example which the reader may keep in mind we mention here potential scattering. The physical system is a non-relativistic particle without spin which is scattered by a potential $V(\underline{x})$. The Hilbert space is $L^2(R^3)$, the unperturbed Hamiltonian is the kinetic energy operator $H_o = K_o = \underline{P}^2$ (we set $h/2\pi = 2m = 1$), and the total Hamiltonian is $H = \underline{P}^2 + V$, where V denotes the maximal multiplication operator in $L^2(R^3)$ determined according to Proposition 2.16 by the real-valued function $V(\underline{x})$ called the potential. When speaking of potential scattering we use the terms free Hamiltonian for H_o and free evolution group for $\{U_t\}$. (It should be borne in mind that in certain physical situations the unperturbed evolution does not describe free particles; as an example we mention scattering from impurities in crystals where $\{U_t\}$ describes the evolution of a particle in a periodic potential, cf. Thomas [1].)

In the presence of the interaction the time evolution of a given state vector $g \in \mathcal{H}$ is governed by the group $\{V_t\}$, i.e. the corresponding state vector at time t is $V_t g$. As already remarked in Chapter 1, in a typical scattering situation one wants to approximate the total evolution in some sense by the free evolution as $t \to \pm\infty$. This is most easily done by supposing that, given the vector $g \in \mathcal{H}$, there exist two state vectors $f_\pm$ such that $V_t g$ converges strongly to $U_t f_\pm$ as $t \to \pm\infty$ respectively, i.e. such that

$$\lim_{t\to-\infty} \|V_t g - U_t f_-\| = 0, \quad \lim_{t\to+\infty} \|V_t g - U_t f_+\| = 0. \qquad (4.1)$$

This means that $V_t g$ is practically indistinguishable from $U_t f_-$ in the remote past and from $U_t f_+$ in the distant future. The requirement of the existence of vectors $f_\pm$ verifying (4.1) is of course a severe restriction on the pair of groups $\{U_t\}$, $\{V_t\}$ and corresponds essentially to imposing the asymptotic condition for the given scattering system.

(4.1) implies in particular that for any projection F

$$\lim_{t\to-\infty} \{\|FV_t g\|^2 - \|FU_t f_-\|^2\} = 0, \qquad (4.2)$$

and similarly for $t \to +\infty$ (Problem 4.2). By considering for instance potential scattering and taking for F the multiplication operator in $L^2(\mathbb{R}^3)$ by the characteristic function of a region Δ, one sees that the difference between the position probability measures over any measurable subset Δ of $\mathbb{R}^3$

for the states $V_t g$ and $U_t f_\pm$ converges to zero as $t \to \pm\infty$ respectively. A similar result holds for the momentum probability measures.

A few points still need further specification in the preceding description of the asymptotic condition. Above all one will have to indicate for what set of vectors g in $\mathcal{H}$ (hereafter called the <u>scattering states</u> of H) the above condition should be required. In many practical situations this may be the set of vectors that describe incoming particles at large negative times and outgoing particles at large positive times. For instance in potential scattering every vector evolving under the free evolution $\{U_t\}$ is localized far away from the scattering center at large times (this follows essentially from Remark 3.15 and will be further discussed in Section 7-1), and then it is clear that (4.1) can be verified only if $V_t g$ has the same property.

The above characterization of scattering states of H requires the notion of position and will be discussed in some detail in Chapter 7. More insight can be obtained by considering a bound state of H. If g is an eigenvector of H, i.e. if $Hg = \mu g$ for some $\mu \in \mathbb{R}$, then the state $V_t g = \exp(-i\mu t)g$ is simply a multiple of g and will therefore not describe a scattering situation. For this reason scattering states are expected to be associated with the continuous spectrum of H (defined in Section 5-6). We shall see in Chapter 5 that the continuous spectrum of H determines a subspace of the Hilbert space $\mathcal{H}$, and for many Hamiltonians this subspace turns out to be identical with the set of scattering states defined with the aid of the position operator.

Here we do not wish to use any particular specification of the scattering states. We simply assume that to each Hamiltonian H one can associate a set of scattering states $M_\infty(H)$ having the following two properties :

(i) $M_\infty(H)$ is a subspace of H,

(ii) $M_\infty(H)$ is invariant under the group $\{\exp(-iHt)\}$, i.e. if $g \in M_\infty(H)$, then $\exp(-iHt)g \in M_\infty(H)$ for all $t \in R$.

(i) means that $M_\infty(H)$ must be a linear set which is also strongly closed. The attribute "scattering state" then corresponds to an observable, which is represented by the self-adjoint operator $E_\infty(H)$ defined as the orthogonal projection with range $M_\infty(H)$, cf. (2.36). (ii) is an expression of the homogeneity in time of the scattering systems considered here : the definition of the scattering states should not depend on time (this need not be verified for time-dependent Hamiltonians). The condition (ii) may be rewritten as (cf. (5.33)-(5.34) or Problem 4.2) :

$$E_\infty(H)\exp(-iHt) = \exp(-iHt)E_\infty(H). \tag{4.3}$$

We now reformulate the asymptotic condition more precisely and in a way which is closer to the physical picture. In a practical scattering situation one starts at large negative times by preparing a state, the time-evolution of which is assumed to be governed by the unperturbed evolution group. This corresponds to giving the state f_- in (4.1). Notice that in (4.1) f_- is interpreted as the initial state <u>at time t = 0</u>, i.e. the prepared state at a negative time t would be $U_t f_-$. f_- should belong to $M_\infty(H_0)$. The first part of the asymptotic condition then requires the existence of a vector $g \varepsilon M_\infty(H)$ such that the first equation in (4.1) is verified. The second part of the asymptotic condition now demands the existence of a vector f_+ in $M_\infty(H_o)$ verifying the second equation in (4.2), where g is the vector obtained from the given f_- by the first equation in (4.1).

Thus the effect of scattering in this picture is to associate with each initial state vector f_- in $M_\infty(H_o)$ a final state vector f_+ in $M_\infty(H_o)$, both being interpreted as states at time $t = 0$. If there is no interaction, i.e. if $V_t = U_t$, one clearly has $f_+ = f_-$. We shall see below that the correspondence $f_- \mapsto f_+$ defines a linear operator on the subspace $M_\infty(H_o)$ of scattering states of the unperturbed Hamiltonian. This operator is called the scattering operator S and is the central object of scattering theory. The difference between S and the identity operator will be an expression for the effect of the interaction, for, if there is no interaction, this difference is zero.

Since V_t is unitary, one may use (2.46) to rewrite the first equation in (4.1) as

$$\lim_{t\to-\infty} \|V_t g - U_t f_-\| = \lim_{t\to-\infty} \|g - V_t^* U_t f_-\| = 0. \tag{4.4}$$

In the first part of the asymptotic condition it was assumed that for $f_- \varepsilon M_\infty(H_o)$ there exists a vector g such that (4.4) holds. Therefore this condition is equivalent to the requirement of the existence of the strong limit of $V_t^* U_t$ as $t\to-\infty$ on the subspace $M_\infty(H_o)$. This strong limit will be called the <u>wave operator</u> Ω_- :

$$\Omega_- \equiv \operatorname*{s-lim}_{t\to-\infty} V_t^* U_t E_\infty(H_o). \tag{4.5}$$

Its interpretation is as follows : when applied to a vector f in $M_\infty(H_0)$, it gives the actual state vector g at time t = 0 that will evolve from the prepared state f in the remote past (in the sense of the first equation in (4.1)). Notice that according to (4.5) Ω_- is defined on all of H and not only on $M_\infty(H_0)$. We have simply set $\Omega_- f = \theta$ on all vectors f in the orthogonal complement of $M_\infty(H_0)$.

Similarly one sees that the second part of the asymptotic condition is equivalent to the requirement of the existence of the strong limit of $U_t^* V_t$ on the range of Ω_- as $t \to +\infty$. For the sake of convenience one usually adopts a more symmetrical treatment with respect to the sign of time. One defines a second wave operator similarly to (4.5) by

$$\Omega_+ \equiv \underset{t\to+\infty}{\text{s-lim}}\, V_t^* U_t E_\infty(H_0), \tag{4.6}$$

if the limit exists. The asymptotic condition is then expressed by the requirement that both of the limits (4.5) and (4.6) exist. The interpretation of Ω_+ is the following : when applied to a vector f in $M_\infty(H_0)$, it gives the actual state vector at time t = 0 which will converge to f in the distant future (in the sense of the second equation in (4.1)).

One sees as in (4.4) that $g = \text{s-lim}\, V_t^* U_t f$ as $t\to\infty$ if and only if $f = \text{s-lim}\, U_t^* V_t g$ as $t\to\infty$. Thus the existence of Ω_+ implies the existence of the strong limit of $U_t^* V_t$ on the range of Ω_+. Hence the second part of the asymptotic condition is verified provided that the range of Ω_- is contained in the range of Ω_+. This postulate has to be added to (4.5) and (4.6) in order that the mathematical description corresponds to the physical picture given above. The situation is depicted in the figure below.

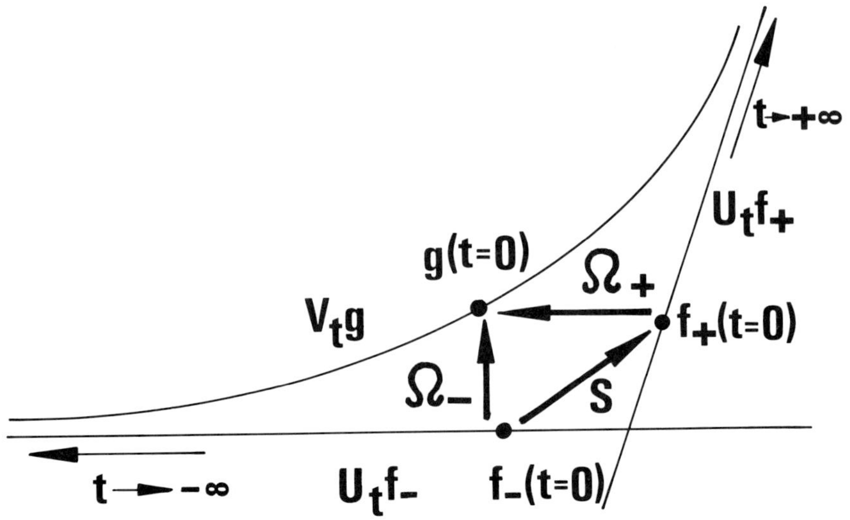

Figure 4.1 : Pictorial representation of the asymptotic condition. Points of the plane represent vectors in Hilbert space. A straight line depicts the trajectory in H of an unperturbed evolution and a curve that of the total evolution of a state vector.

We shall show below that the wave operators are partial isometries with initial set $M_\infty(H_0)$. According to Proposition 2.11 the operators*)

$$F_\pm \equiv \Omega_\pm \Omega_\pm^* \qquad (4.7)$$

are the projections onto the range of the corresponding wave operator. We thus arrive at the final formulation of the asymptotic condition :

(A1) There exist $\text{s-}\lim\limits_{t\to\pm\infty} V_t^* U_t E_\infty(H_0) \equiv \Omega_\pm$ **),

*) An equation bearing the symbol ± means that the equation is valid separately for the upper and for the lower sign.

**) In the physics literature the wave operator at $t\to-\infty$ is often denoted by Ω_+ and that at $t\to+\infty$ by Ω_-.

(A2) $\quad F_-\mathcal{H} \subseteq F_+\mathcal{H}$.

The subspaces $F_\pm\mathcal{H}$ will be contained in the subspace $\mathcal{M}_\infty(H)$ of scattering states of H. The simplest situation is that where

(A3) $\quad F_+\mathcal{H} = F_-\mathcal{H} = \mathcal{M}_\infty(H)$.

If (A3) is verified, then the theory may be called <u>asymptotically complete</u>*), since the evolution $V_t g$ of any scattering state g of H is asymptotically described by the free evolution as in (4.1). Therefore the simplest type of a scattering system is that where both wave operators exist and have range equal to $\mathcal{M}_\infty(H)$, i.e. where (A1) and (A3) are verified. In this situation we speak of a <u>simple scattering system</u>**).

More general situations can now readily be imagined. It may happen that (A1) is not satisfied, as for instance in potential scattering with a Coulomb potential (Chapter 13). One will then have to look for a weaker mathematical formulation of the asymptotic condition which might apply to such situations (cf. Section 4-3). It may happen that the wave operators exist but (A3) does not hold. In such cases one will have to find a different means of giving an asymptotic description of certain scattering states of H (those orthogonal to $F_\pm\mathcal{H}$). This may involve for instance additional free

---

*) It should be said that in the literature one encounters various different definitions of asymptotic completeness. This point will be treated in Chapter 9.

**) Actually this term was introduced by Jauch [1] in a somewhat less general context.

evolution groups as in multichannel scattering (Chapter 14) or the notion of absorption of states (Section 4-6).

To complete this section, we establish some properties verified by the wave operators and define the scattering operator.

<u>PROPOSITION 4.1</u> : The wave operators are partial isometries with initial set $M_\infty(H_o)$.

<u>Proof</u> (for Ω_+) : From the definition of Ω_+ one has $D(\Omega_+) = H$. Furthermore it follows from (4.6), Proposition 2.1 and the unitarity of V_t and U_t that for any $f \varepsilon H$

$$||\Omega_+ f|| = \lim_{t\to\infty} ||V_t^* U_t E_\infty(H_o) f|| = ||E_\infty(H_o) f||. \tag{4.8}$$

The result now follows from Proposition 2.10. #

In many physically interesting cases the set $M_\infty(H_o)$ of scattering states of the unperturbed Hamiltonian is the entire Hilbert space. In that case the wave operators are isometries. In particular this is true for potential scattering which will be discussed in Chapters 8 and 9.

<u>PROPOSITION 4.2</u> : The adjoints of $\Omega_\pm$ are given by

$$\Omega_\pm^* = \text{s-}\lim_{t\to\pm\infty} U_t^* V_t F_\pm . \tag{4.9}$$

<u>Proof</u> (for Ω_+) : By virtue of (2.46) we have for any $f \varepsilon H$

$$\lim_{t\to\infty} ||U_t^* V_t \Omega_+ f - E_\infty(H_o) f|| = \lim_{t\to\infty} ||\Omega_+ f - V_t^* U_t E_\infty(H_o) f|| = 0,$$

i.e.

$$\text{s-}\lim_{t\to+\infty} U_t^* V_t \Omega_+ = E_\infty(H_o). \tag{4.10}$$

Upon multiplying (4.10) on the right by Ω_+^* and using (4.7) and (2.42), one obtains

$$\underset{t\to\infty}{\text{s-lim}}\ U_t^* V_t F_+ = E_\infty(H_0)\Omega_+^* = \Omega_+^*. \#$$

We saw towards the end of Section 2-2 that the existence of $\Omega_\pm$ implies that the adjoint of the sequence $\{V_t^* U_t E_\infty(H_0)\}$ converges weakly to $\Omega_\pm^*$ as $t\to\pm\infty$. By Proposition 4.2 the convergence of the adjoint sequence is strong on the range of $\Omega_\pm$. Let us set $E_\infty(H_0) = I$ for simplicity. We wish to show that the adjoint sequence $\{U_t^* V_t\}$ then cannot converge strongly on the orthogonal complement of $F_\pm H$. Indeed, by repeating the proof of Proposition 4.1, one finds that this strong limit would be isometric on $(F_\pm H)^\perp$. On the other hand it would have to be equal to the weak limit which is zero. Since the zero operator is not isometric, strong convergence of $\{U_t^* V_t\}$ on $(F_\pm H)^\perp$ as $t\to\pm\infty$ is impossible.

PROPOSITION 4.3 : If (A1) is verified, one has for any $\tau\in R$

$$V_\tau \Omega_\pm = \Omega_\pm U_\tau, \qquad (4.11)$$

$$U_\tau \Omega_\pm^* = \Omega_\pm^* V_\tau. \qquad (4.12)$$

These equations signify that the wave operators intertwine the groups $\{U_\tau\}$ and $\{V_\tau\}$. They are often called the intertwining relations.

Proof (for Ω_+) : The proof is based on (A1), Corollary 2.19, the unitarity and the group property of $\{V_t\}$ and $\{U_t\}$, and on (4.3) for the unperturbed Hamiltonian :

$$V_\tau \Omega_+ = V_\tau \ \underset{t\to\infty}{\text{s-lim}}\ V_t^* U_t E_\infty(H_0) = \underset{t\to\infty}{\text{s-lim}}\ V_\tau V_t^* U_t E_\infty(H_0)$$

$$= \underset{t'\to\infty}{\text{s-lim}}\ V_{t'}^* U_{t'+\tau} E_\infty(H_0) = \underset{t'\to\infty}{\text{s-lim}}\ V_{t'}^* U_{t'} E_\infty(H_0) U_\tau = \Omega_+ U_\tau .$$

(4.12) is obtained by taking the adjoint of (4.11), using (2.32) and $U_\tau^* = U_{-\tau}$. #

PROPOSITION 4.4 :

(a) If $f \varepsilon D(H_0)$, then $\Omega_\pm f \varepsilon D(H)$ and $H\Omega_\pm f = \Omega_\pm H_0 f$.

(b) If $g \varepsilon D(H)$, then $\Omega_\pm^* g \varepsilon D(H_0)$ and $H_0 \Omega_\pm^* g = \Omega_\pm^* H g$.

Proof : Let Ω be one of the wave operators. One has

$$\| i\tau^{-1}\Omega(U_\tau - I)f - \Omega H_0 f\| \leq \|\Omega\| \, \| i\tau^{-1}(U_\tau - I)f - H_0 f\| ,$$

which converges to zero as $\tau \to 0$ by Proposition 3.1. Thus $id/d\tau\ \Omega U_\tau f = \Omega H_0 f$. Now (4.11) implies that we must have

$$id/d\tau\ V_\tau \Omega f = id/d\tau\ \Omega U_\tau f = \Omega H_0 f.$$

It then follows from Proposition 3.1 that $\Omega f \varepsilon D(H)$ and $H\Omega f = \Omega H_0 f$. This proves (a). (b) is proved similarly by using (4.12) instead of (4.11). #

PROPOSITION 4.5 : If (A1) is verified, one has for any $\tau \varepsilon \mathcal{R}$

$$F_\pm V_\tau = V_\tau F_\pm . \tag{4.13}$$

This result means that the ranges of the wave operators reduce the group $\{V_t\}$ or equivalently the total Hamiltonian. If (A3) also holds, this result is already contained in (4.3).

Proof : We use (4.7) and Proposition 4.3 :

$$F_\pm V_\tau = \Omega_\pm \Omega_\pm^* V_\tau = \Omega_\pm U_\tau \Omega_\pm^* = V_\tau \Omega_\pm \Omega_\pm^* = V_\tau F_\pm . \#$$

It remains to define the scattering operator S. From Figure 4.1 one sees that it is given by the expression

$$S = \Omega_+^{-1} \Omega_- . \tag{4.14}$$

This definition makes sense if condition (A2) is verified. Indeed the range of Ω_- is then contained in that of Ω_+, and Ω_+ is invertible on its range by Proposition 2.11. Furthermore Ω_+^{-1} may be replaced by Ω_+^*, again as a consequence of Proposition 2.11. Therefore we adopt the following definition of the <u>scattering operator</u>, valid if (A1) and (A2) hold :

$$S = \Omega_+^* \Omega_- . \tag{4.15}$$

<u>PROPOSITION 4.6</u> : S is a partial isometry with initial set $M_\infty(H_0)$, i.e.

$$S^*S = E_\infty(H_0) . \tag{4.16}$$

The range of S is a subspace of $M_\infty(H_0)$ invariant under $\{U_t\}$.

<u>Proof</u> : Since $D(\Omega_-) = D(\Omega_+^*) = \mathcal{H}$, we have $D(S) = \mathcal{H}$. Furthermore one has from (4.15) and (2.32) $S^*S = \Omega_-^*\Omega_+\Omega_+^*\Omega_- = \Omega_-^*F_+\Omega_-$. By condition (A2), the range of Ω_- is contained in that of F_+, so that $F_+\Omega_- = \Omega_-$. Hence $S^*S = \Omega_-^*\Omega_- = E_\infty(H_0)$, which proves that S is a partial isometry as well as (4.16).

The range of S is a subspace of the range of Ω_+^*, and the range of Ω_+^* is $M_\infty(H_0)$ by Proposition 2.11. To see that the range of S is invariant under $\{U_t\}$, one proceeds as in the proof of Proposition 4.5 (most conveniently by using (4.17) below.) #

PROPOSITION 4.7 :

(a) For any $\tau \in \mathcal{R}$ one has $SU_\tau = U_\tau S.$ (4.17)

(b) S leaves $D(H_o)$ invariant and $SH_o \subseteq H_o S.$ (4.18)

Proof : (4.17) follows from (4.15) and Proposition 4.3 :

$$SU_\tau = \Omega_+^*\Omega_- U_\tau = \Omega_+^* V_\tau \Omega_- = U_\tau \Omega_+^* \Omega_- = U_\tau S.$$

To prove (b), one repeats the argument of the proof of Proposition 4.4 by using (4.17) in place of (4.11). #

Actually one can show that the equality sign holds in (4.18) if one restricts oneself to the subspace $\mathcal{M}_\infty(H_o)$ (Problem 4.3), i.e.

$$SH_o E_\infty(H_o) = H_o S\,. \qquad (4.19)$$

Equation (4.18) is a very fundamental property of the scattering operator. It says that S commutes with the unperturbed Hamiltonian H_o, or in more physical terms that the unperturbed energy is conserved in the scattering process, or that the scattering is elastic. It follows for instance from (4.18) that the expectation value of H_o in the final state f_+ is the same as that in the corresponding initial state $f_- \in \mathcal{M}_\infty(H_o) \cap D(H_o)$:

$$(f_+, H_o f_+) = (Sf_-, H_o Sf_-) = (S^*Sf_-, H_o f_-) = (f_-, H_o f_-).$$

The question of the relation between S and physical observables will be discussed in more detail in Section 4-3. (4.18) also means that S and H_o can be diagonalized simultaneously. This aspect will be developed further in Section 5-7 after the introduction of the spectral representation of H_o.

All the properties of S deduced so far depend on the validity of (A1) and (A2) only. One sees that the restriction of S to $M_\infty(H_0)$ is an isometry in the Hilbert space $M_\infty(H_0)$. In particular, if $M_\infty(H_0) = H$ (e.g. in potential scattering), S is an isometry in H. However (A1) and (A2) do not imply that S is unitary in $M_\infty(H_0)$ (Problem 4.8). The unitarity of S holds if and only if the range of S equals $M_\infty(H_0)$, or equivalently if and only if $SS^* = E_\infty(H_0)$. Now $SS^* = \Omega_+^*\Omega_-\Omega_-^*\Omega_+ = \Omega_+^*F_-\Omega_+$. If the range of Ω_+ is contained in that of Ω_-, it follows as in the proof of Proposition 4.6 that $F_-\Omega_+ = \Omega_+$ and hence $SS^* = \Omega_+^*\Omega_+ = E_\infty(H_0)$. If however the range of Ω_- is strictly smaller than that of Ω_+, $F_-\Omega_+$ will be zero on a non-trivial subset of $M_\infty(H_0)$. Hence SS^* will be a strictly smaller projection than $E_\infty(H_0)$ and S will not be unitary. Thus we have shown the following result :

PROPOSITION 4.8 : The scattering operator is a unitary operator in $M_\infty(H_0)$ if and only if the range of Ω_+ is identical with that of Ω_-. In particular S is unitary if the theory is asymptotically complete.

Summary : Let us summarize the principal contents of this section for the most usual case where $M_\infty(H_0) = H$. We have seen that the asymptotic condition leads in a natural way to postulating the existence of the wave operators $\Omega_\pm =$ s-lim $V_t^*U_t$ as $t\to\pm\infty$. They are isometric operators intertwining H and H_0. The scattering operator is simply related to $\Omega_\pm$ by the formula $S = \Omega_+^*\Omega_-$. S is an isometry that commutes with the free Hamiltonian. The unitarity of S does not follow from the existence of the wave operators alone. It is an additional hypothesis which is however verified if one has asymptotic completeness.

To conclude this section we state in rough terms to what extent the above theory is capable of handling potential scattering. In that case the proof of (A1) is usually simple, whereas that of (A3) is quite involved. Methods for establishing (A1) and (A3) and references to the literature will be given in Chapters 8 and 9. The proof of (A1) involves the behaviour of the potential $V(\underline{x})$ at large x. (A1) is verified if the potential tends to zero as $x \to \infty$ faster than 1/x. For Coulomb potentials and potentials that go to zero even more slowly (so-called long range potentials), (A1) does not hold. An alternative method for dealing with such potentials will be indicated in Section 4-3. The validity of (A3) depends on the local behaviour of $V(\underline{x})$. (A3) holds true if the potential is locally less singular than the inverse-square function. For locally more singular potentials, (A3) need not be verified (cf. Sections 4-6C and 9-2E for details).

4-2 SYMMETRIES IN SCATTERING THEORY

One of the most fundamental concepts of modern theoretical physics is that of symmetry. In this short section we indicate how symmetry considerations are incorporated into the general framework given in the preceding section. Certain particular symmetry groups will play an important role in later chapters.

A symmetry may be regarded as a transformation acting

on a physical system or on the parameter space used to describe that system in such a way that the laws of motion remain unchanged. In quantum mechanics a symmetry transformation is represented by a unitary or anti-unitary*) operator acting in the relevant Hilbert space $\mathcal{H}$ [P, § 3-2] which commmutes with the evolution group.

The natural mathematical structure to describe a set of symmetries is that of a group. Thus we say that a scattering system is invariant under a group G if there exists a representation of G by unitary or anti-unitary operators in $\mathcal{H}$ which commutes with both U_t and V_t and which leaves $\mathcal{M}_\infty(H_o)$ and $\mathcal{M}_\infty(H)$ invariant. Thus with each element γ of G there is associated a unitary or anti-unitary operator $U(\gamma)$ such that

$$U(\gamma_1\gamma_2) = U(\gamma_1)U(\gamma_2), \tag{4.20}$$

$$U(\gamma)E_\infty(H_o) = E_\infty(H_o)U(\gamma), \quad U(\gamma)E_\infty(H) = E_\infty(H)U(\gamma), \tag{4.21}$$

$$U(\gamma)U_t = U_{\pm t}U(\gamma), \qquad U(\gamma)V_t = V_{\pm t}U(\gamma). \tag{4.22}$$

In (4.22) the upper sign has to be taken if $U(\gamma)$ is unitary and the lower sign if $U(\gamma)$ is anti-unitary. In either case (4.22) implies that $U(\gamma)H_o = H_oU(\gamma)$ and $U(\gamma)H = HU(\gamma)$. (If $U(\gamma)$ is unitary, this follows from Corollary 3.2; for an anti-unitary $U(\gamma)$ the proof is similar to that of Corollary 3.2; the change of the sign of time is necessary since $U(\gamma)\, i g = -iU(\gamma)g$, cf. (4.30).)

PROPOSITION 4.9 : Suppose that a scattering system is in-

*) The definition of an anti-unitary operator will be given in (4.29) and (4.30) below.

variant under a group G. Then

$$SU(\gamma) = U(\gamma)S \qquad \text{if } U(\gamma) \text{ is unitary,} \tag{4.23}$$

$$S^*U(\gamma) = U(\gamma)S \qquad \text{if } U(\gamma) \text{ is anti-unitary.} \tag{4.24}$$

Thus a unitary representation of a symmetry group reduces the scattering operator. If the representation contains anti-unitary operators, the latter transform S into S^* :

$$U(\gamma)SU(\gamma)^* = S^* \quad \text{if } U(\gamma) \text{ is anti-unitary.} \tag{4.25}$$

Proof : (i) Suppose $U(\gamma)$ is unitary. It follows from (4.21) and (4.22) that

$$U(\gamma)V_t^*U_tE_\infty(H_o) = V_t^*U_tE_\infty(H_o)U(\gamma). \tag{4.26}$$

By taking the strong limit as $t\to\pm\infty$ in (4.26), it follows from Corollary 2.19 that

$$U(\gamma)\Omega_\pm = \Omega_\pm U(\gamma). \tag{4.27}$$

Together with (2.32) this implies that $\Omega_\pm^* U(\gamma)^* = U(\gamma)^*\Omega_\pm^*$. Since $U(\cdot)$ is a unitary representation of G, we have

$$U(\gamma)U(\gamma)^* = I = U(\gamma\gamma^{-1}) = U(\gamma)U(\gamma^{-1}),$$

whence $U(\gamma)^* = U(\gamma^{-1})$. Because each element of G has an inverse in G, one may replace γ by γ^{-1} in the above considerations, which gives

$$\Omega_\pm^* U(\gamma) = U(\gamma)\Omega_\pm^*. \tag{4.28}$$

(4.23) now follows from (4.15), (4.27) and (4.28).

(ii) Suppose $U(\gamma)$ is anti-unitary, i.e.

$$U(\gamma)^*U(\gamma) = U(\gamma)U(\gamma)^* = I \quad ^{*)} \tag{4.29}$$

---

*) To define the adjoint of an anti-linear operator A, one uses instead of (2.30) the formula $\overline{(g,Af)} = (A^*g,f)$, so that A^* is also anti-linear.

and $\quad U(\gamma)\,\alpha f = \bar{\alpha}\,U(\gamma) f \qquad$ for all $\alpha \in C$ and $f \in H$. (4.30)

By repeating the argument of (i), one finds in this case that

$$U(\gamma)\Omega_{\pm} = \Omega_{\mp} U(\gamma) \tag{4.31}$$

and

$$\Omega_{\pm}^{*} U(\gamma) = U(\gamma)\Omega_{\mp}^{*}, \tag{4.32}$$

from which (4.24) follows easily. #

It should be remarked that in the most general case it is only possible to find a projective representation of a symmetry group, i.e. the equation (4.20) is replaced by

$$U(\gamma_1\gamma_2) = \omega(\gamma_1,\gamma_2)U(\gamma_1)U(\gamma_2)$$

with $|\omega(\gamma_1,\gamma_2)| = 1$ (cf. [J, Ch. 9] or [P, § 3-2]). It is easy to check that Proposition 4.9 remains valid for a projective representation of G.

Some particularly important symmetries in potential scattering are the spherical symmetry (Chapter 11) which corresponds to a unitary representation of the rotation group in $L^2(R^3)$ and time-reversal which is represented by an anti-unitary operator Θ (Problem 8.8, [T, Section 6e]). Θ is nothing but the operation of complex-conjugation of the state vectors, i.e. $(\Theta f)(\underline{x}) = \bar{f}(\underline{x})$.

Time-reversal invariance implies that the scattering operator is unitary in $M_\infty(H_o)$ (provided that (A2) holds). In fact the following more general result is true :

PROPOSITION 4.10 : Suppose that (A1) and (A2) hold and that there exists an anti-unitary symmetry transformation, i.e.

an anti-unitary operator U verifying (4.21) and (4.22). Then $F_+ = F_-$ and S is unitary in $M_\infty(H_0)$.

Proof : By (A2) we have $F_- \mathcal{H} \subseteq F_+ \mathcal{H}$. Thus in view of Proposition 4.8 it suffices to show that $F_+ \mathcal{H} \subseteq F_- \mathcal{H}$. Let $g \in F_+ \mathcal{H}$. By (4.31) and (4.32) we have $F_+ U = U F_-$, hence $U^* F_+ = F_- U^*$. Consequently $U^* g = F_- U^* g \in F_- \mathcal{H} \subseteq F_+ \mathcal{H}$, i.e. $F_+ U^* g = U^* g$. Now by (4.31)

$$\|g - \Omega_- U \Omega_+^* U^* g\| = \|g - U \Omega_+ \Omega_+^* U^* g\| = \|U^* g - F_+ U^* g\| = 0.$$

This shows that $g \in (\text{range } \Omega_-) = F_- \mathcal{H}$, hence that $F_+ \mathcal{H} \subseteq F_- \mathcal{H}$. #

4-3 OBSERVABLES AND SCATTERING

As we have described in Section 3-2, a physical observable is represented by a self-adjoint operator in Hilbert space. If A is such an operator, we may consider for instance its expectation value as a function of time for a scattering state g of H evolving under the total evolution group, i.e. the quantity $(V_t g, A V_t g) = (g, V_t^* A V_t g)$. For certain observables these expectation values will become constant as $t \to \pm\infty$. Such observables are very interesting for the description of a scattering process, since they can be used to characterize the asymptotic properties of the scattering states. If one measures the expectation value of such an observable at a finite but sufficiently large time, one has essentially measured its asymptotic value.

We now wish to relate the asymptotic values of observables to the wave operators. Since the latter have been defined as strong limits, we shall consider also the strong limits of $\{V_t^* A V_t\}$ on $M_\infty(H)$ rather than the limits of the expectation values. Of course the existence of these strong limits implies that of the expectation values. Notice that $\{V_t^* A V_t\}$ represents simply the time evolution of the operator A in the Heisenberg picture.

Now one usually characterizes the asymptotic behaviour of scattering states by quantities pertaining to the unperturbed evolution in $M_\infty(H_o)$. For this reason one will admit only those observables A for which $U_t^* A U_t E_\infty(H_o)$ also converges strongly as $t \to \pm\infty$ and which act only in the subspace $M_\infty(H_o)$, i.e. which are such that

$$A = AE_\infty(H_o) = E_\infty(H_o)A. \tag{4.33}$$

<u>PROPOSITION 4.11</u> : Suppose that (A1) and (A3) are verified. Let $A \in \mathcal{B}(H)$ be such that (4.33) holds and such that $U_t^* A U_t$ converges strongly as $t \to \pm\infty$. Then $\{V_t^* A V_t E_\infty(H)\}$ converges strongly as $t \to \pm\infty$.

<u>Proof</u> : One has with (4.33) and (4.3)

$$V_t^* A V_t E_\infty(H) = V_t^* U_t E_\infty(H_o) U_t^* A U_t U_t^* V_t E_\infty(H).$$

Now $V_t^* U_t E_\infty(H_o) \to \Omega_\pm$, $U_t^* V_t E_\infty(H) \to \Omega_\pm^*$ by (4.9), and $U_t^* A U_t$ converges strongly to $A_\pm$, say. It follows from Proposition 2.18 that

$$\underset{t \to \pm\infty}{\text{s-lim}}\, V_t^* A V_t E_\infty(H) = \Omega_\pm A_\pm \Omega_\pm^*. \quad \# \tag{4.34}$$

Proposition 4.11 gives a hint how to obtain a mathematical formulation of the asymptotic condition which is less restrictive than the one given in Section 4-1. First of all it is physically very natural to consider a set of observables (or their expectation values) rather than the states themselves. In fact the outcome of scattering experiments is usually given by the measured values of certain observables. In addition these observables are among those that characterize the unperturbed system, i.e. they commute with $\{U_t\}$. In potential scattering for instance one considers momentum, angular momentum and spin polarizations or helicities if the particle has a non-zero spin.

Thus one may for instance consider the set A_o of all observables that verify (4.33) and commute with $\{U_t\}$. For $A\varepsilon A_o$ one has $A_\pm = A$, and (4.34) becomes

$$\underset{t\to\pm\infty}{\text{s-lim}}\; V_t^* A V_t E_\infty(H) = \Omega_\pm A \Omega_\pm^* . \tag{4.35}$$

Instead of requiring the existence of the wave operators, one could simply require the strong convergence of $\{V_t^* A V_t E_\infty(H)\}$ for all $A\varepsilon A_o$. Physically this means that all constants of the unperturbed motion in $M_\infty(H_o)$ are also asymptotically constants of the perturbed motion, and in this sense the total evolution is asymptotically characterized by quantities pertaining to the unperturbed system.

It is an interesting fact that, if one imposes some relatively weak additional conditions, the existence of the strong limits of $\{V_t^* A V_t E_\infty(H)\}$ as $t\to\pm\infty$ for all $A\varepsilon A_o$ implies the existence of two partial isometries $\Omega_\pm$ with initial set $M_\infty(H_o)$ and range $M_\infty(H)$ such that (4.35) holds for all $A\varepsilon A_o$. Since the strong limits of $\{V_t^* U_t E_\infty(H_o)\}$ as $t\to\pm\infty$ need not exist, these partial isometries can be viewed as generalized wave operators. The proof of this result is rather long and involves the theory of von Neumann algebras (cf. Amrein,

Martin and Misra [1] or Amrein, Georgescu and Martin in [EN].) We shall give some additional details about this method in Section 13-2.

The generalized wave operators verify an intertwining relation $H\Omega_\pm = \Omega_\pm K_\pm$, where $K_\pm$ are self-adjoint functions of H_o which may be different from or equal to H_o (Problem 4.4). An example where $K_\pm \neq H_o$ is that where the perturbation contains a function of H_o. A simple case is that where this function is just a constant, i.e. where $V = W + cI$ for some self-adjoint operator W and some real c. In this case $K_\pm = K = H_o + cI$, and $\{V_t^* \exp(-iKt)E_\infty(H_o)\}$ converges strongly to the generalized wave operators as $t\to\pm\infty$ provided that $\{\exp[i(H_o+W)t] \exp(-iH_o t)E_\infty(H_o)\}$ is strongly convergent as $t\to\pm\infty$. Thus it suffices here to replace H_o by a suitably adjusted unperturbed Hamiltonian K. In more complex situations the strong limits of $\{V_t^* \exp(-iK_\pm t)E_\infty(H_o)\}$ need not exist. The replacement of H_o by a couple of adjusted unperturbed Hamiltonians $K_\pm$ is an example of what physicists call <u>energy renormalization</u>.

An example where $K_\pm = H_o$ is potential scattering with a Coulomb or long range potential for which the above theory is applicable. Thus for Coulomb scattering the usual wave operators do not exist but the generalized ones do. In this case it is also possible to write the generalized wave operators as strong limits of a sequence $\{V_t^* Y_t\}$, where the operators Y_t are functions of H_o but do not form a group in t (i.e. Y_t is not of the form $\exp(-iBt)$ with $B = B^*$). One sees that in this case the long range effect of the potential requires asymptotically an adjustment of the free evolution group which is more complex than a simple renormalization, and the picture of the asymptotic convergence of states given in Section 4-1 will need a new interpretation (cf. Chapter 13 for additional details).

If the generalized wave operators exist, one may define a scattering operator by (4.15). One then has for $f \epsilon M_\infty(H_0)$

$$(Sf,ASf) = (\Omega_- f,\Omega_+ A\Omega_+^* \Omega_- f) = \lim_{t\to+\infty} (V_t \Omega_- f, AV_t \Omega_- f), \quad (4.36)$$

i.e. the expectation value of A in the state Sf is equal to the asymptotic limit as $t\to+\infty$ of the expectation values of A for the scattering state $\Omega_- f$ of H.

In the language of observables one may view the scattering operator as a mapping which assigns to each observable A in A_o another observable S*AS (which is again in A_o provided that S commutes with $\{U_t\}$), and A-S*AS represents the change of the observable A as a result of the perturbation. In particular (f,Af) - (f,S*ASf) is the change of the expectation value of A in the state $f \epsilon M_\infty(H_o)$.

An inspection of (4.35) shows that the generalized wave operators are not completely determined (Problem 4.5); however the correspondence $A \mapsto S^*AS$ is defined uniquely if S commutes with $\{U_t\}$ (Problem 13.4).

There are of course numerous other possibilities of mathematically formulating the asymptotic condition. A physically very natural one is to require convergence of the expectation values of the observables of A_o rather than strong convergence. This hypothesis leads to mathematical complications since weak operator convergence is not compatible with operator multiplication (i.e. w-lim $A_n = A$, w-lim $B_n = B$ does not entail that w-lim $A_nB_n = AB$). One could also think of using weak convergence in the framework of Section 4-1. However the requirement of weak convergence in (4.1) does not determine $f_\pm$ (Problem 4.6) and the requirement of weak convergence in (4.5) is physically not well motivated, and the limits need not be isometries (for a Coulomb potential in potential scattering the weak limit in (4.5) is in fact 0, see Proposition 13.6).

Yet another approach is to use only projections as observables. These are very basic quantities in quantum mechanics, as was explained in Section 3-2. This idea has been exploited by Jauch, Misra and Gibson [1] in a way which is still close to that of Section 4-1. Instead of (4.1) it is required that for every state in $M_\infty(H)$ (represented by a density operator ρ) there exist a pair of density operators $\rho_\pm$ in $M_\infty(H_0)$ such that

$$\lim_{t\to\pm\infty} \sup_F |\mathrm{Tr}\, F(V_t^*\rho V_t - U_t^*\rho_\pm U_t)| = 0, \qquad (4.37)$$

where the supremum is taken over all projections. (4.37) essentially says that at large times the state ρ evolving under V_t is close to a state $\rho_\pm$ evolving under U_t in the sense that the measurement of an arbitrary projection in one or the other of these two states leads to practically the same result. Again it is possible to deduce the existence of a kind of wave operator.

The above form of the asymptotic condition is equivalent to the one where the supremum in (4.37) is taken over all operators A with $\|A\| = 1$ instead of all projections. This theory can accomodate simple energy renormalizations but is too restrictive to be applicable to potential scattering by long range potentials. A more general approach would be to take the supremum in (4.37) only over a subset of projections, for example all projections belonging to A_o. But the scope of such a scheme has not been worked out in detail (see also Mourre [1]).

4-4 RIEMANN-STIELTJES INTEGRATION IN HILBERT SPACE

In this book we shall use two types of integrals of vector-valued or operator-valued functions in Hilbert space. The first one is the analogue of the Riemann integral of complex-valued functions. Its definition and basic properties will be explained in the present section. The second type of integration, which will be introduced in Section 6-1, involves an operator-valued measure.

In principle it is possible to adapt various methods of defining an integral in standard analysis to vector-valued and operator-valued functions. Since we shall usually have continuous functions as integrands, we shall restrict ourselves to Riemann integrals. This means that our integrals will be defined as limits of finite sums in the usual way. The limits are taken in Hilbert space in the strong topology. (For more general integrals the reader may consult e.g. [HP, Chapter 3] or [D].)

Let [a,b] be a finite interval in $\mathcal{R}$. A partition Π of [a,b] is a set of numbers $\{s_0, s_1, \ldots, s_n; u_1, \ldots, u_n\}$ such that $a = s_0 < s_1 < \cdots < s_n = b$ and $u_i \varepsilon (s_{i-1}, s_i]$. We shall denote by $|\Pi|$ the length of the largest subinterval $(s_{i-1}, s_i]$, i.e. $|\Pi| = \max |s_i - s_{i-1}|$ $(i = 1, \ldots, n)$.

Let $f : [a,b] \to \mathcal{H}$ be a function defined on [a,b] whose values are vectors in $\mathcal{H}$. Given any partition Π of [a,b], we may then define the following vector in $\mathcal{H}$:

$$\Sigma_\Pi(f) = \sum_{i=1}^{n} (s_i - s_{i-1}) f(u_i). \tag{4.38}$$

One now takes a sequence of partitions $\{\Pi_r\}$ such that $\lim |\Pi_r| = 0$ as $r \to \infty$, and one defines

$$\int_a^b f(s)ds = \text{s-lim}\, \Sigma_{\Pi_r}(f) \tag{4.39}$$

if this limit exists and is the same for each sequence $\{\Pi_r\}$ with $|\Pi_r| \to 0$ as $r \to \infty$. Instead of (4.39) we shall write

$$\int_a^b f(s)ds = \underset{|\Pi|\to 0}{\text{s-lim}}\, \Sigma_\Pi(f). \tag{4.40}$$

Improper integrals will always be defined as strong limits of integrals over finite sets; for instance

$$\int_a^\infty f(s)ds = \underset{b\to\infty}{\text{s-lim}} \int_a^b f(s)ds. \tag{4.41}$$

In the next proposition we collect some simple properties of these integrals. The domains of integration may be finite or infinite, and it is assumed that all the occuring integrals exist.

PROPOSITION 4.12 :

(a) $\int_a^b f(s)ds + \int_b^c f(s)ds = \int_a^c f(s)ds,$ (4.42)

(b) $\int_a^b \{\alpha f_1(s) + f_2(s)\}ds = \alpha \int_a^b f_1(s)ds + \int_a^b f_2(s)ds,$ (4.43)

(c) $\| \int_a^b f(s)ds \| \leq \int_a^b \| f(s) \| \, ds.$ (4.44)

(d) If $A \varepsilon \mathcal{B}(\mathcal{H})$, then $A \int_a^b f(s)ds = \int_a^b Af(s)ds.$ (4.45)

Proof : (a) and (b) follow immediately from the definitions. To prove (c), we first assume that [a,b] is finite. Then by the triangle inequality (2.9)

$$\| \Sigma_{\Pi}(f) \| \leq \sum_{i=1}^{n} \| f(u_i) \| \, (s_i - s_{i-1}), \tag{4.46}$$

and by Proposition 2.1

$$\| \int_a^b f(s)ds \| = \lim_{|\Pi|\to 0} \| \Sigma_{\Pi}(f) \| \leq \int_a^b \| f(s) \| \, ds.$$

Another application of Proposition 2.1 allows one to pass to infinite intervals. For example if $b = \infty$ and $|a| < \infty$, then

$$\| \int_a^{\infty} f(s)ds \| = \lim_{b\to\infty} \| \int_a^b f(s)ds \| \leq \lim_{b\to\infty} \int_a^b \| f(s) \| \, ds = \int_a^{\infty} \| f(s) \| ds.$$

Similarly (d) is proved first for finite and then for infinite intervals by using each time the fact that $A \cdot \text{s-lim } g_n = \text{s-lim } Ag_n$ as $n \to \infty$ (Problem 1.7). #

We next give sufficient conditions for the existence of such Riemann-Stieltjes integrals. We remind the reader that the function $f : [a,b] \to \mathcal{H}$ is strongly continuous if $\text{s-lim } [f(s + \tau) - f(s)] = \theta$ as $\tau \to 0$ for each $s \varepsilon [a,b]$ and $(s + \tau) \varepsilon [a,b]$.

PROPOSITION 4.13 : Suppose that $f : [a,b] \to \mathcal{H}$ is strongly continuous. If [a,b] is finite, the integral $\int_a^b f(s)ds$ exists. If [a,b] is infinite, this integral exists provided that in addition $\int_a^b \| f(s) \| \, ds < \infty$.

The proof will be given in Section 4-6. In the following proposition [a,b] may be finite or infinite, and all occuring integrals are assumed to exist.

PROPOSITION 4.14 : Suppose $f : [a,b] \to \mathcal{H}$ is strongly differentiable with strongly continuous and integrable derivative f'. Then

$$\int_a^b f'(s)ds = f(b) - f(a). \tag{4.47}$$

If $\phi : [a,b] \to \mathbb{C}$ is a continuously differentiable function, then*)

$$\int_a^b \phi(s)f'(s)ds = \phi(b)f(b) - \phi(a)f(a) - \int_a^b \phi'(s)f(s)ds. \tag{4.48}$$

Proof : We prove the proposition for a finite interval [a,b]. The extension to infinite intervals is straightforward.

(i) Let $c\varepsilon(a,b)$ and define

$$g(x) = \int_c^x f'(s)ds.$$

We first show that $g'(x) = f'(x)$. For this, let $\tau \neq 0$ and let $\Pi = \{s_i;u_i\}$ be a partition of $[x,x+\tau]$. Then

$$g'(x) - f'(x) = \underset{\tau\to 0}{\text{s-lim}} \frac{1}{\tau} \underset{|\Pi|\to 0}{\text{s-lim}} [\sum_i \{f'(u_i) - f'(x)\}(s_i - s_{i-1})]$$

$$\equiv \underset{\tau\to 0}{\text{s-lim}}\, h_\tau .$$

Now fix $\eta > 0$. Since f' is strongly continuous, there exists $\tau_0 > 0$ such that $\|f'(u) - f'(x)\| < \eta$ for all $u \varepsilon [x-\tau_0, x+\tau_0]$. Hence by Proposition 2.1 and the triangle inequality we have for any τ with $|\tau| \leq \tau_0$

---

*) If for instance $b = \infty$, then $\phi(\infty)f(\infty) \equiv \underset{b\to\infty}{\text{s-lim}}\, \phi(b)f(b)$.

$$\|h_\tau\| = \frac{1}{\tau} \lim_{|\Pi|\to 0} \|\sum_i [f'(u_i) - f'(x)](s_i - s_{i-1})\|$$
$$\leq \frac{1}{\tau} \lim_{|\Pi|\to 0} \sum_i (s_i - s_{i-1}) \|f'(u_i) - f'(x)\| < \eta.$$

This shows that s-lim $h_\tau = \theta$ as $\tau \to 0$, i.e. $g'(x)$ exists and $g'(x) = f'(x)$.

(ii) Let $h_o(x) = g(x) - f(x)$ and let $\{e_i\}$ be an orthonormal basis of H. Then for each i, $d/dx\ (e_i, h_o(x)) = 0$, i.e. the components of $h_o(x)$ in the basis $\{e_i\}$ are independent of x. Hence $h_o(x) = h_o$ is a constant vector in H. Consequently

$$\int_a^b f'(s)ds = g(b) - g(a) = f(b) + h_o - f(a) - h_o = f(b) - f(a),$$

which proves (4.47).

(4.48) is an easy consequence of (4.47). It suffices to replace $f(x)$ by $\phi(x)f(x)$ in the latter equation. #

The preceding proposition leads to one of the methods for proving the existence of the wave operators (i.e. of the strong limits in (A1)). The idea is to show that the strong derivative of $V_t^* U_t f$ is integrable on $(-\infty,\infty)$ for a dense set of vectors f in $M_\infty(H_o)$. The details will be given in Chapters 8, 13 and 15. Here we just point out a preliminary result which is a direct consequence of Propositions 4.12 and 4.14.

PROPOSITION 4.15 : Let $\{W_t\}_{t\geq a}$ be a family of operators and $f \in D(W_t)$ for each t. Suppose that $\{W_t f\}$ is strongly continuously differentiable and that

$$\int_a^\infty \| \frac{d}{d\tau} W_\tau f\| \, d\tau < \infty.$$

Then $\{W_t f\}$ converges strongly as $t \to \infty$.

Proof : By applying (4.47) to the vector-valued function $\{W_t f\}$, one gets with (4.44)

$$||W_t f - W_s f|| = || \int_s^t \frac{d}{d\tau} W_\tau f \, d\tau || \leq \int_s^t || \frac{d}{d\tau} W_t f|| \, d\tau. \tag{4.49}$$

By hypothesis the last integral in (4.49) converges to zero as $s,t \to \infty$, which proves that $\{W_t f\}$ is strongly Cauchy. #

We conclude this section with a few remarks about the integration of operator-valued functions. The definition of the integral of an operator-valued function follows naturally from that of the integral of a vector-valued function.

Let $\{A_t\}$ be a family of linear operators defined for t belonging to an interval [a,b]. One may define a new operator denoted by $\int_a^b A_t dt$ as follows :

$$(\int_a^b A_t dt)f = \int_a^b (A_t f)dt. \tag{4.50}$$

Its domain is the set of vectors f such that $f \varepsilon D(A_t)$ for all $t \varepsilon [a,b]$ and such that the integral on the right-hand side of (4.50) exists.

Some properties of this type of integral which follow immediately from Propositions 4.12 and 4.13 are collected below :

PROPOSITION 4.16 : Let $A_t \varepsilon \mathcal{B}(\mathcal{H})$ for each $t \varepsilon [a,b]$. Then

(a) $$||\int_a^b A_t dt|| \leq \int_a^b ||A_t|| \, dt. \tag{4.51}$$

(b) If $B \varepsilon \mathcal{B}(\mathcal{H})$, then $$B \int_a^b A_t dt = \int_a^b BA_t dt. \tag{4.52}$$

(c) Suppose that $\{A_t\}$ is strongly continuous on a finite interval [a,b]. Then $\{A_t\}$ is integrable on [a,b] and the domain of $\int_a^b A_t dt$ is all of H.

(d) If [a,b] is infinite, then the conclusions of (c) hold provided that in addition $\int_a^b \|A_t\| dt < \infty$.

4-5 INTEGRAL REPRESENTATIONS FOR THE WAVE OPERATORS

In this short section we establish an expression for the wave operators in the form of an integral over time. This integral representation of the wave operators is the basic expression in the derivation of the equations of time-independent scattering theory in Chapter 6.

PROPOSITION 4.17 :

(a) If (A1) is satisfied, then

$$\Omega_\pm = \underset{\eta\to+0}{\text{s-lim}}\, \Omega_{\pm\eta} E_\infty(H_o), \tag{4.53}$$

where

$$\Omega_{+\eta} = \eta \int_0^\infty e^{-\eta t} V_t^* U_t dt \tag{4.54}$$

and

$$\Omega_{-\eta} = \eta \int_{-\infty}^0 e^{\eta t} V_t^* U_t dt. \tag{4.55}$$

(b) If (A1) and (A3) are satisfied, then

$$\Omega_\pm^* = \underset{\eta\to+0}{\text{s-lim}}\, \Omega_{\pm\eta}^* E_\infty(H), \tag{4.56}$$

where

$$\Omega_{+\eta}^* = \eta \int_0^\infty e^{-\eta t} U_t^* V_t dt \tag{4.57}$$

and

$$\Omega_{-\eta}^* = \eta \int_{-\infty}^0 e^{\eta t} U_t^* V_t dt. \tag{4.58}$$

Proof : For any $\eta > 0$, the operator-valued function $t \mapsto \eta \exp(-\eta t) V_t^* U_t$ is strongly continuous, and

$$\eta \int_0^\infty e^{-\eta t} \|V_t^* U_t\| \, dt = \eta \int_0^\infty e^{-\eta t} dt = 1. \tag{4.59}$$

Hence the integral in (4.54) defining $\Omega_{+\eta}$ exists by virtue of Proposition 4.16(d).

Now let $f \in M_\infty(H_0)$ and $\delta > 0$. By using (4.44) we get

$$\|\Omega_{+\eta} E_\infty(H_0) f - \Omega_+ f\| = \|\eta \int_0^\infty e^{-\eta t} V_t^* U_t f \, dt - \eta \int_0^\infty e^{-\eta t} \Omega_+ f \, dt\|$$

$$\leq \eta \int_0^\infty e^{-\eta t} \|V_t^* U_t f - \Omega_+ f\| \, dt$$

$$= \eta \int_0^T e^{-\eta t} \|V_t^* U_t f - \Omega_+ f\| \, dt + \eta \int_T^\infty e^{-\eta t} \|V_t^* U_t f - \Omega_+ f\| \, dt.$$

By (A1) we may choose $0 < T < \infty$ such that $\|V_t^* U_t f - \Omega_+ f\| < \delta/2$ for all $t > T$. Hence for $\eta < \eta_0 \equiv \delta(4T \|f\|)^{-1}$ and $t > T$:

$$\|\Omega_{+\eta} E_\infty(H_0) f - \Omega_+ f\| \leq \eta \int_0^T e^{-\eta t} 2\|f\| \, dt + \frac{\delta}{2} \eta \int_T^\infty e^{-\eta t} dt$$

$$< 2\eta T \|f\| + \delta/2 < \delta.$$

This shows that $\text{s-lim}\, \Omega_{+\eta} E_\infty(H_0) = \Omega_+$ as $\eta \to +0$.

The proof of (4.53) for Ω_- is analogous. Similarly part (b) can be established by noticing that $\Omega_\pm^* = \text{s-lim}_{t \to \pm\infty} U_t^* V_t E_\infty(H)$ by virtue of (4.9) and (A3). #

4-6 NOTES AND SUPPLEMENTARY MATERIAL

A. The wave operators were first introduced by Møller [1] in a time-independent formalism and are often called Møller operators in the literature. The time-dependent theory of Section 4-1 with $M_\infty(H_0) = \mathcal{H}$ is essentially due to Friedrichs [1], Jauch [1] and Cook [1]. The definition of the wave operators using certain abstract properties of $M_\infty(H_0)$ was proposed by Wilcox [1].

B. We wish to add some comments concerning the unitarity of the scattering operator, since this concept plays a central role in many theoretical developments (in the so-called S-matrix theory it is even used as an axiom, cf. [E]). In order to simplify the notation we set $M_\infty(H_0) = H$.

Unitarity is sometimes identified with conservation of probability. For S to conserve all transition probabilities, it is sufficient that it be isometric, and this is the case if and only if (A2) holds. So unitarity is not needed in order to have conservation of probability. It should also be said that unitarity does not follow from time reversal invariance and (A1). According to (4.31) and (4.32) time reversal invariance implies only that $F_- = \Theta^{-1}F_+\Theta$, where Θ is the antiunitary time reversal operator. Unitarity is however equivalent to the more restrictive equation $F_- = F_+$. An example of a time reversal invariant scattering system with a non-isometric scattering operator will be mentioned below. An example of a scattering system with isometric but non-unitary S-operator is indicated in Problem 4.8.

C. When introducing the set of scattering states of a Hamiltonian in Section 4-1 we have tacitly assumed that this set is independent of the sign of time. While this is true in numerous situations, a general theory should admit that the states that behave like scattering states as $t\to-\infty$ may not be the same as those that have this property as $t\to+\infty$. One should therefore use two such sets denoted by $M_\infty^\pm(H)$.

In the simple picture indicated in Section 4-1 the states in $M_\infty^-(H)$ would represent incoming particles at large negative times and those in $M_\infty^+(H)$ outgoing particles at large positive times. It may happen that an incoming state will be trapped in the scattering region at all positive times. Also there may be states that are trapped at all negative times and that will escape from the scattering region as $t\to+\infty$. An example of such a Hamiltonian in potential scattering has been given by Pearson [3]. The potential $V(\underline{x})$ is of compact support, singular at $\underline{x} = 0$ and rapidly oscillating near this point. The scattering system is time reversal invariant, the wave operators exist but their ranges are unequal and (A2) does not hold. Thus, if the scattering operator is defined by (4.15), it is not even an isometry. The physical picture is as follows : Take any state vector $f\varepsilon H$, define $g = \Omega_- f$ and $g_+ = F_+ g$. Then $Sf = \Omega_+^* g_+$ and $||Sf|| = ||g_+||$. Thus if g is not in the range of Ω_+, $||Sf||<||f||$. The part g_+ of g will become asymptotically free as $t\to+\infty$, i.e. it describes an outgoing state. The part $g_0 = g - g_+$ of g moves closer and closer to the point $\underline{x} = 0$ as $t\to+\infty$ in the sense that for any a > 0

$$\lim_{t\to\infty} \int_{|\underline{x}|>a} d^3x |(e^{-iHt} g_0)(\underline{x})|^2 = 0. \tag{4.60}$$

In the above example one has a generalized form of a complete asymptotic description in that each scattering state g in $M_\infty^-(H)$ can be split into $g = g_+ \oplus g_0$ such that, as $t\to+\infty$, g_+ becomes asymptotically free and g_0 is absorbed at the singularity (and similarly for $g\varepsilon M_\infty^+(H)$ and $t\to-\infty$). A theory of potential scattering in this general framework has been elaborated by Pearson [4] and by Combescure and Ginibre [2] who proved that under certain assumptions on the potential away from its singularities, all states in H_{ac}(H) orthogonal to the range of Ω_+ (resp. Ω_-) are absorbed as $t\to+\infty$ (resp. $t\to-\infty$). By combining this result with the Gelfand - Levitan solution of the inverse Sturm - Liouville problem [NA, Chapter VIII], one can see that there exist a great many potentials which produce the phenomenon of absorption at local singularities of states belonging to the absolutely continuous subspace of H. It suffices to start from a spectral function for an ordinary second order differential operator $-d^2/dr^2 + V(r)$ whose positive absolutely continuous part has spectral multiplicity 2 on a set of positive Lebesgue measure. Asymptotic completeness will then be violated in the partial wave subspace H_{oo} of angular momentum $\ell = 0$. Indeed the restriction of K_o to H_{oo} has simple spectrum (Proposition 11.5), hence by (4.11) the restriction of H to $\Omega_\pm H_{oo}$ has simple spectrum (Section 5-7), so that $\Omega_\pm H_{oo}$ is strictly smaller than $H_{ac}(H)\cap H_{oo}$ (cf. also Amrein and Georgescu [2]).

It should be stressed that this way of describing absorption and decay has nothing to do with the method of introducing complex potentials [JO, Chapter 20] or contraction semi-groups (Nelson [1], Horwitz, La Vita and Marchand [1],[DA]). H is always a self-adjoint operator, so that one obtains a description of absorption by means of a unitary evolution group.

D. The wave operators between two self-adjoint operators H and H_o have also played an important role in the study of the unitary equivalence of H and H_o. If the limit in (4.5) or (4.6) exists with $E_\infty(H_o) = E_{ac}(H_o)$ and its range is equal to $\mathcal{H}_{ac}(H)$, then the absolutely continuous parts of H and H_o are unitarily equivalent (cf. Proposition 5.21). If $E_{ac}(H_o) \neq I$, the term "generalized wave operators" is used in the mathematical literature for these wave operators, and the reader should keep in mind that the sense of the word "generalized" in this context is different from the one we introduced in Section 4-3.

E. In the integral representations for $\Omega_\pm$ and $\Omega_\pm^*$ given in Proposition 4.17, we used the weight function $\eta \exp(\pm\eta t)$. It is seen from the proof that similar integral representations are valid for other weight functions g_η. For instance one has $\Omega_+ = \text{s-lim}\ \Omega_+(g_\eta)E_\infty(H_o)$ as $\eta\to+0$, where

$$\Omega_+(g_\eta) = \int_0^\infty g_\eta(t)V_t^*U_t dt, \tag{4.61}$$

provided that the following conditions hold :

$$g_\eta \varepsilon C^1([0,\infty)),\ g_\eta \geq 0,\ \int_0^\infty g_\eta(t)dt = 1$$

and $\lim_{\eta\to+0} \int_0^T g_\eta(t)dt = 0$ for each $T < \infty$.

In potential scattering one has $E_\infty(H_o) = I$, so that (4.53) becomes $\Omega_\pm = \text{s-lim}\ \Omega_{\pm\eta}$ as $\eta\to+0$. One can then also show that $\Omega_\pm^* = \text{s-lim}\ \Omega_{\pm\eta}^*$ as $\eta\to+0$, i.e. the projection $E_\infty(H)$ in (4.56) may be dropped. This is a non-trivial extension of Proposition 4.17(b), since in general $U_t^*V_t$ will not be strongly convergent on $\mathcal{M}_\infty(H)^\perp$, so that the strong convergence of $\Omega_{\pm\eta}^* f$ to zero for $f\varepsilon\mathcal{M}_\infty(H)^\perp$ has to be established by some other method. Such a proof has been given by Jauch [1] who uses as additional input certain spectral assumptions on H and H_o.

F. Proof of Proposition 4.13 :

(i) Suppose [a,b] is finite. Let $\delta > 0$. Since f is strongly continuous, it is uniformly continuous on [a,b] in the strong topology, which implies that there exists $\eta > 0$ such that $\|f(s) - f(t)\| < \delta/(b-a)$ whenever $|s-t| < \eta$ (this is proved as for ordinary functions, cf. e.g. [R, Prop. 2.19]).

Let $\{\Pi_r\}$ be a sequence of partitions with $|\Pi_r| \to 0$ as $r \to \infty$, and choose M such that $|\Pi_r| < \eta/2$ for $r > M$. Let $m,n > M$ and let $\Pi_m = \{s_i^{(m)};u_i^{(m)}\}$, $\Pi_n = \{s_j^{(n)};u_j^{(n)}\}$ be the associated partitions. Define the sequence $\{t_k\}$ such that $a_o = t_0 < t_1 < \cdots < t_N = b$ and such that the set $\{s_i^{(m)}, s_j^{(n)}\}$ is identical with the set $\{t_k\}$. Then clearly $t_k - t_{k-1} < \frac{1}{2}\eta$ for all $k = 1, \ldots, N$. Now

$$\Sigma_{\Pi_m}(f) - \Sigma_{\Pi_n}(f) = \sum_{k=1}^{N}\{f(v_k^{(m)}) - f(v_k^{(n)})\}(t_k - t_{k-1}),$$

where each $v_k^{(m)}$ is identical with one of the $u_i^{(m)}$ and each $v_k^{(n)}$ with one of the $u_j^{(n)}$, and $|v_k^{(m)} - v_k^{(n)}| < \eta$ for all k. From the triangle inequality one obtains

$$\|\Sigma_{\Pi_m}(f) - \Sigma_{\Pi_n}(f)\| \leq \sum_{k=1}^{N} \|f(v_k^{(m)}) - f(v_k^{(n)})\| (t_k - t_{k-1})$$

$$\leq \sum_{k=1}^{N} \delta(b-a)^{-1}(t_k - t_{k-1}) = \delta. \qquad (4.62)$$

Hence the sequence of vectors $\Sigma_{\Pi_r}(f)$ is a strong Cauchy sequence, i.e. it has a strong limit denoted by $f_0(\{\Pi_r\})$.

It remains to show that this limit is the same for any pair of such sequences $\{\Pi_r^{(1)}\}$, $\{\Pi_r^{(2)}\}$. This follows from the triangle inequality :

$$\|f_o(\{\Pi_r^{(1)}\}) - f_o(\{\Pi_r^{(2)}\})\| \leq \|f_o(\{\Pi_r^{(1)}\}) - \Sigma_{\Pi_n^{(1)}}(f)\|$$

$$+ \|\Sigma_{\Pi_n^{(1)}}(f) - \Sigma_{\Pi_n^{(2)}}(f)\| + \|\Sigma_{\Pi_n^{(2)}}(f) - f_o(\{\Pi_r^{(2)}\})\|.$$

The second term is less than δ by (4.62) provided that n is large enough. Under the same condition the first and the third term are also less than δ by the definition of $f_o(\{\Pi_r^{(i)}\})$. Hence $\|f_o(\{\Pi_r^{(1)}\}) - f_o(\{\Pi_r^{(2)}\})\|$ is arbitrarily small, which shows that f(s) is integrable over [a,b].

(ii) Suppose for instance $|a| < \infty$, $b = \infty$. Since f(s) is integrable over [a,c] for any $c < \infty$ by part (i), we have from Proposition 4.12

$$\|\int_a^c f(s)ds - \int_a^d f(s)ds\| \leq \int_c^d \|f(s)\| ds.$$

If $\|f(s)\|$ is integrable on $[a,\infty)$, the latter integral converges to zero as $c,d \to \infty$, i.e. $\int_a^c f(s)ds$ is strongly Cauchy as $c \to \infty$. Thus f(s) is integrable on $[a,\infty)$ in this case. #

G. Two mathematical properties of wave operators that are sometimes helpful in proofs are the invariance principle and the chain rule. For their description it is necessary to specify the dependence of $\Omega_\pm$ on H and H_o. We thus write

$$\Omega_\pm(H,H_o) = \underset{t\to\pm\infty}{\text{s-lim}} \exp(iHt)\exp(-iH_o t)E_\infty(H_o). \qquad (4.63)$$

The invariance principle deals with the wave operators for functions of H and H_o. One first defines a class of admissible functions. $\phi : R \to R$ is admissible if the real axis can be divided into a countable number of subintervals Δ_k of length $|\Delta_k|$ such that $|\Delta_k| > \eta > 0$ for all k and such that in each open subinterval, ϕ is differentiable with ϕ' continuous, locally of bounded variation and strictly positive. In all versions of the invariance principle one identifies $M_\infty(H)$ with $H_{ac}(H)$. A typical result is the following :

PROPOSITION 4.18 (Wollenberg [1]) : Let ϕ be an admissible function. If $\Omega_\pm(H,H_o)$ and $\Omega_\pm(\phi(H),\phi(H_o))$ exist and $\Omega_\pm(H,H_o)$ are asymptotically complete, then $\Omega_\pm(H,H_o) = \Omega_\pm(\phi(H),\phi(H_o))$.

Thus the wave operators are the same for each admissible function for which they exist. A different version asserts that, if $\psi(H) - \psi(H_o)$ is a trace class operator for a suitable (not necessarily admissible) function ψ, then for each admissible function ϕ, $\Omega_\pm(\phi(H),\phi(H_o))$ exist, are asymptotically complete and independent of ϕ. This gives therefore a method for proving (A1) and (A3) (cf. Section 9-2).

Let $\mu\varepsilon\mathcal{R}$. Then the following is an admissible function : $\phi(\lambda) = (\mu-\lambda)^{-1}$ for $\lambda > \mu$ and $\phi(\lambda) = \lambda$ for $\lambda < \mu$. If $H_o - \mu > 0$ and $H - \mu > 0$ and if the invariance principle applies, one sees from the functional calculus (Section 5-5) that the study of the scattering problem for the pair $\{H,H_o\}$ may be reduced to that of their resolvents $\{(H-\mu)^{-1},(H_o-\mu)^{-1}\}$ which are bounded operators. For further information on the invariance principle the reader may consult [K, Ch. X.4] and Obermann and Wollenberg [1].

The chain rule expresses the wave operators $\Omega_\pm(H,H_o)$ in terms of $\Omega_\pm(H,H_1)$ and $\Omega_\pm(H_1,H_o)$, where H_1 is an auxiliary self-adjoint operator.

PROPOSITION 4.19 : Suppose that $\Omega_-(H,H_1)$ and $\Omega_-(H_1,H_o)$ exist.

Then $\Omega_-(H,H_0)$ exists and

$$\Omega_-(H,H_0) = \Omega_-(H,H_1)\Omega_-(H_1,H_0). \tag{4.64}$$

If in addition $\Omega_-(H,H_1)$ and $\Omega_-(H_1,H_0)$ are asymptotically complete, then so is $\Omega_-(H,H_0)$. Similar statements hold for Ω_+.

Proof : Define $\Omega_t(A,B) = \exp(iAt)\exp(-iBt)$. Then

$$\Omega_t(H,H_0)E_\infty(H_0) = \Omega_t(H,H_1)E_\infty(H_1)\Omega_t(H_1,H_0)E_\infty(H_0)$$

$$+ \Omega_t(H,H_1)\{I-E_\infty(H_1)\}\Omega_t(H_1,H_0)E_\infty(H_0).$$

As $t\to-\infty$, the first term on the right-hand side converges strongly to $\Omega_-(H,H_1)\Omega_-(H_1,H_0)$ by Proposition 2.18, whereas the second one converges strongly to zero, since the range of $\Omega_-(H_1,H_0)$ is contained in $M_\infty(H_1)$ (in principle by hypothesis, and in practice by using an explicit definition of the scattering states; cf. Problem 7.3). This proves (4.64). The statement about asymptotic completeness is easy to verify. #

PROBLEMS

4.1 : Show that A + B is symmetric if A and B are self-adjoint operators and $D(A)\cap D(B)$ is dense in H.

4.2 : Prove the assertion made in (4.3) and verify (4.2).

4.3 : Prove (4.19).

4.4 : Assume the existence of generalized wave operators $\Omega_\pm$ verifying (4.35) for all $A\varepsilon A_0$. Show that there is a pair of continuous one-parameter unitary groups $\{W_t^\pm\}$ on $M_\infty(H_0)$ commuting with each $A\varepsilon A_0$ and such that their infinitesimal generators $K_\pm$ satisfy $H\Omega_\pm = \Omega_\pm K_\pm$. (Hint : Define $W_t^\pm = \Omega_\pm^* V_t \Omega_\pm$).

4.5 : Let $\Omega_\pm$, $\Omega_\pm'$ be a pair of partial isometries with initial set $M_\infty(H_0)$ verifying (4.35) for all $A\varepsilon A_0$. Show that $\Omega_\pm = \Omega_\pm' U_\pm$, where $U_\pm$ are unitary on $M_\infty(H_0)$ and commute with all $A\varepsilon A_0$.

Verify that $S'^*AS' = S^*AS$ for all $A \varepsilon \mathcal{A}_o$, where $S = \Omega_+^*\Omega_-$ and $S' = \Omega_+'^*\Omega_-'$.

<u>4.6</u> : Show that the requirement of weak convergence in (4.2) need not determine $f_\pm$ uniquely. (Hint : Apply Lemma 5.20).

<u>4.7</u> : Prove part (d) of Proposition 4.12.

<u>4.8</u> : <u>A scattering system with discrete time parameter</u> :

Let $\{e_{i,k}\}$ $(i=1,2,3,\ldots;\ k=0,\pm 1,\pm 2,\ldots)$ be an infinite orthonormal basis of $\mathcal{H}$. Let U,V be the unitary operators defined as follows by their action on $\{e_{i,k}\}$: $Ue_{i,k} = e_{i,k+1}$, $Ve_{i,k} = e_{i,k+1}$ if $i \neq -k,-k+1$, $Ve_{-k,k} = e_{-k+1,k+1}$ and $Ve_{-k+1,k} = e_{-k,k+1}$ for $k < 0$, $Ve_{1,0} = e_{1,1}$. (Make a pictorial representation of the action of U and V by representing each $e_{i,k}$ by its coordinates (i,k) in $\mathcal{R}^2$.) For each integer n, let $U_n = U^n$, $V_n = V^n$. Prove that $\Omega_\pm = \text{s-lim}\ V_n^*U_n$ as $n \to \pm\infty$ exist (it suffices to prove convergence on each vector $e_{i,k}$, cf. Proposition 2.17). Calculate $\Omega_\pm e_{i,k}$ and show that range $\Omega_+ = \mathcal{H}$ whereas range $\Omega_- \neq \mathcal{H}$. Verify explictly that S is isometric but not unitary and that $\text{w-lim}\ U_n^*V_n e_{i,k} = \theta$ as $n \to -\infty$ for each $e_{i,k}$ orthogonal to the range of Ω_-.

<u>Remark</u> : One may similarly construct a scattering system with continuous time parameter for which (A1) and (A2) hold but S is not unitary. For this one replaces the index k by a continuous variable τ, i.e one sets $\mathcal{H} = L^2(\mathcal{R},\mathcal{H}_o;d\tau)$ with $\{e_i\}_{i=1}^\infty$ a countably infinite orthonormal basis of $\mathcal{H}_o$. The discrete shift in the variable k is replaced by a continuous shift in τ and the permutation of two indices i,i+1 over a unit time interval by a corresponding continuous rotation in the two-dimensional subspace spanned by e_i and e_{i+1}.

CHAPTER 5 : SPECTRAL THEORY OF SELF-ADJOINT OPERATORS

This chapter contains mostly mathematical material. It is devoted to the study of certain structural properties of self-adjoint operators that are very useful in quantum mechanics (cf. Section 3-2). The basic theorem is the so-called spectral theorem which in an appropriate sense generalizes the diagonalizability theorem of hermitian matrices in finite dimensions. To motivate the spectral theorem, we discuss in Section 5-1 the eigenvalue problem for hermitian matrices in finite dimensions and compact operators in infinite-dimensional Hilbert spaces. Sections 5-2 and 5-3 define spectral families and associated self-adjoint operators respectively. We state the spectral theorem, prove it for a bounded self-adjoint operator in Section 5-4, and apply it in Section 5-5 to form functions of a self-adjoint operator.

Section 5-6 introduces the important notions of resolvent set and spectrum as well as various decompositions of the spectrum. Section 5-7 deals with the spectral representation of certain self-adjoint operators. Finally Section 5-8 consists of notes and supplementary material, containing in particular a brief indication of the proof of the spectral theorem for an unbounded self-adjoint operator. Two specific applications of spectral theory to scattering may be found in Sections 5-6 and 5-7.

5-1 INTRODUCTION

We remind the reader of the concept of an eigenvalue problem that we have already introduced in Section 2-3 for a compact operator. Let A be a linear densely defined operator in $\mathcal{H}$. A complex number λ is called an eigenvalue of A if there is a non-zero vector $f \varepsilon D(A)$ such that

$$Af = \lambda f. \tag{5.1}$$

In such a case, f is called an eigenvector of A associated with the eigenvalue λ. The eigenvalue problem consists of finding all the eigenvalues and eigenvectors of a given operator A.

If the operator A is self-adjoint, then we know from Proposition 2.26 that the eigenvalues are real and that the eigenvectors associated with different eigenvalues are mutually orthogonal. Denoting the (orthogonal) projection onto the (closed) linear manifold of eigenvectors associated with the eigenvalue λ_k by $E_{\{\lambda_k\}}$, this implies that the ranges of

$E_{\{\lambda_k\}}$ are mutually orthogonal. The dimension n_k of the subspace $E_{\{\lambda_k\}}\mathcal{H}$ is called the multiplicity of the eigenvalue λ_k.

If $\dim \mathcal{H} = n < \infty$, then it is well-known [H] that every hermitian operator can be written as

$$A = \sum_{k=1}^{m} \lambda_k E_{\{\lambda_k\}}, \quad \text{where} \quad n = \sum_{k=1}^{m} n_k. \tag{5.2}$$

This is sometimes expressed by saying that a hermitian operator in a Hilbert space of finite dimension is diagonalizable, i.e. it can be put in the form of a diagonal matrix by a suitable unitary transformation. We now enumerate the different eigenvalues in ascending order, i.e.

$$\lambda_1 < \lambda_2 < \lambda_3 \dots < \lambda_m \tag{5.3}$$

and define an operator-valued function of λ by

$$E_\lambda = \begin{cases} 0 & ; \quad \lambda < \lambda_1 \\ \sum_{\ell=1}^{k} E_{\{\lambda_\ell\}} & ; \quad \lambda_k \leq \lambda < \lambda_{k+1} \\ I & ; \quad \lambda \geq \lambda_m . \end{cases} \tag{5.4}$$

Then it is easy to verify that each E_λ is a projection and that the relations (3.7)-(3.9) are satisfied. Also it follows from the definition (5.4) and relation (5.3) that for $f,g\in\mathcal{H}$,

$$(f,Ag) = \sum_{k=1}^{m} \lambda_k (f,E_{\{\lambda_k\}}g) = \int \lambda \, d(f,E_\lambda g), \tag{5.5}$$

where the integral on the right-hand side is a Stieltjes integral [RN, p. 105].

A similar calculation is possible for a compact operator in an infinite-dimensional Hilbert space. We know from Proposition 2.30 that a self-adjoint compact operator is diagonalizable. For the sake of simplicity of notation, we consider here only a positive compact operator. Then all its eigenvalues λ_k are non-negative, and enumerating them such that $\lambda_1 > \lambda_2 > \ldots > 0$ and possibly $\lambda_o = 0$, we can write

$$A = \sum_{k=1}^{\infty} \lambda_k E_{\{\lambda_k\}} \tag{5.6}$$

where $\lambda_k \to 0$ as $k \to \infty$. Denoting by $E_{\{0\}}$ the projection onto the null-space M_o, we also have

$$\sum_{k=1}^{\infty} E_{\{\lambda_k\}} + E_{\{0\}} = I. \tag{5.7}$$

Then as before, writing $E_o = E_{\{0\}}$ and

$$E_\lambda = \begin{cases} 0 & ; \quad \lambda < 0 \\ E_{\{0\}} + \sum_{\ell=k}^{\infty} E_{\{\lambda_\ell\}} ; & \lambda_{k-1} < \lambda \leq \lambda_k \\ I & ; \quad \lambda \geq \lambda_1, \end{cases} \tag{5.8}$$

we can verify all the properties (3.7) - (3.9) of the projection-valued function E_λ and conclude that

$$(f,Ag) = \int \lambda \, d(f,E_\lambda g). \tag{5.9}$$

Thus we observe that a self-adjoint operator in these two special cases does have an integral representation given by relations (5.5) and (5.9). In these cases such a representation is simply a convenient way of re-expressing diagonalizability. Clearly for self-adjoint operators which do not belong to either of the two classes mentioned above, in particular those which do not have any eigenvalues, the usual concept of diagonalizability is no longer meaningful. However, it may be possible to express any self-adjoint operator in the integral form of (5.5) or (5.9) provided that the spectral family $\{E_\lambda\}$ is appropriately defined. In fact that this is indeed so is the content of the spectral theorem (Propositions

5.8 and 5.9). Before this can be verified, we need to define spectral families abstractly and study some of their properties, which will be done in the next section.

5-2 SPECTRAL FAMILIES

In Section 2-2 we have discussed the one-to-one correspondence between (closed) subspaces of the Hilbert space H and (orthogonal) projection operators in H. Now we suppose there is a non-decreasing family $\{M_\lambda\}$ of subspaces of H depending on the real parameter λ, $-\infty < \lambda < +\infty$, such that the intersection of all the M_λ is θ and their union is dense in H. We call the family $\{M_\lambda\}$ non-decreasing if $M_{\lambda'} \subseteq M_\lambda$ for $\lambda' < \lambda$.

Translating these properties into those of the associated family of projection operators $\{E_\lambda\}$, we have that (Problem 5.1)

(i) $\{E_\lambda\}$ is non-decreasing : $E_{\lambda'} \leq E_{\lambda''}$ for $\lambda' \leq \lambda''$ *) (5.10)

or equivalently $$E_\lambda E_\mu = E_{\min\{\lambda,\mu\}}. \tag{5.11}$$

(ii) $$\underset{\lambda\to-\infty}{\text{s-lim}}\, E_\lambda = 0 \text{ and } \underset{\lambda\to+\infty}{\text{s-lim}}\, E_\lambda = I. \tag{5.12}$$

A family of projections satisfying (5.10) and (5.12) is called a <u>spectral family</u> or a <u>resolution of the identity</u>.

It is easy to show using property (5.10) that s-lim $E_{\lambda\pm\eta}$ as $\eta\to+0$ exist (Problem 5.1). We denote these

---

*) $A \leq B$ means $B-A \geq 0$.

limits by $E_{\lambda\pm 0}$ respectively, i.e.

$$E_{\lambda\pm 0} = \underset{\eta\to+0}{\text{s-lim}}\, E_{\lambda\pm\eta}. \tag{5.13}$$

A spectral family $\{E_\lambda\}$ is said to be right (left) continuous if $E_{\lambda+0} = E_\lambda$ ($E_{\lambda-0} = E_\lambda$). A common convention is to assume right continuity, and from this point on by a spectral family we shall mean a right-continuous spectral family.

Now we give a few more definitions which will be useful later. $\{E_\lambda\}$ is said to be <u>bounded below</u> if $E_\mu = 0$ for some finite μ, in which case $E_\lambda = 0$ for all $\lambda < \mu$. If $\{E_\lambda\}$ is bounded below, then the least upper bound of all such μ is the <u>lower bound</u> of $\{E_\lambda\}$. Similarly one defines the <u>upper bound</u> of $\{E_\lambda\}$ as the greatest lower bound of all finite μ such that $E_\mu = I$. Right-continuity implies that, while E_λ equals I when λ equals the upper bound, E_λ need not be zero when λ equals the lower bound.

A real number μ is called a <u>point of constancy</u> with respect to $\{E_\lambda\}$ if E_λ is a constant in a neighborhood of μ, i.e. if $E_{\mu+\eta} = E_{\mu-\eta}$ for some $\eta > 0$. The set of all points which are not points of constancy (i.e. the set of "points of increase") is called the <u>support</u> of $\{E_\lambda\}$. Then it is clear that $\{E_\lambda\}$ is bounded below (above) if and only if the support of $\{E_\lambda\}$ is.

For any half-open interval $\Delta = (\lambda',\lambda'']$ of the real line, we set

$$E_\Delta = E_{\lambda''} - E_{\lambda'}\ , \tag{5.14}$$

so that E_Δ is the projection onto the subspace $M_{\lambda''} \ominus M_{\lambda'}$, i.e.

the orthogonal complement of $M_{\lambda'}$ in $M_{\lambda''}$. If two such intervals Δ_1 and Δ_2 are disjoint, then by (5.10) and (5.14) the ranges of E_{Δ_1} and E_{Δ_2} are mutually orthogonal, i.e.

$$E_{\Delta_1} E_{\Delta_2} = E_{\Delta_2} E_{\Delta_1} = 0. \tag{5.15}$$

Let us also define

$$E_{\{\lambda\}} = E_\lambda - E_{\lambda-0}. \tag{5.16}$$

Then $E_{\{\lambda\}}$, a strong limit of a sequence of self-adjoint operators, is self-adjoint. Also, by virtue of Proposition 2.18, $E_{\{\lambda\}}^2 = E_{\{\lambda\}}$, thus verifying that $E_{\{\lambda\}}$ defined by (5.16) is a projection. Similarly one derives the property that

$$E_{\{\lambda\}} E_{\{\mu\}} = E_{\{\mu\}} E_{\{\lambda\}} = 0 \qquad \text{if } \lambda \neq \mu. \tag{5.17}$$

$E_{\{\lambda\}}$ is non-zero if and only if $\{E_\lambda\}$ is not strongly continuous at λ. Since in a separable Hilbert space a set of mutually orthogonal projections is at most countable, the relation (5.17) ensures that there are at most countably many such points of discontinuity of $\{E_\lambda\}$.

Definition (5.14) can be extended to closed and open intervals. If $\Delta = [\lambda',\lambda'']$, then we set $E_\Delta = E_{(\lambda',\lambda'']} + E_{\{\lambda'\}}$. On the other hand if $\Delta = (\lambda',\lambda'')$ we first note that $\Delta = \cup_n (\lambda',\lambda''-1/n]$ and then set $E_\Delta = \text{s-lim}\, E_{(\lambda',\lambda''-1/n]}$ as $n \to \infty$, which exists and is a projection. If Δ is the union of a finite number of intervals (open, closed or half-open) then it can be expressed as the finite union of disjoint sets of the same type (open, closed or half-open). If we define E_Δ in such a case to be the sum of the corresponding projections, then it is easy to verify that for two such sets Δ and Δ'

$$E_\Delta E_{\Delta'} = E_{\Delta \cap \Delta'} \ . \tag{5.18}$$

The family $\{E_\Delta\}$ thus defined is called a <u>spectral measure</u> on the class of all sets Δ of the kind described. This measure can then be extended to a countably additive projection-valued measure on the class of all Borel sets of the real real line by standard methods [R]. We shall use the same symbol Δ for a general Borel set as well as for intervals. The spectral measure so defined is a special case of more general <u>positive operator-valued measures</u>, for a discussion of which the interested reader is referred to [AG].

We shall use this (operator-valued) spectral measure for integration in Chapter 6 to study the time-independent scattering theory. Here we restrict ourselves to integration with respect to the numerical-valued measure generated by the function $(f,E_\lambda g)$.

<u>LEMMA 5.1</u> : For any $f,g \in \mathcal{H}$, the complex-valued function $\lambda \mapsto (f,E_\lambda g)$ is of bounded variation. Also, for any unit vector $f \in \mathcal{H}$ the positive function $\lambda \mapsto \|E_\lambda f\|^2$ is normalized and of bounded variation (the definitions will become clear in the proof).

<u>Proof</u> : Let $\Delta = (\lambda',\lambda'']$ and let $\lambda' = \lambda_0 < \lambda_1 < \ldots < \lambda_n = \lambda''$ be an arbitrary partition of Δ. Setting $\Delta_i = (\lambda_{i-1},\lambda_i]$, we note that Δ is the union of all $\{\Delta_i\}$ where the Δ_i's are mutually disjoint. By using also the Schwarz inequality, we get

$$\sum_{i=1}^n |(f,E_{\lambda_i} g) - (f,E_{\lambda_{i-1}} g)| = \sum_{i=1}^n |(f,E_{\Delta_i} g)|$$

$$= \sum_{i=1}^{n} |(E_{\Delta_i} f, E_{\Delta_i} g)| \leq \sum_{i=1}^{n} \|E_{\Delta_i} f\| \, \|E_{\Delta_i} g\|$$

$$\leq (\sum_i \|E_{\Delta_i} f\|^2)^{\frac{1}{2}} (\sum_i \|E_{\Delta_i} g\|^2)^{\frac{1}{2}} = (\sum_i (f, E_{\Delta_i} f))^{\frac{1}{2}} (\sum_i (g, E_{\Delta_i} g))^{\frac{1}{2}}$$

$$= \|E_\Delta f\| \, \|E_\Delta g\| \leq \|f\| \, \|g\| \,. \tag{5.19}$$

Thus the total variation of $(f, E_\lambda g)$ over any interval Δ does not exceed $\|f\| \, \|g\|$, proving the first half of the lemma. The second half follows from the first half by setting $f = g$ and by observing that the property (5.12) ensures that

$$\lim_{\lambda \to -\infty} (f, E_\lambda f) = 0 \text{ and } \lim_{\lambda \to +\infty} (f, E_\lambda f) = 1 \text{ whenever } \|f\| = 1. \quad \#$$

By standard methods ([RN, p. 110]; cf. also Section 4-4), the result of Lemma 5.1 can be used to define Riemann-Stieltjes integrals over any finite interval with respect to the measure generated by the function $(f, E_\lambda g)$. If $\phi(\lambda)$ is a complex-valued continuous function of λ, then the integral $\int_a^b \phi(\lambda) d(f, E_\lambda g)$, defined as the limit of Riemann-Stieltjes sums, exists for every finite a,b and all $f, g \in \mathcal{H}$. Improper integrals over $\mathcal{R}$ are defined as

$$\int_{-\infty}^{\infty} \phi(\lambda) d(f, E_\lambda g) = \lim \int_a^b \phi(\lambda) d(f, E_\lambda g)$$

as $a \to -\infty$, $b \to +\infty$ whenever they exist.

5-3 THE SELF-ADJOINT OPERATOR ASSOCIATED WITH A SPECTRAL FAMILY

To every spectral family $\{E_\lambda\}$ one can associate a self-adjoint operator A formally given by

$$A = \int_{-\infty}^{\infty} \lambda \, dE_\lambda \, . \tag{5.20}$$

The meaning of the above expression is made precise in the

PROPOSITION 5.2 : Let $D(A) = \{g \in H \mid \int_{-\infty}^{\infty} \lambda^2 d(g,E_\lambda g) < \infty\}$. (5.21)
Then there eixsts a self-adjoint operator A with domain D(A) such that for all $f \in H$ and $g \in D(A)$

$$(f,Ag) = \int_{-\infty}^{\infty} \lambda \, d(f,E_\lambda g) . \tag{5.22}$$ *)

Proof : (i) Let $g \in H$ and define $g_n = (E_n - E_{-n})g$ for $n = 1,2,\ldots$ Then g_n converges strongly to g as $n \to \infty$ by virtue of (5.12). On the other hand, by Lemma 5.1

$$\int_{-\infty}^{\infty} \lambda^2 d(g_n,E_\lambda g_n) = \int_{-n}^{n} \lambda^2 d(g,E_\lambda g) \leq n^2 \|g\|^2 , \tag{5.23}$$

i.e. $g_n \in D(A)$. This proves that D(A) is dense.

(ii) For every fixed $g \in D(A)$, consider the linear functional $\phi_g(f) = \int_{-\infty}^{\infty} \lambda d(g,E_\lambda f)$. Then a calculation similar to the one in the proof of Lemma 5.1 shows that

$$|\phi_g(f)| \leq \|f\| \, [\int_{-\infty}^{\infty} \lambda^2 d(g,E_\lambda g)]^{\frac{1}{2}} . \tag{5.24}$$

Therefore by the theorem of Riesz (Proposition 2.3), there exists a vector $g^* \in H$ such that $\phi_g(f) = (g^*,f)$ for each $g \in D(A)$. Setting $g^* = Ag$ for all $g \in D(A)$, we easily see that A is a densely defined linear operator and that A is a symmetric operator satisfying relation (5.22).

(iii) To prove the self-adjointness of A, it suffices to show that the ranges of $(A \pm i)$ are equal to H (Proposition 2.14). To this end we consider the complex-valued functions

---

*) More generally, $\int_{-\infty}^{\infty} \lambda dE_\lambda$ can be defined as a strong Riemann-Stieltjes integral. See [HP, Section 3-3] and Lemma 6.2.

$$\phi_{\pm} : \lambda \mapsto (\lambda \pm i)^{-1}. \tag{5.25}$$

It is clear that $\phi_{\pm}$ are bounded continuous functions on R and $\|\phi_{\pm}\|_{\infty} \equiv \sup |\phi_{\pm}(\lambda)| \leq 1$. Therefore, as in part (ii), one can define two operators $\phi_{\pm}(A)$ such that

$$(f,\phi_{\pm}(A)g) = \int_{-\infty}^{\infty} \phi_{\pm}(\lambda)d(f,E_{\lambda}g),$$

with $\phi_{\pm}(A)\varepsilon B(H)$ and $\|\phi_{\pm}(A)\| \leq 1$. We claim that

$$\phi_{\pm}(A) = (A \pm i)^{-1}. \tag{5.26}$$

To see this, we compute for $g\varepsilon D(A)$ and $f\varepsilon H$

$$\begin{aligned}(f,\phi_{\pm}(A)(A \pm i)g) &= \int_{-\infty}^{\infty} \phi_{\pm}(\lambda)d(f,E_{\lambda}(A \pm i)g)\\ &= \int_{-\infty}^{\infty} \phi_{\pm}(\lambda)d_{\lambda}[\int_{-\infty}^{\infty}(\mu \pm i)d_{\mu}(f,E_{\lambda}E_{\mu}g)]\\ &= \int_{-\infty}^{\infty} \phi_{\pm}(\lambda)d_{\lambda}[\int_{-\infty}^{\lambda}(\mu \pm i)d_{\mu}(f,E_{\mu}g)]\\ &= \int_{-\infty}^{\infty} \phi_{\pm}(\lambda)(\lambda \pm i)d(f,E_{\lambda}g) = (f,g),\end{aligned} \tag{5.27}$$

where we have used (5.11) and the definition (5.25). A similar calculation shows that $\phi_{\pm}(A)H \subseteq D(A)$ and that $(f,(A \pm i)\phi_{\pm}(A)g) = (f,g)$, verifying our claim (5.26) (cf. Lemma 2.27). Thus any vector $f\varepsilon H$ is given as $f = (A \pm i)(\phi_{\pm}(A)f)$, proving that the ranges of $(A \pm i)$ are equal to H. #

If the support of $\{E_{\lambda}\}$ is Δ, a subset of R, then clearly (5.20) can be rewritten as $A = \int_{\Delta} \lambda\, dE_{\lambda}$. If moreover the support is bounded, then it follows that the associated self-adjoint operator A is defined everywhere and bounded. If the lower and upper bounds of $\{E_{\lambda}\}$ are m and M respectively, then one concludes as in (5.23) that $D(A) = H$ and for all $f\varepsilon H$

$$\|Af\|^2 = \int_m^M \lambda^2 d(f,E_{\lambda}f) \leq \max\,\{m^2,M^2\}\|f\|^2. \tag{5.28}$$

In fact it can be shown (Problem 5.2) that the equality holds, establishing the fact that

$$||A|| = \max(|m|,|M|). \qquad (5.29)$$

The relation between eigenvalues and points of discontinuity is shown in the following

LEMMA 5.3 : $f \neq \theta$ and $(A-\mu)f = \theta$ if and only if $||E_\lambda f||^2$ is constant except for $\lambda = \mu$. Also f is an eigenvector associated with the eigenvalue μ if and only if $E_{\{\mu\}}f = f$.

Proof : One gets as in (5.27) that

$$\begin{aligned} ||(A-\mu)f||^2 &= (A-\mu)f,(A-\mu)f) = \int(\lambda-\mu)d_\lambda((A-\mu)f,E_\lambda f) \\ &= \int(\lambda-\mu)^2 d(f,E_\lambda f). \end{aligned} \qquad (5.30)$$

Then it is clear that $(A-\mu)f = \theta$ if and only if the function $(f,E_\lambda f)$ is constant except for a discontinuity at $\lambda = \mu$. Because of the right continuity and (5.12), we conclude that

$$E_\mu f = f \text{ and } E_{\mu-o}f = \theta. \qquad (5.31)$$

Hence μ is an eigenvalue of A if and only if $E_{\{\mu\}} \equiv E_\mu - E_{\mu-o} \neq 0$, as is evident from (5.31), and f is an associated eigenvector if and only if $E_{\{\mu\}}f = f$. #

We conclude this section by presenting some material on the commutativity, decomposition and reduction of (possibly unbounded) operators. These concepts will be independently useful in this book and are essential in proving the spectral theorem.

Two operators $A,B\varepsilon\mathcal{B}(\mathcal{H})$ are said to commute if AB = BA.

It is not possible to extend this notion to unbounded operators in H. However, if one of the two operators, say B, is bounded then we say that A commutes with B when

$$BA \subseteq AB. \tag{5.32}$$

It means that whenever $f \in D(A)$, Bf also belongs to D(A) and ABf = BAf. It is clear that (5.32) is equivalent to the usual definition when both A and B belong to $\mathcal{B}(H)$.

Let $H = M \oplus M^{\perp}$ be a decomposition of H, and F be the projection onto the subspace M. An (unbounded) operator A is said to be decomposed according to $H = M \oplus M^{\perp}$ if

$$FD(A) \subseteq D(A),\ AFD(A) \subseteq M \text{ and } A(I-F)D(A) \subseteq M^{\perp}. \tag{5.33}$$

It is easy to show that (5.33) is equivalent to the condition that A commutes with the projection F, viz.

$$FA \subseteq AF. \tag{5.34}$$

In fact, (5.33) implies that for any $f \in D(A)$, $Ff \in D(A)$, $AFf \in M$ and $A(I-F)f \in M^{\perp}$. Hence $(I-F)AFf = FA(I-F)f = \theta$, leading to the conclusion that AFf = FAf for $f \in D(A)$, which is (5.34). The verification of the converse is left as an exercise (Problem 5.3).

If one of the two equivalent conditions (5.33) and (5.34) is satisfied, the restriction of A to $M \cap D(A)$ can be viewed as an operator in the Hilbert space M. This operator will be called the part of A in M and denoted by A/M.

If the operator A is symmetric, then the following lemma shows that the third condition in (5.33), namely that $A(I-F)D(A) \subseteq M^{\perp}$, is redundant. We say that A is reduced by M

if $FD(A) \subseteq D(A)$ and $AFD(A) \subseteq M$.

LEMMA 5.4 : (a) The symmetric operator A is reduced by a subspace M if and only if A commutes with F, the projection onto M.

(b) If moreover A is selfadjoint, then A/M is also self-adjoint.

(c) Let $B \varepsilon B(H)$ and $[B,F] = 0$. Then $B = B/FH \oplus B/(I-F)H$.

Proof : (a) (i) Suppose A is reduced by M, and let $f \varepsilon D(A)$. Then $Ff \varepsilon D(A)$ and $AFf \varepsilon M$, so that $FAFf = AFf$. Then $(g,FAFf) = (FAFg,f) = (AFg,f) = (g,FAf)$ for all $f,g \varepsilon D(A)$. Since $D(A)$ is dense, we conclude from Proposition 2.2 that $FAFf = FAf$. Thus $AFf = FAFf = FAf$, which says that A commutes with F.

(ii) Conversely, let $FA \subseteq AF$ and $f \varepsilon D(A)$. Then by the definition (5.32) $Ff \varepsilon D(A)$ and $AFf = FAf$. Hence $FAFf = F^2Af = FAf$, which implies with the preceding identity that $AFf = FAFf$, i.e. $AFD(A) \subseteq M$.

(b) Let $f \varepsilon D((A/M)^*)$ and $g \varepsilon D(A)$. Since A is reduced by M, we have that $Fg \varepsilon M \cap D(A)$ and $A(I-F)g = (I-F)Ag \varepsilon M^{\perp}$. Therefore $(f,Ag) = (f,AFg) + (f,A(I-F)g) = (f,AFg) = ((A/M)^* f,g)$. Thus $f \varepsilon M \cap D(A^*) = M \cap D(A) = D(A/M)$, implying that the symmetric operator A/M is self-adjoint.

(c) We have $B = BF + B(I-F) = FBF + (I-F)B(I-F)$. Then by the definition of direct sum of operators in Section 2-4, we get the result. #

5-4 THE SPECTRAL THEOREM

So far we have shown that any spectral family determines a self-adjoint operator by the relation (5.20). In this section we prove the converse, i.e. every self-adjoint operator A admits a representation (5.20) with a spectral family $\{E_\lambda\}$, uniquely determined by A. There are a large number of methods of arriving at the above result, known as the spectral theorem. In this section we give one of these methods, but restrict ourselves to bounded self-adjoint operators, and we relegate to Section 5-8 some indications of the proof of the theorem for unbounded self-adjoint operators.

Before we proceed to the proof of the spectral theorem for a bounded self-adjoint operator we need some preliminary material. We remind the reader that bounded positive operators were defined in Section 3-2. An important property of positive operators is that they have a unique positive square-root. This is the content of the next lemma, the proof of which is given in Section 5-8.

LEMMA 5.5 (Square root lemma) : Let $A \in \mathcal{B}(H)$ and $A \geq 0$. Then there is a unique bounded self-adjoint operator B with $B \geq 0$ and $B^2 = A$. Furthermore, B commutes with every bounded operator that commutes with A. One writes $B = A^{\frac{1}{2}}$.

Next we seek a representation of a bounded operator A similar to the polar form of a complex number z, namely $z = |z| \exp(i \arg z)$ (see Problem 2.29 for a special case). For that we have to first define $|A|$. For this we notice that the bounded operator A*A is positive, because $(f, A^*Af) = \|Af\|^2 \geq 0$

for all $f \in H$. We now define the absolute value of A by $|A| \equiv (A^*A)^{\frac{1}{2}}$, the square root being given by Lemma 5.5.

PROPOSITION 5.6 (Polar decomposition) : Let $A \in B(H)$. Then there is a partial isometry U with initial set $\overline{|A|H}$ such that $A = U|A|$. Such a decomposition is unique in the following sense : If $A = U'B$ where $B \geq 0$ and U' is a partial isometry with initial set $\overline{BH}$, then $B = |A|$ and $U' = U$.

Proof : (i) Since $\| |A|f\|^2 = (f, |A|^2 f) = \|Af\|^2$ for all $f \in H$, the correspondence $|A|f \to Af$ defines an isometric mapping U of the range of $|A|$ onto the range of A : $Af = U|A|f$. By continuity (Proposition 2.6) U can be extended to an isometric operator from the closure of the range of $|A|$ onto the closure of the range of A. U can be further extended to an operator in $B(H)$, which we shall again denote by U, by setting $Ug = 0$ for all g in the orthogonal complement of the range of $|A|$, i.e. in the null space $N(|A|)$ *) of $|A|$. Thus defined, U is partially isometric with its initial set equal to the closure of the range of $|A|$ and its final set equal to the closure of the range of A, and we have

$$A = U|A|. \tag{5.35}$$

(ii) If $A = U'B$, then $A^* = B^*U'^*$ and hence $A^*A = BU'^*U'B = B^2$, because $U'^*U'f = f$ for f in the initial set of U'. Thus B must be a non-negative square root of A^*A, which by Lemma 5.5 is uniquely $|A|$. Then (5.35) determines U' on the range of $|A|$. Since $U' = 0$ on the orthogonal complement of the range of $|A|$, we have $U = U'$. #

*) The null space $N(B)$ of an operator B is $\{f \in D(B) \mid Bf = \theta\}$ (see Problem 5.5).

Now we specialize to the case when A is a bounded self-adjoint operator. In this case $|A| = (A^2)^{\frac{1}{2}} = |A^*|$ and we have

PROPOSITION 5.7 : Let A be a bounded self-adjoint operator and let $A = U|A|$ be its polar decomposition. Then (a) the range of A is equal to the range of $|A|$ and $U = U^*$. (b) Both A and $|A|$ commute with U and any bounded operator that commutes with A commutes with U. (c) $\mathcal{H}$ admits a decomposition :

$$\mathcal{H} = \mathcal{M}_+ \oplus \mathcal{M}_- \oplus \mathcal{M}_0 \tag{5.36}$$

such that A is reduced by each of these subspaces. Furthermore, $A/\mathcal{M}_+ > 0$, $A/\mathcal{M}_- < 0$ and $A/\mathcal{M}_0 = 0$. (d) The decomposition (5.36) is unique in the following sense. If $\mathcal{M}'$ is a subspace that reduces A and if for all $f \in \mathcal{M}'$, $(f,Af) \geq 0$ ($(f,Af) \leq 0$), then $\mathcal{M}' \subseteq \mathcal{M}_-^{\perp}$ ($\mathcal{M}' \subseteq \mathcal{M}_+^{\perp}$).

Proof : (a) Taking the adjoint of the relation $A = U|A|$ we have that

$$A = U|A| = A^* = |A|U^*. \tag{5.37}$$

Consider the bounded operator C given by

$$C = U|A|U^*. \tag{5.38}$$

It is easy to see that C is self-adjoint positive, since $(f,Cf) = (U^*f,|A|U^*f)$ and $|A| \geq 0$. From (5.38) we obtain that $C^2 = U|A|U^*U|A|U^* = U|A|^2U^* = A^2$, where we have used the fact that the initial set of U contains the range of $|A|$ and relation (5.37). Therefore by the uniqueness of the positive square-root (Lemma 5.5) we conclude that $C = |A|$. This combined with (5.37) and (5.38) gives

$$|A| = U|A|U^* = AU^* \tag{5.39}$$

and $$A = |A|U^* = U^*U|A|U^* = U^*|A| = U|A|. \qquad (5.40)$$

The relations (5.39) and (5.37) imply that the ranges of A and $|A|$ are equal, and (5.40) along with the uniqueness property of the polar decomposition gives the second result, viz. $U = U^*$.

(b) Relations (5.37), (5.40) and the fact that $U = U^*$ together imply that both A and $|A|$ commute with U. Now let $B \in \mathcal{B}(\mathcal{H})$ and let B commute with A. By Lemma 5.5, B commutes with $|A|$, the positive square-root of A^2, and

$$BA = BU|A| = AB = U|A|B = UB|A|. \qquad (5.41)$$

Relation (5.41) is equivalent to saying that BU equals UB on the range of $|A|$. Since the null space of any self-adjoint operator is equal to the orthogonal complement of its range (Problem 5.5), it remains to prove that $UB = BU$ on the null space $N(|A|)$ of $|A|$.

Let $f \in N(|A|) = N(A)$. Then $|A|UBf = ABf = BAf = \theta$, where we have used (5.37) and the fact that $U = U^*$. Therefore $UBf \in N(|A|) = N(A)$. On the other hand, the vector UBf clearly belongs to the range of U which by the definition of U (Proposition 5.6) is orthogonal to N(A), implying that $UBf = \theta$. Since U is defined to be zero on $N(|A|) = N(A)$, we have $BUf = \theta$, hence $UB = BU$ on $N(|A|)$.

(c) Define $\mathcal{M}$ to be the closure of the range of A (= the closure of the range of $|A|$). Then $U^2 f = U^*Uf = f$ for $f \in \mathcal{M}$, $Uf = \theta$ for $f \in \mathcal{M}^{\perp}$. Any $f \in \mathcal{M}$ can then be written as $f = f_+ + f_-$ where $Uf_+ = f_+$ and $Uf_- = -f_-$ by setting $f_{\pm} = \frac{1}{2}(I \pm U)f$. It is easy to see that such a decomposition of $\mathcal{M}$ is unique since

$\frac{1}{2}(I \pm U)/M$ are two mutually orthogonal projections in M. Denoting by $M_{\pm}$ the ranges of these two projections in M, we have the decomposition

$$H = M_+ \oplus M_- \oplus M_o, \text{ where } M_o = M^{\perp}.$$

In H, the projections onto $M_{\pm}$ and M_o are given respectively by $E_{\pm} = \frac{1}{2}(U^2 \pm U)$ and $E_{\{0\}} = I-U^2$. Since A commutes with U, it follows from Lemma 5.4(a) that A is reduced by $M_{\pm}$ and M_o.

Since $Af = U|A|f = |A|Uf = \theta$ for all $f \in M_o$, we have that $A/M_o = 0$. On the other hand $Af = |A|f$ for $f \in M_+$ and $Af = -|A|f$ for $f \in M_-$, showing that $A/M_+ \geq 0$ and $A/M_- \leq 0$. However if $(f,Af) = (f,|A|f) = 0$ for any $f \in M_+$, then, since $|A|^{\frac{1}{2}}$ exists by Lemma 5.5, it follows that $|A|^{\frac{1}{2}}f = \theta$. This implies that such a vector f belongs to $N(|A|) = M_o$ as well as to M_+, which is impossible except when $f = \theta$, establishing the result that $A/M_+ > 0$. Similarly one can prove that $A/M_- < 0$.

(d) Let F' be the projection with range M'. Since M' reduces A, it follows from Lemma 5.4 that F' commutes with A. Then, by virtue of (b) of this proposition, F' commutes with U and hence with $E_{\pm}$ and $E_{\{0\}}$. Thus $F'E_- = E_-F'$ is a projection and for any $g \in F'E_-H$, $(g,Ag) \geq 0$. But $(g,Ag) < 0$ which is not possible except when $g = \theta$. Therefore $F'E_- = E_-F' = 0$, implying $M' \subseteq M_-^{\perp}$. The other conclusion follows similarly. #

Though we have proven the polar decomposition theorem for a bounded operator only, an identical result is also true for a closed unbounded operator. Also one has similar results as in Proposition 5.7 for such an operator. We shall discuss this briefly in Section 5-8.

Now we are in a position to prove the spectral theorem for a bounded self-adjoint operator.

PROPOSITION 5.8 (Spectral Theorem) : Let A be a self-adjoint operator in $\mathcal{B}(\mathcal{H})$. Then there exists a unique spectral family $\{E_\lambda\}$ such that the representation (5.20) is valid. Also $\{E_\lambda\}$ commutes with all bounded operators that commute with A.

Proof : (i) Let λ be any real number and let $A-\lambda I = U(\lambda)|A-\lambda I|$ be the polar decomposition of the bounded self-adjoint operator $A-\lambda I$. Then, using the notation of Proposition 5.7 (c), we set

$$E_\lambda = I - \tfrac{1}{2}[U(\lambda) + U(\lambda)^2] = I - E_+(\lambda) = E_-(\lambda) + E_{\{0\}}(\lambda). \quad (5.42)$$

(ii) Now we show that $\{E_\lambda\}$ is a spectral family. Denoting by $\mathcal{M}_\lambda$ the range of E_λ, we have by Proposition 5.7(c) that

$$(f,(A-\lambda I)f) \leq 0 \text{ for all } f \in \mathcal{M}_\lambda. \quad (5.43)$$

Since $\mathcal{M}_\lambda$ clearly reduces A and since $(f,(A-\mu I)f) \leq 0$ for all $f\in\mathcal{M}_\lambda$ and $\mu \geq \lambda$, it follows by Proposition 5.7(d) and the definition (5.42) that $\mathcal{M}_\lambda \subseteq \mathcal{M}_\mu$ for $\lambda \leq \mu$. This shows that $\{\mathcal{M}_\lambda\}$ is a non-decreasing family of subspaces of $\mathcal{H}$, or equivalently $\{E_\lambda\}$ is a non-decreasing family of projections. Therefore, by Problem 5.1 the strong limits $E_{\pm\infty} = \text{s-lim}\, E_\lambda$ as $\lambda \to \pm\infty$ exist and are projections. Each E_λ commute with A since $U(\lambda)$ does, and thus $E_{\pm\infty}$ also commute with A. Let $\lambda < -\|A\|$ and $f \neq \theta$. Then by Lemma 5.30, it follows that $(f,(A-\lambda I)f) > (f,Af) + \|A\|\,\|f\|^2 \geq (f,Af) + |(f,Af)| \geq 0$, so that $A - \lambda I > 0$. Thus Proposition 5.7 (d) implies that $E_-(\lambda) = 0$. On the other hand $A - \lambda I > 0$ also means that

$E_{\{0\}}(\lambda) = 0$. Therefore $E_\lambda = 0$ for $\lambda < -\|A\|$ by virtue of (5.42). Similarly one can show that for $\lambda > \|A\|$, $A - \lambda I < 0$ and hence $E_\lambda = I$. This verifies relation (5.12), viz. $E_{-\infty} = 0$ and $E_{+\infty} = I$.

(iii) Let $\lambda < \mu$ and $E_{(\lambda,\mu]} = E_\mu - E_\lambda$ as in (5.14). Then

$$E_{(\lambda,\mu]}E_\mu = E_{(\lambda,\mu]} = E_{(\lambda,\mu]}(I-E_\lambda), \tag{5.44}$$

and we have that $(A-\mu I)E_\mu \leq 0$ by (5.43) and $(A-\lambda I)(I-E_\lambda) = (A-\lambda I)E_+(\lambda) \geq 0$ by the definition (5.42) and Proposition 5.7 (c). Thus it follows that $(A-\mu I)E_{(\lambda,\mu]} = E_{(\lambda,\mu]}(A-\mu I)E_\mu \cdot E_{(\lambda,\mu]} \leq 0$ and $(A-\lambda I)E_{(\lambda,\mu]} = E_{(\lambda,\mu]}(A-\lambda I)(I-E_\lambda)E_{(\lambda,\mu]} \geq 0$. These two inequalities can be combined to yield for $\lambda < \mu$

$$\lambda E_{(\lambda,\mu]} \leq AE_{(\lambda,\mu]} \leq \mu E_{(\lambda,\mu]}. \tag{5.45}$$

By taking the strong limits $\mu \to \lambda+0$ in (5.45), we obtain that

$$(A-\lambda I)(E_{\lambda+0}-E_\lambda) = 0. \tag{5.46}$$

Since $(E_{\lambda+0}-E_\lambda)H$ reduces A, (5.46) implies by virtue of Proposition 5.7(d) that $(E_{\lambda+0}-E_\lambda)E_+(\lambda) = 0$ or equivalently $(E_{\lambda+0}-E_\lambda)E_\lambda = E_{\lambda+0}-E_\lambda$. On the other hand, (5.11) implies that $E_{\lambda+0}E_\lambda = E_\lambda$, allowing us to conclude that $E_{\lambda+0} = E_\lambda$, or that the family $\{E_\lambda\}$ is right continuous.

(iv) Finally we must show that the self-adjoint operator $A' = \int\lambda dE_\lambda$, associated as in Proposition 5.2 with the spectral family $\{E_\lambda\}$ of definition (5.42), coincides with the given operator A. From part (ii) of the proof, it is clear that $\{E_\lambda\}$ has bounded support with lower bound m and upper bound M, say. We assume for the moment that $E_{\{m\}} = 0$ and consider a partition Π of the interval $[m,M]$ given by $m = \lambda_0 < \lambda_1 < \lambda_2 \ldots < \lambda_{n-1} < \lambda_n = M$ and sum the individual inequalities (5.45) with $\lambda = \lambda_{k-1}$ and $\mu = \lambda_k$ to arrive at

$$\sum_{k=1}^{n} \lambda_{k-1}(E_{\lambda_k}-E_{\lambda_{k-1}}) \leq A \sum_{k=1}^{n} (E_{\lambda_k}-E_{\lambda_{k-1}})$$

$$\leq \sum_{k=1}^{n} \lambda_k(E_k-E_{\lambda_{k-1}}). \tag{5.47}$$

The middle member of (5.47) is clearly equal to A. With $f \varepsilon \mathcal{H}$ and $\lambda_k' \varepsilon (\lambda_{k-1}, \lambda_k]$, we can rewrite the above as

$$-\sum_{k=1}^{n}(\lambda_k'-\lambda_{k-1})(f,(E_{\lambda_k}-E_{\lambda_{k-1}})f) \leq (f,[A-\sum_{k=1}^{n} \lambda_k'(E_{\lambda_k}-E_{\lambda_{k-1}})]f)$$

$$\leq \sum_{k=1}^{n} (\lambda_k-\lambda_{k'})(f,(E_{\lambda_k}-E_{\lambda_{k-1}})f).$$

Denoting by $|\Pi|$ the length of the partition, i.e. $|\Pi| = \max|\lambda_k-\lambda_{k-1}|$, we obtain that

$$-|\Pi| \, \|f\|^2 \leq (f,[A-\sum_{k=1}^{n} \lambda_k'(E_{\lambda_k}-E_{\lambda_{k-1}})]f) \leq |\Pi| \, \|f\|^2, \tag{5.48}$$

where we have used the fact $\sum_{k=1}^{n} (E_{\lambda_k}-E_{\lambda_{k-1}}) = I$.

Thus we have shown that the Stieltjes integral $\int \lambda d(f,E_\lambda f)$ exists and also that $(f,Af) = \int \lambda d(f,E_\lambda f)$ for all $f \varepsilon \mathcal{H}$. Since by Problem 2.22 we can write (f,Ag) as a linear combination of four terms of the type (h,Ah), we can conclude that $(f,Ag) = \int \lambda d(f,E_\lambda g)$ for all $f,g \,\varepsilon\, \mathcal{H}$, or symbolically $A = \int_{m-o}^{M} dE_\lambda$.*) Here we have shown the weak convergence of $\sum \lambda_k'(E_{\lambda_k}-E_{\lambda_{k-1}})$ to the operator A. However, as is evident from the relation (5.48), the same estimates are sufficient to show strong and even uniform convergence. That E_λ commutes with every bounded operator which commutes with A follows

*) The notation m-0 is necessary only if $E_{\{m\}} \neq 0$. The preceding proof can easily be adapted to this case by replacing [m,M] by $[m-\eta,M]$ with $\eta > 0$.

from the same property of $U(\lambda)$ and the definition (5.42). The proof of the uniqueness of the family $\{E_\lambda\}$ is deferred till the end of Section 5-8. #

There is a similar theorem for any (unbounded) self-adjoint operator, which we state without proof. For a brief sketch of its proof, the reader is referred to Section 5-8.

<u>PROPOSITION 5.9</u> (Spectral theorem) : Let A be any self-adjoint operator in H. Then there exists a unique spectral family $\{E_\lambda\}$ such that the representation (5.20) is valid. Also each E_λ commutes with all bounded operators that commute with A.

In fact, if we assume the polar decomposition result for an arbitrary closed operator (indeed there is one, see e.g. [K, page 334]), then the proof of Proposition 5.8 goes through in this case. Also since for an unbounded operator the support of $\{E_\lambda\}$ is necessarily unbounded, the convergence of $\int \lambda dE_\lambda$ will not be in the uniform operator topology but rather only in the strong sense. Another possibility is to write a bounded function of the given unbounded self-adjoint operator and conclude the spectral theorem for the unbounded one from the knowledge of that for the bounded one. This is what we shall do briefly in Section 5-8. It is the spectral theorem for an unbounded self-adjoint operator which is of practical interest, since most operators in the applications (e.g. the free Hamiltonian K_o in Section 3-3) are unbounded.

5-5 FUNCTIONAL CALCULUS

So far we have studied a self-adjoint operator A and its spectral representation given by (5.20), where the spectral family $\{E_\lambda\}$ is uniquely determined by A. Now we want to form functions of such an operator. If A belongs to $\mathcal{B}(\mathcal{H})$, then it is easy to guess the natural definition for functions like polynomials and exponentials. For example $\exp(A) \equiv \sum_{n=0}^{\infty} A^n/n!$, since the right hand side converges in the operator norm. However, functions which are not so simple cannot be defined in the above manner, and when A is unbounded, all such natural definitions are of no help. But the spectral theorem allows us to form a large class of functions of a self-adjoint operator A.

Let ϕ be a complex-valued continuous function of the real variable λ. As in Proposition 5.2, one may define an operator, denoted $\phi(A)$, by writing

$$D(\phi(A)) = \{f\varepsilon\mathcal{H} \Big| \int_{-\infty}^{\infty} |\phi(\lambda)|^2 d(f,E_\lambda f) < \infty\} \tag{5.49}$$

and for $f\varepsilon D(\phi(A))$, $g \varepsilon \mathcal{H}$

$$(g,\phi(A)f) = \int_{-\infty}^{\infty} \phi(\lambda) d(g,E_\lambda f). \tag{5.50}$$

We shall again write formally

$$\phi(A) = \int_{-\infty}^{\infty} \phi(\lambda) dE_\lambda. \tag{5.51}$$

For $\phi(\lambda) = \lambda$, we get back the operator A as expected. The denseness of $D(\phi(A))$ follows as in the proof of Proposition 5.2 by considering vectors in $\cup_n (E_n - E_{-n})\mathcal{H}$. If $\phi(\lambda)$ is furthermore bounded on the support of $\{E_\lambda\}$, then one sees by reasoning as in (2.49) that $D(\phi(A)) = \mathcal{H}$ and that $\phi(A)$ is bounded with

$$||\phi(A)|| = \sup_{\lambda\varepsilon\text{supp}\{E_\lambda\}} |\phi(\lambda)| . \qquad (5.52)$$

The following proposition summarizes some properties of $\phi(A)$.

PROPOSITION 5.10 : Let ϕ, ϕ_1 and ϕ_2 be complex-valued, continuous functions, bounded on the support of $\{E_\lambda\}$.

(a) $\phi(A)^* = \bar{\phi}(A)$ where $\bar{\phi}(\lambda) = \overline{\phi(\lambda)}$.

(b) If $\phi(\lambda) = \phi_1(\lambda)\phi_2(\lambda)$, then $\phi(A) = \phi_1(A)\phi_2(A)$.

(c) If $\phi(\lambda) = \alpha_1\phi_1(\lambda) + \alpha_2\phi_2(\lambda)$, then $\phi(A) = \alpha_1\phi_1(A) + \alpha_2\phi_2(A)$.

(d) $\phi(A)$ is normal, i.e. $\phi(A)^*\phi(A) = \phi(A)\phi(A)^*$.

(e) $\phi(A)$ commutes with all bounded operators that commute with A.

(f) If A is reduced by a projection F, then $\phi(A)/F\mathcal{H} = \phi(A/F\mathcal{H})$.

The proof of this simple proposition is left as an exercise for the reader (Problem 5.6). One can seek generalizations in two directions. Firstly, one can consider the integrals in relations (5.20) and (5.51) in a sense other than that of Riemann-Stieltjes and thereby enlarge the class of allowable functions. In fact one can consider Lebesgue-Stieltjes integrals [R] and admit functions $\phi(\lambda)$ (not necessarily continuous) that are measurable and finite almost everywhere with respect to each of the measures $||E_\lambda f||^2$ for all $f\varepsilon\mathcal{H}$. Readers interested in details are referred to [RN]. Secondly, if ϕ, though continuous, is not bounded on the support of $\{E_\lambda\}$, then properties (a), (d), (e) and (f) of Proposition 5.10 are still true, and (b) and (c) are replaced by $\phi_1(A)\phi_2(A) \subseteq \phi(A)$ and $\alpha_1\phi_1(A) + \alpha_2\phi_2(A) \subseteq \phi(A)$.

The proof of these statements is simple (Problem 5.6). In particular it is clear that $\phi(A)$ is self-adjoint if ϕ is a real function and is unitary if $|\phi(\lambda)| = 1$ for all λ. The latter follows from the fact that under this hypothesis $\phi(A)$ is bounded and $\phi(A)^*\phi(A) = \phi(A)\phi(A)^* = (\phi\bar{\phi})(A) = I$.

Two particularly interesting functions of a self-adjoint operator are the resolvent and the unitary group. A particular case of the former has already been introduced in the proof of Proposition 5.2 where we used the functions $\phi_\pm(\lambda) = (\lambda \pm i)^{-1}$. Here we replace $\pm i$ by a general complex number z, $\operatorname{Im} z \neq 0$.

Let $\phi_z(\lambda) = (\lambda-z)^{-1}$ and $\psi_t(\lambda) = \exp(-it\lambda)$, where $\operatorname{Im} z \neq 0$ and $-\infty < t < \infty$. Since both ϕ_z and ψ_t are bounded continuous functions of λ, we can define as in (5.51)

$$\phi_z(A) \equiv R_z(A) = \int (\lambda-z)^{-1}\, dE_\lambda \tag{5.53}$$

and

$$\psi_t(A) \equiv e^{-itA} = \int e^{-it\lambda}\, dE_\lambda . \tag{5.54}$$

The next proposition gives some properties of these two families of bounded operators.

<u>PROPOSITION 5.11</u> : Let ϕ_z and ψ_t be as above. Then

(a) $\phi_z(A) = (A-z)^{-1} \in \mathcal{B}(\mathcal{H})$ and $\|\phi_z(A)\| \leq \dfrac{1}{|\operatorname{Im} z|}$. (5.55)

(b) $\{\psi_t(A)\}$ forms a strongly continuous one-parameter unitary group whose infinitesimal generator is the self-adjoint operator A.

Proof : The first part of (a) is established by the same argument as that in part (iii) of the proof of Proposition 5.2, and the inequality in (5.55) follows from (5.52). That $\{\psi_t(A)\}$ forms a one-parameter unitary group follows from Proposition 5.10(b). Since $\|\psi_t(A)f-f\|^2 = \int |\exp(-it\lambda)-1|^2 d\,\|E_\lambda f\|^2$, $|\exp(-it\lambda)-1| \leq 2$ and $\int d\,\|E_\lambda f\|^2 = \|f\|^2 < \infty$, we conclude strong continuity at $t = 0$ after an application of the Lebesgue dominated convergence theorem.

Now for all $f\varepsilon D(A)$ and $t \neq 0$, by using (3.34), we obtain $\|it^{-1}(\psi_t(A)-I)f-Af\|^2 =$

$$\int_{-\infty}^{\infty}|it^{-1}(e^{-it\lambda}-1)-\lambda|^2 d\,\|E_\lambda f\|^2 \leq 4\int_{-\infty}^{\infty}\lambda^2 d\,\|E_\lambda f\|^2. \quad (5.56)$$

The last integral is finite by (5.21). Another application of the Lebesgue dominated convergence theorem to (5.56) and the definition (3.4) show that the self-adjoint operator A is the infinitisimal generator of $\{\psi_t(A)\}$. #

5-6 RESOLVENT SET, SPECTRUM, DECOMPOSITION OF THE SPECTRUM

We have seen in the last section that for all complex z ($\operatorname{Im} z \neq 0$), $(A-z)^{-1}$ defines a normal operator in $\mathcal{B}(\mathcal{H})$, called the resolvent of the self-adjoint operator A. A moment's reflexion should convince the reader that the same is true for real z outside the support of $\{E_\lambda\}$, the spectral family associated with A. The set of complex numbers z for which $(A-z)^{-1}\varepsilon\mathcal{B}(\mathcal{H})$ is called the resolvent set of A, denoted $\rho(A)$. The complementary set in the complex plane is called the spectrum of A, written as $\sigma(A)$. Hence $\sigma(A) \equiv \mathcal{C}-\rho(A)$.*) We

*) If Δ,Δ' are two subsets of $\mathcal{C}$ or $\mathcal{R}^n$, then we denote by $\Delta-\Delta'$ the set of all points in Δ which are not in Δ'.

have seen in the last section that all complex z with $\mathrm{Im}\, z \neq 0$ belong to $\rho(A)$. That $\rho(A)$ is open in the complex plane is shown in the next lemma.

LEMMA 5.12 : Let A be a self-adjoint operator.

(a) $AR_z \in \mathcal{B}(\mathcal{H})$ for all $z \in \rho(A)$, (5.57)

where we have suppresed the explicit dependence of R_z on A.

(b) The resolvent R_z satisfies the first resolvent equation

$R_z - R_{z'} = (z-z')R_z R_{z'}$, for $z, z' \in \rho(A)$. (5.58)

(c) $\rho(A)$ is open in the complex plane and R_z is uniformly holomorphic*) in each connected component of $\rho(A)$.

Proof : (a) By the definition of $R_z(A)$, we know that R_z maps $\mathcal{H}$ onto $D(A)$ and therefore AR_z is defined on all of $\mathcal{H}$. Also, $AR_z = (A-zI)R_z + zR_z = I + zR_z \in \mathcal{B}(\mathcal{H})$.

(b) (5.58) is an easy consequence of (5.57) and the relation $R_z - R_{z'} = R_z(A-z'I)R_{z'} - R_z(A-zI)R_{z'}$. Note that (5.58) implies that R_z and $R_{z'}$ commute.

(c) Let ξ be a fixed point in $\rho(A)$. Then for $|z-\xi| < \|R_\xi\|^{-1}$, the Neumann series (2.51) for the operator $[I-(z-\xi)R_\xi]^{-1}$ is uniformly convergent and we define

$$R'_z \equiv R_\xi[I-(z-\xi)R_\xi]^{-1} = \sum_{n=0}^{\infty} (z-\xi)^n R_\xi^{n+1}. \qquad (5.59)$$

Since $R_\xi(A-zI)f = f-(z-\xi)R_\xi f$ for every $f \in D(A)$, we have that

*) A vector-valued function f_z is said to be strongly (weakly) holomorphic in the domain $\mathcal{O}$ if f_z has a strong (weak) derivative in $\mathcal{O}$. It can be shown [K, p. 139] that strong and weak holomorphy are equivalent. A $\mathcal{B}(\mathcal{H})$-valued function A is said to be strongly (uniformly) holomorphic in $\mathcal{O}$ if A_z has strong (uniform) derivatives in $\mathcal{O}$.

$$R'_z(A-zI)f = \sum_{n=0}^{\infty}[(z-\xi)^n R_\xi^n f - (z-\xi)^{n+1}R_\xi^{n+1} f] = f.$$

Similarly, for all $f \varepsilon H$

$$(A-zI)R'_z f = \underset{n\to\infty}{s\text{-}\lim} \sum_{k=0}^{n} (z-\xi)^k (A-zI)R_\xi^{k+1} f = f ,$$

where we have used the closedness of $(A-zI)$ to deduce that $R'_z f \varepsilon D(A)$ for $z \varepsilon \rho(A)$. This shows that $R'_z = R_z$ and that for every $\xi \varepsilon \rho(A)$ there is a small open disc around ξ all points of which are in $\rho(A)$. Thus $\rho(A)$ is open.

Conversely, if z and ξ are two points in $\rho(A)$ satisfying the inequality $|z-\xi| < \|R_\xi\|^{-1}$, then R_z is given by the uniformly convergent series (5.59). This establishes the uniform holomorphy of R_z, with the Neumann series (5.59) as the Taylor series for R_z. #

As $\rho(A)$ is open, it is clear that the spectrum $\sigma(A)$ is closed. If μ is an eigenvalue of A, then μ is a point of increase of E_λ and $A-\mu$ is not invertible. On the other hand if μ is a point of increase of E_λ but not an eigenvalue, then $(A-\mu)^{-1}$ exists and is unbounded. Therefore, it follows that $\sigma(A)$ consists of the points of increase of E_λ. Thus the integrals in (5.20) and (5.51) need be taken only over $\sigma(A)$.

The notions of resolvent set and spectrum are not restricted to self-adjoint operators but can be defined for any closed operator A in H as follows : $\rho(A) = \{z \varepsilon C \,|\, (A-zI)^{-1} \varepsilon B(H)\}$ and $\sigma(A) = C - \rho(A)$. Then a reasoning similar to that in the proof of Lemma 5.12 shows that $\rho(A)$ is an open set while $\sigma(A)$ is closed, though not in general a subset of the real axis.

As we have seen in Lemma 5.3, for a self-adjoint A with associated $\{E_\lambda\}$, $E_{\{\lambda\}} \equiv E_\lambda - E_{\lambda-o} \neq 0$ if and only if λ is an eigenvalue of A. In this case $E_{\{\lambda\}}$ is the orthogonal projection onto the associated eigenspace. Also it follows from Proposition 2.26 that the projections $E_{\{\lambda\}}$ for different values of λ are mutually orthogonal, i.e. $E_{\{\lambda\}}E_{\{\mu\}} = 0$ for $\lambda \neq \mu$. The set of all eigenvalues of A, called the point spectrum of A and denoted $\sigma_p(A)$, is therefore at most a countable set if $\mathcal{H}$ is separable.

Let $\mathcal{H}_p$ be the (closed) subspace spanned by all the $E_{\{\lambda\}}\mathcal{H}$. If $\mathcal{H}_p = \mathcal{H}$, A is said to have pure point spectrum. In general $\sigma_p(A)$ is not a closed set. For example, the spectrum of an operator A with pure point spectrum consists of the point spectrum and its accumulation points : $\sigma(A) = \overline{\sigma_p(A)}$. $\mathcal{H}_p$ reduces A since each $E_{\{\lambda\}}\mathcal{H}$ does. Let A_p be the part of A in $\mathcal{H}_p$. Then A_p has pure point spectrum.

If on the other hand $\mathcal{H}_p = \{\theta\}$, i.e. if A has no eigenvalues at all, A is said to have purely continuous spectrum. In general, the part A_c of A in $\mathcal{H}_c \equiv \mathcal{H}_p^\perp$ has purely continuous spectrum. $\sigma(A_c)$, the spectrum of A_c, is called the continuous spectrum of A and is denoted by $\sigma_c(A)$. According to this definition $\sigma_c(A)$ is the spectrum of a self-adjoint operator, i.e. it is a closed set, unlike $\sigma_p(A)$.

It should be emphasized that the above decomposition is more like a decomposition of $\mathcal{H}$ than one of the spectrum, since $\sigma_p(A)$ and $\sigma_c(A)$ are not necessarily disjoint sets. In fact, there are differential operators with eigenvalues in the continuous spectrum (see e.g. von Neumann and Wigner [1]).

However, there is another kind of decomposition which decomposes the spectrum of a self-adjoint operator into two disjoint sets, also called pure point and continuous spectrum respectively [HP,AG]. By this definition, a real number λ belongs to the continuous spectrum if and only if $(A-\lambda I)^{-1}$ exists and is unbounded with dense domain. We shall have no occasion to use this latter definition in this book.

A_p and A_c will be called the discontinuous and continuous parts of A, respectively. The subspaces $\mathcal{H}_p \equiv \mathcal{H}_p(A)$ and $\mathcal{H}_c \equiv \mathcal{H}_c(A)$ are called the <u>subspaces of discontinuity</u> and <u>continuity</u>, respectively, with respect to A. The following lemma characterizes $\mathcal{H}_c$ completely.

<u>LEMMA 5.13</u> : (a) $f \in \mathcal{H}_c(A)$ if and only if $(f, E_\lambda f)$ is a continuous function of λ. (b) Let Δ be a countable subset of $\mathcal{R}$. Then the part of A in $E_\Delta \mathcal{H}$ has pure point spectrum, or equivalently $E_\Delta \mathcal{H}_c(A) = \{\theta\}$.

<u>Proof</u> : (a) Assume $(f, E_\lambda f)$ to be continuous. Then $(f, E_{\{\lambda\}} f) = 0$ for all $\lambda \in \mathcal{R}$. Since $E_{\{\lambda\}}$ is a projection,

$$|(f, E_{\{\lambda\}} g)|^2 \leq (f, E_{\{\lambda\}} f)(g, E_{\{\lambda\}} g) = 0 \text{ for all } g \in \mathcal{H}, \qquad (5.60)$$

so that f is orthogonal to the ranges of all $E_{\{\lambda\}}$ and therefore to $\mathcal{H}_p$. Hence $f \in \mathcal{H}_c$.

Conversely, if $f \in \mathcal{H}_c = \mathcal{H}_p^\perp$, then f is orthogonal to $E_{\{\lambda\}} f$ for all λ, so that by (5.16), $(f, E_\lambda f)$ is continuous.

(b) Δ is a countable union of points, say $\Delta = \cup \{\lambda_i\}$. By the countable additivity of the spectral measure (see (5.64)), $E_\Delta f = \sum_i E_{\{\lambda_i\}} f$, for each $f \in \mathcal{H}$. Either $E_{\{\lambda_i\}} f = \theta$ or

$E_{\{\lambda_i\}}f$ is an eigenvector of A by Lemma 5.3. Hence $E_\Delta \mathcal{H} \subseteq \mathcal{H}_p(A)$.#

Now we consider some examples to illustrate the above concepts.

Example 5.14 : Let $\mathcal{H} = L^2[0,1]$ and Q be the position operator defined in (3.12); $(Qf)(x) = xf(x)$. Then, as we have remarked before, Q is a bounded self-adjoint operator, defined everywhere. We have also seen that the associated spectral family $\{E_y\}$ is given by : $(E_y f)(x) = \chi_{(-\infty,y]}(x)f(x)$. Then

$$(f,Qg) = \int_0^1 \overline{f(x)}xg(x)dx = \int_0^1 y\, d(\int_0^y \overline{f(x)}g(x)dx)$$

$$= \int_0^1 y\, d(f,E_y g). \tag{5.61}$$

Since $(f,E_y f) = \int_0^y |f(x)|^2 dx$ is a continuous (even differentiable) function of y and E_y is not constant anywhere in [0,1], it follows that $\sigma(Q)$ is the whole interval [0,1] and that Q has a purely continuous spectrum.

Example 5.15 : Consider $\mathcal{H} = C \oplus L^2[0,1]$, the direct sum of a one-dimensional space and $L^2[0,1]$ (cf. Section 2-1). Let the operator A be given by $A(\alpha,f) = (\frac{1}{2}\alpha,Qf)$, where $\alpha\varepsilon C$, $f\varepsilon L^2[0,1]$ and Q is the operator defined in the previous example. Then it is obvious that A is a bounded self-adjoint operator. As is clear from the previous example, the equation $Qf = \lambda f$ has no non-zero solution in $L^2[0,1]$ and hence the vector $1 \oplus \theta$ is the normalized eigenvector in $\mathcal{H}$ corresponding to the only eigenvalue $\frac{1}{2}$ of A. Thus $\mathcal{H}_p = C \oplus \theta$, $\mathcal{H}_c \equiv \mathcal{H}_p^\perp = \theta \oplus L^2[0,1]$, and the continuous spectrum of A is $\sigma_c(A) = \sigma(Q) = [0,1]$. So this example illustrates the case of an eigenvalue embedded in the continuum.

An eigenvalue of A need not be an isolated point of $\sigma(A)$ even when A has only pure point spectrum. In fact $\sigma_p(A)$ can be a countable set everywhere dense in $\mathcal{R}$, as is illustrated by the following example.

Example 5.16 : Let $\{e_n\}_1^\infty$ be an orthonormal basis in $L^2[0,1]$, and $\{\gamma_n\}$ be an enumeration of all rationals in [0,1]. Set $Ae_n = \gamma_n e_n$ and extend it by linearity to all of $L^2[0,1]$. A is then clearly a bounded self-adjoint operator with $\{\gamma_n\}$ as its eigenvalues. Since these are everywhere dense in [0,1], one has $\sigma(A) = \overline{\sigma_p(A)} = [0,1]$.

As is clear from the spectral theorem for any self-adjoint operator A, the self-adjoint operator UAU* (U unitary) has the same spectrum as A and its spectral family is $\{UE_\lambda U^*\}$ (see also Proposition 5.21). However, a comparison of the two operators Q and A in Examples 5.14 and 5.16, respectively, shows that though $\sigma(Q) = \sigma(A) = [0,1]$, the two operators are quite different and not unitarily equivalent.

It is useful, particularly in the context of scattering theory (see Section 7-6), to further subdivide $\mathcal{H}_c$ into two parts. We have remarked in Section 5-2 that the spectral family $\{E_\lambda\}$ determines a spectral measure E_Δ with the properties ($\emptyset$ denotes the empty set),

$$E_{\Delta\cap\Delta'} = E_\Delta E_{\Delta'} \tag{5.62}$$

$$E_{\Delta\cup\Delta'} = E_\Delta + E_{\Delta'} \quad \text{if } \Delta\cap\Delta' = \emptyset \tag{5.63}$$

$$E_{\cup\Delta_n} = \sum_1^\infty E_{\Delta_n} \quad \text{if } \Delta_m \cap \Delta_n = \emptyset \text{ for } m \neq n. \tag{5.64}$$

Thus for any fixed $f\varepsilon\mathcal{H}$, one constructs a non-negative countably additive Borel measure [R] by setting $m_f(\Delta) = (f,E_\Delta f) = \|E_\Delta f\|^2$. Now the Lebesgue decomposition theorem [R] states that any measure m on $\mathcal{R}$ has a unique decomposition $m = m_{ac} + m_s$, where m_{ac} is absolutely continuous and m_s is singular with respect to the Lebesgue measure $|\cdot|$ on $\mathcal{R}$. Hence we expect a similar decomposition of the Hilbert space $\mathcal{H}$ relative to the family of measures $\{m_f\}$. This is the content of the next proposition, for the proof of which the reader is referred to [K, Thm. X.1.5]. Before stating this proposition we give the relevant definitions in the present context. A vector f is said to be <u>absolutely continuous with respect to A</u> if m_f is absolutely continuous with respect to the Lebesgue measure on $\mathcal{R}$, i.e. if $|\Delta| = 0$ implies $m_f(\Delta) = \|E_\Delta f\|^2 = 0$. Similarly if m_f is singular with respect to the Lebesgue measure on $\mathcal{R}$, i.e. if there is a Borel set Δ_0 with $|\Delta_0| = 0$ such that $m_f(\Delta) = m_f(\Delta \cap \Delta_0)$ for all Borel sets $\Delta \subseteq \mathcal{R}$, then f is said to be <u>singular with respect to A</u>. The set of all vectors in $\mathcal{H}$ which are absolutely continuous (singular) with respect to A is denoted by $\mathcal{H}_{ac}(A)$ $(\mathcal{H}_s(A))$ or simply by $\mathcal{H}_{ac}$ $(\mathcal{H}_s)$ and called the <u>subspace of absolute continuity (singularity)</u> with respect to A (anticipating the results of the following proposition).

<u>PROPOSITION 5.17</u> : $\mathcal{H}_{ac}$ and $\mathcal{H}_s$ are (closed) subspaces of $\mathcal{H}$, orthogonal complements to each other and reduce A.

Since the point set $\{\lambda\}$ has Lebesgue measure zero, $(f,E_\lambda f) = (f,E_{\lambda-0}f)$ for all $f \in \mathcal{H}_{ac}$ and all $\lambda \in \mathcal{R}$. Therefore $\mathcal{H}_{ac} \subseteq \mathcal{H}_c$ by Lemma 5.13, and $\mathcal{H}_p \subseteq \mathcal{H}_s$. Setting $\mathcal{H}_{sc} = \mathcal{H}_c \ominus \mathcal{H}_{ac}$, we get for each self-adjoint operator A the following decompositions of the Hilbert space :

$$\mathcal{H} = \mathcal{H}_{ac} \oplus \mathcal{H}_{s} = \mathcal{H}_{c} \oplus \mathcal{H}_{p} = \mathcal{H}_{ac} \oplus \mathcal{H}_{sc} \oplus \mathcal{H}_{p}. \qquad (5.65)$$

We shall denote by $E_{ac}(A)$ the projection with range $\mathcal{H}_{ac}(A)$.

If $\mathcal{H}_{ac} = \mathcal{H}$ (or equivalently $\mathcal{H}_s = \{\theta\}$), A is said to be (spectrally) absolutely continuous. Similarly, if $\mathcal{H} = \mathcal{H}_s$, A is (spectrally) singular; if $\mathcal{H} = \mathcal{H}_{sc}$, A is (spectrally) singularly continuous. In general, the part $A_{ac}(A_s, A_{sc})$ of A in the reducing subspace $\mathcal{H}_{ac}(\mathcal{H}_s, \mathcal{H}_{sc})$ is called the (spectrally) absolutely continuous (singular, singularly continuous) part of A. Then $\sigma_{ac}(A)(\sigma_s(A), \sigma_{sc}(A))$, the absolutely continuous (singular, singularly continuous) spectrum of A, is given by $\sigma(A_{ac})(\sigma(A_s), \sigma(A_{sc}))$. As with the decomposition of the spectrum into continuous and point spectrum, in this case also the same λ can belong to both $\sigma_{ac}(A)$ and $\sigma_s(A)$. However,

$$\sigma(A) = \sigma_{ac}(A) \cup \sigma_s(A).$$

In Example 5.14, Q is absolutely continuous. This is so because $\|E_\Delta f\|^2 = \int_\Delta |f(x)|^2 dx$, and if the Lebesgue measure of Δ is zero, then clearly $\|E_\Delta f\|^2 = 0$, for all $f \in L^2[0,1]$. Similary, the free Hamiltonian K_0 defined in (3.29) is also absolutely continuous. In fact, both these examples are special cases of the following result. Let $\mathcal{H} = L^2(\Delta)$, Δ a measurable subset of $\mathcal{R}^n$ and ψ be a real-valued measurable function on Δ. Setting $(Af)(x) = \psi(x)f(x)$, we observe by Proposition 2.16 that A is a self-adjont operator. It can be shown (Problem 5.7) that if ψ is continuously differentiable and if grad $\psi \neq 0$ almost everywhere in Δ, then A is absolutely continuous.

A useful property of H_{ac} is given in the next proposition.

PROPOSITION 5.18 : Let $g \in H$ and $f \in H_{ac}$. Then the function $\lambda \mapsto (g, E_\lambda f)$ is absolutely continuous and

$$\left|\frac{d}{d\lambda}(g, E_\lambda f)\right|^2 \leq \frac{d}{d\lambda}(g_{ac}, E_\lambda g_{ac}) \frac{d}{d\lambda}(f, E_\lambda f) \tag{5.66}$$

almost everywhere, where g_{ac} is the projection of g on H_{ac}.

Proof : For the definition and some properties of absolutely continuous functions, the reader is referred to [R]. Observe that the right hand side of (5.66) is non-negative and finite almost everywhere, for $(g_{ac}, E_\lambda g_{ac})$ and $(f, E_\lambda f)$ are absolutely continuous and non-decreasing in λ.

Since $E_\Delta E_\lambda f = E_\lambda E_\Delta f = \theta$ for $|\Delta| = 0$, it follows that $E_\lambda f \in H_{ac}$ for all λ. By the decomposition (5.65), we then have $(g, E_\lambda f) = (g_{ac}, E_\lambda f)$. By Problem 2.22, we can write $(g_{ac}, E_\lambda f)$ as the sum of four absolutely continuous functions and conclude that $(g, E_\lambda f)$ is absolutely continuous.

Now by the Schwarz inequality, it follows that for any $\Delta \subseteq R$

$$|(g, E_\Delta f)|^2 = |(g_{ac}, E_\Delta f)|^2 \leq (g_{ac}, E_\Delta g_{ac})(f, E_\Delta f). \tag{5.67}$$

Taking $\Delta \equiv (\lambda, \lambda+\tau]$ and dividing (5.67) by τ^2, (5.66) follows because the derivatives exist almost everywhere by virtue of the absolute continuity of all the functions involved. #

The physical relevance of the decomposition (5.65) will be discussed in Chapter 7. Here we only mention that, if H is the Hamiltonian of a quantum-mechanical system, $\mathcal{H}_p(H)$ usually contains the bound states of H while $\mathcal{H}_c(H)$ (in fact $\mathcal{H}_{ac}(H)$) consists of scattering states. For most self-adjoint operators in applications, $\mathcal{H}_{sc} = \{\theta\}$ or equivalently $\mathcal{H}_c = \mathcal{H}_{ac}$. But one can construct second order differential operators which have a non-trivial singularly continuous part, cf. Aronszajn [1]. Some attempts have been made recently to characterize $\mathcal{H}_{ac}$ and $\mathcal{H}_{sc}$, see Gustafson and Johnson [1], Sinha [1] and the contribution of Gustafson in [LM].

There is still another useful way of decomposing the spectrum of a self-adjoint operator A. The isolated eigenvalues of finite multiplicity, i.e. all isolated λ in $\sigma(A)$ with the range of $E_{\{\lambda\}}$ finite-dimensional, form the discrete spectrum $\sigma_d(A)$ of A. The set complementary to $\sigma_d(A)$ in $\sigma(A)$ is said to constitute the essential spectrum $\sigma_e(A)$ of A. Unlike the earlier decompositions, this one breaks up the spectrum $\sigma(A)$ into two disjoint sets $\sigma_e(A)$ and $\sigma_d(A)$. From the definition, it is also clear that $\sigma_e(A)$ consists of the continuous spectrum, the accumulation points of the point spectrum and eigenvalues of infinite multiplicity. A point λ of $\sigma_e(A)$ is characterized by (Problem 5.11)

$$\dim E_{(\lambda-\eta,\lambda+\eta]}\mathcal{H} = \infty \text{ for all } \eta > 0. \tag{5.68}$$

If A is compact and self-adjoint in an infinite-dimensional $\mathcal{H}$, then the essential spectrum $\sigma_e(A)$ of A consists of one point $\{0\}$, and the non-zero eigenvalues (which are necessarily isolated and of finite multiplicity by Proposition 2.25) form the discrete spectrum $\sigma_d(A)$. The next lemma gives a useful characterization of the essential spectrum of a self-adjoint operator.

LEMMA 5.19 : Let A be a selfadjoint operator in $\mathcal{H}$. Then $\sigma_e(A)$ is closed and $\mu \in \sigma_e(A)$ if and only if there exists a sequence $\{f_n\} \in D(A)$ such that $\|f_n\| = 1$, w-lim $f_n = \theta$ and s-lim$(A-\mu I)f_n = \theta$ as $n \to \infty$.

Proof : By (5.68) $\mu \in \sigma_e(A)$ if and only if E_{Δ_n} is an infinite dimensional projection for any interval Δ_n containing μ in its interior. The closedness of $\sigma_e(A)$ follows from its definition.

(i) Consider a sequence of nested open intervals Δ_n containing μ and contracting to μ as $n \to \infty$. Since each subspace $\mathcal{H}_n \equiv E_{\Delta_n}\mathcal{H}$ is infinite dimensional, it is possible to construct an infinite orthonormal sequence $\{f_n\}$ such that f_n belongs to $\mathcal{H}_n (n = 1,2,\ldots)$. Then by Bessel's inequality (2.11) f_n converges weakly to zero, and

$$\|(A-\mu I)f_n\|^2 = \|(A-\mu I)E_{\Delta_n}f_n\|^2 =$$

$$\int_{\Delta_n} (\lambda-\mu)^2 d\|E_\lambda f_n\|^2 \leq |\Delta_n|^2 \|f_n\|^2 = |\Delta_n|^2 \to 0,$$

where by $|\Delta_n|$ we mean the length of the interval Δ_n.

(ii) Conversely, let $\{f_n\}$ be a sequence satisfying the hypotheses of the lemma and let $\Delta = (a,b)$ be an arbitrary interval containing μ in its interior. Then we have for all $f \in D(A)$,

$$\|(A-\mu I)f\|^2 = \int_{-\infty}^{\infty} (\lambda-\mu)^2 d\|E_\lambda f\|^2 \geq (b-\mu)^2 \int_b^{\infty} d\|E_\lambda f\|^2 +$$

$$(a-\mu)^2 \int_{-\infty}^{a} d\|E_\lambda f\|^2 = (b-\mu)^2 \|(I-E_b)f\|^2 + (a-\mu)^2 \|E_a f\|^2.$$

This implies that $(I-E_b)f_n$ and $E_a f_n$ converge strongly to zero or that $(I-E_\Delta)f_n$ converges strongly to zero as $n \to \infty$. Consequently, $\|E_\Delta f_n\|$ converges to $\|f_n\| = 1$ as $n \to \infty$. Thus the subspace $E_\Delta \mathcal{H}$ cannot be of finite dimension, because if it were, by Proposition 2.23 the weak convergence of f_n to θ would imply the strong convergence of $E_\Delta f_n$ to zero, which would contradict the fact that $\|E_\Delta f_n\| \to 1$. #

The usefulness of the decomposition of the spectrum into its essential and discrete parts derives from the following result which will be proven in Section 8-1. If A and B are two self-adjoint operators and if A-B is compact, then $\sigma_e(A) = \sigma_e(B)$. We end this section by giving some applications.

LEMMA 5.20 : Let U_t be a strongly continuous one-parameter unitary group with infinitesimal generator A and let $f \in \mathcal{H}_{ac}(A)$. Then $U_t f$ converges weakly to zero as $t \to \pm\infty$.

Proof : We have $U_t = \exp(-iAt) = \int \exp(-i\lambda t) dE_\lambda$. Let $g \in \mathcal{H}$. Then by Proposition 5.18 $(g, E_\lambda f)$ is an absolutely continuous function. Hence its derivative $d(g,E_\lambda f)/d\lambda$ is defined a.e. and is integrable by (5.66) and Schwarz'inequality. Thus

$$(g, U_t f) = \int e^{-i\lambda t} d(g, E_\lambda f) = \int e^{-i\lambda t} \frac{d(g, E_\lambda f)}{d\lambda} d\lambda.$$

An application of the Riemann-Lebesgue lemma [R] to the above expression yields $\lim (g, U_t f) = 0$ as $t \to \pm\infty$. #

PROPOSITION 5.21 : Let U_t, V_t, $\Omega_\pm$ and $F_\pm$ be as in Section 4-1. Then (a) the subspaces $F_\pm \mathcal{H}$ reduce H and the self-adjoint operators $H/F_\pm \mathcal{H}$ have the same spectrum as $H_0/\mathcal{M}_\infty(H_0)$. (b) If ϕ

is a complex-valued, continuous and bounded function on $\mathcal{R}$, one has

$$\phi(H)\Omega_\pm = \Omega_\pm\phi(H_o). \tag{5.69}$$

(c) $$E_\lambda\Omega_\pm = \Omega_\pm E^o_\lambda \quad \text{for all } \lambda\varepsilon\mathcal{R}. \tag{5.70}$$

Proof : From Proposition 4.4(b) we have that if $g\varepsilon D(H)$ then $\Omega^*_\pm g\varepsilon D(H_o)$ and $H_o\Omega^*_\pm g = \Omega^*_\pm Hg$. Multiplying on the left of the last relation by $\Omega_\pm$, we obtain

$$\Omega_\pm H_o\Omega^*_\pm g = \Omega_\pm\Omega^*_\pm Hg = F_\pm Hg.$$

On the other hand, since $\Omega^*_\pm g\varepsilon D(H_o)$, we can apply Proposition 4.4(a) to conclude that $\Omega_\pm\Omega^*_\pm g = F_\pm g\varepsilon D(H)$ and

$$\Omega_\pm H_o\Omega^*_\pm g = H\Omega_\pm\Omega^*_\pm g = HF_\pm g. \tag{5.71}$$

The two preceding relations together with the result of Lemma 5.4 show that $F_\pm\mathcal{H}$ reduce H.

Since $E_\infty(H_o)\Omega^*_\pm = \Omega^*_\pm$, we also get $F_\pm Hg = HF_\pm g = \Omega_\pm H_oE_\infty(H_o)\Omega^*_\pm g$, and we say that $HF_\pm$ are unitarily equivalent to $H_oE_\infty(H_o)$. Denoting by E_λ and E^o_λ the spectral family of H and H_o respectively, it follows that

$$HF_\pm = \int\lambda dE_\lambda F_\pm = \int\lambda d(\Omega_\pm E^o_\lambda E_\infty(H_o)\Omega^*_\pm).$$

On the other hand by the spectral theorem (Proposition 5.9) the spectral families $\{E_\lambda F_\pm\}$ of the self-adjoint operators $HF_\pm$ are unique, and $\Omega_\pm E^o_\lambda E_\infty(H_o)\Omega^*_\pm$ are clearly two spectral families in $F_\pm\mathcal{H}$. Therefore

$$E_\lambda F_\pm = \Omega_\pm E^o_\lambda E_\infty(H_o)\Omega^*_\pm, \tag{5.72}$$

which gives (5.70) upon multiplication by $\Omega_\pm$ from the right and using Proposition 2.11. (5.70) implies that for all $f,g\varepsilon\mathcal{H}$

$$(f,\phi(H)\Omega_{\pm}g) = \int\phi(\lambda)d(f,E_{\lambda}\Omega_{\pm}g)$$
$$= \int\phi(\lambda)d(\Omega_{\pm}^{*}f,E_{\lambda}^{o}g) = (\Omega_{\pm}^{*}f,\phi(H_o)g) = (f,\Omega_{\pm}\phi(H_o)g).$$

This proves (5.69). By taking in particular $\phi(\lambda) = (\lambda-z)^{-1}$, we get from (5.69) and Proposition 5.10(f) that

$$(H/F_{\pm}\mathcal{H} - z)^{-1}\Omega_{\pm} = (H-z)^{-1}\Omega_{\pm} = \Omega_{\pm}(H_o-z)^{-1} = \Omega_{\pm}(H_o/\mathcal{M}_{\infty}(H_o) - z)^{-1},$$

which shows that $H/F_{\pm}\mathcal{H}$ and $H_o/\mathcal{M}_{\infty}(H_o)$ have identical resolvent sets, proving the second assertion of (a). #

Remark 5.22 : It should be pointed out that (5.70) contains more information than the simple statement that $HF_{\pm}$ and $H_oE_{\infty}(H_o)$ have the same spectrum. Indeed, by using the definition of absolute continuity, one easily deduces from (5.72) that $\Omega_{\pm}\mathcal{H}_{ac}(H_o) \subseteq \mathcal{H}_{ac}(H)$ (Problem 5.11). Thus, if $H_oE_{\infty}(H_o)$ is absolutely continuous, as is the case with the free Hamiltonian K_o, then (5.70) leads to the conclusion that $HF_{\pm}$ are absolutely continuous, or equivalently that $F_{\pm}\mathcal{H} \subseteq \mathcal{H}_{ac}(H)$. This consideration also illustrates why in general the wave operators in potential scattering will be only isometric and not unitary. In fact, if the point spectrum of H is not void, $\Omega_{\pm}$ cannot be unitary since $\mathcal{H}_p(H)$ is orthogonal to $F_{\pm}\mathcal{H}$ by the preceding inclusion.

5-7 SPECTRAL REPRESENTATIONS

In this section we introduce a particular representation of the Hilbert space $\mathcal{H}$ with respect to a given self-adjoint operator. The material of this section will be useful in defining the so-called "on-shell S-matrix" which will be used in deriving an expression for the scattering cross section in Section 7-3.

As discussed in Section 2-1, the space $L^2(\mathcal{R}^n)$ consists of (equivalence classes of) square-integrable functions with domain $\mathcal{R}^n$ and with values in the one-dimensional space $\mathcal{C}$. A natural generalization of such a structure would be to consider functions taking values in a more general vector space, for example in a fixed Hilbert space $\mathcal{H}_o$ of arbitrary dimension. For this, let Λ be a subset of the real line, measurable with respect to Lebesgue measure.*) Suppose that f is a <u>vector-valued function</u> on Λ taking values in a fixed separable Hilbert space $\mathcal{H}_o$. We shall use the notation f_λ for the value (in $\mathcal{H}_o$) of the function f at the point λ and denote the function f by the collection $\{f_\lambda\}$. Such a function f is said to be <u>measurable</u> if for each vector $h\varepsilon\mathcal{H}_o$, the complex-valued function $(h,f_\lambda)_o$ is Lebesgue measurable, where $(\cdot,\cdot)_o$ is the scalar product in $\mathcal{H}_o$. If f and g are two measurable vector-valued functions, then $(f_\lambda, g_\lambda)_o$ is also a measurable function. This follows from the expression

$$(f_\lambda,g_\lambda)_o = \sum_{k=1}^{\infty} (f_\lambda,e_k)_o(e_k,g_\lambda)_o,$$

where $\{e_k\}$ is any orthonormal basis in $\mathcal{H}_o$ and from the fact that the limit of a sequence of measurable functions is measurable if it exists [R].

Two such (vector-valued) measurable functions f and g

---

*) In a more general definition, Λ has to be a measure space with an associated non-negative measure μ. For the application that we have in mind (Chap. 7), however, the restriction to the Lebesgue measure suffices.

are said to be equivalent if they differ only on a set of measure zero. As with the space $L^2(\mathcal{R}^n)$, by the space $L^2(\Lambda,\mathcal{H}_o)$ we mean the set of all equivalence classes of measurable vector-valued functions satisfying

$$\|f\|^2 = \int_\Lambda \|f_\lambda\|_o^2 \, d\lambda < \infty. \tag{5.73}$$

The operations of vector addition, of multiplication of a vector by a scalar and of forming a scalar product of two vectors in $L^2(\Lambda,\mathcal{H}_o)$ are given by the relations :

$$f + g = \{f_\lambda + g_\lambda\}, \quad \alpha f = \{\alpha f_\lambda\}, \tag{5.74}$$

and

$$(f,g) = \int_\Lambda (f_\lambda, g_\lambda)_o d\lambda, \tag{5.75}$$

where $f = \{f_\lambda\}$ and $g = \{g_\lambda\}$. The existence of the integral in (5.75) follows from the Schwarz inequality in $\mathcal{H}_o$ and the condition (5.73) as in (2.14). All the axioms of a Hilbert space including that of completeness can be checked for $L^2(\Lambda,\mathcal{H}_o)$ in the same way as in the case of $L^2(\mathcal{R}^n)$, and one can also show that $L^2(\Lambda,\mathcal{H}_o)$ is separable. Let $\{e_k\}$ be an orthonormal basis of $\mathcal{H}_o$ and set

$$\alpha_k(\lambda) = (e_k, f_\lambda)_o. \tag{5.76}$$

Then it is often convenient to express all the properties of $L^2(\Lambda,\mathcal{H}_o)$ in terms of $\alpha_k(\lambda)$, the "coordinate functions". In particular, a function $f = \{f_\lambda\}$ belongs to $L^2(\Lambda,\mathcal{H}_o)$ if and only if $\Sigma_k \int_\Lambda |\alpha_k(\lambda)|^2 d\lambda < \infty$.

Remark 5.23 : An important property of the Hilbert space $L^2(\Lambda,\mathcal{H}_o)$ that we shall use often in the sequel is the following. There exists a sequence $\{f_n\}$, $f_n \varepsilon L^2(\Lambda,\mathcal{H}_o)$ such that for every fixed λ, $\{f_{n,\lambda}\}$ is a fundamental set in $\mathcal{H}_o$. This can be seen by writing $f_{n,\lambda} = e_n t(\lambda)$, where $\{e_n\}$ is an orthonormal basis in $\mathcal{H}_o$ and $t(\lambda)$ is a measurable square-integrable function which does not vanish anywhere in Λ.

In the notation introduced above we may rewrite $L^2(\mathcal{R}^n)$ as $L^2(\mathcal{R}^n,\mathcal{C})$. The next example deals with a case where $\mathcal{H}_o$ is infinite dimensional.

Example 5.24 : Let $S^{(2)}$ be the two-dimensional unit sphere embedded in $\mathcal{R}^3$. We coordinatize $S^{(2)}$ by the polar angle θ $(0 \le \theta \le \pi)$ and the azimuthal angle ϕ $(0 \le \phi \le 2\pi)$. Let $L^2(S^{(2)})$ be the space of equivalence classes of functions $f(\theta,\phi)$ such that

$$\int\int|f(\theta,\phi)|^2 \sin\theta \, d\theta \, d\phi < \infty. \tag{5.77}$$

By introducing spherical coordinates, we have for any $f\varepsilon L^2(\mathcal{R}^3)$

$$||f||^2 = \int|f(\underline{x})|^2 d^3x = \int\int\int|f(r,\theta,\phi)|^2 r^2 \sin\theta \, dr \, d\theta \, d\phi,$$

where we have written

$$f(x_1,x_2,x_3) = f(r\sin\theta\cos\phi,\ r\sin\theta\sin\phi,\ r\cos\theta) \equiv f(r,\theta,\phi).$$

Clearly $rf(r,\cdot,\cdot)\,\varepsilon\,L^2(S^{(2)})$ for almost all r with respect to the Lebesgue measure on the half-line $0 \le r < \infty$. Therefore, $f \equiv \{rf(r,\cdot,\cdot)\}\,\varepsilon\,L^2([0,\infty),\ L^2(S^{(2)}))$ and we may write

$$L^2(\mathcal{R}^3) = L^2([0,\infty),\ L^2(S^{(2)})). \tag{5.78}$$

Next we consider some bounded operators in the space $\mathcal{H} \equiv L^2(\Lambda,\mathcal{H}_o)$. Let $\phi(\cdot)$ be an essentially bounded (with respect to the Lebesgue measure on Λ) complex-valued function on Λ. Setting

$$L_\phi f = \{\phi(\lambda)f_\lambda\}, \tag{5.79}$$

where $f = \{f_\lambda\}$, one observes that

$$||L_\phi f||^2 = \int_\Lambda |\phi(\lambda)|^2 ||f_\lambda||_o^2 d\lambda \le \text{ess sup}\,|\phi(\lambda)|^2\, ||f||^2.$$

Thus we conclude that L_ϕ defined by (5.79) belongs to $\mathcal{B}(\mathcal{H})$, and one sees as in Proposition 2.16(b) that

$$\|L_\phi\| = \operatorname*{ess\,sup}_{\lambda\in\Lambda}|\phi(\lambda)|. \tag{5.80}$$

An immediate generalization suggests itself. For each $\lambda\in\Lambda$, let $A(\lambda)$ belong to $\mathcal{B}(\mathcal{H}_o)$. Such an operator-valued function A is said to be measurable if for every vector $f\in L^2(\Lambda,\mathcal{H}_o)$ the vector-valued function given by $\{A(\lambda)f_\lambda\}$ is measurable. Then for this measurable $\mathcal{B}(\mathcal{H}_o)$-valued function, we define an operator A in the Hilbert space $L^2(\Lambda,\mathcal{H}_o)$ by giving its domain as

$$D(A) = \{f\in L^2(\Lambda,\mathcal{H}_o) \mid \textstyle\int_\Lambda \|A(\lambda)f_\lambda\|_o^2\, d\lambda < \infty\}$$

and setting

$$Af = \{A(\lambda)f_\lambda\}. \tag{5.81}$$

Let $\{g_n\}$ be a countable dense set in $\mathcal{H}_o$. Then (Problem 5.13)

$$\|A(\lambda)\|_o = \sup\|g_n\|_o^{-1}\|A(\lambda)g_n\|_o = \sup\|g_n\|_o^{-1}\|g_m\|_o^{-1}|(g_m,A(\lambda)g_n)_o|,$$

where the supremum is taken over all $m,n = 1,2,\ldots$ Since $(g_m,A(\lambda)g_n)_o$ is a measurable function of λ for fixed n,m and since the supremum of a countable set of measurable functions is measurable, it follows that $\|A(\lambda)\|$[*)] is a measurable non-negative function. It can be shown furthermore (Problem 5.8) that the operator A defined by (5.81) has dense domain and is closed. An operator A defined by (5.81) is called decomposable. If moreover $A(\lambda) = \phi(\lambda)I_o$, where ϕ is a measurable function finite almost everywhere and I_o is the identity operator in $\mathcal{H}_o$, then A is said to be diagonalizable. From Problem 5.8 it is easy to see that a diagonalizable operator A given by a

---

*) When no confusion is possible, we shall omit the subscript o referring to quantities in $\mathcal{H}_o$.

function ϕ is self-adjoint if and only if the function ϕ is real-valued. A decomposable operator A is sometimes written as

$$A = \{A(\lambda)\}. \tag{5.82}$$

Next we prove two propositions which characterize the class of bounded decomposable operators.

PROPOSITION 5.25 : A decomposable operator A is bounded if and only if $\|A(\lambda)\|$ is essentially bounded, and in that case

$$\|A\| = \operatorname*{ess\,sup}_{\lambda\varepsilon\Lambda} \|A(\lambda)\|. \tag{5.83}$$

Proof : (i) We have already remarked that $\|A(\lambda)\|$ is a measurable function defined almost everywhere. Let M = ess sup $\|A(\lambda)\|$ and suppose $M < \infty$. Then for all vectors $f = \{f_\lambda\}$, the vector $\{A(\lambda)f_\lambda\}$ is in $L^2(\Lambda, H_o)$ and one has

$$\|Af\|^2 = \int \|A(\lambda)f_\lambda\|_o^2 d\lambda \leq M^2\|f\|^2, \text{ or equivalently } \|A\| \leq M. \tag{5.84}$$

(ii) Conversely, let A be bounded and χ_Δ be the characteristic function of the measurable set $\Delta \subseteq \Lambda$. Then for any vector $f = \{f_\lambda\} \varepsilon H$,

$$\int_\Lambda \chi_\Delta(\lambda)\|A(\lambda)f_\lambda\|_o^2 d\lambda = \|A\chi_\Delta f\|^2 \leq \|A\|^2 \int_\Lambda \chi_\Delta(\lambda)\|f_\lambda\|_o^2 d\lambda.$$

Therefore $\int_\Delta (\|A\|^2\|f_\lambda\|_o^2 - \|A(\lambda)f_\lambda\|_o^2)\, d\lambda \geq 0$

for all measurable sets $\Delta \subseteq \Lambda$, which implies that

$$\|A(\lambda)f_\lambda\|_o \leq \|A\|\,\|f_\lambda\|_o \text{ for a.a.}\lambda. \tag{5.85}$$

Let $\{f_{n,\lambda}\}$ be a countable fundamental set in H_o (Remark 5.23). Then (5.85) implies that for all n and almost all λ

$$\|A(\lambda)f_{n,\lambda}\|_o \leq \|A\|\,\|f_{n,\lambda}\|_o , \tag{5.86}$$

since the set where the inequality (5.86) is violated is a countable union of null sets which is of measure zero. Thus (5.86) together with Problem 5.13 leads to the conclusion that $\|A(\lambda)\| \leq \|A\|$ a.e. and hence $M \leq \|A\|$. By combining this with (5.84) we arrive at the equality (5.83). #

PROPOSITION 5.26 : Let A and B be two bounded decomposable operators in $L^2(\Lambda, H_o)$, given as $A = \{A(\lambda)\}$ and $B = \{B(\lambda)\}$. Then

(a) $A + B = \{A(\lambda) + B(\lambda)\}$; $\alpha A = \{\alpha A(\lambda)\}$,

(b) $AB = \{A(\lambda)B(\lambda)\}$,

(c) $A^* = \{A(\lambda)^*\}$.

Proof : We prove only the last assertion. Let f and g be any two vectors in $L^2(\Lambda, H_o)$ and let A' be the bounded decomposable operator given by $A' = \{A(\lambda)^*\}$. Then $(f, A'g) = \int (f_\lambda, A(\lambda)^* g_\lambda)_o d\lambda = \int (A(\lambda) f_\lambda, g_\lambda)_o d\lambda = (Af, g)$. Thus $A' = A^*$. #

Now let us consider a special diagonalizable operator A, viz. the one for which $\phi(\lambda) = \lambda$. Then we write

$$A = \{\lambda I_o\}, \tag{5.87}$$

and by virtue of what we have remarked before, it is self-adjoint. Setting

$$E_\mu = \{\chi_\mu(\lambda) I_o\}, \tag{5.88}$$

where χ_μ is the characteristic function the set $\Lambda_\mu \equiv \Lambda \cap (-\infty, \mu]$, we conclude that

$$(f, E_\mu g) = \int_{\Lambda_\mu} (f_\lambda, g_\lambda)_o d\lambda. \tag{5.89}$$

Then for all $g \in D(A)$,

$$(f,Ag) = \int_\Lambda \lambda(f_\lambda,g_\lambda)_o d\lambda = \int_\Lambda \lambda d(f,E_\lambda g),$$

which shows that we can associate a self-adjoint operator of the type given by (5.87) (in fact, a certain class of mutually commuting operators, viz. the diagonalizable operators) with the space $L^2(\Lambda,H_o)$. The converse of this statement, which is equivalent to the spectral theorem, will be stated later after we have introduced more general L^2-spaces of vector-valued functions. Incidentally, by virtue of (5.89) the operator A defined in (5.87) is absolutely continuous. This is of course due to our choice of the Lebesgue measure in defining $L^2(\Lambda,H_o)$. Any decomposable operator $A = \{A(\lambda)\}$ commutes with all the operators L_ϕ defined in (5.79). This can be seen as follows. Since ϕ is essentially bounded,

$$\|AL_\phi g\|^2 = \int|\phi(\lambda)|^2\|A(\lambda)g_\lambda\|_o^2 d\lambda \leq \text{ess sup}|\phi(\lambda)|^2\int\|A(\lambda)g_\lambda\|_o^2 d\lambda < \infty$$

for all $g\varepsilon D(A)$. Also for all such g,

$$AL_\phi g = \{A(\lambda)(L_\phi g)_\lambda\} \equiv \{\phi(\lambda)A(\lambda)g_\lambda\} = L_\phi Ag,$$

which implies $L_\phi A \subseteq AL_\phi$. The converse of this property is the content of the next proposition which is of some interest in scattering theory.

<u>PROPOSITION 5.27</u> : Let B be a bounded operator in $L^2(\Lambda,H_o)$ that commutes with the self-adjoint operator A given by (5.87). Then B is a decomposable operator given by $B = \{B(\lambda)\}$ and

$$\|B\| = \underset{\lambda\varepsilon\Lambda}{\text{ess sup}}\,\|B(\lambda)\| < \infty. \tag{5.90}$$

<u>Proof</u> : Let $\{f_k\}$ be a sequence in $L^2(\Lambda,H_o)$ such that $f_{k,\lambda} = e_k t(\lambda)$, where $\{e_k\}$ and $t(\lambda)$ are as introduced in Remark 5.23. Define an operator $C(\lambda)$ in H_o by setting

$$C(\lambda)f_{k,\lambda} = (B^*f_k)_\lambda \quad \text{for all } k, \tag{5.91}$$

and extending it by linearity to the countable dense set $\mathcal{D}_\lambda$ of $\mathcal{H}_o$ generated by $\{f_{k,\lambda}\}$. Denoting by Γ_k the null set where (B^*f_k) is not defined, we see that $\Gamma = \cup\,\Gamma_k$ has measure zero and $C(\lambda)$ given by (5.91) is densely defined for each $\lambda \notin \Gamma$.

Since $E_\Delta = L\chi_\Delta$ and since by the spectral theorem $E_\Delta B^* = B^*E_\Delta$ for all Borel sets $\Delta \subseteq \Lambda$, we get as in the proof of (5.85) that $\|C(\lambda)g_\lambda\|_o \leq \|B^*\|\,\|g_\lambda\|_o$ for all $\lambda \notin \Gamma$ and all $g_\lambda \varepsilon \mathcal{D}_\lambda$. Thus for such λ, $C(\lambda)$ is bounded and hence by Proposition 2.6 it has a unique bounded extension which we also denote by $C(\lambda)$. We have for every $h \varepsilon \mathcal{H}_o$ and $g = \{g_\lambda\} \varepsilon \mathcal{H}$

$$\begin{aligned}(h,C(\lambda)g_\lambda)_o &= \textstyle\sum_k (h,C(\lambda)e_k)_o(e_k,g_\lambda)_o \\ &= t(\lambda)^{-1} \textstyle\sum_k (h,(B^*f_k)_\lambda)_o(e_k,g_\lambda)_o.\end{aligned}$$

Since each factor in the last term is measurable, $\{C(\lambda)\}$ is measurable. Define $B(\lambda) = C(\lambda)^*$ for $\lambda \notin \Gamma$ and $B(\lambda) = 0$ for $\lambda \varepsilon \Gamma$. Then ess sup $\|B(\lambda)\| \leq \|B\|$, so that the decomposable operator $B' \equiv \{B(\lambda)\}$ is bounded by Proposition 5.25.

Finally, let $g \varepsilon L^2(\Lambda,\mathcal{H}_o)$. By (5.89) we have that $(f_k,BE_\mu g) = (B^*f_k,E_\mu g) = \int_{-\infty}^{\mu} (f_{k,\lambda},B(\lambda)g_\lambda)_o\,\chi_\Lambda(\lambda)d\lambda$ and $(f_k,E_\mu Bg) = \int_{-\infty}^{\mu} (f_{k,\lambda},(Bg)_\lambda)_o\,\chi_\Lambda(\lambda)d\lambda$. Since $E_\mu B = BE_\mu$ and and since μ is arbitrary, we obtain $(f_{k,\lambda},B(\lambda)g_\lambda)_o = (f_{k,\lambda},(Bg)_\lambda)_o$ for all k and almost all λ. Hence by Proposition 2.2

$$B(\lambda)g_\lambda = (Bg)_\lambda \quad \text{for almost all } \lambda. \tag{5.92}$$

(5.92) and the definition of B' show that $B = B'$. #

COROLLARY 5.28 : A bounded decomposable operator $B = \{B(\lambda)\}$ is bounded invertible (respectively self-adjoint, unitary, partially isometric) if and only if $B(\lambda)$ is bounded invertible (respectively self-adjoint, unitary, partially isometric) almost everywhere.

Proof : Let $A = \{\lambda I_o\}$. Then B commutes with A. If $B^{-1} \varepsilon \mathcal{B}(\mathcal{H})$, then it follows that B^{-1} also commutes with A and therefore, by the preceeding proposition, is decomposable. We write $B^{-1} = \{B'(\lambda)\}$. By Proposition 5.26(b) we have for all $f \varepsilon L^2(\Lambda, \mathcal{H}_o)$

$$f = BB^{-1}f = \{B(\lambda)B'(\lambda)f_\lambda\} = B^{-1}Bf = \{B'(\lambda)B(\lambda)f_\lambda\},$$

which implies that $B(\lambda)B'(\lambda) = I_o = B'(\lambda)B(\lambda)$ for a.a.λ (Problem 5.9) and proves that $B(\lambda)$ is bounded invertible for almost all λ.

Similarly, let $U = \{U(\lambda)\}$ be unitary. By Proposition 5.26(c), $U^* = \{U(\lambda)^*\}$. Then again for all $f \varepsilon L^2(\Lambda, \mathcal{H}_o)$, $f = U^*Uf = UU^*f = \{U(\lambda)^*U(\lambda)f_\lambda\} = \{U(\lambda)U(\lambda)^*f_\lambda\}$, which as before implies that $U(\lambda)U^*(\lambda) = U(\lambda)^*U(\lambda) = I_o$ for a.a.λ. The other assertions can be established the same way. #

Now we give an example, namely that of the Hilbert space $L^2(\mathcal{R}^3)$ and K_o, the free Hamiltonian discussed in Section 3-3. We consider the space $L^2(\mathcal{R}^3)$ as the Hilbert space $L^2([o,\infty), L^2(S^{(2)}))$ as we have done in Example 5.24, using spherical coordinates in the momentum representation instead of the position representation (i.e. replacing the functions f in Example 5.24 by their Fourier transforms $\tilde{f}$). Setting $\lambda = \underline{k}^2 = \sum_{i=1}^{3} k_i^2$, we define for each f in $L^2(\mathcal{R}^3)$ and for almost all $\lambda \varepsilon [0,\infty)$ a vector $f_\lambda \varepsilon L^2(S^{(2)})$ by $f_\lambda(\theta,\phi) = 2^{-\frac{1}{2}}\lambda^{\frac{1}{4}}\cdot \tilde{f}(\lambda^{\frac{1}{2}},\theta,\phi)$, and conclude as in Example 5.24 that

$$\|f\|^2 = \|\tilde{f}\|^2 = \int\int\int k^2|\tilde{f}(k,\theta,\phi)|^2 dk\ \sin\theta d\theta\, d\phi$$
$$= \int_0^\infty \|f_\lambda\|^2_{L^2(S^{(2)})} d\lambda. \tag{5.93}$$

Let G_o be the Hilbert space defined in the above manner with norm given by (5.93). It is clear that

$$G_o = L^2([0,\infty), L^2(S^{(2)})) \tag{5.94}$$

and that $L^2(R^3)$ is unitarily isomorphic to G_o, the isomorphism U_o being given by

$$(U_of)_\lambda(\underline{\omega}) = 2^{-\frac{1}{2}}\lambda^{\frac{1}{4}}\tilde{f}(\lambda^{\frac{1}{2}}\underline{\omega}), \quad \underline{\omega}\,\varepsilon\, S^{(2)}. \tag{5.95}$$

Remembering the definition of the free Hamiltonian K_o in (3.29) (with $h = 2\pi$, $2m = 1$), we have for all $g\varepsilon D(K_o)$

$$(U_oK_og)_\lambda = \lambda(U_og)_\lambda, \tag{5.96}$$

or equivalently $U_oK_oU_o^{-1} = \{\lambda I_o\}$. It also follows that each essentially bounded function ϕ of K_o is given by $U_o\phi(K_o)U_o^{-1} = \{\phi(\lambda)I_o\}$. The above representation (5.94) of $L^2(R^3)$ will be called the spectral representation of the Hilbert space $L^2(R^3)$ relative to K_o or simply the <u>spectral representation</u> of K_o. For the sake of simplicity of notation we shall often identify $f\varepsilon L^2(R^3)$ with $U_of\varepsilon G_o$ and write for example instead of (5.96), $(K_og)_\lambda = \lambda g_\lambda$.

We know from Section 4-1 that the S-operator defined by (4.15) commutes with $H_oE_\infty(H_o)$. When H_o is the free Hamiltonian K_o, $E_\infty(H_o) = I$ and therefore in such a case S commutes with K_o. Thus by Proposition 5.27 and Corollary 5.28 we conclude that S is a bounded decomposable operator in the representation (5.94) of $L^2(\mathcal{R}^3)$, i.e.

$$\mathcal{U}_0 S \mathcal{U}_0^{-1} = \{S(\lambda)\} \quad \text{and} \quad S(\lambda)^* S(\lambda) = I_0 \text{ a.e.} \tag{5.97}$$

If S is unitary, so is $S(\lambda)$ for almost all λ. The operator $S(\lambda)$ is often called the <u>on-shell S-operator</u> or simply the <u>S-matrix</u> at energy λ.

So far we have talked about the space $L^2(\Lambda, \mathcal{H}_0)$ where the functions take values in one fixed Hilbert space $\mathcal{H}_0$. We now introduce a generalization of this, the knowledge of which is not required for the understanding of potential scattering. Let $\{\Lambda_j\}$ (j = 0,1,2,...,) be a partition of $\mathcal{R}$ into disjoint Borel sets. For each j, let $\mathcal{K}_j$ be a separable Hilbert space with dim $\mathcal{K}_j = j$. For each j we may construct, as before, the Hilbert space $L^2(\Lambda_j, \mathcal{K}_j)$ of square-integrable (with respect to the Lebesgue measure) vector-valued functions taking values in $\mathcal{K}_j$. In this notation the space $L^2(\Lambda_0, \mathcal{K}_0)$ consists of the vector θ. Then we form the direct sum (cf. Section 2-4) of these Hilbert spaces and write

$$\mathcal{G} = \oplus_j \; L^2(\Lambda_j, \mathcal{K}_j). \tag{5.98}$$

The next Proposition gives the spectral theorem for an absolutely continuous operator in this language. A similar theorem is true for an arbitrary self-adjoint operator provided that one admits more general measures in the definition of $L^2(\Lambda_j, \mathcal{K}_j)$.

<u>PROPOSITION 5.29</u> : Let A be an absolutely continuous self-adjoint operator in $\mathcal{H}$. Then there exists a Hilbert space $\mathcal{G}$ of the form (5.98) and a unitary map $\mathcal{U}$ from $\mathcal{H}$ onto $\mathcal{G}$ such that

$$(\mathcal{U}Af)_{j,\lambda} = \lambda(\mathcal{U}f)_{j,\lambda} \; ; \; \lambda\varepsilon\Lambda_j, \tag{5.99}$$

for each $f\varepsilon D(A)$. Furthermore, for each bounded Borel function ϕ defined on $\mathcal{R}$, we have

$$(\mathcal{U}\phi(A)f)_{j,\lambda} = \phi(\lambda)(\mathcal{U}f)_{j,\lambda} \; ; \; \lambda\varepsilon\Lambda_j. \tag{5.100}$$

The Hilbert space $\mathcal{G}$ of the form (5.98) in which A acts like multiplication by λ is called the <u>spectral representation</u> of A (more precisely, the spectral representation of $\mathcal{H}$ relative to A). The unitary operator $\mathcal{U}$ is called the <u>spectral transformation</u> of A. Each point λ in $\mathcal{R}$ clearly belongs to one of the sets Λ_j for some j. We call the corresponding non-negative integer j the <u>spectral multiplicity</u> of the point λ. This defines a function $p : \mathcal{R} \to \{0,1,2,\ldots,\infty\}$. The function $p(\lambda)$ is measurable, defined almost everywhere and is equal to zero if $\lambda\not\varepsilon\sigma(A)$. Note that $\sigma(A)$ is equal to the closure of the union of all Λ_j with $j \neq 0$. For a detailed discussion of these questions, see the article of Brown in [PE].

We should point out that, given a self-adjoint operator A, it is the spectral multiplicity function p rather than the spectral transformation which is uniquely determined by A. In fact, if a given spectral transformation $\mathcal{U}$ is composed with any decomposable unitary operator in $\mathcal{G}$, one arrives at a new spectral transformation. In the literature the spectral representation (5.98) is sometimes written in the language of <u>direct integrals of Hilbert spaces</u> and denoted as $\int^{\oplus}\mathcal{H}_\lambda d\lambda$ with $\mathcal{H}_\lambda = \mathcal{K}_{j(\lambda)}$, where $j(\lambda)$ is the integer j determined by $p(\lambda) = j$.

An absolutely continuous self-adjoint operator A is

said to have simple spectrum if its multiplicity function $p(\lambda)$ assumes only the values 0 and 1. This notion of simple spectrum is the continuous analogue of that of a hermitian matrix without repeated eigenvalues. If A has simple spectrum, then any bounded operator B that commutes with A is not only decomposable but also diagonalizable in the spectral representation of A. This follows from the fact that $\mathcal{B}(\mathcal{K}) = \{\alpha I \mid \alpha \in \mathcal{C}\}$ when $\dim \mathcal{K} = 1$. Another way of expressing the same thing is to say that B is a function of the self-adjoint operator A, where the associated function $\phi_B(\lambda)$ is measurable and essentially bounded. It is clear that the operator appearing in Example 5.14 has simple spectrum (Problem 5.10).

If $\mathcal{U}$ is a spectral transformation of A and if $B = WAW^{-1}$ with W unitary, then $\mathcal{U}W^{-1}$ is a spectral transformation of B. Hence A and B have identical spectral multiplicity functions. For a scattering system, this and (5.71) imply that $H/F_{\pm}\mathcal{H}$ has the same spectral multiplicity as $H_o/\mathcal{M}_\infty(H_o)$. In particular, if $H_o/\mathcal{M}_\infty(H_o)$ has simple spectrum, then the spectrum of $H/F_{\pm}\mathcal{H}$ is also simple.

We have shown in Propositions 4.6 and 4.7 that for a general scattering system verifying (A1) and (A2), the S-operator is isometric on $\mathcal{M}_\infty(H_o)$ and commutes with H_o. Assume that $\mathcal{M}_\infty(H_o) = \mathcal{H}_{ac}(H_o)$ and consider the spectral representation $\mathcal{G}$ of $H_{o,ac}$, given by (5.98). Then the projection onto the space $L^2(\Lambda_j, \mathcal{K}_j)$ is equal to $\mathcal{U}E^o_{\Lambda_j}\mathcal{U}^*$, where $\{E^o_\lambda\}$ is the spectral family of H_o. Since S commutes with H_o, it commutes with $E^o_{\Lambda_j}$ and hence $\mathcal{U}S\mathcal{U}^*/L^2(\Lambda_j, \mathcal{K}_j)$ commutes with $\mathcal{U}\, H_{o,ac}\mathcal{U}^*/L^2(\Lambda_j, \mathcal{K}_j)$. Thus by applying Proposition 5.27 in

$L^2(\Lambda_j, K_j)$ for a fixed j, we infer that $USU^*/L^2(\Lambda_j, K_j)$ is decomposable and is given by $\{S_j(\lambda)\}$. Finally we set

$$S(\lambda) = \oplus_j S_j(\lambda) \tag{5.101}$$

and observe that USU^* is the operator $\{S(\lambda)\}$ in the spectral representation G. (Note that $U^* = U^{-1}$ by Remark 10.2.)

5-8 NOTES AND SUPPLEMENTARY MATERIAL

A. For the proof of Lemma 5.5, we need the following preliminary result.

LEMMA 5.30 : (a) Let A be a bounded self-adjoint operator. Then

$$\sup_{\|f\|=1} |(f,Af)| = \|A\| . \tag{5.102}$$

(b) Let $B \in \mathcal{B}(H)$. Then $\|B^*B\| = \|B\|^2$.

Proof : (a) Since $\text{Re}(f,Ag) = \frac{1}{2}[(f,Ag)+(g,Af)] = \frac{1}{4}(f+g,A(f+g)) - \frac{1}{4}(f-g,A(f-g))$, it follows that

$$|\text{Re}(f,Ag)| \leq \tfrac{1}{4}(\|f+g\|^2 + \|f-g\|^2) \sup_{\|h\|=1} |(h,Ah)|, \tag{5.103}$$

where we have used the simple fact that

$$|(f,Af)| \leq \|f\|^2 \sup_{\|h\|=1} |(h,Ah)|.$$

By the parallelogram law : $\|f+g\|^2 + \|f-g\|^2 = 2\|f\|^2 + 2\|g\|^2$, the inequality (5.103) reduces to

$$|\text{Re}(f,Ag)| \leq \sup_{\|h\|=1} |(h,Ah)|,$$

provided that $\|f\| = \|g\| = 1$. Now it is possible to choose a real number α so that $(e^{i\alpha}f,Ag) = |(f,Ag)|$, and hence

$$|(f,Ag)| \leq \sup_{||h||=1} |(h,Ah)|, \text{ if } ||f|| = ||g|| = 1. \tag{5.104}$$

By Proposition 2.3 and (2.24),

$$\sup_{||f||=||g||=1} |(f,Ag)| = \sup_{||g||=1} ||Ag|| = ||A||. \tag{5.105}$$

Therefore (5.104) reduces to

$$||A|| \leq \sup_{||h||=1} |(h,Ah)| \leq \sup_{||h||=1} ||h|| \, ||Ah|| = ||A||,$$

showing that these inequalities must in fact be equalities.

(b) $||B^*B|| = \sup(f,B^*Bf) = \sup ||Bf||^2 = ||B||^2$, with $||f|| = 1$. #

Proof of Lemma 5.5 :

(i) It is no loss of generality to assume $||A|| \leq 1$, for otherwise we consider $A' = A/||A||$. Then by Lemma 5.30, we conclude that $||I-A|| = \sup |(f,(I-A)f)| \leq 1$, with $||f|| = 1$. Now we use the fact (Problem 5.12) that the power series for $\sqrt{1-z}$ converges absolutely for $|z| \leq 1$. Let this power series be given as $\sqrt{1-z} = \sum_{n=0}^{\infty} c_n z^n$. Then the series $\sum_{n=0}^{\infty} c_n (I-A)^n$ converges in norm and defines a bounded self-adjoint operator B, since each c_n is clearly real. Since $0 \leq I-A \leq I$, it follows that $0 \leq (f,(I-A)^n f) \leq 1$ with $||f|| = 1$. Thus $(f,Bf) = 1 + \sum_{n=1}^{\infty} c_n (f,(I-A)^n f) \geq 1 + \sum_{n=1}^{\infty} c_n = \lim \sqrt{1-x} = 0$ as $x \to 1-0$, where we have used the fact that each c_n is negative for $n \geq 1$ (Problem 5.12). This shows that $B \geq 0$. Since the series converges absolutely, a rearrangement of the terms shows that $1-z = (\sum_{n=0}^{\infty} c_n z^n)^2$, and an identical rearrangement of the terms of the series $(\sum_{n=0}^{\infty} c_n (I-A)^n)^2$ proves that $B^2 = A$. Also since the series $\sum_{n=0}^{\infty} c_n (1-A)^n$ converges in norm, any bounded operator that commutes with A will commute with B.

(ii) Suppose that $B' \geq 0$ is such that $B'^2 = A$. Since $B'A = B'^3 = AB'$, B' commutes with B. Therefore, setting $C = B^{\frac{1}{2}}$, $C' = B'^{\frac{1}{2}}$ (the square roots being defined as in (i)) and $g = (B-B')f$ with $f \varepsilon H$, we get that

$$\|Cg\|^2 + \|C'g\|^2 = (f,\{(B-B')B+(B-B')B'\}g) = (f,(B^2-B'^2)g) = 0,$$

implying that $Cg = C'g = \theta$. This relation leads to the conclusion that $C^2 g = B(B-B')f = \theta$ and $C'^2 g = B'(B-B')f = \theta$ for all $f \varepsilon H$. Thus $\|(B-B')f\|^2 = (f,(B-B')^2 f) = 0$, and hence $B = B'$. #

B. As we have mentioned in Section 5-4, if one has a proof of the polar decomposition of a closed (unbounded) operator, then the proofs of Propositions 5.7 and 5.8 remain essentially identical, if one is careful about domains. The proof of the polar decomposition depends on the possibility of defining a unique positive square root of a positive (unbounded) operator. However the method employed in Lemma 5.5 for a bounded operator is not applicable for the general case and one needs to use more powerful methods [K, p. 281]. Here we shall by-pass this difficulty and indicate briefly the proof of the spectral theorem for bounded normal and unitary operators. Finally we shall relate an (unbounded) self-adjoint operator with a unitary operator and obtain the general spectral theorem, also obtaining as a by-product the polar decomposition for a self-adjoint operator.

PROPOSITION 5.31 : Let A be a bounded normal operator, i.e. $A^*A = AA^*$. Then there exists a unique projection-valued measure E defined on the Borel sets of the plane $\mathcal{R}^2$ such that

$$A = \int_{-\infty}^{\infty} \int_{-\infty}^{\infty} z\, E\,(dxdy), \quad A^* = \int_{-\infty}^{\infty} \int_{-\infty}^{\infty} \bar{z}\, E\,(dxdy), \tag{5.106}$$

where $z = x + iy$, $\bar{z} = x - iy$. Also the spectral measure E commutes with all bounded operators that commute with A.

Sketch of the proof : (i) Every normal operator can be written in the form $A = A_1 + iA_2$, where A_1 and A_2 are bounded self-adjoint operators. In fact, we need only write

$$A_1 = \tfrac{1}{2}(A+A^*) \text{ and } A_2 = \tfrac{1}{2}i(A^*-A). \tag{5.107}$$

Then the property of normality of A is expressed by the fact that A_1 and A_2 commute, i.e. $A_1A_2 = A_2A_1$. Also it is clear that $||A_1|| \leq ||A||$ and $||A_2|| \leq ||A||$.

(ii) Let $E_{1,x}$ and $E_{2,y}$ $(-||A|| \leq x,y \leq ||A||)$ be the spectral families associated with the self-adjoint operators A_1 and A_2 respectively. Then by Proposition 5.8 they are given uniquely by the operator A and they commute. Then,

$$A = A_1 + iA_2 = \int x dE_{1,x} \cdot \int dE_{2,y} + i\int dE_{1,x} \cdot \int y dE_{2,y}$$

$$= \int\int (x + iy) dE_{1,x}\, dE_{2,y}, \tag{5.108}$$

where the last integration is over the square of side $2\,||A||$. The extension of the product measure $E_{1,\Delta_1} \cdot E_{2,\Delta_2}$, to a measure E_I on rectangles I is standard, and then it is easy to show that such a measure E satisfies the relations (5.62)-(5.64). #

PROPOSITION 5.32 : Let U be a unitary operator. Then there exists a unique spectral family $\{E_\phi\}$ defined on the unit circle $\{0 \leq \phi \leq 2\pi\}$ such that

$$U = \int_0^{2\pi} e^{i\phi} dE_\phi. \tag{5.109}$$

Moreover, $\{E_\phi\}$ commute with all bounded operators that commute with U.

Sketch of the proof : A unitary operator is clearly normal and hence admits a spectral representation as in (5.106). Since $\|Uf\| = \|f\|$, E_I for such an operator is 0 for every rectangle I which lies entirely outside or entirely inside the unit circle $x^2+y^2 = |z|^2 = 1$. In other words, the support of the spectral family E_I is a subset of the unit circle. Let $I_1, I_2, \dots, I_n$ be (open, half-open or closed) rectangles covering the arc $0 < \psi \leq \phi$ of the unit circle in a simple manner and having no other points in common with the circle. Denoting by E_ϕ the sum $\sum_{k=1}^n E_{I_k}$, which depends only on the arc in question, it is easy to arrive at (5.109) from (5.106). #

Proof of Proposition 5.9 :

(i) We know from Proposition 2.14 and Section 5-5 that the ranges of $(A \pm i)$ are H and that $(A \pm i)^{-1}$ are bounded operators. Therefore the operator V given by

$$V = (A-i)(A+i)^{-1} \tag{5.110}$$

is defined everywhere. Also since $\|(A-i)f\|^2 = \|Af\|^2 + \|f\|^2 = \|(A+i)f\|^2$, and since the range of V is H, it follows that V is unitary. In the case when A is closed symmetric but not self-adjoint, the corresponding V is isometric and not unitary. V given by (5.110) is called the Cayley transform of A. It is easy to see that $(I-V)^{-1}$ exists, though is not necessarily bounded and we can recover A from V by writing

$$A = i(I+V)(I-V)^{-1}. \tag{5.111}$$

(ii) Let F_ϕ be the spectral family of V defined in the preceding proposition, with the requirement that $F_o = 0$ and $F_{2\pi} = I$. F_ϕ is continuous not only at $\phi = 0$ (by virtue of the above choice) but also at $\phi = 2\pi$, for otherwise I-V would not be invertible, contradicting our observation in (i). Now setting $\lambda = -\cot(\phi/2)$

and
$$E_\lambda = F_\phi = F_{-2\,\mathrm{arccot}\,\lambda}, \tag{5.112}$$

we have that

$$A = i(I+V)(I-V)^{-1} = \int_0^{2\pi} i(1+e^{i\phi})(1-e^{i\phi})^{-1} dF_\phi = \int_{-\infty}^{\infty} \lambda\, dE_\lambda .$$

To have a complete justification of the above procedure one has to decompose the interval $[0,2\pi]$ into an infinite number of intervals with endpoints accumulating to 0 and 2π and discuss the integrals involved on such partitions. For the details the reader is referred to [RN; Section 121].

If a bounded operator B commutes with A in the sense of (5.32), then it follows easily that B commutes with V and hence with $\{F_\phi\}$ by Proposition 5.32. Thus by (5.112) B also commutes with the spectral family $\{E_\lambda\}$. #

If A is self-adjoint and positive, then, writing $(f,Af) = \int_{-\infty}^{\infty} \lambda d\|E_\lambda f\|^2 \geq 0$ for all $f \in D(A)$, it is clear that the support of $\{E_\lambda\}$ is the right half line, $\lambda \geq 0$. On the right half-line the function $\lambda \mapsto \lambda^{\frac{1}{2}}$ is single-valued and continuous if we choose the positive branch. Then by the functional calculus (cf. Section 5-5) we can define the unique square root of A as

$$A^{\frac{1}{2}} = \int_0^{\infty} \lambda^{\frac{1}{2}} dE_\lambda, \tag{5.113}$$

and verify easily that $A^{\frac{1}{2}}$ is positive self-adjoint and

$(A^{\frac{1}{2}})^2 = A.$

Next, let A be any self-adjoint operator with spectral representation (5.20). Then the operator A^2 given by $\int_{-\infty}^{\infty}\lambda^2 dE_\lambda$ is clearly positive self-adjoint and its square root, denoted $|A|$ as before, is given by

$$|A| = \int_{-\infty}^{\infty}|\lambda|dE_\lambda = \int_{0}^{\infty}\lambda dE_\lambda - \int_{-\infty}^{0}\lambda dE_\lambda, \tag{5.114}$$

and $D(|A|) = D(A)$. It is also easy to verify that $N(A) = N(|A|)$ = the range of $E_{\{0\}}$, and since $E_\lambda(I-E_o) = 0$ for $\lambda \leq 0$, $E_\lambda E_{-o} = E_{-o}$ for $\lambda \geq 0$, it follows that

$$A = (I-E_o-E_{-o})|A|. \tag{5.115}$$

Writing $U = I-E_o-E_{-o}$, we see that U is self-adjoint, partially isometric and that it commutes with A and $|A|$. Thus (5.115) is exactly the polar decomposition of the (unbounded) self-adjoint operator A. This incidentally also demonstrates the uniqueness of the spectral family $\{E_\lambda\}$ in the spectral theorem. In fact, let $A = \int\lambda dE'_\lambda$, where $\{E'_\lambda\}$ is another spectral family. Then by the above discussion, we know that the operator $(A-\lambda I)$ has the polar decomposition $A-\lambda I = U(\lambda)|A-\lambda I|$ where $U(\lambda) = I-E'_\lambda-E'_{\lambda-o}$. On the other hand the family E_λ has been defined in (5.42) in terms of $U(\lambda)$ and a simple computation shows that $E_\lambda = E'_\lambda$.

PROBLEMS

5.1 : (a) Using the properties of $\{M_\lambda\}$, verify (5.10)-(5.12) for $\{E_\lambda\}$. (b) Show by using only (5.10) that the limits (5.13) as well as s-lim E_μ as $\mu \to \pm\infty$ exist.

5.2 : Let $\{E_\lambda\}$ be the spectral family of a self-adjoint operator A and let m and M be the lower and upper bound of $\{E_\lambda\}$.

Show that $||A|| = \max\{|m|, |M|\}$.

5.3 : Verify that (5.34) implies (5.33).

5.4 : Let $A \in \mathcal{B}(\mathcal{H})$ and (f, Af) real for all $f \in \mathcal{H}$. Then A is self-adjoint. In particular, a bounded positive operator is self-adjoint.

5.5 : Show that the null space of any closed, densely defined operator is a subspace of $\mathcal{H}$ and is the orthogonal complement of the range of its adjoint.

5.6 : (a) Prove Proposition 5.10. (b) If A is any self-adjoint operator and if ϕ_1, ϕ_2 are continuous (not necessarily bounded) functions on $\mathcal{R}$, then show that $\phi_1(A)\phi_2(A) \subseteq \phi(A)$ if $\phi = \phi_1\phi_2$ and $\alpha_1\phi_1(A) + \alpha_2\phi_2(A) \subseteq \phi(A)$ if $\phi = \alpha_1\phi_1 + \alpha_2\phi_2$.

5.7 : Let $\mathcal{H} = L^2(\Delta)$, where Δ is a measurable subset of $\mathcal{R}^n$ and let ψ be a real-valued measurable function on Δ. Set $(Af)(\underline{x}) = \psi(\underline{x})f(\underline{x})$. Show that, if ψ is continuously differentiable and if $\text{grad}\,\psi \neq 0$ a.e. in Δ, then A is (spectrally) absolutely continuous. (Hint : Consider the hypersurface $S_\mu = \{\underline{x} | \psi(\underline{x}) = \mu\}$ and use the fact that $d^n\underline{x} = |\text{grad}\psi|^{-1} dS_\mu d\mu$, where dS_μ is the surface element.)

5.8† : Let A be defined as in (5.81). Show that (a) A is densely defined and closed, and (b) A is self-adjoint if and only if $A(\lambda)$ is self-adjoint for almost all λ.

5.9 : Let A and B be two bounded decomposable operators : $A = \{A(\lambda)\}$ and $B = \{B(\lambda)\}$, and let $A = B$. Then $A(\lambda) = B(\lambda)$ a.e.

5.10 : (a) The operators in Examples 5.14 - 5.16 have simple spectrum. (b) A projection has pure point spectrum.

5.11 : Prove that $\Omega_\pm \mathcal{H}_{ac}(H_0) \subseteq \mathcal{H}_{ac}(H)$. Also verify (5.68).

5.12 : Show that the power series for $(1-z)^{\frac{1}{2}}$ about $z = 0$ converges absolutely not only for $|z| < 1$ but also for $|z| = 1$. (Hint : All derivatives of $(1-z)^{\frac{1}{2}}$ at $z = 0$ are negative.)

5.13 : (a) Let $f \in \mathcal{H}$ and $\mathcal{D}$ be a dense set. Show that $||f|| = \sup|(g,f)|/||g||$, with g varying over $\mathcal{D}$ and $g \neq \theta$. (b) Let $A \in \mathcal{B}(\mathcal{H})$ and $\{g_n\}$ be a countable dense set in $\mathcal{H}$. Show that $||A|| = \sup_n (||Ag_n|| \, ||g_n||^{-1}) = \sup_{n,m}(|(g_n, Ag_m)| \, ||g_n||^{-1} ||g_m||^{-1})$.

5.14† : Prove Proposition 5.17.

5.15† : Write out in detail the proofs of Propositions 5.31 and 5.32.

CHAPTER 6 : TIME-INDEPENDENT SCATTERING THEORY

Time-independent or stationary state scattering theory was introduced long before the time-dependent one. In potential scattering it arose by separation of variables in the time-dependent Schrödinger equation $i\partial f_t/\partial t = Hf_t$, leading to the time-independent Schrödinger equation $-\Delta f(\underline{x}) + V(\underline{x})f(\underline{x}) = \lambda f(\underline{x})$. The solution describing a stationary scattering situation was characterized by a suitable boundary condition at infinity (plane wave plus outgoing spherical wave). By transforming this partial differential equation together with the boundary condition into an integral equation, one arrived at integral expressions for the solution $f(\underline{x})$ involving the potential and certain Green's functions. (Green's functions are kernels of resolvent operators when these are written as integral operators in $L^2(\mathcal{R}^3)$.)

In later years similar equations were formally written for "eigenfunctions" of an abstract pair of Hamiltonians H and H_o (cf. Lippmann and Schwinger [1]). Unfortunately such formal equations are devoid of precise mathematical meaning. A well-known example of such an equation is $\psi = \phi-(H_o-\lambda-io)^{-1}V\psi$, where ϕ and ψ are eigenfunctions with eigenvalue λ of H_o and H respectively. Since scattering is associated with the continuous spectrum of the Hamiltonian (cf. Section 7-1), ϕ and ψ cannot belong to the Hilbert space. Thus the action of the Hilbert space operators V and $(H_o-\lambda-i\eta)^{-1}$ on ψ is not defined, and the designation of ψ as an "eigenfunction" of H does not even make sense unless H is represented in a particular form (e.g. as a differential operator as in potential scattering, in which case ψ may be a twice continuously differentiable solution of the corresponding eigenvalue differential equation which is not absolutely square-integrable).

We shall return to the integral equations for potential scattering in Chapter 10. The purpose of the present chapter is to derive equations which are analogous to the above-mentioned formal equations but which have a precise mathematical meaning. These equations hold for an abstract scattering system, i.e. H and H_o are not required to be operators of any particular type. The important point is that the equations of time-independent scattering theory are valid as a consequence of the asymptotic condition. They will be written as operator equations in Hilbert space. The designation "time-independent scattering theory" stems from the fact that the one-parameter unitary groups V_t and U_t appearing in the definition of the wave and scattering operators will be eliminated and replaced by resolvents or spectral projections.

The importance of the time-independent theory resides in the fact that these equations are much more appropriate for actual calculations than the time-dependent expressions of Chapter 4. Also one of the methods for proving asymptotic completeness is based on these equations, cf. Section 9-2.

The time-independent scattering theory involves a new type of integral in Hilbert space, namely the integral of an operator-valued function with respect to a spectral measure. This so-called spectral integral will be defined in Section 6-1. The remainder of Section 6-1 contains some theorems which will be needed in the derivation of the equations of time-independent scattering theory in Section 6-2. In particular we prove in Proposition 6.3 the existence of all the spectral integrals that will appear in these equations. For this reason it is unnecessary to study abstract sufficient conditions for the existence of such integrals. Some relevant results will however be collected in Section 6-3.

6-1 SPECTRAL INTEGRALS

In Section 5-5 we discussed integrals of the form $\int\phi(\lambda)dE_\lambda$, where ϕ is a continuous function from R to C and $\{E_\lambda\}$ a spectral family. These integrals defined an operator the existence of which was established by considering the associated sesquilinear form $\int\phi(\lambda)d(g,E_\lambda f)$ and using the Riesz representation theorem. In this chapter we need a generalization of such integrals. It consists of admitting for $\phi(\lambda)$ also operator-valued functions. A convenient definition for this case is to use a Riemann-Stieltjes integral similar to that defined in Section 4-4.

Let $(a,b]$ be a finite interval in $\mathcal{R}$ and $\Pi = \{\lambda_0, \ldots, \lambda_n; \mu_1, \ldots, \mu_n\}$ a partition of $(a,b]$. For each $\lambda \varepsilon (a,b]$, let A_λ and B_λ be operators in $\mathcal{B}(\mathcal{H})$, and let $\{E_\lambda\}$ be a spectral family. If $f \varepsilon \mathcal{H}$, we define

$$\int_a^b A_\lambda dE_\lambda B_\lambda f = \underset{|\Pi|\to 0}{\text{s-lim}} \sum_{k=1}^n A_{\mu_k} E_{(\lambda_{k-1},\lambda_k]} B_{\mu_k} f \tag{6.1}$$

if the limit exists and is independent of the chosen sequence of partitions. The set of all vectors f for which this holds is a linear manifold which is the domain of the linear opera- $\int_a^b A_\lambda dE_\lambda B_\lambda$ defined by (6.1).

The spectral integral over an infinite interval is defined as the strong limit of that over a sequence of finite intervals as in (4.41). The following identity which follows immediately from (5.62) and the above definitions will frequently be used in the sequel :

$$\int_{\mathcal{R}} A_\lambda dE_\lambda E_{(a,b]} B_\lambda = \int_a^b A_\lambda dE_\lambda B_\lambda . \tag{6.2}$$

As already mentioned in the introduction, little knowledge of the properties of spectral integrals is required to deduce the equations of time-independent scattering theory. For this reason we postpone the discussion of various properties of spectral integrals till Section 6-3 and proceed to establish the basic theorem permitting the passage from time-dependent to time-independent scattering theory (Proposition

6.3 below). All the spectral integrals appearing in this chapter will be defined everywhere in H, even though in some cases the integrand will contain an unbounded operator. Consequently these integrals will be written down without specifying their domains.

LEMMA 6.1 : Let A be a self-adjoint operator, $\{E_\lambda\}$ its spectral family, $[a,b]$ a finite or infinite interval and $\{B_\lambda\}$ a family of bounded operators such that $\|B_\lambda\| \leq M < \infty$ for all $\lambda \varepsilon [a,b]$. Then

$$\int_a^b B_\lambda (A-\lambda) dE_\lambda = 0. \tag{6.3}$$

Proof : It suffices to prove (6.3) for the case where $[a,b]$ is a finite interval. Let $\Pi = \{\lambda_k ; \mu_k\}$ be a partition of $[a,b]$ and set $\Delta_k = (\lambda_{k-1}, \lambda_k]$, $k = 1,\ldots,n$. (5.30) implies that

$$\|(A-\mu_k)E_{\Delta_k} f\|^2 = \int_{\lambda_{k-1}}^{\lambda_k} (\lambda-\mu_k)^2 d\|E_\lambda E_{\Delta_k} f\|^2 \leq (\lambda_k-\lambda_{k-1})^2 \|E_{\Delta_k} f\|^2. \tag{6.4}$$

We get from this by applying successively the triangle inequality, (6.4) and the Schwarz inequality for sums that

$$\begin{aligned} \|\textstyle\sum_{k=1}^n B_{\mu_k} (A-\mu_k)E_{\Delta_k} f\| &\leq M \textstyle\sum_{k=1}^n \|(A-\mu_k)E_{\Delta_k} f\| \\ &\leq M \textstyle\sum_{k=1}^n |\Pi|^{\frac{1}{2}} (\lambda_k-\lambda_{k-1})^{\frac{1}{2}} \|E_{\Delta_k} f\| \\ &\leq M|\Pi|^{\frac{1}{2}} \{\textstyle\sum_{k=1}^n (\lambda_k-\lambda_{k-1})\}^{\frac{1}{2}} \{\textstyle\sum_{k=1}^n \|E_{\Delta_k} f\|^2\}^{\frac{1}{2}}. \end{aligned} \tag{6.5}$$

Now $$\textstyle\sum_{k=1}^n \|E_{\Delta_k} f\|^2 = \|\sum_{k=1}^n E_{\Delta_k} f\|^2 = \|E_{(a,b]} f\|^2 \tag{6.6}$$

(cf. (5.19) or (2.10)), so that the last term in (6.5) is equal to $M|\Pi|^{\frac{1}{2}}(b-a)^{\frac{1}{2}}\|E_{(a,b]} f\|$, which converges to zero as $|\Pi| \to 0$. #

If A is a self-adjoint operator with associated spectral family $\{E_\lambda\}$, then the spectral integral $\int\phi(\lambda)dE_\lambda$, if it exists, will coincide with the operator $\phi(A)$ defined in Section 5-5. This is the content of the next lemma.

<u>LEMMA 6.2</u> : Let A be self-adjoint, $\{E_\lambda\}$ its spectral family, $\phi : \mathcal{R} \to \mathcal{C}$ a continuous function and $f\varepsilon D(\phi(A))$. Then

$$\phi(A)f = \int_{\mathcal{R}}\phi(\lambda)dE_\lambda f, \tag{6.7}$$

where the integral in (6.7) is a spectral integral.

<u>Proof</u> : Let $\Delta = (a,b]$ be a finite interval and define $f_\Delta = E_\Delta f$. We have $f_\Delta \varepsilon D(\phi(A))$ by (5.49). Let $\Pi = \{\lambda_k;\mu_k\}$ be a partition of Δ and set $\Delta_k = (\lambda_{k-1},\lambda_k]$, k=1,...,n. By (5.50) we have

$$\phi(A)f_\Delta = \underset{|\Pi|\to 0}{\text{w-lim}} \sum_{k=1}^{n}\phi(\mu_k)E_{\Delta_k}f_\Delta \equiv \underset{|\Pi|\to 0}{\text{w-lim}}\ \Sigma_\Pi(f).$$

Since $E_{\Delta_j}E_{\Delta_k} = 0$ for $j \neq k$, we have by (2.10)

$$\|\sum_{k=1}^{n}\phi(\mu_k)E_{\Delta_k}f_\Delta\|^2 = \sum_{k=1}^{n}|\phi(\mu_k)|^2(f_\Delta,E_{\Delta_k}f_\Delta), \tag{6.8}$$

which by (5.50) and Proposition 5.10(a,b) converges to $(f_\Delta,\phi(A)^*\phi(A)f_\Delta)$ as $|\Pi|\to 0$. This means that $\|\Sigma_\Pi(f)\|\to\|\phi(A)f_\Delta\|$ as $|\Pi|\to 0$. Hence by Proposition 2.1, $\Sigma_\Pi(f)$ converges strongly to $\phi(A)f_\Delta$, which proves (6.7) on the interval Δ. The general case now follows by letting Δ tend to $\mathcal{R}$, since $\phi(A)f - \phi(A)f_\Delta = (I-E_\Delta)\phi(A)f$ converges strongly to zero as Δ tends to $\mathcal{R}$. #

The following proposition contains the identities relating time-dependent and time-independent equations. Their proof amounts essentially to replacing the unitary evolution groups either by a spectral integral as in Lemma 6.2 or by a resolvent, using the formulae (3.58).

PROPOSITION 6.3 : Let A,B and C be self-adjoint operators, $X_t = \exp(-iAt)$, $Y_t = \exp(-iBt)$ and $Z_t = \exp(-iCt)$ the associated one-parameter unitary groups, and $\{E_\lambda\}$ the spectral family of C. Let $T \in \mathcal{B}(\mathcal{H})$, $\eta > 0$ and $\delta > 0$. Then all of the integrals below exist (as strong limits), and one has the identities

$$J_- \equiv \int_{-\infty}^0 dt\, e^{\eta t} X_t^* \int_{-\infty}^0 ds\, e^{\delta s} Y_s^* T Z_s Z_t = -\int_{\mathbb{R}} R_{\lambda+i\eta}(A) R_{\lambda+i\delta}(B) T dE_\lambda, \quad (6.9)$$

$$J_+ \equiv \int_0^\infty dt\, e^{-\eta t} X_t^* \int_{-\infty}^0 ds\, e^{\delta s} Y_s^* T Z_s Z_t = \int_{\mathbb{R}} R_{\lambda-i\eta}(A) R_{\lambda+i\delta}(B) T dE_\lambda, \quad (6.10)$$

$$\int_{-\infty}^0 dt\, e^{\eta t} X_t^* T Z_t = -i \int_{\mathbb{R}} R_{\lambda+i\eta}(A) T\, dE_\lambda, \quad (6.11)$$

$$\int_0^\infty dt\, e^{-\eta t} X_t^* T Z_t = i \int_{\mathbb{R}} R_{\lambda-i\eta}(A) T dE_\lambda. \quad (6.12)$$

Proof : The function $s \mapsto \exp(\delta s) Y_s^* T Z_s$ is strongly continuous, and $\int_{-\infty}^0 ds \exp(\delta s) \| Y_s^* T Z_s \| = \delta^{-1} \|T\|$. Hence the integral over the variable s in $J_\pm$ exists by Proposition 4.16(d) and defines an operator in $\mathcal{B}(\mathcal{H})$. A similar argument shows that the integrals over t exist. Hence $J_\pm$ are well defined and in $\mathcal{B}(\mathcal{H})$.

Fix $0 < m < \infty$ and define $E(m) \equiv E_{(-m,m]}$. Let $f \in \mathcal{H}$, let $\Pi = \{\lambda_k; \mu_k\}$ be a partition of $[-m,m]$, set $\Delta_k = (\lambda_{k-1}, \lambda_k]$ and define

$$J_\Pi(f) = \int_{-\infty}^0 dt\, e^{\eta t} X_t^* \int_{-\infty}^0 ds\, e^{\delta s} Y_s^* T \sum_{k=1}^n \exp[-i\mu_k(s+t)] E_{\Delta_k} f.$$

We get from (4.44) that $\| J_- E(m) f - J_\Pi(f) \| \leq$

$$\|T\| \int_{-\infty}^0 dt\, e^{\eta t} \int_{-\infty}^0 ds\, e^{\delta s} \| Z_{s+t} E(m) f - \sum_{k=1}^n \exp[-i\mu_k(s+t)]\, E_{\Delta_k} f \|. \quad (6.13)$$

Now the integrand of the double integral in (6.13) converges pointwise to zero as $|\Pi| \to 0$ by Lemma 6.2. Also, by (6.8), it is majorized uniformly in Π by the integrable function $2\exp(\eta t)\exp(\delta s)\|f\|$. Thus it follows from the Lebesgue dominated convergence theorem that

$$\underset{|\Pi|\to 0}{\text{s-lim}}\, J_\Pi(f) = J_- E(m) f. \tag{6.14}$$

Now by a change of variables and by applying (3.58) we obtain

$$J_\Pi(f) = \sum_{k=1}^n \int_{-\infty}^0 dt\, e^{-i(\mu_k+i\eta)t} X_t^* \int_{-\infty}^0 ds\, e^{-i(\mu_k+i\delta)s} Y_s^* T E_{\Delta_k} f$$

$$= \sum_{k=1}^n \int_0^\infty d\tau e^{i(\mu_k+i\eta)\tau} X_\tau \int_0^\infty d\sigma\, e^{i(\mu_k+i\delta)\sigma} Y_\sigma T E_{\Delta_k} f$$

$$= - \sum_{k=1}^n R_{\mu_k+i\eta}(A) R_{\mu_k+i\delta}(B) T E_{\Delta_k} f. \tag{6.15}$$

Since $J_\Pi(f)$ is strongly convergent to $J_- E(m) f$ as $|\Pi| \to 0$, (6.15) implies that

$$J_- E(m) f = - \int_{-m}^{m} R_{\lambda+i\eta}(A) R_{\lambda+i\delta}(B)\, T\, dE_\lambda f. \tag{6.16}$$

Now $J_- E(m) f$ converges strongly to $J_- f$ as $m \to \infty$. This establishes the existence of the spectral integral over the entire real line in (6.9) and completes the proof of the identity (6.9). The other equations (6.10) - (6.12) can be obtained in the same way. #

6-2 STATIONARY STATE SCATTERING THEORY

In this section we derive various time-independent equations satisfied by the wave operators and by the scattering operator. Some comments about the relation of these equations with the expressions found in the physics literature will be appended. The notation in this section is as follows : H and H_o denote the total and unperturbed Hamiltonians respectively, $\{E_\lambda\}$ the spectral family of H, $\{E^o_\lambda\}$ that of H_o, $V_t = \exp(-iHt)$, $U_t = \exp(-iH_o t)$, $R_z = (H-z)^{-1}$ and $R^o_z = (H_o-z)^{-1}$.

PROPOSITION 6.4 : (a) If (A1) is satisfied, then

$$\Omega_+ = \underset{\eta\to+0}{\text{s-lim}}\, i\eta \int_{\mathbb{R}} R_{\lambda-i\eta}\, dE^o_\lambda\, E_\infty(H_o), \tag{6.17}$$

$$\Omega_- = \underset{\eta\to+0}{\text{s-lim}}(-i\eta) \int_{\mathbb{R}} R_{\lambda+i\eta}\, dE^o_\lambda\, E_\infty(H_o). \tag{6.18}$$

(b) If (A1) and (A3) are satisfied, then

$$\Omega^*_\pm = \underset{\eta\to+0}{\text{s-lim}} \pm i\eta \int_{\mathbb{R}} R^o_{\lambda\mp i\eta}\, dE_\lambda E_\infty(H). \tag{6.19}$$

Proof : These equations follow easily from Propositions 4.17 and 6.3. For (6.17) for instance we use (4.53), (4.54) and (6.12) where we set $X_t = V_t$, $Z_t = U_t$ and $T = I$:

$$\Omega_+ = \underset{\eta\to+0}{\text{s-lim}}\, \Omega_{+\eta} E_\infty(H_o) = \text{s-lim}\, \eta\int_0^\infty e^{-\eta t} V^*_t U_t dt\, E_\infty(H_o)$$

$$= \underset{\eta\to+0}{\text{s-lim}}\, i\eta \int_{\mathbb{R}} R_{\lambda-i\eta}\, dE^o_\lambda\, E_\infty(H_o).\ \# \tag{6.20}$$

In order to deduce time-independent equations involving the interaction V, it is necessary to impose a relation between the domains of H and H_o. We assume in this section that H is a self-adjoint extension of the symmetric operator $\hat{H} = H_o + V$ defined on some dense set $D(\hat{H}) \subseteq D(H_o) \cap D(V)$. One condition which is verified in numerous applications and which will be discussed in detail in Section 8-1 is the following :

$$H^* = H = H_o + V \text{ with } D(H) = D(H_o) \subseteq D(V). \tag{6.21}$$

PROPOSITION 6.5 (Lippmann-Schwinger equations) :

(a) If (A1) and (6.21) are satisfied, then

$$\Omega_{\pm} = E_{\infty}(H_o) - \operatorname*{s-lim}_{\eta\to+0} \int_{R} R_{\lambda\mp i\eta} V\, dE^o_{\lambda}\, E_{\infty}(H_o), \tag{6.22}$$

and

$$\Omega_{\pm} = E_{\infty}(H_o) - \operatorname*{s-lim}_{\eta\to+0} \int_{R} R^o_{\lambda\mp i\eta} V\Omega_{\pm} dE^o_{\lambda}. \tag{6.23}$$

(b) If (A1), (A3) and (6.21) are satisfied, then

$$\Omega^*_{\pm} = E_{\infty}(H) + \operatorname*{s-lim}_{\eta\to+0} \int_{R} R^o_{\lambda\mp i\eta} V dE_{\lambda}\, E_{\infty}(H). \tag{6.24}$$

Proof : By (5.55) we have $\|R_{\lambda\pm i\eta}\| \leq \eta^{-1}$. Hence, if $\eta > 0$, one obtains from Lemma 6.1 that

$$\int R_{\lambda\pm i\eta}(H_o-\lambda) dE^o_{\lambda} = 0. \tag{6.25}$$

Now let $0 < n < \infty$ and define $E^o(n) = E^o_{(-n,n]}$. Since $E^o(n)\mathcal{H} \subseteq D(H_o)$, (6.21) implies that

$$HE^o(n) = H_oE^o(n) + VE^o(n). \tag{6.26}$$

We then get by using successively (6.20), (6.25) and (6.26):

$$\begin{aligned}
\Omega_{+\eta}E^o(n)E_{\infty}(H_o) &= i\eta \int_{R} R_{\lambda-i\eta}E^o(n) dE^o_{\lambda}\, E_{\infty}(H_o) \\
&= \int_{R} R_{\lambda-i\eta}(H_o-\lambda+i\eta)E^o(n) dE^o_{\lambda}\, E_{\infty}(H_o) \\
&= \int_{R} R_{\lambda-i\eta}\, [(H-\lambda+i\eta)-V]\, E^o(n) dE^o_{\lambda}\, E_{\infty}(H_o) \\
&= E^o(n)E_{\infty}(H_o) - \int_{-n}^{n} R_{\lambda-i\eta} V dE^o_{\lambda}\, E_{\infty}(H_o).
\end{aligned} \tag{6.27}$$

Since $E^o(n)E_{\infty}(H_o) \to E_{\infty}(H_o)$ strongly as $n \to \infty$, (6.27) implies the existence of the spectral integral over R in (6.22) and

$$\Omega_{+\eta}E_\infty(H_o) = E_\infty(H_o) - \int_{\mathbb{R}} R_{\lambda-i\eta}VdE_\lambda^o E_\infty(H_o). \qquad (6.28)$$

(6.22) follows from (6.28) since $\Omega_{+\eta}E_\infty(H_o) \to \Omega_+$ strongly as $\eta \to +0$.

To prove (6.23), one starts from

$$E_\infty(H_o) = \Omega_+^*\Omega_+ = \text{s-lim}_{\eta\to+0}\, \eta\int_0^\infty dt\, e^{-\eta t}U_t^*V_t\Omega_+ = \text{s-lim}_{\eta\to+0}\, i\eta\int_{\mathbb{R}} R_{\lambda-i\eta}^o dE_\lambda\Omega_+$$

and proceeds as above, and similarly one obtains (6.24) (Problem 6.1). #

Remark 6.6 : (a) The preceding proof shows that (6.22) remains true if the condition (6.21) is replaced by the weaker hypothesis that (6.26) holds for each $n < \infty$.

(b) Although V may be an unbounded operator, the spectral integrals appearing in Proposition 6.5 are defined everywhere in $\mathcal{H}$. This is a consequence of the hypothesis (6.21), as can be seen from the above proof.

We next deduce some time-independent equations for the scattering operator. These will be useful later on for expressing the differential scattering cross section in terms of the potential.

PROPOSITION 6.7 : (a) Suppose that (A1), (A3) and (6.21) hold. Then

$$S = E_\infty(H_o) + \text{s-lim}_{\eta\to+0}\int_{\mathbb{R}} (R_{\lambda-i\eta}^o - R_{\lambda+i\eta}^o)V\Omega_- dE_\lambda^o. \qquad (6.29)$$

(b) If (A1), (A2) and (6.21) hold, then

$$S = E_\infty(H_o) + \text{s-lim}_{\eta\to+0}\int_{\mathbb{R}} (R_{\lambda-i\eta}^o - R_{\lambda+i\eta}^o)\Omega_+^*VdE_\lambda^o E_\infty(H_o). \qquad (6.30)$$

Proof : We have from (4.15) and Proposition 4.1 that $S-E_\infty(H_o) = \Omega_+^*\Omega_- - \Omega_-^*\Omega_- = (\Omega_+^* - \Omega_-^*)\Omega_-$. Hence by Proposition 6.5(b) we obtain

$$S - E_\infty(H_o) = \text{s-}\lim_{\eta\to+0} \int_R (R^o_{\lambda-i\eta} - R^o_{\lambda+i\eta}) V dE_\lambda E_\infty(H)\Omega_- ,$$

and (6.29) follows from the fact that $E_\infty(H)\Omega_- = \Omega_-$ by (A3) and from the intertwining relation (5.70). The proof of (6.30) is left as an exercise (Problem 6.2). #

Another important expression for the S-operator will be given in Proposition 6.11 below. Its proof is considerably longer and will be prepared in the next Lemma.

LEMMA 6.8 : If (A1) and (A3) are satisfied, then

$$\Omega_+^*\Omega_- = \text{s-}\lim_{\eta\to+0}\ \text{s-}\lim_{\delta\to+0} (\eta+\delta)\delta \int_0^\infty dt\, e^{-\eta t} U_t^* \int_{-\infty}^{o} ds\, e^{\delta s} V_s^* U_{s+t} E_\infty(H_o), \tag{6.31}$$

$$\Omega_-^*\Omega_- = \text{s-}\lim_{\eta\to+0}\ \text{s-}\lim_{\delta\to+0} (\eta-\delta)\delta \int_{-\infty}^{o} dt\, e^{\eta t} U_t^* \int_{-\infty}^{o} ds\, e^{\delta s} V_s^* U_{s+t} E_\infty(H_o). \tag{6.32}$$

Proof : (i) One gets from Proposition 4.17(b), the fact that $E_\infty(H)\Omega_- = \Omega_- E_\infty(H_o)$, the intertwining relation $V_t\Omega_- = \Omega_- U_t$ and (4.3) that

$$\Omega_+^*\Omega_- = \text{s-}\lim_{\eta\to+0} \eta \int_0^\infty dt\, e^{-\eta t} U_t^* \Omega_- U_t E_\infty(H_o). \tag{6.33}$$

(ii) Let $\Omega_{-\delta}$ be defined by (4.55). We have as in (4.59) that

$$\|\Omega_{-\delta}\| \le 1. \tag{6.34}$$

Now let $f\in M_\infty(H_o)$ and $\eta > 0$. Then

$$\left\|\int_0^\infty dt\, e^{-\eta t} U_t^* (\Omega_- - \Omega_{-\delta}) U_t f\right\| \le \int_0^\infty dt\, e^{-\eta t} \|(\Omega_- - \Omega_{-\delta}) U_t f\|.$$

The integrand in the last integral converges to zero as $\delta\to+0$ for each t by (4.53) and is majorized uniformly in δ by the integrable function $\exp(-\eta t)\{\|\Omega_- U_t f\| + \|\Omega_{-\delta} U_t f\|\} \leq 2\exp(-\eta t)\|f\|$. Thus it follows from the Lebesgue dominated convergence theorem that for each $\eta > 0$

$$\underset{\delta\to+0}{\text{s-lim}} \int_0^\infty dt\, e^{-\eta t} U_t^* (\Omega_- - \Omega_{-\delta}) U_t E_\infty(H_o) = 0. \tag{6.35}$$

By inserting (6.35) into (6.33) and using (4.55), we obtain

$$\Omega_+^* \Omega_- = \underset{\eta\to+0}{\text{s-lim}}\ \underset{\delta\to+0}{\text{s-lim}}\, \eta\delta \int_0^\infty dt\, e^{-\eta t} U_t^* \int_{-\infty}^0 ds\, e^{\delta s} V_s^* U_{s+t} E_\infty(H_o).$$

(iii) To complete the proof of (6.31), it remains to show that

$$\underset{\delta\to+0}{\text{s-lim}}\, \delta\, I(\delta) \equiv \underset{\delta\to+0}{\text{s-lim}}\, \delta \int_0^\infty dt\, e^{-\eta t} U_t^* \Omega_{-\delta} U_t E_\infty(H_o) = 0.$$

This is true because the integral $I(\delta)$ exists for each $\delta > 0$ and converges strongly to a bounded operator as $\delta\to+0$ by (6.35).

The other equation (6.32) is proved in the same way. #

We shall also need some relations between the resolvents of H and H_o which we give in the following proposition and its corollary.

<u>PROPOSITION 6.9</u> <u>(Second Resolvent Equation)</u> : Let H_o be closed and V such that $D(H_o) \subseteq D(V)$.

(a) If H is a closed extension of $\hat{H} = H_o + V$ (with $D(H) = D(H_o)$), one has for each $z\in\rho(H)\cap\rho(H_o)$

$$(H-z)^{-1} = (H_o-z)^{-1} - (H-z)^{-1} V (H_o-z)^{-1}. \tag{6.36}$$

(b) If in addition $D(H) = D(H_o)$, then

$$(H-z)^{-1} = (H_o-z)^{-1} - (H_o-z)^{-1}V(H-z)^{-1}. \qquad (6.37)$$

Remark : Notice that VR_z^o and VR_z are everywhere defined operators under our assumptions on D(V) and D(H), so that (6.36) and (6.37) will hold when applied to any vector $f\varepsilon H$. We shall use this proposition only in the case where H_o and H are self-adjoint.

Proof : Using the definition of $\hat{H}$ and the fact that H and $\hat{H}$ coincide on $D(H_o)$, we get

$$V(H_o-z)^{-1} = \{(\hat{H}-z)-(H_o-z)\}(H_o-z)^{-1} = (H-z)(H_o-z)^{-1} - I.$$

(6.36) follows from this upon multiplying by $(H-z)^{-1}$ from the left. (6.37) is proved similarly. #

COROLLARY 6.10 : Assume the hypotheses of Proposition 6.9(b), and let $z\varepsilon\rho(H)\cap\rho(H_o)$, $\xi\varepsilon\rho(H)\cap\rho(H_o)$. Then

$$(\xi-z)R_\xi^oR_z = R_\xi^o - R_\xi^oVR_z - R_z. \qquad (6.38)$$

Proof : We apply successively the resolvent equations (6.37), (5.58) and again (6.37)

$$(\xi-z)R_\xi^oR_z = (\xi-z)R_\xi^oR_z^o(I-VR_z) = (R_\xi^o-R_z^o)(I-VR_z) = R_\xi^o-R_\xi^oVR_z-R_z. \#$$

PROPOSITION 6.11 : Suppose that (A1), (A3) and (6.21) hold. Then

$$S = E_\infty(H_o) + \underset{\eta\to+0}{\text{s-lim}}\ \underset{\delta\to+0}{\text{s-lim}} \int (R^o_{\lambda-i\eta}-R^o_{\lambda+i\eta})(V-VR_{\lambda+i\delta}V)dE^o_\lambda\, E_\infty(H_o). \qquad (6.39)$$

$$S^* = E_\infty(H_o) - \underset{\eta\to+0}{\text{s-lim}}\ \underset{\delta\to+0}{\text{s-lim}} \int (R^o_{\lambda-i\eta}-R^o_{\lambda+i\eta})(V-VR_{\lambda-i\delta}V)dE^o_\lambda E_\infty(H_o). \qquad (6.40)$$

Proof : First notice that, as a consequence of (6.21), Proposition 6.9 and its corollary may be applied. Also, the operators $VR_{\lambda+i\delta}$ and $V E^o(n) \equiv VE^o_{(-n,n]}$ are defined everywhere if $n < \infty$, so that the integrand of the spectral integral in (6.39) over any finite interval $(-n,n]$ is defined on every vector $f \in H$. It will be seen that the spectral integral over the entire real line exists as a consequence of Proposition 6.3.

We have $S - E_\infty(H_o) = \Omega_+^* \Omega_- - \Omega_-^* \Omega_-$. Upon inserting the expressions of Lemma 6.8 and transforming the double integral over time into a spectral integral by using Proposition 6.3, one obtains

$$S - E_\infty(H_o) = \text{s-}\lim_{\eta\to+0} \text{s-}\lim_{\delta\to+0} \delta \int_{\mathcal{R}} X_\lambda(\eta,\delta)\, dE_\lambda^o\, E_\infty(H_o) \tag{6.41}$$

with $X_\lambda(\eta,\delta) = (\eta+\delta) R^o_{\lambda-i\eta} R_{\lambda+i\delta} + (\eta-\delta) R^o_{\lambda+i\eta} R_{\lambda+i\delta}$.

By using (6.38), this expression may be transformed into

$$\begin{aligned} X_\lambda(\eta,\delta) &= i\{R^o_{\lambda-i\eta} - R^o_{\lambda-i\eta} VR_{\lambda+i\delta} - R_{\lambda+i\delta}\} \\ &\quad - i\{R^o_{\lambda+i\eta} - R^o_{\lambda+i\eta} VR_{\lambda+i\delta} - R_{\lambda+i\delta}\} \\ &= i(R^o_{\lambda-i\eta} - R^o_{\lambda+i\eta}) - i(R^o_{\lambda-i\eta} - R^o_{\lambda+i\eta}) VR_{\lambda+i\delta}. \end{aligned} \tag{6.42}$$

Now by Proposition 6.3

$$i \int R^o_{\lambda\mp i\eta}\, dE^o_\lambda = \int_0^{\pm\infty} dt\, e^{\mp\eta t} U_t^* U_t = \pm\eta^{-1}. \tag{6.43}$$

It follows that for any $\eta > 0$

$$\text{s-}\lim_{\delta\to+0} \delta \int_{\mathcal{R}} (R^o_{\lambda+i\eta} - R^o_{\lambda-i\eta})\, dE^o_\lambda = 0.$$

Therefore, by inserting (6.42) into (6.41) and then using Lemma 6.1, we arrive at

$$S-E_\infty(H_o) = \underset{\eta\to+0}{\text{s-lim}}\ \underset{\delta\to+0}{\text{s-lim}}(-i\delta)\int_R (R^o_{\lambda-i\eta}-R^o_{\lambda+i\eta})VR_{\lambda+i\delta}dE^o_\lambda E_\infty(H_o)$$

$$= \underset{\eta\to+0}{\text{s-lim}}\ \underset{\delta\to+0}{\text{s-lim}}\int_R (R^o_{\lambda-i\eta}-R^o_{\lambda+i\eta})VR_{\lambda+i\delta}(H-V-\lambda-i\delta)dE^o_\lambda E_\infty(H_o)$$

$$= \underset{\eta\to+0}{\text{s-lim}}\ \underset{\delta\to+0}{\text{s-lim}}\int_R (R^o_{\lambda-i\eta}-R^o_{\lambda+i\eta})(V-VR_{\lambda+i\delta}V)dE^o_\lambda E_\infty(H_o).$$

This proves (6.39). The proof of (6.40) is similar (Problem 6.2). #

We end this section with some comments on the preceding formulae.

Remark 6.12 : If, as is often the case, $M_\infty(H_o) = H_{ac}(H_o)$, then all the propositions of this section remain true if one replaces the condition (A3) by $F_+H = F_-H = H_{ac}(H)$ and $E_\infty(H)$ by $E_{ac}(H)$ in the time-independent equations.

The Lippmann-Schwinger equations (6.22) furnish expressions for $\Omega_\pm$ in terms of the interaction V, the resolvent of the total Hamiltonian H and the spectral family of the unperturbed Hamiltonian H_o. In practical situations, e.g. in potential scattering, all quantities referring to H_o can be written down explicitly, and the interaction V is also given. On the other hand quantities pertaining to H are usually not known explicitly. It follows that (6.22) will be of practical use only insofar as one may replace R_z by some approximate expression.

The second type of Lippmann-Schwinger equations, viz. (6.23), is an integral equation for $\Omega_\pm$ involving only the interaction and quantities pertaining to the unperturbed evolution. As an integral equation it may in principle be solved by iteration, setting for instance for the first approximation $\Omega_{\pm} = E_{\infty}(H_0)$ in the integral on the right-hand side of (6.23) (this is called the Born approximation for $\Omega_{\pm}$). The iteration leads to an expression for $\Omega_{\pm}$ as a power series (the Born series for $\Omega_{\pm}$) in the coupling constant γ, if V is replaced by γV. The convergence of this series under suitable assumptions has to be studied separately. The Born series for related operators is discussed in Section 12-1.

The operators $V\Omega_-$ and

$$T_z \equiv V - VR_zV \tag{6.44}$$

appearing in the expressions (6.29) and (6.39) for the S-operator are often used by physicists to calculate the scattering amplitude, which will be defined in Section 7-3. Let us analyze in some detail equation (6.29). For the sake of simplicity we set $E_{\infty}(H_0) = I$ and assume that H_0 is absolutely continuous and has constant spectral multiplicity, i.e. that its spectral representation is given by $G = L^2(\Lambda, H_0)$ with $\Lambda \subseteq \sigma(H_0)$.

We have seen in Section 5-7 that the operator $R \equiv S-I$ is decomposable, i.e. $R = \{R(\lambda)\}$. It will be shown in Section 7-3 that $R(\lambda)$ determines the scattering cross section at energy λ. (6.29) expresses S-I in terms of $V\Omega_-$. Now $V\Omega_-$ is in general not decomposable in the spectral representation of H_0. However (6.29) signifies that $R(\lambda)$ should be proportional to the on-shell part*) of $V\Omega_-$, if the latter makes

---

*) It may happen that a non-decomposable operator B can be written as an integral operator in $L^2(\Lambda, H_0)$, i.e. that there exists a measurable family $\{B(\lambda,\lambda')\}$ of operators in H_0 such

sense. Formally this can be seen as follows. Assume that $V\Omega_-$ is an integral operator in $L^2(\Lambda, \mathcal{H}_o)$, i.e. that there exists a family $\{(V\Omega_-)(\lambda,\mu)\}$, $\lambda,\mu\in\Lambda$, of operators in $\mathcal{H}_o$ such that $(V\Omega_- f)_\lambda = \int d\mu (V\Omega_-)(\lambda,\mu) f_\mu$. Now $(R^o_{\mu+i\eta} f - R^o_{\mu-i\eta} f)_\lambda =$ $2i\eta[(\mu-\lambda)^2+\eta^2]^{-1} f_\lambda$, so that by writing (6.29) in the spectral representation of H_o we find

$$\begin{aligned}(f,Rg) &= \int d\lambda (f_\lambda, R(\lambda) g_\lambda)_o \\ &= \lim_{\eta\to+0} \iint d\lambda d\mu \frac{2i\eta}{(\mu-\lambda)^2+\eta^2} (f_\lambda, (V\Omega_-)(\lambda,\mu) g_\mu)_o \\ &= 2\pi i \int d\lambda (f_\lambda, (V\Omega_-)(\lambda,\lambda) g_\lambda)_o . \end{aligned} \tag{6.45}$$

The last equality follows formally from the fact that $\lim \eta[(\mu-\lambda)^2+\eta^2]^{-1} = \pi\delta(\mu-\lambda)$ as $\eta\to 0$. Thus, provided that $V\Omega_-$ can be suitably written in the above form and that the last identity in (6.45) can be rigourously justified, we have

$$R(\lambda) = 2\pi i (V\Omega_-)(\lambda,\lambda). \tag{6.46}$$

A slightly different way of interpreting (6.29) is as follows. By Propositions 4.7 and 5.9 one has $SE^o_\Delta = E^o_\Delta S$ for each interval Δ. Since E^o_Δ is a projection, this implies also that $SE^o_\Delta = E^o_\Delta S E^o_\Delta$. We thus obtain from (6.29)

$$(S-I)E^o_\Delta = E^o_\Delta (S-I) E^o_\Delta = \text{s-}\lim_{\eta\to+0} \int_\Delta (R^o_{\lambda-i\eta} - R^o_{\lambda+i\eta}) E^o_\Delta V\Omega_- E^o_\Delta dE^o_\lambda, \tag{6.47}$$

where we have taken the projection E^o_Δ appearing on the left inside the strong limit as $\eta \to +0$ and inside the strong limit defining the spectral integral, which is justified by Corol-

---

*) Ctd.

that $(Bf)_\lambda = \int d\lambda' B(\lambda,\lambda') f_{\lambda'}$. Then the on-shell part of B at energy λ is formally defined to be $B(\lambda,\lambda)$. This makes sense only under suitable assumptions on B (cf. Birman and Entina [1]), since the set $\{(\lambda,\lambda') | \lambda=\lambda'\}$ has measure zero in $\Lambda\times\Lambda$.

lary 2.19. Equation (6.47) means that the action of the operator S-I on states having support in an arbitrarily small interval Δ in the spectral representation of H_o is completely determined by the part $E^o_\Delta V\Omega_- E^o_\Delta$ of $V\Omega_-$ in the subspace $E^o_\Delta \mathcal{H}$ corresponding to the energy interval Δ. If one can somehow perform the limit where Δ shrinks to a single point, one will again find that $R(\lambda)$ is proportional to the on-shell part of $V\Omega_-$. Of course this limiting procedure cannot be carried through without further assumptions. One such assumption is that $|V|^{\frac{1}{2}}(H_o+i)^{-1}$ and $|V|^{\frac{1}{2}}(H+i)^{-1}$ be Hilbert-Schmidt; see Birman and Entina [1] for details.

In a similar way (6.39) means that $R(\lambda)$ is proportional to the on-shell part of $T_{\lambda+io} = V-VR_{\lambda+io}V$, i.e. $R(\lambda) = 2\pi i\ T_{\lambda+io}(\lambda,\lambda)$, provided that the limit of $T_{\lambda+i\eta}$ as $\eta\to+0$ is suitably definable and that its on-shell part makes sense. We shall refrain from going into rigorous derivations of these properties on an abstract level. The case of potential scattering will be treated in Chapter 10 and separable potentials in Section 8-3.

6-3 NOTES AND SUPPLEMENTARY MATERIAL

A. We have derived the stationary equations of Section 6-2 as a consequence of the asymptotic condition which is the basic physical principle of scattering theory. From the purely mathematical point of view it is sometimes useful to invert this procedure. Thus in numerous mathematical papers one proves the existence of the time-independent wave operators, e.g. eqns. (6.17)-(6.19). In this form these operators may exist under more general conditions than the time-dependent ones; one then shows under some additional assumptions that the time-dependent limits also exist and are equal to the stationary ones. As already pointed out, the stationary equations are useful in proving asymptotic completeness. Some relevant references are Birman and Entina [1], Kato and Kuroda [1] and Kuroda [5]. It should be remarked that these authors work with the spectral representations rather than with the spectral families of H and H_o.

B. The Lippmann-Schwinger equations (Proposition 6.5) can also be deduced from Proposition 6.4 by using the second resolvent equation. Other integral equations involving the interaction V may be obtained similarly by inserting into (6.17) and (6.18) any expression for R_z involving the interaction V. In this way it is possible to obtain Lippmann-Schwinger type equations for scattering systems for which (6.21) does not hold. An example is potentials of the Rollnik class in potential scattering [SM].

C. The following theorem gives sufficient conditions for the existence of spectral integrals. For the proof the reader is referred to Birman and Solomjak [1].

PROPOSITION 6.13 : Let [a,b] be a finite interval and assume that $\{A_\lambda\}$ is Hölder continuous in operator norm with Hölder index $\alpha > \frac{1}{2}$, i.e.

$$||A_\lambda - A_\mu|| \leq M|\lambda-\mu|^\alpha \text{ for all } \lambda,\mu\in[a,b]. \qquad (6.48)$$

Then $\int_a^b A_\lambda dE_\lambda$ exists as the uniform limit of the corresponding Riemann-Stieltjes sums.

One occasionally has to consider double spectral integrals. Since $E_\Delta E_{\Delta'} = 0$ if $\Delta \cap \Delta' = \emptyset$, one expects that such integrals can be reduced to single ones. The following result is easy to prove (Problem 6.4) :

PROPOSITION 6.14 : Assume that $\int_R A_\lambda dE_\lambda$ and $\int_R dE_\lambda B_\lambda$ exist. Then $\int_R A_\lambda dE_\lambda B_\lambda$ exists, and

$$\int_R A_\lambda dE_\lambda \int_R dE_\mu B_\mu = \int_R A_\lambda dE_\lambda B_\lambda . \qquad (6.49)$$

Similarly one can prove that $\int A_\lambda (\int B_\mu dE_\mu) dE_\lambda = \int A_\lambda B_\lambda dE_\lambda$ under more stringent conditions (e.g. conditions of the type (6.48)). For this and other information about spectral integrals in scattering theory the reader may consult Amrein, Georgescu and Jauch [1], Birman and Solomjak [1], Pearson [1] and Thomas [3]. Other types of integral representations are also given in Galindo [1].

PROBLEMS

6.1 : Prove the equations (6.23) and (6.24).

6.2 : Prove the equations (6.30) and (6.40).

6.3 : (a) Let $D \in \mathcal{B}(\mathcal{H})$ and assume the conditions of Proposition 6.3. Show that the integrals below exist and that one has

$$i \int_{-\infty}^{0} dt\, e^{\eta t} DX_t^* TZ_t = \int_R DR_{\lambda+i\eta}(A) T dE_\lambda = D \int_R R_{\lambda+i\eta}(A) T dE_\lambda . \qquad (6.50)$$

(b) Assume (A1), (6.21), $V(H_o+i)^{-1} \in \mathcal{B}(\mathcal{H})$ and $V(H+i)^{-1} \in \mathcal{B}(\mathcal{H})$. (The last two requirements in fact follow from (6.21), see Section 8-1). Show that for $f \in D(H_o)$

$$V\Omega_- f = \text{s-}\lim_{\eta \to +0} \int_R (V - VR_{\lambda+i\eta} V) dE_\lambda^o E_\infty(H_o) f . \qquad (6.51)$$

(Hint : Use $D = V(H+i)^{-1}$, $T = I + V(H_o+i)^{-1}$ in (6.50) to show that $\int VR_{\lambda+i\eta} dE_\lambda^o f$ exists. Then use the second identity in (6.50) to prove that $-i\eta \int VR_{\lambda+i\eta} dE_\lambda^o E_\infty(H_o) f$ converges strongly to $V\Omega_{-}f$, and show that this sequence also converges to the right-hand side of (6.51).)

6.4 : Prove Proposition 6.14.

6.5 : Prove (6.10) - (6.12).

CHAPTER 7 : POSITION IN SCATTERING THEORY

The scattering theory that we have developed so far is very general in the sense that it presupposes only the knowledge of two one-parameter unitary groups and the existence of certain limits in a Hilbert space. No spectral properties of the Hamiltonians have been used, and the position operator, which is one of the basic observables in quantum mechanics, has played no role in our arguments. In this Chapter we introduce other observables besides the total and the unperturbed energy and develop various physical features of scattering theory within this more specific structure. As pointed out in Section 3-2, the two basic observables are position and momentum. Position is essential since physical events take place in space, and momentum appears naturally because in many instances the unperturbed Hamiltonian is a function of the momentum operators.

In Section 7-1 we study the behaviour of the position probability measure of states under a given evolution group, which leads in a natural way to the definition of the bound states and the scattering states of its infinitesimal generator H. We shall see how these two sets of states may be related to spectral subspaces of H. In Section 7-2 we introduce the notion of time delay and show for a simple example how it can be related to the scattering operator. Section 7-3 contains the calculation of the differential cross section in terms of the scattering operator under certain additional assumptions on the scattering system. In Section 7-4 we take up some related questions such as the optical theorem, conditions for the finiteness of the total cross section and estimates of its high energy behaviour. Finally in Section 7-5 we consider the scattering of two particles by a translation invariant local interaction and show how it can be reduced to potential scattering by removing the movement of the center of mass. Some topics of this Chapter will be analyzed in the specific context of potential scattering.

7-1 BOUND STATES AND SCATTERING STATES

In the discussion of the asymptotic condition in Section 4-1 we had to assume that to each Hamiltonian H one may associate a subspace $M_\infty(H)$ of scattering states. The purpose of the present section is to give a specific definition of this subspace and to establish some of its properties.

Roughly speaking, a physical system is in a scattering

state if at large negative and at large positive times the system splits up into at least two subsystems which are localized far from each other. In the same spirit one may visualize a bound state as being such that all subsystems stay close together at all times. This picture involves, in addition to the one-parameter unitary group describing the time-evolution of the system, the notion of subsystems and that of the relative position of subsystems. A simple scattering system is usually composed of no more than two subsystems which are elementary in the sense that they cannot be further split into smaller parts (e.g. two elementary particles). Furthermore the case of two subsystems with a translation invariant total Hamiltonian may be reduced to that of a single particle moving in an external field (e.g. a potential) by removing the center-of-mass motion (cf. Sections 7-5 and 7-6 for details). We shall therefore illustrate the considerations of this section by potential scattering and specify what mathematical developments are valid in a more general setting.

In potential scattering the Hilbert space is $L^2(\mathcal{R}^3)$, and according to Example 3.4 the variable $\underline{x} \in \mathcal{R}^3$ is interpreted as the position variable of the particle (or in the case of a pair of particles as the relative position between them). The relevant quantity for our present purposes is the position probability measure, or equivalently the family of projections $\{F_\Delta\}$, where F_Δ is the multiplication operator in $L^2(\mathcal{R}^3)$ by the characteristic function χ_Δ of the Borel set $\Delta \subseteq \mathcal{R}^3$. More particularly we shall work with the projections F_r $(r \geq 0)$ obtained by taking Δ to be the ball

$S_r = \{\underline{x} \mid |\underline{x}| \leq r\}$:

$$(F_r f)(\underline{x}) = \chi_r(\underline{x}) f(\underline{x}), \tag{7.1}$$

where χ_r is the characteristic function of S_r. Thus, if $\|f\| = 1$, the quantity $\|F_r f\|^2$ represents the probability that the particle in the state f is localized in the ball S_r. The family $\{F_r\}$ has the following properties :

$$\left.\begin{aligned} F_r &= F_r^* = F_r^2 , \\ \underset{r\to\infty}{\text{s-lim}}\, F_r &= I. \end{aligned}\right\} \tag{7.2}$$

Let H be a Hamiltonian operator and $V_t = \exp(-iHt)$. A vector f will be called a <u>scattering state of H</u> if it disappears from every finite ball S_r as $t\to+\infty$ and as $t\to-\infty$, i.e. if for every $r < \infty$, $\|F_r V_t f\|^2/\|f\|^2$ converges to zero as $|t|\to\infty$. Hence we define $M_\infty(H)$ by

$$M_\infty(H) = \{f\varepsilon \mathcal{H} \mid \underset{|t|\to\infty}{\text{s-lim}}\, F_r \exp(-iHt)f = \theta \text{ for each } r < \infty\}. \tag{7.3}$$

Similarly one may define the bound states of H. Such a state is characterized by the property that the probability of finding the particle outside some suitably chosen finite ball remains arbitrarily small at <u>all</u> times. Hence f is a <u>bound state of H</u> if, given any $\eta > 0$, there exists $r < \infty$ such that $\|(I-F_r)V_t f\|^2/\|f\|^2 < \eta$ for all $t\varepsilon R$. Hence we may define a subset $M_0(H)$ of $\mathcal{H}$ by

$$M_0(H) = \{f\varepsilon\mathcal{H} \mid \lim_{r\to\infty} \sup_{t\varepsilon R} \|(I-F_r)\exp(-iHt)f\|^2 = 0\}. \tag{7.4}$$

We proceed to derive some simple consequences of these definitions. In Propositions 7.1 through 7.4 the specific form of F_r and of H is irrelevant; it suffices to impose (7.2).

PROPOSITION 7.1 : $M_\infty(H)$ and $M_o(H)$ are (closed) subspaces of H which are mutually orthogonal.

Proof : If $f,g \varepsilon H$, one has

$$||f+g||^2 \leq ||f+g||^2 + ||f-g||^2 = 2||f||^2 + 2||g||^2. \qquad (7.5)$$

Consequently

$$||F_rV_t(f_1+\alpha f_2)||^2 \leq 2||F_rV_tf_1||^2 + 2|\alpha|^2 ||F_rV_tf_2||^2.$$

Hence, if $f_1,f_2 \varepsilon M_\infty(H)$, then $||F_rV_t(f_1+\alpha f_2)||^2 \to 0$ as $t \to \pm\infty$, which shows that $M_\infty(H)$ is a linear manifold.

Now suppose $f_n \varepsilon M_\infty(H)$ and s-lim $f_n = f$ as $n \to \infty$. Then by (7.5) and since $||F_r|| = ||V_t|| = 1$:

$$||F_rV_tf||^2 \leq 2||F_rV_t(f-f_n)||^2 + 2||F_rV_tf_n||^2$$
$$\leq 2||f-f_n||^2 + 2||F_rV_tf_n||^2. \qquad (7.6)$$

Given $\eta > 0$, one may choose n sufficiently large that $||f-f_n||^2 < \eta/4$. Since $f_n \varepsilon M_\infty(H)$, $||F_rV_tf_n||^2 < \eta/4$ provided that $|t| > T$ for some finite T. This shows that $||F_rV_tf||^2 < \eta$ provided that $|t| > T$, i.e. $f \varepsilon M_\infty(H)$. Hence $M_\infty(H)$ is closed.

The proof that $M_o(H)$ is a subspace is similar and left as an exercise (Problem 7.1). The orthogonality of these two subspaces is obtained as follows :

$$|(f,g)|^2 = |(V_tf,V_tg)|^2 = |(F_rV_tf,V_tg) + (V_tf,(I-F_r)V_tg)|^2$$
$$\leq 2|(F_rV_tf,V_tg)|^2 + 2|(V_tf,(I-F_r)V_tg)|^2$$
$$\leq 2||g||^2 ||F_rV_tf||^2 + 2||f||^2 ||(I-F_r)V_tg||^2.$$

Let $\eta > 0$. If $g \varepsilon M_o(H)$, one may choose $r < \infty$ such that the second term is less than $\eta/2$ for all $t \varepsilon R$. If $f \varepsilon M_\infty(H)$, there exists $t\in\mathbb{R}$ such that the first term is less than $\eta/2$. Hence $|(f,g)|^2 < \eta$ for any $\eta > 0$, i.e. $(f,g) = 0$. #

PROPOSITION 7.2 : $\mathcal{H}_p(H) \subseteq \mathcal{M}_o(H)$ and $\mathcal{M}_\infty(H) \subseteq \mathcal{H}_c(H)$.

Proof : Suppose f is an eigenvector of H, i.e. $Hf = \mu f$ for some $\mu\in\mathbb{R}$. Then

$$\|(I-F_r)V_t f\|^2 = \|(I-F_r)e^{-i\mu t}f\|^2 = \|(I-F_r)f\|^2.$$

This quantity is independent of t and converges to zero as $r\to\infty$ by (7.2). Hence $f\in\mathcal{M}_o(H)$. The first inclusion now follows because $\mathcal{H}_p(H)$ is defined to be the subspace spanned by all the eigenvectors of H and because $\mathcal{M}_o(H)$ is also a closed subspace. It implies that $\mathcal{M}_o(H)^\perp \subseteq \mathcal{H}_p(H)^\perp = \mathcal{H}_c(H)$, cf. (5.65). Since $\mathcal{M}_\infty(H) \subseteq \mathcal{M}_o(H)^\perp$ by Proposition 7.1, the second inclusion is also established. #

One sees from Proposition 7.2 that eigenvectors and their linear combinations are always bound states and that scattering states are always associated with the continuous spectrum of the Hamiltonian. The most usual situation is that where one has equality in both statements of Proposition 7.2, i.e. where each bound state is a linear combination of eigenvectors and where the set of scattering states is identical with the subspace of continuity of H. The next proposition gives sufficient conditions for this to be true. It should be stressed though that in the general case $\mathcal{M}_o(H)$ and $\mathcal{M}_\infty(H)$ need not be orthogonal complements of each other. Also there may be bound states which lie in $\mathcal{H}_c(H)$ (i.e. $\mathcal{M}_o(H)\cap\mathcal{H}_c(H) \neq \{\theta\}$, cf. Section 4-6). Throughout the remainder of this section we shall tacitly assume that the continuous part of H is absolutely continuous, i.e. that $H_{sc}(H) = \{\theta\}$. The arguments can easily be adapted to the case where this assumption is not made, cf. Section 7-6.

We first give a lemma which shows one aspect of the importance of compact operators in scattering theory.

LEMMA 7.3 : Let $H = H^*$. If A is a compact operator and $f\varepsilon H_{ac}(H)$, then s-lim $A\exp(-iHt)f = \theta$ as $t \to \pm\infty$.

Proof : $\{\exp(-iHt)f\}$ converges weakly to zero as $t \to \pm\infty$ by Lemma 5.20, and the result follows from Proposition 2.23. #

PROPOSITION 7.4 : For $n > 0$, let $E(n) = E_n - E_{-n}$ be the spectral projection of H associated with the interval $(-n,n]$. Suppose that for each $r < \infty$ and each $n < \infty$, $F_r E(n)$ is a compact operator. Then $M_0(H) = H_p(H)$ and $M_\infty(H) = H_c(H)$ (H_c assumed absolutely continuous).

Proof : Let $f\varepsilon H_c(H) = H_{ac}(H)$ and $\eta > 0$. We have as in (7.6)

$$\|F_r V_t f\|^2 \leq 2\,\|[I-E(n)]f\|^2 + 2\|F_r E(n)\, V_t f\|^2 .$$

Since s-lim $E(k) = I$ as $k \to \infty$, we may choose a number $n < \infty$ such that $\|[I-E(n)]f\|^2 < \eta/4$. By Lemma 7.3 there exists $T < \infty$ such that $\|F_r E(n)\, V_t f\|^2 < \eta/4$ provided that $|t| > T$. Hence $\|F_r V_t f\|^2 < \eta$ for all t verifying $|t| > T$, which means that $f\varepsilon M_\infty(H)$.

We have thus shown that $H_c(H) \subseteq M_\infty(H)$ under the hypotheses of the proposition, and combining with Proposition 7.2 we arrive at $M_\infty(H) = H_c(H)$. Since $M_0(H)$ is orthogonal to $M_\infty(H)$, we must have $M_0(H) \subseteq H_p(H)$, and again the equality sign must hold by Proposition 7.2. #

COROLLARY 7.5 : Suppose that $F_r(H-z)^{-\alpha}$ is compact for some $z\varepsilon\rho(H)$, some $\alpha > 0$ and all $r < \infty$. Then the conclusions of Proposition 7.4 hold.

Proof : This follows from the identity $F_rE(n) = F_r(H-z)^{-\alpha}(H-z)^{\alpha}E(n)$, the fact that $(H-z)^{\alpha}E(n)\varepsilon\mathcal{B}(\mathcal{H})$ if $n < \infty$ and Proposition 2.22(b). #

We now return to potential scattering. It follows directly from Remark 3.15 that all states in $L^2(\mathcal{R}^3)$ are evanescent under the free evolution, i.e. $\mathcal{M}_\infty(K_0) = L^2(\mathcal{R}^3)$. We shall give an alternative proof below based on Corollary 7.5.

We use the notation of Example 3.4. We shall often encounter operators of the form $\psi(\underline{Q})\phi(\underline{P})$, where $\psi(\underline{Q})$ is a function of the position operators Q_1,Q_2,Q_3, i.e. a multiplication operator in $L^2(\mathcal{R}^3)$ by a function $\psi(\underline{x})$, and $\phi(\underline{P})$ is a function of the momentum operators P_1,P_2,P_3, i.e. a multiplication operator in the space of the Fourier transforms of the vectors of $L^2(\mathcal{R}^3)$ by a function $\phi(\underline{k})$. A particular operator of the form $\phi(\underline{P})$ is the resolvent of the free Hamiltonian K_0; as is easily checked, it is given by

$$[\mathcal{F}(K_0-z)^{-1}f](\underline{k}) = (\underline{k}^2-z)^{-1}\tilde{f}(\underline{k}). \qquad (7.7)$$

If the function ψ is not essentially bounded, the corresponding operator $\psi(\underline{Q})$ is unbounded. The same is true for ϕ and $\phi(\underline{P})$. It turns out that under certain integrability assumptions on ψ and ϕ, the product $\psi(\underline{Q})\phi(\underline{P})$ is bounded on its domain of definition, so that its closure is in $\mathcal{B}(\mathcal{H})$ and sometimes even compact. With some abuse of notation, we shall denote this closure simply by $\psi(\underline{Q})\phi(\underline{P})$.

LEMMA 7.6 : Suppose that the functions ψ and ϕ are in $L^2(\mathcal{R}^3)$. Then $\psi(\underline{Q})\phi(\underline{P})$ and $\phi(\underline{P})\psi(\underline{Q})$ are compact (in fact Hilbert-Schmidt) operators, and

$$\|\psi(\underline{Q})\phi(\underline{P})\| \leq (2\pi)^{-3/2}\|\psi\|_2\|\phi\|_2. \tag{7.8}$$

Proof : (i) Let $f \in L^2(\mathcal{R}^3)$. Then by the Schwarz inequality $\|\phi(\underline{k})\tilde{f}(\underline{k})\|_1 \leq \|\phi\|_2 \|f\| < \infty$, i.e. $\phi(\underline{k})\tilde{f}(\underline{k}) \in L^1(\mathcal{R}^3)$. Fix $\underline{x} \in \mathcal{R}^3$. Then

$$[\phi(\underline{P})f](\underline{x}) = (2\pi)^{-3/2} \int d^3k e^{i\underline{k}\cdot\underline{x}} \phi(\underline{k})\tilde{f}(\underline{k}).$$

As both $\bar{\phi}$ and f are in $L^2(\mathcal{R}^3)$, the last integral may be viewed as a scalar product which, by the unitarity of F, is equal to the scalar product between the inverse Fourier transforms of $\exp(-i\underline{k}\cdot\underline{x})\bar{\phi}(\underline{k})$ and f in the position representation. Thus, since

$$(2\pi)^{-3/2} \int e^{i\underline{k}\cdot\underline{y}} e^{-i\underline{k}\cdot\underline{x}} \bar{\phi}(\underline{k})d^3k = \tilde{\bar{\phi}}(\underline{y}-\underline{x}) ,$$

we have $$[\phi(\underline{P})f](\underline{x}) = (2\pi)^{-3/2} \int d^3y\, \tilde{\phi}(\underline{y}-\underline{x})f(\underline{y}). \tag{7.9}$$

This shows that $\phi(\underline{P})$ is an integral operator in $L^2(\mathcal{R}^3)$ with kernel $(2\pi)^{-3/2}\tilde{\phi}(\underline{y}-\underline{x})$.

(ii) It follows that $\psi(\underline{Q})\phi(\underline{P})$ is an integral operator with kernel $(2\pi)^{-3/2}\psi(\underline{x})\tilde{\phi}(\underline{y}-\underline{x})$. Its Hilbert-Schmidt norm is finite, since

$$\|\psi(\underline{Q})\phi(\underline{P})\|_{HS}^2 = (2\pi)^{-3} \int d^3x\, d^3y |\psi(\underline{x})|^2 |\tilde{\phi}(\underline{y}-\underline{x})|^2$$

$$= (2\pi)^{-3} \int d^3x\, d^3z |\psi(\underline{x})|^2 |\tilde{\phi}(\underline{z})|^2 = (2\pi)^{-3}\|\psi\|_2^2 \|\phi\|_2^2. \tag{7.10}$$

The estimate (7.8) now follows from (2.62).

(iii) The preceding results imply that $\bar{\psi}(\underline{Q})\bar{\phi}(\underline{P})\varepsilon\mathcal{B}_2$. By applying Proposition 2.32(a), we see that $\phi(\underline{P})\psi(\underline{Q}) = [\bar{\psi}(\underline{Q})\bar{\phi}(\underline{P})]^*\varepsilon\mathcal{B}_2$. #

<u>PROPOSITION 7.7</u> : If $W\varepsilon L^2(\mathcal{R}^3)$ and $z\varepsilon\rho(K_0)$, then $W(\underline{Q})(K_0-z)^{-1}$ and $(K_0-z)^{-1}W(\underline{Q})$ are Hilbert-Schmidt operators, and

$$||W(\underline{Q})(K_0-z)^{-1}|| \leq 2^{-3/2}\pi^{-\frac{1}{2}}(\operatorname{Im} z^{\frac{1}{2}})^{-\frac{1}{2}}||W||_2 \tag{7.11}$$

with $\operatorname{Im} z^{\frac{1}{2}} \geq 0$.

<u>Proof</u> : Since the function $\phi(\underline{k}) = (\underline{k}^2-z)^{-1}$ is square-integrable, the first two assertions follow from Lemma 7.6. The bound (7.11) is obtained from (7.8) by calculating the following integral

$$||\phi||_2^2 = 4\pi \int_0^\infty k^2 dk|k^2-z|^{-2} = \pi^2(\operatorname{Im} z^{\frac{1}{2}})^{-1}. \ \#$$

<u>COROLLARY 7.8</u> : Let $r < \infty$ and $z\varepsilon\rho(K_0)$. Then

$$F_r(K_0-z)^{-1}\varepsilon\mathcal{B}_0 \tag{7.12}$$

and

$$\mathcal{M}_\infty(K_0) = \mathcal{H}_{ac}(K_0) = L^2(\mathcal{R}^3). \tag{7.13}$$

<u>Proof</u> : The characteristic function χ_r is in $L^2(\mathcal{R}^3)$, whence (7.12). The first identity in (7.13) now follows from Corollary 7.5, and the second one was established in Problem 5.7 and in (5.96). #

We shall now specify the scattering states of the total Hamiltonian $H = K_0 + V$ for a certain class of potentials. The proof is based on the relation between the resolvents of H and K_0 proved in Proposition 6.9. We use the notation $R_z = (H-z)^{-1}$ and $R_z^o = (K_0-z)^{-1}$.

PROPOSITION 7.9 : Suppose that the potential $V(\underline{x})$ may be written as $V(\underline{x}) = V_1(\underline{x}) + V_2(\underline{x})$ with $V_1 \varepsilon L^2(\mathcal{R}^3)$ and V_2 essentially bounded, and let $H = K_o + V$. Then $\mathcal{M}_\infty(H) = \mathcal{H}_c(H)$ and $\mathcal{M}_o(H) = \mathcal{H}_p(H)$ (H_c assumed absolutely continuous).

Remark : It will be shown in Section 8-1 that $H = K_o + V$ is self-adjoint on $D(K_o)$.

Proof : Since $V_2(\underline{x})$ is essentially bounded, the multiplication operator V_2 is bounded and defined everywhere by Proposition 2.16. Let $f \varepsilon D(K_o)$. By Proposition 3.10 there exists a number c such that $|f(\underline{x})| \leq c$ almost everywhere. Hence $\|V_1 f\|^2 \leq c^2 \|V_1\|_2^2$, which is finite since $V_1(\cdot) \varepsilon L^2(\mathcal{R}^3)$. Thus $D(K_o) \subseteq D(V_1) = D(V)$.

We may now use (6.36) to write

$$R_z F_r = R_z^o F_r - R_z V_1 R_z^o F_r - R_z V_2 R_z^o F_r . \tag{7.14}$$

By Lemma 7.6 we have $R_z^o F_r \varepsilon \mathcal{B}_2$ and $V_1 R_z^o \varepsilon \mathcal{B}_2$. Since all the other operators appearing on the right-hand side of (7.14) are in $\mathcal{B}(\mathcal{H})$, $R_z F_r \varepsilon \mathcal{B}_2$ by Proposition 2.32. Hence $F_r(H-\bar{z})^{-1} = (R_z F_r)^* \varepsilon \mathcal{B}_2$, and the result follows from Corollary 7.5. #

Example 7.10 : Proposition 7.9 applies in particular to Coulomb potentials $V(\underline{x}) = \gamma|\underline{x}|^{-1}, \gamma \varepsilon \mathcal{R}$. It suffices to set $V_1(\underline{x}) = V(\underline{x})\chi_r(\underline{x})$ and $V_2(\underline{x}) = V(\underline{x}) - V_1(\underline{x})$ with $0 < r < \infty$.

Some additional comments about the characterization of scattering states may be found in Section 7-6.

7-2 TIME DELAY

Time delay is an important theoretical concept although it is usually such a small quantity that it is not directly experimentally measurable. Roughly speaking, the time delay is the excess time that the scattered particle spends in the scattering region when compared to a free particle subject to the same initial conditions. Thus a positive time delay means that a particle takes more time to pass through the region where it is influenced by the potential than a particle which is allowed to propagate freely through this same region. A negative time delay means that on the average the scattered particles are accelerated by the effects of the potential.

In the present section we show how the concept of time delay may be mathematically formulated in terms of the position observable. Since time delay pertains to scattering, it ought to be expressible in terms of the scattering operator. This relation will be established in the following form : there exists a self-adjoint operator ΔT commuting with the free evolution group $\{U_t\}$ which is constructed from the scattering operator and whose expectation values give the correct time delay.

To define the time delay, we fix a ball S_r of sufficiently large radius. For $f \varepsilon \mathcal{H}$, we consider the integral

$$I_r(f,H) = \int_{-\infty}^{\infty} \| F_r \exp(-iHt) f \|^2 \, dt. \qquad (7.15)$$

To find its interpretation, we look at an approximating sum $\sum_{k=1}^{n} \| F_r \exp(-iu_k H) f \|^2 (t_k - t_{k-1})$, where $\{t_k; u_k\}$ is a partition of the interval $[-\tau,\tau]$. This is a weighted sum of small time intervals, the weight being the probability that in the respective time interval the state f is localized in the ball S_r (provided that $\|f\| = 1$). The sum may therefore be interpreted as the total time spent in the ball S_r by the particle in the state f, if the time evolution is governed by $V_t = \exp(-iHt)$. This will be our interpretation of (7.15).

To obtain the time delay corresponding to an initial state f, we have to consider the difference between $I_r(\Omega_- f,H)$ and $I_r(f,H_0)$, i.e. between the time spent in S_r by the scattering state $\Omega_- f$ evolving under the total evolution and the initial state f evolving freely. In order to arrive at a quantity which is independent of the radius of S_r, one will finally take the limit $r\to\infty$. Thus, if we assume that this limit exists, the time delay $(\Delta T)(f)$ for the initial state f is given by

$$(\Delta T)(f) = \lim_{r\to\infty} \int_{-\infty}^{\infty} dt\, \{\|F_r V_t \Omega_- f\|^2 - \|F_r U_t f\|^2\}$$

$$= \lim_{r\to\infty} \int_{-\infty}^{\infty} dt (f, U_t^* [\Omega_-^* F_r \Omega_- - F_r] U_t f), \quad (7.16)$$

where we have used the intertwining relation (4.11). Our aim is now to define a linear operator ΔT such that $(\Delta T)(f) = (f,\Delta T f)$.

The mathematical issues related to (7.16) are as follows. Firstly one will have to show that the improper integrals $I_r(\Omega_- f,H)$ and $I_r(f,H_0)$ are finite. In potential scattering the latter is finite by Corollary 3.13, since $\|F_r U_t f\|^2$ behaves as $|t|^{-3}$ at large t provided that $f\in L^1(R^3)\cap L^2(R^3)$ (the condition $f\in L^1(R^3)$ is not necessary, cf. Sinha [1]). In order to prove that $I_r(\Omega_- f,H) < \infty$ for certain vectors f, one will have to impose some restrictions on the interaction. Next, the existence of the limit $r \to \infty$ for sufficiently many vectors f has to be established. In this way one will obtain a quadratic form on some dense set $\mathcal{D}$. Finally one has to see whether this quadratic form determines a linear operator or not.

These mathematical questions are quite delicate. In our discussion below we shall make an additional hypothesis on the scattering system which considerably simplifies matters. We shall then prove the existence of the infinite integrals over time and of the limit $r \to \infty$ on a suitable dense set $\mathcal{D}$. The problem of when these limits determine a linear operator will not be discussed. However we shall define a certain self-adjoint operator in terms of the scattering operator and verify that its expectation values on $\mathcal{D}$ coincide with the right-hand side of (7.16). It is in this sense that the relation between the time delay and the scattering operator will be determined. The reader will find further comments as well as references in Section 7-6.

The additional hypothesis mentioned above is as follows : We have seen in Section 5-7 that the scattering operator is decomposable in the spectral representation of $H_0 E_\infty(H_0)$. We shall assume in addition that S is diagonalizable, i.e. that S is multiplication by a function $S(\lambda)$ in the spectral representation of $M_\infty(H_0)$ relative to H_0, which means that S is a function of $H_0/M_\infty(H_0)$. This hypothesis is always verified if H_0 has simple spectrum. The theory is also applicable to potential scattering with spherically symmetric potentials, a point that will be discussed in Chapter 11, which is devoted to spherical symmetry. To simplify the proofs, we also assume that the spectral representation of $H_0/M_\infty(H_0)$ has the form $G = L^2(\Lambda, H_0)$, where Λ is an interval.

We assume the asymptotic condition in the form (A1) and (A2) and also that $M_\infty(H_0) \subseteq H_{ac}(H_0)$. According to our hypothesis and (5.50), we have for $f, g \varepsilon M_\infty(H_0)$

$$(f, Sg) = \int_R S(\lambda) d(f, E_\lambda^0 g). \tag{7.17}$$

If the function $S(\cdot)$ is continuously differentiable, we may also define an operator S' as the multiplication operator by the derivative $S'(\lambda) \equiv dS(\lambda)/d\lambda$ in the spectral representation of $M_\infty(H_0)$ relative to H_0, i.e.

$$(f, S'g) = \int_R S'(\lambda) d(f, E_\lambda^0 g). \tag{7.18}$$

We now prove the above-mentioned results. We begin by transforming (7.16) into an expression containing the S-operator, which is possible under an additional hypothesis on the rate of convergence of the wave operators.

PROPOSITION 7.11 : Let $f \varepsilon M_\infty(H_0)$. Assume (A1), (A2) and

(α) $\|F_r U_t f\| \varepsilon L^1(0,\infty)$, $\|F_r U_t Sf\| \varepsilon L^1(0,\infty)$ for each $r < \infty$,

(β) $\|V_t \Omega_- f - U_t f\| \varepsilon L^1(-\infty, 0)$, $\|V_t \Omega_- f - U_t Sf\| \varepsilon L^1(0,\infty)$.

Then the integral in (7.16) exists for each $r < \infty$. If the limit as $r \to \infty$ in (7.16) exists, then

$$(\Delta T)(f) = \lim_{r\to\infty} \int_0^\infty dt (SU_t f, [F_r S - SF_r] U_t f). \tag{7.19}$$

Proof : We define

$$J_r^{\pm}(f,t) \equiv \|F_r V_t \Omega_{\pm} f\|^2 - \|F_r U_t f\|^2$$
$$= (V_t\Omega_{\pm}f - U_t f, F_r U_t f) + (V_t\Omega_{\pm}f, F_r[V_t\Omega_{\pm} - U_t]f). \qquad (7.20)$$

We may then write

$$\int_{-\infty}^{\infty} dt\{\|F_r V_t \Omega_- f\|^2 - \|F_r U_t f\|^2\} = \int_{-\infty}^{0} J_r^-(f,t)dt +$$
$$+ \int_0^{\infty} J_r^+(Sf,t)dt + \int_0^{\infty} dt\{\|F_r U_t Sf\|^2 - \|F_r U_t f\|^2\}. \qquad (7.21)$$

It follows by applying the Schwarz inequality to (7.20) that

$$|J_r^{\pm}(f,t)| \leq 2\,\|f\|\,\|V_t\Omega_{\pm}f - U_t f\|.$$

Thus by the assumption (β), $J_r^-(f,t)$ is majorized on $-\infty < t \leq 0$ uniformly in r by a function belonging to $L^1(-\infty,0)$. Therefore the first integral on the righ-hand side of (7.21) is finite. Also, since $J_r^-(f,t) \to 0$ as $r \to \infty$ for each t, we get from the dominated convergence theorem that the first integral in (7.21) converges to zero as $r \to \infty$.

Similarly, the second part of assumption (β) implies that the second integral in (7.21) exists and converges to zero as $r \to \infty$. The third integral is finite by assumption (α), since e.g. $\|F_r U_t f\|^2 \leq \|f\|\,\|F_r U_t f\|$. Thus

$$(\Delta T)(f) = \lim_{r\to\infty} \int_0^{\infty} dt\{\|F_r U_t Sf\|^2 - \|F_r U_t f\|^2\}. \qquad (7.22)$$

By applying (4.17), (4.16) and (4.3), we get

$$\|F_r U_t Sf\|^2 - \|F_r U_t f\|^2 = (SU_t f, F_r SU_t f) - (SU_t f, SF_r U_t f), \quad (7.23)$$

and (7.19) follows by inserting (7.23) into (7.22). #

We next give two auxiliary lemmas. The proof of the first one is given at the end of Section 7-6.

LEMMA 7.12 : Let $\phi : \mathcal{R} \to \mathcal{C}$ be a twice continuously differentiable function of bounded support. Then the Fourier transforms of ϕ and ϕ' are in $L^1(\mathcal{R})$.

LEMMA 7.13 : Let $f \in \mathcal{H}_{ac}(H_o)$ such that $E^o_\Delta f = f$ for some finite interval Δ. Let $\rho \in C_o^\infty(\mathcal{R})$ such that $\rho(\lambda) = 1$ for all $\lambda \in \Delta$. Define a function $S_\rho : \mathcal{R} \to \mathcal{C}$ by $S_\rho(\lambda) = S(\lambda)\rho(\lambda)$, let $\tilde{S}_\rho$ be the Fourier transform of S_ρ and suppose that S_ρ is twice continuously differentiable. Then for each $g \in \mathcal{H}$

$$(g,Sf) = (2\pi)^{-\frac{1}{2}} \int_{\mathcal{R}} dt\ \tilde{S}_\rho(t)(g,U_{-t}f), \tag{7.24}$$

$$(g,S'f) = (2\pi)^{-\frac{1}{2}} i \int_{\mathcal{R}} dt\ t\ \tilde{S}_\rho(t)(g,U_{-t}f). \tag{7.25}$$

Proof : We have by (7.17) and our assumption on ρ

$$\begin{aligned}(g,Sf) &= \int S(\lambda)\rho(\lambda)d(g,E^o_\lambda f)\\ &= (2\pi)^{-\frac{1}{2}} \int d(g,E^o_\lambda f) \int \tilde{S}_\rho(t)e^{it\lambda}dt,\end{aligned} \tag{7.26}$$

which gives (7.24) provided that one may interchange the order of integration in (7.26). For this, we notice that by Proposition 5.18 the absolute value of the double integral in (7.26) is majorized by

$$\int d\lambda \left\{\frac{d}{d\lambda}(f,E^o_\lambda f)\right\}^{\frac{1}{2}} \left\{\frac{d}{d\lambda}(g_{ac},E^o_\lambda g_{ac})\right\}^{\frac{1}{2}} \int dt|\tilde{S}_\rho(t)|,$$

which is finite since $[d/d\lambda(f,E^o_\lambda f)]^{\frac{1}{2}}$ and $[d/d\lambda(g_{ac},E^o_\lambda g_{ac})]^{\frac{1}{2}}$ are in $L^2(\mathcal{R})$ and since $\tilde{S}_\rho \in L^1(\mathcal{R})$ by Lemma 7.12. Hence the interchange of the order of integration in (7.26) is justified by Fubini's theorem.

(7.25) is proved in the same way by noticing that the Fourier transform of S'_ρ is $(it\ \tilde{S}_\rho(t))$. #

We now relate the time delay to the S-operator.

PROPOSITION 7.14 : Let $f \in M_\infty(H_0) \subseteq H_{ac}(H_0)$ such that $E^0_\Delta f = f$ for some finite interval Δ. Assume all the hypotheses of Proposition 7.11 and that $S(\cdot)$ is twice continuously differentiable. Then the limit in (7.16) exists, and

$$\lim_{r\to\infty} \int_{-\infty}^{\infty} dt\{\|F_r V_t \Omega_- f\|^2 - \|F_r U_t f\|^2\} = -i(f, S^* S' f). \tag{7.27}$$

Proof : Let ρ be as in Lemma 7.13. By using (7.19) and (7.24), we get

$$(2\pi)^{\frac{1}{2}} (\Delta T)(f) = \lim_{r\to\infty} \int_0^\infty dt \int_R d\tau\, \tilde{S}_\rho(\tau) \cdot \{(SU_t f, F_r U_{t-\tau} f) - (SU_t f, U_{-\tau} F_r U_t f)\}. \tag{7.28}$$

By the Schwarz inequality, the absolute value of the integrand in (7.28) is majorized by $|\tilde{S}_\rho(\tau)|\,\|f\|\,(\|F_r U_t Sf\| + \|F_r U_t f\|)$, which is integrable as a consequence of Lemma 7.12 and assumption (α). Thus Fubini's theorem allows us to interchange the order of integration in (7.28), which after a change of variables in the second term leads to

$$(2\pi)^{\frac{1}{2}} (\Delta T)(f) = \lim_{r\to\infty} \int_R d\tau\, \tilde{S}_\rho(\tau) \int_0^\tau dt (SU_t f, F_r U_{t-\tau} f). \tag{7.29}$$

The absolute value of the integrand in (7.29) is majorized by $|\tilde{S}_\rho(\tau)|\,\|f\|^2$, which is integrable over $R \times [0,\tau]$ since $\tau \tilde{S}_\rho(\tau) \in L^1(R)$ by Lemma 7.12. As the integrand converges pointwise to $\tilde{S}_\rho(\tau)(Sf, U_{-\tau} f)$ as $r \to \infty$, we may apply the dominated convergence theorem to interchange the limit with the double integral. Hence by using also (7.25)

$$(\Delta T)(f) = (2\pi)^{-\frac{1}{2}} \int d\tau\, \tilde{S}_\rho(\tau) \int_0^\tau dt (Sf, U_{-\tau} f)$$

$$= (2\pi)^{-\frac{1}{2}} \int \tau\, \tilde{S}_\rho(\tau)(Sf, U_{-\tau} f) d\tau = -i(Sf, S'f) = -i(f, S^* S' f). \quad \#$$

In view of Proposition 7.14 it is natural to define

the time delay operator ΔT as the following diagonalizable operator in the spectral representation of $M_\infty(H_0)$ relative to H_0 :

$$\Delta T = \{(\Delta T)(\lambda)\} = \{-i\bar{S}(\lambda) \frac{d}{d\lambda} S(\lambda) I_0\}.$$

If S is unitary in $M_\infty(H_0)$, then by (5.101) and Corollary 5.28 $S(\lambda)$ is unitary for almost all λ. Since S was assumed to be diagonalizable, this implies the existence of a real-valued function δ on Λ such that $S(\lambda) = \exp[2i\delta(\lambda)]$, so that

$$(\Delta T)(\lambda) = 2 \frac{d\delta(\lambda)}{d\lambda} I_0. \tag{7.30}$$

Thus if $\delta(\lambda)$ is piecewise continuously differentiable, the operator ΔT defined in the above manner is self-adjoint and commutes with $H_0/M_\infty(H_0)$. Furthermore its expectation value $(f,\Delta Tf)$ coincides with the time delay $(\Delta T)(f)$ whenever f verifies the hypotheses of Proposition 7.14. For potential scattering with a spherically symmetric potential decreasing sufficiently rapidly at infinity, it can be shown that there is a dense set of vectors f in H verifying all these assumptions. (The hypotheses (α) and (β) of Proposition 7.11 can be verified by the method developed in Section 13-1 provided that the partial wave S-matrix S_ℓ is three times continuously differentiable. The required differentiability of S_ℓ can be established using (11.39) and Proposition 11.16(d). See Problem 13.3 and Martin [3]).

The essential result of this section is expressed by equation (7.30) which signifies that the time delay at energy λ is given by the derivative of the phase of the on-shell scattering operator $S(\lambda)$ with respect to the energy. This fact is often used in relation with scattering resonances, a subject that will be mentioned in Chapter 11.

7-3 SCATTERING INTO CONES. DIFFERENTIAL CROSS SECTIONS

An important quantity in relation with scattering experiments is the probability that the scattered particles will emerge in a given cone C with apex at the origin. This probability was computed in Section 3-3 for the case of free particles in potential scattering and shown to be equal to the probability that their momentum lies in the cone C. One expects a similar result to be true also if the potential is non-zero. In this case the emerging particles associated with the initial state f are asymptotically described by $U_t Sf$ at large positive times, and therefore the probability that they emerge in the cone C should be equal to the probability that the final state Sf has momentum in C. This result is the content of Proposition 7.15 below.

The differential scattering cross section is obtained by taking for C an infinitesimal cone and for the initial state f a monochromatic wave. Since monochromatic or plane waves are not square-integrable functions over $\mathbb{R}^3$, one will have to do some kind of limit to deduce the differential cross section from Hilbert space objects. In the course of this calculation it is necessary to make certain approximations. Some of these refer to mathematical properties of the S-operator; they can be proved by imposing suitable conditions on the potential. A second type of approximation can be justified by invoking specific properties of the experimental setup, for instance that the initial state is practically monoenergetic and well collimated, or whether the incident particles are scattered coherently or incoherently by the target. Therefore it should be borne in mind that the expression for the scattering cross section in terms of the S-operator may vary with the type of experiment.

These remarks will be illustrated in detail in this section where we specify a set of hypotheses that are often verified and deduce the formula for the differential cross section under these assumptions. The individual assumptions will be introduced at the point where they are first used in the derivation, and a short summary will be given at the end (Proposition 7.18). Throughout this section we stay within the framework of potential scattering, although this restriction is unimportant (cf. Section 7-6).

The discussion of scattering into cones is similar to that given in Section 3-3 for free particles. Let C be a cone in $\mathcal{R}^3$ with apex at the origin. We may define two projections F_C and G_C in $L^2(\mathcal{R}^3)$ by

$$(F_C g)(\underline{x}) = \chi_C(\underline{x})g(\underline{x}), \quad (\mathcal{F}G_C g)(\underline{k}) = \chi_C(\underline{k})\tilde{g}(\underline{k}). \qquad (7.31)$$

The range of F_C consists of the states localized in the cone C, whereas that of G_C are the states the momentum of which lies in C.

<u>PROPOSITION 7.15</u> : Assume the asymptotic condition in the form (A1) and (A2). Let $g \in \mathcal{M}_\infty(H_0)$, $\|g\| = 1$, and let P(g,C) be the probability that the corresponding scattered state will be found in the cone C as $t \to +\infty$, i.e.

$$P(g,C) \equiv \lim_{t\to+\infty} \|F_C V_t \Omega_- g\|^2. \qquad (7.32)$$

Then $$P(g,C) = ||G_C Sg||^2 = \int_C d^3k |(\mathcal{F}Sg)(\underline{k})|^2. \quad (7.33)$$

Proof : By applying successively (2.88), (2.37), (A1) and (A2) we get

$$\lim_{t\to\infty} \Big| ||F_C V_t \Omega_- g|| - ||F_C U_t Sg|| \Big|$$
$$\leq \lim_{t\to\infty} ||F_C V_t \Omega_- g - F_C U_t Sg|| \leq \lim_{t\to\infty} ||\Omega_- g - V_t^* U_t Sg||$$
$$= ||\Omega_- g - \Omega_+ \Omega_+^* \Omega_- g|| = ||\Omega_- g - \Omega_- g|| = 0. \quad (7.34)$$

This implies with (7.32) and (3.51) that

$$P(g,C) = \lim_{t\to\infty} ||F_C U_t Sg||^2 = P^+_{free}(Sg,C) = ||G_C Sg||^2. \ \#$$

Proposition 7.15 gives the expected result for the probability that the scattered state will finally lie in a given cone C. In order to get the scattering cross section, it is useful to introduce the following auxiliary quantity :

$$P'(g,C) = \lim_{t\to+\infty} ||F_C(V_t\Omega_- - U_t)g||^2. \quad (7.35)$$

As in the proof of Proposition 7.15, one sees that one may replace $F_C V_t \Omega_- g$ asymptotically by $F_C U_t Sg$, i.e.

$$P'(g,C) = \lim_{t\to\infty} ||F_C(U_t Sg - U_t g)||^2, \quad (7.36)$$

whence by (3.51)

$$P'(g,C) = ||G_C(Sg-g)||^2 = \int_C d^3k \ |(\mathcal{F}Sg)(\underline{k}) - \tilde{g}(\underline{k})|^2. \ (7.37)$$

The difference between the scattering state and the initial state used in (7.35), i.e. $V_t\Omega_- g - U_t g$, is not a state vector that is actually realized in an experiment, but it is much used in theoretical considerations. The idea behind this is that, in order to extract from the final state the part that is really due to scattering, one should remove from it the part that would be there if there was no interaction. The following lemma relates $P'(g,C)$ and $P(g,C)$.

LEMMA 7.16 : Let $g \in M_\infty(H_0)$. Then

$$|P'(g,C) - P(g,C)| \leq P^{+}_{\text{free}}(g,C) + 2\{P^{+}_{\text{free}}(g,C)P(g,C)\}^{\frac{1}{2}}.$$

Proof : We have from (7.33), (7.37) and (2.8)

$$|P'(g,C) - P(g,C)| = \left| \|G_C(Sg-g)\|^2 - \|G_C Sg\|^2 \right|$$

$$= \left| \|G_C Sg\|^2 + \|G_C g\|^2 - (G_C Sg, G_C g) - (G_C g, G_C Sg) - \|G_C Sg\|^2 \right|$$

$$\leq \|G_C g\|^2 + 2\|G_C g\| \, \|G_C Sg\| ,$$

which is the desired inequality. #

Suppose now that the initial state is well collimated, e.g. that the support of $\tilde{g}(\underline{k})$ lies in a cone C_0 of small solid angle. The "axis" of C_0 may be interpreted as the forward direction or the direction of motion of the incident particle. If the intersection of C with C_0 consists only of the point $\underline{x} = 0$, i.e. if one will not observe the scattered particles inside the forward cone, then $P^{+}_{\text{free}}(g,C) = 0$ by (3.51). Hence by Lemma 7.16 we have $P'(g,C) = P(g,C)$, i.e. the probability of finding the final state in the cone C at large positive times is also given by $P'(g,C)$. It is only in this sense, i.e. by replacing $P(g,C)$ by $P'(g,C)$, that we shall subtract the initial state from the scattering state. We see that the relevant operator in (7.37) is

$$R \equiv S - E_\infty(H_0). \tag{7.38}$$

Under the above hypotheses, we then have

$$P(g,C) = P'(g,C) = \int_C d^3k |(\mathcal{F}Rg)(\underline{k})|^2. \qquad (7.39)$$

We now choose the x_3-coordinate axis along the axis of C_o, i.e. in the forward direction. To define a cross section, we consider a large (in fact infinite) number of independent scattering events. The initial state of each of them is obtained by translating the state g by some vector $\underline{a}$ in the $\{x_1, x_2\}$-plane (the "impact parameter plane"). We denote the translated state by $g_{\underline{a}}$. By taking a uniform distribution over this plane of the parameter $\underline{a}$ of these translations, this may be interpreted as a uniform beam of uncorrelated incoming particles.

The hypothesis that the incoming particles are uncorrelated and scattered independently means that, in order to get the total number of particles scattered into the cone C, we may add the individual probabilities for scattering into the cone C, i.e.

$$N_{tot}(g,C) = \sum_{\underline{a}} P_{\underline{a}}(g,C) \equiv \sum_{\underline{a}} P(g_{\underline{a}},C). \qquad (7.40)$$

From the definition of the scattering cross section given in Chapter 1, we see that this quantity should be divided by the number N_o of points $\underline{a}$ lying in any unit square of the impact parameter plane. This is equivalent to giving each point the weight of a small surface $\Delta_{\underline{a}} = N_o^{-1}$ in the right-hand sum of (7.40). In the limit wnere N_o tends to infinity, this amounts to replacing the sum in (7.40) by an integral :

$$\sigma(g,C) \equiv \lim_{N_o \to \infty} N_o^{-1} N_{tot}(g,C) = \int d^2a \, P(g_{\underline{a}},C) =$$

$$= \int d^2a \int_C d^3k |(\mathcal{F}Rg_{\underline{a}})(\underline{k})|^2. \qquad (7.41)$$

It remains to transform (7.41) into an expression from which one may read off the formula for the differential cross section under suitable assumptions on R and g. For this we use the spectral representation (5.94) of K_o and the notation introduced in relation with (5.94). We denote by $\underline{\omega}$ the unit vectors in $\mathbb{R}^3$ (i.e. the variable in $S^{(2)}$) and by $d\omega$ the surface element in $S^{(2)}$.

By Propositions 4.7 and 5.27, R is decomposable in the representation (5.94) of $L^2(\mathbb{R}^3)$, i.e. for almost every $\lambda \geq 0$ there exists a bounded operator $R(\lambda)$ in $L^2(S^{(2)})$ such that $Rg = \{R(\lambda)g_\lambda\}$. We shall now make the hypothesis that each $R(\lambda)$ is a Hilbert Schmidt operator in $L^2(S^{(2)})$. This implies as in Proposition 2.33 that $R(\lambda)$ is an integral operator with Hilbert-Schmidt kernel, i.e. that for almost every $\lambda \geq 0$ there exists a function $R(\lambda;\underline{\omega},\underline{\omega}')$ with

$$\|R(\lambda)\|_{HS}^2 = \iint d\omega d\omega' |R(\lambda;\underline{\omega},\underline{\omega}')|^2 < \infty \qquad (7.42)$$

such that

$$[R(\lambda)g_\lambda](\underline{\omega}) = \int d\omega' R(\lambda;\underline{\omega},\underline{\omega}')g_\lambda(\underline{\omega}'). \qquad (7.43)$$

The integral in (7.43) exists for almost all $\underline{\omega}$ since the integrand is in $L^1(S^{(2)})$ for almost all $\underline{\omega}$.

We may now rewrite (7.41) as

$$\sigma(g,C) = \int d^2a \int_{\underline{\omega}\in C} d\lambda d\omega \left| \int d\omega' R(\lambda;\underline{\omega},\underline{\omega}') e^{ik\underline{\omega}'\cdot\underline{a}} g_\lambda(\underline{\omega}') \right|^2, \qquad (7.44)$$

where $k = \lambda^{\frac{1}{2}}$. The next step is to interchange the integrals with respect to the variables $\underline{a}$ and $(\lambda,\underline{\omega})$ and to simplify the resulting integral over d^2a by using the fact that the support of $\tilde{g}(\underline{k})$ is contained in the cone C_o of small solid angle. For this we prove the following lemma in which $(k,\underline{\omega})$ denotes the spherical coordinates of the point $\underline{k} = (k_1,k_2,k_3) \in R^3$.

LEMMA 7.17 : Let $\phi : R^3 \to C$ be a function with support in a closed cone C_0 verifying $k_3 > 0$ for all $\underline{k} \in C_0 - \{0\}$. Let $k > 0$ be fixed and suppose that $\int d\omega |\phi(k,\underline{\omega})|^2 < \infty$. Then

$$I_k \equiv \int d^2a \ |\frac{1}{2\pi} \int d\omega \exp(ik_1a_1 + ik_2a_2)\phi(k,\underline{\omega})|^2$$

$$= \int d\omega \ |\phi(k,\underline{\omega})|^2 (kk_3)^{-1}. \qquad (7.45)$$

Proof : In view of the support condition on ϕ, we may make the change of variables $(k,\underline{\omega}) \to (k,k_1,k_2) = (k, k\sin\theta\cos\phi, k\sin\theta\sin\phi)$. The corresponding Jacobian determinant is easily calculated to be

$$\left| \det \frac{\partial(k_1 , k_2)}{\partial(\cos\theta,\phi)} \right| = k^2\cos\theta = kk_3 = k(k^2-k_1^2-k_2^2)^{\frac{1}{2}}.$$

By rewriting the first integral of (7.45) in the new variables and using the Parseval identity (2.22) for the Fourier transformation, we obtain

$$I_k = \int d^2a \left| \frac{1}{2\pi} \int dk_1dk_2 \exp(ik_1a_1 + ik_2a_2)\phi(k,k_1,k_2)(kk_3)^{-1} \right|^2$$

$$= \int dk_1dk_2 |\phi(k,k_1,k_2)|^2 (kk_3)^{-2} = \int d\omega \ |\phi(k,\underline{\omega})|^2 (kk_3)^{-1},$$

where the last equation is obtained by reintroducing the original variables $(k,\underline{\omega})$. #

We now apply this lemma to the function $\phi(k,\underline{\omega}') \equiv R(\lambda;\underline{\omega},\underline{\omega}')g_\lambda(\underline{\omega}')$ for fixed k and $\underline{\omega}$. This function has the required support properties by the assumption made on g. In order to make sure that it is square-integrable over the unit sphere for almost all values of λ and $\underline{\omega}$, we assume that there exists $M < \infty$ such that $|g_\lambda(\underline{\omega}')| \leq M$ for all λ and $\underline{\omega}'$. Then we have

$$J(\lambda,\underline{\omega}) \equiv \int d^2a \left| \int d\omega' R(\lambda;\underline{\omega},\underline{\omega}') e^{ik\underline{\omega}'\cdot\underline{a}} g_\lambda(\underline{\omega}')\right|^2$$
$$= (2\pi)^2 \int d\omega' |R(\lambda;\underline{\omega},\underline{\omega}')|^2 |g_\lambda(\underline{\omega}')|^2 \lambda^{-1} (\cos\theta')^{-1}, \quad (7.46)$$

where θ' is the angle between the direction $\underline{\omega}'$ and the k_3-axis (Figure 7.1). By the assumption on the support of g, there exists $M_1 < \infty$ such that $|\cos\theta'|^{-1} \leq M_1$ for all $\underline{\omega}'$ in the domain in which the integrand in (7.46) is different from zero. Thus

$$\iint d\lambda d\omega \, J(\lambda,\underline{\omega}) \leq 4\pi^2 M^2 M_1 \int_\Delta d\lambda \, \lambda^{-1} \|R(\lambda)\|^2_{HS},$$

where $\Delta = \{\lambda \geq 0 \mid \|g_\lambda\|_o \neq 0\}$. If the last integral is finite

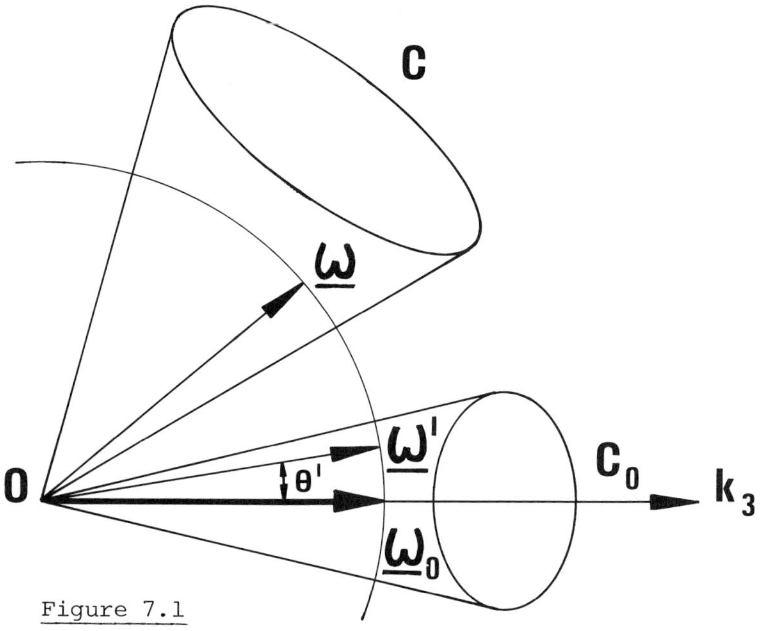

Figure 7.1

(e.g. if g is such that Δ is a finite closed interval not containing $\lambda = 0$ and $R(\lambda)$ is continuous in Hilbert-Schmidt norm on Δ), $\sigma(g,C)$ is finite, and we get from (7.44) and (7.46) that

$$\sigma(g,C) = (2\pi)^2 \int_C d\lambda d\omega \int d\omega' |R(\lambda;\underline{\omega},\underline{\omega}')|^2 |g_\lambda(\underline{\omega}')|^2 \lambda^{-1} (\cos\theta')^{-1}. \tag{7.47}$$

We now define the <u>scattering amplitude</u> $f(\lambda;\underline{\omega}'\to\underline{\omega})$ by

$$f(\lambda;\underline{\omega}'\to\underline{\omega}) = -2\pi i \lambda^{-\frac{1}{2}} R(\lambda;\underline{\omega},\underline{\omega}'). \tag{7.48}$$

Since C is an arbitrary cone (with $C \cap C_0 = \{\underline{0}\}$), the Radon-Nikodym derivative $d\sigma(g;\underline{\omega})/d\omega$ exists for almost all $\underline{\omega}$ [R] and may be interpreted as the differential scattering cross section associated with the state vector g. It is given by

$$\begin{aligned} d\sigma(g,\underline{\omega})/d\omega &= \int_0^\infty d\lambda \int d\omega' |f(\lambda;\underline{\omega}'\to\underline{\omega})|^2 |g_\lambda(\underline{\omega}')|^2 (\cos\theta')^{-1} \\ &= \int_0^\infty d\lambda \int d\omega' \frac{d\sigma}{d\omega}(\lambda;\underline{\omega}'\to\omega) |g_\lambda(\omega')|^2 (\cos\theta')^{-1}, \end{aligned} \tag{7.49}$$

where we have set

$$\frac{d\sigma}{d\omega}(\lambda;\underline{\omega}'\to\underline{\omega}) = |f(\lambda;\underline{\omega}'\to\underline{\omega})|^2. \tag{7.50}$$

Since $|g_\lambda(\underline{\omega}')|^2$ represents the probability of the initial state having energy λ and direction $\underline{\omega}'$, one may interpret $|f(\lambda;\underline{\omega}'\to\underline{\omega})|^2$ as the <u>differential cross section</u> for scattering from the initial direction $\underline{\omega}'$ to the final direction $\underline{\omega}$ at energy λ (the factor $\cos\theta'$ is practically equal to 1 and stems from the fact that the direction of the incoming state has a small spread so that the impact parameter plane is only approximately orthogonal to the vectors $\underline{\omega}'$ in C_0). This interpretation of $d\sigma/d\omega$ has to be taken in the sense of (7.49) which tells us how the differential cross section has to be averaged over the momentum distribution of the incoming state.

In experiments one does not usually know the exact form of $|g_\lambda(\underline{\omega})|^2$. For this reason it would be preferable to have an expression for the scattering cross section which is insensitive to variations in $|g_\lambda(\underline{\omega})|^2$. Such an expression can be deduced from (7.47) under some additional continuity assumptions on $|f(\lambda;\underline{\omega}'\to\underline{\omega})|^2$. Specifically we make the hypothesis that for each $\eta > 0$ there exists a neighbourhood $\mathcal{V}$ of the fixed point $\underline{k}_o = \lambda_o^{\frac{1}{2}}\underline{\omega}_o$ such that for each cone C and all $\lambda,\underline{\omega}'$ with $\lambda^{\frac{1}{2}}\underline{\omega}'\varepsilon\mathcal{V}$

$$\int_{\Omega_C} d\omega|f(\lambda;\underline{\omega}'\to\underline{\omega})|^2 - \int_{\Omega_C} d\omega|f(\lambda_o;\underline{\omega}_o\to\underline{\omega})|^2 < \eta, \tag{7.51}$$

where $\Omega_C \equiv C\cap S^{(2)}$ is the part of $S^{(2)}$ lying in the cone C.

If η and $\mathcal{V}$ are as indicated, one obtains from (7.47) that

$$\sigma(g,C) = \int_{\Omega_C} d\omega|f(\lambda_o;\underline{\omega}_o\to\underline{\omega})|^2 \, ||g||^2 + r(\underline{k}_o;\mathcal{V}) \tag{7.52}$$

with

$$|r(\underline{k}_o;\mathcal{V})| \leq \eta M_1 \, ||g||^2 + \int_{\Omega_C} d\omega \, |f(\lambda_o;\underline{\omega}_o\to\underline{\omega})|^2 (M_1-1)||g||^2.$$

Since $||g||^2 = 1$, these expressions are independent of g (provided that $g_\lambda(\underline{\omega})$ is bounded and has support in $\mathcal{V}$). Since $M_1 \to 1$ as $\mathcal{V}$ shrinks to the point $\underline{k}_o$, the second term in $r(\underline{k}_o;\mathcal{V})$ may be neglected with respect to the first term in (7.52). The first term in $r(\underline{k}_o;\mathcal{V})$ converges to zero as $\mathcal{V}$ shrinks to $\underline{k}_o$. Thus (7.52) shows that, under the hypothesis (7.51), $\sigma(g,C)$ is practically independent of g provided that the support of $\tilde{g}$ lies in a sufficiently small neighbourhood of $\underline{k}_o$ and is given by $\sigma(g,C) = \int_{\Omega_C} d\omega|f(\lambda_o;\underline{\omega}_o\to\underline{\omega})|^2$.

We collect these results in the following proposition and then continue with some additional comments.

PROPOSITION 7.18 : Let $\Delta \subseteq (0,\infty)$ be a closed interval. Assume that for each $\lambda \in \Delta$, $R(\lambda)$ is a Hilbert-Schmidt operator such that $\|R(\lambda)\|_{HS}^{2} \in L^{1}(\Delta)$. Let g be a state vector such that $g_{\lambda}(\underline{\omega})$ is bounded and $g_{\lambda}(\underline{\omega}) = 0$ for all $\lambda \notin \Delta$ and all $\underline{\omega} \notin C_{0}$. Then the differential cross section $d\sigma(g;\underline{\omega})/d\omega$ for the state g and $\underline{\omega} \notin C_{0}$ is given by (7.49). If in addition $R(\lambda)$ verifies (7.51), then the differential cross section for scattering from the initial direction $\underline{k}_{0} = \lambda^{\frac{1}{2}}\underline{\omega}_{0}$ into the final direction $\underline{k} = \lambda^{\frac{1}{2}}\underline{\omega}$ is given as $d\sigma(\lambda;\underline{\omega}_{0} \to \underline{\omega})/d\omega = |f(\lambda;\underline{\omega}_{0} \to \underline{\omega})|^{2}$.

We have seen in Section 5-7 that $R(\lambda)$ is determined only for almost all λ. The hypothesis that for such λ, $R(\lambda)$ be a Hilbert-Schmidt operator, determines $R(\lambda;\underline{\omega},\underline{\omega}')$ only almost every where in $S^{(2)} \times S^{(2)}$. Clearly the replacement of $R(\lambda)$ or $R(\lambda;\underline{\omega},\underline{\omega}')$ by a representative in its equivalence class has no effect either on the formula (7.47) for the cross section or on $S = R + I$ as an operator in $L^{2}(R^{3})$. We shall often choose the most convenient representative. In particular we shall see in Sections 8-3, 10-2 and 16-2 that, under suitable hypotheses on V, one can choose a representative $R(\lambda;\underline{\omega},\underline{\omega}')$ which is continuous in $\underline{\omega}$, $\underline{\omega}'$ and λ when λ is restricted to suitable intervals $\Delta \subseteq (0,\infty)$. Then all the hypotheses of Proposition 7.18 are verified and $d\sigma/d\omega(\lambda;\underline{\omega}_{0} \to \underline{\omega})$ is continuous.

Even if (7.51) holds true, it may not be possible in a practical situation to prepare a sufficiently monoenergetic and well collimated beam such that one may neglect the variations of the scattering amplitude over the range of momenta of this beam. Equation (7.49) then indicates how the scattering amplitude has to be averaged over the momentum distribution of the incoming state.

Equation (7.50) gives the differential cross section for the scattering of an incident beam from a single scattering center under the hypothesis that the individual particles of the beam are uncorrelated and scattered independently. Suppose now that the target consists of a large number N of randomly distributed individual scatterers. If multiple scattering in the target is negligible and if the arrangement of the scatterers does not lead to any coherence effects, the probability of the beam being scattered into a cone C (per unit incident flux) is the sum of the corresponding probabilities for each individual scatterer (this excludes for example crystals as targets; as is well known, one has to add amplitudes rather than probabilities whenever coherence effects are present). Thus the differential cross section in this case is obtained by simply multiplying the right-hand side of (7.49) or (7.50) by N.

Remark 7.19 : The expression giving the scattering amplitude in terms of the kernel of the integral operator $R(\lambda)$ depends on how the operator R is written as an integral operator. This is occasionally done in $\underline{k}$-space in one of the following two ways :

$$(FRg)(\underline{k}) = \int d^3k'\,\delta(k-k')R'(\underline{k},\underline{k}')\tilde{g}(\underline{k}'), \tag{7.53}$$

$$(FRg)(\underline{k}) = \int d^3k'\,\delta(\lambda_k-\lambda_{k'})R^o(\underline{k},\underline{k}')\tilde{g}(\underline{k}'), \tag{7.54}$$

where $\lambda_k = (h/2\pi)^2k^2/2m$ (we reintroduce the factors $h/2\pi$ and $2m$). The scattering amplitude is then given as follows in terms of these new integral kernels (Problem 7.4) :

$$f(\lambda_k;\underline{\omega}_o\to\underline{\omega}) = -2\pi ik^{-1}R(\lambda_k;\underline{\omega},\underline{\omega}_o)$$

$$= -2\pi ik\,R'(k\underline{\omega},k\underline{\omega}_o) = -2\pi im(2\pi/h)^2R^o(k\underline{\omega},k\underline{\omega}_o). \tag{7.55}$$

7-4 THE TOTAL CROSS SECTION

In this section we first derive some relations satisfied by the scattering amplitude that are frequently used in theoretical considerations. They are obtained by rewriting the statement that S is an isometric or unitary operator in terms of the kernel of the integral operators $R(\lambda)$ and remembering that this kernel determines the scattering amplitude. A particular relation of this type is the so-called optical theorem which relates the scattering amplitude in the forward direction to the total cross section. We shall conclude this section by describing a method for establishing the finiteness of the total cross section and its behaviour at large energies, but we shall refer to the literature for a complete proof.

We will need a characterization of the adjoint of an integral operator and of the product of two integral operators. Formally it is simple to check that these are again integral operators and to compute their kernels. We give here the result for the case of Hilbert-Schmidt kernels and defer the proof till Section 7-6.

PROPOSITION 7.20 : Let Δ be a measurable subset of $\mathcal{R}^n$ and $\mathcal{H} = L^2(\Delta, d\mu)$ the Hilbert space of square-integrable functions from Δ into $\mathcal{C}$ with respect to a measure μ on Δ. Suppose that A and B are integral operators in $\mathcal{H}$ with Hilbert-Schmidt kernels $K_A(\underline{x},\underline{y})$ and $K_B(\underline{x},\underline{y})$ respectively. Then A^* and AB are integral operators with Hilbert-Schmidt kernels $K_{A^*}(\underline{x},\underline{y})$ and $K_{AB}(\underline{x},\underline{y})$ respectively given by

$$K_{A*}(\underline{x},\underline{y}) = \overline{K_A(\underline{y},\underline{x})} \equiv \overline{K_A}(\underline{y},\underline{x}), \tag{7.56}$$

$$K_{AB}(\underline{x},\underline{y}) = \int_\Delta d\mu(z) K_A(\underline{x},\underline{z}) K_B(\underline{z},\underline{y}). \tag{7.57}$$

Let us now write (4.16) in terms of the R-operator (we assume $E_\infty(H_o) = I$) :

$$S^*S = (R^*+I)(R+I) = R^*R + R^* + R + I = I,$$

i.e.

$$R^*R = -(R^*+R). \tag{7.58}$$

This implies by Proposition 5.26 that for almost all λ

$$R(\lambda)^*R(\lambda) = -\{R(\lambda)^* + R(\lambda)\}. \tag{7.59}$$

If $R(\lambda)$ is a Hilbert-Schmidt operator in $H_o = L^2(S^{(2)})$, we may apply Proposition 7.20 to (7.59) to obtain

$$\int d\omega' \bar{R}(\lambda;\underline{\omega}',\underline{\omega}) R(\lambda;\underline{\omega}',\underline{\omega}_o) = -\{\bar{R}(\lambda;\underline{\omega}_o,\underline{\omega}) + R(\lambda;\underline{\omega},\underline{\omega}_o)\}. \tag{7.60}$$

In terms of the scattering amplitude this reads

$$\int d\omega' \bar{f}(\lambda;\underline{\omega}\to\underline{\omega}') f(\lambda;\underline{\omega}_o\to\underline{\omega}') = \frac{2\pi i}{\sqrt{\lambda}}\{\bar{f}(\lambda;\underline{\omega}\to\underline{\omega}_o) - f(\lambda;\underline{\omega}_o\to\underline{\omega})\}. \tag{7.61}$$

This relation is satisfied by the scattering amplitude for almost all values of $\underline{\omega}_o$ and $\underline{\omega}$ as a consequence of the fact that S is an isometric operator. In particular, if we set $\underline{\omega} = \underline{\omega}_o$, we get the following result known as the <u>optical theorem</u> :

$$\int d\omega' \frac{d\sigma}{d\Omega}(\lambda;\underline{\omega}_o\to\underline{\omega}') = \sigma_{tot}(\lambda;\omega_o) = \frac{4\pi}{\lambda^{\frac{1}{2}}} \operatorname{Im} f(\lambda;\underline{\omega}_o\to\underline{\omega}_o). \tag{7.62}$$

This equation relates the total cross section for the given initial momentum $\lambda^{\frac{1}{2}}\underline{\omega}_o$ to the imaginary part of the scattering amplitude at the same energy and in the forward direction. It implies that the scattering amplitude must have a non-vanishing imaginary part, at least near the forward direction. (7.60) and (7.61) hold almost everywhere. Therefore, for

(7.62) to make sense, further assumptions are necessary. One such assumption is that the scattering amplitude be a continuous function of its arguments.

The optical theorem was derived as a consequence of the fact that S is an isometric operator. If S is unitary, i.e. if in addition $SS^* = I$, one gets in the same way as above additional relations satisfied by the scattering amplitude, viz.

$$\int d\omega' \bar{f}(\lambda;\underline{\omega}' \to \underline{\omega}) f(\lambda;\underline{\omega}' \to \underline{\omega}_o) = \frac{2\pi i}{\lambda^{\frac{1}{2}}} \{\bar{f}(\lambda;\underline{\omega} \to \underline{\omega}_o) - f(\lambda;\underline{\omega}_o \to \underline{\omega})\} \quad (7.63)$$

and for $\underline{\omega} = \underline{\omega}_o$

$$\int d\omega' \frac{d\sigma}{d\Omega} (\lambda;\underline{\omega}' \to \underline{\omega}_o) = \frac{4\pi}{\lambda^{\frac{1}{2}}} \operatorname{Im} f(\lambda;\underline{\omega}_o \to \underline{\omega}_o). \quad (7.64)$$

By comparing (7.64) with the optical theorem, one sees that the unitarity of S implies that the total cross section for the initial wave vector $\underline{k}_o = \lambda^{\frac{1}{2}}\underline{\omega}_o$ is identical to the integral over all initial directions of the differential cross section for scattering with final wave vector $\underline{k}_o$:

$$\int d\omega' \frac{d\sigma}{d\omega} (\lambda;\underline{\omega}_o \to \underline{\omega}') = \int d\omega' \frac{d\sigma}{d\omega} (\lambda;\underline{\omega}' \to \underline{\omega}_o). \quad (7.65)$$

(7.65) may be viewed as an integral version of the property called "detailed balance" in the physics literature [T, page 354]. It is seen that this identity is not directly related to time-reversal invariance, although the latter may often be useful in proving the unitarity of S (cf. Proposition 4.10). Time-reversal invariance in potential scattering leads to other relations for the scattering amplitude. In fact, by using the definition of the time-reversal operator Θ and transcribing (4.25) into an equation for the scattering amplitude, one sees that time-reversal invariance implies that (Problems 8.8 and 10.6)

$$f(\lambda;\underline{\omega}_o \to \underline{\omega}) = f(\lambda;-\underline{\omega}\to -\underline{\omega}_o), \tag{7.66}$$

which leads to

$$\sigma_{tot}(\lambda;\underline{\omega}_o) = \int d\omega' \frac{d\sigma}{d\omega}(\lambda;\underline{\omega}' \to -\underline{\omega}_o). \tag{7.67}$$

We now consider the question of the finiteness of the total cross section. For this it is useful to define its average $\bar{\sigma}(\lambda)$ over all initial directions :

$$\bar{\sigma}(\lambda) = \frac{1}{4\pi} \int d\omega_o \, \sigma_{tot}(\lambda;\underline{\omega}_o). \tag{7.68}$$

We shall give a criterion for $\bar{\sigma}(\lambda)$ to be finite. If the potential is spherically symmetric, i.e. if the rotation group is a symmetry group, then $\sigma_{tot}(\lambda;\underline{\omega}_o)$ is independent of the initial direction $\underline{\omega}_o$, hence $\bar{\sigma}(\lambda) = \sigma_{tot}(\lambda;\underline{\omega}_o)$, and our criterion is directly applicable to the total cross section.

In view of (7.50), (7.48) and (7.42), one sees that

$$\bar{\sigma}(\lambda) = \pi\lambda^{-1} \|R(\lambda)\|^2_{HS}. \tag{7.69}$$

Thus the hypothesis that $R(\lambda) = S(\lambda)-I_o$ is a Hilbert-Schmidt operator implies that $\bar{\sigma}(\lambda)$ is finite (as regards the point $\lambda = 0$, one has to require in addition that $\lim \lambda^{-1}\|R(\lambda)\|^2_{HS}$ as $\lambda \to 0$ exists and is finite). In general, if $R(\lambda)$ is an integral operator in $L^2(S^{(2)})$, $\bar{\sigma}(\lambda)$ is finite if and only if its kernel is a Hilbert-Schmidt kernel. In particular $\bar{\sigma}(\lambda) < \infty$ if $R(\lambda)$ belongs to the set $\mathcal{B}_1(\mathcal{H}_o)$ of trace class operators in $\mathcal{H}_o$. By using (2.68) and (7.59), $\bar{\sigma}(\lambda)$ may be expressed in terms of traces :

$$\bar{\sigma}(\lambda) = \frac{\pi}{\lambda} \mathrm{Tr}_o[R(\lambda)^*R(\lambda)] = -\frac{\pi}{\lambda} \mathrm{Tr}_o[R(\lambda)^* + R(\lambda)] =$$

$$= -\frac{2\pi}{\lambda} \mathrm{Re}\, \mathrm{Tr}_o R(\lambda), \tag{7.70}$$

where the symbol Tr_0 indicates that the trace is taken in $\mathcal{H}_0$.

The following theorem allows one to derive sufficient conditions for $R(\lambda)$ to belong to $\mathcal{B}_1(\mathcal{H}_0)$. We use the spectral representation of H_0 in the form given in Proposition 5.29. Since there each $\lambda\varepsilon\mathcal{R}$ belongs to one set Λ_j, we may associate with each $\lambda\varepsilon\mathcal{R}$ a non-negative integer $j(\lambda)$ by the requirement that $\lambda\varepsilon\Lambda_{j(\lambda)}$. In the next proposition a subscript λ refers to the Hilbert space $\mathcal{K}_{j(\lambda)}$.

PROPOSITION 7.21 : Let H_0 be absolutely continuous, and let $\mathcal{D}$ be the dense set of functions $f = \{f_{j,\lambda}\}$ such that $\|f_{j,\lambda}\|_\lambda \leq M$ for all j and λ. Let $A\varepsilon\mathcal{B}_1$. Then there exists a family $\{A(\lambda)\}$ of trace class operators, $A(\lambda)\varepsilon\mathcal{B}_1(\mathcal{K}_{j(\lambda)})$, such that for all $f,g\varepsilon\mathcal{D}$

$$\int_{\mathcal{R}} dt(f,U_t^*AU_tg) = \sum_{j=1}^{\infty} \int_{\sigma(H_0)} (f_{j,\lambda},A(\lambda)g_{j,\lambda})_\lambda \, d\lambda < \infty. \quad (7.71)$$

The family $\{A(\lambda)\}$ is essentially unique in the sense that if $\{A'(\lambda)\}$ is another such family, then $A(\lambda) = A'(\lambda)$ for almost all λ. Furthermore

$$\int_{\sigma(H_0)} d\lambda \; Tr_\lambda \; A(\lambda) = 2\pi \; Tr \; A \quad (7.72)$$

and

$$\int_{\sigma(H_0)} d\lambda \, |||A(\lambda)|||_{1,\lambda} \leq 2\pi |||A|||_1 . \quad (7.73)$$

For the proof of this result, we refer to Jauch, Misra and Sinha [1] or Martin and Misra [1]. A part of it is rather easy to understand. Formally the operator $\int dt \, U_t^*AU_t$ commutes with H_0, so that it will be decomposable, which justifies the equality in (7.71). The finiteness of these integrals and (7.72) and (7.73) have to be proved separately. The essential uniqueness of the family $\{A(\lambda)\}$ is shown by reasoning similar to that of the proof of Proposition 5.28.

We now show how this theorem applies in scattering theory. For this we prove the following two propositions. For the sake of simplicity we assume that H_o has uniform spectral multiplicity, i.e. that all K_j with $j \neq 0$ are identical with one Hilbert space H_o.

PROPOSITION 7.22 : Let ϕ and ψ be bounded functions defined on $\mathcal{R}$. Suppose that $H_o\phi(H_o)$, $H\phi(H_o)$ and $V\phi(H_o)$ are in $\mathcal{B}(\mathcal{H})$ and $\psi(H)V\phi(H_o)\varepsilon\mathcal{B}_1$. Then

(a) $R(\lambda) \varepsilon \mathcal{B}_1(\mathcal{H}_o)$ for almost all λ for which $\phi(\lambda)\psi(\lambda) \neq 0$.

(b) $|\mathrm{Re}\int_{\sigma(H_o)} d\lambda\ \phi(\lambda)\psi(\lambda)\mathrm{Tr}_o R(\lambda)| < \infty$. (7.74)

Proof : In view of the assumptions on $\phi(H_o)$, of (5.69) and of Propositions 4.14 and 4.12(d), the following identities hold

$$\psi(H_o)R\phi(H_o) = \psi(H_o)\Omega_+^*(\Omega_- - \Omega_+)\phi(H_o)$$

$$= -\Omega_+^*\psi(H)\int_{-\infty}^{\infty}dt\ \frac{d}{dt}\ V_t^*\phi(H_o)U_t = -i\int_{-\infty}^{\infty}dt\ U_t^*\Omega_+^*\psi(H)V\phi(H_o)U_t.$$

Since $\Omega_+^*\psi(H)V\phi(H_o)\varepsilon\mathcal{B}_1$, we may apply Proposition 7.21 to conclude that there exists a family of trace class operators $\{A(\lambda)\}$ such that for $f,g \varepsilon \mathcal{D}$

$$(f,\psi(H_o)R\phi(H_o)g) = \int d\lambda(f_\lambda,A(\lambda)g_\lambda)_o.$$

Now by Proposition 5.26(b)

$$(f,\psi(H_o)R\phi(H_o)g) = \int d\lambda\ \psi(\lambda)\phi(\lambda)(f_\lambda,R(\lambda)g_\lambda)_o,$$

so that by the essential uniqueness of $\{A(\lambda)\}$ we have $A(\lambda) = \psi(\lambda)\phi(\lambda)R(\lambda)$ a.e., which implies (a). (b) is obtained from (7.72) :

$$\int \mathrm{Tr}_o A(\lambda)d\lambda = \int \psi(\lambda)\phi(\lambda)\mathrm{Tr}_o R(\lambda)d\lambda = 2\pi \ \mathrm{Tr}\ [\Omega_+^* \psi(H)V\phi(H_o)],$$

which is finite. #

PROPOSITION 7.23 : Let $H_o = K_o$, let V be a potential in $L^1(\mathcal{R}^3) \cap L^2(\mathcal{R}^3)$ and $H = K_o + V$. Let $z,\xi \in \rho(H) \cap \rho(H_o)$. Then $(H-z)^{-1}V(K_o-\xi)^{-1}$ is a trace class operator.

Remark : In the remainder of this section we shall use the fact that, under the assumptions of Proposition 7.23, the following operators are in $\mathcal{B}(\mathcal{H})$: VR_ξ^o, HR_ξ^o and $K_o R_z$, where $R_\xi^o = (K_o-\xi)^{-1}$ and $R_z = (H-z)^{-1}$. The proof of this will be given in Section 8-1.

Proof : We introduce the following factorization of V :

$$V = |V|^{\frac{1}{2}} V^{\frac{1}{2}}, \tag{7.75}$$

where $|V|^{\frac{1}{2}}$ is the maximal multiplication operator by $|V(\underline{x})|^{\frac{1}{2}}$ and $V^{\frac{1}{2}}$ that by $|V(\underline{x})|^{\frac{1}{2}}$ sign $V(\underline{x})$. It follows that

$$R_z VR_\xi^o = R_z(K_o-z)R_z^o|V|^{\frac{1}{2}}V^{\frac{1}{2}}R_\xi^o. \tag{7.76}$$

Since $V\epsilon L^1(\mathcal{R}^3)$, $|V(\underline{x})|^{\frac{1}{2}}\epsilon L^2(\mathcal{R}^3)$, so that by Proposition 7.7 both $R_z^o|V|^{\frac{1}{2}}$ and $V^{\frac{1}{2}}R_\xi^o$ belong to $\mathcal{B}_2$. Hence $R_z^o|V|^{\frac{1}{2}}V^{\frac{1}{2}}R_\xi^o \in \mathcal{B}_1$ by Proposition 2.34.

Since $K_o R_{\bar{z}} \in \mathcal{B}(\mathcal{H})$, we have $[(K_o-\bar{z})R_{\bar{z}}]^* \epsilon \mathcal{B}(\mathcal{H})$. Thus the operator $R_z(K_o-z)$ has an extension belonging to $\mathcal{B}(\mathcal{H})$, namely $[(K_o-\bar{z})R_{\bar{z}}]^*$. It now follows from Proposition 2.34 that $R_z VR_\xi^o = [(K_o-\bar{z})R_{\bar{z}}]^* R_z VR_\xi^o \in \mathcal{B}_1$. #

It follows from Propositions 7.22(a) and 7.23 that

$R(\lambda)$ is a trace class operator, i.e. that the total cross section at fixed energy is finite for almost all λ if $V \in L^1(R^3) \cap L^2(R^3)$. Furthermore, if we choose $z = i$ and $\xi = -i$, we obtain by using Proposition 7.22(b) and (7.70)

$$\left| \mathrm{Re} \int_0^\infty d\lambda \frac{\mathrm{Tr}_0 R(\lambda)}{1+\lambda^2} \right| = \left| \frac{1}{2\pi} \int_0^\infty d\lambda \frac{\lambda \bar{\sigma}(\lambda)}{1+\lambda^2} \right| < \infty. \tag{7.77}$$

This inequality restricts the total cross section at high energies. If $\bar{\sigma}(\lambda)$ is assumed to be a monotonic function of λ as $\lambda \to \infty$, then (7.77) implies that $\lim \bar{\sigma}(\lambda) = 0$ as $\lambda \to \infty$. Of course (7.77) does not exclude the possibility of $\bar{\sigma}(\lambda)$ having a sequence of narrow peaks at energies $\{\lambda_k\}$ accumulating at infinity.

One sees that Proposition 7.22 allows one to deduce the finiteness of the total cross section at fixed energy as well as bounds on its high energy behaviour in the form of integrability conditions. This result is applicable to abstract scattering systems; in particular no specific form of H_0 has to be imposed. An application to relativistic potential scattering can be found in Martin and Misra [1]. For non-relativistic potential scattering the results obtainable in this way are not quite as strong as those found by other methods. A further discussion of this point as well as an improvement of the condition (7.77) on the high-energy behaviour of the cross section is given in Sections 7-6 and 12-1.

7-5 SCATTERING OF TWO PARTICLES

The main purpose of this section is to show that the scattering problem for a system consisting of two quantum-mechanical particles may be reduced to that of a single particle in an external field provided that the interaction between the two subsystems is invariant under translations of the entire system. An example of such a two-body interaction is a potential V depending only on the relative distance $\underline{x}_2 - \underline{x}_1$ of the subsystems.

Let $\mathcal{H}_1$ be the Hilbert space of the states of the first particle and $\mathcal{H}_2$ that of the second particle. If the two particles are distinguishable, the Hilbert space $\mathcal{H}$ describing the states of the composite system is the tensor product of $\mathcal{H}_1$ and $\mathcal{H}_2$, i.e. $\mathcal{H} = \mathcal{H}_1 \otimes \mathcal{H}_2$. The observables referring only to the first particle are of the form $A_1 \otimes I$, where A_1 is a self-adjoint operator acting in $\mathcal{H}_1$. Similarly those referring only to the second particle have the form $I \otimes A_2$ with $A_2 = A_2^*$ acting in $\mathcal{H}_2$.

In the case of two identical particles, $\mathcal{H}$ consists of the symmetrized or anti-symmetrized subspace of $\mathcal{H}_1 \otimes \mathcal{H}_2$ according to whether the particles obey Bose-Einstein or Fermi-Dirac statistics. The symmetrized (respectively anti-symmetrized) subspace is that spanned by the set of all vectors of the form $f_1 \otimes f_2 + f_2 \otimes f_1$ (resp. $f_1 \otimes f_2 - f_2 \otimes f_1$) with $f_1, f_2 \in \mathcal{H}_1$ (note that here $\mathcal{H}_1 = \mathcal{H}_2$).

In order to be concrete we discuss the case of two distinguishable non-relativistic spinless particles of masses m_1 and m_2. We then have $\mathcal{H} = L^2(\mathbb{R}^3) \otimes L^2(\mathbb{R}^3)$, which may be identified with $L^2(\mathbb{R}^6)$ by the considerations of Section 2-4. We write $\underline{x}_1 = \{x_{1,1}, x_{1,2}, x_{1,3}\}$ for the variable in the first space $L^2(\mathbb{R}^3)$ defining $\mathcal{H}$ and $\underline{x}_2 = \{x_{2,1}, x_{2,2}, x_{2,3}\}$ for the variable in the second one. Similarly the variables occuring in the Fourier transforms of the functions in H will be denoted by $\underline{k}_1$ and $\underline{k}_2$. The j-th component of the position observable $Q_{1,j}$ for the first particle is the maximal multiplication operator by $x_{1,j}$ in $L^2(R^6)$. The j-th component of momentum observable $P_{1,j}$ of the first particle is the maximal multiplication operator by $k_{1,j}$ acting in the space of the Fourier transforms of the functions in $L^2(R^6)$. Similarly one defines $Q_{2,j}$ and $P_{2,j}$. It should be pointed out that for instance $Q_{1,j}$ may be written as $Q_{1,j} = Q_j \otimes I$, where Q_j is the operator defined in (3.16) and acting here in the first factor $L^2(R^3)$ (see Problem 2.39).

The free Hamiltonian H_o is the sum of the free Hamiltonians $K_{o,1} \otimes I$ and $I \otimes K_{o,2}$ of each particle :

$$D(H_o) = \{f \varepsilon L^2(R^6) \,|\, \underline{k}_1^2 \tilde{f}(\underline{k}_1,\underline{k}_2) \varepsilon L^2(R^6) \text{ and } \underline{k}_2^2 \tilde{f}(\underline{k}_1,\underline{k}_2) \varepsilon L^2(R^6)\}$$

and $$(FH_o f)(\underline{k}_1,\underline{k}_2) = (\underline{k}_1^2/2m_1 + \underline{k}_2^2/2m_2)\tilde{f}(\underline{k}_1,\underline{k}_2). \qquad (7.78)$$

It follows that the free evolution group $U_t = \exp(-iH_o t)$ factorizes into

$$U_t = U_t^{(1)} \otimes U_t^{(2)} \qquad (7.79)$$

with $U_t^{(s)} = \exp(-iK_{o,s}t)$ acting in $L^2(R^3)$, $s = 1,2$. (7.79) expresses the fact that each of the two particles moves independently; if the initial state is a product state $f = f_1 \otimes f_2$, then $U_t f$ remains a product state at all times by virtue of (2.79).

We now introduce the <u>total mass</u> $M = m_1 + m_2$, the <u>reduced mass</u> $m = m_1 m_2/M$ as well as the <u>total momentum</u> observable $\underline{P}_{tot}$ and the <u>relative momentum</u> observable $\underline{P}_{rel}$ as the self-adjoint closures of the following operators (Problem 7.7)

$$\hat{\underline{P}}_{tot} = \underline{P}_1 + \underline{P}_2, \quad \hat{\underline{P}}_{rel} = (m_1 \underline{P}_2 - m_2 \underline{P}_1)/M. \qquad (7.80)$$

It follows from the identity

$$\frac{1}{2M}(\underline{k}_1 + \underline{k}_2)^2 + \frac{1}{2mM^2}(m_1 \underline{k}_2 - m_2 \underline{k}_1)^2 = \underline{k}_1^2/2m_1 + \underline{k}_2^2/2m_2$$

that we may write

$$H_o = K_{o,CM} + K_{o,rel} \qquad (7.81)$$

where $K_{o,CM}$ and $K_{o,rel}$ are the maximal multiplication operators by $(\underline{k}_1^2 + \underline{k}_2^2)/2M$ and by $(2mM^2)^{-1}(m_1 \underline{k}_2 - m_2 \underline{k}_1)^2$ respectively on the Fourier transforms of the functions in $L^2(\mathcal{R}^6)$.

We define the new variables $\underline{k}_{CM} = \underline{k}_1 + \underline{k}_2$, $\underline{k}_{rel} = (m_1 \underline{k}_2 - m_2 \underline{k}_1)/M$ and notice that the Jacobian determinant associated with this change of variables is 1. We may thus write the scalar product in $L^2(\mathcal{R}^6)$ in these new variables :

$$(f,g) = \int d^3k_1 d^3k_2 \; \bar{\tilde{f}}(\underline{k}_1, \underline{k}_2)\tilde{g}(\underline{k}_1, \underline{k}_2)$$

$$= \int d^3k_{CM}\, d^3k_{rel} \; \bar{\tilde{f}}(\underline{k}_{CM}, \underline{k}_{rel})\tilde{g}(\underline{k}_{CM}, \underline{k}_{rel}).$$

This means that we may identify $\mathcal{H} = L^2(\mathcal{R}^6)$ in a different way with $L^2(\mathcal{R}^3) \otimes L^2(\mathcal{R}^3)$ as follows : When taking the Fourier transforms of the functions in this new decomposition of $\mathcal{H}$, the variable in the first factor is $\underline{k}_{CM}$ and that in the second factor $\underline{k}_{rel}$. In the remainder of this section all decompositions of operators as tensor products will refer to this latter tensor product structure of $\mathcal{H}$. In particular the free Hamiltonians may be written as $K_{o,CM} = \mathcal{K}_{o,CM} \otimes I$ and $K_{o,rel} = I \otimes \mathcal{K}_{o,rel}$.

To determine the two position variables $\underline{x}_{CM}$ and $\underline{x}_{rel}$

in this new factorization of H, we must carry out the preceding change of variables in the inverse Fourier transformation, i.e. write

$$\int d^3k_1 d^3k_2 \exp(i\underline{k}_1 \cdot \underline{x}_1 + i\underline{k}_2 \cdot \underline{x}_2)\tilde{f}(\underline{k}_1,\underline{k}_2)$$

$$= \int d^3k_{CM}\, d^3k_{rel} \exp(i\underline{k}_{CM} \cdot \underline{x}_{CM} + i\underline{k}_{rel} \cdot \underline{x}_{rel})\tilde{f}(\underline{k}_{CM},\underline{k}_{rel}).$$

It is seen from this that $\underline{x}_{CM}$ and $\underline{x}_{rel}$ must be such that $\underline{k}_1 \cdot \underline{x}_1 + \underline{k}_2 \cdot \underline{x}_2 = \underline{k}_{CM} \cdot \underline{x}_{CM} + \underline{k}_{rel} \cdot \underline{x}_{rel}$, which is satisfied if $\underline{x}_{CM}$ is the <u>center-of-mass coordinate</u> $\underline{x}_{CM} = (m_1 \underline{x}_1 + m_2 \underline{x}_2)/M$ and $\underline{x}_{rel}$ the <u>relative coordinate</u> $\underline{x}_{rel} = \underline{x}_2 - \underline{x}_1$. It is also seen that, if we denote by $\underline{Q}_{CM}$ and $\underline{Q}_{rel}$ the corresponding maximal multiplication operators, then the following commutation relations hold on $S(R^6)$:

$$[\underline{P}_{tot},\underline{Q}_{rel}] = [\underline{P}_{rel},\underline{Q}_{CM}] \subseteq 0,$$

$$[P_{tot,j},Q_{CM,k}] = [P_{rel,j},Q_{rel,k}] \subseteq -i\delta_{jk}I.$$

(7.81) implies that the free evolution group U_t factorizes also in the new decomposition of H as a tensor product :

$$U_t = U_{t,CM} \otimes U_{t,rel} \tag{7.82}$$

with $U_{t,CM} = \exp(-iK_{0,CM}t)$ and $U_{t,rel} = \exp(-iK_{0,rel}t)$ acting in the respective factor $L^2(R^3)$. Thus under the free evolution the motion of the center of mass and the relative motion are independent. By comparing with (7.79), we see that our system may also be viewed as being composed of two fictitious non-relativistic particles of mass M and m.

Assume now that the total evolution group is also factorized in the form

$$V_t = U_{t,CM} \otimes V_{t,rel}. \tag{7.83}$$

By differentiation (Proposition 3.1) one sees that the total Hamiltonian H must be an extension of $K_{o,CM} \hat{\otimes} I + I \hat{\otimes} H_{rel}$ (which is essentially self-adjoint by Problem 14.8). Conversely, for (7.83) to be verified, it suffices that the difference $H - H_0$ (i.e. the interaction) acts non-trivially only in the space of the relative motion of the two particles.

If the wave operators exist, one finds from (2.81) and (2.82) that

$$\Omega_\pm = \underset{t\to\pm\infty}{\text{s-lim}} (U^*_{t,CM} U_{t,CM} \otimes V^*_{t,rel} U_{t,rel}) = I \otimes \Omega_{\pm,rel}, \quad (7.84)$$

where $\Omega_{\pm,rel} = \text{s-lim}\, V^*_{t,rel} U_{t,rel}$ as $t \to \pm\infty$ in $L^2(R^3)$. Also

$$S = \Omega^*_+ \Omega_- = I \otimes \Omega^*_{+,rel} \Omega_{-,rel} \equiv I \otimes S_{rel}. \quad (7.85)$$

This shows that, under the hypothesis (7.83), it suffices to solve the scattering problem in the space $L^2(R^3)$ of the relative motion. This scattering problem is that of a non-relativistic particle of mass m (the reduced mass) with interaction V. If V is a potential depending only on the distance $\underline{x}_2 - \underline{x}_1$ of the two original particles, the resulting scattering problem is that of potential scattering.

The above elimination of the motion of the center of mass of the two particles is often interpreted as setting the total momentum equal to zero, or as choosing a reference frame moving along with the center-of-mass frame. The formula (7.52) for the differential cross section is then valid in that frame, i.e. one considers cones lying in the space of relative coordinates with apex at the center of mass of the two particles. To relate this cross section to an actually observed one, it is necessary to transform it into the reference frame in which the experiment is carried out. The cross section in any reference frame is always defined as in Chapter 1, with the requirement that all quantities involved (initial data, scattering angles, solid angles) be expressed in that reference frame. It often happens that one of the particles is initially at rest. The corresponding reference frame is then called the laboratory system. The transcription of $d\sigma/d\omega$ from one reference frame into another one is a purely kinematical problem. The details of this calculation will not be given here (Problem 7.5). It should be said, though, that a two-particle system having a sharp value of the total momentum is an idealization in quantum mechanics, since $\underline{P}_{tot}$ has purely continuous spectrum. Thus in principle cross sections for two-particle systems should be defined by working with states in $L^2(\mathbb{R}^6)$ and by a limiting procedure similar to that of Proposition 7.18.

7-6 NOTES AND SUPPLEMENTARY MATERIAL

A. We comment here on the characterization of bound states and scattering states given in Section 7-1. Sufficient conditions in abstract form implying that $\mathcal{H}_p(H) = \mathcal{M}_o(H)$ and $\mathcal{M}_\infty(H) = \mathcal{H}_c(H)$ were given by Amrein and Georgescu [1]. Necessary and sufficient conditions could be obtained by combining their results with those of Baumgärtel [1]. In their paper they do not assume that the continuous part of H is absolutely continuous. However, if H has a singularly continuous part, the definition of $\mathcal{M}_\infty(H)$ has to be modified. Instead of (7.3) one defines a subspace

$$\bar{M}_\infty(H) = \{f\varepsilon\mathcal{H} | \lim_{T\to\infty} \frac{1}{2T}\int_{-T}^{T} dt \, \|F_r e^{-iHt} f\|^2 = 0 \text{ for each } r < \infty\}.$$

All results of Section 7-1 remain true if $M_\infty(H)$ is replaced everywhere by $\bar{M}_\infty(H)$.

The states in $\bar{M}_\infty(H)$ disappear from each bounded region in the time average. However, if $f\varepsilon\mathcal{H}_{sc}(H)$, then $\|F_r\exp(-iHt)f\|$ need not converge to zero as $t\to\pm\infty$. The decay rate of this quantity has been further studied by Sinha [1].

The identities $\mathcal{H}_p(H) = M_o(H)$ and $\bar{M}_\infty(H) = \mathcal{H}_c(H)$ are true under much more general conditions than those of Propositions 7.8 and 7.9. They hold in particular if H is an arbitrary self-adjoint function of the momentum operators (Problem 7.2). In potential scattering it is sufficient to require that $V\varepsilon L^2_{loc}(\mathbb{R}^3)$ *), which can be seen as follows. Suppose that $f\varepsilon D(K_o)$ and $\phi\varepsilon C_o^\infty(\mathbb{R}^3)$. Then

$$K_o\phi(\underline{Q})f = \phi(\underline{Q})K_o f - (\Delta\phi)(\underline{Q})f - 2i\sum_{k=1}^{3}(\frac{\partial\phi}{\partial x_k})(\underline{Q})P_k f. \quad (7.86)$$

Since f is also in $D(P_k)$, this implies that $\phi(\underline{Q})f\varepsilon D(K_o)$. Also $\phi(\underline{Q})f\varepsilon D(V)$ by Proposition 3.10. Let H be any self-adjoint extension of $\hat{H} = K_o + V$, $D(\hat{H}) = \{g | g = \phi(\underline{Q})f$ for some $f\varepsilon D(K_o)$ and some $\phi\varepsilon C_o^\infty(\mathbb{R}^3)\}$. In view of Corollary 7.5, it suffices to show that $(H-i)^{-1}F_r\varepsilon\mathcal{B}_o$ for each $r<\infty$.

Given $r < \infty$, choose $\psi\,\varepsilon\, C_o^\infty(\mathbb{R}^3)$ such that $\psi(\underline{x}) = 1$ for $|\underline{x}| \leq r$. Then, by (7.86)

---

*) $V\varepsilon L^2_{loc}(\Delta)$ means that $V\varepsilon L^2(\Delta')$ for each compact subset Δ' of Δ.

$$(H-i)^{-1}F_r = R_i\psi(\underline{Q})F_r = R_i\psi(\underline{Q})(K_o-i)(K_o-i)^{-1}F_r$$

$$= R_i\{(H-i)\psi(\underline{Q})-V\psi(\underline{Q}) + (\Delta\psi)(\underline{Q}) + 2i(\underline{\nabla}\psi)(\underline{Q})\cdot\underline{P}\}R_i^o F_r.$$

Now $\psi(\underline{Q})R_i^o$, $V\psi(\underline{Q})R_i^o$ and $(\Delta\psi)(\underline{Q})R_i^o$ are in $\mathcal{B}_2$ by Proposition 7.7, and $(\partial\psi/\partial x_k)(\underline{Q})P_kR_i^o$ is in $\mathcal{B}_o$, being the uniform limit of the sequence of Hilbert-Schmidt operators $\{(\partial\psi/\partial x_k)(\underline{Q})\cdot P_kR_i^oG^k(n)\}$ as $n \to \infty$, where $G^k(n)$ is the spectral projection of P_k associated with the interval $(-n,n]$. Hence $(H-i)^{-1}F_r\varepsilon\mathcal{B}_o$ by Proposition 2.22(d).

An example where $\mathcal{H}_p(H)$ is strictly smaller than $\mathcal{M}_o(H)$ (in which there are bound states belonging to the absolutely continuous subspace of H) can be obtained by taking a potential of the form $W(\underline{x}) = V(\underline{x})-\chi_r(\underline{x})$, where $V(\underline{x})$ is the potential given by Pearson [3] which we mentioned in Section 4-6 and S_r is a ball containing the support of V. The absolutely continuous spectrum of $H = K_o + W$ will then contain a negative part. The states associated with the negative absolutely continuous spectrum of H are orthogonal to the ranges of $\Omega_\pm$ by Proposition 5.21, hence they are absorbed at the point $\underline{x} = 0$ (in the sense of (4.60)) as t tends to both $+\infty$ and $-\infty$ (cf. Theorem 3 of Pearson [4]). Since the time evolution is continuous, one then easily shows that these states are in $\mathcal{M}_o(H)$ (cf. also Section 4-6$\underline{C}$ and Amrein and Georgescu [2]).

$\underline{B}$. Lemma 7.6 has the following generalization :

LEMMA 7.24 : Let $2 \leq p < \infty$ and suppose that the functions ψ and ϕ are in $L^p(\mathbb{R}^n)$. Then $\psi(\underline{Q})\phi(\underline{P})$ is a compact operator in

$L^2(\mathbb{R}^n)$ and $\|\psi(\underline{Q})\phi(\underline{P})\| \leq (2\pi)^{-n/p}\ \|\psi\|_p\|\phi\|_p$.

The proof is based on the Hölder and the Hausdorff-Young inequalities (cf. Propositions 6.1 and 6.2 of Faris [1]).

C. The definition (7.15) of the time of sojourn of a state in a ball of radius r could also be used to characterize scattering states. Typically one would expect a scattering state to spend only a finite amount of time in any bounded region. This question has been investigated by Sinha [1] who obtained the following results : If $I_r(f,H) < \infty$, then $\|F_r \exp(-iHt)f\|^2 \to 0$ as $t \to \infty$. Furthermore, if $I_r(f,H) < \infty$ for each $r < \infty$, then $f \varepsilon \mathcal{H}_{ac}(H)$.

One may define a set $\mathcal{N}_\infty(H) = \{f | I_r(f,H) < \infty$ for each $r < \infty\}$. One difficulty that arises from this definition is that in general $\mathcal{N}_\infty(H)$ is not closed. In potential scattering one can however prove that the closure of $\mathcal{N}_\infty(H)$ equals $\mathcal{H}_{ac}(H)$ under certain assumptions, e.g. if $D(K_o) \subseteq D(H)$ or if V is in the Rollnik class (cf. Section 8-4). Furthermore $\mathcal{N}_\infty(H)$ is closed if the potential is a sufficiently small perturbation of K_o, cf. Sinha [1].

D. As regards time delay in potential scattering, we should first say that the quantity $(\Delta T)(f)$ introduced in (7.16) should be viewed as a global time delay in the sense that its calculation involves the scattered wave in all directions. Another possibility is to introduce a notion of time delay from a given direction of the incoming beam into a given direction of observation of the scattered particles, i.e. an angular time delay. The global time delay must then somehow be an average of the angular one over all directions. This latter approach involves two additional difficulties :

(i) in order to get a correct probabilistic interpretation, one has to normalize to unity the part of the initial state which is being scattered into a cone, and therefore the angular time delay will not be obtainable from a linear operator;

(ii) one will have to pass to the limit of plane waves, a problem that we have already encountered in Section 7-3 when deriving the differential cross section. For a discussion of the global and the angular time delay and further references the reader may consult the paper by Bollé and Osborn [1].

Eq. (7.30) is known in potential scattering as the Eisenbud-Wigner relation (cf. Wigner [2]). Its generalization to the case where S is not a diagonalizable operator in the spectral representation of H_o is mathematically rather delicate. The following two investigations of this question should be cited. (a) Jauch, Misra and Sinha [1] assume that $(H-z)^{-1} - (H_o-z)^{-1}$ is a trace class operator and that H_o is absolutely continuous. By applying Proposition 7.21 they show that for each $r < \infty$ there exists a family $\{\Gamma_r(\lambda)\}$ of trace class operators, where $\Gamma_r(\lambda)$ acts in $\mathcal{K}_{j(\lambda)}$, the Hilbert space associated with the point λ in the spectral representation of H_o, such that

$$I_r(\Omega_- f,H) - I_r(f,H_o) = \int d\lambda\ (f_\lambda,\Gamma_r(\lambda)f_\lambda)_\lambda < \infty$$

for each f in some dense set $\mathcal{D}_o$. Finally $\mathrm{Tr}_\lambda\ \Gamma_r(\lambda)$ converges as $r \to \infty$ in the space $\mathcal{D}'$ of distributions, and the limit distribution ΔT verifies a generalization of the Eisenbud-Wigner relation : if we write $S(\lambda) = \exp[2i\Delta(\lambda)]$, then $(\Delta T)(\lambda) = 2\, d/d\lambda\, \mathrm{Tr}_\lambda \Delta(\lambda)$ in $\mathcal{D}'$. (b) Lavine in [LM] considers for each $r < \infty$ the sesquilinear form $(H_o f, \Delta T_r g) \equiv \int_{-\infty}^{\infty} dt (H_o f, U_t^*[\Omega_-^* F_r \Omega_- - F_r] U_t g)$ for $f, g \varepsilon E^o_{(a,b)}\mathcal{H}$ with $0 < a < b < \infty$ (compare with (7.16)). By using the theory of relatively smooth operators (see Section 9-2B), he shows that, under suitable assumptions on V for potential scattering in $L^2(\mathcal{R})$, $(H_o f, \Delta T_r g)$ converges to $\int_{-\infty}^{\infty} (\Omega_- f, V_t^*(V + \frac{1}{2} QV')V_t \Omega_- g) dt$ as $r \to \infty$. Hence, if $V(x) + \frac{1}{2} x V'(x)$ is everywhere positive or negative, then the time delay also has that sign.

E. Equation (7.33) for scattering into cones is due to Green and Lanford [1] and Dollard [3]. In the latter paper a similar expression is derived for scattering by a Coulomb potential. This relation was generalized to other long range potentials by Martin [1].

The restriction of our considerations in Section 7-3 to potential scattering is inessential. The only explicit property of the free Hamiltonian K_o that was used is eq. (3.51), i.e. the formula for the probability that a state evolving under the free evolution $\{\exp(-iK_o t)\}$ will eventually be localized in a cone C. Now this formula has been established for a large class of unperturbed Hamiltonians that are functions of the momentum operators by Jauch, Lavine and Newton [1], so that in principle all the proofs in Section 7-3 can be carried through for scattering systems with such unperturbed Hamiltonians.

Cross sections are often defined in terms of the flux

rather than the number of particles. The relation between these two ways of defining $d\sigma/d\omega$ has been studied by Combes, Newton and Shtokhamer [1].

F. In applying Proposition 7.22 to potential scattering, it is essential that $V \in L^1(\mathbb{R}^3)$ in order to obtain trace class operators. This set of potentials includes those verifying $|V(\underline{x})| \leq \alpha(1+|\underline{x}|)^{-\beta}$ with $\beta > 3$. This result is not optimal, since the finiteness of $\sigma_{tot}(\lambda)$ is known to hold whenever $\beta > 2$ but not necessarily when $\beta < 2$ [LL, no. 106]. A rigorous proof that $\sigma_{tot}(\lambda) < \infty$ if $\beta > 2$ follows from Proposition 3.1 of Kuroda [5] or from Section 10-2 where it is shown that $R(\lambda)$ is Hilbert-Schmidt (for spherically symmetric potentials one may also consult Villarroel [1]). Other abstract conditions ensuring that $R(\lambda) \in \mathcal{B}_1$ as well as a time-independent proof of Proposition 7.21 are given in Birman [1].

As far as the high energy behaviour is concerned, the integrability condition (7.77) can be improved so as to yield essentially the same result as that obtained by means of analyticity [SM, Ch. VI]. For this one chooses in Proposition 7.22 $\phi(\lambda) = \psi(\lambda) = (\lambda-c)^{-\beta}$ with c less than the lower bound of $\sigma(H)$ and replaces (7.76) by $R_c^\beta V R_c^{o\beta} = R_c^\beta (K_o-c)^\beta R_c^{o\beta} |V|^{\frac{1}{2}} V^{\frac{1}{2}} R_c^{o\beta}$. One finds that $R_c^{o\beta} V R_c^{o\beta} \in \mathcal{B}_1$ provided that $\beta > 3/4$. Also $(K_o-c)^\beta R_c^\beta \in \mathcal{B}(\mathcal{H})$ for each $0 \leq \beta \leq 1$, which can be seen by applying the so-called complex method of interpolation to the scales of Hilbert spaces $\{\mathcal{H}_s\}$ and $\{\mathcal{H}_s^o\}$ ($s \in \mathbb{R}$) associated with the self-adjoint operators H and K_o respectively [RS II, p. 44] : $(K_o-c)^\beta$ is a bounded map from $\mathcal{H}_2$ to $\mathcal{H}_{2-2\beta}^o$ (since $K_o R_z \in \mathcal{B}(\mathcal{H})$) and from $\mathcal{H}_o = L^2(\mathbb{R}^3)$ to $\mathcal{H}^o_{-2\beta}$, hence by the Calderón-Lions interpolation theorem it is bounded from $\mathcal{H}_{2t}$ to

$\mathcal{H}^{o}_{2t-2\beta}$ for $0 < t < 1$, in particular from $\mathcal{H}_{2\beta}$ to $\mathcal{H}^{o}_{o} = L^2(\mathbb{R}^3)$. Thus $R_c^{\beta} V R_c^{o\beta} \in \mathcal{B}_1$ if $3/4 < \beta \leq 1$ and $V \in L^1(\mathbb{R}^3) \cap L^2(\mathbb{R}^3)$, and (7.74) gives $\int_0^{\infty} d\lambda\, \lambda \bar{\sigma}(\lambda)(\lambda - c)^{-2\beta} < \infty$. Therefore, if $\bar{\sigma}(\lambda)$ is a monotonic function, then $\bar{\sigma}(\lambda) \leq c_{\eta} \lambda^{-\frac{1}{2}+\eta}$ for any $\eta > 0$. For a proof with $\beta = 1$, see also Jauch and Sinha [1].

G. The total momentum operator $\underline{P}_{tot}$ of a system of two particles defined in (7.80) is the infinitesimal generator of the three-parameter group of the simultaneous translations of these particles :

$$[\exp(-i\underline{a}\cdot\underline{P}_{tot})f](\underline{x}_1,\underline{x}_2) = f(\underline{x}_1 - \underline{a}, \underline{x}_2 - \underline{a}).$$

If instead of (7.83) one requires only that the interaction between the two particles be translation invariant, more precisely that $[V_t, \exp(-i\underline{a}\cdot\underline{P}_{tot})] = 0$ for all t and $\underline{a}$, then V_t is decomposable in the spectral representation $\mathcal{G} = L^2(\mathbb{R}^3, L^2(\mathbb{R}^3))$ of $L^2(\mathbb{R}^6)$ with respect to $\underline{P}_{tot}$, i.e. $V_t = \{V_t(\underline{k}_{CM})\}$, where, for each $\underline{k}_{CM} \in \mathbb{R}^3$, $V_t(\underline{k}_{CM})$ acts in $L^2(\mathbb{R}^3)$. The same is of course true for $U_t = \exp(-iH_o t)$, so that it is again sufficient to consider the scattering problem in the subspace $L^2(\mathbb{R}^3)$, interpreted as the space of the relative motion at total momentum $\underline{k}_{CM}$. This is more general than what we have done in Section 7-5, since here the interaction may also depend on the total momentum $\underline{k}_{CM}$. The scattering operator will be of the form $S = \{S(\underline{k}_{CM})\}$, i.e. it is given by a family of scattering operators acting in the space $L^2(\mathbb{R}^3)$ of the relative motion which may depend explicitly on the total momentum $\underline{k}_{CM}$.

H. <u>Proof of Lemma 7.12</u> : The hypotheses on ϕ imply that both ϕ and ϕ'' are in $L^1(\mathcal{R})\cap L^2(\mathcal{R})$. Upon integrating by parts, one obtains the following results for the Fourier transforms $\tilde{\phi}'$ and $\tilde{\phi}''$ of ϕ' and ϕ'' :

$$\tilde{\phi}'(k) = ik\,\tilde{\phi}(k) \ , \ h(k) \equiv \tilde{\phi}(k)-\tilde{\phi}''(k) = (1+k^2)\tilde{\phi}(k). \tag{7.87}$$

Since $\phi - \phi'' \in L^1(\mathcal{R})$, h is uniformly bounded, i.e. $|h(k)| \leq c$ for some $c < \infty$ and all $k\in\mathcal{R}$. Hence by (7.87) $|\tilde{\phi}(k)| \leq c(1+k^2)^{-1}$, i.e. $\tilde{\phi}\in L^1(\mathcal{R})$. Furthermore, since $\phi - \phi'' \in L^2(\mathcal{R})$, we have $h\in L^2(\mathcal{R})$. Hence we get from (7.87) and the Schwarz inequality

$$\int|\tilde{\phi}'(k)|dk \leq \int k(1+k^2)^{-1}|h(k)|dk \leq \|k(1+k^2)^{-1}\|\,\|h\| < \infty. \ \#$$

Proof of Proposition 7.20 : (i) Let D be the integral operator with kernel $\bar{K}_A(\underline{y},\underline{x})$. Then for $f,g \in \mathcal{H} = L^2(\Delta,d\mu)$ we get from the Schwarz inequality

$$\int|\bar{f}(\underline{x})\bar{K}_A(\underline{y},\underline{x})g(\underline{y})|d\mu(\underline{x})d\mu(\underline{y}) \leq \{\int|f(\underline{x})|^2|g(\underline{y})|^2d\mu(\underline{x})d\mu(\underline{y})\ \cdot$$
$$\cdot \int|K_A(\underline{x},\underline{y})|^2d\mu(\underline{x})d\mu(\underline{y})\}^{\frac{1}{2}} = \|f\|\,\|g\|\,\|A\|_{HS} < \infty. \tag{7.88}$$

This allows us to interchange the order of integration in the following integral by Fubini's theorem

$$(f,Dg) = \int d\mu(\underline{x})\bar{f}(\underline{x}) \int d\mu(\underline{y})\bar{K}_A(\underline{y},\underline{x})g(\underline{y})$$
$$= \int d\mu(\underline{y}) \int d\mu(\underline{x})\bar{K}_A(\underline{y},\underline{x})\bar{f}(\underline{x})g(\underline{y}) = (Af,g).$$

This proves that $D = A^*$.

(ii) Let T be the integral operator with kernel (7.57). One has as above

$$\int d\mu(\underline{x})d\mu(\underline{y})d\mu(\underline{z})\,|\bar{f}(\underline{x})K_A(\underline{x},\underline{z})K_B(\underline{z},\underline{y})g(\underline{y})|$$
$$\leq \|f\|\,\|g\|\,\|A\|_{HS}\|B\|_{HS}\,.$$

Then, by interchanging the integrals with respect to the

variables $\underline{y}$ and $\underline{z}$, one obtains

$$(f,Tg) = \int d\mu(\underline{x})\bar{f}(\underline{x}) \int d\mu(\underline{y}) \int d\mu(\underline{z}) K_A(\underline{x},\underline{z}) K_B(\underline{z},\underline{y}) g(\underline{y})$$

$$= \int d\mu(\underline{x})\bar{f}(\underline{x}) \int d\mu(\underline{z}) K_A(\underline{x},\underline{z}) \int d\mu(\underline{y}) K_B(\underline{z},\underline{y}) g(\underline{y}) = (f,ABg).$$

Hence T = AB by Lemma 2.27. #

PROBLEMS

7.1 : Prove that $M_o(H)$ is a subspace.

7.2 : Assuming the continuous part of H to be absolutely continuous, show that the conclusions of Proposition 7.4 hold (a) if there exists an operator $T \varepsilon \mathcal{B}(\mathcal{H})$ with dense range verifying $TV_t = V_t T$ for all real t and such that $F_r T \varepsilon \mathcal{B}_o$ for each $r < \infty$, (b) if $\mathcal{H} = L^2(\mathcal{R}^3)$ and H is a self-adjoint function of the momentum operators P_1, P_2, P_3. [Hint : apply (a) with $T = (|\underline{P}|^2 + 1)^{-1}$]. (c) Solve the same Problem as in (b) in $L^2(\mathcal{R}^n)$. (Use Lemma 7.24 or the fact that $\chi_s(\underline{k})(k^2+1)^{-1} \varepsilon L^2(\mathcal{R}^n)$ and Proposition 2.22(d) to prove compactness of $F_r T$).

7.3 : Let $H_o = K_o$, let $H = H^*$ be such that the wave operators $\Omega_\pm$ exist and let $M_\infty^\pm(H)$ be defined by (7.3) for $t \to \pm\infty$ respectively. Show that $\Omega_+ \mathcal{H} \subseteq M_\infty^+(H)$ and $\Omega_- \mathcal{H} \subseteq M_\infty^-(H)$.

7.4 : Verify (7.55).

7.5 : Consider the scattering between two particles of mass m_1 and m_2 one of which is initially at rest, and express the differential cross section in the laboratory system in terms of that in the center-of-mass frame (cf. [T, Ch. 4-d]). To what measurement does the differential cross section in the laboratory system correspond ?

7.6 : Generalize the theory of time delay of Section 7-2 to the case where the spectral representation of $H_o/M_\infty(H_o)$ is $\mathcal{G} = \oplus_i L^2(\Lambda_i, \mathcal{K}_i)$ and each Λ_i is an open set. (Hint : Each open set is the union of a countable collection of disjoint open intervals [R, Prop. 2.8]).

7.7 : Show that the operator sum $\hat{\underline{P}}_{tot} = \underline{P}_1 + \underline{P}_2$ is essentially self-adjoint but not self-adjoint and determine its closure.

PART III

SPECIAL TOPICS
IN
POTENTIAL SCATTERING

CHAPTER 8 : SELF-ADJOINTNESS. EXISTENCE OF WAVE OPERATORS

This chapter deals with two fundamental questions in potential scattering. In Section 8-1 we give a general theorem (Theorem of Rellich-Kato) in the perturbation theory of self-adjoint operators ensuring stability of self-adjointness and apply it to potential scattering. In Section 8-2 we use the so-called Cook-Hack criterion to establish the existence of the wave operators for short-range potentials. Further topics in this section are scattering at large distances and screening for short-range potentials. Finally in Section 8-3 we work out in detail the scattering theory for the example of so-called separable interactions.

8-1 SELF-ADJOINTNESS OF THE HAMILTONIAN

We have seen in Section 4-1 that for the description of a scattering process we begin with two evolution groups $\{U_t\}$ and $\{V_t\}$ or equivalently their self-adjoint generators H_o and H respectively. As in classical mechanics, in many cases of physical interest the total Hamiltonian H can be written formally as $H = H_o + V$ where H_o is the unperturbed Hamiltonian and V is the interaction Hamiltonian. The explicit forms of H_o and V will be dictated by physical intuition in the context of the model employed (see also Section 3-2).

Since frequently both H_o and V are unbounded operators, the sum $H_o + V$ defined on $D(H_o) \cap D(V)$ need not be self-adjoint or even essentially self-adjoint. If $H_o + V$ admits more than one self-adjoint extension, then the evolution group $\{V_t\}$ is not uniquely determined and one needs to specify further conditions. However there exist sufficient conditions on V which force $H_o + V$ to be self-adjoint or essentially self-adjoint on $D(H_o) \cap D(V)$. One such condition is the following which is usually called the <u>condition of Rellich-Kato</u>.

Let A and B be linear operators in $\mathcal{H}$. If

$$D(A) \subseteq D(B) \tag{8.1}$$

and there exist $\alpha, \beta \geq 0$ such that for all $f \in D(A)$

$$\|Bf\| \leq \alpha\|f\| + \beta\|Af\|, \tag{8.2}$$

then we say that B is relatively bounded with respect to A or simply <u>A-bounded</u>. The greatest lower bound β_o of all β for which (8.2) is satisfied is called the <u>A-bound</u> of B.

Conditions (8.1) and (8.2) are not entirely independent of each other, as can be seen from the following proposition of which we shall not make use in the sequel.

PROPOSITION 8.1 : Let A be self-adjoint and B be a closed operator in $\mathcal{H}$ with $D(A) \subseteq D(B)$. Then (8.2) is verified.

Proof : Since $D(A) \subseteq D(B)$, the operator $B(A-z)^{-1}$ is defined everywhere in $\mathcal{H}$ for each $z \in \rho(A)$. Next suppose that there exists a sequence $\{f_n\}$ converging strongly to f in $\mathcal{H}$ such that $\{B(A-z)^{-1}f_n\}$ also converges strongly. Since $(A-z)^{-1} \in \mathcal{B}(\mathcal{H})$, it is clear that $\{(A-z)^{-1}f_n\}$ converges strongly to $(A-z)^{-1}f$. Therefore $\{B(A-z)^{-1}f_n\}$ converges to $B(A-z)^{-1}f$ because B is a closed operator. Thus $B(A-z)^{-1}$ is a closed operator defined everywhere, and by the closed graph theorem [K, Theorem III.5.20] we conclude that it belongs to $\mathcal{B}(\mathcal{H})$.

Finally, let $f \in D(A)$. Then for any $z \in \rho(A)$,

$$\|Bf\| = \|B(A-z)^{-1}(A-z)f\|$$
$$\leq \|B(A-z)^{-1}\| \, \|Af\| + |z| \, \|B(A-z)^{-1}\| \, \|f\| \,, \qquad (8.3)$$

and we have the inequality (8.2) with $\alpha = |z| \, \|B(A-z)^{-1}\|$, $\beta = \|B(A-z)^{-1}\|$. (The proposition is also true if A is only closed with non-empty resolvent set.) #

We can rewrite the inequality (8.2) in a different way : there exist $\alpha', \beta' \geq 0$ such that for all $f \in D(A)$,

$$\|Bf\|^2 \leq \alpha'^2 \|f\|^2 + \beta'^2 \|Af\|^2. \qquad (8.4)$$

That (8.4) implies (8.2) is immediate on writing $\alpha = \alpha'$ and $\beta = \beta'$. Conversely, since for $\eta > 0$, $(\eta^{-\frac{1}{2}}\alpha \|f\| - \eta^{\frac{1}{2}}\beta \|Af\|)^2 =$

$\eta^{-1}\alpha^2\|f\|^2 + \eta\beta^2\|Af\|^2 - 2\alpha\beta\,\|f\|\,\|Af\| \geq 0$, one arrives at (8.4) on squaring both sides of (8.2) and writing $\alpha' = \alpha(1+\eta^{-1})^{\frac{1}{2}}$, $\beta' = \beta(1+\eta)^{\frac{1}{2}}$. Thus the A-bound of B may as well be taken as the greatest lower bound of the possible values of β' satisfying (8.4). With this understanding, we shall often use (8.4) instead of (8.2).

The next proposition establishes the relation between an A-bounded operator B and the operator $B(A-z)^{-1}$.

PROPOSITION 8.2 : Let A be self-adjoint, and consider the following statements :

(a) B is A-bounded.

(b) $B(A-z)^{-1}\varepsilon\mathcal{B}(\mathcal{H})$ for some $z\varepsilon\rho(A)$.

(c) B is A-bounded with A-bound less than 1.

(d) There exists a complex $z\varepsilon\rho(A)$ such that $\|B(A-z)^{-1}\| < 1$.

Then (a) is equivalent to (b) and (c) is equivalent to (d).

Proof : That (b) implies (a) is evident from (8.3). To show that (a) implies (b), we combine (8.4) with (2.48) to conclude that $\|Bf\|^2 \leq \|(\beta'A \pm i\alpha')f\|^2$ for each $f\varepsilon D(A)$. Hence upon writing $f = (A \pm i\alpha'/\beta')^{-1}g$, we get that for all $g\varepsilon\mathcal{H}$

$$\|B(A \pm i\alpha'/\beta')^{-1}g\|^2 \leq \beta'^2\|g\|^2. \qquad (8.5)$$

If moreover the A-bound of B is less than 1, then there exists a positive $\beta' < 1$ satisfying (8.4), and we observe for the corresponding pair α',β' that $\|B(A \pm i\alpha'/\beta')^{-1}\| < 1$. The converse implication is immediate from (8.3). #

Remark 8.3 : (a) If B is A-bounded, then $B(A-z)^{-1} \varepsilon \mathcal{B}(\mathcal{H})$ for all $z \varepsilon \rho(A)$. In fact, by Proposition 8.2, $B(A-z_0)^{-1} \varepsilon \mathcal{B}(\mathcal{H})$ for some $z_0 \varepsilon \rho(A)$ and the conclusion follows from the first resolvent equation (5.58) by writing,

$$B(A-z)^{-1} = B(A-z_0)^{-1} + (z-z_0)B(A-z_0)^{-1}(A-z)^{-1}. \quad (8.6)$$

(b) From the A-boundedness of B it also follows that the closure of $(A-z)^{-1}B^*$ is bounded. This is so because by (2.33), $(A-z)^{-1}B^* \subseteq (B(A-\bar{z})^{-1})^* \varepsilon \mathcal{B}(\mathcal{H})$. If furthermore B is self-adjoint, as is often the case in potential scattering, then the closure of $(A-z)^{-1}B$ is in $\mathcal{B}(\mathcal{H})$.

We introduce another concept involving the pair of operators A,B which is useful in applications. We say that a symmetric operator B is relatively compact with respect to A or simply A-compact if $D(A) \subseteq D(B)$ and if the operator $B(A-z)^{-1}$ is compact for some $z \varepsilon \rho(A)$. Clearly, if B is A-compact then it is A-bounded. That the property of A-compactness implies some restriction on the A-bound of B is the content of the next proposition.

PROPOSITION 8.4 : Let A be self-adjoint and B be A-compact. Then the A-bound of B is 0.

Proof : We observe from (8.6) that if $B(A-z_0)^{-1}$ is compact for some $z_0 \varepsilon \rho(A)$, then $B(A-z)^{-1}$ is compact for all $z \varepsilon \rho(A)$. Thus with no loss of generality we can assume that $B(A-i)^{-1}$ is compact. It is easy to see that (Problem 8.1) the sequence of bounded operators $\{(A+i)(A+in)^{-1}\}$ converges strongly to zero as $n \to \infty$ and that $\|(A+i)(A+in)^{-1}\| \leq 1$ for all n. Since

$B(A-in)^{-1} = [B(A-i)^{-1}][(A-i)(A-in)^{-1}]$, it follows by Lemma 8.23 (see Section 8-4) that $\{B(A-in)^{-1}\}$ converges in norm to zero as $n \to \infty$. Therefore, by choosing n large enough, the norm $\|B(A-in)^{-1}\|$ can be made arbitrarily small and we arrive at the conclusion by using (8.3). #

Now we are ready to state the central theorem of this section giving sufficient conditions for A+B to be a self-adjoint operator.

PROPOSITION 8.5 (Theorem of Rellich-Kato) : Let $A = A^*$ and B be a symmetric operator. (a) If B is A-bounded with A-bound less than 1, then A+B is self-adjoint with $D(A+B) = D(A)$. If moreover A is bounded below*), then A+B is also bounded below. (b) If B is A-compact, then A+B is self-adjoint and is bounded below whenever A is. Furthermore, the essential spectra of A + B and of A are the same, i.e. $\sigma_e(A+B) = \sigma_e(A)$.

Proof : Without loss of generality, we may assume that (8.4) holds with $\alpha' > 0$ and $0 < \beta' < 1$. Clearly A+B is symmetric on D(A), and for any $f \in D(A)$ we have

$$[\beta'/\alpha'(A+B) \pm i]f = [I + B(A \pm i\alpha'/\beta')^{-1}](\beta' A/\alpha' \pm i)f. \qquad (8.7)$$

By (8.5), $\|B(A \pm i\alpha'/\beta')\| \leq \beta' < 1$ and thus the operators $I+B(A \pm i\alpha'/\beta')^{-1}$ map H one-to-one onto H by Proposition 2.20.

---

*) A self-adjoint operator A is said to be bounded below if there exists a finite M such that $(f,Af) \geq M\|f\|^2$ for all $f \in D(A)$. The least upper bound $T(A)$ of all such M is called the lower bound of A. $T(A)$ coincides with the lower bound of the spectral family of A, i.e. the lower bound of $\sigma(A)$ (Problem 8.2).

Since by Proposition 2.14, $(\beta' A/\alpha' \pm i)$ map D(A) onto H, we conclude that the ranges of the operators $[\beta'/\alpha'(A+B)\pm i]$ are the whole of H. Therefore by Proposition 2.14 the operator $\beta'/\alpha'(A+B)$ and hence A+B is self-adjoint with its domain $D(A+B) = D(A)$.

In order to show that A+B is bounded below, it suffices to exhibit a real number M such that all real z less than M belong to $\rho(A+B)$. Now for $z < T(A)$, $(A-z)^{-1} \varepsilon B(H)$ (see Problem 8.2), so that by (8.2) and (5.52) we have

$$\begin{aligned} \|B(A-z)^{-1}\| &\leq \alpha\|(A-z)^{-1}\| + \beta\|A(A-z)^{-1}\| \\ &\leq \alpha(T(A)-z)^{-1} + \beta \sup_{\lambda\varepsilon\sigma(A)} |\lambda|(\lambda-z)^{-1} \\ &\leq \alpha(T(A)-z)^{-1} + \beta \max\{1,|T(A)|(T(A)-z)^{-1}\}, \end{aligned}$$

which is less than 1 whenever $z < M$, where $M = T(A) - \max\{\alpha/(1-\beta),\alpha+\beta|T(A)|\}$. Thus for $z < M < T(A)$, $(A-z)^{-1}\varepsilon B(H)$ and $[I+B(A-z)^{-1}]^{-1}\varepsilon B(H)$ by Proposition 2.20, and the conclusion follows by taking the inverse of the identity $A+B-z = [I+B(A-z)^{-1}](A-z)$. We note that the lower bound $T(A+B)$ is greater than or equal to M, determined above.

(b) If B is A-compact, then in particular it is A-bounded with A-bound less than 1, and therefore all the conclusions of (a) follow in this case also. If $\lambda\varepsilon\sigma_e(A)$, then by Lemma 5.19 there exists a sequence of vectors $\{f_n\}$ in D(A) such that $\|f_n\| = 1$, $\{f_n\}$ converges weakly to zero and $\{(A-\lambda)f_n\}$ converges strongly to zero as $n\to\infty$. It follows that for any $z\varepsilon\rho(A)$, $(A-z)f_n = (A-\lambda)f_n + (\lambda-z)f_n$ converges weakly to θ. Therefore, $(A+B-\lambda)f_n = (A-\lambda)f_n + Bf_n =$

$(A-\lambda)f_n + [B(A-z)^{-1}](A-z)f_n$ converges strongly to zero, since $B(A-z)^{-1}$, a compact operator, transforms a weakly convergent sequence into a strongly convergent one (Proposition 2.23). Thus, again by Lemma 5.19, λ belongs to $\sigma_e(A+B)$ and we have $\sigma_e(A) \subseteq \sigma_e(A+B)$.

Furthermore, it follows from (8.7) that for $z = i\alpha'/\beta'$ the resolvents $(A+B-z)^{-1}$ and $(A-z)^{-1}$ satisfy the relation

$$(A+B-z)^{-1} = (A-z)^{-1}(I+B(A-z)^{-1})^{-1}, \tag{8.8}$$

and thus $B(A+B-z)^{-1}$ is also compact. Therefore, by reversing the roles of A and A+B in the above argument, we get the inclusion $\sigma_e(A+B) \subseteq \sigma_e(A)$ and hence the equality $\sigma_e(A) = \sigma_e(A+B)$. #

Remark 8.6 : If B is A bounded with A-bound less than 1, then B is also A+B-bounded. To see this we use the inequality $\|Af\| \leq \|(A+B)f\| + \|Bf\|$ and rewrite (8.2) as $\|Bf\| \leq \alpha\|f\| + \beta\|(A+B)f\| + \beta\|Bf\|$. Since (8.2) is satisfied with $0 < \beta < 1$, we can subtract $\beta\|Bf\|$ to arrive at the inequality $\|Bf\| \leq \alpha(1-\beta)^{-1}\|f\| + \beta(1-\beta)^{-1}\|(A+B)f\|$, which establishes our claim. However the (A+B)-bound of B, in general, will not be less than 1, though its A-bound is. A simple counter-example is obtained by taking $B = -3/4\,A$ where A is unbounded. On the other hand, if the A-bound of B is less than $\frac{1}{2}$, it follows easily from the above considerations that its A+B-bound is less than one. Also, as we have observed in (8.8), if B is A-compact, then it is A+B-compact and thus the A+B-bound of B is 0 in this case. From the identity $(A-z)(A+B-z)^{-1} = I-B(A+B-z)^{-1}$ it is clear that A is (A+B)-bounded.

For applications in potential scattering, we identify $A = K_0$ and $B = V$, where V is the multiplication operator by a real potential $V(\underline{x})$. In all statements concerning potential scattering, we assume V to be a real-valued function without mentioning it. The next proposition contains the implications of various assumptions on $V(\underline{x})$ on the self-adjointness of K_0+V. Operators of the form K_0+V are often called <u>Schrödinger operators</u>.

<u>PROPOSITION 8.7</u> : Let K_0 be the free Hamiltonian defined in (3.29). (a) If $V \in L^2(\mathbb{R}^3)$, then V is K_0-compact and K_0+V is self-adjoint with $D(K_0+V) = D(K_0)$. (b) If $V \in L^\infty(\mathbb{R}^3)$, then V is K_0-bounded with K_0-bound 0 and K_0+V is selfadjoint. If moreover $V(\underline{x})$ converges to zero as $|\underline{x}| \to \infty$ *), then V is K_0-compact. (c) If $V = V_1 + V_2$ where $V_1 \in L^2(\mathbb{R}^3)$, $V_2 \in L^\infty(\mathbb{R}^3)$ and converges to 0 as $|\underline{x}| \to \infty$, then V is K_0-compact and K_0+V is self-adjoint.

<u>Proof</u> : (a) First we need to show that $D(K_0) \subseteq D(V)$. By Proposition 3.10 we know that, if $f \in D(K_0)$, then $f(\underline{x})$ is a bounded uniformly continuous function. Therefore $\int |V(\underline{x}) f(\underline{x})|^2 d^3x \leq \|f\|_\infty^2 \|V\|_2^2 < \infty$ and thus $D(K_0) \subseteq D(V)$. Then by Prop. 7.7 it follows that $VR_z^0 \equiv V(K_0 - z)^{-1}$ is a Hilbert-Schmidt operator and hence compact. Proposition 8.5 (b) leads to the conclusion that K_0+V is self-adjoint with domain $D(K_0)$.

(b) Since in this case $D(V) = \mathcal{H}$, it is clear that $D(K_0) \subseteq D(V)$. Hence for all $f \in \mathcal{H}$, $\|Vf\| \leq \|V\|_\infty \|f\|$. Thus V has K_0-bound 0, and by Proposition 8.5(a) K_0+V is self-adjoint.

---

*) It is understood that the convergence is uniform with respect to the angular variables.

If moreover $V(\underline{x}) \to 0$ as $|\underline{x}|\to\infty$, then for any $\eta > 0$ we can find $r > 0$ such that $|V(\underline{x})| < \eta$ for all $|\underline{x}| > r$. Let F_r be the projection defined in (7.1). Then $VF_rR_z^o$ is Hilbert-Schmidt by Lemma 7.6 for $z\varepsilon\rho(K_o)$ and $\|V(I-F_r)\| < \eta$. Since $\|VR_z^o - VF_rR_z^o\| \le \|V(I-F_r)\|\,\|R_z^o\| < \eta\,\|R_z^o\|$, we conclude that VR_z^o is the uniform limit of the sequence of compact operators $\{VF_rR_z^o\}$ and hence compact (cf. Proposition 2.22(d)).

(c) Since $VR_z^o = V_1R_z^o + V_2R_z^o$, the conclusion follows from (a) and (b). #

The following well-known potentials serve as examples of possible applications of Proposition 8.7.

Example 8.8 : The Yukawa potential : $V(\underline{x}) = \gamma|\underline{x}|^{-1}e^{-\alpha|\underline{x}|}$, $\gamma\varepsilon R$ and $\alpha > 0$, and the square-well potential :

$$V(\underline{x}) = \begin{cases} V_o & \text{for } |\underline{x}| \le r_o,\ V_o < 0 \\ 0 & \text{for } |\underline{x}| > r_o. \end{cases} \tag{8.9}$$

Both these potentials belong to $L^2(R^3)$, and thus by Proposition 8.7(a) V is K_o-compact, yielding that the total Hamiltonian $H = K_o+V$ is a self-adjoint operator.

Example 8.9 : The Coulomb potential : $V(\underline{x}) = \gamma|\underline{x}|^{-1}$, $\gamma\varepsilon R$. By splitting up $V(\underline{x})$ as in Example 7.10, one concludes from Proposition 8.7(c) that V is K_o-compact and that $H_c = K_o+V$, the Coulomb Hamiltonian, is self-adjoint.

Remark 8.10 : It is worth noticing that in Examples 8.8 and 8.9, V is K_o-compact and hence by Proposition 8.5(b) the essential spectrum of the total Hamiltonian is the same as that of the free Hamiltonian K_o. In other words, $\sigma_e(K_o+V) = [0,\infty)$. Also since the lower bound $T(K_o) = 0$, it follows from

Proposition 8.5(a) that K_o+V is bounded below with lower bound $T(K_o+V)$. These two facts, together with the definition of the essential spectrum, lead to the conclusion that the discrete spectrum (corresponding to bound states in the interpretation of Section 7-1) lies in the finite interval $[T(K_o+V),0)$ with 0 being the only possible accumulation point. A more detailed analysis shows that, whereas in Example 8.8 the discrete spectrum consists of at most a finite number of points, it is infinite in number and accumulates at 0 in Example 8.9 if $\gamma < 0$ ([LL, Section 16] or [NE, equation (14.49)]).

The invariance of the essential spectrum under a relatively compact perturbation implies that the part of the spectrum in $\mathcal{R}-\sigma_e$ consists of only isolated eigenvalues of finite multiplicity (for both the unperturbed and the perturbed operator). On the other hand, in such a situation the nature of the essential spectrum may change. For example, one may start with an absolutely continuous operator and end up with an operator having pure point spectrum after a relatively compact perturbation. This is contained in the following proposition.

PROPOSITION 8.11 (Weyl-Von Neumann) : Let A be a self-adjoint operator in a separable Hilbert space $\mathcal{H}$. Then given any $\eta > 0$, there exists a self-adjoint Hilbert-Schmidt operator B with $\|B\|_{HS} < \eta$ such that the self-adjoint operator A+B has pure point spectrum.

Since the proof of this proposition is not of particular interest for this book, we refer the reader to

[K, Thm. X.2.1]. More information on the spectrum will be available in the next section when we study the existence of wave operators.

Remark 8.12 : If in relation (8.2), β is chosen very close to β_o, the A-bound of B, then the other constant α will in general have to be chosen very large. Thus it is usually impossible to set $\beta = \beta_o$ in (8.2). This can be explicitly seen in the case when $A = K_o$ in $L^2(R^3)$ and $V \in L^2(R^3)$ (Problem 8.2).

As far as perturbation theory is concerned the main result of this section, Proposition 8.7, fails to cover a certain number of cases of physical interest, e.g. the harmonic oscillator potential : $V(\underline{x}) = \kappa|\underline{x}|^2/2$ ($\kappa > 0$) and the Stark potential : $V(\underline{x}) = Ex_1$ (see however Problem 8.11 and Proposition 8.24). In both these examples the interaction V is not small compared to the free Hamiltonian K_o in the sense that no inequality of the type (8.2) is satisfied by V. However it is intuitively clear that, in order to have any scattering, the potential must be small at infinity in a suitable sense, and therefore these two examples are not relevant in the context of potential scattering.

Proposition 8.5 needs the assumption that the A-bound of B is strictly less than 1. This restriction cannot be dropped in general, as is evident from the simple example in which A is (unbounded) self-adjoint and B = -A; here the A-bound of B is 1, but A+B, being a proper restriction of the operator 0, is only essentially self-adjoint. However, one has the following result which we state without proof (see [K, Thm. V.4.6] and Wüst [1]).

PROPOSITION 8.13 : Let A be essentially self-adjoint and B be symmetric. If B is A-bounded satisfying (8.4) with $\beta' = 1$ or (8.2) with $\beta = 1$, then A + B is essentially self-adjoint.

8-2 EXISTENCE OF WAVE OPERATORS

In this section we verify part (A1) of the asymptotic condition discussed in Section 4-1. We first give a relatively abstract condition ensuring the existence of wave operators and then apply it to potential scattering. According to Section 7-5, this covers the scattering of two particles interacting via a translation-invariant potential. The free Hamiltonian for such a system is K_o (with m = the reduced mass = $\frac{1}{2}$). If V has suitable properties, then, as we have seen in the last section, $H = K_o + V$ is a self-adjoint operator with $D(H) = D(K_o)$. Thus we have two evolution groups, the free evolution group $U_t = \exp(-iK_o t)$ and total evolution group $V_t = \exp(-iHt)$, which are defined as in Proposition 5.11.

Though by Proposition 8.7, $K_o + V$ is self-adjoint when $V \in L^2(R^3) + L^\infty(R^3)$, it is intuitively clear that in general such a potential need not give rise to scattering in the sense that (A1) need not be satisfied. For example, consider a constant real potential, $V(\underline{x}) = V_o$ for all $\underline{x}$. Then $V_t^* U_t = \exp(-iV_o t)$ has no limit as $t \to \pm\infty$. From a physical point of view, in order that the wave operators $\Omega_\pm$ exist, the potential function $V(\underline{x})$ should in some sense tend to zero at infinity. The following proposition and its corollary describe one sufficient condition for the existence of the wave operators $\Omega_{\pm}$.

PROPOSITION 8.14 : Let $U_t = \exp(-iH_o t)$ and let V be H_o-bounded with H_o-bound less than 1. Then the total evolution group V_t is well-defined. If furthermore $|\int_a^{\pm\infty} \|VU_t E_\infty(H_o)f\| dt| < \infty$ for some finite a and all f belonging to some fundamental set $\mathcal{D}_o \subseteq D(H_o)$, then (A1) is verified, i.e. the wave operators $\Omega_{\pm}$ exist.

Proof : By Proposition 8.5 we have that $H = H_o + V$ is self-adjoint with domain $D(H_o)$ and by the functional calculus (Proposition 5.11) we can then define the total evolution group $V_t = \exp(-iHt)$.

Set $\Omega(t) = V_t^* U_t E_\infty(H_o)$. It can be shown (Problem 8.3) that, if V is H_o-bounded, then $\Omega(t)f$ is strongly continuously differentiable for all $f \in D(H_o)$ and then $d/dt[\Omega(t)f] = iV_t^* VU_t E_\infty(H_o)f$. Thus $\int_a^\infty \|\frac{d}{dt}\Omega(t)f\| dt = \int_a^\infty \| iV_t^* VU_t E_\infty(H_o)f\| dt = \int_a^\infty \|VU_t E_\infty(H_o)f\| dt < \infty$ for all $f \in \mathcal{D}_o \subseteq D(H_o)$. Therefore by Proposition 4.15, s-lim $\Omega(t)f$ as $t \to \infty$ exists for each $f \in \mathcal{D}_o$, so so that Ω_+ exists by Proposition 2.17. Similarly one proves the existence of Ω_-. #

COROLLARY 8.15 : Let $H_o = K_o$, the free Hamiltonian in $L^2(\mathcal{R}^3)$, and assume that the potential function V belongs to $L^2(\mathcal{R}^3)$. Then $\Omega_{\pm}$ exist.

Proof : First note that $E_\infty(K_o) = I$. By Proposition 8.7, V is K_o-bounded with K_o-bound 0. We choose $\mathcal{D}_o$ in the above proposition to be $\mathcal{S}(\mathcal{R}^3)$, which is dense in $L^2(\mathcal{R}^3)$. Since $\mathcal{S}(\mathcal{R}^3) \subseteq L^1(\mathcal{R}^3) \cap L^2(\mathcal{R}^3)$, it follows from (3.42) that for all

$f \varepsilon S(R^3)$ and $t \neq 0$, $|(U_t f)(\underline{x})| \leq c\|f\|_1 |t|^{-3/2}$, where c is a constant independent of f and t. Thus for $f \varepsilon S(R^3)$

$$\|VU_t f\|^2 = \int |V(\underline{x})|^2 |(U_t f)(\underline{x})|^2 d^3x \leq c^2 \|f\|_1^2 |t|^{-3} \|V\|_2^2, \quad (8.10)$$

hence $|\int_{\pm a}^{\pm\infty} \|VU_t f\| \, dt| \leq c\|f\|_1 \|V\|_2 \int_a^{\infty} t^{-3/2} dt < \infty$ if $a > 0$. Therefore all the hypotheses of Proposition 8.14 are satisfied and we have the desired conclusion. #

If we assume that the potential $V(\underline{x})$ decreases at infinity faster than an inverse power of $|\underline{x}|$, i.e. if there exist $R, c < \infty$ such that

$$|V(\underline{x})| \leq c|\underline{x}|^{-\beta}, \text{ for all } |\underline{x}| > R, \quad (8.11)$$

then V belongs to $L^2(R^3 - S_R)$ provided that the exponent β is larger than 3/2. In Section 8-4 we shall see how we can push this result (the existence of $\Omega_\pm$) to all exponents β greater than 1 by using the estimate (3.62) instead of (3.42). That we cannot achieve more, in particular that the wave operators $\Omega_\pm$ as defined in Section 4-1 do not exist when V is a Coulomb potential, will be seen in Chapter 13. Thus it is convenient to introduce the following classes of potential functions. Let $V \varepsilon L^2_{loc}(R^3)$, i.e. $V\chi_\Delta \varepsilon L^2(R^3)$ for each bounded Borel set $\Delta \subseteq R^3$. Then V is said to be <u>short range</u> if V satisfies (8.11) with β greater than 1. On the other hand potentials for which $\lim |V(\underline{x})| = 0$ and $\{|\underline{x}||V(\underline{x})|\}$ converges to a non-zero (possibly infinite) limit as $|\underline{x}| \to \infty$ are called <u>long range</u>. It is then clear that the Yukawa and the square-well potentials belong to the first category while the Coulomb potential belongs to the second.

With some adjustments the proof of Corollary 8.15

goes through for potentials that may have strong local singularities such that $V \notin L^2_{loc}(\mathbb{R}^3)$ (cf. Section 8-4). In anticipation of this result, we can say that the existence of $\Omega_\pm$ (or more precisely, the applicability of Proposition 8.14) depends crucially on the behaviour of the potential V near infinity but hardly at all on its local behaviour.

Next we prove some results on scattering at large distances and screening for short range potentials. The following abstract proposition is useful for this purpose.

PROPOSITION 8.16 : Let $\{V_n\}$, $n = 0,1,2,3,\ldots$ be a sequence of symmetric operators satisfying

(a) $D(H_o) \subseteq D(V_n)$ for all n,

(b) $\|V_n(H_o-z)^{-1}\| \leq M < 1$ for all n and some $z \in \rho(H_o)$,

(c) $\{V_n(H_o-z)^{-1}\}$ converges strongly to $V_o(H_o-z)^{-1}$ as $n \to \infty$,

(d) The improper integrals $\int_a^{\pm\infty} \|V_n e^{-iH_o t} E_\infty(H_o) f\| \, dt$ converge uniformly in n for all f in some fundamental set $\mathcal{D}_o \subseteq D(H_o)$ and for some finite a.

Set $H_n = H_o + V_n$ for $n = 1,2,\ldots$ and $H = H_o + V_o$. Then $\Omega_\pm^{(n)} \equiv$ s-lim $\exp(iH_n t)\exp(-iH_o t)E_\infty(H_o)$ and $\Omega_\pm \equiv$ s-lim $\exp(iHt)\exp(-iH_o t)E_\infty(H_o)$ exist as $t \to \pm\infty$. Furthermore

$$\text{s-lim } \Omega_\pm^{(n)} = \Omega_\pm \text{ as } n \to \infty. \tag{8.12}$$

Proof (for the + sign) : (i) (a) and (b) together with Propositions 8.2 and 8.5 allow us to conclude that H_n, H are self-adjoint with $D(H_n) = D(H) = D(H_o)$ for all n. On the other hand the convergence of the integrals in (d) ensures the existence of $\Omega_\pm^{(n)}$ and $\Omega_\pm$ by Proposition 8.14.

(ii) Set $\Omega^{(n)}(t) = \exp(iH_n t)\exp(-iH_0 t)E_\infty(H_0)$ and $\Omega(t) = \exp(iHt)\exp(-iH_0 t)E_\infty(H_0)$. We observe as in the proof of Proposition 8.14 that for any $f\varepsilon\mathcal{D}_0$, $\|\Omega^{(n)}f-\Omega^{(n)}(t)f\| \le \int_t^\infty \|\frac{d}{ds}\Omega^{(n)}(s)f\|\,ds \le \int_t^\infty \|V_n\exp(-iH_0 s)E_\infty(H_0)f\|ds$. Therefore it follows from (d) that $\{\Omega^{(n)}(t)f\}$ converges strongly to $\Omega_+^{(n)}f$ as $t\to+\infty$, uniformly in n.

(iii) Let $f\varepsilon D(H_0) = D(H)$. Then

$$\exp(iHt)f-\exp(iH_n t)f = \exp(iH_n t)[\exp(-iH_n t)\exp(iHt)f-f]$$

$$= -i\ \exp(iH_n t)\int_0^t \exp(-iH_n s)(V_n-V_0)\exp(iHs)f\ ds. \qquad (8.13)$$

We may write $f = (H-z)^{-1}g$ for some $g\varepsilon\mathcal{H}$ and $z\varepsilon\rho(H)$ and obtain from (8.13) that

$$\|e^{iHt}f-e^{iH_n t}f\| \le \int_0^t \|[V_n(H-z)^{-1} - V_0(H-z)^{-1}]e^{iHs}g\|\ ds. \quad (8.14)$$

Now $\exp(iHs)$ leaves $D(H) = D(H_0)$ invariant (Corollary 3.2(b)) and $V_n(H-z)^{-1}g-V_0(H-z)^{-1}g = [V_n(H_0-z)^{-1}-V_0(H_0-z)^{-1}](H_0-z)\cdot(H-z)^{-1}g$. By Remark 8.6 we know that $(H_0-z)(H-z)^{-1}\varepsilon\mathcal{B}(\mathcal{H})$ and therefore by (c) of the hypotheses $\{V_n(H-z)^{-1}\}$ converges strongly to $V_0(H-z)^{-1}$ as $n\to\infty$. On the other hand (b) implies that the integrand in (8.14) is bounded by $2\|(H_0-z)(H-z)^{-1}\|\,\|g\|$, which is integrable in $[0,t]$. Thus by the Lebesgue dominated convergence theorem we have that $\exp(iH_n t)f$ converges strongly to $\exp(iHt)f$ for all $f\varepsilon D(H_0)$. Since the sequence $\{\exp(iH_n t)\}$ is uniformly bounded, the convergence is in fact true for all $f\varepsilon\mathcal{H}$ (see Proposition 2.17). Then $\Omega^{(n)}(t)f-\Omega(t)f = [\exp(iH_n t)-\exp(iHt)]\ \exp(-iH_0 t)E_\infty(H_0)f$ also converges strongly to zero for each positive t and all $f\varepsilon\mathcal{H}$ as $n\to\infty$.

(iv) Finally, by the triangle inequality, we have that for all $f \in \mathcal{D}_0$, $\|\Omega_+^{(n)} f - \Omega_+ f\| \leq$

$$\|\Omega_+^{(n)} f - \Omega^{(n)}(t)f\| + \|\Omega^{(n)}(t)f - \Omega(t)f\| + \|\Omega_+ f - \Omega(t)f\|. \quad (8.15)$$

Given any $\eta > 0$, by (ii) the first term on the right hand side of (8.15) can be made less than $\eta/3$ by choosing t large enough independently of n, and the same choice of t renders the last term also less than $\eta/3$. Then, because of (iii), we can choose n large enough to make the second term less than $\eta/3$, thus proving (8.12) on $\mathcal{D}_0$. Since $\Omega_+^{(n)}$ is isometric for all n, the convergence can then be extended to the whole of $\mathcal{H}$ by Proposition 2.17. #

As an application of Proposition 8.16 we consider a sequence of scattering events between two particles where their relative distance in the initial state (at t = 0) becomes larger and larger. Intuitively it is clear that if the potential is sufficiently short range, the particles will fail to be affected by their mutual interaction and will not be scattered at all in the limit of infinite relative distance. To express this idea mathematically, let $\underline{a}$ be a vector in $\mathcal{R}^3$ and consider the transformation : $\underline{x}_{rel} \to \underline{x}_{rel} + \underline{a}$. This induces in $\mathcal{H}_{rel}$ (cf. Section 7-5) a (continuous) unitary representation $U(\underline{a})$ of the translation group of $\mathcal{R}^3$, given by :

$$(U(\underline{a})f)(\underline{x}) = f(\underline{x}-\underline{a}). \quad (8.16)$$

We expect that for each fixed $f \in \mathcal{H}_{rel}$ the S-operator will act like the identity operator on $U(\underline{a})f$ for large enough $|\underline{a}|$, or more precisely $\|SU(\underline{a})f - U(\underline{a})f\| \to 0$ as $|\underline{a}| \to \infty$. The following corollary in fact proves a stronger result, viz. that

$\|\Omega_{\pm}U(\underline{a})f - U(\underline{a})f\| \to 0$ as $|\underline{a}| \to \infty$ (Problem 8.9).

COROLLARY 8.17 : Let $H_0 = K_0$ in $L^2(R^3)$ and $V \varepsilon L^2(R^3)$. Then $\Omega_{\pm}(\underline{a}) \equiv U(\underline{a})\Omega_{\pm}U(-\underline{a})$ converges strongly to I as $|\underline{a}| \to \infty$.

Proof : Set $V_a(\underline{x}) = V(\underline{x}-\underline{a})$ for each $\underline{a} \varepsilon R^3$. Since $V_a \varepsilon L^2(R^3)$ for each $\underline{a}$ and $\|V_a R_z^0\|_{HS}^2 = \|VR_z^0\|_{HS}^2 \leq c\|V\|_2^2(\text{Im}\sqrt{z})^{-1} < 1$ (cf. (7.11)) uniformly in $\underline{a}$, provided that we choose Im z large enough, we have verified the hypotheses (a) and (b) of Proposition 8.16 for the sequence $\{V_a\}$.

Next we claim that $V_a(K_0-z)^{-1}$ converges strongly to 0 as $|\underline{a}| \to \infty$. For this it suffices to show that $V_a f$ converges strongly to θ for all $f \varepsilon D(K_0)$. By Proposition 3.10 f is the inverse Fourier transform of an L^1-function, so that by the Riemann-Lebesgue lemma $f(\underline{x}+\underline{a})$ converges to zero as $|\underline{a}| \to \infty$. On the other hand in the expression $\|V_a f\|^2 = \int |V(\underline{x})|^2 |f(\underline{x}+\underline{a})|^2 d^3x$, the integrand is majorized by $\|f\|_\infty^2 \cdot |V(\underline{x})|^2$, which is integrable. Thus by the Lebesgue dominated convergence theorem $\|V_a f\|^2 \to 0$ as $|\underline{a}| \to \infty$, which proves (c) with $V_0 = 0$.

For all $f \varepsilon S(R^3) \subseteq D(K_0)$, we have from (8.10) that for $|t| \neq 0$, $\|V_a \exp(-iK_0 t)f\| \leq c\|f\|_1 \|V_a\|_2 |t|^{-3/2} = c\|f\|_1 \|V\|_2 |t|^{-3/2}$, and thus (d) is also established. Hence by Proposition 8.16, we conclude that $\Omega_{\pm}(K_0+V_a,K_0)$ converges strongly to $\Omega_{\pm}(K_0,K_0) = I$ as $|\underline{a}| \to \infty$. At the same time we observe that $\Omega_{\pm}(K_0+V_a,K_0) \equiv$ s-lim $\exp[i(K_0+V_a)t]\exp(-iK_0 t) =$ s-lim $U(\underline{a})\exp[i(K_0+V)t]\exp(-iK_0 t)U(-\underline{a}) = U(\underline{a})\Omega_{\pm}(K_0+V,K_0)U(-\underline{a})$ as $t \to \pm\infty$, where we have used the simple facts that $U(\underline{a})$

commutes with K_0 and that $V_a = U(\underline{a})VU(-\underline{a})$. Thus we have shown that the translated wave operators $\Omega_\pm(\underline{a}) = U(\underline{a})\Omega_\pm U(-\underline{a})$ converge strongly to I as $|\underline{a}|\to\infty$. #

COROLLARY 8.18 (Screened short range potentials) : Let $H_0 = K_0$ in $L^2(R^3)$ and $V\in L^2(R^3)$. Also let $\{\phi_\eta(\underline{x})\}$ $(\eta > 0)$ be a set of real-valued functions defined on R^3 such that $|\phi_\eta(\underline{x})| \leq 1$ and $\phi_\eta(\underline{x})$ converges to 1 pointwise as $\eta\to +0$. Define the function V_η by setting $V_\eta(\underline{x}) = V(\underline{x})\phi_\eta(\underline{x})$. Then $\Omega_\pm(K_0+V_\eta,K_0)$ converges strongly to $\Omega_\pm(K_0+V,K_0)$ as $\eta\to +0$.

Proof : Since $\|V_\eta\|_2^2 \leq \|V\|_2^2$ and $\|V_\eta - V\|_2^2 \to 0$ as $\eta\to+0$, all the hypotheses of Proposition 8.16 can be verified easily for the sequence of symmetric operators $\{V_\eta\}$. #

In the above, if we choose $\phi_\eta(\underline{x}) = \exp(-\eta|\underline{x}|)$, then this represents the exponential screening often used in the physics literature. Corollary 8.18 says that for short range potentials, screening does not give any surprising results as the screening parameter is made smaller and smaller. On the other hand for long range potentials the situation is not so simple. In fact when V is a Coulomb potential, $\Omega_\pm(K_0+V_\eta,K_0)$ converges weakly to 0 as $\eta\to +0$ (see Dollard [2]).

8-3 SCATTERING BY SEPARABLE INTERACTIONS

A very simple model in which one can not only show the existence and completeness of the wave operators $\Omega_\pm$, but also calculate explicitly the on-shell S operator or equivalently the scattering cross section, is the so-called separable potential model. In such a model the interaction Hamiltonian V is assumed to be a (bounded) operator of rank one, given as

$$Vf = (h,f)h, \tag{8.17}$$

where h is a fixed vector in $\mathcal{H}$. It is easy to verify that such an operator is Hilbert-Schmidt, in fact trace class (cf. (2.69)). If $\mathcal{H} = L^2(\mathcal{R}^n)$, then by Proposition 2.33 V is an integral operator in $L^2(\mathcal{R}^n)$ with kernel

$$V(\underline{x},\underline{y}) = h(\underline{x})\overline{h(\underline{y})}. \tag{8.18}$$

Since the action of V on a vector $f \in L^2(\mathcal{R}^n)$ is given as $(Vf)(\underline{x}) = \int V(\underline{x},\underline{y})f(\underline{y})d^n y = h(\underline{x})\int \overline{h(y)}f(y)d^n y$ and not as multiplication by a function, one sometimes refers to such interactions as non-local. Such a model has been found useful in some physically interesting cases, particularly in low-energy nuclear problems ([JO, Chapter 19.3]).

The next proposition shows the existence of $\Omega_\pm$ under a restrictive hypothesis on the function h in (8.17). However a more general result is valid, viz. $\Omega_\pm$ exist for any H_0 with $E_\infty(H_0) = E_{ac}(H_0)$ and for any $h \in \mathcal{H}$ (see [K, Theorem X.4.3]). In any case, since V given by (8.17) is in $\mathcal{B}(\mathcal{H})$, it is clear that $H \equiv H_0 + \gamma V$ is self-adjoint for any real γ. The following lemma is useful in proving the existence of $\Omega_\pm$ and obtaining an expression for the scattering cross section.

<u>LEMMA 8.19</u> : Let U_t be the free evolution group $\exp(-iK_0 t)$ in $L^2(\mathcal{R}^3)$ and let f and g be in $L^1(\mathcal{R}^3) \cap L^2(\mathcal{R}^3)$. Then $(f,U_t g) \in L^1(-\infty,\infty) \cap L^2(-\infty,\infty)$, as a function of t.

<u>Proof</u> : Since $|(f,U_t g)| \leq \|f\|\,\|g\|$, we need to show integra-

bility only at infinity. But by (3.42), $|(U_t g)(\underline{x})| \le c|t|^{-3/2} \|g\|_1$. Hence $|(f,U_t g)| \le c|t|^{-3/2} \|f\|_1 \|g\|_1$, which is integrable and square-integrable at infinity. #

PROPOSITION 8.20 : Let $H = L^2(R^3)$ and $H_0 = K_0$, the free Hamiltonian. Furthermore, assume that h in (8.17) is in $L^1(R^3) \cap L^2(R^3)$. Then the wave operators for the pair (H,K_0) exist.

Proof : Let $g \varepsilon L^1(R^3) \cap L^2(R^3)$, which is dense in $L^2(R^3)$. Since $\|VU_t g\| = \|h\| \, |(h,U_t g)|$ is integrable by Lemma 8.19, we arrive at the result by Proposition 8.14. #

One can also verify asymptotic completeness in the sense that $\Omega_\pm \Omega_\pm^* = E_{ac}(H)$ for this restricted class of h by using the methods of Section 9-1 (Problem 9.4). For the purposes of this section, we shall assume this and go on to calculate the on-shell S-operator using the time-independent formulae developed in Chapter 6. In the course of this proof we shall invoke the following lemma which will be proven in Section 8-4.

LEMMA 8.21 : Let Φ be a non-constant holomorphic function in the upper complex half-plane. Assume that for each real λ, $\Phi(\lambda+i\eta)$ converges to a limit $\Phi_+(\lambda)$ as $\eta \to +0$, the convergence being uniform in λ on each compact subset of R. Then for any complex number α, the set $\Gamma_0 = \{\lambda \varepsilon R | \Phi_+(\lambda) = \alpha\}$ is a closed set of Lebesgue measure zero. A similar statement holds for the lower half-plane.

In the next Proposition we use the spectral representation of K_o introduced in (5.94).

PROPOSITION 8.22 : Under the hypotheses of Proposition 8.20 the S-matrix in $L^2(S^{(2)})$ is given by

$$R(\lambda)g \equiv [S(\lambda)-I_o]g = -2\pi i\gamma[1+\gamma(h,R^o_{\lambda+io}h)]^{-1}(h_\lambda,g)_o h_\lambda \tag{8.19}$$

for almost all $\lambda\varepsilon[0,\infty)$, where $g\varepsilon L^2(S^{(2)})$. $R(\lambda)$ is an integral operator in $L^2(S^{(2)})$. Furthermore there exists a closed set in $[0,\infty)$ of Lebesgue measure zero, in the complement of which the kernel $R(\lambda;\underline{\omega},\underline{\omega}')$ of $R(\lambda)$ is continuous in all three variables.

Proof : We begin by setting $\gamma = 1$ and observe that the result for $\gamma \neq 1$ is obtained by substituting $\gamma^{\frac{1}{2}}h$ (or $\gamma^{\frac{1}{2}}h_\lambda$) for h (or h_λ) in the final expressions.

(i) Since $D(H) = D(H_o) \subseteq D(V) = H$, we have by the second resolvent equation (Proposition 6.9) that

$$(h,R_z h) = [1+(h,R^o_z h)]^{-1}(h,R^o_z h). \tag{8.20}$$

Thus the operator $T_z \equiv (V-VR_zV)$ (see (6.44)) in this case takes the simple form

$$T_z f = [1-(h,R_z h)](h,f)h = [1+(h,R^o_z h)]^{-1}(h,f)h. \tag{8.21}$$

(ii) We know from (3.58) that for $\delta > 0$, $(h,R^o_{\lambda+i\delta}h) = i\int_0^\infty \exp(i\lambda t-\delta t)(h,U_t h)dt$. Note that $|\int_0^\infty \exp(i\lambda t)(h,U_t h)dt| < \infty$ by Lemma 8.19 and that $|(h,R^o_{\lambda+i\delta}h)-i\int_0^\infty \exp(i\lambda t)(h,U_t h)dt| \leq \int_0^\infty |e^{-\delta t}-1||(h,U_t h)|dt$. The last integral converges to 0 as $\delta\to+0$ by the Lebesgue dominated convergence theorem. This shows that the limit of $(h,R^o_{\lambda+i\delta}h)$ exists uniformly in λ as $\delta\to+0$. We denote this limit by $(h,R^o_{\lambda+io}h)$ and have

$$(h,R^{o}_{\lambda+io}h) = i\int_{0}^{\infty} e^{i\lambda t}(h,U_t h)dt, \qquad (8.22)$$

which is bounded and continuous in λ.

Now $(h,R^{o}_{z}h)$ is holomorphic in the upper half complex plane and has boundary values as $z\to\lambda+io$, uniformly in λ. Hence by Lemma 8.21 there exists a closed set Γ_o of Lebesgue measure zero in the complement of which $1+(h,R^{o}_{\lambda+io}h)$ is invertible, and hence $[1+(h,R^{o}_{\lambda+io}h)]^{-1}$ is bounded continuous in every compact subset of $\mathcal{R}-\Gamma_o$. It follows by (8.20) that $(h,R_{\lambda+i\delta}h)$ also has boundary values for all $\lambda\varepsilon\mathcal{R}-\Gamma_o$.

(iii) Let $\mathcal{S}_o$ be a countable subset of $\mathcal{S}(\mathcal{R}^3)$ such that for every $\lambda>0$, $\{f_\lambda | f\varepsilon\mathcal{S}_o\}$ is fundamental in $L^2(S^{(2)})$ (Problem 8.6). Let $f\varepsilon\mathcal{S}_o$ and $\eta>0$. Then as in (ii), $(f,R^{o}_{\lambda-i\eta}h)$ has a limit as $\eta\to+0$, viz. $-i\int_{-\infty}^{0}\exp(i\lambda t)(f,U_t h)dt$. Thus $(f,[R^{o}_{\lambda-i\eta}-R^{o}_{\lambda+i\eta}]h)$ has a limit (uniformly in λ), denoted by $(f,\Delta R^{o}_{\lambda}h)$, which is equal to $-i\int_{-\infty}^{\infty}\exp(i\lambda t)(f,U_t h)dt$. By Lemma 8.19, $(f,U_t h)\varepsilon L^1(-\infty,\infty)\cap L^2(-\infty,\infty)$. Thus $(f,\Delta R^{o}_{\lambda}h)$ is $-i(2\pi)^{\frac{1}{2}}$ times the Fourier transform of the L^2-function $(f,U_t h)$. On the other hand by (5.96) and (5.75) we know that

$$(f,U_t h) = \int_{-\infty}^{\infty} e^{-i\lambda t}(f_\lambda,h_\lambda)_o\, d\lambda\,, \qquad (8.23)$$

where the index o refers to the space $L^2(S^{(2)})$ to which f_λ and h_λ belong. Since $f\varepsilon\mathcal{S}(\mathcal{R}^3)$, it follows that $(f_\lambda,h_\lambda)_o\,\varepsilon$ $L^2(\mathcal{R})\cap L^1(\mathcal{R})$. Thus by the unitarity of the Fourier transformation we obtain that for almost all λ :

$$(f,\Delta R^{o}_{\lambda}h) = -i\int_{-\infty}^{\infty}e^{i\lambda t}(f,U_t h)dt = -2\pi i(f_\lambda,h_\lambda)_o. \qquad (8.24)$$

(iv) Let Σ be a compact interval in $(0,\infty)-\Gamma_o$, $\Delta\subseteq\Sigma$ and $f,g\varepsilon\mathcal{S}_o$. Using (7.38), (6.39) and (8.21), we get

$$(f,RE^o_\Delta g) = \lim_{\eta\to+0}\ \lim_{\delta\to+0} \int_\Delta (f,[R^o_{\lambda-i\eta} - R^o_{\lambda+i\eta}]T_{\lambda+i\delta}dE^o_\lambda g) =$$

$$\lim_{\eta\to+0}\ \lim_{\delta\to+0} \int_\Delta (f,[R^o_{\lambda-i\eta}-R^o_{\lambda+i\eta}]h)[1+(h,R^o_{\lambda+i\delta}h)]^{-1}d(h,E^o_\lambda g).\ (8.25)$$

Since by (ii) $[1+(h,R^o_{\lambda+i\delta}h)]^{-1}$ has a bounded continuous limit as $\delta\to+0$ for $\lambda\in\Sigma$, we can use the Lebesgue dominated convergence theorem to interchange the first limit and the integral and get

$$(f,RE^o_\Delta g) = \lim_{\eta\to+0} \int_\Delta (f,[R^o_{\lambda-i\eta}-R^o_{\lambda+i\eta}]h)\,[1+(h,R^o_{\lambda+io}h)]^{-1}d(h,E^o_\lambda g).$$

An identical reasoning allows us to take the limit $\eta\to+0$ inside the integral to conclude

$$(f,RE^o_\Delta g) = (-2\pi i)\int_\Delta (f_\lambda,h_\lambda)_o[1+(h,R^o_{\lambda+io}h)]^{-1}d(h,E^o_\lambda g). \qquad (8.26)$$

Finally, since K_o has absolutely continuous spectrum, we have from (5.89) that $d(h,E^o_\lambda g) = (h_\lambda,g_\lambda)_o d\lambda$. Thus

$$(f,RE^o_\Delta g) = \int_\Delta (f_\lambda,h_\lambda)_o\{-2\pi i[1+(h,R^o_{\lambda+io}h)]^{-1}(h_\lambda,g_\lambda)_o\}d\lambda.$$

Since Δ is arbitrary, $(f_\lambda,R(\lambda)g_\lambda)_o$ equals the integrand above for $\lambda\varepsilon\Sigma$ except on a null set $N(f,g)$. One can find a null set N_o such that equality holds for all $f,g\varepsilon S_o$ and all $\lambda\varepsilon\Sigma - N_o$ (Problem 8.13). As $\{f_\lambda|f\varepsilon S_o\}$ is fundamental in $L^2(S^{(2)})$, we obtain (8.19) for all such λ. The right-hand side of (8.19) is defined for all $\lambda\varepsilon\Sigma$, hence (8.19) holds for all $\lambda\varepsilon\Sigma$ after suitably modifying $R(\lambda)$ on N_o. Since $(0,\infty)-\Gamma_o$ is open, it is a countable union of disjoint open intervals $\{\Sigma_k\}_{k=1}^\infty$ [R, Proposition 2.8]. Therefore every $\lambda\varepsilon(0,\infty)-\Gamma_o$ belongs to some compact interval in some Σ_k and we have (8.19) for every $\lambda\varepsilon(0,\infty)-\Gamma_o$.

(v) Since $R(\lambda)$ is an operator of rank one in $L^2(S^{(2)})$, it is Hilbert-Schmidt and hence an integral operator with kernel given by (with the coupling constant γ reintroduced)

$$R(\lambda;\underline{\omega},\underline{\omega}') = -2\pi i\gamma[1+\gamma(h,R_{\lambda+io}h)]^{-1}h_\lambda(\underline{\omega})\overline{h_\lambda(\underline{\omega}')}, \qquad (8.27)$$

for all $\lambda\varepsilon(0,\infty)-\Gamma_o$. Now, by equation (5.95), the vector h_λ is given by $h_\lambda(\underline{\omega}) = 2^{-\frac{1}{2}}\lambda^{\frac{1}{4}}\tilde{h}(k\underline{\omega})$. Since $h\varepsilon L^1(R^3)\cap L^2(R^3)$ by hypothesis, $\tilde{h}$ is bounded and uniformly continuous. Therefore $h_\lambda(\underline{\omega})$ is continuous in λ and $\underline{\omega}$. From this and (ii) we conclude that $R(\lambda;\underline{\omega},\underline{\omega}')$ given in (8.27) is continuous in $\lambda,\underline{\omega},\underline{\omega}'$ for λ in any compact subset of $(0,\infty)$ not intersecting Γ_o. #

The scattering amplitude is defined in (7.48) as $f(\lambda;\underline{\omega}_o\to\underline{\omega}) = -2\pi i\ \lambda^{-\frac{1}{2}}R(\lambda;\underline{\omega},\underline{\omega}_o)$, and thus (8.27) with the expression for $h_\lambda(\underline{\omega})$ leads to the result that

$$f(\lambda;\underline{\omega}_o\to\underline{\omega}) = -2\pi^2\gamma[1+\gamma(h,R^o_{\lambda+io}h)]^{-1}\tilde{h}(\lambda^{\frac{1}{2}}\underline{\omega})\bar{\tilde{h}}(\lambda^{\frac{1}{2}}\underline{\omega}_o)\ . \qquad (8.28)$$

Then, since the kernel $R(\lambda;\underline{\omega},\underline{\omega}_o)$ satisfies all the required conditions of continuity (cf. Section 7-3), the differential scattering cross section is given by (7.50) and we write it as

$$\frac{d\sigma}{d\omega}(\lambda;\underline{\omega}_o\to\underline{\omega}) = |f(\lambda;\underline{\omega}_o\to\underline{\omega})|^2$$

$$= 4\pi^4\gamma^2|[1+\gamma(h,R^o_{\lambda+io}h)]|^{-2}|\tilde{h}(\lambda^{\frac{1}{2}}\underline{\omega})|^2|\tilde{h}(\lambda^{\frac{1}{2}}\underline{\omega}_o)|^2, \qquad (8.29)$$

which is finite for all $\lambda\varepsilon(0,\infty)-\Gamma_o$. Similarly the relation (7.69) gives an expression for the average total cross section which becomes: $\bar{\sigma}(\lambda) = \pi\lambda^{-1}\|R(\lambda)\|^2_{HS} =$

$$\pi^3\gamma^2|[1+\gamma(h,R^o_{\lambda+io}h)]|^{-2}\left|\int|\tilde{h}(\lambda^{\frac{1}{2}}\underline{\omega})|^2d\omega\right|^2. \qquad (8.30)$$

This shows that, if $\lambda = 0$ is not in Γ_o, then $\bar{\sigma}(0)$ is finite and that $\bar{\sigma}(\lambda)\to 0$ as $\lambda\to\infty$ (Problem 8.7). More about separable potentials can be found in Problems 9.4 and 10.10 and in Section 12-1.

8-4 NOTES AND SUPPLEMENTARY MATERIAL

A. We begin with a result used in the Proposition 8.4.

LEMMA 8.23 : Let $A \in \mathcal{B}_0(\mathcal{H}), \mathcal{B}_2(\mathcal{H})$ or $\mathcal{B}_1(\mathcal{H})$ and let $\{B_n\}$ be a sequence in $\mathcal{B}(\mathcal{H})$ such that $\{B_n^*\}$ converges strongly to B^*. Then $\{AB_n\}$ converges to AB in operator norm, Hilbert-Schmidt norm or trace norm respectively.

Proof : (i) Let $f \in \mathcal{H}$ and let $Af = \sum_1^\infty \lambda_k(e_k,f)h_k$ be the canonical expansion of the compact operator A [see (2.58)]. For every integer $N \geq 1$, define a finite rank operator A_N by setting $A_N f = \sum_{k=1}^N \lambda_k(e_k,f)h_k$. Then by Proposition 2.31 and Problem 2.37(a), $\{A_N\}$ converges to A in operator, Hilbert-Schmidt or trace norm respectively. We denote these norms by adjoining subscripts 0,2,1 respectively. Since $A_N B_n f = \sum_{k=1}^N \lambda_k(B_n^* e_k,f)h_k$, we have by Problem 2.37(b) that

$$\|A_N B_n\|_{0,2,1} \leq |||A_N B_n|||_1 \leq \sum_{k=1}^N \lambda_k \|B_n^* e_k\|. \quad (8.31)$$

We may now assume without loss of generality that $B^* = 0$ and infer from (8.31) that for fixed N, $A_N B_n$ converges to zero as $n \to \infty$ in operator, Hilbert-Schmidt or trace norm respectively.

(ii) By the principle of uniform boundedness, $\|B_n\| = \|B_n^*\| \leq M < \infty$ for all n. On the other hand

$$\|AB_n\|_{0,2,1} \leq \|A-A_N\|_{0,2,1}\|B_n\| + \|A_N B_n\|_{0,2,1}.$$

Thus given $\eta > 0$, we may first choose $N_0(\eta)$ so that for

$N > N_0(\eta)$, $\|A-A_N\|_{0,2,1} < \eta/2M$ respectively and then choose n_0 to ensure that $\|A_N B_n\| < \eta/2$ for $n > n_0$. This implies that $\|AB_n\|_{0,2,1} < \eta$ for $n > n_0$ respectively. #

B. The question of self-adjointness or essential self-adjointness of the formal sum $K_0 + V$ has been of great interest in recent times. A review on this subject can be found in [K, supplementary notes] and the lecture of Simon in [TU].

If $V = V_1 + V_2$ where $V_1 \varepsilon L^2(R^3)$ and $V_2 \varepsilon L^\infty(R^3)$, then by Proposition 8.7 we have self-adjointness of $K_0 + V$ on $D(K_0)$. If instead V is only locally square-integrable, $K_0 + V$ is still defined on $C_0^\infty(R^3)$. Since K_0 is essentially self-adjoint on $C_0^\infty(R^3)$ (see the remark following Proposition 3.9), one expects that $K_0 + V$ is also essentially self-adjoint on $C_0^\infty(R^3)$. The next proposition gives the relevant result (Simander [1]).

PROPOSITION 8.24 : Let $V_1 \varepsilon L^2_{loc}(R^n)$, $V_1(\underline{x}) \geq -\alpha|\underline{x}|^2$ for some $\alpha \varepsilon R$ and let V_2 be K_0-bounded in $L^2(R^n)$ with K_0-bound less than 1. Then $\hat{H} = K_0 + V_1 + V_2$ is essentially self-adjoint on $C_0^\infty(R^n)$.

We shall call a potential strongly singular if it violates the condition of local square-integrability. If V is locally square-integrable everywhere except at one point, say $\underline{x} = 0$, $K_0 + V$ is clearly defined on $C_0^\infty(R^n-\{0\})$. As regards essential self-adjointness, the potential $V(\underline{x}) = \gamma|\underline{x}|^{-2}$ acts as a reference case. For example if $\gamma \geq 3/4$ then the ordinary differential operator $-d^2/dx^2 + \gamma x^{-2}$ is essentially self-adjoint on $C_0^\infty((0,\infty))$ [DS, Theorem XIII 6.23].

For $\gamma < 3/4$ the same differential operator has an infinite number of self-adjoint extensions in $L^2(0,\infty)$, though each of them remains bounded below if $-1/4 < \gamma < 3/4$. In general one has the following result (Simander [1]).

<u>PROPOSITION 8.25</u> : Let $V_1 \varepsilon L^2_{1oc}(R^n-\{0\})$, $V_1(\underline{x}) \geq \gamma|\underline{x}|^{-2} - \alpha|\underline{x}|^2$, for some $\alpha\varepsilon R$ and $\gamma = 1-4^{-1}(n-2)^2$. Also let $V_2 \varepsilon L^\infty(R^n)$. Then $\hat{H} = K_o + V_1 + V_2$ is essentially self-adjoint on $C_o^\infty(R^n-\{0\})$.

If the potential is strongly singular on a closed set Σ of measure zero instead of a point, $K_o + V$ defines a symmetric operator on the dense domain $C_o^\infty(R^n-\Sigma)$. In a more general situation, it may happen that $D(K_o)\cap D(V) = \{\theta\}$ [SM]. In some cases it is still possible to define a "sum" of K_o and V by using quadratic forms, as we shall see below.

It is instructive to consider the potential $V(x) = \gamma x^{-\beta}$, $x \geq 0$. By [DS, Theorem XIII 6.23] $-d^2/dx^2 + V$ is essentially self-adjoint on $C_o^\infty((0,\infty))$ if $\gamma > 0$ and $\beta > 2$. However if the potential is attractive, i.e. if $\gamma < 0$ and $\beta > 2$, $-d^2/dx^2 + V$ has an infinite number of self-adjoint extensions and none of them is bounded below. On the other hand if $3/2 < \beta < 2$, $-d^2/dx^2 + V$ is not essentially self-adjoint on $C_o^\infty((0,\infty))$ for any γ. But in this case one can define a distinguished extension with the help of quadratic forms.

Schrödinger operators with magnetic vector potentials have also been considered in the literature. For results on essential self-adjointness and scattering theory, the reader may consult Simon [2], Kato and Kuroda [1] and references cited there.

C. A quadratic form is a map $q : Q(q) \times Q(q) \to \mathbb{C}$, where $Q(q)$, its form domain, is a dense linear manifold in $\mathcal{H}$, such that q is linear in the second argument and antilinear in the first. If $q(f,g) = \overline{q(g,f)}$, q is said to be symmetric, and if $q(f,f) \geq M\|f\|^2$ for some M, q is semibounded. Let A be a self-adjoint operator and $A = |A|^{\frac{1}{2}}U|A|^{\frac{1}{2}}$ its polar decomposition (see (5.115)). Define Q(A), the form domain of the operator A, by $Q(A) = D(|A|^{\frac{1}{2}})$ and

$$q(f,g) = (|A|^{\frac{1}{2}}f, U|A|^{\frac{1}{2}}g), \quad f,g \in Q(A). \tag{8.32}$$

Then q given by (8.32) is called the quadratic form associated with the operator A.

A semibounded quadratic form q is said to be closed if $Q(q)$ is complete in the norm

$$\|f\|_q = [q(f,f) + (1-M)\|f\|^2]^{\frac{1}{2}}. \tag{8.33}$$

If A is a self-adjoint operator with lower bound M (see the footnote on page 319) and if q is the associated quadratic form, then the closedness of $|A|^{\frac{1}{2}}$ implies that $Q(A) = D(|A|^{\frac{1}{2}})$ is complete in the norm (8.33). That the converse is also true is the content of the next proposition [K, Thm. VI.2.6].

PROPOSITION 8.26 : If q is a closed (symmetric) semibounded quadratic form, then it is the quadratic form of a unique self-adjoint operator.

If A is symmetric and bounded below, then one can associate a symmetric quadratic form by writing $Q(A) = D(A)$ and $q(f,g) = (f,Ag)$ for $f,g \in D(A)$. Then this form has a closed extension with the same lower bound and thus by Proposition

8.26 determines a self-adjoint operator A_F, called the Friedrichs extension of A. Amongst all the self-adjoint extensions of A with the same lower bound as A, A_F is distinguished by the fact that it has the smallest form domain of all [K, Chapter VI.2.3].

Next we consider perturbations of quadratic forms. Let a be the quadratic form of a semibounded self-adjoint operator and b be a symmetric quadratic form. We say b is relatively a-form bounded if $Q(a) \subseteq Q(b)$ and there exist positive numbers α and β so that

$$|b(f,f)| \leq \beta a(f,f) + \alpha\|f\|^2, \text{ for all } f \in Q(a). \qquad (8.34)$$

In such a case the greatest lower bound of all β satisfying (8.34) is called the a-form bound of b, and one has a result similar to the Rellich-Kato theorem (see Simon in [TU]) :

PROPOSITION 8.27 : Let a be the quadratic form of a semibounded self-adjoint operator and b a symmetric form which is a-form bounded with bound less than 1. Then the form $a+b$ defined on $Q(a)$ is the quadratic form of a self-adjoint operator.

If furthermore the form b is the quadratic form of a self-adjoint operator B, then the form boundedness of b relative to the form of a positive operator A can be shown to be equivalent to the (operator) boundedness of $|B|^{\frac{1}{2}}$ relative to $A^{\frac{1}{2}}$. Thus even in the case where $D(A) \cap D(B) = \{\theta\}$ it may happen that $Q(A) \cap Q(B)$ is dense and we may be able to define a self-adjoint "sum" of A and B by the quadratic form technique. As applications we consider the two following examples.

Example 8.28 : Let $A = -d^2/dx^2$ in $L^2(\mathcal{R})$ and $b(f,g) = \bar{f}(0)g(0)$, for $f,g\in C(\mathcal{R})$, the class of continuous bounded functions on $\mathcal{R}$. Taking $Q(A) = D(A)$ we note for the same reasons as in Proposition 3.10 that $Q(A) \subseteq Q(b)$ and that for $f\in Q(A)$, $|b(f,f)| = |f(0)|^2 \leq \|f\|_\infty^2 \leq \beta(f,Af) + \alpha\|f\|^2$, where β can be chosen as small as we want. Thus by Proposition 8.27 we can give sense to the operator $-d^2/dx^2 + \delta(x)$, where δ is the Dirac delta function. In general, as in this example, the domain of the self-adjoint operator (so defined) may contain vectors which are neither in $D(A)$ nor in $D(B)$, so that in some sense cancellations take place in the formal sum $Af + Bf$.

Example 8.29 : We shall call a real measurable function V on $\mathcal{R}^3$ a Rollnik potential if

$$\|V\|_R^2 \equiv \int \frac{|V(\underline{x})|\,|V(\underline{y})|}{|\underline{x} - \underline{y}|^2}\, d^3x\, d^3y < \infty. \tag{8.35}$$

This class contains all functions in $L^{3/2}(\mathcal{R}^3)$; in particular if $\beta < 2$ the function $\gamma|\underline{x}|^{-\beta}$ belongs to $L^{3/2}(\mathcal{R}^3) + L^\infty(\mathcal{R}^3)$. For an exhaustive study of such potentials the reader is referred to [SM]. It can be shown that $C_0^\infty(\mathcal{R}^3) \subseteq D(|V|^{\frac{1}{2}})$ and thus $((K_0+\kappa^2)^{-1}|V|^{\frac{1}{2}}f)(\underline{x}) = (4\pi)^{-1}\int|\underline{x}-\underline{y}|^{-1}\exp(-\kappa|\underline{x}-\underline{y}|)|V(\underline{y})|^{\frac{1}{2}}\cdot f(\underline{y})d^3y$ where $\kappa > 0$. On $C_0^\infty(\mathcal{R}^3)$ the operator $|V|^{\frac{1}{2}}(K_0+\kappa^2)^{-1}|V|^{\frac{1}{2}}$ is well defined by (8.35) and is an integral operator with Hilbert-Schmidt kernel $(4\pi)^{-1}|V(\underline{x})|^{\frac{1}{2}}|\underline{x}-\underline{y}|^{-1}\exp(-\kappa|\underline{x}-\underline{y}|)^{-1}\cdot|V(\underline{y})|^{\frac{1}{2}}$. Furthermore, (8.35) shows that

$$\| |V|^{\frac{1}{2}}(K_0+\kappa^2)^{-1}|V|^{\frac{1}{2}}\|^2 \leq \| |V|^{\frac{1}{2}}(K_0+\kappa^2)^{-1}|V|^{\frac{1}{2}}\|_{HS}^2$$

$$= (4\pi)^{-2}\int|\underline{x}-\underline{y}|^{-2}|V(\underline{x})|\exp(-2\kappa|\underline{x}-\underline{y}|)|V(\underline{y})|d^3x\,d^3y \leq \|V\|_R^2. \tag{8.36}$$

The integral in (8.36) converges to zero as $\kappa \to \infty$ by the Lebesgue dominated convergence theorem. Thus given $\beta < 1$, we can find $\kappa > 0$ so that $\|(K_0+\kappa^2)^{-\frac{1}{2}}|V|^{\frac{1}{2}}f\|^2 = (f, |V|^{\frac{1}{2}}(K_0+\kappa^2)^{-1}|V|^{\frac{1}{2}}f) \leq \beta^2\|f\|^2$, or equivalently $|V|^{\frac{1}{2}}$ is $K_0^{\frac{1}{2}}$-bounded with $K_0^{\frac{1}{2}}$-bound less than one. By virtue of previous remarks and Proposition 8.27 this implies that we can define $H = K_0 + V$ as a self-adjoint operator with form domain $Q(H) = Q(K_0) = D(K_0^{\frac{1}{2}})$.

D. We first mention an additional result about relative compactness and then give two further theorems about the existence of wave operators. Assume that $V \in L^p(R^3)$ with $2 \leq p < \infty$. By Lemma 7.24, $V(K_0+i)^{-1}$ is compact since the function $(\underline{k}^2+i)^{-1}$ is in $L^p(R^3)$. Hence by Proposition 8.5 $H = K_0 + V$ is self-adjoint on $D(K_0)$.

The criterion for the existence of the wave operators described in Proposition 8.14 is essentially due to Cook [1] and Hack [1]. A generalization of Corollary 8.15 is given in the next proposition.

PROPOSITION 8.30 : Let $H_0 = K_0$ in $L^2(R^3)$ and $V \in L^p(R^3)$ with $2 \leq p < 3$. Then $\Omega_\pm$ exist.

Proof : Define q and s such that $p^{-1} + q^{-1} = \frac{1}{2}$ and $s^{-1} + q^{-1} = 1$. Then we have that $6 < q \leq \infty$ and $1 \leq s < 6/5$. Let $f \in L^s(R^3) \cap L^2(R^3)$. We know from (3.62) that $\|U_t f\|_q \leq c|t|^{-3(s^{-1}-\frac{1}{2})}\|f\|_s$. Since $1/3 < s^{-1}-\frac{1}{2} \leq 1/2$, we can rewrite the above inequality as

$$\|U_t f\|_q \leq c|t|^{-1-\eta}\|f\|_s \text{ with } \eta > 0. \tag{8.37}$$

V is K_0-compact and thus by Problem 8.3 $d/dt\, V_t^* U_t g$ is strongly continuous for each $g \varepsilon C_0^\infty(R^3)$. Furthermore by the Hölder inequality (3.61) and (8.37) we get that $\|VU_t g\| \leq \|V\|_p \|U_t g\|_q \leq c\|V\|_p \|g\|_s |t|^{-1-\eta}$. This verifies the integrability hypothesis of Proposition 8.14 and leads to the desired result. #

The above Proposition roughly includes the results of Jauch and Zinnes [1] in the sense that if $V \varepsilon L^p(R^3)$ $(2 \leq p < 3)$ then V satisfies their integrability condition, viz. for some $\delta > 0$,

$$\int (1+|\underline{x}|)^{-1+\delta} |V(\underline{x})|^2 d^3x < \infty. \tag{8.38}$$

Proposition 8.30 covers essentially the whole class of short-range potentials, i.e. all $V \varepsilon L^2_{loc}(R^3)$ which decrease at infinity as $|\underline{x}|^{-1-\eta}$ for some $\eta > 0$.

As we have mentioned in Section 8-2, the local behaviour of the potential V has no influence on the existence of the wave operators. A more precise statement, essentially due to Kupsch and Sandhas [1], is contained in the following

PROPOSITION 8.31 : Suppose there exists $r < \infty$ such that $[1-\chi_r(\underline{x})]V(\underline{x}) \varepsilon L^p(R^3)$ with $2 \leq p < 3$. Let H be an arbitrary self-adjoint extension of $\hat{H} = H_0 + V$ with $D(\hat{H}) = \{f \varepsilon S(R^3) | f(\underline{x}) = 0 \text{ for } |\underline{x}| \leq r\}$. Then $\Omega_\pm = \text{s-lim} \exp(iHt)\exp(-iK_0 t)$ exist as $t \to \pm\infty$.

Proof : (i) We choose a function $\phi \varepsilon C^\infty(R^3)$ with the following properties : $0 \leq \phi(\underline{x}) \leq 1$, $\phi(\underline{x}) = 0$ for $|\underline{x}| \leq r$ and $\phi(\underline{x}) = 1$ for $|\underline{x}| \geq 2r$. We write for $f \varepsilon S(R^3)$,

$$V_t^* U_t f = V_t^* \phi(\underline{Q}) U_t f + V_t^* [I-\phi(\underline{Q})] U_t f. \tag{8.39}$$

Since $\|[I-\phi(\underline{Q})]g\| \leq \|F_{2r}g\|$ for all $g\epsilon L^2(R^3)$, we have that $\|V_t^*[I-\phi(\underline{Q})]U_tf\| = \|[I-\phi(\underline{Q})]U_tf\| \leq \|F_{2r}U_tf\|$, which converges to zero as $t\to\pm\infty$ by Remark 3.15. Thus the convergence of $V_t^*U_tf$ is equivalent to the convergence of $V_t^*\phi(\underline{Q})U_tf$.

(ii) It is clear from (3.33) that $U_tf\epsilon S(R^3)$ and that $\phi(\underline{Q})U_tf\epsilon S(R^3)$ and is 0 for $|\underline{x}| \leq r$. Thus $\phi(\underline{Q})U_tf \in D(\hat{H}) \subseteq D(H)$. Therefore $V_t^*\phi(\underline{Q})U_tf$ is strongly differentiable and

$$d/dt[V_t^*\phi(\underline{Q})U_tf] = iV_t^*(\hat{H}\phi(\underline{Q}) - \phi(\underline{Q})K_o)U_tf$$

$$= iV_t^*(V\phi(\underline{Q}) + [K_o,\phi(\underline{Q})])\,U_tf. \qquad (8.40)$$

The hypothesis of the Proposition implies that $V\phi \in L^p(R^3)$ with $2 \leq p < 3$. Thus $V\phi(\underline{Q})$ is K_o-compact. Also one sees from (7.86) that $[K_o,\phi(\underline{Q})]$ is K_o-bounded and by Problem 8.3, we conclude that $V_t^*\phi(\underline{Q})U_tf$ has a strongly continuous derivative.

By (7.86) we have for all $f\epsilon S(R^3)$ that $\|V\phi(\underline{Q})U_tf + [K_o,\phi(\underline{Q})]U_tf\| \leq \|V\phi(\underline{Q})U_tf\| + \|(\Delta\phi)(\underline{Q})U_tf\| + 2\|(\underline{\nabla}\phi)(\underline{Q})\cdot U_t\underline{P}f\|$, since $\underline{P}$ commutes with U_t. Since $P_kf\epsilon S(R^3)$, one can use the argument of the proof of Proposition 8.30 to show that $\|d/dt[V_t^*\phi(\underline{Q})U_tf]\|$ is integrable, so that $\Omega_\pm$ exist by Proposition 8.14 (if $p = 2$, one may alternatively use the proof of Corollary 8.15). #

By a similar argument, the existence of the wave operators can also be shown for potentials having a hard core. A <u>hard core</u> means that a compact subset Δ_o of R^3 is inaccessible to the particle. If for example $V \in L^2(R^3-\Delta_o)$, one defines the total Hamiltonian H as the Friedrichs extension of $-\Delta+V$ on $C_o^\infty(R^3-\Delta_o)$. Then $\Omega_\pm \equiv$ s-lim $\exp(iHt)\,F\exp(-iK_ot)$

as $t \to \pm\infty$ exist, where F is the projection of $L^2(\mathbb{R}^3)$ onto $L^2(\mathbb{R}^3-\Delta_0)$, and $\Omega_\pm$ map $L^2(\mathbb{R}^3)$ into $L^2(\mathbb{R}^3-\Delta_0)$. For details, see Hunziker [4].

The corollaries 8.17 and 8.18 are also valid when $V \in L^p(\mathbb{R}^3)$, $2 \le p < 3$. We have already seen that the wave operators $\Omega_\pm$ exist in this case. Furthermore we know from Lemma 7.24 that for $\operatorname{Im} z \neq 0$, $V(K_0-z)^{-1}$ is compact and also that $\|V_{\underline{a}}(K_0-z)^{-1}\| \le (2\pi)^{-3/p}\,\|V_{\underline{a}}\|_p \|(k^2-z)^{-1}\|_p = (2\pi)^{-3/p}\,\|V\|_p \cdot \|(k^2-z)^{-1}\|_p$, where we have used the translation invariance of Lebesgue measure. To show that $V_{\underline{a}}(K_0-z)^{-1}$ converges strongly to zero as $|\underline{a}| \to \infty$, we note that $\|V_{\underline{a}}(K_0-z)^{-1}f\| = \|V(K_0-z)^{-1}U(-\underline{a})f\|$, that $U(-\underline{a})f$ converges weakly to zero, and use Proposition 2.23. Finally by the estimates in the proof of Proposition 8.30, we have that $\int_1^\infty \|V_{\underline{a}}U_t f\|\,dt \le c\|V_{\underline{a}}\|_p\|f\|_s = c\|V\|_p\|f\|_s$. This shows how to prove Corollary 8.17 when $V \in L^p(\mathbb{R}^3)$, $2 \le p < 3$. The proof of the other corollary in this case is still simpler (Problem 8.10).

E. We now prove a theorem about boundary values of holomorphic functions and Lemma 8.21 as a corollary. These results will play an essential role in the proof of asymptotic completeness in Chapter 9.

PROPOSITION 8.32 : Let $f(z)$ be a non-constant complex-valued function, holomorphic in the open unit disc $\{z = re^{i\theta} \mid r < 1\}$. Assume that $f(re^{i\theta})$ has boundary values $f^*(e^{i\theta})$ as $r \to 1$, uniformly in θ. Then for any complex number α the set $\Gamma_o = \{\theta \in [0,2\pi] \mid f^*(e^{i\theta}) = \alpha\}$ is a closed set of Lebesgue measure zero.

Proof : Without loss of generality we may assume that $f(z)$ is not identically zero, $|f(z)| < 1$ for $|z| < 1$ and $\alpha = 0$. Set $\phi(z) = \log |f(z)|$. Then $\phi(z)$ is negative and harmonic in the open unit disc except at the zeros of f where it assumes the value $-\infty$.

It is possible to choose a sequence of numbers $\rho < 1$ converging to 1 such that no zeros of f lie on the circle of radius ρ, because there are at most finitely many zeros of any holomorphic function in any compact subset of its domain of holomorphy (cf. [RU]). For each such ρ, we define the Poisson-integral

$$\phi_\rho(re^{i\theta}) \equiv 1/2\pi \int_0^{2\pi} \phi(\rho e^{i\theta'}) \frac{\rho^2-r^2}{\rho^2+r^2-2\rho r\cos(\theta-\theta')} d\theta', \qquad (8.41)$$

and note that $\phi_\rho(z)$ is harmonic in the open disc of radius ρ and continuous in its closure.

For each such ρ, also set $D_\rho(z) = \phi(z) - \phi_\rho(z)$. Since $\phi_\rho(z)$, defined in (8.41), is finite for $|z| < \rho$, it is evident that $D_\rho(z)$ is harmonic there except at the finite set of zeros of $f(z)$, where $D_\rho(z)$ assumes the value $-\infty$. Furthermore $\phi_\rho(re^{i\theta})$ converges to $\phi(\rho e^{i\theta})$ as $r \to \rho$ and hence $D_\rho(\rho e^{i\theta}) = 0$. Thus by the maximum modulus principle for harmonic functions, $D_\rho(z) \leq 0$ for $|z| \leq \rho$.

Consider in the open unit disc a fixed point $z_0 = r_0 e^{i\theta}$ at which f does not vanish. Since $\phi(\rho e^{i\theta'}) \leq 0$ and since for $\rho > r_0$, $(\rho^2-r_0^2)[\rho^2+r_0^2-2\rho r_0\cos(\theta-\theta')]^{-1} \geq (\rho-r_0)(\rho+r_0)^{-1} > 0$, we have from (8.41) that

$$\phi_\rho(z_0) < \frac{\rho-r_0}{\rho+r_0} \int_0^{2\pi} \phi(\rho e^{i\theta'})d\theta' \leq \frac{\rho-r_0}{\rho+r_0} \int_{\Gamma_0} \phi(\rho e^{i\theta'})d\theta'. \qquad (8.42)$$

We now assume that the Lebesgue measure of Γ_o is positive. Then, considering a sequence of $\rho < 1$ converging to 1 with the above-mentioned properties and such that $\rho > r_o$, we observe from (8.42) that $\phi_\rho(z_o)$ converges to $-\infty$ as $\rho \to 1$. This is so because in (8.42), $\phi(\rho e^{i\theta'})$ converges to $-\infty$ as $\rho \to 1$ uniformly in θ' for θ' in Γ_o. On the other hand $\phi(z_o) = \phi_\rho(z_o) + D_\rho(z_o)$, with $D_\rho(z_o) \leq 0$. Thus the sum $\phi_\rho(z_o) + D_\rho(z_o)$ converges to $-\infty$ as $\rho \to 1$, while $\phi(z_o)$ is finite. This is a contradiction and hence the Lebesgue measure of Γ_o must be 0.

Finally, since the limit function f^* is clearly continuous, it follows that the set Γ_o on which it vanishes must also be closed. #

A part of the conclusion of the above Proposition, namely that Γ_o is of Lebesgue measure zero, can be obtained with fewer assumptions. Let f be a holomorphic function in the open unit disc U satisfying $\|f\|_\infty \equiv \sup|f(z)| < \infty$, where the supremum is taken over all $z \in U$. With this norm the set of bounded holomorphic functions in U forms a complete normed linear space (i.e. a Banach space) called H^∞. Then Fatou's theorem [RU, Theorem 11.21] states that every $f \in H^\infty$ has a radial limit f^* at almost all points of the unit circle $S^{(1)}$. Moreover, if $f \in H^\infty$ and is not identically zero, then $\log|f^*| \in L^1(S^{(1)})$ [RU, Theorem 17.17]. From this it follows that the limit function f^* can be zero only on a set Γ_o of Lebesgue measure zero. H^∞ is one of the Hardy spaces H^p, $0 < p \leq \infty$, and both Fatou's theorem and a result similar to that of Proposition 8.32 except the closedness of Γ_o can be derived for each f in any of the H^p-spaces. The closedness of Γ_o in Proposition 8.32 is due to the continuity of f* and does not follow from H^p-considerations.

Proof of Lemma 8.21 : It suffices to show that $\Gamma_o \cap [a,b]$ is a closed set of measure zero for each finite closed interval $[a,b]$. For this, we set $\ell = \frac{1}{2}|b-a|$ and consider the closed half-disc V in the upper half plane of radius ℓ and base $[a,b]$, i.e. $V = \{z | \mathrm{Im}\, z \geq 0, |z-\frac{1}{2}(a+b)| \leq \ell\}$. The assumptions made on Φ imply that it is continuous on V (where it is understood that $\Phi(\lambda) = \Phi_+(\lambda)$ for real λ).

For $z \varepsilon V$, let $\xi = \ell^{-1}[z-\frac{1}{2}(a+b)]$ and $w = (\xi^2+i\xi+1) \cdot (\xi^2-i\xi+1)^{-1}$. The correspondence $\Psi : z \mapsto w$ maps V one-to-one and continuously onto the closed unit disc $\{w \mid |w| \leq 1\}$ and is a conformal mapping. The image of the interval $[a,b]$ is the segment $\Sigma = \{e^{i\theta} | -4/3 \leq \tan\theta \leq 4/3\}$ of the unit circle. If we define $f(w) = \Phi(z) \equiv \Phi(\Psi^{-1}(w))$, then f is continuous on the closed unit disc and holomorphic in its interior. In fact f is uniformly continuous, since the disc is a compact set. It follows that f verifies all the hypotheses of Proposition 8.32, so that the set $\Gamma_{oo} = \{w | w \varepsilon \Sigma \text{ and } f(w) = \alpha\}$ is a closed set of measure zero . Since Ψ is continuous, $\Gamma_o = \Psi^{-1}(\Gamma_{oo})$ is a closed set in $[a,b]$.

To prove that $|\Gamma_o| = 0$, we show that the measure of $\Psi(\Delta)$ is positive whenever $|\Delta| > 0$, $\Delta \subseteq [a,b]$. For $\lambda \varepsilon [a,b]$, we write $\Psi(\lambda) = \exp[i\theta(\lambda)]$ and easily obtain that $d\theta/d\lambda = 2\ell^{-1}(1-\mu^2)[(\mu^2+1)^2 + \mu^2]^{-1}$, where $\mu = \ell^{-1}[\lambda-\frac{1}{2}(a+b)] \varepsilon [-1,+1]$. Since $d\theta/d\lambda$ is positive for all $\lambda \varepsilon (a,b)$ and integrable over $[a,b]$, we get for the measure of $\Psi(\Delta)$:

$$|\Psi(\Delta)| = \int_{\Psi(\Delta)} d\theta = \int_{\Delta} \frac{d\theta}{d\lambda}\, d\lambda > 0 \text{ if } |\Delta| > 0.$$

The proof for the lower half-plane follows similarly. #

PROBLEMS

8.1 : Show that if A is self-adjoint, then $(A+i)(A+in)^{-1} \varepsilon \mathcal{B}(\mathcal{H})$ for every integer $n \geq 1$ and $\{(A+i)(A+in)^{-1}\}$ converges strongly to zero as $n \to \infty$.

8.2 : Prove the statements made (a) in the footnote on page 319, (b) in Remark 8.12.

8.3 : Let V be H_0-bounded. Show that for every $f \varepsilon D(H_0)$, $V_t^* U_t f$ is strongly continuously differentiable.

8.4 : Assume the hypotheses of Proposition 8.5(a), replacing the self-adjointness of A by essential self-adjointness. Show that A+B is essentially self-adjoint.

8.5 : In $\mathcal{H} \equiv L^2(\mathcal{R})$, let H_0 and H be the self-adjoint operators determined by $H_0 = -id/dx$ and $H = H_0+V$, where V is assumed to be a bounded real function in $L^1(\mathcal{R})$. Prove (A1) and (A3) and give explicit expressions for $\Omega_\pm$ and S in terms of V. (Hint : Show that $(U_t f)(x) = f(x-t)$ and $(V_t f)(x) = \exp[-iW(x)+iW(x-t)] \cdot f(x-t)$ for a suitable function W.)

8.6 : There is a countable subset $\mathcal{S}_0$ of $\mathcal{S}(\mathcal{R}^3)$ such that for each $\lambda > 0$, the set $\{f_\lambda | f \varepsilon \mathcal{S}_0\}$ is dense in $L^2(S^{(2)})$. (Hint : Use Remark 5.23 and the spherical harmonics, Section 11-5A.)

8.7 : Verify the statements made after (8.30) and show in addition that $\bar{\sigma}(\lambda) = o(\lambda^{-1})$ as $\lambda \to \infty$. (Hint : $(h, R^0_{\lambda \pm io} h) \to 0$ as $\lambda \to \infty$.)

8.8 : Define Θ in $L^2(\mathcal{R}^3)$ by $(\Theta f)(\underline{x}) = \bar{f}(\underline{x})$. (a) Show that Θ is an anti-unitary operator. (b) Let $H = K_0+V$, where V is a real potential such that V has K_0-bound less than 1. Show that $\Theta U_t \Theta = U_t^*$ and $\Theta V_t \Theta = V_t^*$ and verify (7.66). (Hint : Verify that $\Theta U_t \Theta$ and $\Theta V_t \Theta$ are strongly continuous one-parameter unitary groups and that $\Theta K_0 \Theta = K_0$ and $\Theta H \Theta = H$ on $D(K_0)$. Note the definition of Θ^* from the footnote on page 150.)

8.9 : Show that if s-lim $U(\underline{a})\Omega_{\pm}U(-\underline{a}) = I$ as $|\underline{a}| \to \infty$, then $U(\underline{a})\Omega_{\pm}^{*}U(-\underline{a}) \to I$ and $U(\underline{a})SU(-\underline{a}) \to I$ as $|\underline{a}| \to \infty$. (Hint : Problem 2.4(b).)

8.10[†]: (a) Write out in detail the proofs of Corollaries 8.17 and 8.18 under the hypothesis $V \in L^p(\mathcal{R}^3)$, $2 \leq p < 3$. (Hint : See Section 8-4D.) (b) Show that under the same assumptions $U(\underline{a})V_t U(-\underline{a}) \to U_t$ uniformly in t as $|\underline{a}| \to \infty$.

8.11 : Let $H = K_o + \kappa^2 \underline{Q}^2$ in $L^2(\mathcal{R}^3)$ be the harmonic oscillator Hamiltonian. (a) Show that H is essentially self-adjoint on $\mathcal{S}(\mathcal{R}^3)$ and find the domain of self-adjointness of H. (Hint : H has pure point spectrum.) (b) Show that $\exp(iHt)\exp(-iK_o t)$ converges weakly to 0 as $t \to \pm\infty$ and hence cannot converge strongly.

8.12 : Suppose $V = V_1 + V_2$ with $V_1 \in L^2(\mathcal{R}^3)$, $V_2(\underline{x}) = 0$ for $|x| < R$ and $|V_2(\underline{x})| \leq c|\underline{x}|^{-\beta}$ for $|x| \geq R$ and $0 < \beta \leq 3/2$. Show that for $f \in \mathcal{S}(\mathcal{R}^3)$ and $|t| > 1$, $\|VU_t f\| \leq c'|t|^{-\beta}$ if $0 < \beta < 3/2$, $\|VU_t f\| \leq c''|t|^{-3/2}\log|t|$ for $\beta = 3/2$.

8.13 : (a) Let Σ be an interval and $A = \{A(\lambda)\}$ a bounded operator. Assume that $\int_\Delta (f_\lambda, A(\lambda)g_\lambda)_o d\lambda = 0$ for all $f,g \in \mathcal{S}_o$ and all subintervals Δ of Σ. Then there is a null set N_o in Σ such that $(f_\lambda, A(\lambda)g_\lambda) = 0$ for all $\lambda \in \Sigma - N_o$ and all $f,g \in \mathcal{S}_o$. (Hint : Proof of Proposition 5.27).
(b) Under the above hypothesis, show that $A(\lambda) = 0$ for all $\lambda \in \Sigma - N_o$.

CHAPTER 9 : ASYMPTOTIC COMPLETENESS

As we have seen in Chapter 4, asymptotic completeness means that the time evolution of every scattering state of the total Hamiltonian H approaches asymptotically a state vector evolving under the unperturbed evolution group. According to the discussion following Proposition 4.2, asymptotic completeness is equivalent to the property that $U_t^*V_t f$ converges strongly as $t\to\pm\infty$ for each $f\in M_\infty(H)$. One sees from this that the task of proving asymptotic completeness is formally the same as that of proving the existence of the wave operators, the only difference being an interchange of the roles played by $\{U_t\}$ and $\{V_t\}$.

We have seen in Section 8-2 that the existence of the wave operators in potential scattering can be established with simple methods. It suffices for instance to show that, for a fundamental set of vectors f in $M_\infty(H_0)$, $\|VU_t f\| \in L^1(R)$ as a function of t, and this is readily done by using the explicit form (3.37) for $(U_t f)(\underline{x})$. Similarly asymptotic completeness is implied by the property that $\|VV_t f\| \in L^1(R)$ for all f belonging to a fundamental set $\mathcal{D}$ in $M_\infty(H)$. However, since no explicit expression like (3.37) is available for $(V_t f)(\underline{x})$, the method of Section 8-2 cannot be applied here. These remarks illustrate the fact that the proof of asymptotic completeness in potential scattering and the proof of the existence of the wave operators for more general unperturbed evolution groups require a considerable amount of mathematical machinery and are quite lengthy.

Section 9-1 is devoted to a detailed proof of asymptotic completeness in potential scattering for two simple classes of potentials. To treat more general classes of potentials and to go beyond potential scattering, it is convenient to develop abstract methods for proving the existence of strong limits of sequences of the form $\{Y_t^* Z_t f\}$, where $\{Y_t\}$ and $\{Z_t\}$ are one-parameter continuous unitary groups of operators. Some of these methods together with applications to potential scattering will be described in Section 9-2, but we shall refer to the literature for complete proofs.

9-1 A COMPLETENESS PROOF IN POTENTIAL SCATTERING

We have seen in Section 5-7 that the free Hamiltonian of potential scattering K_0 is spectrally absolutely continuous. Therefore by Remark 5.22 the ranges of $\Omega_\pm$ are contained in $\mathcal{H}_{ac}(H)$. Now if the potential is absolutely square-integrable (at least locally), then $\mathcal{H}_{ac}(H) = \mathcal{M}_{\infty}(H)$ by Proposition 7.9 (or Section 7-6A), provided that H has no singularly continuous spectrum. In this case asymptotic completeness is equivalent to the requirement that the ranges of $\Omega_{\pm}$ are equal to $\mathcal{H}_{ac}(H)$.

The main result of the present section is the equality of the ranges of $\Omega_{\pm}$ with $\mathcal{H}_{ac}(H)$ for two classes of potentials, and asymptotic completeness will be understood here in this sense*). In principle one should also show that $\mathcal{H}_{sc}(H) = \{\theta\}$ (or at least that no vector in $\mathcal{H}_{sc}(H)$ is a scattering state in the sense of Section 7-1). This stronger property (i.e. $F_{\pm} = E_{ac}(H)$ and $\mathcal{H}_{sc}(H) = \{\theta\}$) is sometimes called strong asymptotic completeness. More will be said about $\mathcal{H}_{sc}(H)$ in Section 10-3.

We shall first present the chain of arguments leading to asymptotic completeness on an abstract level. In this part the explicit form of H_0 is irrelevant. The hypotheses in these theorems are such that the results are easily applicable to the classes of potentials that we have in mind; however we should point out that similar results can be established under weaker assumptions. An essential feature of our method is a factorization of the interaction as V = AB. The principal results are Propositions 9.4 and 9.6. The second part of this section is devoted to a verification of the hypotheses of these propositions for two classes of potentials

---

*) One also says that the wave operators $\Omega_{\pm}$ are asymptotically complete or simply complete.

belonging to $L^2(\mathbb{R}^3)$. Finally we shall mention a simpler method of proving asymptotic completeness which is applicable to spherically symmetric potentials.

The notation is the following. We always assume that H, H_o, A and B are self-adjoint operators such that $H = H_o + AB$ with $D(H) = D(H_o)$. B will be H-bounded and A will belong to $\mathcal{B}(\mathcal{H})$. We set $V_t = \exp(-iHt)$, $U_t = \exp(-iH_ot)$ and denote by $\{E_\lambda\}$ the spectral family of H. We begin with the following convergence criterion.

<u>PROPOSITION 9.1</u> : Let $f\varepsilon D(H)$ and assume that

$$h(t) \equiv \|AU_tA\| \varepsilon L^1(\mathbb{R}) \tag{9.1}$$

and

$$g(t) \equiv \|BV_tf\| \varepsilon L^2(\mathbb{R}) \tag{9.2}$$

as functions of the variable t. Then $U_t^*V_tf$ converges strongly as $t \to \pm\infty$.

<u>Proof</u> : We have as in the proof of Proposition 8.14 that $U_t^*V_tf - U_\tau^*V_\tau f = -i\int_\tau^t ds\, U_s^*VV_sf$. By using the Schwarz inequality and (2.25), we obtain from this

$$\|U_t^*V_tf-U_\tau^*V_\tau f\|^2 = \int_\tau^t ds \int_\tau^t d\sigma\, (U_s^*ABV_sf, U_\sigma^*ABV_\sigma f)$$

$$= \int_\tau^t ds \int_\tau^t d\sigma (BV_sf, AU_{s-\sigma}ABV_\sigma f)$$

$$\leq \int_\tau^t ds \int_\tau^t d\sigma\, \|BV_sf\|\, \|AU_{s-\sigma}A\|\, \|BV_\sigma f\|$$

$$= \int_\tau^t ds \int_\tau^t d\sigma\, g(s)h(s-\sigma)g(\sigma). \tag{9.3}$$

We get by a change of variables and by applying the Schwarz inequality to the integral over the variable s that

$$\int_{-\infty}^{\infty} ds \int_{-\infty}^{\infty} d\sigma\; g(s)h(s-\sigma)g(\sigma) = \int_{-\infty}^{\infty} dx\; h(x) \int_{-\infty}^{\infty} ds\; g(s)g(s-x)$$

$$\leq \int_{-\infty}^{\infty} dx\, h(x) \{ \int_{-\infty}^{\infty} ds|g(s)|^2 \int_{-\infty}^{\infty} dy|g(y)|^2\}^{\frac{1}{2}} = \|h\|_1 \; \|g\|^2.$$

The last term is finite, since $g\varepsilon L^2(\mathcal{R})$ and $h\varepsilon L^1(\mathcal{R})$ by hypothesis. Together with (9.3), this implies that $\{U_t^* V_t f\}$ is strongly Cauchy as $t \to \pm\infty$. #

In potential scattering the hypothesis (9.1) of the above proposition is easy to verify (Lemma 9.8). The hypothesis (9.2) involves the total evolution group and is usually handled by transforming it into a time-independent form. For this one has to introduce a factor $\exp(\mp\eta t)$ with $\eta > 0$ (for $t \gtrless 0$ respectively), which will bring in the resolvents $R_{\lambda\pm i\eta}$ of H in the time-independent form. These operators become unbounded as $\eta \to 0$, since $\|R_{\lambda\pm i\eta}\| = \eta^{-1}$ if $\lambda \varepsilon \sigma(H)$. However in certain cases it is possible to show that $\|BR_{\lambda\pm i\eta}C\|$ is bounded uniformly in $\eta > 0$ and in $\lambda\varepsilon\Delta$ if B and C are suitably chosen operators and Δ is a suitable interval, so that the limit of $BR_{\lambda\pm i\eta}C$ as $\eta \to 0$ may exist. In this way one may prove that $\int\|BV_tE_\Delta Cf\|^2 dt < \infty$ for all $f\varepsilon D(C)$, if the operator C and the interval Δ are suitably chosen. This is the content of the next two lemmas.

<u>LEMMA 9.2</u> : Let $\Delta = (a,b]$ be a half-open bounded interval, $B = B^*$ an H-bounded operator and $C\varepsilon\mathcal{B}(\mathcal{H})$. Assume that $\|B(R_{\lambda+i\eta} - R_{\lambda-i\eta})C\| \leq M < \infty$ for all $0 < \eta \leq 1$ and all $\lambda\varepsilon\Delta$. Then $\|BE_{\Delta_0}C\| \leq (2\pi)^{-1} M|\Delta_0|$ for each half-open subinterval $\Delta_0 = (c,d]$ of Δ and $\|B\phi(H)E_\Delta C\| \leq (2\pi)^{-1} M \int_\Delta |\phi(\lambda)|d\lambda$ for each continuous function $\phi : \mathcal{R} \to \mathcal{C}$.

Proof : We use the following relation between the spectral projections E_Δ and the resolvent of H, valid for each finite interval $\Delta = (a,b]$ (see Section 9-2 for its proof) :

$$E_{(a,b]} = \underset{\delta\to+0}{\text{s-lim}}\ \underset{\eta\to+0}{\text{s-lim}}\ (2\pi i)^{-1}\int_{a+\delta}^{b+\delta} d\lambda\,(R_{\lambda+i\eta} - R_{\lambda-i\eta}). \qquad (9.4)$$

Since B is H-bounded, $BE_{\Delta_0}f$ is defined for all $f\varepsilon\mathcal{H}$. For $g\varepsilon D(B)$, we obtain by using (9.4) for the interval $\Delta_0 = (c,d]$

$$2\pi i(g,BE_{\Delta_0}Cf) = \lim_{\delta\to+0}\ \lim_{\eta\to+0}\int_{c+\delta}^{d+\delta} d\lambda\,(Bg,[R_{\lambda+i\eta} - R_{\lambda-i\eta}]Cf)$$

$$= \lim_{\delta\to+0}\ \lim_{\eta\to+0}\int_{c+\delta}^{d+\delta} d\lambda\,(g,B[R_{\lambda+i\eta} - R_{\lambda-i\eta}]Cf). \qquad (9.5)$$

Now by the Schwarz inequality

$$\left|\int_{c+\delta}^{d+\delta} d\lambda\,(g,B[R_{\lambda+i\eta} - R_{\lambda-i\eta}]Cf)\right| \le (d-c)\,M\|g\|\,\|f\|. \qquad (9.6)$$

(9.5) and (9.6) imply $|(g,BE_{\Delta_0}Cf)| \le (2\pi)^{-1}M|\Delta_0|\,\|f\|\|g\|$. The first assertion of the lemma now follows from Problem 5.13. For the second one, we use Proposition 6.2 and $BE_\Delta\varepsilon\mathcal{B}(\mathcal{H})$ to get $B\phi(H)E_\Delta C = \int_\Delta B\phi(\lambda)dE_\lambda C$. Let $\Pi = \{\lambda_j;\mu_k\}$ be a partition of (a,b]. Then, by using the first part of the lemma, we obtain

$$\|B\phi(H)E_\Delta Cf\| \le \lim_{|\Pi|\to 0}\sum_{k=1}^{n(\Pi)}\|B\phi(\mu_k)E_{(\lambda_{k-1},\lambda_k]}Cf\|$$

$$\le \lim_{|\Pi|\to 0}\sum_{k=1}^{n(\Pi)}|\phi(\mu_k)|\,\frac{M}{2\pi}\,(\lambda_k-\lambda_{k-1})\,\|f\| = \frac{M}{2\pi}\int_a^b|\phi(\lambda)|d\lambda\,\|f\|. \#$$

Remarks : (a) The lemma may also be proved if C is unbounded by replacing the operators that involve C by their closures. Thus for example the hypothesis on $B(R_{\lambda+i\eta} - R_{\lambda-i\eta})C$ would then be that its closure is bounded uniformly in $0 < \eta \le 1$ and $\lambda\varepsilon\Delta$.

(b) The use of half-open intervals in the lemma is

not essential; the result also holds for open and closed intervals. We have taken half-open ones in order to simplify the notation in the proof, since in Chapters 4 and 6 partitions were defined in terms of half-open intervals.

LEMMA 9.3 : If the assumptions of Lemma 9.2 are verified, then $\|BV_tE_\Delta Cf\| \in L^2(\mathbb{R})$ for each $f \in \mathcal{H}$.

Proof : Let $0 < \eta \le 1$. Then

$$I_\eta \equiv \int_0^\infty \|BV_tE_\Delta Cf\|^2 e^{-\eta t}dt + \int_{-\infty}^0 \|BV_tE_\Delta Cf\|^2 e^{\eta t}dt$$

$$= \int_0^\infty (Cf, V_t^*(BE_\Delta)^*BE_\Delta V_tCf)e^{-\eta t}dt$$

$$+ \int_{-\infty}^0 (Cf, V_t^*(BE_\Delta)^*BE_\Delta V_tCf)e^{\eta t}dt. \tag{9.7}$$

Since B is H-bounded, we have $BE_\Delta = B(H+i)^{-1}(H+i)E_\Delta \in \mathcal{B}(\mathcal{H})$. Let $\Pi = \{\lambda_k;\mu_j\}$ be a partition of (a,b]. We obtain by applying Proposition 6.3 and the Schwarz inequality that

$$I_\eta = i \lim_{|\Pi|\to 0} \sum_{k=1}^{n(\Pi)} (Cf, (R_{\mu_k-i\eta} - R_{\mu_k+i\eta})(BE_\Delta)^*BE_\Delta E_{(\lambda_{k-1},\lambda_k]}Cf)$$

$$\le \lim_{|\Pi|\to 0} \sum_{k=1}^{n(\Pi)} \|BE_\Delta(R_{\mu_k+i\eta} - R_{\mu_k-i\eta})C\| \, \|f\| \, \|BE_{(\lambda_{k-1},\lambda_k]}Cf\|. \tag{9.8}$$

From Lemma 9.2 one gets the following estimates

$$\|BE_{(\lambda_{k-1},\lambda_k]}Cf\| \le (2\pi)^{-1}M|\lambda_k-\lambda_{k-1}| \, \|f\|, \tag{9.9}$$

$$\|BE_\Delta(R_{\lambda+i\eta}-R_{\lambda-i\eta})C\| \le (2\pi)^{-1}M \int_\Delta \frac{2\eta}{(\lambda-\mu)^2+\eta^2}d\mu$$

$$< \pi^{-1}M \int_{\mathbb{R}} \eta(x^2+\eta^2)^{-1}dx = \pi^{-1}M\int_{\mathbb{R}}(1+y^2)^{-1}dy = M. \tag{9.10}$$

By combining (9.8)-(9.10), one arrives at

$$I_\eta \leq \lim_{|\Pi|\to 0} \sum_{k=1}^{n(\Pi)} (2\pi)^{-1} M^2 |\lambda_k - \lambda_{k-1}| \, \|f\|^2$$

$$= (2\pi)^{-1} M^2 \|f\|^2 (b-a) \equiv \alpha_0 .$$

This shows that the integrals I_η are bounded uniformly in $0 < \eta \leq 1$. As $\eta \to +0$, the integrands in (9.7) are monotone increasing and converge to $\|BV_tE_\Delta Cf\|^2$. Hence $\|BV_tE_\Delta Cf\|^2 \in L^1(R)$ and $\int_R \|BV_tE_\Delta Cf\|^2 dt \leq \alpha_0$ by the monotone convergence theorem [R, Theorem 4.9]. #

The hypothesis of the preceding lemma is a bound on $\|B(R_{\lambda+i\eta} - R_{\lambda-i\eta})C\|$ as $\eta \to 0$. One method of obtaining such a bound is based on the second resolvent equation (6.37). By remembering that $V = AB$, we obtain $BR_zC = BR_z^oC - BR_z^oA \cdot BR_zC$, which implies that

$$BR_zC = (I+BR_z^oA)^{-1} BR_z^oC, \tag{9.11}$$

provided that the inverse exists. Suppose that $\|BR_z^oA\| \leq M < \infty$ for all $z \in \rho(H_o)$. The inverse in (9.11) will then exist for <u>weak coupling</u>. For this one introduces a <u>coupling constant</u> $\gamma \in R$, i.e. one considers the Hamiltonian $H_\gamma = H_o + \gamma AB$ for a sufficiently small value of $|\gamma|$. The following result then holds true.

<u>PROPOSITION 9.4</u> : Assume that $\|AU_tA\| \in L^1(R)$ and that $\|BR^o_{\lambda\pm i\eta}A\| \leq M_1$ for all $\lambda \in R$ and all $0 < \eta \leq 1$. Suppose there exists a bounded operator C with dense range in H such that $\|BR^o_{\lambda\pm i\eta}C\| \leq M_2 < \infty$ for all $\lambda \in R$ and all $0 < \eta \leq 1$. Let $H_\gamma = H_o + \gamma AB$ with γ real and $|\gamma| < M_1^{-1}$. Assume in addition that H_o is spectrally absolutely continuous and that the wave

operators $\Omega_\pm$ = s-lim $\exp(iH_\gamma t)\exp(-iH_0 t)$ as $t \to \pm\infty$ exist. Then H_γ is spectrally absolutely continuous and the theory is asymptotically complete, i.e. $\Omega_\pm \mathcal{H} = \mathcal{H}$ and $\Omega_\pm$ are unitary.

Proof : We set $R_z^{(\gamma)} = (H_\gamma - z)^{-1}$. Since $\|\gamma B R_z^0 A\| \leq |\gamma| M_1 < 1$, $(I+\gamma B R_z^0 A)^{-1}$ exists and $\|(I+\gamma B R_z^0 A)^{-1}\| \leq (1-|\gamma| M_1)^{-1}$ for each $z = \lambda + i\eta$ with $0 < \eta \leq 1$ by Proposition 2.20. (9.11) now implies that

$$\|BR_{\lambda\pm i\eta}^{(\gamma)}C\| \leq \|(I+\gamma BR_{\lambda\pm i\eta}^{0}A)^{-1}\| \, \|BR_{\lambda\pm i\eta}^{0}C\| \leq \frac{M_2}{1-|\gamma|M_1} < \infty$$

for all $\lambda \in \mathcal{R}$ and all $0 < \eta \leq 1$. Thus by Lemma 9.3 $\|B\exp(-iH_\gamma t)E_\Delta Cf\| \in L^2(\mathcal{R})$ for each half-open interval $\Delta = (a,b]$ and each $f \in \mathcal{H}$, and by Proposition 9.1, $U_t^* \exp(-iH_\gamma t)E_\Delta Cf$ converges strongly as $t \to \pm\infty$. Since the range of C is dense and s-lim $E_{(a,b]} = I$ as $a \to -\infty$, $b \to +\infty$, this means that $\{U_t^* \exp(-iH_\gamma t)\}$ converges strongly on a dense set in $\mathcal{H}$. Hence the ranges of $\Omega_\pm$ contain a dense set, which means that they are equal to $\mathcal{H}$ since they are closed subspaces. This shows that $\Omega_\pm$ are unitary. The fact that H_γ is spectrally absolutely continuous follows from Remark 5.22. #

The completeness proof becomes considerably more delicate for large values of the coupling constant. If $BR_z^0 A$ is a compact operator for each $z \in \rho(H_0)$, if $(I+BR_z^0 A)^{-1} \in \mathcal{B}(\mathcal{H})$ whenever Im $z \neq 0$ and if the limits $BR_{\lambda\pm io}^0 A \equiv \lim BR_{\lambda\pm i\eta}^0 A$ as $\eta \to +0$ exist uniformly in λ, one can show by using Lemma 8.21 that $(I+BR_{\lambda\pm io}^0 A)^{-1}$ exists and is bounded for all $\lambda \in \mathcal{R}$ with the possible exception of a closed set of measure zero. This is the content of the next lemma, which will then lead to an abstract theorem about asymptotic completeness.

LEMMA 9.5 : Let $\{W_z\}$ be a holomorphic family of compact operators in $\mathcal{C}-\mathcal{R}$. Suppose that $I+W_z$ is invertible for each $z\varepsilon\mathcal{C}-\mathcal{R}$ and that $W_{\lambda\pm io} \equiv \lim W_{\lambda\pm i\eta}$ as $\eta \to +0$ exist in operator norm, the convergence being uniform in λ on each compact subset of $\mathcal{R}$. Define $\Gamma_o^{\pm} = \{\lambda\varepsilon\mathcal{R} | I+W_{\lambda\pm io}$ is not invertible$\}$ and $\Gamma_o = \Gamma_o^{+} \cup \Gamma_o^{-}$.

(a) Let Δ be a closed interval such that $\Delta\cap\Gamma_o = \emptyset$. Then there exists a constant $M_\Delta < \infty$ such that $\|(I+W_{\lambda\pm i\eta})^{-1}\| \leq M_\Delta$ for all $\lambda \varepsilon \Delta$ and all $0 \leq \eta \leq 1$. Furthermore $(I+W_z)^{-1}$ is continuous in operator norm on each of the rectangles $\Delta_{\pm} = \{z = \lambda\pm i\eta \mid \lambda\varepsilon\Delta,\ 0 \leq \eta \leq 1\}$.

(b) Γ_o is a closed set of Lebesgue measure zero called the exceptional set associated with $\{W_z\}$.

Proof : $W_{\lambda\pm io} \varepsilon \mathcal{B}_o$ by Proposition 2.22(d), and by Proposition 2.24, $(I+W_z)^{-1} \varepsilon \mathcal{B}(\mathcal{H})$ if $\text{Im}\, z \neq 0$ and $(I+W_{\lambda\pm io})^{-1} \varepsilon \mathcal{B}(\mathcal{H})$ if $\lambda \notin \Gamma_o^{\pm}$. Now let $z,\xi \varepsilon \Delta_+$. Then $(I+W_z)^{-1}$, $(I+W_\xi)^{-1} \varepsilon \mathcal{B}(\mathcal{H})$ and one has the identity

$$I + W_z = [I-(W_\xi-W_z)(I+W_\xi)^{-1}](I+W_\xi). \tag{9.12}$$

It implies that

$$(I+W_z)^{-1}-(I+W_\xi)^{-1} = (I+W_\xi)^{-1}\{[I-(W_\xi-W_z)(I+W_\xi)^{-1}]^{-1}-I\}. \tag{9.13}$$

Since $\|W_z-W_\xi\| \to 0$ as $z \to \xi$, the inverse of the square bracket in (9.13) exists if $|z-\xi|$ is sufficiently small. We thus get by using (2.89) and Proposition 2.20 :

$$\begin{aligned} \Big| \|(I+W_z)^{-1}\| - \|(I+W_\xi)^{-1}\| \Big| &\leq \|(I+W_z)^{-1} - (I+W_\xi)^{-1}\| \\ &\leq \|(I+W_\xi)^{-1}\| \textstyle\sum_{n=1}^{\infty} [\|W_\xi-W_z\|\,\|(I+W_\xi)^{-1}\|]^n \\ &= \|(I+W_\xi)^{-1}\|^2 \|W_\xi-W_z\|\,[1-\|W_\xi-W_z\|\,\|(I+W_\xi)^{-1}\|]^{-1}. \end{aligned}$$

This shows that $\|(I+W_z)^{-1} - (I+W_\xi)^{-1}\| \to 0$ as $z \to \xi$ and that $z \mapsto \|(I+W_z)^{-1}\|$ is a continuous function on Δ_+. Since Δ_+ is compact, this function is bounded [R, Proposition 9.9]. Similarly $\|(I+W_z)^{-1}\|$ is bounded on Δ_-. This completes the proof of (a). That of (b) is given in Section 9-2. #

PROPOSITION 9.6 : Let H, H_o, A and B be self-adjoint operators such that $A\varepsilon\mathcal{B}(\mathcal{H})$, B is H_o-bounded and $H = H_o + AB$ with $D(H) = D(H_o)$. Assume that $\|AU_tA\| \varepsilon L^1(\mathcal{R})$, that $W_z \equiv BR_z^oA$ verifies the hypotheses of Lemma 9.5 and that there exists an operator $C\varepsilon\mathcal{B}(\mathcal{H})$ with dense range such that $\|BR^o_{\lambda\pm i\eta}C\| \leq M < \infty$ for all $\lambda\varepsilon\mathcal{R}$ and all $0 < \eta \leq 1$. Then, if the wave operators $\Omega_\pm$ = s-lim $V_t^*U_tE_{ac}(H_o)$ as $t \to \pm\infty$ exist, their ranges are equal to $\mathcal{H}_{ac}(H)$.

Proof : (i) Let Γ_o be the closed set of measure zero determined by Lemma 9.5 and define $\Gamma = \mathcal{R}-\Gamma_o$. Since the Lebesgue measure of Γ_o is zero, $E_{\Gamma_o}\mathcal{H}$ is orthogonal to $\mathcal{H}_{ac}(H)$, cf. Section 5-6. Thus each $g\varepsilon\mathcal{H}_{ac}(H)$ is in the range of E_Γ. We shall prove in (ii) below that $E_\Gamma\mathcal{H} \subseteq \Omega_\pm\mathcal{H}$. Since $\Omega_\pm\mathcal{H} \subseteq \mathcal{H}_{ac}(H)$ by Remark 5.22, we arrive at the following chain of inclusions : $\mathcal{H}_{ac}(H) \subseteq E_\Gamma\mathcal{H} \subseteq \Omega_\pm\mathcal{H} \subseteq \mathcal{H}_{ac}(H)$, which implies that

$$\Omega_\pm\mathcal{H} = \mathcal{H}_{ac}(H) = E_\Gamma\mathcal{H} = E_{\mathcal{R}-\Gamma_o}\mathcal{H}. \tag{9.14}$$

(ii) We now prove that $E_\Gamma\mathcal{H} \subseteq \Omega_\pm\mathcal{H}$. Since Γ_o is closed, Γ is open. Thus Γ is a countable union of disjoint open intervals $\{\Delta_k\}_{k=1}^\infty$ [R, Proposition 2.8]. Hence $E_\Gamma = \sum_{k=1}^\infty E_{\Delta_k}$ by (5.64), and it suffices to prove that $E_{\Delta_k}\mathcal{H} \subseteq \Omega_\pm\mathcal{H}$ for each k. So let $\Delta = (a,b)$ be one of the intervals $\{\Delta_k\}$. We choose a sequence of half-open intervals $(a_i,b_i]$ contained in Δ such that $a < a_i < b_i < b$ and $\lim a_i = a$, $\lim b_i = b$ as $i \to \infty$. It follows from the right-continuity of the spectral family $\{E_\lambda\}$ (Section 5-2) that s-lim $E_{(a,a_i]} = 0$ and s-lim $E_{(b_i,b)} = 0$ as $i \to \infty$, or equivalently that $E_{(a,b)} = \text{s-lim}\, E_{(a_i,b_i]}$ as $i \to \infty$. Thus the union of the ranges of $\{E_{(a_i,b_i]}\}_{i=1}^{\infty}$ is dense in $E_{(a,b)}\mathcal{H}$, and consequently it suffices to show that $E_{(a_i,b_i]}\mathcal{H} \subseteq \Omega_\pm\mathcal{H}$ for each i.

Now by Lemma 9.5, $\|(I-BR^o_{\lambda\pm i\eta}A)^{-1}\| \le M_{[a_i,b_i]} < \infty$ for all $\lambda\varepsilon[a_i,b_i]$ and all $0 \le \eta \le 1$, hence $\|BR_{\lambda\pm i\eta}C\| \le M_{[a_i,b_i]}M$ by (9.11) for all such λ and η. Since $D(H) = D(H_o)$ and B is H_o-bounded, it is also H-bounded (cf. e.g. Proposition 8.1), so that $\|BV_tE_{(a_i,b_i]}Cf\| \varepsilon L^2(\mathbb{R})$ for each $f\varepsilon\mathcal{H}$ by Lemma 9.3. Thus s-lim $U_t^*V_tE_{(a_i,b_i]}Cf$ as $t \to \pm\infty$ exist for each $f\varepsilon\mathcal{H}$ by Proposition 9.1. Since the set $\{Cf | f\varepsilon\mathcal{H}\}$ is dense in $\mathcal{H}$, it follows from Proposition 2.17 that s-lim $U_t^*V_tE_{(a_i,b_i]}$ as $t \to \pm\infty$ exist on all of $\mathcal{H}$. This means that $E_{(a_i,b_i]}\mathcal{H}$ belongs to the ranges of $\Omega_\pm$, as we wished to prove. #

Remark 9.7 : It also follows from part (i) of this proof that $\mathcal{H}_s(H) = E_{\Gamma_o}\mathcal{H}$. Thus, as Γ_o is a closed set, the singular spectrum of H is contained in Γ_o : $\sigma_s(H) \subseteq \Gamma_o$.

In the remainder of this section we take $H_o = K_o = \underline{P}^2$, the free Hamiltonian of potential scattering. U_t is the free evolution group given by (3.33) or (3.37). We begin with an auxiliary result and then prove asymptotic completeness for a simple class of potentials.

LEMMA 9.8 : Let $\phi_1, \phi_2 : \mathcal{R}^3 \to \mathcal{C}$ be two bounded and absolutely square-integrable functions, and denote by A_1 and A_2 the multiplication operators by ϕ_1 and ϕ_2 respectively in $L^2(\mathcal{R}^3)$. Then

(a) $\|A_1 U_t A_2\| \in L^1(\mathcal{R})$ as a function of t,

(b) $\|A_1 R^o_{\lambda \pm i\eta} A_2\| \leq M < \infty$ for all $\lambda \in \mathcal{R}$ and all $\eta > 0$.

(c) For each $\lambda \in \mathcal{R}$, the sequence $\{A_1 R^o_{\lambda \pm i\eta} A_2\}$ converges in operator norm as $\eta \to +0$, the convergence being uniform in $\lambda \in \mathcal{R}$.

Proof : (i) Since ϕ_1 and ϕ_2 are bounded, we have $A_1, A_2 \in \mathcal{B}(\mathcal{H})$ by Proposition 2.16. Hence $\|A_1 U_t A_2\| \leq \|A_1\| \|A_2\| \equiv M_{12}$.

(ii) Let $f \in L^2(\mathcal{R}^3)$. Then $A_2 f \in L^1(\mathcal{R}^3) \cap L^2(\mathcal{R}^3)$, since by the Schwarz inequality $\|A_2 f\|_1 \leq \|\phi_2\|_2 \|f\|$. Thus by (3.42) $|(U_t A_2 f)(\underline{x})| \leq |4\pi t|^{-3/2} \|A_2 f\|_1 \leq |4\pi t|^{-3/2} \|\phi_2\|_2 \|f\|$. Hence

$$\|A_1 U_t A_2 f\|^2 = \int d^3x |\phi_1(\underline{x})|^2 |(U_t A_2 f)(\underline{x})|^2 \leq |4\pi t|^{-3} \|\phi_1\|_2^2 \|f\|^2 \|\phi_2\|_2^2,$$

which implies that $\|A_1 U_t A_2\| \leq |4\pi t|^{-3/2} \|\phi_1\|_2 \|\phi_2\|_2$. From this and (i) we infer that $\|A_1 U_t A_2\| \leq \max\{M_{12}, |4\pi t|^{-3/2} \|\phi_1\|_2 \|\phi_2\|_2\}$, which is integrable. (The fact that $\|A_1 U_t A_2\|$ is measurable may be proved by reasoning as in the remark preceding Proposition 5.25, and by using Problem 5.13.)

(iii) To prove (b), we use (3.58) and Proposition 4.16(a) :

$$\|A_1 R^o_{\lambda + i\eta} A_2\| = \|\int_0^\infty dt\, e^{i\lambda t} e^{-\eta t} A_1 U_t A_2\| \leq \int_{-\infty}^{\infty} dt \|A_1 U_t A_2\| < \infty, \quad (9.15)$$

and similarly for $\|A_1 R^o_{\lambda - i\eta} A_2\|$.

(iv) By Proposition 4.16(d), $i \int_0^\infty dt\, e^{i\lambda t} A_1 U_t A_2$ defines an operator in $\mathcal{B}(\mathcal{H})$ which we denote symbolically by $A_1R^o_{\lambda+io}A_2$. We get as above that for all $\lambda\varepsilon\mathcal{R}$ and $\eta > 0$

$$\|A_1R^o_{\lambda+i\eta}A_2 - A_1R^o_{\lambda+io}A_2\| \leq \int_0^\infty dt\, |e^{-\eta t}-1|\ \|A_1U_tA_2\|. \qquad (9.16)$$

As $\eta \to +0$, the integrand converges pointwise to zero. Also it is majorized uniformly in $\eta > 0$ by the integrable function $2\,\|A_1U_tA_2\|$. Hence the integral converges to zero as $\eta \to +0$ by the Lebesgue dominated convergence theorem, showing that u-lim $A_1R^o_{\lambda+i\eta}A_2 = A_1R^o_{\lambda+io}A_2$ as $\eta\to+0$. Since the integral in (9.16) is independent of λ, the convergence is uniform in $\lambda\varepsilon\mathcal{R}$. Similarly one obtains the boundary values of $A_1R^o_zA_2$ in the lower half-plane. #

PROPOSITION 9.9 : Let V be a real-valued measurable function defined on $\mathcal{R}^3$ belonging to $L^1(\mathcal{R}^3)\cap L^\infty(\mathcal{R}^3)$, and let $H = K_o + V$. Then

(a) The scattering theory determined by U_t and $V_t = \exp(-iHt)$ is asymptotically complete.

(b) If $H_\gamma = K_o + \gamma V$ with $\gamma\varepsilon\mathcal{R}$, then for sufficiently small values of $|\gamma|$, H_γ is spectrally absolutely continuous and the wave operators $\Omega_{\pm,\gamma}$ = s-lim $\exp(iH_\gamma t)U_t$ as $t\to\pm\infty$ are unitary.

We remark that the hypothesis of this proposition is verified for all potentials satisfying $|V(\underline{x})| \leq M(1+|\underline{x}|)^{-3-\eta}$ for some $\eta > 0$ and some $M < \infty$.

Proof : (i) Since $V\varepsilon\mathcal{B}(\mathcal{H})$ by Proposition 2.16, we have $D(H) = D(K_o)$. Furthermore

$$\int d^3x|V(\underline{x})|^2 \leq \|V\|_\infty \int d^3x|V(\underline{x})| = \|V\|_\infty\|V\|_1 < \infty. \qquad (9.17)$$

Thus $V(\cdot) \in L^2(R^3)$, and the wave operators exist by Corollary 8.15.

(ii) We use the factorization (7.75), i.e. we take $A = V^{\frac{1}{2}}$ and $B = |V|^{\frac{1}{2}}$. Notice that the functions $V^{\frac{1}{2}}(\cdot)$ and $|V|^{\frac{1}{2}}(\cdot)$ are in $L^2(R^3) \cap L^\infty(R^3)$. We choose C to be the multiplication operator by $\psi(\underline{x}) = (1+\underline{x}^2)^{-1}$. The range of C is dense since it contains $S(R^3)$.

It follows from Lemma 9.8 that $\|AU_tA\| \in L^1(R^3)$ and that there exists $M < \infty$ such that $\|BR^o_{\lambda\pm i\eta}A\| \leq M$ and $\|BR^o_{\lambda\pm i\eta}C\| \leq M$ for all $\lambda\in R$ and all $\eta > 0$. Hence all the hypotheses of Proposition 9.4 are verified, and we have asymptotic completeness and unitarity of $\Omega_{\pm,\gamma}$ for weak coupling (cf. Problem 9.1 for a simple upper bound for $|\gamma|$.)

(iii) Asymptotic completeness for an arbitrary coupling constant follows from Proposition 9.6 provided that we show that $W_z \equiv BR^o_zA$ verifies the hypotheses of Lemma 9.5. This is not difficult to do. (α) W_z is holomorphic by Lemma 5.12. (β) Since $V^{\frac{1}{2}}(\cdot) \in L^2(R^3)$, BR^o_z is a Hilbert-Schmidt operator by Proposition 7.7 if $\text{Im}\, z \neq 0$. Since $A\in B(H)$, BR^o_zA is Hilbert-Schmidt and hence compact. (γ) For $\text{Im}\, z \neq 0$, we obtain from Proposition 6.9 that

$$(I+BR^o_zA)(I-BR_zA) = I-B(R_z-R^o_z+R^o_zVR_z)A = I \tag{9.18}$$

and

$$(I-BR_zA)(I+BR^o_zA) = I-B(R_z-R^o_z+R_zVR^o_z)A = I. \tag{9.19}$$

Hence $I + BR^o_zA$ is invertible and

$$(I+BR^o_zA)^{-1} = (I-BR_zA) \in B(H). \tag{9.20}$$

(δ) The convergence of $W_{\lambda\pm i\eta}$ as $\eta\to+0$, uniformly in $\lambda\varepsilon\mathcal{R}$, follows from Lemma 9.8(c). #

In the above proof the estimates on the free resolvent near the real axis were easily obtainable from the fact that $\|BU_tA\|$ and $\|BU_tC\|$ belong to $L^1(\mathcal{R})$. To arrive at asymptotic completeness for a larger class of potentials, one writes the free resolvent as an integral operator in $L^2(\mathcal{R}^3)$ (with respect to the position variable) and uses estimates on the kernel of this integral operator. This is the content of the next proposition and the ensuing lemmas. In Proposition 9.10 we take into account the factor 2m appearing in (3.29).

<u>PROPOSITION 9.10</u> : Let $K_o = \underline{P}^2/2m$ and $z\;\varepsilon\;\rho(K_o)$. Then $(K_o-z)^{-1}$ is an integral operator in $L^2(\mathcal{R}^3)$ with kernel

$$G_z^o(\underline{x},\underline{y}) = \frac{2m}{4\pi|\underline{x}-\underline{y}|}\exp\{i(2mz)^{\frac{1}{2}}|\underline{x}-\underline{y}|\}, \tag{9.21}$$

where the branch of the square root is given by Im $[(2mz)^{\frac{1}{2}}]>0$.

<u>Proof</u> : Since the function $\underline{k}\mapsto\phi_z(\underline{k})\equiv[(\underline{k}^2/2m)-z]^{-1}$ is in $L^2(\mathcal{R}^3)$, it follows from part (i) of the proof of Lemma 7.6 that $(K_o-z)^{-1}$ is an integral operator in the position variable with kernel $(2\pi)^{-3/2}\tilde{\phi}_z(\underline{y}-\underline{x})$. (9.21) is obtained by evaluating the Fourier transform of ϕ_z. The details of this calculation are given in Section 9-2. #

<u>Remark 9.11</u> : The integral kernel $G_z(\underline{x},\underline{y})$ is called the <u>free Green's function</u>. For $\lambda\;\varepsilon\;\sigma(K_o) = [0,\infty)$, the limits of $G^o_{\lambda\pm i\eta}(\underline{x},\underline{y})$ as $\eta\to+0$ exist for almost all $\underline{x},\underline{y}$ and define the <u>retarded</u> and the <u>advanced free Green's functions</u> $G^o_{\lambda+io}$ and

$G^o_{\lambda-io}$ respectively :

$$G^o_{\lambda\pm io}(\underline{x},\underline{y}) = \frac{2m}{4\pi|\underline{x}-\underline{y}|}\exp\{\pm i|\sqrt{2m\lambda}|\,|\underline{x}-\underline{y}|\},\ \lambda \geq 0. \tag{9.22}$$

Thus $R^o_{\lambda\pm io}$ are also given as integral operators, but of course the kernels (9.22) do not define bounded operators. The Green's functions verify the equation $(-\Delta_x/2m - z)G^o_z(\underline{x}-\underline{y}) = \delta^{(3)}(\underline{x}-\underline{y})$, which is formally obtained by writing the identity $(K_o-z)(K_o-z)^{-1} = I$ in terms of integral kernels or by differentiating (9.21) and (9.22). The Green's functions for the total Hamiltonian will be discussed in Section 10-4$\underline{B}$.

LEMMA 9.12 : Define $I(\underline{x}) = \int d^3y|\underline{x}-\underline{y}|^{-2}(1+|\underline{y}|)^{-2\alpha}$, where $\alpha > \frac{1}{2}$. Then there exists $M_\alpha < \infty$ such that

$$I(\underline{x}) \leq M_\alpha(1+\underline{x}^2)^{-1} \quad \text{if } \alpha > 3/2, \tag{9.23}$$

$$I(\underline{x}) \leq M_\alpha(1+\underline{x}^2)^{\frac{1}{2}-\alpha} \quad \text{if } \tfrac{1}{2} < \alpha < 3/2. \tag{9.24}$$

Proof : Clearly $I(\underline{x})$ depends only on $r \equiv |\underline{x}|$, so that we may choose $\underline{x} = r\underline{e}_3$, where $\underline{e}_3$ is the unit vector along the third coordinate axis. By setting $\underline{u} = \underline{x} - \underline{y}$ and introducing spherical coordinates for $\underline{u}$, one gets

$$\begin{aligned} I(\underline{x}) &= 2\pi\int_0^\infty u^2du\int_{-1}^{+1}d\cos\theta\, u^{-2}[1+(r^2+u^2-2r\,u\cos\theta)^{\frac{1}{2}}]^{-2\alpha} \\ &\leq 2\pi\int_0^\infty du\int_{-1}^{+1}d\cos\theta(1+|u-r|)^{-2\alpha} \leq 4\pi\int_{-\infty}^\infty du(1+|u|)^{-2\alpha} \\ &= 8\pi(2\alpha-1)^{-1}. \end{aligned}$$

Thus $I(\underline{x})$ is uniformly bounded, and (9.23) and (9.24) are verified for $|\underline{x}| \leq 1$. For $|\underline{x}| > 1$, we divide the domain of integration in the definition of $I(\underline{x})$ into $D_1 \cup D_2$ with $D_1 = \{\underline{y}|\ |\underline{y}| \leq r/2\}$, $D_2 = \{\underline{y}|\ |\underline{y}| > r/2\}$, which leads to a splitting of $I(\underline{x})$ into $I(\underline{x}) = I_1(\underline{x}) + I_2(\underline{x})$. One gets the following estimates for $\alpha \neq 3/2$:

$$I_1(\underline{x}) \leq 4r^{-2}\int_{D_1} d^3y(1+|\underline{y}|)^{-2\alpha} < 16\pi r^{-2}\int_0^{r/2} dy(1+y)^{2-2\alpha}$$

$$= \frac{16\pi}{3-2\alpha} r^{-2} [(1+r/2)^{3-2\alpha}-1] < c_\alpha r^{-2} + d_\alpha r^{1-2\alpha}. \qquad (9.25)$$

To estimate $I_2(\underline{x})$, we set $\underline{v} = r^{-1}\underline{y}$ and obtain

$$I_2(\underline{x}) \leq \int_{|\underline{v}|>\frac{1}{2}} r^3 d^3v[r^{-2}|\underline{e}_3-\underline{v}|^{-2}]|r\underline{v}|^{-2\alpha}$$

$$= r^{1-2\alpha}\int_{|\underline{v}|>\frac{1}{2}} d^3v|\underline{e}_3-\underline{v}|^{-2}|\underline{v}|^{-2\alpha} = m_\alpha r^{1-2\alpha}, \qquad (9.26)$$

since the last integral in (9.26) is finite if $\alpha > \frac{1}{2}$. (9.25) and (9.26) imply that for $|\underline{x}| > 1$, $I(\underline{x}) \leq c_\alpha r^{-2} + (d_\alpha + m_\alpha)r^{1-2\alpha}$ $\leq 2c_\alpha(1+r^2)^{-1} + 2^{\alpha-\frac{1}{2}}(d_\alpha+m_\alpha)(1+r^2)^{\frac{1}{2}-\alpha}$, which implies (9.23) and (9.24). #

We shall now introduce a class of potentials that will be denoted by $\mathcal{V}$ in the sequel. A real-valued measurable function V defined on $\mathbb{R}^3$ belongs to $\mathcal{V}$ if it can be factorized as

$$V(\underline{x}) = (1+|\underline{x}|)^{-\nu}[B_1(\underline{x}) + B_2(\underline{x})] \qquad (9.27)$$

with $B_1(\cdot) \in L^2(\mathbb{R}^3)$, $|B_2(\underline{x})| \leq \beta(1+|\underline{x}|)^{-\frac{1}{2}-\delta}$ for some $\nu > 3/2$, $\delta > 0$ and $0 \leq \beta < \infty$. These conditions together with (7.5) imply that

$$\int d^3x|V(\underline{x})|^2 \leq 2\int d^3x|B_1(\underline{x})|^2 + 2\beta^2\int d^3x(1+|\underline{x}|)^{-2\nu} < \infty,$$

so that each function of $\mathcal{V}$ belongs to $L^2(\mathbb{R}^3)$. Roughly speaking, $\mathcal{V}$ consists of potentials that are absolutely square-integrable and decrease to zero more rapidly than $|\underline{x}|^{-2-\eta}$ for some $\eta > 0$ as $|\underline{x}| \to \infty$. It is easy to see that $\mathcal{V}$ is a linear set. By using Hölder's inequality in the proof of

Lemma 9.13 below, one may relax the condition imposed on B_2 by requiring only that $B_2 \varepsilon L^p(\mathcal{R}^3)$ for some $2 < p < 6$ (Kato and Kuroda [1]).

We add some other simple properties of potentials of the class $\mathcal{V}$ that will be used in the proofs below and in later chapters. We define A, B_1, B_2 and B to be the maximal multiplication operators in $L^2(\mathcal{R}^3)$ by $A(\underline{x}) \equiv (1+|\underline{x}|)^{-\nu}$, $B_1(\underline{x})$, $B_2(\underline{x})$ and $B(\underline{x}) \equiv B_1(\underline{x}) + B_2(\underline{x})$ respectively. A and B_2 are in $\mathcal{B}(\mathcal{H})$ by Proposition 2.16. B and V = AB are K_0-compact as a consequence of Proposition 8.7, so that $BR_z^0 \varepsilon \mathcal{B}(\mathcal{H})$ for all $z \varepsilon \rho(K_0)$ (see Proposition 8.2 and Remark 8.3). Furthermore, since $BR_z = BR_z^0(K_0 - z)R_z = BR_z^0(I - VR_z)$ and $VR_z \varepsilon \mathcal{B}(\mathcal{H})$ by Remarks 8.6 and 8.3 for $\mathrm{Im}\, z \neq 0$, BR_z is in $\mathcal{B}(\mathcal{H})$ for all such z, so that B is H-bounded.

<u>LEMMA 9.13</u> : Let V belong to $\mathcal{V}$ with associated factorization (9.27). For $z \varepsilon \rho(K_0)$ set $W_z = BR_z^0 A$, and for $\lambda \geq 0$ define $W_{\lambda \pm io}$ to be the integral operator with kernel

$$W_{\lambda \pm io}(\underline{x},\underline{y}) \equiv B(\underline{x}) G^o_{\lambda \pm io}(\underline{x},\underline{y}) A(\underline{y}). \tag{9.28}$$

Then (a) $W_z \varepsilon \mathcal{B}_2$ for $z \varepsilon \rho(K_0)$ and $z = \lambda \pm io$, and $\{W_z\}$ is continuous in Hilbert-Schmidt norm in $\{z |\ \mathrm{Im}\, z \geq 0\}$ and in $\{z |\ \mathrm{Im}\, z \leq 0\}$, (b) $\{W_z\}$ verifies all the hypotheses of Lemma 9.5, (c) $\|W_{\lambda \pm io}\|^2_{HS}$ is independent of λ and given by

$$\|W_{\lambda \pm io}\|^2_{HS} = (4\pi)^{-2} \iint d^3x\, d^3y |B(\underline{x})|^2 |\underline{x}-\underline{y}|^{-2} |A(\underline{y})|^2.$$

<u>Proof</u> : (a) For $z \varepsilon \rho(K_0)$, W_z belongs to $\mathcal{B}(\mathcal{H})$ by the preceding remarks. By Propositions 2.33 and 9.10 we have

$$\|W_z\|_{HS}^2 = \iint d^3x\, d^3y |B(\underline{x})|^2 |G_z^o(\underline{x},\underline{y})|^2 (1+|\underline{y}|)^{-2\nu}$$

$$\leq (4\pi)^{-2} \iint d^3x\, d^3y |B_1(\underline{x}) + B_2(\underline{x})|^2 |\underline{x}-\underline{y}|^{-2} (1+|\underline{y}|)^{-2\nu}$$

$$\leq M_\nu (8\pi^2)^{-1} \{\int d^3x |B_1(\underline{x})|^2 (1+|\underline{x}|^2)^{-1} + \int d^3x |B_2(\underline{x})|^2 (1+|\underline{x}|^2)^{-1}\},$$

where we have used (7.5) and (9.23) to obtain the last inequality. By using the hypotheses made on B_1 and B_2, we obtain

$$8\pi^2 \|W_z\|_{HS}^2 \leq M_\nu \|B_1\|_2^2 + \beta M_\nu \int d^3x (1+|\underline{x}|^2)^{-1} (1+|\underline{x}|)^{-1-2\delta} < \infty.$$

Thus W_z is Hilbert-Schmidt for each $z \varepsilon \rho(K_o)$ as well as for $z = \lambda \pm io$, $\lambda \geq 0$. We also get from (9.21), (9.22) and (2.67) that

$$\|W_z - W_\xi\|_{HS} = (4\pi)^{-2} \iint d^3x\, d^3y |B(\underline{x})|^2 |\underline{x}-\underline{y}|^{-2} \cdot$$

$$\cdot (1+|\underline{y}|)^{-2\nu} \left|\exp[(iz^{\frac{1}{2}} - i\xi^{\frac{1}{2}}) |\underline{x}-\underline{y}|] - 1\right|^2. \qquad (9.29)$$

It follows from the preceding estimates that the integrand in (9.29) is majorized uniformly in z and ξ by an integrable function. Since it also converges pointwise to zero as $\xi \to z$ in the closed upper or lower half-plane, we arrive at the continuity of $\{W_z\}$ in Hilbert-Schmidt norm by applying the Lebesgue dominated convergence theorem.

(b) We have already shown in (a) that W_z is compact and that $\lim W_{\lambda \pm i\eta} = W_{\lambda \pm io}$ as $\eta \to +0$ in Hilbert-Schmidt norm. Let Δ be a compact subset of R. Since the restriction of a continuous function to a compact set is uniformly continuous [R, Proposition 9.16], it follows from (a) that $\{W_z\}$ is uniformly continuous in Hilbert-Schmidt norm on each of the two compact subsets $\{z = \lambda \pm i\eta |\ \lambda \varepsilon \Delta,\ 0 \leq \eta \leq \eta_o\}$ of C, which by (2.62) implies in particular that $\|W_{\lambda \pm i\eta} - W_{\lambda \pm io}\| \to 0$ as

$\eta \to +0$ uniformly in $\lambda \varepsilon \mathcal{R}$.

By (5.58) one has $(z-\xi)^{-1}(BR_z^oA - BR_\xi^oA) = BR_z^oR_\xi^oA$, which leads to

$$\frac{d}{dz} BR_z^oA = B(R_z^o)^2 A, \tag{9.30}$$

with the limit existing in operator norm. Thus $\{W_z\}$ is holomorphic in $\mathcal{C}-[0,\infty)$. The fact that $(I+BR_z^oA)^{-1} \varepsilon \mathcal{B}(\mathcal{H})$ for $z \varepsilon \mathcal{C}-\mathcal{R}$ follows from (9.20), since $BR_zA \varepsilon \mathcal{B}(\mathcal{H})$.

(c) This is an immediate consequence of (9.28) and (9.22). #

PROPOSITION 9.14 : Let $V \varepsilon \mathcal{V}$ and $H = K_o + V$. Then the scattering theory determined by H and K_o is asymptotically complete. Moreover, if $H_\gamma = K_o + \gamma V$ with $\gamma \varepsilon \mathcal{R}$, then for sufficiently small values of $|\gamma|$, H_γ is spectrally absolutely continuous and the wave operators associated with the pair $\{H_\gamma, K_o\}$ are unitary.

Proof : Since $V \varepsilon L^2(\mathcal{R}^3)$, we have $D(H) = D(K_o)$ by Proposition 8.7. Also the wave operators exist by Corollary 8.15.

Let A and B be as indicated before Lemma 9.13 and define C = A. The range of C is dense since it contains all functions in $L^2(\mathcal{R}^3)$ of bounded support. Lemma 9.8 implies that $\|AU_tA\| \varepsilon L^1(\mathcal{R})$. B is K_o-bounded as well as H-bounded. By Lemma 9.13, $W_z \equiv BR_z^oA$ verifies the hypotheses of Lemma 9.5, so that the assertions of the proposition follow immediately from Propositions 9.4 and 9.6. #

The results about asymptotic completeness given so far can be extended to a larger class of potentials by applying more refined inequalities than those used in Lemmas 9.8 and 9.12. Some details about this can be found in Section 9-2. We end this section with another theorem about asymptotic completeness which is applicable when the unperturbed Hamiltonian H_o has simple spectrum, in particular in potential scattering with a spherically symmetric potential if one considers the scattering problem in each partial wave subspace (cf. Chapter 11). The theorem states that asymptotic completeness holds provided that one is able to verify certain spectral properties of the total Hamiltonian H.

<u>PROPOSITION 9.15</u> : Suppose that the absolutely continuous part of H_o is unitarily equivalent to the multiplication operator by the independent variable in $L^2(a,b)$, where (a,b) is a finite or infinite interval (i.e. the spectral representation of the part of H_o in $\mathcal{H}_{ac}(H_o)$ is $\mathcal{G}_o = L^2(a,b)$.) Assume that $\Omega_\pm = \text{s-lim} \exp(iHt)\exp(-iH_o t)E_{ac}(H_o)$ as $t \to \pm\infty$ exist, that H has simple spectrum and that $\sigma_e(H) \subseteq [a,b]$. Then the ranges of $\Omega_\pm$ are equal to $\mathcal{H}_{ac}(H)$.

<u>Proof</u> (for Ω_+) : By the assumptions made on H, it follows that the spectral representation space of H_{ac} is $\mathcal{G} = L^2(\Lambda)$ for some $\Lambda \subseteq [a,b]$. Set $F_+ = \Omega_+\Omega_+^*$. By Remark 5.22 and Proposition 5.21, $F_+\mathcal{H} \subseteq \mathcal{H}_{ac}(H)$ and F_+ reduces H. By Proposition 5.27 and since H has simple spectrum, F_+ is diagonalizable in $\mathcal{G}$, i.e. $F_+ = \{\phi(\lambda)I_o\}$. As a consequence of Proposition 5.26 and Problem 5.9, $F_+(\lambda) \equiv \phi(\lambda)I_o$ is a projection for almost all $\lambda\varepsilon\Lambda$, in particular $[\phi(\lambda)]^2 = \phi(\lambda)$ a.e. Hence $\phi(\lambda) = 0$ or 1 a.e. Thus, if we set $\Delta = \{\lambda\varepsilon\Lambda | \phi(\lambda) = 1\}$, then $F_+ = \{\chi_\Delta(\lambda)I_o\}$, so that by (5.88) $F_+ = E_\Delta E_{ac}(H)$.

Now define $\Delta' = [a,b]-\Delta$. One has from (5.18)

$$E_{\Delta'}F_+ = E_{\Delta'}E_\Delta E_{ac}(H) = E_{\Delta'\cap\Delta}E_{ac}(H) = E_\emptyset E_{ac}(H) = 0.$$

This, together with the intertwining relation (5.70), implies that

$$E_{ac}(H_o)E^o_{\Delta'} = \Omega_+^*\Omega_+E^o_{\Delta'} = \Omega_+^*E_{\Delta'}\Omega_+ = \Omega_+^*E_{\Delta'}F_+\Omega_+ = 0.$$

Since $E_{ac}(H_o)E^o_{\Delta'}$ is the multiplication operator by $\chi_{\Delta'}(\lambda)$ in $G_o = L^2(a,b)$, the Lebesgue measure of Δ' must be zero. Thus $E_{\Delta'}E_{ac}(H) = 0$ by the definition of the subspace of absolute continuity. Also, since $\sigma_{ac}(H) \subseteq [a,b]$, we have $E_{R-[a,b]}E_{ac}(H) = 0$, so that we arrive at $E_{ac}(H) = E_\Delta E_{ac}(H) + E_{\Delta'}E_{ac}(H) + E_{R-[a,b]}E_{ac}(H) = E_\Delta E_{ac}(H) = F_+$. #

In the above proof we used in an essential way the hypothesis that $\Delta' \subseteq [a,b]$ and $E^o_{\Delta'}E_{ac}(H_o) = 0$ imply $|\Delta'| = 0$. It would not be sufficient to require in Proposition 9.15 only that H_o has simple spectrum and that $\sigma_{ac}(H_o) = [a,b]$, as can be seen from the following example : Let $H = L^2(0,1)$, let Δ be a measurable subset of $[0,1]$ which is everywhere dense in $[0,1]$ and such that $0 < |\Delta| < 1$, and set $\Delta' = [0,1]-\Delta$. Define H and H_o by $(Hf)(x) = xf(x)$ for $0 \leq x \leq 1$, $(H_of)(x) = xf(x)$ for $x\in\Delta$ and $(H_of)(x) = -f(x)$ for $x\in\Delta'$. Then $H_{ac}(H_o) = L^2(\Delta)$, $\sigma_{ac}(H_o) = [0,1] = \sigma_{ac}(H)$ (σ_{ac} is a closed set !), $E^o_{\Delta'} = 0$ but $|\Delta'| = 1-|\Delta| > 0$. The wave operators exist and are equal to E^o_Δ, so that their range is strictly smaller than $H_{ac}(H) = L^2(0,1)$ (see also Problem 13.11).

In the preceding example the spectral representation of $H_{o,ac}$ is $L^2(\Delta,C)$; in particular each point in $R-\Delta$ has spectral multiplicity 0. A more general theorem than Proposition 9.15 can be obtained by using spectral multiplicity theory. In fact H_{ac} and $H_{o,ac}$ can be unitarily equivalent (in particular the wave operators, if they exist, can be asymptotically complete) if and only if H_{ac} and $H_{o,ac}$ have identical multiplicity functions (see the article of Brown in [PE].)

9-2 NOTES AND SUPPLEMENTARY MATERIAL

In the first part of this section we describe briefly some other methods for proving asymptotic completeness.

A. We first mention a possible generalization of Proposition 9.9. Let $\phi_1, \phi_2 \in L^{3+\delta}(\mathbb{R}^3) \cap L^{3-\delta}(\mathbb{R}^3)$ for some $\delta > 0$ and let A_1, A_2 be as in Lemma 9.8. By using the Hölder and the Hausdorff-Young inequalities, one may prove that $\|A_1 U_t A_2\| \leq \text{const } |t|^{-3/p} \|\phi_1\|_p \|\phi_2\|_p$ for $p = 3 \pm \delta$, see Theorem 6.4 of Kato [2]. Hence $\|A_1 U_t A_2\| \in L^1(\mathbb{R})$ (use the bound with $p = 3+\delta$ for $|t| \leq 1$ and that with $p = 3-\delta$ for $|t| > 1$), and the conclusions of Lemma 9.8 are verified.

Now assume that $V \in L^{3/2+\eta}(\mathbb{R}^3) \cap L^{3/2-\eta}(\mathbb{R}^3)$ with $\eta > 0$. Then by Example 8.29, V is a Rollnik potential, i.e. H is defined by quadratic forms, $V^{\frac{1}{2}} E_\Delta \in \mathcal{B}(\mathcal{H})$ for each compact Δ and $W_z = |V|^{\frac{1}{2}} R_z^0 V^{\frac{1}{2}} \in \mathcal{B}_2$ for each $z \in \mathbb{C}$, including $z = \lambda \pm i0$. Let A,B, C be as in the proof of Proposition 9.9. One sees that parts (ii) and (iii) of that proof are valid for this case, so that s-lim $U_t^* V_t E_{ac}(H)$ as $t \to \pm\infty$ exist by the proof of Proposition 9.6.

It remains to establish the existence of the wave operators. Since $\|AU_tC\| \in L^1(\mathcal{R})$, one has $\|E_\Delta V_t^* V U_t C f\| \leq \|V^{\frac{1}{2}} E_\Delta\| \cdot \|AU_tC\| \, \|f\| \in L^1(\mathcal{R})$, so that s-lim $E_\Delta V_t^* U_t C$ as $t \to \pm\infty$ exist by Proposition 4.15 if Δ is compact. Since C has dense range, s-lim $E_\Delta V_t^* U_t$ as $t \to \pm\infty$ exist. Now let $f \in D(K_o^{\frac{1}{2}})$, i.e. $f = (K_o^{\frac{1}{2}} + 1)^{-1} g$ with $g \in \mathcal{H}$, and let E(n) be as in Proposition 7.4. Then

$$V_t^* U_t f = E(n) V_t^* U_t f + [I - E(n)](|H|^{\frac{1}{2}}+1)^{-1}(|H|^{\frac{1}{2}}+1)(K_o^{\frac{1}{2}}+1)^{-1} U_t g.$$

Now $\|(|H|^{\frac{1}{2}}+1)(K_o^{\frac{1}{2}}+1)^{-1}\| \equiv M < \infty$ since $Q(H) = Q(K_o)$, and $\|[I-E(n)](|H|^{\frac{1}{2}}+1)^{-1}\| \leq n^{-\frac{1}{2}}$, so that

$$\|V_t^* U_t f - V_s^* U_s f\| \leq \|E(n)(V_t^* U_t - V_s^* U_s) f\| + 2n^{-\frac{1}{2}} M \|g\|. \quad (9.31)$$

Given $\delta > 0$, one first chooses n such that the second term is less than $\delta/2$ and then T such that the first term is less than $\delta/2$ if $s,t > T$ (or $s,t < -T$), which proves that $\{V_t^* U_t\}$ is strongly convergent on a dense set and hence everywhere.

The preceding proof of asymptotic completeness applies in particular for all potentials in $L^1(\mathcal{R}^3) \cap L^2(\mathcal{R}^3)$, a class which played an important role in the early days of scattering theory (cf. D below). It should also be mentioned that existence of $\Omega_\pm$ and asymptotic completeness have been proved by Simon for all Rollnik potentials [SM].

B. Our proof of asymptotic completeness in Section 9-1 bears a certain resemblance to the method of relatively smooth operators introduced by Kato [2,3]. Relative smoothness is a somewhat stronger form of condition (9.2) in the sense that the bound should be uniform on the set $\{f \mid \|f\| = 1\}$.

More precisely, if H is self-adjoint and A closed with $D(H) \subseteq D(A)$, then A is said to be H-smooth if

$$\|A\|_H^2 \equiv \sup_{\theta \neq f \in D(H)} (2\pi)^{-1} \|f\|^{-2} \int_{-\infty}^{\infty} \|Ae^{-iHt} f\|^2 dt < \infty. \tag{9.32}$$

Further consequences of this definition are stated in Problem 9.10. A look at Proposition 9.1 makes it clear that smoothness will be useful in proving asymptotic convergence. In this context we cite the following theorem by Lavine [2].

PROPOSITION 9.16 : Let H_1, H_2 be self-adjoint, $\{E_\lambda^1\}$, $\{E_\lambda^2\}$ their respective spectral families, $T \in \mathcal{B}(\mathcal{H})$ and Δ an open interval. Suppose that there exist two operators A and B such that A is H_1-bounded, AE_Δ^1 is H_1-smooth, B is H_2-bounded, BE_Δ^2 is H_2-smooth and such that for all $f_1 \in D(H_1)$, $f_2 \in D(H_2)$

$$(Tf_1, H_2 f_2) - (H_1 f_1, Tf_2) = (Af_1, Bf_2). \tag{9.33}$$

Then $\quad T_\pm(\Delta) \equiv \text{s-lim} \exp(iH_2 t)\, T \exp(-iH_1 t) E_\Delta^1$

and $\quad T'_\pm(\Delta) \equiv \text{s-lim} \exp(iH_1 t) T^* \exp(-iH_2 t) E_\Delta^2$

as $t \to \pm\infty$ exist and $T'_\pm(\Delta) = [T_\pm(\Delta)]^*$.

This theorem may be used to prove the asymptotic convergence of observables (take $H_1 = H_2 = H$ and $T \in \mathcal{A}_o$, the set of free observables defined in Section 4-3) or to prove the existence and asymptotic completeness of wave operators. For the latter proof one sets for example $H_1 = H_o$, $H_2 = H$ and $T = I$, so that formally AB represents a factorization of $V = H - H_o$ (e.g. $A = V^{\frac{1}{2}}$, $B = |V|^{\frac{1}{2}}$). If the hypotheses of the proposition are verified with $\Delta = \mathbb{R}$, then the wave operators exist and one has asymptotic completeness (in fact $\Omega_\pm$ are unitary), since the roles of H and H_o are interchangeable in the theorem (which differentiates this method from that of Section 9-1). In more general situations one may try to verify the smoothness conditions in the proposition for sufficiently many intervals, so that asymptotic completeness will follow as in the proof of Proposition 9.6.

Various methods for proving relative smoothness have been developed. One of them is that of positive commutators. If $A \in \mathcal{B}(\mathcal{H})$ and $i(HA-AH) \geq 0$, then $[i(HA-AH)]^{\frac{1}{2}}$ is H-smooth (Problem 9.7). Results about relative smoothness based on modifications of this simple theorem were given by Lavine [2,3]. Alternatively, one may express $||A||_H$ in terms of the resolvent R_z or the spectral family $\{E_\lambda\}$ of H, e.g. as (see Kato [2])

$$||A||_H^2 = \sup_{f,z} (2\pi)^{-1} ||f||^{-2} |(A^*f, [R_z - R_{\bar{z}}]A^*f)|$$
$$= \sup_{f,\Delta} ||E_\Delta A^*f||^2 / (||f||^2 |\Delta|), \qquad (9.34)$$

where f varies over $D(A^*)$, z over $\mathcal{C}-\mathcal{R}$ and $\Delta = (a,b]$ over all finite half-open intervals. Another way of arriving at relative smoothness therefore consists in obtaining estimates on the resolvent near the real axis. It can easily be seen for instance from the considerations of Section 9-1 and 9-2A that, if V is as in Section 9-2A, then $V^{\frac{1}{2}}$ is H_0-smooth and $V^{\frac{1}{2}}E_\Delta$ is H-smooth for each open interval Δ such that $\bar{\Delta} \cap \Gamma_0 = \emptyset$, where $\bar{\Delta}$ is the closure of Δ and Γ_0 the closed set of measure zero defined in Lemma 9.5 for $W_z = |V|^{\frac{1}{2}} R_z^0 V^{\frac{1}{2}}$. Stronger results on smoothness and asymptotic completeness in potential scattering can be found in Lavine [3]. See also Rejto [2].

Smoothness is also related to absolute continuity. If

AE_Δ is H-smooth, then $E_\Delta A^*D(A^*) \subseteq \mathcal{H}_{ac}(H)$, which follows immediately from the second identity in (9.34). In particular, if A^* has dense range, then Δ contains no singular spectrum of H. If A is H-smooth and $f \varepsilon \mathcal{H}_s(H)$, then for each $g \varepsilon D(A^*)$, $(Af,g) = (f,A^*g) = 0$ since $A^*g \varepsilon \mathcal{H}_{ac}(H)$. Hence $Af = \theta$ for each $f \varepsilon \mathcal{H}_s(H)$.

It is also interesting to note that, if A is H-smooth, then it is $|H|^\alpha$-bounded for each $\alpha > \frac{1}{2}$, cf. Theorem 5.8 of Kato [2]. For related results about commutators and wave operators the reader may also consult the book by Putnam [PU].

C. We next make some comments on the stationary method of proving asymptotic completeness which is based on the equations of time-independent scattering theory given in Section 6-2. To illustrate the typical steps that it involves, we indicate the proof of a theorem similar to Proposition 9.6 but where the hypothesis that $\|AU_tA\| \varepsilon L^1(\mathbb{R})$ is replaced by the weaker time-independent assumption that $A(R^o_{\lambda+i\eta} - R^o_{\lambda-i\eta})A$ converges as $\eta \to +0$, uniformly in λ (see the estimates in Lemma 9.8(c)), and where we set $C = A$ for simplicity.

PROPOSITION 9.17 : Let H, H_o, A and B be self-adjoint operators such that $E_\infty(H_o) = I$, A has dense range and $R_z = R^o_z - R^o_z B(I+AR^o_z B)^{-1} AR^o_z$ for $z \varepsilon \rho(H_o) \cap \rho(H)$. Assume that the stationary wave operators $\Omega_\pm = \text{w-lim}(\pm i\eta) \int R_{\lambda\mp i\eta} dE^o_\lambda$ as $\eta \to +0$ exist, that the closure of $AR^o_z B$ verifies the hypotheses of Lemma 9.5 and that the closure of $A(R^o_{\lambda+i\eta} - R^o_{\lambda-i\eta})A$ converges in operator norm as $\eta \to +0$, uniformly in λ on compact subsets of $\mathbb{R}$. Then $\Omega_\pm \Omega^*_\pm = E_{ac}(H)$.

It should be said that the above relation between R_z and R_z^o is verified under the assumptions of Proposition 9.6 by (9.20). Also, the condition of the existence of the stationary wave operators is redundant, since it can be deduced from the other hypotheses and Proposition 6.13.

Proof : The idea is to show that $\|\Omega_\pm^* E_\Delta Af\| = \|E_\Delta Af\|$ for f in D(A) and all intervals $\Delta = [a,b]$ such that a and b are not eigenvalues of H and $\Delta \cap \Gamma_o = \emptyset$, where Γ_o is the closed set of measure zero determined in Lemma 9.5. It follows that $\Omega_\pm^*$ are isometric on a dense set in $E_\Delta \mathcal{H}$, hence on $E_\Delta \mathcal{H}$. By using the fact that $\Gamma \equiv \mathcal{R}-\Gamma_o$ is a countable union of open intervals, one concludes similarly as in the proof of Proposition 9.6 that $\Omega_\pm^*$ are isometric on $E_\Gamma \mathcal{H}$. Hence $E_\Gamma \mathcal{H} \subseteq \Omega_\pm \mathcal{H}$ by Proposition 2.11, which leads again to (9.14)

To show that $\|\Omega_+^* E_\Delta Af\| = \|E_\Delta Af\|$ if Δ is as indicated, we insert the expression for R_z into that for Ω_+, which, together with Lemma 6.1, leads to

$$\Omega_+ E_\Delta^o = \text{w-}\lim_{\eta\to+0} \int_\Delta [I-R^o_{\lambda-i\eta}B(I+AR^o_{\lambda-i\eta}B)^{-1}A]dE_\lambda^o$$

$$\equiv \text{w-}\lim_{\eta\to+0} \int_\Delta L_{\lambda-i\eta}dE_\lambda^o. \tag{9.35}$$

Consequently $E_\Delta^o\Omega_+^* = \text{w-}\lim \int_\Delta dE_\lambda^o L^*_{\lambda-i\eta}$ as $\eta \to +0$, where the limit defining the spectral integral is now also a weak limit. But this integral exists as a strong limit, since the integrand verifies (6.48). Thus

$$\kappa \equiv \|\Omega_+^* E_\Delta Af\|^2 = (f, A\Omega_+ E_\Delta^o \Omega_+^* Af)$$

$$= \lim_{\sigma\to+0} \lim_{\eta\to+0} (f, \int_\Delta AL_{\lambda-i\eta}dE_\lambda^o \int_\Delta dE_\mu^o L^*_{\mu-i\sigma}Af)$$

By Proposition 6.14, the double spectral integral may be replaced by a single one. We write this spectral integral as a limit of Riemann-Stieltjes sums by using partitions $\Pi = \{\lambda_k;\mu_k\}$ such that $\lambda_k \notin \sigma_p(H)$ and then apply (9.42) to each term :

$$\kappa = (2\pi i)^{-1} \lim_{\sigma\to+0} \lim_{\eta\to+0} \lim_{|\Pi|\to 0} \lim_{\delta\to+0} \sum_{k=1}^{n(\Pi)} \int_{\lambda_{k-1}}^{\lambda_k} d\lambda \cdot$$

$$(f, AL_{\mu_k - i\eta}[R^o_{\lambda+i\delta} - R^o_{\lambda-i\delta}]L^*_{\mu_k - i\sigma}Af).$$

Now $AL_z = D_zA$ with $D_z = I - AR^o_zB(I+AR^o_zB)^{-1}$. $D_{\lambda-i\eta}$ converges in operator norm as $\eta \to +0$, uniformly in $\lambda\varepsilon\Delta$. Since $A(R^o_{\lambda+i\eta} - R^o_{\lambda-i\eta})A$ has the same convergence properties by hypothesis, we may interchange the limits as $|\Pi|\to 0$ and $\delta \to 0$. Since the scalar product in the last integral is uniformly continuous in μ_k and λ, the limit $|\Pi|\to 0$ may be taken and gives

$$\kappa = (2\pi i)^{-1} \lim_{\sigma\to+0} \lim_{\eta\to+0} \lim_{\delta\to+0} \int_a^b d\lambda (f, D_{\lambda-i\eta}A[R^o_{\lambda+i\delta} - R^o_{\lambda-i\delta}]AD^*_{\lambda-i\sigma}f). \tag{9.36}$$

The integrand in (9.36) is uniformly continuous in all the variables (up to the real axis), so that by the Lebesgue dominated convergence theorem one may replace the multiple limit by a single one by putting $\delta = \eta = \sigma$.

Now using Proposition 6.9 and (9.20), one easily sees that $L_zR^o_z = R_z$, so that by (5.58), $L_z(R^o_{\bar z} - R^o_z)L^*_z = -2i(\operatorname{Im} z)\cdot L_zR^o_zR^{o*}_zL^*_z = -2i(\operatorname{Im} z)R_zR_{\bar z} = R_{\bar z} - R_z$. By setting $D_zA = AL_z$ in (9.36) and then using the preceding identity and (9.42), one arrives at the result

$$\|\Omega^*_+E_\Delta Af\|^2 = (2\pi i)^{-1} \lim_{\eta\to+0} \int_a^b d\lambda (Af, [R_{\lambda+i\eta} - R_{\lambda-i\eta}]Af) = \|E_\Delta Af\|^2. \ \#$$

Some examples that are covered by Proposition 9.17 are potentials in $\mathcal{V}$ (choose A and B as in Proposition 9.14), Rollnik potentials [take $B = V^{\frac{1}{2}}$ and A the multiplication operator by $\phi(\underline{x})$ with $\phi(\underline{x}) = |V(\underline{x})|^{\frac{1}{2}}$ if $V(\underline{x}) \neq 0$ and $\phi(\underline{x}) = (1+|\underline{x}|^2)^{-2}$ if $V(\underline{x}) = 0$] and potentials verifying $|V(\underline{x})| \leq \beta(1+|\underline{x}|)^{-1-\eta}$ with $\eta > 0$ (see Section 10-4C). The above form of the stationary method is due to Thomas [3] who used it to prove asymptotic completeness in the three-body problem. A slightly different abstract stationary theory, based on the spectral representations of H and H_o rather than spectral integrals, has been developed by Kato and Kuroda [1]. See also Birman and Entina [1] and Schechter [1].

It is sometimes useful to have a stationary theory that is local in the energy variables. Let Δ be a Borel subset of $\mathbb{R}$ and define $\Omega_{\pm,\Delta} = \Omega_{\pm}E^o_{\Delta}$. If in Proposition 9.17 the convergence of $AR^o_{\lambda\pm i\eta}B$ and $A(R^o_{\lambda+i\eta}-R^o_{\lambda-i\eta})A$ as $\eta \to +0$ is imposed only for $\lambda\varepsilon\Delta$, one will obtain the result that $\Omega_{\pm,\Delta}\Omega^*_{\pm,\Delta} = E_{ac}(H)E_{\Delta}$; see e.g. Kato and Kuroda [1].

The abstract stationary theory is more general than the time-dependent one in the following sense : It may happen that the limits $W_{\pm} = \text{s-lim } (\pm i\eta) \int R_{\lambda\mp i\eta}\, dE^o_{\lambda}E_{ac}(H_o)$ as $\eta \to +0$ exist and define partially isometric operators but that $\{V^*_tU_tE_{ac}(H_o)\}$ is not strongly convergent as $t\to\pm\infty$. The existence of $W_{\pm}$ implies that of the Abelian time-dependent limits $W_{\pm} = \text{s-lim } (\pm\eta) \int_0^{\pm\infty} dt \exp(\mp\eta t)V^*_tU_tE_{ac}(H_o)$ as $\eta \to +0$. Sufficient conditions in addition to the existence of $W_{\pm}$ ensuring strong convergence of $\{V^*_tU_tE_{ac}(H_o)\}$ as $t\to\pm\infty$ can be found in Kato and Kuroda [1] and in Baumgärtel [2].

D. Another method for proving the existence and the asymptotic completeness of wave operators involves the use of trace class operators. Formally a trace criterion is as follows : If $\psi(H)-\psi(H_o)$ is a trace class operator for a suitable function ψ, then $\Omega_\pm(H,H_o)$ = s-lim $V_t^* U_t E_{ac}(H_o)$ and $\Omega_\pm(H_o,H)$ = s-lim $U_t^* V_t E_{ac}(H)$ as $t\to\pm\infty$ exist. A relatively simple result is the following ([K, Theorem X.4.12], [PU, Corollary 5.7.1]) :

PROPOSITION 9.18 : Let H and H_o be self-adjoint and suppose that $(H-z)^{-M} - (H_o-z)^{-M} \in \mathcal{B}_1$ for some positive integer M and some $z\in\mathcal{C}-\mathcal{R}$. Then $\Omega_\pm(H,H_o)$ and $\Omega_\pm(H_o,H)$ exist.

This theorem immediately implies asymptotic completeness in potential scattering with $V\in L^1(\mathcal{R}^3)\cap L^2(\mathcal{R}^3)$, since then $(H-z)^{-1}-(H_o-z)^{-1}\in\mathcal{B}_1$ by Proposition 7.23. (This was the first completeness result in potential scattering, given by Kuroda [1]).

More general results can be obtained by restricting the considerations to finite spectral intervals, see Birman [1]. In this context we cite the following theorem given by Pearson [5].

PROPOSITION 9.19 : Let H and H_o be self-adjoint and $T\in\mathcal{B}(\mathcal{H})$ such that $TD(H_o) \subseteq D(H)$. Let $A = HT-TH_o$ with $D(A) = D(H_o)$ and assume that the closure of A is of trace class. Then the following limits exist :

$$\Omega_\pm(H,H_o;T) = \underset{t\to\pm\infty}{\text{s-lim}} \exp(iHt)T\exp(-iH_o t)E_{ac}(H_o).$$

As an application one may take $T = E_\Delta E_\Delta^o$, where Δ is an interval. The condition on A then essentially means that $E_\Delta VE^o_\Delta \in \mathcal{B}_1$ and leads to the existence of the local wave operators s-lim $E_\Delta V^*_t U_t E^o_\Delta E_{ac}(H_o)$ and s-lim $E^o_\Delta U^*_t V_t E_\Delta E_{ac}(H)$ as $t\to\pm\infty$. Trace conditions of this type have been useful to prove asymptotic completeness for certain locally strongly singular potentials (Pearson [2,4], Deift and Simon [1], Combescure and Ginibre [2]) and existence and asymptotic completeness for certain potentials that are very rapidly oscillating near infinity with the amplitude of the oscillations diverging to $+\infty$ as $|\underline{x}|\to\infty$ (Matveev and Skriganov [1]).

E. In potential scattering there is yet another way of arriving at asymptotic completeness, namely that of eigenfunction expansions. The idea is to determine a set of non-normalizable eigenfunctions of H such that each vector in $\mathcal{H}_{ac}(H)$ can be expanded in terms of these eigenfunctions, and then to relate $\Omega_\pm$ to these eigenfunctions. Details and references will be given in the next chapter.

We summarize in rough terms the conditions under which asymptotic completeness has been established in potential scattering. As far as the behaviour of $V(\underline{x})$ at infinity is concerned, it suffices that $|V(\underline{x})| \leq M(1+|\underline{x}|)^{-1-\eta}$ for some $\eta > 0$ and all $|\underline{x}| \geq R$ (Lavine [3], Agmon [1]). Locally V may be strongly singular provided that the singularities are positive (Pearson [4], Combescure and Ginibre [2] and references given therein). Asymptotic completeness has also been proved for some potentials with strong negative singularities (Amrein and Georgescu [2], Pearson [2]), but for such potentials the phenomenon of local absorption cannot in general be excluded (Section 4-6C).

F. We end by giving the proofs that were omitted in Section 9-1.

Proof of (9.4) : Let $\delta,\eta > 0$ and define

$$\psi_{\delta,\eta}(\lambda) = \frac{1}{2\pi i}\int_{a+\delta}^{b+\delta} [(\mu-\lambda-i\eta)^{-1} - (\mu-\lambda+i\eta)^{-1}]d\lambda =$$

$$\frac{1}{2\pi i}\int_{a+\delta}^{b+\delta} 2i\eta[(\mu-\lambda)^2+\eta^2]^{-1}d\mu = \frac{1}{\pi}[\arctan\frac{b+\delta-\lambda}{\eta} - \arctan\frac{a+\delta-\lambda}{\eta}].$$

Clearly $\psi_{\delta,\eta}$ is a continuous function of λ and $|\psi_{\delta,\eta}(\lambda)| \leq 1$. Also, if $\chi_{(a,b]}$ denotes the characteristic function of $(a,b]$, then

$$\lim_{\delta\to+0}\lim_{\eta\to+0}\psi_{\delta,\eta}(\lambda) = \chi_{(a,b]}(\lambda). \tag{9.37}$$

As in (6.8), we have that

$$\|\int_{(a,b]}[1-\psi_{\delta,\eta}(\lambda)]dE_\lambda f\|^2 = \int_{(a,b]}|1-\psi_{\delta,\eta}(\lambda)|^2 d(f,E_\lambda f). \tag{9.38}$$

The integrand converges pointwise to zero as $\delta,\eta\to 0$ and is bounded by a constant uniformly in δ and η. Thus by the Lebesgue dominated convergence theorem the norm in (9.38) converges to zero as $\delta,\eta\to+0$ (in the correct order). Similarly

$$\|\int_{R-(a,b]}\psi_{\delta,\eta}(\lambda)dE_\lambda f\|^2 \to 0 \text{ as } \delta,\eta\to +0. \tag{9.39}$$

We thus have shown that

$$E_{(a,b]} \equiv \int_{a+o}^{b+o} dE_\lambda = \operatorname*{s-lim}_{\delta\to+0}\ \operatorname*{s-lim}_{\eta\to+0}\int_R \psi_{\delta,\eta}(\lambda)dE_\lambda. \tag{9.40}$$

It remains to show that for fixed $\delta,\eta > 0$

$$\int_R\psi_{\delta,\eta}(\lambda)dE_\lambda \equiv \frac{1}{2\pi i}\int_R\{\int_{a+\delta}^{b+\delta}d\mu[(\mu-\lambda-i\eta)^{-1} - (\mu-\lambda+i\eta)^{-1}]\}dE_\lambda$$

$$= \frac{1}{2\pi i}\int_{a+\delta}^{b+\delta} d\mu(R_{\mu+i\eta} - R_{\mu-i\eta})$$

$$\equiv \frac{1}{2\pi i}\int_{a+\delta}^{b+\delta}d\mu\int_R[(\mu-\lambda-i\eta)^{-1} - (\mu-\lambda+i\eta)^{-1}]dE_\lambda, \tag{9.41}$$

which amounts to justifying the interchange of the order of integration. This may be done by applying once more the Lebesgue dominated convergence theorem or Fubini's theorem (Problem 9.2). #

One sees from this proof that one may set $\delta = 0$ in (9.4) provided that a and b are not eigenvalues of H. In that case

$$E_{(a,b)} = E_{[a,b]} = \underset{\eta\to+0}{\text{s-lim}}\ (2\pi i)^{-1}\int_a^b d\lambda(R_{\lambda+i\eta} - R_{\lambda-i\eta}). \quad (9.42)$$

Proof of Lemma 9.5(b) : (i) It suffices to show that $\Gamma_o^+ \cap [a,b]$ and $\Gamma_o^- \cap [a,b]$ are closed sets of measure zero for each finite closed interval [a,b] (Problem 9.8). We consider for instance Γ_o^+. Since $\{W_z\}$ is continuous in the operator norm in the domain $\{z|\text{Im } z > 0\}$ and $W_{\lambda+i\eta}$ converges to $W_{\lambda+io}$ uniformly in $\lambda\varepsilon[a,b]$, $\{W_z\}$ is uniformly continuous in operator norm on the compact set $\{z = \lambda+i\eta|\lambda \varepsilon [a,b],\ 0 \le \eta \le 1\}$. Thus one may subdivide [a,b] into a finite union of closed intervals $[a_k,b_k]$ of length $|b_k-a_k| = \eta_o \le 2$ such that $||W_z-W_\xi|| < \frac{1}{2}$ whenever z and ξ both belong to one of the semi-circular domains $\Delta_k \equiv \{z|\text{Im } z \ge 0$ and $|z-\frac{1}{2}(a_k+b_k)| \le \frac{1}{2}\eta_o\}$, and it is sufficient to show that $\Gamma_o^+ \cap [a_k,b_k]$ is closed and of measure zero for each such subinterval.

To establish this latter fact, we shall construct a function $\phi : \Delta_k \to C$ having the following properties :

(α) ϕ is holomorphic in the interior of Δ_k,

(β) ϕ is continuous on Δ_k,

(γ) $\phi(z) = 0$ if and only if $(I+W_z)$ is not invertible.

Since $(I+W_z)^{-1}$ exists if Im $z \neq 0$, ϕ will not be identically zero. One then sees from the proof of Lemma 8.21 (Section 8-4) that the set $\{\lambda\varepsilon[a_k,b_k] | \phi(\lambda) = 0\}$ is closed and of measure zero, which means that $\Gamma_o^+ \cap [a_k,b_k]$ is a closed set of measure zero.

(ii) It remains to find ϕ. We first notice that W_z is compact for each $z\varepsilon\Delta_k$ by Proposition 2.22(d). We now fix $\zeta\,\varepsilon\,\Delta_k$ and choose a finite rank operator T of the form (2.52) such that $||W_\zeta - T|| < \frac{1}{2}$. We denote by M the range of T and by F the orthogonal projection with range M. It follows that for any $z\varepsilon\Delta_k$, $||W_z-T|| \leq ||W_z-W_\zeta|| + ||W_\zeta-T|| < \frac{1}{2} + \frac{1}{2} = 1$, so that $(I+W_z-T)^{-1}\varepsilon\mathcal{B}(\mathcal{H})$ for each $z\varepsilon\Delta_k$ by Proposition 2.20. If we define $Y_z = T(I+W_z-T)^{-1}$, we obtain as in (2.86) that

$$I+W_z = (I+Y_z)(I+W_z-T). \tag{9.43}$$

Y_z is holomorphic in the interior of Δ_k and continuous on Δ_k (Problem 9.3).

(9.43) implies that $(I+W_z)$ is invertible if and only if $(I+Y_z)$ is. Now by the proof of Proposition 2.24 (Section 2-5), $(I+Y_z)$ is invertible if and only if its restriction to M is, i.e. iff $F+Y_zF$ is invertible as an operator in M. This operator may be represented as a $n\times n$ matrix $\{a_{ik}(z)\}$ ($n = \dim M < \infty$) by fixing an orthonormal basis in M. The determinant $\phi(z) = \det\{a_{ik}(z)\}$ verifies (α), (β) and (γ) (Problem 9.3). #

<u>Calculation of the Green's function (9.21)</u> : It remains to show that

$$J \equiv (2\pi)^{-3}\int d^3k\, e^{-i\underline{k}\cdot\underline{x}}(\underline{k}^2/2m - z)^{-1} = \frac{2m}{4\pi r}\exp\{i(2mz)^{\frac{1}{2}}r\}, \tag{9.44}$$

with $r = |\underline{x}|$. By introducing spherical coordinates and integrating over the angles, one gets

$$J = \frac{1}{(2\pi)^2} \int_0^\infty k^2 dk (k^2/2m - z)^{-1} \cdot \frac{1}{ikr} (e^{ikr} - e^{-ikr})$$

$$= \frac{2m}{2(2\pi)^2 ir} \int_{-\infty}^{\infty} k\, dk\, (k^2 - 2mz)^{-1} (e^{ikr} - e^{-ikr}) \equiv I_1 - I_2, \qquad (9.45)$$

where in the second step we have used the fact that the integrand is an even function of k. The integrals I_1 and I_2 can be evaluated by contour integration. The integrand is holomorphic in the complex k-plane except for the two singular points $k = \pm(2mz)^{\frac{1}{2}}$. One may close the contour by a semicircle at infinity in the upper (lower) half plane for I_1 (respectively I_2) without changing the value of the integrals. These can now be evaluated by the residue theorem, giving

$$I_1 = \frac{2m}{2(2\pi)^2 ir} \frac{2\pi i}{2} \exp\{i(2mz)^{\frac{1}{2}} r\}, \quad I_2 = -I_1.$$

(9.44) follows by inserting these expressions into (9.45). #

In the above proof we have evaluated the Fourier transform of $\phi_z(\underline{k}) = (\underline{k}^2/2m - z)^{-1}$ pointwise. Since ϕ_z is not in $L^1(R^3)$ but only in $L^2(R^3)$, this Fourier transform should in principle be calculated as a limit in the mean. The consistency of the above procedure can however easily be checked as follows : the function $\underline{x} \mapsto |\underline{x}|^{-1} \exp\{i(2mz)^{\frac{1}{2}} |\underline{x}|\}$ is in $L^1(R^3) \cap L^2(R^3)$, so that its inverse Fourier transform may be calculated pointwise. This can be done by evaluating elementary integrals, and the result is, apart from a numerical factor, the function $\phi_z(\underline{k})$.

PROBLEMS

9.1 : Show that the operators $\Omega_{\pm,\gamma}$ in Proposition 9.9 are unitary whenever $|\gamma| < \gamma_o = 4\pi/3 \|V\|_\infty^{-1/3} \|V\|_1^{-2/3}$. (Hint : Use the bound on $\|AU_t B\|$ given in the proof of Lemma 9.8).

9.2 : Prove equation (9.41).

9.3 : Show that the function $\phi(z)$ defined at the end of the proof of Lemma 9.5(b) (Section 9-2) verifies (α), (β) and (γ) of part (i) of that proof (establish the holomorphy and continuity of Y_z, $F + Y_z F$ and $a_{ik}(z)$ in the appropriate domains and use the definition of the determinant).

9.4 : Prove asymptotic completeness for the separable potentials considered in Proposition 8.20. (Hint : Set $A = V^{1/2}$, $B = \gamma V^{1/2}$. Prove that $(h, V_t E_\Delta f) \in L^2(\mathcal{R})$ for $\Delta \in \mathcal{R} - \Gamma_o$ by using the results of the proof of Proposition 8.22, or else use Proposition 9.6 with $Cf = \sum_i \alpha_i (e_i, f) e_i$, where $\{e_i\}$ is an orthonormal basis of $L^2(\mathcal{R}^3)$, each e_i is in $L^1(\mathcal{R}^3)$, $\alpha_i > 0$, $\sum_i \alpha_i < \infty$ and $\sum \alpha_i \|e_i\|_1 < \infty$. Note that the exceptional sets $\Gamma_o^\pm$ given in Lemma 9.5 coincide with the sets $\{\lambda \in \mathcal{R} | 1 + \gamma (h, R^o_{\lambda \pm io} h) = 0\}$ used in Proposition 8.22).

9.5† : Write out in detail the stationary proof of asymptotic completeness of Proposition 9.17.

9.6 : Let $H = H^*$ and $\Delta \subseteq \mathcal{R}$ an interval. Assume there exists a dense set $\mathcal{D}$ in $\mathcal{H}$ such that for each $f \in \mathcal{D}$, $|(f, R_{\lambda + i\eta} f)| \leq M = M(f) < \infty$ for all $\lambda \in \Delta$, $0 < \eta \leq 1$. Show that the part of H in $E_\Delta \mathcal{H}$ is absolutely continuous. (Hint : Use (9.4)).

9.7 : If $A \in \mathcal{B}(\mathcal{H})$, $H = H^*$ and $i(HA - AH) \geq 0$, then $[i(HA-AH)]^{\frac{1}{2}}$ is H-smooth. (Hint : Use (9.32) and the monotone convergence theorem).

9.8 : (a) Let Δ be a measurable set in $\mathcal{R}$. Prove that, if $\Delta \cap [a,b]$ is closed and of measure zero for each finite $[a,b]$, then Δ is closed and $|\Delta| = 0$. (b) Let Δ be a measurable set in $(0,\infty)$. Assume that $\Delta \cap [a,b]$ is closed and of measure zero for each finite $[a,b] \subseteq (0,\infty)$. Show that $\Delta \cup \{0\}$ is closed and of measure zero.

9.9 : Let $W \in \mathcal{Y}$ and $V \in \mathcal{Y}$ with factorization $V = AB$, and define $H = K_o + W + \gamma V \equiv H_1 + \gamma V$, $\gamma \in \mathcal{R}$. (a) Prove that, in the notation (4.63), $\Omega_\pm(H,K_o)$, $\Omega_\pm(H,H_1)$ and $\Omega_\pm(H_1,K_o)$ exist and are complete. Also show that the corresponding R-operators verify

$$R(H,K_o) = R(H_1,K_o) + \Omega_+(H_1,K_o)^* R(H,H_1)\Omega_-(H_1,K_o). \qquad (9.46)$$

(Hint : Proposition 4.19). (b) Let $\Gamma_o(H_1,K_o)$ be the exceptional set associated with H_1. Under the assumption that $\Gamma_o(H_1,K_o) \cap [0,\infty) = \emptyset$, verify the hypotheses of Lemma 9.5 for the family $\{B(H_1-z)^{-1}A\}$ for Re $z \geq 0$. (c) What happens if the condition $\Gamma_o(H_1,K_o) \cap [0,\infty) = \emptyset$ is dropped ?

9.10†: Let A be closed, H self-adjoint and $D(H) \subseteq D(A)$. If A is H-smooth, the map $f \mapsto \{A \exp(-iHt)f\}$ defines a bounded operator from $D(H) \subseteq \mathcal{H}$ into $L^2(\mathcal{R},\mathcal{H};dt)$. Show that this implies that, for each $g \in \mathcal{H}$, $\exp(-iHt)g \in D(A)$ for almost all t, that $\{A \exp(-iHt)g\} \in L^2(\mathcal{R},\mathcal{H};dt)$ and that

$$\int_{-\infty}^{\infty} dt\, \|A \exp(-iHt)g\|^2 \leq 2\pi \|A\|_H^2 \|g\|^2.$$

(Hint : A strong Cauchy sequence in $L^2(\mathcal{R},\mathcal{H};dt)$ has a subsequence which converges pointwise for almost all t, see [L II, Theorem XI.2]).

CHAPTER 10 : EIGENFUNCTION EXPANSIONS

This chapter deals with some spectral properties of Hamiltonians of the form $H = K_0 + V$, $V \varepsilon \mathcal{V}$ and expressions for the scattering amplitude. Section 10-1 begins with a heuristic description of scattering phenomena in terms of eigenfunctions of H and K_0. To make this rigorous, we start with the construction of the spectral transformation of H_{ac} in terms of that of K_0 (Proposition 10.5). The associated improper eigenfunctions are then obtained from this spectral transformation in the following proposition. In Proposition 10.8 we prove that these eigenfunctions satisfy the Lippmann-Schwinger integral equation and form a complete set in a generalized sense. Finally in Proposition 10.11 we establish the boundedness and continuity of the eigenfunctions and their asymptotic behaviour at large distances. Section 10-2 is concerned with the on-shell R-operator and the scattering amplitude.

We verify in particular the continuity assumptions made on these quantities in Section 7-3 in order to define the scattering cross section. In Section 10-3 we show that there is no singularly continuous spectrum, thereby establishing strong asymptotic completeness of the theory. Section 10-4 contains amongst other things generalizations of the mathematical concepts introduced in Section 10-1, a discussion of the total Green's function and a study of the scattering problem for potentials verifying $|V(\underline{x})| \leq c(1+|\underline{x}|)^{-1-\delta}$ $(\delta > 0)$.

We shall use the following convention : Let $f = \{f_\lambda\}$ be in $L^2(\Lambda, \mathcal{K})$. If $f_\lambda = \theta$ (in $\mathcal{K}$) for almost all λ, then we shall choose the representative of f such that $f_\lambda = \theta$ for all λ, without explicitly mentioning it. This choice will lead to an expression for $S(\lambda)$ for $\lambda \notin \Gamma_o$ which is continuous in λ.

Notation : In this and the next two chapters, some quantities will carry a wave-vector $\underline{k}$ as a subscript. For typographical reasons we shall replace $\underline{k}$ by $\mathcal{k}$ whenever it appears as a subscript, but not otherwise.

10-1 EIGENFUNCTIONS

We begin with an outline of the traditional approach to the description of scattering phenomena, in which one considers the so-called improper eigenfunctions of the given pair of Hamiltonians (H, H_o) and their asymptotic behaviour at large distances from the scattering centre (see e.g. [LL, Section 105]). For $H_o = K_o$, consider the corresponding time-independent Schrödinger equation :

$$(\Delta + \lambda)\psi(\underline{x}) = 0, \tag{10.1}$$

where the positive number λ is the energy. One can easily find solutions of (10.1), e.g. the plane waves $\exp(\pm i\underline{k}\cdot\underline{x})$, labelled by the wave-vectors $\underline{k}$ such that $\underline{k}^2 \equiv k^2 = \lambda$. These solutions do not belong to the Hilbert space $L^2(R^3)$ and this is typical whenever one talks about an eigenvalue problem derived from a self-adjoint operator corresponding to a point in its continuous spectrum. On the other hand an eigenvector associated with an eigenvalue of a self-adjoint operator belongs by definition to the Hilbert space. Physicists often require box normalization or δ-function normalization for the improper eigenfunctions which do not belong to the Hilbert space ([LL, Section 5]). For instance, letting $\psi^o_k(\underline{x}) = (2\pi)^{-3/2} \cdot \exp(i\underline{k}\cdot\underline{x})$, one writes that

$$\int \overline{\psi^o_k(\underline{x})}\psi^o_{k'}(\underline{x})d^3x = \delta^{(3)}(\underline{k}-\underline{k}'). \tag{10.2}$$

We shall avoid such equations which make no sense in the mathematical language we have introduced, namely that of a Hilbert space. We adopt the convention that a non-normalizable (improper) eigenfunction like ψ^o_k, corresponding to a point in the continuous spectrum, will simply be called an eigen-function.

The eigenfunctions $\{\psi^o_k\}$ form a complete set leading to an eigenfunction expansion in the following sense : by Section 2-5A every function $f \in L^2(R^3)$ can be expressed as a continuous superposition of ψ^o_k with coefficients $\tilde{f}(\underline{k})$ given by (2.17) and satisfying $\|f\|^2 = \int |\tilde{f}(\underline{k})|^2 d^3k$.

For the perturbed Hamiltonian $H = K_o + V$, the associated time-independent Schrödinger equation is

$$(\Delta - V(\underline{x}) + \lambda)\psi(\underline{x}) = 0. \tag{10.3}$$

Though the number of potentials for which (10.3) is explicitly solvable in terms of elementary functions is small (see [N, page 417]), one can nevertheless make certain general statements about the nature of the solutions. In most cases of interest, the solutions of (10.3) that correspond to scattering are the ones with positive energy $\lambda = k^2$. These we denote by $\psi_{k}^{\pm}$. More precisely, for every wave-vector $\underline{k}$ with $k^2 = \lambda$, one looks for solutions $\psi_{k}^{\pm}$ of (10.3) having the following asymptotic behaviour as $|\underline{x}| \equiv r \to \infty$:

$$\psi_{k}^{\pm}(\underline{x}) \simeq \psi_{k}^{0}(\underline{x}) + r^{-1} e^{\mp ikr}(2\pi)^{-3/2} f_{\pm}(\lambda;\underline{\omega}_k,\underline{\omega}_x), \tag{10.4}$$

where we have written $\underline{\omega}_x = \underline{x}/r$.

To understand the asymptotic behaviour (10.4) in physical terms we imagine a picture of scattering of a wave from a fixed target. Then the first term on the right describes the incident wave or the unscattered part and the second corresponds to the scattered part of the wave. If the potential is very short range (we refer to Problem 10.4 for details), then at large distances from the target no further influence of the target on the scattered wave is expected and hence the total flux across any sphere of sufficiently large radius R should be independent of R, leading to the r^{-1} behaviour of the second term. In cases where the asymptotic behaviour (10.4) can be verified, the functions $f_{\pm}$ determine the scattering amplitude; e.g. $f_{-}(\lambda;\underline{\omega}_k,\underline{\omega}_x) = f(\lambda;\underline{\omega}_k \to \underline{\omega}_x)$. The eigenfunctions $\psi_{k}^{\pm}$ once more do not belong to the Hilbert space $L^2(R^3)$, and one often writes down box or δ-function normalization conditions as in (10.2).

In the above discussions the eigenfunctions of the Hamiltonians could be obtained as solutions of the associated partial differential equations. However, there is an abstract definition of eigenfunctions and eigenfunction expansion for any self-adjoint operator A and this definition will be given in Section 10-4. In general this definition involves a linear topological space, in a certain sense bigger than $\mathcal{H}$, in which the eigenfunctions must lie. The main difficulty in eigenfunction expansion is to find such a space. One can often consider the situation as a perturbation problem in the following sense. Assuming that the eigenfunction expansion of H_0 is known, one can construct such a space and the eigenfunctions for the perturbed operator $H = H_0 + V$ by using the perturbation V itself. In this section we carry this through for potential scattering and give some remarks about abstractions in Section 10-4. We shall only consider the class $\mathcal{V}$ of potentials and use the existence of $\Omega_\pm$ and the asymptotic completeness of the theory which have already been established for this class in Section 9-1. Usually one approaches the problem from the opposite direction, i.e. by making an eigenfunction expansion without using scattering theory and then constructing $\Omega_\pm$ from the eigenfunctions.

At first we study a more general problem, namely that of finding the spectral transformation $\mathcal{U}$ for the perturbed operator H when that of H_0 is known (see Section 5-7). We remind the reader that if $H_0 = K_0$, then the spectral transformation $\mathcal{U}_o$ is given as in (5.96) by*)

$$(\mathcal{U}_o f)_\lambda(\underline{\omega}) = 2^{-\frac{1}{2}}\lambda^{\frac{1}{4}}\,\tilde{f}(\underline{k}). \tag{10.5}$$

In the above we have used the notation that $\underline{\omega}$ is the unit vector in the direction of $\underline{k}$ and $\underline{k}^2 = \lambda \geq 0$, or equivalently $\underline{k} = \sqrt{\lambda}\underline{\omega}$. It is clear that for almost all λ, $(\mathcal{U}_o f)_\lambda \varepsilon L^2(S^{(2)})$.

We have seen in earlier chapters that some quantities appearing in scattering theory are associated with a fixed value of the kinetic energy. To each such value λ there is associated a Hilbert space $L^2(S^{(2)})$ in the spectral representation of K_o. While the theory is defined in $L^2(\mathcal{R}^3)$, certain operators such as the S-matrix act in $L^2(S^{(2)})$. To relate quantities acting in these two different Hilbert spaces it is very useful to have some operators mapping $L^2(\mathcal{R}^3)$ into $L^2(S^{(2)})$ or $L^2(S^{(2)})$ into $L^2(\mathcal{R}^3)$. We now define a class of such operators. Let ϕ be a measurable function finite almost everywhere and set for all $f \varepsilon D(\phi(\underline{Q}))$,

$$M_\phi(\lambda)f = (\mathcal{U}_o \phi(\underline{Q})f)_\lambda. \tag{10.6}$$

$M_\phi(\lambda)$ is interpreted as an operator from $L^2(\mathcal{R}^3)$ into $L^2(S^{(2)})$. Its properties are contained in the

LEMMA 10.1 : Let $\phi \varepsilon L^2(\mathcal{R}^3)$ and $\lambda \geq 0$. Then the closure of $M_\phi(\lambda)$ (also denoted by $M_\phi(\lambda)$) is a Hilbert-Schmidt operator from $L^2(\mathcal{R}^3)$ into $L^2(S^{(2)})$. Moreover the function $\lambda \mapsto M_\phi(\lambda)$ is continuous in Hilbert-Schmidt norm.

---

*) For the sake of clarity and in order to bring out the analogy between $\mathcal{U}$ and $\mathcal{U}_o$, we deviate in this chapter from the convention of not mentioning explicitly the free spectral transformation $\mathcal{U}_o$ in (5.96).

Remark 10.2 : In Chapter 2 we have introduced only operators acting in a single Hilbert space. One can also have an operator A from one Hilbert space $\mathcal{H}$ into another $\mathcal{H}'$, i.e. a linear mapping from a manifold $D(A) \subseteq \mathcal{H}$ into $\mathcal{H}'$. All the notions of bounded, closed, compact, Hilbert-Schmidt and trace class operators carry over if one uses the $\mathcal{H}$-norm in D(A) and the $\mathcal{H}'$-norm in the range of A. The only important change that takes place is in the definition of the adjoint. It is not difficult to see that if A is densely defined, then A* is a well defined linear operator from $\mathcal{H}'$ to $\mathcal{H}$ (Problem 10.1). In particular, if U is an unitary operator from $\mathcal{H}$ onto $\mathcal{H}'$ then $U^* = U^{-1}$. The canonical expansion (2.58) of a compact operator remains unchanged provided that the orthonormal sets $\{e_k\}$ and $\{h_k\}$ are in $\mathcal{H}$ and $\mathcal{H}'$ respectively. Similarly one can show that $\|A\|_{HS}^2 = \sum_k \|Ae_k\|^2 = \sum_k \|A^*h_k\|^2$, where $\{e_k\}$ and $\{h_k\}$ are any two orthonormal bases in $\mathcal{H}$ and $\mathcal{H}'$ respectively (Problem 10.1). Also Proposition 2.33 remains valid, e.g. if $\mathcal{H} = L^2(\mathcal{R}^n, d\mu)$ and $\mathcal{H}' = L^2(\mathcal{R}^m, d\nu)$, then A is Hilbert-Schmidt if and only if A is an integral operator with kernel $K_A(x,y)$ such that $\|A\|_{HS}^2 = \int\int |K_A(x,y)|^2 d\mu(x) d\nu(y) < \infty$. We shall denote by $\mathcal{B}(\mathcal{H},\mathcal{H}')$, $\mathcal{B}_o(\mathcal{H},\mathcal{H}')$, $\mathcal{B}_2(\mathcal{H},\mathcal{H}')$ and $\mathcal{B}_1(\mathcal{H},\mathcal{H}')$ the set of all bounded everywhere defined, compact, Hilbert-Schmidt and trace class operators from $\mathcal{H}$ to $\mathcal{H}'$ respectively.

Proof of Lemma 10.1 : Since ϕ and f are in $L^2(\mathcal{R}^3)$, it follows that $\phi f \in L^1(\mathcal{R}^3)$ and that $\widetilde{\phi f}$ is a bounded function. Then by (10.5) and (10.6) we have $|(M_\phi(\lambda)f)(\underline{\omega})| \le (16\pi^3)^{-\frac{1}{2}}\lambda^{\frac{1}{4}} \|\phi\|_2 \|f\|_2$ for all $\underline{\omega} \in S^{(2)}$, and thus $M_\phi(\lambda)f \in L^2(S^{(2)})$, because the Lebesgue measure of $S^{(2)}$ is finite. Therefore (10.6) defines a bounded operator $M_\phi(\lambda)$ from $D(\phi(\underline{Q}))$ into $L^2(S^{(2)})$ and by a result similar to Proposition 2.6 its closure [also denoted $M_\phi(\lambda)$] belongs to $\mathcal{B}(L^2(\mathbb{R}^3), L^2(S^{(2)}))$. It is clear that $\|M_\phi(\lambda)\| \le (2\pi)^{-1}\lambda^{\frac{1}{4}}\|\phi\|_2$. Now

$$(M_\phi(\lambda)f)(\underline{\omega}) = (16\pi^3)^{-\frac{1}{2}}\lambda^{\frac{1}{4}} \int e^{-i\lambda^{\frac{1}{2}}\underline{\omega}\cdot\underline{x}}\phi(\underline{x})f(\underline{x})d^3x, \qquad (10.7)$$

the integral being defined for all $f\varepsilon L^2(\mathbb{R}^3)$. This shows that $M_\phi(\lambda)$ is an integral operator with kernel

$$M_\phi(\lambda)(\underline{\omega},\underline{x}) = (16\pi^3)^{-\frac{1}{2}}\lambda^{\frac{1}{4}} e^{-i\lambda^{\frac{1}{2}}\underline{\omega}\cdot\underline{x}}\phi(\underline{x}). \qquad (10.8)$$

By Remark (10.2),

$$\|M_\phi(\lambda)\|_{HS}^2 = (4\pi^2)^{-1}\lambda^{\frac{1}{2}}\|\phi\|_2^2 < \infty, \qquad (10.9)$$

proving that $M_\phi(\lambda)$ is a Hilbert-Schmidt operator for every fixed λ.

We observe that

$$\|M_\phi(\lambda) - M_\phi(\mu)\|_{HS}^2 = \int\int |M_\phi(\lambda)(\underline{\omega},\underline{x}) - M_\phi(\mu)(\underline{\omega},\underline{x})|^2 d\omega d^3x$$

$$= (16\pi^3)^{-1}\int\int |\lambda^{\frac{1}{4}}e^{-i\lambda^{\frac{1}{2}}\underline{\omega}\cdot\underline{x}} - \mu^{\frac{1}{4}}e^{-i\mu^{\frac{1}{2}}\underline{\omega}\cdot\underline{x}}|^2 \; |\phi(\underline{x})|^2 d\omega d^3x. \qquad (10.10)$$

For fixed $\underline{\omega},\underline{x}$ and λ, the integrand converges to 0 as $\mu \to \lambda$. Also for μ sufficiently close to λ, it is majorized by the integrable function $8\lambda^{\frac{1}{2}}|\phi(\underline{x})|^2$, uniformly in μ. Therefore the Lebesgue dominated convergence theorem gives the continuity of $M_\phi(\lambda)$ in Hilbert-Schmidt norm. #

Next we want to relate the operators $M_\phi(\lambda)$, defined above, to the spectral representation of K_o.

<u>LEMMA 10.3</u> : Let $\mathcal{U}_o$ be the spectral transformation of K_o given in (10.5) and $\phi\varepsilon L^2(\mathbb{R}^3)$. Then

(a) $\phi(\underline{Q})^*g = \int M_\phi(\lambda)^*(\mathcal{U}_o g)_\lambda d\lambda$ for all $g\varepsilon \mathcal{S}(\mathbb{R}^3)$, (10.11) where $M_\phi(\lambda)^*$, the adjoint of $M_\phi(\lambda)$, maps $L^2(S^{(2)})$ into $L^2(\mathcal{R}^3)$ for every fixed λ and the above integral converges as an improper Riemann integral.

(b) For each bounded continuous function Φ on $\mathcal{R}$, any interval $\Delta \subseteq \mathcal{R}$ and all $g \in \mathcal{S}(\mathcal{R}^3)$,

$$\phi(\underline{Q})^*\Phi(K_o)g = \int \Phi(\lambda) M_\phi(\lambda)^*(\mathcal{U}_o g)_\lambda d\lambda \tag{10.12}$$

$$\phi(\underline{Q})^* E^o_\Delta g = \int_\Delta M_\phi(\lambda)^*(\mathcal{U}_o g)_\lambda d\lambda, \tag{10.13}$$

(c) for every $g \in \mathcal{S}(\mathcal{R}^3)$, the sequence $\{\phi(\underline{Q})^*(R^o_{\lambda+i\eta} - R^o_{\lambda-i\eta})g\}$ converges weakly to $2\pi i M_\phi(\lambda)^*(\mathcal{U}_o g)_\lambda$ as $\eta \to +0$.

(d) Let Δ be a compact interval in $\mathcal{R}$ and $h : \Delta \to L^2(\mathcal{R}^3)$ be strongly continuous. Then for every $g \in \mathcal{S}(\mathcal{R}^3)$,

$$\int_\Delta (h(\lambda), \phi(\underline{Q})^* dE^o_\lambda g) = \int_\Delta (h(\lambda), M_\phi(\lambda)^*(\mathcal{U}_o g)_\lambda) d\lambda.$$

<u>Proof</u> : We first note that $M_\phi(\lambda)^*$ is a Hilbert-Schmidt operator from $L^2(S^{(2)})$ to $L^2(\mathcal{R}^3)$, continuous in Hilbert-Schmidt norm and $\|M_\phi(\lambda)^*\| = \|M_\phi(\lambda)\| \leq (2\pi)^{-1}\lambda^{\frac{1}{4}}\|\phi\|_2$. Since $(\mathcal{U}_o g)_\lambda(\underline{\omega}) = 2^{-\frac{1}{2}}\lambda^{\frac{1}{4}}\tilde{g}(\underline{k})$ and since $g \in \mathcal{S}(\mathcal{R}^3)$, we obtain that $M_\phi(\lambda)^*(\mathcal{U}_o g)_\lambda$ is a strongly continuous function of λ. Furthermore, $\|M_\phi(\lambda)^*(\mathcal{U}_o g)_\lambda\| \leq (2\pi)^{-1}\|\phi\|_2 \lambda^{\frac{1}{4}} \|(\mathcal{U}_o g)_\lambda\|_o$ and is integrable in $[0,\infty)$ (Problem 10.2), establishing the strong integrability of $M_\phi(\lambda)^*(\mathcal{U}_o g)_\lambda$ by Proposition 4.13. On the other hand, for every $g \in \mathcal{S}(\mathcal{R}^3)$ and all $f \in D(\phi(\underline{Q}))$, we have that

$$(g,\phi(\underline{Q})f) = (\mathcal{U}_o g, \mathcal{U}_o \phi(\underline{Q})f) = \int ((\mathcal{U}_o g)_\lambda, (\mathcal{U}_o \phi(\underline{Q})f)_\lambda)_o d\lambda =$$

$$\int ((\mathcal{U}_o g)_\lambda, M_\phi(\lambda)f)_o d\lambda = \int (M_\phi(\lambda)^*(\mathcal{U}_o g)_\lambda, f) d\lambda = (\int M_\phi(\lambda)^*(\mathcal{U}_o g)_\lambda d\lambda, f),$$

since the integral defines a vector in $L^2(\mathcal{R}^3)$. Thus $g \in D(\phi(\underline{Q})^*)$ (which alternatively follows since $g \in \mathcal{S}(\mathcal{R}^3)$) and we have (a).

The improper integral $\int \Phi(\lambda) M_\phi(\lambda)^*(\mathcal{U}_o g)_\lambda d\lambda$ exists and thus defines a vector in $L^2(\mathcal{R}^3)$ for every $g \in \mathcal{S}(\mathcal{R}^3)$. The same reasoning as above allows one to infer that $\Phi(K_o)g \in D(\phi(\underline{Q})^*)$ and that we have (10.12). Noting that $\int_\Delta M_\phi(\lambda)^*(\mathcal{U}_o g)_\lambda d\lambda$ is a well-defined Riemann integral, one similarly gets (10.13).

Setting $\Phi(\lambda) = e^{-i\lambda t}$ in (10.12) with $t \in \mathcal{R}$, we have for all $f \in L^2(\mathcal{R}^3)$

$$(f, \phi(\underline{Q})^* U_t g) = \int e^{-i\lambda t}(f, M_\phi(\lambda)^*(\mathcal{U}_o g)_\lambda) d\lambda. \tag{10.14}$$

On the other hand $\|\phi(\underline{Q})^* U_t g\| \leq \|\phi(\underline{Q})^*(K_o+i)^{-1}\| \, \|(K_o+i)g\|$, which is finite by Proposition 7.7 and because $g \in \mathcal{S}(\mathcal{R}^3) \subseteq D(K_o)$. Also as in the proof of Corollary 8.15 we see that $\|\phi(\underline{Q})^* U_t g\| \leq$ constant $\cdot |t|^{-3/2}$ for $t \neq 0$. Thus $\|\phi(\underline{Q})^* U_t g\| \in L^1(\mathcal{R}) \cap L^2(\mathcal{R})$ for all $g \in \mathcal{S}(\mathcal{R}^3)$, and the unitarity of the Fourier transformation and (10.14) lead to the result (for $\lambda \geq 0$) :

$$2\pi(f, M_\phi(\lambda)^*(\mathcal{U}_o g)_\lambda) = \int e^{i\lambda t}(f, \phi(\underline{Q})^* U_t g) dt. \tag{10.15}$$

Finally, using (3.58) as well as the fact that $\|\phi(\underline{Q})^* U_t g\| \in L^1(\mathcal{R})$ and proceeding as in part (iii) of the proof of Proposition 8.22, we conclude that $(f, \phi(\underline{Q})^*[R^o_{\lambda+i\eta} - R^o_{\lambda-i\eta}]g)$ converges to $i \int_{-\infty}^{\infty}(f, \phi(\underline{Q})^* U_t g) e^{i\lambda t} dt$ as $\eta \to +0$. This along with (10.15) leads to (c). The proof of (d) is left as an exercise (Problem 10.3). #

<u>Remark 10.4</u> : Let $\mu \mapsto f(\mu) \in D(\phi(\underline{Q}))$ be a strongly continuous function on a subset Δ of $(0,\infty)$. Since $M_\phi(\lambda) f(\mu)$ is strongly continuous in both variables and equal to $(\mathcal{U}_o \phi(\underline{Q}) f(\mu))_\lambda$ for a.a.λ, we may assume the equality for all λ by the convention of page 395, and then $M_\phi(\mu) f(\mu) = (\mathcal{U}_o \phi(\underline{Q}) f(\mu))_\mu$ for all $\mu \in \Delta$.

Operators like $M_\phi(\lambda)$ can be defined abstractly. Given a self-adjoint operator H, there exists a class of operators C to each of which one may associate a mapping $M_C(\lambda)$ from $\mathcal{H}_{ac}(H)$ into $\mathcal{H}_\lambda$ (see page 227) for $\lambda\varepsilon\sigma_{ac}(H)$. $M_C(\lambda)$ has properties similar to those given in Lemma 10.3. One such abstraction will be discussed in Section 10-4C.

The next proposition gives the spectral transformation $\mathcal{U}$ for $H = K_0 + V$, where V is a potential in class $\mathcal{V}$, introduced in (9.27). We use the notations of Lemma 9.13 and also denote by Γ_0 the closed set of measure zero on which one of the operators $(I+W_{\lambda\pm io})$ is not invertible (see Lemma 9.5). We set $W_\lambda^\pm \equiv W_{\lambda\pm io} = \lim BR^o_{\lambda\pm i\eta}A$ in Hilbert-Schmidt norm as $\eta \to +0$.

PROPOSITION 10.5 : Let $H_0 = K_0$ in $L^2(\mathbb{R}^3)$ and $H = K_0 + V$ with $V\varepsilon\mathcal{V}$. Define two operators $\mathcal{U}_\pm$ by

$$\mathcal{U}_\pm = \mathcal{U}_0\, \Omega^*_\pm. \tag{10.16}$$

Then $\mathcal{U}_\pm$ are two spectral transformations of H_{ac} mapping $\mathcal{H}_{ac}(H)$ onto $\mathcal{G}_0 \equiv L^2([0,\infty),L^2(S^{(2)}))$. Furthermore, if f is in the range of A and $\lambda\varepsilon(0,\infty)-\Gamma_0$, they are given by

$$(\mathcal{U}_\pm f)_\lambda = M_A(\lambda)(I+W_\lambda^\pm)^{-1}A^{-1}f = (\mathcal{U}_0 A(I+W_\lambda^\pm)^{-1}A^{-1}f)_\lambda . \tag{10.17}$$

Proof : The existence of $\Omega_\pm$ and the asymptotic completeness of the theory have been established in Proposition 9.14. It is easy to see that $\mathcal{U}_\pm\mathcal{U}^*_\pm = I$ and $\mathcal{U}^*_\pm\mathcal{U}_\pm = F_\pm = E_{ac}(H)$. Thus $\mathcal{U}_\pm$ are partial isometries with initial set $\mathcal{H}_{ac}(H)$. Also it follows from Proposition 4.4(b) that for any $f\varepsilon D(H_{ac})$,

$(\mathcal{U}_\pm H_{ac}f)_\lambda = (\mathcal{U}_\pm HF_\pm f)_\lambda = (\mathcal{U}_o\Omega_\pm^* HF_\pm f)_\lambda = (\mathcal{U}_o K_o \Omega_\pm^* f)_\lambda =$
$\lambda(\mathcal{U}_o\Omega_\pm^* f)_\lambda = \lambda(\mathcal{U}_\pm f)_\lambda$. Therefore $\mathcal{U}_\pm H_{ac}\mathcal{U}_\pm^*$ is the multiplication operator by λ in $\mathcal{G}_o$, showing that $\mathcal{U}_\pm$ are indeed spectral transformations of H_{ac}.

From (6.22) and the fact that $E_\infty(K_o) = I$, we get for f in the range of A and $g\varepsilon S_o \subseteq S(R^3)$ (see Problem 8.6)

$$(f,\Omega_\pm g) = (f,g) - \lim_{\eta\to+0} \int (f,R_{\lambda\mp i\eta}VdE_\lambda^o g).$$

By (9.20) and since V = AB = BA, the above expression can be rewritten as

$$(f,\Omega_\pm g) = \lim_{\eta\to+0} \int (A^{-1}f,[I+BR_{\lambda\pm i\eta}^o A]^{-1*}AdE_\lambda^o g).$$

Now we note that $A(\cdot)\,\varepsilon\, L^2(R^3)\cap L^\infty(R^3)$ and we can apply Lemma 10.3(d) with $h(\lambda) = [1+BR_{\lambda\pm i\eta}^o A]^{-1}A^{-1}f$. Thus for any compact interval $\Delta \subseteq [0,\infty)$,

$$(f,\Omega_\pm E_\Delta^o g) = \lim_{\eta\to+0} \int_\Delta ([I+BR_{\lambda\pm i\eta}^o A]^{-1}A^{-1}f, M_A(\lambda)^*(\mathcal{U}_o g)_\lambda)d\lambda.$$

We know from Lemmas 9.5 and 9.13 that $[I+BR_{\lambda\pm i\eta}^o A]^{-1}$ converges in norm to $(I+W_\lambda^\pm)^{-1}$ whenever $\lambda\varepsilon[0,\infty)-\Gamma_o$. Restricting Δ to be a compact interval in $(0,\infty)-\Gamma_o$ (which we can do since Γ_o is closed), we see that $\|[I+BR_{\lambda\pm i\eta}^o A]^{-1}\|$ is uniformly bounded in η for all $\lambda\varepsilon\Delta$. Since $\|M_A(\lambda)^*(\mathcal{U}_o g)_\lambda\|$ is integrable (see the proof of Lemma 10.3(a)), we apply the Lebesgue dominated convergence theorem to obtain

$$(f,\Omega_\pm E_\Delta^o g) = \int_\Delta ((I+W_\lambda^\pm)^{-1}A^{-1}f, M_A(\lambda)^*(\mathcal{U}_o g)_\lambda)d\lambda. \qquad (10.18)$$

The relation (10.18), combined with the fact that $(f,\Omega_\pm E_\Delta^o g)$ $= \int_\Delta ((\mathcal{U}_o\Omega_\pm^* f)_\lambda\ ,(\mathcal{U}_o g)_\lambda)_o d\lambda$ for any compact Δ, leads to the result that for almost all $\lambda\varepsilon(0,\infty)-\Gamma_o$

$$((\mathcal{U}_\pm f)_\lambda, (\mathcal{U}_o g)_\lambda)_o = ((I+W_\lambda^\pm)^{-1}A^{-1}f, M_A(\lambda)^*(\mathcal{U}_o g)_\lambda)$$
$$= (M_A(\lambda)(I+W_\lambda^\pm)^{-1}A^{-1}f, (\mathcal{U}_o g)_\lambda)_o \qquad (10.19)$$

Since the set $\{(\mathcal{U}_o g)_\lambda | g\varepsilon \mathcal{S}_o\}$ is fundamental in $L^2(S^{(2)})$ for each $\lambda\varepsilon(0,\infty)$ (Problem 8.6), (10.19) yields the first part of (10.17) for almost all $\lambda\varepsilon(0,\infty)-\Gamma_o$ as in Problem 8.13. But $\lambda\mapsto M_A(\lambda)(I+W_\lambda^\pm)^{-1}A^{-1}f$ is strongly continuous on $(0,\infty)-\Gamma_o$, and hence by the convention adopted on page 395, the first part of (10.17) is true for all $\lambda\varepsilon(0,\infty)-\Gamma_o$. The second equality in (10.17) follows from Remark 10.4. #

One may see from the proof of Proposition 10.5 that it is valid under more general circumstances once we have a more general definition of $M_A(\lambda)$. An abstract method of finding the spectral transformation for the perturbed operator given that of the unperturbed one has been discussed by Rejto [1] and Kuroda [2]. From Section 5-7 it is clear that all the information about the absolutely continuous part of H is contained in the spectral transformations $\mathcal{U}_\pm$. However texts on quantum mechanics usually employ eigenfunctions rather than the spectral transformations. So we shall now relate the spectral transformations with what we shall call the "evaluation functionals", which are also eigenfunctions under certain conditions.

We now introduce two auxiliary Hilbert spaces derived from the self-adjoint operator A, defined in Lemma 9.13. It is clear that the null space of A is $\{\theta\}$ and that its range is dense in $L^2(R^3)$. We denote this range by E and define a scalar product on E as follows :

$$(f,g)_E \equiv \int (1+|\underline{x}|)^{2\nu}\bar{f}(\underline{x})g(\underline{x})d^3x, \text{ for } f,g\varepsilon E. \tag{10.20}$$

This scalar product makes E into a Hilbert space, the norm being given by $\|f\|_E^2 = (f,f)_E$ (Problem 10.9). The completeness of E is proven as for $L^2(R^3)$. We define another linear space E' (called the dual of E, see Section 10-4A) as follows :

$$E' = \{f: R^3 \to C \,|\, f \text{ measurable}, \|f\|_{E'}^2 \equiv \int (1+|\underline{x}|)^{-2\nu}|f(\underline{x})|^2 d^3x < \infty\}. \tag{10.21}$$

For an abstract definition of dual spaces, we refer the reader to [GS, Vol. 2] and to Section 10-4. For our purposes the above concrete definition suffices. We note that all essentially bounded measurable functions are in this space E', i.e. $L^\infty(R^3) \subseteq E'$. In particular, the functions ψ_k^0 (given for every vector $\underline{k}\varepsilon R^3$ by $\psi_k^0(\underline{x}) = (2\pi)^{-3/2}\exp(i\underline{k}\cdot\underline{x})$) belong to E'. E' is also a Hilbert space with the obvious scalar product suggested by (10.21).

There is a linear one-to-one correspondence between E and E'. Indeed let $f\varepsilon E$. Then,

$$\|f\|_E^2 \equiv \int (1+|\underline{x}|)^{2\nu}|f(\underline{x})|^2 d^3x = \int (1+|\underline{x}|)^{-2\nu}|(1+|\underline{x}|)^{2\nu}f(\underline{x})|^2 d^3x$$

$$\equiv \|A^{-2}f\|_{E'}^2, \tag{10.22}$$

where we have written $(A^{-2}f)(\underline{x}) = (1+|\underline{x}|)^{2\nu}f(\underline{x})$. Conversely, if $g\varepsilon E'$ then $\|g\|_{E'}^2 = \|A^2 g\|_E^2$. Thus multiplication by $(1+|\underline{x}|)^{2\nu}$ and by $(1+|\underline{x}|)^{-2\nu}$ define one-to-one linear mappings from E onto E' and from E' onto E respectively. We also define the evaluation of $g\varepsilon E'$ at $f\varepsilon E$ by

$$\langle g,f\rangle \equiv \int \bar{g}(\underline{x})f(\underline{x})d^3x = (A^2 g,f)_E. \tag{10.23}$$

Since $\psi_k^o \varepsilon E'$, we may associate with it a vector $\Psi_k^o \varepsilon E$ given by $\Psi_k^o = A^2\psi_k^o$. Then the spectral transformation U_o of K_o can be written in terms of Ψ_k^o as follows. Setting as before $\underline{k} = \lambda^{\frac{1}{2}}\underline{\omega}$ with $\lambda\varepsilon[0,\infty)$ and $\underline{\omega}\varepsilon S^{(2)}$ and using (10.5), we get for every $f\varepsilon E \subseteq L^1(R^3)\cap L^2(R^3)$,

$$2^{\frac{1}{2}}\lambda^{-\frac{1}{4}}(U_o f)_\lambda(\underline{\omega}) = (Ff)(\underline{k}) \equiv (2\pi)^{-3/2}\int \exp(-i\underline{k}\cdot\underline{x})f(\underline{x})d^3x$$
$$= \int \overline{\psi_k^o(\underline{x})}f(\underline{x})d^3x = \langle\psi_k^o,f\rangle = (\Psi_k^o,f)_E \;. \qquad (10.24)$$

The properties of the Fourier transformation discussed in Section 2-2 show that $(Ff)(\underline{k})$ is square integrable and $\int|(Ff)(\underline{k})|^2 d^3k = \|f\|^2$. Thus (10.24) proves that for $f\varepsilon E$, $\langle\psi_k^o,f\rangle \equiv (\Psi_k^o,f)_E \varepsilon L^2(R^3,d^3k)$ and

$$\int|(Ff)(\underline{k})|^2 d^3k = \int|\langle\psi_k^o,f\rangle|^2 d^3k = \|f\|^2 = \|E_{ac}(K_o)f\|^2. \qquad (10.25)$$

Similarly for each $\underline{k}$, we look for vectors $\psi_k^\pm \varepsilon E'$ to which we associate $\Psi_k^\pm = A^2\psi_k^\pm \varepsilon E$ such that for all $f\varepsilon E$,

$$2^{\frac{1}{2}}\lambda^{-\frac{1}{4}}(U_\pm f)_\lambda(\underline{\omega}) = \langle\psi_k^\pm,f\rangle = (\Psi_k^\pm,f)_E \;, \qquad (10.26)$$

where $U_\pm$ are the spectral transformations of H_{ac} established in Proposition 10.5. The functions ψ_k^o and $\psi_k^\pm$ are called the <u>evaluation functionals</u> for the self-adjoint operators K_o and H_{ac} respectively.

By analogy with the first equality of (10.24) we define a pair of operators $F_\pm$, called the <u>generalized Fourier transformations</u>, by setting for every $f\varepsilon L^2(R^3)$,

$$(F_\pm f)(\underline{k}) \equiv 2^{\frac{1}{2}}\lambda^{-\frac{1}{4}}(U_\pm f)_\lambda(\underline{\omega}). \qquad (10.27)$$

Then it follows from (10.16) that

$$\int|(F_\pm f)(\underline{k})|^2 d^3k = \|U_\pm f\|^2 = \|U_o\Omega_\pm^* f\|^2 = (\Omega_\pm\Omega_\pm^* f,f) = \|F_\pm f\|^2. \qquad (10.28)$$

If furthermore the theory is asymptotically complete in the sense that $F_{\pm} = E_{ac}(H)$ and if the evalutation functionals $\psi_{k}^{\pm}$ verifying (10.26) exist then we obtain the analogue of (10.25) for $F_{\pm}$ and $\psi_{k}^{\pm}$, viz. that for all $f \varepsilon E$,

$$\int |(F_{\pm}f)(\underline{k})|^2 d^3k \equiv \int |\langle \psi_{k}^{\pm}, f \rangle|^2 d^3k = \|E_{ac}(H)f\|^2. \qquad (10.29)$$

This also shows that $F_{\pm}$ satisfying (10.29) can be extended to partial isometries with $H_{ac}(H)$ as their initial set and obeying (10.27).

We shall now show the existence of the evaluation functionals $\psi_{k}^{\pm}$ for H with $V \varepsilon Y$ for almost all $\underline{k} \varepsilon R^3$.

<u>PROPOSITION 10.6</u> : Assume the hypotheses of Proposition 10.5. Then for every $\underline{k} \varepsilon \Gamma^{(3)} \equiv \{\underline{k} \varepsilon R^3 | k^2 = \lambda \varepsilon (0,\infty) - \Gamma_0\}$, the evaluation functionals $\psi_{k}^{\pm}$ for the absolutely continuous part of H, verifying (10.26) and (10.29) exist. The associated vectors $\Psi_{k}^{\pm} \equiv A^2 \psi_{k}^{\pm} \varepsilon E$ are given by

$$\Psi_{k}^{\pm} = A(I+W_{\lambda}^{\pm})^{-1*} A^{-1} \Psi_{k}^{o}. \qquad (10.30)$$

Moreover, $\Psi_{k}^{\pm}$ is the unique solution in E of the equation

$$\Psi_{k}^{\pm} = \Psi_{k}^{o} - AW_{\lambda}^{\pm *} A^{-1} \Psi_{k}^{\pm}. \qquad (10.31)$$

<u>Proof</u> : Let $f \varepsilon E$. Then from Proposition 10.5 we know that for every $\lambda \varepsilon (0,\infty) - \Gamma_0$ the spectral transformations $U_{\pm}$ of H_{ac} are given as $(U_{\pm}f)_{\lambda} = (U_o A(I+W_{\lambda}^{\pm})^{-1} A^{-1} f)_{\lambda}$. Using (10.27) and (10.24) to write this in terms of $F_{\pm}$ we have

$$(F_{\pm}f)(\underline{k}) = (FA(I+W_{\lambda}^{\pm})^{-1} A^{-1} f)(\underline{k}) = (\Psi_{k}^{o}, A(I+W_{\lambda}^{\pm})^{-1} A^{-1} f)_E$$

$$= (A^{-1}\Psi_{k}^{o}, (I+W_{\lambda}^{\pm})^{-1} A^{-1} f) = (A(I+W_{\lambda}^{\pm})^{-1*} A^{-1} \Psi_{k}^{o}, f)_E \ , \qquad (10.32)$$

where we have used the simple facts that the vectors $A(I+W_\lambda^\pm)^{-1}A^{-1}f$ and $A(I+W_\lambda^\pm)^{-1*}A^{-1}\Psi_k^o$ belong to E if $\underline{k}\varepsilon\Gamma^{(3)}$ and also that $(g,f)_E = (A^{-1}g,A^{-1}f)$ (see Problem 10.9). Comparing (10.32) with (10.27) and (10.26), we arrive at the existence of vectors $\Psi_k^\pm \varepsilon E$ verifying (10.30). Therefore by the discussion preceding this proposition, we also have the existence of the evaluation functionals $\psi_k^\pm$ in E'. Since the theory is asymptotically complete we get (10.29) by using (10.28).

Since $A^{-1}\Psi_k^\pm \varepsilon H = D(W_\lambda^{\pm *})$, we obtain from (10.30) that

$$\Psi_k^\pm - \Psi_k^o = A[(I+W_\lambda^\pm)^{-1}-I]^*A^{-1}\Psi_k^o$$

$$= -AW_\lambda^{\pm *}(I+W_\lambda^\pm)^{-1*}A^{-1}\Psi_k^o = -A\,W_\lambda^{\pm *}A^{-1}\Psi_k^\pm .$$

The uniqueness of the solution follows easily from the fact that $(I+W_\lambda^\pm)^{-1}\varepsilon B(H)$ whenever $\lambda\notin\Gamma_o$. #

The results of Proposition 10.6 establish the existence of the evaluation functionals $\psi_k^\pm$ for H_{ac} as vectors in E'. From the definition of the space E' it is clear however that these are actually functions of $\underline{x}$. The next two propositions give some properties of these functions. The following lemma is useful for this purpose.

<u>LEMMA 10.7</u> : Assume the hypotheses of Proposition 10.5 and let $\Delta^{(3)} \equiv \{\underline{k}\varepsilon R^3 | k^2 = \lambda\varepsilon\Delta\}$, Δ a compact subset of $(0,\infty)-\Gamma_o$. Then $\psi_k^\pm$ are bounded and continuous in the E'-norm for $\underline{k}\,\varepsilon\Delta^{(3)}$.

<u>Proof</u> : By (10.22), (10.30) and Problem 10.9 we have for each $\underline{k}\varepsilon\,\Delta^{(3)}$

$$\|\psi_k^\pm\|_{E'} = \|\Psi_k^\pm\|_E = \|A^{-1}\Psi_k^\pm\| \leq \|(I+W_\lambda^\pm)^{-1}\|\,\|A^{-1}\Psi_k^o\| . \quad (10.33)$$

Since $\|A^{-1}\Psi^o_k\| = \|A\|_2$, Lemmas 9.5 and 9.13 and (10.33) lead to the boundedness of $\|\psi^\pm_k\|_{E'} = \|\Psi^\pm_k\|_E$.

Similary for $\underline{k}$ and $\underline{k}' \in \Delta^{(3)}$ one has

$$\|\psi^\pm_k - \psi^\pm_{k'}\|_{E'} = \|\Psi^\pm_k - \Psi^\pm_{k'}\|_E \leq \|(I+W^\pm_\lambda)^{-1}\| \, \|A^{-1}\Psi^o_k - A^{-1}\Psi^o_{k'}\| +$$
$$\|(I+W^\pm_\lambda)^{-1} - (I+W^\pm_{\lambda'})^{-1}\| \, \|A^{-1}\Psi^o_{k'}\|. \tag{10.34}$$

Now

$$\|A^{-1}\Psi^o_k - A^{-1}\Psi^o_{k'}\|^2 = (2\pi)^{-3}\int(1+|\underline{x}|)^{-2\nu}\left|e^{i\underline{k}\cdot\underline{x}} - e^{i\underline{k}'\cdot\underline{x}}\right|^2 d^3x. \tag{10.35}$$

Since the integrand in (10.35) converges pointwise to zero as $\underline{k}' \to \underline{k}$ and since the integrand is majorized by the integrable function $4(1+|\underline{x}|)^{-2\nu}$, we can apply the Lebesgue dominated convergence theorem to conclude that $A^{-1}\Psi^o_k$ is continuous in $\underline{k}$ in the H-norm. On the other hand $(1+W^\pm_\lambda)^{-1}$ is continuous in operator norm (see Lemma 9.5) for $\lambda\varepsilon\Delta$ and thus (10.34) implies the continuity of $\psi^\pm_k$ in the E'-norm. #

<u>PROPOSITION 10.8</u> : Assume the hypotheses of Proposition 10.5. Then (a) for every $\underline{k}\varepsilon\Gamma^{(3)}$ the integral equation, called the <u>Lippmann-Schwinger equation</u> :

$$\psi^\pm_k(\underline{x}) = \psi^o_k(\underline{x}) - (4\pi)^{-1}\int|\underline{x}-\underline{y}|^{-1}e^{\mp ik|\underline{x}-\underline{y}|}V(\underline{y})\psi^\pm_k(\underline{y})d^3y \tag{10.36}$$

has a unique solution in E'.

(b) Each of the two families $\{\psi^\pm_k\}$ form a complete orthonormal system of evaluation functionals of H_{ac} in the following sense :

Let $\{\Gamma_n\}$ be an increasing sequence of compact subsets of $\Gamma \equiv (0,\infty)-\Gamma_0$ such that $\cup_n\Gamma_n = \Gamma$ and let $\Gamma_n^{(3)} = \{\underline{k}\varepsilon R^3 | k^2 = \lambda\varepsilon\Gamma_n\}$. Then for every n and each $f\varepsilon H_{ac}(H)$, the integral $\int_{\Gamma_n^{(3)}}(F_\pm f)(\underline{k})\psi_k^\pm d^3k$ exists as a vector-valued integral in E'. Moreover, the vector defined by the integral belongs to H and

$$f = \text{s-lim}_{n\to\infty} \int_{\Gamma_n^{(3)}} (F_\pm f)(\underline{k})\, \psi_k^\pm \, d^3k. \tag{10.37}$$

The generalized Fourier coefficients $(F_\pm f)(\underline{k})$ are given by

$$(F_\pm f)(\underline{k}) = \underset{n\to\infty}{\ell.\text{i.m.}} \int_{|\underline{x}|\leq n} \overline{\psi_k^\pm}(\underline{x}) f(\underline{x}) d^3x. \tag{10.38}$$

Remark 10.9 : The integral in (10.37) can be given a pointwise meaning if for instance $\psi_k^\pm(\underline{x})$ is bounded for $\underline{x}\varepsilon R^3$ and $\underline{k}\varepsilon\Gamma_n^{(3)}$. This is because $F_\pm f \,\varepsilon\, L^1_{loc}(R^3)$ for every $f\varepsilon H$. In the next proposition we shall verify the boundedness of $\psi_k^\pm(\underline{x})$ for fixed $V\varepsilon Y$ under some further restrictions on B_1. However using L^∞-techniques this fact can be established for the whole class Y (see Alsholm and Schmidt [1]). Since we have not developed analysis in L^∞-spaces, we have chosen to define the integral in (10.37) as a vector-valued integral. This point is discussed further in the proof below. Note on the other hand that the integral in (10.38) is an ordinary Lebesgue integral.

Proof of Proposition 10.8 : (a) By (9.22), (9.28) and Proposition 7.20 we have that $W_\lambda^{\pm *}(\underline{x},\underline{y}) = \overline{W}_{\lambda\pm io}(\underline{y},\underline{x}) = A(\underline{x})G^o_{\lambda\mp io}(\underline{x},\underline{y})B(\underline{y})$. Since we are looking for solutions $\psi_k^\pm$ in E', it follows that $A\psi_k^\pm\varepsilon L^2(R^3)$ for every $\underline{k}\varepsilon\Gamma^{(3)}$. The integral in (10.36) is equal to $-A(\underline{x})^{-1}\int W_\lambda^{\pm *}(\underline{x},\underline{y})A(\underline{y})\psi_k^\pm(\underline{y})\, d^3y$ which is finite for almost all $\underline{x}$ and for any $\psi_k^\pm\varepsilon E'$ because $A(\underline{x}) \neq 0$ and $\int|W_\lambda^{\pm *}(\underline{x},\underline{y})|^2 d^3y < \infty$ for almost all $\underline{x}$. Remembering that $\Psi^o_k(\underline{x}) = A(\underline{x})^2\psi^o_k(\underline{x})$ and setting $\Psi^\pm_k(\underline{x}) = A(\underline{x})^2\psi^\pm_k(\underline{x})$, we may write (10.31) as

$$A(\underline{x})^2\psi^\pm_k(\underline{x}) = A(\underline{x})^2\psi^o_k(\underline{x}) -$$

$$(4\pi)^{-1}A(\underline{x})^2\int|\underline{x}-\underline{y}|^{-1}e^{\mp ik|\underline{x}-\underline{y}|}B(\underline{y})A(\underline{y})\psi^\pm_k(\underline{y})d^3y. \qquad (10.39)$$

Dividing both sides of equation (10.39) by $A(\underline{x})^2$ we arrive at (10.36). Similarly it can be shown that (10.31) follows from (10.36). The existence and uniqueness of the solution of (10.36) is then a consequence of the results of Proposition 10.6.

(b) By Lemma 10.7 and Proposition 4.13 we note that the integral $\int_{\Gamma_n^{(3)}}(F_\pm f)(\underline{k})\psi^\pm_k d^3k$ exists as a Riemann integral for every $f\epsilon H_{ac}(H)$ such that $(F_\pm f)(\underline{k})$ is continuous in $\underline{k}$, and defines a vector in E'. The set D of such vectors f is dense in $H_{ac}(H)$ since $F_\pm$ is unitary from $H_{ac}(H)$ onto $L^2(R^3)$ and since the set of continuous functions in $L^2(R^3)$ is dense. It is possible to define the above integral for all $f\epsilon H_{ac}(H)$, either by admitting a more general definition of vector-valued integrals [HP, Section 3.7], or defining it as the limit in the E'-norm of a sequence of integrals with vectors $f_m \epsilon D$ such that s-lim $f_m = f$ as $m \to \infty$. For the rest of the proof we shall assume that the integral in (10.37) makes sense not only for this dense set but also for all $f\epsilon H_{ac}(H)$.

By (9.14), $E_{\Gamma_n} = E_{ac}(H)E_{\Gamma_n}$ and for all $g\epsilon E$, we get from (10.16) and (10.27) that

$$(E_{\Gamma_n}f,g) = (U_\pm E_{\Gamma_n}f,U_\pm g) = \int_{\Gamma_n}((U_\pm f)_\lambda,(U_\pm g)_\lambda)_o d\lambda$$
$$= \int_{\Gamma_n^{(3)}}\overline{(F_\pm f)(\underline{k})}(F_\pm g)(\underline{k})d^3k. \qquad (10.40)$$

On writing $(F_{\pm}g)(\underline{k}) = \langle\psi^{\pm}_{k},g\rangle$ as in (10.26), (10.40) becomes

$$(E_{\Gamma_n}f,g) = \langle\int_{\Gamma_n^{(3)}}(F_{\pm}f)(\underline{k})\psi^{\pm}_{k}d^3k,g\rangle \ . \tag{10.41}$$

Since (10.41) is true for all $g\varepsilon E$ and since $H \subseteq E'$, we conclude $E_{\Gamma_n}f = \int_{\Gamma_n^{(3)}}(F_{\pm}f)(\underline{k})\psi^{\pm}_{k}d^3k$ in E'. This shows that the integral in fact defines a vector in H. But E_{Γ_n} converges strongly to $E_{\Gamma} = E_{ac}(H)$ as $n \to \infty$, proving (10.37).

Let $f\varepsilon H_{ac}(H) \subseteq L^2(R^3)$ and let $f_n = \chi_n f$, where χ_n is the characteristic function of the set $\{\underline{x}\varepsilon R^3 | |\underline{x}| \leq n\}$. Then $f_n\varepsilon E$ for every n, $\{f_n\}$ converges strongly to f in $L^2(R^3)$ as $n \to \infty$ and the integral

$$\int_{|\underline{x}|\leq n}\overline{\psi^{\pm}_{k}}(\underline{x})f(x)d^3x = \int\overline{\psi^{\pm}_{k}}(\underline{x})f_n(\underline{x})d^3x = \langle\psi^{\pm}_{k},f_n\rangle \equiv (F_{\pm}f_n)(\underline{k})$$

is finite for each $\underline{k}\varepsilon\Gamma^{(3)}$. Since $F_{\pm}$ are bounded operators (in fact partial isometries with initial set $H_{ac}(H)$), $\{F_{\pm}f_n\}$ converges strongly to $F_{\pm}f$, proving (10.38). #

As we mentioned in the beginning of this section, the function ψ^{o}_{k} is a solution of the differential equation (10.1). In general, however, the evaluation functionals need not satisfy the equation : $H\psi_k = \lambda(\underline{k})\psi_k$ in any reasonable sense. It can be shown that ψ_k are actually eigenfunctions if there exists a suitable linear manifold $N \subseteq E$ such that $N \subseteq D(H)$ and $HN \subseteq E$. For the class of potentials considered in Proposition 10.8, $\psi^{\pm}_{k}$ are indeed eigenfunctions of $H = K_o + V$ (see Section 10-4$\underline{A}$ for details), so that Proposition 10.8(b) establishes what is called an <u>eigenfunction expansion</u> associated with H_{ac}.

The relations (10.37) and (10.38) can be looked upon

as the continuous analogue of the expansion of a vector in terms of a complete orthonormal set in H and as the expression of the coefficients of the expansion (see page 26). On the other hand if the integral in (10.37) can be given a pointwise meaning (see Remark 10.9) then these two relations are similar to the reciprocity relations (2.17) and (2.20) of the Fourier transformation. This justifies the terminology generalized Fourier transformations for $F_{\pm}$. These equations are the mathematically rigorous counterpart of the common statement that the eigenfunctions $\psi_k^{\pm}$ are δ-function normalized as in (10.2) and that they are complete.

Now we study some detailed properties of the evaluation functionals $\psi_k^{\pm}$. Though the results of the next proposition are true for the whole class V of potentials (see Section 10-4), we make one restrictive assumption for simplicity of presentation. Before we proceed to the main set of results, we prove a lemma.

LEMMA 10.10 : Let $\phi \in L^2(R^3) \cap L^\infty(R^3)$. Then

$$\int |\underline{x}-\underline{y}|^{-2} |\phi(\underline{y})|^2 d^3y \leq 4\pi \|\phi\|_\infty^2 + \|\phi\|_2^2.$$

Proof :

$$\int_{|\underline{x}-\underline{y}|\leq 1} |\phi(\underline{y})|^2 |\underline{x}-\underline{y}|^{-2} d^3y \leq \|\phi\|_\infty^2 \int_{|\underline{x}-\underline{y}|\leq 1} |\underline{x}-\underline{y}|^{-2} d^3y = 4\pi \|\phi\|_\infty^2,$$

and $\int_{|\underline{x}-\underline{y}|>1} |\phi(\underline{y})|^2 |\underline{x}-\underline{y}|^{-2} d^3y \leq \int_{|\underline{x}-\underline{y}|>1} |\phi(\underline{y})|^2 d^3y \leq \|\phi\|_2^2$. #

PROPOSITION 10.11 : Assume the hypotheses of Proposition 10.5, and that $B_1 \in L^2(R^3) \cap L^\infty(R^3)$ (see the discussion following (9.27)). (a) If Δ is an arbitrary compact subset of $\Gamma \equiv (0,\infty)-\Gamma_0$, then the evaluation functionals $\psi_k^{\pm}(\underline{x})$ are bounded for $(\underline{x},\underline{k})$ in $R^3 \times \Delta^{(3)}$ and $\psi_k^{\pm}(\underline{x}) - \psi_k^{0}(\underline{x})$ are uniformly continuous in the same domain. (b) For $|\underline{x}| = r \to \infty$ and $\underline{k}$ in any compact subset of $\Gamma^{(3)}$,

$$\psi_k^{\pm}(\underline{x}) - \psi_k^{0}(\underline{x}) = 0(r^{-\delta}) + 0(r^{-1})^{*)}, \text{ uniformly in } \underline{k}, \tag{10.42}$$

where $\delta > 0$ is the number appearing in the bound of B_2 in (9.27). (c) $\psi_k^{\pm}$ are the solutions of the equation (10.3), i.e. $(\Delta-V(\underline{x})+k^2)\psi_k^{\pm}(\underline{x}) = 0$. Furthermore $\Delta\psi_k^{\pm}$ is locally square-integrable in R^3.

Here we shall prove only (a) and (b) of the above Proposition and give some indication of the proof of (c) in Section 10-4A.

<u>Proof</u> (for ψ_k^-) : Set $T_k(\underline{x},\underline{y}) = -\dfrac{e^{ik|\underline{x}-\underline{y}|}}{4\pi|\underline{x}-\underline{y}|} B(\underline{y})$. (10.43)

Then the Lippmann-Schwinger equation (10.36) can be rewritten as

$$\psi_k^-(\underline{x}) = \psi_k^0(\underline{x}) + \int T_k(\underline{x},\underline{y})A(\underline{y})\psi_k^-(\underline{y})d^3y \equiv \psi_k^0(\underline{x}) + v_k(\underline{x}). \tag{10.44}$$

Now

$$\int|T_k(\underline{x},\underline{y})|^2d^3y = (4\pi)^{-2}\int|\underline{x}-\underline{y}|^{-2}|B(\underline{y})|^2d^3y \le$$

$$2(4\pi)^{-2}\int|\underline{x}-\underline{y}|^{-2}|B_1(\underline{y})|^2d^3y + 2(4\pi)^{-2}\int|\underline{x}-\underline{y}|^{-2}|B_2(\underline{y})|^2d^3y,$$

which is uniformly bounded in $\underline{x}$ by Lemmas 10.10 and 9.12. Since $|\psi_k^0(\underline{x})| = (2\pi)^{-3/2}$, we have by the Schwarz inequality

---

*) $f(\underline{x}) = 0(g(\underline{x}))$ as $|\underline{x}|\to a$ means that there exists a non-zero constant c such that $|f(\underline{x})|\le c|g(\underline{x})|$ as $|\underline{x}|\to a$. $f(\underline{x}) = o(g(\underline{x}))$ as $|\underline{x}|\to a$ means that $|g(\underline{x})|^{-1}|f(\underline{x})| \to 0$ as $|\underline{x}| \to a$.

$$|\psi_{\underline{k}}^{-}(\underline{x})|^2 \leq 2(2\pi)^{-3} + 2\|\Psi_{\underline{k}}^{-}\|_E^2 \int |T_k(\underline{x},\underline{y})|^2 d^3y. \tag{10.45}$$

By Lemma 10.7, $\|\Psi_{\underline{k}}^{-}\|_E$ is bounded in $\underline{k}$ for $\underline{k}\varepsilon\Delta^{(3)}$ and hence the boundedness of $\psi_{\underline{k}}^{-}(\underline{x})$ for $(\underline{x},\underline{k}) \varepsilon R^3 \times \Delta^{(3)}$ is a consequence of (10.45).

Let $\underline{k}\varepsilon\Delta^{(3)}$. Then

$$|v_{\underline{k}}(\underline{x}_1)-v_{\underline{k}}(\underline{x}_2)|^2 \leq \|\Psi_{\underline{k}}^{-}\|_E^2 \int |T_k(\underline{x}_1,\underline{y})-T_k(\underline{x}_2,\underline{y})|^2 d^3y \leq$$

$$2(4\pi)^{-2}\|\Psi_{\underline{k}}^{-}\|_E^2 |\underline{x}_1-\underline{x}_2|^2 k^2 \int |\underline{x}_1-\underline{y}|^{-2}|B(\underline{y})|^2 d^3y +$$

$$2(4\pi)^{-2}\|\Psi_{\underline{k}}^{-}\|_E^2 |\underline{x}_1-\underline{x}_2|^2 \int |\underline{x}_1-\underline{y}|^{-2}|\underline{x}_2-\underline{y}|^{-2}|B(\underline{y})|^2 d^3y. \tag{10.46}$$

In the above we have used the inequalities

$$\left|\exp(ik|\underline{x}_1-\underline{y}|)-\exp(ik|\underline{x}_2-\underline{y}|)\right| \leq k|\underline{x}_1-\underline{x}_2|$$

and $\left||\underline{x}_1-\underline{y}|^{-1} - |\underline{x}_2-\underline{y}|^{-1}\right| \leq |\underline{x}_1-\underline{y}|^{-1}|\underline{x}_2-\underline{y}|^{-1}|\underline{x}_1-\underline{x}_2|$,

both following from the triangle inequality: $\left||\underline{x}_1-\underline{y}| - |\underline{x}_2-\underline{y}|\right| \leq |\underline{x}_1-\underline{x}_2|$. As was shown in the preceding paragraph, the first integral in (10.46) is uniformly bounded in $\underline{x}_1$. Since $B\varepsilon L^\infty(R^3)$, the second integral in (10.46) is majorized by $\|B\|_\infty^2 \int |\underline{x}_1-\underline{y}|^{-2}|\underline{x}_2-\underline{y}|^{-2}d^3y$. The integral $\int |\underline{x}_1-\underline{y}|^{-2}|\underline{x}_2-\underline{y}|^{-2}d^3y$ is clearly homogeneous in $\underline{x}_1-\underline{x}_2$ and a calculation similar to that of Lemma 9.12 shows that the integral is equal to $c|\underline{x}_1-\underline{x}_2|^{-1}$ where $c = \int y^{-2}|\underline{y}-\underline{e}_3|^{-2}d^3y < \infty$. Thus (10.46) leads to the result that $|v_{\underline{k}}(\underline{x}_1)-v_{\underline{k}}(\underline{x}_2)| \leq c_1 k|\underline{x}_1-\underline{x}_2| + c_2|\underline{x}_1-\underline{x}_2|^{\frac{1}{2}}$, establishing (Hölder) continuity of $v_{\underline{k}}(\underline{x})$ for every fixed $\underline{k}\varepsilon\Delta^{(3)}$.

Similarly,

$$|v_{k_1}(\underline{x})-v_{k_2}(\underline{x})|^2 \le 2\,\|\Psi^-_{k_1}\|^2_E \int |T_{k_1}(\underline{x},\underline{y}) - T_{k_2}(\underline{x},\underline{y})|^2 d^3y +$$

$$2\,\|\Psi^-_{k_1} - \Psi^-_{k_2}\|^2_E \int |T_{k_2}(\underline{x},\underline{y})|^2 d^3y. \tag{10.47}$$

Observing that $\left|\exp(ik_1|\underline{x}-\underline{y}|) - \exp(ik_2|\underline{x}-\underline{y}|)\right| \le |k_1-k_2|\,|\underline{x}-\underline{y}|$, we can write for every $R > 0$ that

$$\int |T_{k_1}(\underline{x},\underline{y}) - T_{k_2}(\underline{x},\underline{y})|^2 d^3y$$

$$\le |k_1-k_2|^2\{c_3\int|B_1(\underline{y})|^2 d^3y + c_4\int_{|y|<R}|B_2(\underline{y})|^2 d^3y\}$$

$$+ c_5\int_{|y|>R}|B_2(\underline{y})|^2|\underline{x}-\underline{y}|^{-2}d^3y.$$

The first two integrals are finite. For the third one, we choose $0 < \eta < \delta$ and note that

$$\int_{|\underline{y}|>R}|B_2(\underline{y})|^2|\underline{x}-\underline{y}|^{-2}d^3y \le \beta^2R^{-2\eta}\int d^3y(1+|\underline{y}|)^{-1-2(\delta-\eta)}|\underline{x}-\underline{y}|^{-2}$$

$$\le c_6R^{-2\eta}(1+|\underline{x}|^2)^{-(\delta-\eta)} \le c_6R^{-2\eta} \text{ for all } \underline{x}\in R^3,$$

where we have used (9.24). We can choose R sufficiently large to make this term arbitrarily small, and then the other two integrals can be made small by choosing $|\underline{k}_1-\underline{k}_2|$ small, thus establishing the convergence of the first term in (10.47) to zero as $|\underline{k}_1-\underline{k}_2| \to 0$ with $\underline{k}_1$, $\underline{k}_2\in\Delta^{(3)}$. The convergence of the second term in (10.47) to zero as $|\underline{k}_1-\underline{k}_2| \to 0$ follows from Lemma 10.7 and the fact that $\int|T_k(\underline{x},\underline{y})|^2d^3y$ is independent of k for $\underline{k}$ in $\Delta^{(3)}$ and finite, which was established in the first part of this proof. This proves (a).

(b) Since $\psi_k^{\pm}(\underline{x})$ is uniformly bounded for all $\underline{x}\in R^3$ and $\underline{k}$ in some compact subset $\Delta^{(3)}$ of $\Gamma^{(3)}$, we can make an estimate of $v_k(\underline{x})$ for large values of $|\underline{x}|$. In fact by (10.44) and the Schwarz inequality, we have

$$|v_k(\underline{x})| \le c_7 \int |\underline{x}-\underline{y}|^{-1} |V(\underline{y})| d^3y \le$$

$$c_7\{\|B_1\|_2 (\int |\underline{x}-\underline{y}|^{-2} |A(\underline{y})|^2 d^3y)^{\frac{1}{2}} + \|A\|_2 (\int |\underline{x}-\underline{y}|^{-2} |B_2(\underline{y})|^2 d^3y)^{\frac{1}{2}}\}.$$

This, by Lemma 9.12, yields the result that there exists finite constants c_1' and c_2' such that $|v_k(\underline{x})| \le c_1'|\underline{x}|^{-1} + c_2'|\underline{x}|^{-\delta}$ for every $\underline{x}$, which leads to (10.42). For an asymptotic expansion see Problem 10.4. #

We know from the result of Problem 10.4 that if $B_2 = 0$ [or more generally if $V\in L^1(R^3)\cap L^2(R^3)$] the asymptotic behaviour of the evaluation functional $\psi_k^-(\underline{x})$ is what has been traditionally used by physicists for the description of the the scattering process. On the other hand if $B_2 \ne 0$ [e.g. the potential function $V(\underline{x}) \sim |\underline{x}|^{-2-\delta}$ as $|\underline{x}| \to \infty$ with $0 < \delta < 1$], then Proposition 10.11(b) indicates that this description may break down. In such a situation the predominant term in the asymptotic expression for $\psi_k^-(\underline{x}) - \psi_k^0(\underline{x})$ is $O(|x|^{-\delta})$ as $|\underline{x}| \to \infty$, which need not correspond to the usual picture of an expanding spherical wave. Yet, as we have seen in Section in 9-1, the wave operators exist and are complete, leading to a well-defined unitary S-operator. We shall see in the next section that the on-shell S-operator also in this case has all the desirable properties. It should be mentioned, however, that we are not aware of an example where the exact asymptotic behaviour of $v_k(\underline{x})$ has been computed for this class.

The physics of the above situation may be understood as follows. If $V(\underline{x}) \sim |\underline{x}|^{-3-\delta}$ (i.e. $B_2 = 0$) then outside a sphere of sufficiently large radius, no further scattering takes place and thus the outgoing total scattered flux across spheres of radius R remains independent of R for large R. This leads to the inverse square law for the flux density or equivalently the asymptotic behaviour of Problem 10.4. However if $V(\underline{x}) \sim |x|^{-2-\delta}$ $(0 < \delta < 1)$, there may exist no such finite R and scattering may continue at all distances from the target adding non-zero though decreasing total flux across every sphere, however large.

10-2 THE S-MATRIX

In this section we derive some useful expressions for the on-shell S-operator or S-matrix for potentials in the class $\mathcal{V}$ and study its properties. In particular we want to know the detailed behaviour of the kernel $R(\lambda;\underline{\omega},\underline{\omega}')$ of $R(\lambda)$ so that we may verify the assumptions made in Sections 7-3 and 7-4, which enabled us to define differential and total cross sections. Before we proceed, we remind the reader that the operators $M_A(\lambda)$ and $M_{B_1}(\lambda)$ are well-defined operators from $L^2(\mathcal{R}^3)$ into $L^2(S^{(2)})$ and have the properties stated in Lemmas 10.1 and 10.3. Since B_2 is not square-integrable, these lemmas do not apply immediately to B_2. However we shall see in Lemma 10.21 that for every λ in a compact interval of $(0,\infty)$, $M_{B_2}(\lambda)$ is well-defined and is a $\mathcal{B}_o(L^2(\mathcal{R}^3), L^2(S^{(2)}))$-valued continuous function. It also satisfies (10.13) and Lemma 10.3(d), see Lemma 10.21. For the purposes of the next proposition we shall assume this result. Note also that for the results of this section we dispense with the assumption that $B_1 \varepsilon L^{\infty}(R^3)$, which was made occasionally in Section 10-1.

PROPOSITION 10.12 : Let $V \varepsilon \mathcal{Y}$. Then

(a) $$R(\lambda) \equiv S(\lambda)-I_0 = -2\pi i\, M_B(\lambda)(I+\bar{W}_\lambda^-)^{-1*} M_A(\lambda)^* \tag{10.48}$$

$$= -2\pi i\, M_A(\lambda)(1+W_\lambda^+)^{-1} M_B(\lambda)^*, \tag{10.49}$$

for all $\lambda \varepsilon (0,\infty)-\Gamma_0$. Furthermore $R(\lambda)$ is Hilbert-Schmidt in $L^2(S^{(2)})$ and is continuous in Hilbert-Schmidt norm, for all λ in $(0,\infty)-\Gamma_0$.

(b) The scattering amplitude is given by

$$f(\lambda;\underline{\omega}'\to\underline{\omega}) = -\,2\pi^2 \int \overline{\psi^0_{\sqrt{\lambda}\underline{\omega}}(\underline{x})}\, V(\underline{x})\, \psi^-_{\sqrt{\lambda}\underline{\omega}'}(\underline{x}) d^3x, \tag{10.50}$$

where the integral exists as a limit in the mean in $L^2(S^{(2)})$ for every fixed $\lambda \varepsilon (0,\infty)-\Gamma_0$ and $\underline{\omega}' \varepsilon S^{(2)}$. If moreover $B_2 = 0$, i.e. $V \varepsilon L^1 \cap \mathcal{Y}$, then the integral in (10.50) exists as a Lebesgue integral.

Proof : (a) We start with the expression (6.40) for R^* and write (with $E_\infty(K_0) = I$)

$$(f,R^*g) = -\lim_{\eta\to+0}\ \lim_{\delta\to+0} \int (f,[R^0_{\lambda-i\eta} - R^0_{\lambda+i\eta}](V-VR_{\lambda-i\delta}V)dE^0_\lambda g).$$

For V in $\mathcal{Y}$, $(V-VR_{\lambda-i\delta}V) = A(I-BR_{\lambda-i\delta}A)B = A(I+W_{\lambda-i\delta})^{-1}B$, where we have used (9.20). Now let $f,g \varepsilon \mathcal{S}_0$ and let Δ be a compact interval in $(0,\infty)-\Gamma_0$. Then by Lemma 10.3(d) the above expression for R^* can be rewritten as

$$(f,R^*E^0_\Delta g) = -\lim_{\eta\to+0}\ \lim_{\delta\to+0} \int_\Delta (A[R^0_{\lambda+i\eta} - R^0_{\lambda-i\eta}]f\ ,\ (I+W_{\lambda-i\delta})^{-1}BdE^0_\lambda g)$$

$$= -\lim_{\eta\to+0}\lim_{\delta\to+0}\int_\Delta (A[R^o_{\lambda+i\eta} - R^o_{\lambda-i\eta}]f\,,\,(I+W_{\lambda-i\delta})^{-1}M_B(\lambda)^*(\mathcal{U}_o g)_\lambda)d\lambda. \tag{10.51}$$

By Lemmas 9.5 and 9.13, the integrand converges pointwise as $\delta\to+0$, the limit being obtained on replacing $(I+W_{\lambda-i\delta})^{-1}$ by $(I+W^-_\lambda)^{-1}$. Remember that A is bounded, $\|(I+W_{\lambda-i\delta})^{-1}\|$ is bounded uniformly in δ for λ in Δ, $\|M_B(\lambda)^*\|$ is uniformly bounded in Δ, and that $(\mathcal{U}_o g)_\lambda$ is square-integrable, hence locally integrable. Therefore we may apply the Lebesgue dominated convergence theorem to interchange the limit $\delta\to+0$ and the integral in (10.51) to obtain

$$(f,R^*E^o_\Delta g) = -\lim_{\eta\to+0}\int_\Delta (A[R^o_{\lambda+i\eta} - R^o_{\lambda-i\eta}]f,(I+W^-_\lambda)^{-1}M_B(\lambda)^*(\mathcal{U}_o g)_\lambda)d\lambda.$$

By Lemma 10.3(c) $A[R^o_{\lambda+i\eta} - R^o_{\lambda-i\eta}]f$ converges weakly to $2\pi i\, M_A(\lambda)^*(\mathcal{U}_o f)_\lambda$ as $\eta\to+0$ and it also follows from the proof of Lemma 10.3(c) that $\|AR^o_{\lambda\pm i\eta}f\| \le \int\|AU_t f\|\,dt$, which is bounded in λ and η. Thus another application of the Lebesgue dominated convergence theorem leads to the result

$$(f,R^*E^o_\Delta g) = 2\pi i\int_\Delta (M_A(\lambda)^*(\mathcal{U}_o f)_\lambda,(I+W^-_\lambda)^{-1}M_B(\lambda)^*(\mathcal{U}_o g)_\lambda)d\lambda$$

$$= \int_\Delta ((\mathcal{U}_o f)_\lambda,[2\pi i\, M_A(\lambda)(I+W^-_\lambda)^{-1}M_B(\lambda)^*](\mathcal{U}_o g)_\lambda)_o d\lambda. \tag{10.52}$$

On the other hand we know from (5.97) that R* is a decomposable operator in the spectral representation of K_o, and thus $(f,R^*E^o_\Delta g) = (\mathcal{U}_o f,\mathcal{U}_o R^*E^o_\Delta \mathcal{U}^*_o\mathcal{U}_o g) = \int_\Delta((\mathcal{U}_o f)_\lambda,\ R(\lambda)^*(\mathcal{U}_o g)_\lambda)_o d\lambda$, where we have also used Proposition 5.26(c). Since Δ is an arbitrary compact interval in $(0,\infty)-\Gamma_o$ and since for any $\lambda>0$, $\{(\mathcal{U}_o f)_\lambda\}$ and $\{(\mathcal{U}_o g)_\lambda\}$ are fundamental in $L^2(S^{(2)})$ as f,g vary over $\mathcal{S}_o$, the relation (10.52) implies as in Problem 8.13 that $R(\lambda)^* = 2\pi i\, M_A(\lambda)(I+W^-_\lambda)^{-1}\, M_B(\lambda)^*$. Then (10.48) follows by taking the adjoint on both sides.

By Lemmas 10.3 and 10.21, $M_A(\lambda)^*$ is Hilbert-Schmidt and $M_B(\lambda)$ is bounded, and both are continuous in their respective norms. Thus by (2.63) $R(\lambda)$ is a Hilbert-Schmidt operator in $L^2(S^{(2)})$ and is continuous in λ in Hilbert-Schmidt norm. The alternative expression (10.49) for $R(\lambda)$ follows on using (6.39) instead of (6.40) and its proof is left as an exercise (Problem 10.5).

(b) Let $f\varepsilon L^2(S^{(2)})$. Then by the expression (10.48) and Proposition 10.5 we have that $R(\lambda)^*f =$ $2\pi i\, M_A(\lambda)(I+W_\lambda^-)^{-1}M_B(\lambda)^*f = 2\pi i\,(\mathcal{U}_- AM_B(\lambda)^*f)_\lambda$. Using (10.26) and the relation between the spaces $\mathcal{E}$, $\mathcal{H}$ and $\mathcal{E}'$, this leads to

$$(R(\lambda)^*f)(\underline{\omega}') = 2\pi i\; 2^{-\frac{1}{2}}\,\lambda^{\frac{1}{4}}(\Psi^-_{\underline{k}'},AM_B(\lambda)^*f)_{\mathcal{E}}$$

$$= 2\pi i\; 2^{-\frac{1}{2}}\lambda^{\frac{1}{4}}(A\psi^-_{\underline{k}'},M_B(\lambda)^*f) = 2\pi i\; 2^{-\frac{1}{2}}\,\lambda^{\frac{1}{4}}(M_B(\lambda)A\psi^-_{\underline{k}'},f)_o, \quad (10.53)$$

where we have written $\underline{k}' = \lambda^{\frac{1}{2}}\underline{\omega}'$ and also noted that $A\psi^-_{\underline{k}'}\varepsilon L^2(\mathcal{R}^3)$. From Proposition 7.20 we have that $R(\lambda;\underline{\omega},\underline{\omega}') =$ $R(\lambda)(\underline{\omega},\underline{\omega}') = \overline{R(\lambda)^*(\underline{\omega}',\underline{\omega})}$. Thus (10.53) gives us an expression for the kernel of $R(\lambda)$:

$$R(\lambda;\underline{\omega},\underline{\omega}') = -2\pi i\; 2^{-\frac{1}{2}}\lambda^{\frac{1}{4}}(M_B(\lambda)A\psi^-_{\underline{k}'})(\underline{\omega}), \quad (10.54)$$

where for each fixed $\lambda\varepsilon(0,\infty)-\Gamma_o$ and $\underline{\omega}'\varepsilon S^{(2)}$, the right hand side is defined for almost all $\underline{\omega}\varepsilon S^{(2)}$. Now (7.48) and (10.54) lead to the following expression for the scattering amplitude

$$f(\lambda;\underline{\omega}'\to\underline{\omega}) = -4\pi^2\; 2^{-\frac{1}{2}}\;\lambda^{-\frac{1}{4}}(M_B(\lambda)A\psi^-_{\underline{k}'})(\underline{\omega}). \quad (10.55)$$

For fixed λ and $\underline{\omega}'$, $f(\lambda;\underline{\omega}'\to\cdot)\;\varepsilon\, L^2(S^{(2)})$ and $\|f(\lambda;\underline{\omega}'\to\cdot)\|_o =$ $4\pi^2 2^{-\frac{1}{2}}\lambda^{-\frac{1}{4}}\;\|M_B(\lambda)A\psi^-_{\underline{k}'}\|_o$. Set $f_N(\lambda;\underline{\omega}'\to\underline{\omega}) = -4\pi^2 2^{-\frac{1}{2}}\lambda^{-\frac{1}{4}}\cdot$ $(M_B(\lambda)F_NA\psi^-_{\underline{k}'})(\underline{\omega})$, where F_N is the projection defined in (7.1).

Then $\| f(\lambda;\underline{\omega}' \to \cdot)-f_N(\lambda;\underline{\omega}' \to \cdot)\|_0 \leq 4\pi^2 2^{-\frac{1}{2}}\lambda^{-\frac{1}{4}} \|M_B(\lambda)\| \cdot$ $\|(I-F_N)A\psi^-_{k'}\|$ and thus $f_N(\lambda;\underline{\omega}' \to \cdot)$ converges strongly to $f(\lambda;\underline{\omega}' \to \cdot)$ in $L^2(S^{(2)})$ as $N \to \infty$.

On the other hand $M_B(\lambda)F_N A\psi^-_{k'} = M_{B_1}(\lambda)F_N A\psi^-_{k'} + M_{B_2}(\lambda)F_N A\psi^-_{k'}$, and the kernel of $M_{B_1}(\lambda)$ is given by (10.8) as $M_{B_1}(\lambda)(\underline{\omega},\underline{x}) = 2^{-\frac{1}{2}}\lambda^{\frac{1}{4}}\overline{\psi^o_k}(\underline{x})B_1(\underline{x})$. Since $B_2 \varepsilon \mathcal{B}(\mathcal{H})$, we have by (10.5) and (10.6) that $(M_{B_2}(\lambda)F_N A\psi^-_{k'})(\underline{\omega}) = 2^{-\frac{1}{2}}\lambda^{\frac{1}{4}} \cdot$ $\int_{|\underline{x}|<N} \overline{\psi^o_k}(\underline{x})B_2(\underline{x})A(\underline{x})\psi^-_{k'}(\underline{x})d^3x$. Putting these together we get that $f_N(\lambda;\underline{\omega}' \to \underline{\omega}) = -2\pi^2\int_{|\underline{x}|<N} \overline{\psi^o_k}(\underline{x})V(\underline{x})\psi^-_{k'}(\underline{x})d^3x$, and since $f_N(\lambda;\underline{\omega}' \to \cdot)$ converges to $f(\lambda;\underline{\omega}' \to \cdot)$ strongly, we arrive at (10.50).

From the above discussion it is easy to see that, if $B_2 = 0$ then the integral in (10.50) can be written as $(B_1\psi^o_k, A\psi^-_{k'})$, since $\|B_1\psi^o_k\|_2 = (2\pi)^{-3/2}\|B\|_2$ and $A\psi^-_{k'} \varepsilon L^2(\mathcal{R}^3)$ by Proposition 10.6. #

The expression (10.50) for the scattering amplitude is not unique. In fact starting with (10.49) instead of (10.48) one arrives at (Problem 10.5)

$$f(\lambda;\underline{\omega}' \to \underline{\omega}) = -2\pi^2 \int \overline{\psi^+_k}(\underline{x})V(\underline{x})\,\psi^o_{k'}(\underline{x})d^3x, \qquad (10.56)$$

where the integral has to be interpreted in the same way as in Proposition 10.12. Certain symmetry relations exist for the scattering amplitude. These follow from the symmetry relations of $\psi^{\pm}$ and ψ^o. It is easy to see that $\overline{\psi^o_{k'}} = \psi^o_{-k'}$ and then it follows from the Lippmann-Schwinger equation (10.36) that $\overline{\psi^+_k} = \psi^-_{-k}$ (Problem 10.6). This and the expressions

(10.50) and (10.56) yield the symmetry relation (7.66), viz. $f(\lambda;\underline{\omega}'\to\underline{\omega}) = f(\lambda;-\underline{\omega}\to-\underline{\omega}')$.

A comparison of (10.50) and the asymptotic expansion of ψ_k^- in Problem 10.4 (with $B_2 = 0$) allows us to conclude that as $|\underline{x}| = r\to\infty$,

$$\psi^-_{\sqrt{\lambda}\underline{\omega}}(\underline{x}) = \psi^o_{\sqrt{\lambda}\underline{\omega}}(\underline{x}) + \frac{e^{i\lambda^{\frac{1}{2}}r}}{r}(2\pi)^{-3/2}f(\lambda;\underline{\omega}\to\underline{\omega}_x)+o(r^{-1}),$$

which was given in (10.4). The above asymptotic expansion is often the starting point of a scattering theory in the physics literature. Thus if $B_2 = 0$, the coefficient of the expanding spherical wave gives the scattering amplitude up to a constant multiple.

Next we prove certain continuity properties of the scattering amplitude that were necessary to define differential and total scattering cross sections in Sections 7-3 and 7-4.

PROPOSITION 10.13 : Let $V\varepsilon\mathcal{V}$. Then (a) the scattering amplitude satisfies the continuity condition (7.51), i.e. $\int_{\Omega_C} d\omega|f(\lambda;\underline{\omega}'\to\underline{\omega})|^2$ is a continuous function of λ and $\underline{\omega}'$ for $\lambda\varepsilon(0,\infty)-\Gamma_0$ and $\underline{\omega}'\varepsilon S^{(2)}$, uniformly with respect to Ω_C.

(b) If furthermore $B_2 = 0$, then $f(\lambda;\underline{\omega}'\to\underline{\omega})$ is continuous in all three variables for $\lambda\varepsilon(0,\infty)-\Gamma_0$ and $(\underline{\omega},\underline{\omega}')\,\varepsilon\, S^{(2)}\times S^{(2)}$.

Proof : (a) Let $\lambda,\lambda_o\varepsilon(0,\infty)-\Gamma_0$ and $\underline{\omega}',\underline{\omega}_o\,\varepsilon\, S^{(2)}$. By Lemmas 10.7, 10.1 and 10.21, $M_B(\lambda)A\psi^-_{k'}$ converges strongly in $L^2(S^{(2)})$ to $M_B(\lambda_o)A\psi^-_{k_o}$ as $\lambda\to\lambda_o$ and $\underline{\omega}'\to\underline{\omega}_o$. Together with (10.55), this allows us to conclude that $G_C f(\lambda;\underline{\omega}'\to\cdot)$ converges strongly to $G_C f(\lambda_o;\underline{\omega}_o\to\cdot)$ as $\lambda\to\lambda_o$ and $\underline{\omega}'\to\underline{\omega}_o$, uniformly in Ω_C, where G_C is the projection operator in $L^2(S^{(2)})$ corresponding to the set Ω_C in $S^{(2)}$. Thus $\int_{\Omega_C} d\omega|f(\lambda;\underline{\omega}'\to\underline{\omega})|^2 = \|G_C f(\lambda;\underline{\omega}'\to\cdot)\|_o^2$ is a continuous function of λ and $\underline{\omega}'$, uniformly in Ω_C (see Proposition 2.1).

(b) As we have seen in the proof of part (b) of the last proposition, in this case the scattering amplitude can be written as $f(\lambda;\underline{\omega}'\to\underline{\omega}) = -2\pi^2(B_1\psi^o_{\underline{k}}, A\psi^-_{\underline{k}'})$. Since $\psi^o_{\underline{k}}$ is a continuous function of $\underline{k}$ and $B_1\varepsilon L^2$, it follows by the Lebesgue dominated convergence theorem that $B_1\psi^o_{\underline{k}}$ is strongly continuous in $\underline{k}$. As $A\psi^-_{\underline{k}'}$ is strongly continuous by Lemma 10.7, we arrive at the stated continuity of $f(\lambda;\underline{\omega}'\to\underline{\omega})$. #

10-3 THE SINGULAR SPECTRUM

We have stated in the beginning of Section 9-1 that, in order to prove asymptotic completeness in the sense of (A3), we need to establish two things : first that the ranges of $\Omega_\pm$ are equal to $\mathcal{H}_{ac}(H)$ and second, that $\mathcal{H}_{sc}(H) = \{\theta\}$. The first objective was reached in Section 9-1 for potentials in the class $\mathcal{V}$. In this section we show that the total Hamiltonian $H = K_o + V$ does not have any singularly continuous spectrum if $V\varepsilon\mathcal{V}$. For a Hamiltonian H, the states belonging to $\mathcal{H}_{sc}(H)$ do not yield to any simple physical interpretation. For example, though $\|F_r \exp(-iHt)f\|^2$ may converge to 0 as $t\to\pm\infty$, the total time spent in a sphere in $\mathbb{R}^3$ of radius r, given by $\int\|F_r \exp(-iHt)f\|^2 dt$, is infinite for $f\varepsilon\mathcal{H}_{sc}(H)$ (see e.g. Sinha [1]). From this point of view also, it is desirable to have $\mathcal{H}_{sc}(H) = \{\theta\}$.

Since every potential V in $\mathcal{V}$ is also in $L^2(\mathcal{R}^3)$, we know from Proposition 8.7 that V is K_o-compact, and thus by Proposition 8.5, $\sigma_e(H) = \sigma_e(K_o) = [0,\infty)$. Therefore the discrete spectrum of H is a subset of the negative half-line. If $\mathcal{H}_{sc}(H)$ is empty, then $\sigma_e(H)$ consists of $\sigma_{ac}(H)$ and the positive eigenvalues of H along with their limit points. These eigenvalues are embedded in the continuum, since they are clearly not isolated points in $\sigma(H)$. By the description of the bound states given in Section 7-1, the positive energy eigenstates as well as states belonging to $\mathcal{H}_d(H)$ are candidates for bound states. However from a physical point of view the negative energy eigenstates, i.e. the states in $\mathcal{H}_d(H)$, are the bound states. This is because the positive energy eigenstates are not stable under small perturbations which couple them to the states in $\mathcal{H}_c(H)$ and which are often present in physical situations. Still it is desirable to prove that the positive eigenvalues are absent.

The classical example of von Neumann and Wigner [1] provides a potential V such that $H = K_o + V$ has positive eigenvalues. Intuitively speaking the situation is as follows. If the potential tends to zero very slowly (e.g. as $|\underline{x}|^{-1}$ or more slowly) and if it is rapidly oscillating at infinity, then the waves may get reflected to interfere constructively, and build up a positive energy eigenstate. Conversely, the method of proving the absence of positive eigenvalues of H usually consist of two steps. First, one shows that the associated eigenvector vanishes outside a large sphere in $\mathcal{R}^3$.

Then one applies the unique continuation theorem for elliptic differential equations to conclude that such a vector is θ. For details on this point see Jansen and Kalf [1] and references cited therein. We add that for a satisfactory scattering theory this information, namely the absence of positive energy bound states, does not play any significant role. In view of this, we shall not attempt to discuss this point for the class $\mathcal{V}$. We want to add however, that if $B_1 = 0$, so that $|V(\underline{x})| \leq \beta(1+|\underline{x}|)^{-2-\delta}$ $(\delta > 0)$ for all $\underline{x}\varepsilon R^3$, Kato [1] has established the absence of positive eigenvalues.

For the purposes of this section we set $\tilde{\Gamma}_o = (0,\infty)\cap\Gamma_o$. Then from Remark 9.7 and the above discussion we know that $\sigma_s(H)\cap(0,\infty) \subseteq \tilde{\Gamma}_0$. It is clear then that all the positive eigenvalues are in $\tilde{\Gamma}_0$. What we prove next will show that in fact $\tilde{\Gamma}_0$ consists of only positive eigenvalues and that $\sigma_{sc}(H)$ is empty. For this we shall first need two lemmas.

Let A and B be the multiplication operators in $L^2(R^3)$ defined in connection with the class $\mathcal{V}$ in (9.27). Let $\lambda\varepsilon\Gamma_0$ be an <u>exceptional point</u> with <u>exceptional vector</u> $f_\pm$, i.e. let

$$(I+W_\lambda^\pm)f_\pm = \theta \text{ with } f_\pm \neq \theta. \tag{10.57}$$

<u>LEMMA 10.14</u> : Let $\lambda\varepsilon\tilde{\Gamma}_0$ and $f_\pm$ be as in (10.57). Then $(\mathcal{U}_o Af_\pm)_\lambda = \theta$ in $L^2(S^{(2)})$, i.e. $M_A(\lambda)f_\pm = \theta$.

<u>Proof</u> : Since $A\varepsilon L^2(R^3)\cap L^\infty(R^3)$, the relation (10.6) defining $M_A(\mu)$ is valid for all $f\varepsilon\mathcal{H}$, and $(\mathcal{U}_o Af)_\mu = M_A(\mu)f$ is a continuous $L^2(S^{(2)})$-valued function by Lemma 10.1. Thus it makes sense to talk of $(\mathcal{U}_o Af)_\mu$ for all $\mu\varepsilon[0,\infty)$.

Remembering that $W_\lambda^\pm$ is a Hilbert-Schmidt operator with kernel given by (9.28) and denoting by $Z_\lambda^\pm$ the Hilbert-Schmidt operator with kernel

$$Z_\lambda^\pm(\underline{x},\underline{y}) = A(\underline{x})G^o_{\lambda\pm io}(\underline{x},\underline{y})A(\underline{y}), \tag{10.58}$$

we have that $W_\lambda^\pm(\underline{x},\underline{y}) = B(\underline{x})A^{-1}(\underline{x})Z_\lambda^\pm(\underline{x},\underline{y}) = C(\underline{x})Z_\lambda^\pm(\underline{x},\underline{y})$, where we have written $C(\underline{x}) = B(\underline{x})A^{-1}(\underline{x})$. Let C be the self-adjoint multiplication operator by the function $C(\underline{x})$. Since $\|CZ_\lambda^\pm f\|^2 = \|W_\lambda^\pm f\|^2$ for each $f\varepsilon H$, one sees from the definition of D(C) in Proposition 2.16 that D(C) contains the range of $Z_\lambda^\pm$. Hence $W_\lambda^\pm = CZ_\lambda^\pm$.

From Lemmas 10.1, 10.22 and by (10.57) we get

$$((U_oAf_\pm)_\lambda,(U_oAf_\pm)_\lambda)_o = (M_A(\lambda)f_\pm, M_A(\lambda)f_\pm)_o =$$

$$(M_A(\lambda)^* M_A(\lambda)f_\pm, f_\pm) = (2\pi i)^{-1} \lim_{\eta\to+0} (A[R^o_{\lambda+i\eta}-R^o_{\lambda-i\eta}]Af_\pm, f_\pm) =$$

$$\pm\pi^{-1} \lim_{\eta\to+0} \mathrm{Im}\ (AR^o_{\lambda\pm i\eta}Af_\pm, f_\pm) = \pm\pi^{-1}\ \mathrm{Im}\ (Z_\lambda^\pm f_\pm, f_\pm) =$$

$$\mp\pi^{-1}\ \mathrm{Im}\ (Z_\lambda^\pm f_\pm, W_\lambda^\pm f_\pm) = \mp\pi^{-1}\ \mathrm{Im}\ (Z_\lambda^\pm f_\pm, CZ_\lambda^\pm f_\pm) = 0,$$

since C is symmetric. This leads to the result that $\|(U_oAf)_\lambda\|_o^2 = 0$, i.e. $(U_oAf)_\lambda = \theta$ in $L^2(S^{(2)})$. #

<u>LEMMA 10.15</u> : Let $V\varepsilon Y$ with $B_2 = 0$ and let $f_\pm$ be the exceptional vector defined in (10.57), associated with $\lambda\varepsilon\Gamma_o$.

(a) $|\underline{Q}|\,Af \varepsilon L^1(R^3)$, where $(|\underline{Q}|h)(\underline{x}) = |\underline{x}|h(\underline{x})$.

(b) Let $\lambda > 0$. For each $\mu > 0\ (\mu \neq \lambda)$, define $g_\mu^\pm \equiv (\mu-\lambda)^{-1}(U_oAf_\pm)_\mu \varepsilon L^2(S^{(2)})$. Then $\|g_\mu^\pm\|_o$ belongs to $L^2([0,\infty))$. Thus $g^\pm \equiv U_o^{-1}\{g_\mu^\pm\}$ defines a non-zero vector in H.

Proof : (a) Since $B_2 = 0$, $f_\pm \varepsilon H$ and $Af_\pm = -AW_\lambda^\pm f_\pm$, we get from the estimate (9.23) that

$$|(Af_\pm)(\underline{x})| \le c|B_1(\underline{x})A(\underline{x})|(\int|\underline{x}-\underline{y}|^{-2}|A(\underline{y})|^2 d^3y)^{\frac{1}{2}} \|f_\pm\|$$

$$\le c|B_1(\underline{x})A(\underline{x})|(1+|\underline{x}|^2)^{-\frac{1}{2}} \|f_\pm\|. \qquad (10.59)$$

Hence $|\underline{x}||(Af_\pm)(\underline{x})| \le c\|f_\pm\| \, |B_1(\underline{x})A(\underline{x})|$, which is integrable since both B_1 and A belong to $L^2(R^3)$.

(b) Divide the interval $[0,\infty)$ into two parts, $N(\lambda) = \{\mu \big| |\mu-\lambda| \le \lambda/2\}$ and its complement $D(\lambda) = \{\mu \big| |\mu-\lambda| > \lambda/2\}$. Then it is clear that $\|g_\mu^\pm\|_0 \,\varepsilon\, L^2(D(\lambda))$. Hence it suffices to show that $\|g_\mu^\pm\|_0 \,\varepsilon\, L^2(N(\lambda))$. Since by Lemma 10.14, $(U_0 Af_\pm)_\lambda = \theta$ in $L^2(S^{(2)})$, this follows if we can prove

$$\|(U_0 Af_\pm)_\mu\|_0 \equiv \|(U_0 Af_\pm)_\mu - (U_0 Af_\pm)_\lambda\|_0 \le c|\mu-\lambda|^\delta \text{ with } \delta > \tfrac{1}{2}, \qquad (10.60)$$

where c may depend on λ, A and $f_\pm$. For then $\int_{N(\lambda)}\|g_\mu^\pm\|_0^2 \, d\mu \le c^2\int_{N(\lambda)} |\mu-\lambda|^{2\delta-2}d\mu \le c^2 \int_{|\mu|<\lambda/2} \mu^{2\delta-2}d\mu < \infty$, proving that $\|g_\mu^\pm\|_0 \,\varepsilon\, L^2(N(\lambda))$ and hence the result that $g^\pm \,\varepsilon\, H$. Next we show that when $B_2 = 0$, we have the inequality (10.60) with $\delta = 1$.

Note that

$$(U_0 Af_\pm)_\mu(\underline{\omega}) - (U_0 Af_\pm)_\lambda(\underline{\omega}) = 2^{-\frac{1}{2}}\{(\mu^{\frac{1}{4}}-\lambda^{\frac{1}{4}})(FAf_\pm)(\mu^{\frac{1}{2}}\underline{\omega})$$

$$+ \lambda^{\frac{1}{4}}[(FAf_\pm)(\mu^{\frac{1}{2}}\underline{\omega}) - (FAf_\pm)(\lambda^{\frac{1}{2}}\underline{\omega})]\}. \qquad (10.61)$$

Since $Af_\pm \varepsilon L^1(R^3)$, $(FAf_\pm)(\mu^{\frac{1}{2}}\underline{\omega})$ is uniformly bounded. For $\mu \varepsilon N(\lambda)$, we have $|\mu^{\frac{1}{4}}-\lambda^{\frac{1}{4}}| = 4^{-1}|\int_\lambda^\mu s^{-3/4}ds| \le \text{const} \cdot \lambda^{-3/4}|\mu-\lambda|$. Thus the first term in (10.61) is majorized by $c_1\|FAf_\pm\|_\infty \lambda^{-3/4}|\mu-\lambda|$, uniformly in $\underline{\omega}$. Next using the inequality

$|\exp(-i\alpha s) - \exp(-i\beta s)| \leq |\alpha-\beta||s|$, we obtain similarly that

$$|(FAf_\pm)(\mu^{\frac{1}{2}}\underline{\omega})-(FAf_\pm)(\lambda^{\frac{1}{2}}\underline{\omega})|$$
$$\leq (2\pi)^{-3/2}\int|\exp(-i\mu^{\frac{1}{2}}\underline{\omega}\cdot\underline{x}) - \exp(-i\lambda^{\frac{1}{2}}\underline{\omega}\cdot\underline{x})||A(\underline{x})f_\pm(\underline{x})|d^3x$$
$$\leq c_2\lambda^{-\frac{1}{2}}|\mu-\lambda| \int |\underline{x}||A(\underline{x})f_\pm(\underline{x})|d^3x = c_3\lambda^{-\frac{1}{2}}|\mu-\lambda|,$$

by virtue of (a). Since all these estimates are uniform in $\underline{\omega}$, we obtain the inequality (10.60) with $\delta = 1$ for any $\lambda > 0$. #

Remark 10.16 : Part (b) of Lemma 10.15 is actually true without the assumption $B_2 = 0$. Indeed the validity of (10.60) follows from the fact that $M_A(\mu)$ is locally Hölder continuous in operator norm, i.e. $\|M_A(\mu) - M_A(\mu')\| \leq c_\Delta|\mu-\mu'|$ for all μ,μ' belonging to a compact interval Δ in $(0,\infty)$. This result will be given in Lemma 10.21.

PROPOSITION 10.17 : Let $V\varepsilon\mathcal{V}$. Let $\lambda > 0$ be an exceptional point and $f_\pm$ be an associated exceptional vector as defined in (10.57). Then (a) $g^\pm$, defined in Lemma 10.15(b), belongs to D(H) and satisfies the eigenvalue equation

$$Hg^\pm = \lambda g^\pm, \tag{10.62}$$

and (b) $\mathcal{H}_{sc}(H) = \{\theta\}$.

Proof : From Lemma 10.15 and Remark 10.16 we know that $g^\pm\varepsilon\mathcal{H}$ and is given in the K_0-spectral representation by $(\mathcal{U}_0 g^\pm)_\mu \equiv g^\pm_\mu = (\mu-\lambda)^{-1}(\mathcal{U}_0 Af_\pm)_\mu$. A simple computation shows that $(\mathcal{U}_0 K_0 g^\pm)_\mu = (\mathcal{U}_0 Af_\pm)_\mu + \lambda(\mathcal{U}_0 g^\pm)_\mu$, and thus $g^\pm\varepsilon D(K_0)$ which, as we have observed before, is equal to D(H). We see at the same time that $(K_0-\lambda)g^\pm = Af_\pm$.

Consider the sequence of vectors $\{g^{\pm}(\eta)\}$ given by $g^{\pm}(\eta) = R^{o}_{\lambda\pm i\eta}Af_{\pm}$ $(\eta > 0)$. Clearly $g^{\pm}(\eta)\varepsilon D(V)$ for every $\eta > 0$, since V is K_o-compact. Also $\{g^{\pm}(\eta)\}$ converges strongly to $g^{\pm}$ as $\eta\to 0$. This follows from the fact that $\|g^{\pm}(\eta)-g^{\pm}\|^2 = \int|(\mu-\lambda\mp i\eta)^{-1} - (\mu-\lambda)^{-1}|^2\|(U_o Af_{\pm})_{\mu}\|_o^2\, d\mu$ and the Lebesgue dominated convergence theorem. That $\{Vg^{\pm}(\eta)\}$ also converges strongly to $AW^{\pm}_{\lambda}f_{\pm}$ is clear from the identity $Vg^{\pm}(\eta) = ABR^{o}_{\lambda\pm i\eta}Af_{\pm}$ and Lemma 9.13. V is a closed operator and therefore $g^{\pm}\varepsilon D(V)$ [which we already know since $D(K_o) \subseteq D(V)$] and $Vg^{\pm} = AW^{\pm}_{\lambda}f_{\pm}$. Thus $(H-\lambda)g^{\pm} = (K_o-\lambda)g^{\pm} + Vg^{\pm} = Af_{\pm} + AW^{\pm}_{\lambda}f_{\pm} = \theta$, by (10.57).

Thus each point of $\tilde{\Gamma}_o$ is an eigenvalue of H. In view of the separability of $H = L^2(R^3)$, $\tilde{\Gamma}_o$ is a countable set. Let $\Delta = \{0\}\cup\tilde{\Gamma}_o$. Then by Lemma 5.13(b), $E_{\Delta}H \subseteq H_p(H)$. Since V is K_o-compact, Proposition 8.5(b) and relation (9.14) imply that $E_{[0,\infty)-\Delta} = E_{ac}(H)$, so that $E_{[0,\infty)} = E_{\Delta} + E_{ac}(H)$. Combining these facts with the relation : $H_{sc}(H) \subseteq E_{[0,\infty)}H$, we conclude that the singular continuous spectrum of H is absent, i.e. $H_{sc}(H) = \{\theta\}$. #

<u>Remark 10.18</u> : (a) Using Lemma 10.14 it is not hard to see that for any λ, the null spaces of $I + W^{+}_{\lambda}$ and of $I + W^{-}_{\lambda}$ are identical, implying that $\Gamma^{+}_{o} = \Gamma^{-}_{o} \equiv \Gamma_o$. Also by Proposition 2.25, these null spaces are finite dimensional. Furthermore, if λ is an eigenvalue of H, then B maps the corresponding eigenspace $M_{\{\lambda\}}$ injectively into the null space of $I+W^{\pm}_{\lambda}$, showing that each subspace $M_{\{\lambda\}}$ is finite dimensional. If $\lambda \neq 0$, the above map is in addition surjective. That B is injective on $M_{\{\lambda\}}$ is seen as follows : if $Bg = \theta$, then

$(K_0-\lambda)g = -Vg = \theta$ implying $g = \theta$ since K_0 is absolutely continuous. The verification of the other statements is given as Problem 10.8.

(b) The result of Proposition 10.17 need not be true for $\lambda = 0$. If $\lambda = 0$ is an exceptional point, then the associated vector g defined in Lemma 10.15(b) is not necessarily in $L^2(R^3)$. If $g \varepsilon L^2(R^3)$, then it is an eigenvector, i.e. $Hg = \theta$. If $g \notin L^2(R^3)$, it still verifies $Hg = 0$ where H is interpreted as the formal differential expression : $H = -\Delta + V$. In the second case g is sometimes called a virtual eigenvector or a quasi-bound state. More about this for spherically symmetric potentials can be found in Remark 11.17(c) and in Problems 11.11 and 11.12. The property that $\lambda = 0$ is an exceptional point is unstable under small changes of the potential. In fact, suppose for a certain value γ_0 of the coupling constant, $I + \gamma_0 W_0$ is not invertible. Then by Proposition 2.24, $-\gamma_0^{-1}$ is an eigenvalue of W_0. Since by Proposition 2.25 the non-zero eigenvalues of a compact operator are isolated, $I + \gamma W_0$ must be invertible for all γ ($\neq \gamma_0$) in some interval (a,b) containing γ_0 in its interior. Also we know from Proposition 9.4 that $I + \gamma W_0$ is invertible for $|\gamma|$ small enough and therefore it follows that for fixed $V \varepsilon \mathcal{V}$, the set $\{\gamma \varepsilon R | I+\gamma W_0$ is not invertible$\}$ is a discrete set.

Proposition 10.17 together with Proposition 9.14 proves the asymptotic completeness for the class $\mathcal{V}$ in the sense of (A3). As far as scattering theory is concerned (as described in Section 4-1), this is sufficient. However, there are some other questions about the spectrum of H that one may be interested in. Is the set of positive eigenvalues in any compact subset of $(0,\infty)$ finite ? More ambitiously, is it true that there are no positive eigenvalues ? Thirdly, is the set of negative energy eigenstates or the bound states of H (including multiplicities) finite ? The first question will be briefly discussed in Section 10-4E. For the second, no definite answer is known for the class $\mathcal{V}$, though it is believed that indeed there are no positive eigenvalues of H. It is well-known that for the class $\mathcal{V}$, there are only finitely many negative energy bound states. For this the reader is referred to the review article of Simon in [LI]. Stronger results about the absence of the singularly continuous spectrum can be found in Agmon [1], Aguilar and Combes [1] and Lavine [3].

10-4 NOTES AND SUPPLEMENTARY MATERIAL

A. Here we give an abstract definition of generalized eigenfunctions. For that we need the notion of Gelfand triplets [GV]. First we remind the reader of the definition of bounded linear functionals on a Hilbert space H as introduced in Section 2-1. Similarly one may define a continuous linear functional on any linear topological vector space as a continuous linear map from it to C, where continuity means continuity with respect to the given topology of the space and the standard topology on C. The collection of all continuous linear functionals (equipped with the so-called weak topology) also forms a linear topological vector space and is called the (topological) dual of the first one. From the Riesz representation theorem (Proposition 2.3), we see that the dual $\mathcal{H}'$ of a Hilbert space $\mathcal{H}$ is itself a Hilbert space and is isometrically anti-isomorphic to $\mathcal{H}$. Let $\mathcal{E}$ be a linear manifold, dense in $\mathcal{H}$ and suppose that $\mathcal{E}$ is equipped with its own topology which is assumed to be stronger than that of $\mathcal{H}$. Then clearly the canonical injection J from $\mathcal{E}$ into $\mathcal{H}$ is continuous. Let $\mathcal{E}'$ be the dual of $\mathcal{E}$. If $f' \varepsilon \mathcal{E}'$, we write $\langle f', g\rangle$ for the value in C of the linear functional f' at $g \varepsilon \mathcal{E}$. Since $\mathcal{H}' \subseteq \mathcal{E}'$, we can define J', the dual of J, by writing $\langle J'f, g\rangle \equiv \langle f, Jg\rangle$ where $g \varepsilon \mathcal{E}$ and $f \varepsilon \mathcal{H}'$. Thus J' is one-to-one and is the canonical injection of $\mathcal{H}'$ into $\mathcal{E}'$. If Z is the (anti-linear) isometry mapping $\mathcal{H}$ onto $\mathcal{H}'$, then, setting $L = J'Z$, we observe that L is the (anti-linear) embedding of $\mathcal{H}$ into $\mathcal{E}'$ given by $\langle Lf, g\rangle = (f,g)$, $g \varepsilon \mathcal{E}$. A triplet $\mathcal{E}$, $\mathcal{H}$ and $\mathcal{E}'$ with the above properties is called a Gelfand triplet and often written symbolically as

$$\mathcal{E} \underset{\to}{\subseteq} \mathcal{H} \underset{\to}{\subseteq} \mathcal{E}'. \tag{10.63}$$

In our application, the topologies of $\mathcal{E}$ and $\mathcal{E}'$ were given by relations (10.20) and (10.21) respectively, and it is clear that both $\mathcal{E}$ and $\mathcal{E}'$ are Hilbert spaces.

Let H be a self-adjoint operator in $\mathcal{H}$. Let (X, Σ, μ) be a σ-finite measure space and $\psi_\tau \varepsilon \mathcal{E}'$ for almost all $\tau \varepsilon X$. Assume that $\tau \mapsto \psi_\tau$ defines a mapping from X into $\mathcal{E}'$ such that $\langle \psi_\tau, g\rangle$ is μ-measurable for every $g \varepsilon \mathcal{E}$. Furthermore suppose that the following two conditions are satisfied :

(i) for every $g \varepsilon \mathcal{E}$, the relation

$$\int_X |\langle \psi_\tau, g\rangle|^2 d\mu(\tau) = \|E_{ac}(H)g\|^2 \tag{10.64}$$

holds and the mapping $g \mapsto \langle \psi_\tau, g\rangle$ from $\mathcal{E}$ into $L^2(\mu)$ can be extended to a partial isometry F from $\mathcal{H}$ onto $L^2(\mu)$ with initial set $\mathcal{H}_{ac}(H)$,

(ii) there exists a real-valued measurable function λ defined on X such that $F H_{ac} F^{-1}$ is the operator of multiplication by $\lambda(\tau)$ in $L^2(\mu)$.

Then (X,Σ,μ), E, ψ and λ are said to determine a spectral representation of H with the evaluation functional ψ_τ. The mapping F is called the generalized Fourier transformation and differs from the spectral transformation U, introduced in Section 5-7, only because in general the measures are chosen differently.

In addition, if we assume that there exists a dense linear manifold N of E (N separable in its own topology) such that $N \subseteq D(H)$ and $HN \subseteq E$, then μ, E, ψ and λ are said to determine an eigenfunction expansion associated with the self-adjoint operator H. In this connection the following lemma is instructive.

LEMMA 10.19 : Let μ, E, ψ and λ determine an eigenfunction expansion associated with H. Regarding H as an operator from N into E, let $H' : E' \to N'$ be the dual of H. Embed E' naturally in N' as we embedded H' in E'. Then

$$H'\psi_\tau = \lambda(\tau)\psi_\tau \,, \quad \mu\text{-a.e.} \qquad (10.65)$$

Proof : Observe that from the definition of ψ_τ and F we have $\langle\psi_\tau, f\rangle = (Ff)_\tau$, for all $f \in E$. Then with $f \in N$, $\langle H'\psi_\tau, f\rangle = \langle\psi_\tau, Hf\rangle = (FHf)_\tau = \lambda(\tau)(Ff)_\tau$ from the property (ii). Thus $\langle H'\psi_\tau, f\rangle = \lambda(\tau)\langle\psi_\tau, f\rangle$ for μ-a.a.τ. Using the separability of N, one can remove the dependence of the null-set on f, and then the denseness of N in E leads to (10.65). #

For the operator $H = K_o + V$ with $V \varepsilon \mathcal{Y}$, we have $\mathcal{R}^3$, equipped with the Lebesgue measure, as the measure space (X,μ), with the wave-vector $\underline{k}$ replacing τ. We can take for $\mathcal{N}$ the space $\mathcal{S} \equiv \mathcal{S}(\mathcal{R}^3)$, which is not a Hilbert space. Then, since $\mathcal{E}' \subseteq \mathcal{S}'$, we can look upon the equation (10.65) as an equation in $\mathcal{S}'$, viz.

$$(-\Delta+V)\psi_k(\underline{x}) = \lambda(\underline{k})\psi_k(\underline{x}), \text{ with } \lambda(\underline{k}) = \underline{k}^2. \qquad (10.66)$$

The same equation can be derived by taking $\mathcal{N}$ to be $C_o^\infty(\mathcal{R}^3)$, and in that case the functions ψ_k are called the <u>weak solutions</u> of (10.66). If we assume furthermore that $V \varepsilon \mathcal{Y}$ with $B_1 \varepsilon L^2(\mathcal{R}^3) \cap L^\infty(\mathcal{R}^3)$, then by Proposition 10.11(a) we observe that for $\underline{k} \varepsilon \Gamma^{(3)}$, ψ_k satisfying equation (10.66) is twice differentiable in an ordinary sense and $\Delta\psi_k \varepsilon L^2_{loc}(\mathcal{R}^3)$. For more general results of this type, see Alsholm and Schmidt [1].

There exist schemes which are in a sense different from the one described above. For these, the reader is referred to [GS, Vol. III] and [B].

Mathematicians often prove the eigenfunction expansion first and then establish the scattering theory. We have done the opposite. Titchmarsh in his book [TI] studies equations of the type (10.66) and uses function theoretic methods to study the solutions $\psi_k(\underline{x})$ and their completeness properties, e.g. (10.37) and (10.38). The eigenfunction expansion for the case when $V \varepsilon \mathcal{Y}$ or V in some related class has been established in 3 dimensions by Povzner [1] and Ikebe [1], and in n dimensions ($n \geq 3$) by Alsholm and Schmidt [1]. The basic technique used in all these works is to show the compactness of the operators $T_\lambda^\pm$ in $L^\infty(\mathcal{R}^n)$, where $T_\lambda^\pm$ are defined by giving the kernels as $T^{\pm}_{\lambda}(\underline{x},\underline{y}) = |\underline{x}-\underline{y}|^{-1}\exp(\pm i\lambda^{\frac{1}{2}}|\underline{x}-\underline{y}|)V(\underline{y})$, and then to study the solutions of the corresponding Lippman-Schwinger equation. Then the completeness relations (10.37) and (10.38) are established for these eigenfunctions. The wave operators $\Omega_{\pm}$ are defined by writing $\Omega_{\pm} = F^*_{\pm}F$, where $F_{\pm}$ are the generalized Fourier transformations associated with the two sets of generalized eigenfunctions $\psi^{\pm}_{k}$ and F is the ordinary Fourier transformation in $L^2(R^n)$. Finally one verifies that these $\Omega_{\pm}$ coincide with the ones defined from the time-dependent scattering theory. This gives the completeness of the wave operators in the sense that the ranges of $\Omega_{\pm}$ are equal to $H_{ac}(H)$, since the initial sets of $F_{\pm}$ are equal to $H_{ac}(H)$ by construction. In relation to Proposition 10.11, we would like to add that the uniform boundedness of $\psi^{\pm}_{k}(\underline{x})$ and the uniform continuity of $\psi^{\pm}_{k}(\underline{x}) - \psi^{o}_{k}(\underline{x})$ for $(\underline{x},\underline{k}) \in R^3 \times \Delta^{(3)}$ without the restriction of boundedness on B_1 has been obtained by Alsholm and Schmidt [1]. Agmon [1] has shown that, for potentials V which verify $V \in L^p_{loc}(R^n)$ with $p = 2$ when $n \geq 3$ and $p > n/2$ when $n \geq 4$ and $|V(\underline{x})| = 0(|x|^{-1-\delta})$, $\delta > 0$, at infinity, there exists an eigenfunction expansion for $H = K_0 + V$ in $L^2(R^n)$ and that the same sort of argument as above establishes the completeness of $\Omega_{\pm}$. In order to deal with such a general problem, however, one has to use the complex method of interpolation for Sobolev spaces, and we refrain from discussing this method in this book.

One can also establish eigenfunction expansions for potentials which are locally more singular than what we have discussed so far, though the behaviour of V at infinity remains $0(|\underline{x}|^{-2-\delta})$. For these results, the reader is referred to [SM] and Davies [1]. The methods of Kuroda [2] and Kato-Kuroda [1] are similar in spirit to those of this book but differ by the use of the stationary method to prove the completeness of $\Omega_\pm$ before going on to establish the eigenfunction expansion.

B. We next study the total Green's function for potentials verifying the hypotheses of Proposition 10.11. We have seen in Proposition 9.10 that R_z^o is an integral operator with kernel $G_z^o(\underline{x},\underline{y})$ given by (9.21). Since V is K_0-compact we obtain from (6.37) that $R_z = R_z^o - R_z^o V R_z$. Moreover $V \in L^2(R^3)$ and hence $R_z^o V R_z$ is Hilbert-Schmidt. Thus by Proposition 2.33, $R_z^o V R_z$ and therefore R_z is an integral operator. In conformity with the Remark 9.11, we shall call the kernel $G_z(\underline{x},\underline{y})$ of R_z the <u>total Green's function</u>. Then it is easy to establish an integral equation similar to the Lippmann-Schwinger equation (10.36), namely

$$G_z(\underline{x},\underline{y}) = G_z^o(\underline{x},\underline{y}) - \int G_z^o(\underline{x},\underline{u})V(\underline{u})G_z(\underline{u},\underline{y})d^3u. \qquad (10.67)$$

Next, by (9.20) we have $AR_zA = (I+W_{\bar{z}})^{-1*}AR_z^oA$. Therefore it follows from Lemmas 9.13 and 9.5 that AR_zA has boundary values (written symbolically $AR_{\lambda\pm io}A$) as $z\to\lambda\pm i0$ in Hilbert-Schmidt norm whenever $\lambda\notin\Gamma_0$. Let $G_{\lambda\pm io}(\underline{x},\underline{y}) = A(\underline{x})^{-1}\cdot(AR_{\lambda\pm io}A)(\underline{x},\underline{y})A(\underline{y})^{-1}$. Then clearly $G_{\lambda\pm io}$ are two measurable functions on $R^3\times R^3$ and we get an integral equation similar to (10.67) :

$$G_{\lambda\pm io}(\underline{x},\underline{y}) = G^o_{\lambda\pm io}(\underline{x},\underline{y}) - \int G^o_{\lambda\pm io}(\underline{x},\underline{u})V(\underline{u})G_{\lambda\pm io}(\underline{u},\underline{y})d^3u. \quad (10.68)$$

We use the notation $[AG_z^o(\cdot,\underline{y})](\underline{x}) = A(\underline{x})G_z^o(\underline{x},\underline{y})$ and $[AG_z(\cdot,\underline{y})](\underline{x}) = A(\underline{x})G_z(\underline{x},\underline{y})$. Then for fixed $\underline{y}$, $AG_z^o(\cdot,\underline{y})\in L^2(R^3)$

for all z including $z = \lambda \pm i0$, because by Lemma 9.12, $\int A^2(\underline{x})|\underline{x}-\underline{y}|^{-2}d^3x < \infty$. This along with the equation $(1+W_{\bar{z}})^* \cdot AG_z(\cdot,\underline{y}) = AG_z^o(\cdot,\underline{y})$ and the invertibility of $(I+W_{\bar{z}})$ yields the result that $AG_z(\cdot,\underline{y}) \in L^2(R^3)$ for all $z \in C-\Gamma_o$. Also $\|AG_z^o(\cdot,\underline{y}) - AG_{\lambda\pm io}^o(\cdot,\underline{y})\|_2 \to 0$ as $z \to \lambda\pm i0$, and thus $\|AG_z(\cdot,\underline{y})-AG_{\lambda\pm io}(\cdot,\underline{y})\|_2 \to 0$, once we use the continuity of $(I+W_z)^{-1}$ up to the real axis for all $\lambda\notin\Gamma_o$. These considerations combined with the inequality $|G_z(\underline{x},\underline{y}) - G_{\lambda-io}(\underline{x},\underline{y})|^2 \leq 2|G_z^o(\underline{x},\underline{y}) - G_{\lambda-io}^o(\underline{x},\underline{y})|^2 + 4\|AG_z(\cdot,\underline{y})\|_2^2 \int |T_z(\underline{x},\underline{u}) - T_k(\underline{x},\underline{u})|^2 d^3u + 4\|AG_z(\cdot,\underline{y})-AG_{\lambda-io}(\cdot,\underline{y})\|_2^2 \int |T_k(\underline{x},\underline{u})|^2 d^3u$, and estimates similar to those of Proposition 10.11 lead to the conclusion that for each $\lambda \notin \Gamma_o$, $G_z(\underline{x},\underline{y})$ converges to $G_{\lambda\pm io}(\underline{x},\underline{y})$ pointwise almost everywhere. The limit functions $G_{\lambda+io}(\underline{x},\underline{y})$ and $G_{\lambda-io}(\underline{x},\underline{y})$ are called the retarded and advanced total Green's functions respectively, in analogy with the free case. The details of the above discussion constitute the Problem 10.11.

Formally we can compute $(-\Delta_x-\lambda)G_{\lambda\pm io}^o(\underline{x},\underline{y})$, where Δ_x is the Laplacian with respect to the variable $\underline{x}$. As in Remark 9.11 we write down this equation as $(-\Delta_x-\lambda)G_{\lambda\pm io}^o(\underline{x},\underline{y}) = \delta^{(3)}(\underline{x}-\underline{y})$, where $\delta^{(3)}$ is the Dirac delta function in 3 dimensions. For a rigorous justification of this equation, one has to view $G_{\lambda\pm io}^o$ as a distribution and the above differential equation as a distributional differential equation. Proceeding formally it is easy to derive the (distribution) differential equation satisfied by $G_{\lambda\pm io}(\underline{x},\underline{y})$, namely $(-\Delta_x+V-\lambda)G_{\lambda\pm io}(\underline{x},\underline{y}) = \delta^{(3)}(\underline{x}-\underline{y})$. It is common to define the Green's function of a differential expression as the distribution solution of the above type of equations.

C. Now we prove a lemma which shows how operators like $M_\phi(\lambda)$ can be defined in an abstract way.

LEMMA 10.20 : Let H be an absolutely continuous self-adjoint operator in a separable Hilbert-space $\mathcal{H}$ with spectral transformation $\mathcal{U}$ and spectral representation $L^2(\Lambda,\mathcal{H}_0)$ where Λ is an interval, and let A be a self-adjoint H-bounded operator. Furthermore assume that the bounded operator $\overline{AE_\Delta A}$ satisfies the inequality $\|\overline{AE_\Delta A}\| \leq c(A)^2|\Delta|$ for every finite interval $\Delta \subseteq \Lambda$, where $\{E_\Delta\}$ is the spectral measure of H and $|\Delta|$ denotes the Lebesgue measure of Δ. Then there exists a $\mathcal{B}(\mathcal{H},\mathcal{H}_0)$-valued measurable function $M_A(\lambda)$ defined on Λ such that

(a) $(\mathcal{U}Af)_\lambda = M_A(\lambda)f$ for all $f\varepsilon D(H)$,

(b) $\|M_A(\lambda)\| \leq c(A)$ for all $\lambda \varepsilon \Lambda$,

(c) $M_A(\lambda)^*\varepsilon\mathcal{B}(\mathcal{H}_0,\mathcal{H})$ for all λ and $AE_\Delta f = \int_\Delta M_A(\lambda)^*(\mathcal{U}f)_\lambda d\lambda$, for all $f\varepsilon\mathcal{H}$ and all compact Δ.

We shall henceforth write f_λ for $(\mathcal{U}f)_\lambda$, where $\mathcal{U}$ is the spectral transformation of H.

Proof : Let $f\varepsilon D(A)$. Set $Af = \{g_\lambda\}$, where $g_\lambda\varepsilon\mathcal{H}_0$ for a.a.λ, and set $h(\Delta) = \int_\Delta g_\lambda d\lambda$. Since $\|g_\lambda\|_0 \varepsilon L^2(\Lambda)$, it follows that $h(\cdot)$ is an $\mathcal{H}_0$-valued set-function. Then by the Schwarz inequality,

$$\begin{aligned}\|h(\Delta)\|_0^2 &\leq |\Delta|\int_\Delta\|g_\lambda\|_0^2 d\lambda = |\Delta|\ \|E_\Delta g\|^2\\ &\leq |\Delta|\ \|\overline{AE_\Delta A}\|\ \|f\|^2 \leq c(A)^2|\Delta|^2\|f\|^2. \qquad (10.69)\end{aligned}$$

$h(\Delta)$ is an indefinite (Bochner) integral of a locally integrable $\mathcal{H}_0$-valued function g_λ and thus has a strong Radon-Nikodym derivative almost everywhere [HP, Corollary of Theorem 3.8.5] in the sense that $h'(\lambda) = \lim |\Delta'|^{-1}h(\Delta')\varepsilon\mathcal{H}_0$ exists and equals g_λ for a.a.λ when $\Delta' = [\lambda-\delta,\lambda+\delta]$ and $\delta \to +0$ (which we shall express by saying that Δ' shrinks to $\{\lambda\}$).

Choose a countable dense set $\{f_n\}$ in D(A) and define $g_{n,\lambda}$, $h_n(\Delta)$ and $h_n'(\lambda)$ as above. There exists a null set $N \subseteq \Lambda$ such that $h_n'(\lambda)$ exists for all n and $\lambda \notin N$. For any $\eta > 0$ and any $f\varepsilon D(A)$, choose f_n such that $||f-f_n|| < \eta$ and observe from (10.69) that $||h(\Delta)-h_n(\Delta)|| \leq c(A)|\Delta|\eta$. From this one may deduce that, if $\lambda\notin N$, $\lim\sup|| \, |\Delta_1|^{-1}h(\Delta_1)-|\Delta_2|^{-1}h(\Delta_2)|| \leq 2\,c(A)\eta$ as both Δ_1 and Δ_2 shrink to $\{\lambda\}$, showing that $h'(\lambda)$ exists for $\lambda\notin N$ for all $f\varepsilon D(A)$. Clearly h' is linear in f and $||h'(\lambda)|| \leq c(A)||f||$ by (10.69). For $\lambda\notin N$ define $M_A(\lambda)$ by $M_A(\lambda)f = h'(\lambda)$ and set $M_A(\lambda) = 0$ if $\lambda\varepsilon N$. Then $||M_A(\lambda)|| \leq c(A)$ and since D(A) is dense in $\mathcal{H}$, we can extend $M_A(\lambda)$ to an operator in $\mathcal{B}(\mathcal{H},\mathcal{H}_0)$ with the same norm. Using the same notation for the extended operator, we arrive at (b). The proof of the measurability of $\{M_A(\lambda)\}$ in norm is as in the discussion following (5.81).

Clearly $M_A(\lambda)^* \varepsilon \mathcal{B}(\mathcal{H}_0,\mathcal{H})$ if $\lambda\varepsilon\Lambda$, so that, for every $f = \{f_\lambda\}\varepsilon\mathcal{H}$, $M_A(\lambda)^*f_\lambda \varepsilon \mathcal{H}$ for almost all $\lambda\varepsilon\Lambda$, and $M_A(\lambda)^*f_\lambda$ is measurable in both the weak and strong sense since $\mathcal{H}$ is separable. Thus $M_A(\lambda)^*f_\lambda$ is integrable on each compact Δ since $||M_A(\lambda)^*f_\lambda|| \leq c(A)\,||f_\lambda||_0$ a.e. and $||f_\lambda||_0$ is locally integrable. Then (c) follows easily. #

For a presentation of the above lemma without the assumptions that H has constant spectral multiplicity and that Λ is an interval, see Kato [3]. We shall not apply this lemma to obtain $M_{B_2}(\lambda)$, where B_2 is the multiplication operator in $L^2(R^3)$ defined for the class V, but rather exploit the spherical symmetry of K_0 to derive this directly.

LEMMA 10.21 : Let $\phi(\underline{Q})$ be the multiplication operator in $L^2(R^3)$ by the function $\phi(\underline{x})$ satisfying $|\phi(\underline{x})| \leq \beta(1+|\underline{x}|)^{-\frac{1}{2}-\delta}$ for some $\delta > 0$, $0 < \beta < \infty$ and all $\underline{x}\varepsilon R^3$. Let K_0 be the free Hamiltonian in $L^2(R^3)$. (a) For all $\lambda\varepsilon(0,\infty)$ there exists an operator $M_\phi(\lambda)\varepsilon B_0(L^2(R^3),L^2(S^{(2)}))$ verifying (10.6). $M_\phi(\lambda)$ is a continuous function of λ in the operator norm, satisfying $\|M_\phi(\mu)-M_\phi(\lambda)\| \leq c_\Delta|\mu-\lambda|^\rho$ for all μ,λ belonging to a compact interval Δ in $(0,\infty)$, with $\rho = \delta$ if $\delta < 1$ and $\rho = 1$ if $\delta \geq 1$.

(b) $M_\phi(\lambda)$ satisfies (10.13) and Lemma 10.3(d), provided that Δ is a compact interval in $(0,\infty)$.

Proof : (a) $\phi(\underline{Q})$ is in $B(H)$ and we can write $\phi(\underline{Q}) = CD$, where C and D are multiplication operators by $C(\underline{x}) = (1+|\underline{x}|)^{-\frac{1}{2}-\delta}$ and $D(\underline{x}) = (1+|\underline{x}|)^{\frac{1}{2}+\delta}\phi(\underline{x})$ respectively. Then $\|D\| \leq \beta$ and we note that $M_\phi(\lambda)$ exists if $M_C(\lambda)$ exists and $M_\phi(\lambda)f = M_C(\lambda)Df$ for all $f\varepsilon H \equiv L^2(R^3)$.

We exploit the spherical symmetry of C and use some of the material to be presented in Chapter 11. Write $H = \oplus H_{\ell m}$, with $\ell = 0,1,2,\ldots$ and $m = 0, \pm 1, \ldots, \pm\ell$, where $H_{\ell m}$ is isomorphic to $L^2([0,\infty))$ for each ℓ and m. By introducing spherical polar coordinates in $L^2(R^3)$ (as we did in Example 5.24)), each $f\varepsilon L^2(R^3)$ may be expressed as $f = \sum_{\ell m} r^{-1} f_{\ell m}(r)\cdot Y_{\ell m}(\underline{\omega})$, where $f_{\ell m} \varepsilon L^2([0,\infty))$ and $Y_{\ell m}$ are the spherical harmonics (see Section 11-5A). We have $\|f\|_2^2 = \sum_{\ell m}\int_0^\infty |f_{\ell m}(r)|^2 dr$.

Each $H_{\ell m}$ reduces K_o, and the reduced K_o is the ordinary differential operator $-d^2/dr^2 + \ell(\ell+1)r^{-2}$ in $L^2([0,\infty))$ (an additional boundary condition is needed for $\ell = 0$, for details see Section 11-3). The reduced K_o has generalized eigenfunctions $u^o_\ell(\lambda,r) = (2/\pi\lambda)^{\frac{1}{2}} i^\ell \hat{j}_\ell(\lambda^{\frac{1}{2}} r)$ $(\lambda > 0)$, where $\hat{j}_\ell$ are the Riccati-Bessel functions (see Section 11-5A).

Set $s_{\ell m}(\lambda,r) = 2^{-\frac{1}{2}}\lambda^{\frac{1}{4}}(1+r)^{-\frac{1}{2}-\delta} u^o_\ell(\lambda,r)$, note that $s_{\ell m}(\lambda,\cdot) \in L^2([0,\infty))$ since $u^o_\ell(\lambda,r)$ is bounded, and we can compute $(Cf)_\lambda$ for any $f \in L^2(R^3)$ to find that (see Section 11-1) $(Cf)_\lambda(\underline{\omega}) = \sum_{\ell m}(s_{\ell m}(\lambda,\cdot),f_{\ell m})_+ Y_{\ell m}(\underline{\omega})$, where $(\cdot,\cdot)_+$ denotes the scalar product in $L^2([0,\infty))$. Now we define

$$M_C(\lambda) = \bigoplus_{\ell m} M_{C,\ell m}(\lambda), \tag{10.70}$$

where $M_{C,\ell m}(\lambda) \in \mathcal{B}(H_{\ell m}, H_o)$ is given by $M_{C,\ell m}(\lambda) f_{\ell m} = (s_{\ell m}(\lambda,\cdot),f_{\ell m})_+ Y_{\ell m}$. We have $\|M_{C,\ell m}(\lambda)\| = \|s_{\ell m}(\lambda,\cdot)\|_+$ and hence (see Lemma 11.8) $\|M_C(\lambda)\| = \sup_{\ell m}\|M_{C,\ell m}(\lambda)\| = \sup_{\ell m}\|s_{\ell m}(\lambda,\cdot)\|_+$. Similarly $\|M_C(\mu)-M_C(\lambda)\| \leq \sup_{\ell m}\|s_{\ell m}(\mu,\cdot)-s_{\ell m}(\lambda,\cdot)\|_+$. A detailed analysis involving Bessel functions (see Kuroda [3]) shows that for every μ,λ in a compact interval Δ of $(0,\infty)$, $\|s_{\ell m}(\mu,\cdot)\|_+ \leq c(\Delta)$ and $\|s_{\ell m}(\mu,\cdot)-s_{\ell m}(\lambda,\cdot)\|_+ \leq N(\Delta)\,|\mu-\lambda|^\rho$, where the constants $c(\Delta)$ and $N(\Delta)$ depend only on Δ and not on ℓ, m, and ρ is as indicated in the statement of the lemma. This completes the proof of (a) except for the compactness which will be established in Proposition 10.23 (see also Problem 12.4 for an alternative proof).

(b) If $g \in S(R^3)$, then g_λ is a $L^2(S^{(2)})$-valued continuous function of λ and hence $\int_\Delta M_\phi(\lambda)^* g_\lambda d\lambda$ exists as a Riemann integral. A straightforward calculation as in the proof of Lemma 10.3(a) leads to (10.13) in this case also. For the verification of the result of Lemma 10.3(d), see Problem 10.3. #

Notice that, as in Remark 10.4, $(B_2 f(\lambda))_\lambda = M_{B_2}(\lambda)f(\lambda)$ makes sense whenever $f(\cdot)$ is a $L^2(R^3)$-valued continuous function. We also remark that by virtue of Lemmas 10.1 and 10.21, one may define $M_\phi(\lambda)$ for a function ϕ verifying $\phi \varepsilon L^2_{loc}(R^3)$ and $|\phi(\underline{x})| \leq \beta(1+|\underline{x}|)^{-\frac{1}{2}-\delta}$ for $|\underline{x}| \geq R$ with $\delta > 0$. This is done by writing $\phi_1(\underline{x}) = \phi(\underline{x})\chi_R(\underline{x})$, $\phi_2(\underline{x}) = \phi(\underline{x})-\phi_1(\underline{x})$ and setting $M_\phi(\lambda) = M_{\phi_1}(\lambda) + M_{\phi_2}(\lambda)$. The next lemma gives some additional properties of $M_A(\lambda)$ and $M_B(\lambda)$ associated with the class $\mathcal{Y}$, and was used in Section 10-3 and will be invoked again in Chapter 12.

<u>LEMMA 10.22</u> : Let A,B and $W_\lambda^\pm$ be as in Lemma 9.13. Then $W_\lambda^+ - W_\lambda^- = 2\pi i\, M_B(\lambda)^* M_A(\lambda)$ for each $\lambda\varepsilon(0,\infty)$. In particular, $AR^o_{\lambda+io}A - AR^o_{\lambda-io}A = 2\pi i\, M_A(\lambda)^* M_A(\lambda)$ for all $\lambda\varepsilon[0,\infty)$, where by $AR^o_{\lambda\pm io}A$ we mean the limits of $AR^o_{\lambda\pm i\eta}A$ in Hilbert-Schmidt norm as $\eta \to +0$ (these limits exist by Lemma 9.13).

<u>Proof</u> : Let $g\varepsilon S(R^3)$. Then $Ag\varepsilon S(R^3)$ and we have from Lemma 10.3(c) that w-lim $B_1[R^o_{\lambda+i\eta} - R^o_{\lambda-i\eta}]Ag = 2\pi i\, M_{B_1}(\lambda)^*(Ag)_\lambda = 2\pi i\, M_{B_1}(\lambda)^* M_A(\lambda)g$ as $\eta \to +0$ for all $\lambda \geq 0$. On the other hand, by Lemma 9.13, $B_1 R^o_{\lambda\pm i\eta}A$ converges in Hilbert-Schmidt norm to $B_1 R^o_{\lambda\pm io}A$ as $\eta\to +0$. Thus $[B_1 R^o_{\lambda+io}A - B_1 R^o_{\lambda-io}A]g = 2\pi i\, M_{B_1}(\lambda)^* M_A(\lambda)g$ for all $\lambda \geq 0$ and all $g\varepsilon S(R^3)$, and the equality of the two operators follows from Lemma 2.27. Since $A\varepsilon L^2(R^3)$, we can replace B_1 by A and obtain that

$$AR^o_{\lambda+io}A-AR^o_{\lambda-io}A = 2\pi i\; M_A(\lambda)^*M_A(\lambda) \text{ for all } \lambda \geq 0.$$

We also know from Lemma 9.13 that $B_2R^o_{\lambda\pm i\eta}A$ converges in Hilbert-Schmidt norm to $B_2R^o_{\lambda\pm io}A$ as $\eta\to+0$ and that the limit operators are continuous in Hilbert-Schmidt norm. Therefore by an application of the Lebesgue dominated convergence theorem, one has that $\lim_{\eta\to+0}\int_\Delta d\lambda(B_2[R^o_{\lambda+i\eta}-R^o_{\lambda-i\eta}]Ag,f) = \int_\Delta d\lambda([B_2R^o_{\lambda+io}A-B_2R^o_{\lambda-io}A]g,f)$ for every compact interval $\Delta \subseteq [0,\infty)$. Since K_o has no eigenvalues, the left hand side of the above equality reduces to $2\pi i(B_2E^o_\Delta Ag,f)$ by (9.42). Since $Ag\varepsilon S(R^3)$, it follows by Lemma 10.21 that $(B_2E^o_\Delta Ag,f) = (\int_\Delta d\lambda M_{B_2}(\lambda)^*(Ag)_\lambda,f) = \int_\Delta d\lambda(M_{B_2}(\lambda)^* M_A(\lambda)g,f)$, provided that $\Delta \subseteq (0,\infty)$. Note that in the above sets of equalities Δ is an arbitrary compact interval in $(0,\infty)$ and, for fixed f and g the integrands are continuous functions. Therefore $([B_2R^o_{\lambda+io}A - B_2R^o_{\lambda-io}A]g,f) = 2\pi i\;(M_{B_2}(\lambda)^*M_A(\lambda)g,f)$ for all $\lambda\varepsilon(0,\infty)$. Since f is arbitrary and $S(R^3)$ is dense in H, we have $B_2R^o_{\lambda+io}A - B_2R^o_{\lambda-io}A = 2\pi i\; M_{B_2}(\lambda)^*M_A(\lambda)$ for all $\lambda\varepsilon(0,\infty)$. Since $W^\pm_\lambda = B_1R^o_{\lambda\pm io}A + B_2R^o_{\lambda\pm io}A$, we have the required result. #

D. By Lemma 10.21 we can actually obtain the on-shell S-operator $S(\lambda)$ for potentials V satisfying $|V(\underline{x})| \leq c(1+|x|)^{-1-\delta}$, $\delta > 0$. Write $V = CDC$, where C and D are the multiplication operators in $L^2(R^3)$ given by $C(\underline{x}) = (1+|x|)^{-\frac{1}{2}-\delta/2}$ and $D(\underline{x}) = (1+|x|)^{1+\delta}V(\underline{x})$ respectively. In order to be able to apply Lemma 9.5, we need to show that the boundary values of the analytic $B(H)$-valued function $W_z = DCR^o_zC$ exist in operator norm, since by Proposition 8.7(c) CR^o_zC is compact for $\text{Im}\, z \neq 0$. The following proposition sums up the relevant results.

PROPOSITION 10.23 : Let the potential V satisfy $|V(\underline{x})| \leq c(1+|\underline{x}|)^{-1-\delta}$ for some $\delta > 0$. Define C and D as above. Then

(a) for each $\lambda > 0$, $\{W_{\lambda\pm i\eta}\}$ converges in operator norm to a continuous $\mathcal{B}_0$-valued function $W_\lambda^\pm$ as $\eta\to+0$.

(b) Let $\Gamma_0 = \{0\}\cup\{\lambda\varepsilon\mathcal{R}-\{0\}\,|\,(I+W_\lambda^+)$ or $(I+W_\lambda^-)$ is not invertible$\}$. Then Γ_0 is a closed set of Lebesgue measure zero,

(c) The wave operators $\Omega_\pm$ exist and $\Omega_\pm\Omega_\pm^* = E_{ac}(H)$.

(d) One has $R(\lambda) = -2\pi i\, M_C(\lambda)D(1+W_\lambda^-)^{-1*}M_C(\lambda)^*$. Also $R(\lambda)\varepsilon\mathcal{B}_0(L^2(S^{(2)}))$ for all $\lambda\varepsilon(0,\infty)-\Gamma_0$ and is continuous in the operator norm.

Proof : (a) Let Δ be a compact interval in $(0,\infty)$ and Δ' be another such interval containing Δ in its interior. Let $z = \lambda \pm i\eta$ with $\lambda\varepsilon\Delta$ and $\eta > 0$. Write

$$\begin{aligned} CR_z^oC &= CR_z^oE_{\Delta'}^oC + CR_z^oE_{\mathcal{R}-\Delta'}^oC \\ &= CR_z^oE_{\Delta'}^oC + C(H_oE_{\mathcal{R}-\Delta'}^o-z)^{-1}E_{\mathcal{R}-\Delta'}^oC. \end{aligned} \tag{10.71}$$

The second term in (10.71) is analytic in operator norm in a complex domain containing a neighbourhood of Δ. Hence its boundary values as $z\to\lambda\pm io$ exist, are continuous in operator norm and identical.

By Lemma 10.21, Problem 10.2(b), relations (10.12) and (10.6) the first term in (10.71) can be expressed as

$$CR_z^oE_{\Delta'}^oC = \int_{\Delta'}(\mu-z)^{-1}M_C(\mu)^*M_C(\mu)d\mu, \tag{10.72}$$

where the integral exists in operator norm. Its boundary values as $z\to\lambda\pm io$ exist and are continuous (in fact Hölder continuous) in operator norm, since $M_C(\mu)$ is locally Hölder continuous by Lemma 10.21. This follows by adapting the proof of the theorem of Plemelj-Privalov [MU, § 16] to $\mathcal{B}(\mathcal{H})$-valued functions (see also Theorem A.3 of Thomas [3]). Hence by Proposition 2.22(d), $CR^o_{\lambda\pm io}C$ are compact, and the result of (a) follows by noting that $W_z = DCR^o_z C$.

(b) As in Lemma 9.5, $\Gamma_o \cap \Delta$ is a closed set of measure zero if Δ is any compact set in $(0,\infty)$, from which (b) follows (see Problem 9.8).

(c) The existence of $\Omega_\pm$ was established in Proposition 8.30. Asymptotic completeness follows by applying the stationary proof of Proposition 9.17. Notice that the convergence required in Lemma 9.5 has been verified only on $\mathcal{R}-\{0\}$. By the definition of Γ_o in (b), the conclusions of the Lemma 9.5 remain valid for this Γ_o.

(d) By using (a) and a reasoning identical to that in the proof of Lemma 10.22, one finds

$$2\pi i\ M_C(\lambda)^* M_C(\lambda) = CR^o_{\lambda+io}C - CR^o_{\lambda-io}C \varepsilon \mathcal{B}_o(\mathcal{H}).$$

Therefore by Problem 2.34, $M_C(\lambda)$ belongs to $\mathcal{B}_o(\mathcal{H},\mathcal{H}_o)$.

The expression for $R(\lambda)$ is obtained as in Proposition 10.12. Since $M_C(\lambda)$ is compact and since $(I+W^-_\lambda)^{-1}\varepsilon\mathcal{B}(\mathcal{H})$ for all $\lambda\varepsilon(0,\infty)-\Gamma_o$, we conclude that $R(\lambda)$ is compact. The continuity of $R(\lambda)$ follows from the continuity of each factor. #

Agmon [1] has given the following expression for the distribution kernel of $R(\lambda)$ when $|V(\underline{x})| = 0(|\underline{x}|^{-1-\delta})$, $\delta > 0$ and $V\varepsilon L^p_{loc}(\mathcal{R}^n)$ with $p = 2$ for $n \leq 3$ and $p > n/2$ for $n \geq 4$:

$$R(\lambda;\underline{\omega},\underline{\omega}') = -\pi i\lambda^{n/2-1} \int \psi^-_{k\underline{\omega}}(\underline{x})V(\underline{x})\overline{\psi^o_{k\underline{\omega}'}(\underline{x})}d^n x, \qquad (10.73)$$

where $\psi^{-}_{k\underline{\omega}}$ is one of the generalized eigenfunctions that he obtains for this class of potentials and $k = \lambda^{\frac{1}{2}}$. The integral in (10.73), though formally similar to that in (10.50) which we derived for the class $\mathcal{V}$, does not converge in any reasonable sense and has to be understood as a distribution in a suitable way.

Let B be a compact operator with $\{\mu_\ell\}$ as its singular values (see Section 2-3). Then one defines the <u>Schatten - von Neumann ideal</u> $\mathcal{B}_\eta(\mathcal{H})$ to be the subset of compact operators satisfying $\sum_\ell \mu_\ell^\eta < \infty$. Then Hilbert-Schmidt and trace class operators are in $\mathcal{B}_\eta$ with $\eta = 2$ and 1 respectively. Kuroda [4] has shown that if $(1+|x|)^\alpha V \in L^\infty(\mathcal{R}^n)$ with $\alpha > 1$, then for all $\lambda \in (0,\infty)-\Gamma_0$, $R(\lambda) \in \mathcal{B}_\eta(L^2(S^{(2)}))$, where $\eta > (n-1)(\alpha-1)^{-1}$. On the other hand Birman and Entina [1] have established that $R(\lambda) \in \mathcal{B}_1(L^2(S^{(2)}))$ for $V \in \mathcal{B}_1(\mathcal{H})$ or for V relatively trace-class (e.g. for $V \in L^1(\mathcal{R}^3) \cap L^2(\mathcal{R}^3)$).

<u>E</u>. We end this chapter by mentioning some results on the finiteness of the set of positive eigenvalues. The proof of such results is often based on the following lemma.

<u>LEMMA 10.24</u> : Assume that V can be factorized as V = AB such that $(I+|\underline{Q}|)^\delta A \in \mathcal{B}(\mathcal{H})$ for some $\delta > 0$ and B is K_0-compact. Let Δ be a compact interval in $\mathcal{R}$ and assume that

$$\|(I+|\underline{Q}|)^\delta g\| \leq c_\Delta \|g\| \tag{10.74}$$

for all eigenvectors g of H associated with eigenvalues $\lambda \in \Delta$. Then the set of eigenvalues (multiplicities counted) in Δ is finite.

The proof of this lemma is given as an exercise (Problem 10.13). The verification of (10.74) for any $\Delta = [a,b]$ with $0 < a < b < \infty$ can be done for the class $\mathcal{V}$ (with $B_2 = 0$) by the methods developed in this chapter. The same result has been obtained by Agmon [1] for the class of potentials mentioned before equation (10.73).

PROBLEMS

10.1 : Verify the statements made in Remark 10.2.

10.2 : (a) Let $M_\phi(\lambda)$ be as in Lemma 10.1 and $g \in \mathcal{S}(\mathbb{R}^3)$. Show that $M_\phi(\lambda)^*(\mathcal{U}_0 g)_\lambda$ is a strongly continuous $\mathcal{H}$-valued function of $\lambda \in [0,\infty)$, and that $M_\phi(\lambda)^*(\mathcal{U}_0 g)_\lambda$ is Riemann integrable in the sense of Proposition 4.13. (b)† For every $g \in L^2(\mathbb{R}^3)$, $M_\phi(\lambda)^*(\mathcal{U}_0 g)_\lambda$ is locally Bochner integrable (see [HP, Section 3-7]).

10.3 : Prove Lemma 10.3(d). Verify the same result if ϕ is as in Lemma 10.21 and Δ is a compact interval in $(0,\infty)$.

10.4† : Suppose that $V \in \mathcal{V}$ with $B_2 = 0$ and let $\underline{k} \in \Gamma^{(3)}$. Assume furthermore that either $\psi^{\pm}_{\underline{k}}(\underline{x})$ is bounded or $|B_1(\underline{x})| \leq c(1+|\underline{x}|)^{-3/2-\eta}$ for all $|\underline{x}| > r_0$ and some $\eta > 0$. Show that $\psi^{\pm}_{\underline{k}}(\underline{x}) = \psi^{o}_{\underline{k}}(\underline{x}) - (4\pi r)^{-1}\exp(\mp ikr)\int \exp(\pm ik\underline{\omega}_x\cdot\underline{y})V(\underline{y})\psi^{\pm}_{\underline{k}}(\underline{y})d^3y + o(r^{-1})$ as $r \to \infty$. (Hint : For $|y| < r$, one has $|\underline{x}-\underline{y}|^{-1} = r^{-1}+\eta_1$ and $\exp(ik|\underline{x}-\underline{y}|) = \exp(ikr-ik\underline{\omega}_x\cdot\underline{y}+ikr\eta_2)$ with $\eta_1 = O(|y|r^{-2})$ and $\eta_2 = O(|y|^2 r^{-2})$. Use (10.36) and divide the domain into $\{\underline{y} \mid |\underline{y}| < \sqrt{r}\}$ and its complement. For the second integral apply the Schwarz inequality and Lemma 9.13. Use that $A\psi^{\pm}_{\underline{k}} \in L^2(\mathbb{R}^3)$ if $\psi^{\pm}_{\underline{k}}$ is not assumed to be bounded. Reference : Ikebe [1, Lemma 3.2]).

10.5 : Derive (10.49) and (10.56).

10.6 : Derive the symmetry relation $\overline{\psi^{+}_{\underline{k}}} = \psi^{-}_{-\underline{k}}$ and (7.66). (See also Problem 8.8.)

10.7† : Prove the results of Proposition 10.12 for the potenials considered in Proposition 9.9.

10.8 : (a) Prove the assertions of Remark 10.18(a). (Hint : Use Lemma 10.22 and note s-lim $\eta R^o_{\lambda\pm i\eta} = 0$ as $\eta \to 0$.) (b) Verify that an eigenvector g satisfies the homogeneous version of (10.36). (Hint : $Ag = (AR^o_{\lambda\pm io}A)Bg$.)

10.9 : Verify that the spaces E and E' are Hilbert spaces. Show that both are isomorphic to $H \equiv L^2(R^3)$ and find the corresponding isomorphisms. Prove that $(g,f)_E = (A^{-1}g, A^{-1}f)_H$.

10.10†: Consider a separable potential as in Proposition 8.20 and assume in addition that $|\underline{Q}|h \varepsilon L^1(R^3)$. (a) Show that $H = K_o + \gamma F_h$ has no singularly continuous spectrum and that each eigenvalue has spectral multiplicity one. (Hint. Show that $h_\lambda = \theta$ if $\lambda \varepsilon \Gamma_o \cap (0,\infty)$. See also Problem 9.4.) (b) Let $\lambda_o \varepsilon \Gamma_o \cap (0,\infty)$. Then $g \equiv \{(\lambda-\lambda_o)^{-1}h_\lambda\}$ is a corresponding eigenvector and $d(h, R^o_{\lambda\pm io}h)/d\lambda\Big|_{\lambda=\lambda_o} = [\lim_{\delta\to+0} d(h, R^o_{\lambda\pm i\delta}h)/d\lambda]_{\lambda=\lambda_o} = \|g\|^2$. (Hint : Use the dominated convergence theorem in the $\underline{x}$- and the spectral representations.) (c) Let Δ be a compact subset of $(0,\infty)$. Show that $\Delta \cap \sigma_p(H)$ is a finite set. (Hint : Use (b) and the following fact. If $f(\lambda)$ is continuously differentiable and if λ_o is an accumulation point of the zeros of f in Δ, then λ_o is also an accumulation point of the zeros of f'.) (d) Let $\lambda_o > 0$ be an eigenvalue of H. Prove that $\lim \bar{\sigma}(\lambda)$ as $\lambda \to \lambda_o$ exists and is equal to zero.

10.11 : Verify in detail the statements made in Section 10-4$\underline{B}$.

10.12†: In Problem 9.9, assume in addition the hypotheses of Proposition 10.11 for V, and that $\Gamma_o(H_1, K_o) \cap [0,\infty) = \emptyset$. (a) Define $M^\pm_\phi(\lambda)$ by $M^\pm_\phi(\lambda)f = (U_\pm \phi(\underline{Q})f)_\lambda$, where $U_\pm$ are the spectral transformations of $H_{1,ac}$. Prove that $M^\pm_\phi(\lambda)$ have properties similar to those of $M_\phi(\lambda)$ in Lemmas 10.1 and 10.3. (b) Assume $B_2 = 0$ for V. Derive expressions for $R^\pm(H,H_1)(\lambda)$ in the spectral representations of $H_{1,ac}$ analogous to those in Proposition 10.12. (c) Show that

$$R(H,K_o)(\lambda) = R(H_1,K_o)(\lambda) + R^+(H,H_1)(\lambda)S(H_1,K_o)(\lambda). \qquad (10.75)$$

Remark : The restriction $B_2 = 0$ can be removed using Lemmas 10.20 and 10.21, and the method employed in Proposition 10.23. The usefulness of (10.75) will become apparent in Problem 12.7.

10.13 : Prove Lemma 10.24. (Hint : First show that $\|(I+|\underline{Q}|)^\delta (K_o+I)g\| \leq \alpha_\Delta \|g\|$. Then assume that the set of eigenvalues in Δ is infinite, and let $\{g_n\}$ be an infinite orthonormal sequence of associated eigenvectors. Show that $\text{w-lim}(I+|\underline{Q}|)^\delta (K_o+I)g_n = \theta$ and use Proposition 8.7(b) to arrive at a contradiction.)

CHAPTER 11 : SPHERICAL SYMMETRY IN SCATTERING THEORY

Thus far we have verified the basic requirements of scattering theory, described in Chapter 4, for potentials in class $\mathcal{V}$. In many physical situations, one need not consider the scattering problem in its full generality. This is because of various symmetries that are present in nature. One such symmetry, often called spherical symmetry, is that of rotation in the physical 3-dimensional Euclidean space. The hypothesis of spherical symmetry simplifies the mathematical structure of the theory. This is exhibited in Section 11-1 where we show that the Hilbert space can be split up into what are called the partial wave subspaces and that it suffices to solve the scattering problem in each such subspace. In Section 11-3 we prove that in each partial wave subspace the free and the total Hamiltonians are ordinary differential operators and discuss their essential self-adjointness.

Section 11-4 is devoted to a discussion of their eigenfunctions and of the phase shifts. As an application of the partial wave analysis, we study in Section 11-2 the spin-orbit interaction.

11-1 PARTIAL WAVE ANALYSIS

We begin by parametrizing the rotations in 3 dimensional Euclidean space. For this, we set $\underline{x} \equiv (x_1,x_2,x_3)$, i.e. $x_i = \underline{e}_i \cdot \underline{x}$ $(i = 1,2,3)$, where $\underline{e}_i$ are the three cartesian orthonormal basis vectors. Then the operation of rotation is a linear map : $\underline{x} \mapsto \underline{x}'$ such that $|\underline{x}| = |\underline{x}'|$. From this it easily follows that the rotations in R^3 are in one-to-one correspondence with the set of 3×3 real orthogonal matrices R. This correspondence also includes reflections about the origin, e.g. $\underline{x} \mapsto -\underline{x}$. In order to isolate the set of proper rotations (i.e. rotations without reflections) one needs to require further that the determinant of the associated 3×3 matrices be equal to +1. In the following we shall identify each proper rotation in R^3 with a 3×3 real orthogonal matrix R of determinant 1, i.e.

$$R^T R = I \text{ and } \det R = +1, \tag{11.1}$$

where we have written R^T for the transpose matrix of R. This set of matrices forms a group, i.e. it is closed under matrix multiplication and inversion, and the identity matrix is its identity element. This group is called the rotation group and is denoted by SO(3). There exist other parametrizations of the rotations, for instance in terms of the three Euler angles associated with a given direction in R^3 (see [HA]). Note, however, that the parametrization of the rotations by the orthogonal matrices of determinant 1 has the advantage of admitting a natural extension to dimensions other than 3.

Since here rotations are assumed to be symmetries, it follows from Wigner's theorem (see [P]) on symmetries in quantum mechanics that each rotation R can be represented by either a unitary or an anti-unitary operator in the Hilbert space of states. If one considers only a proper rotation given by a matrix $R \in SO(3)$, then using the connectedness of the Lie group SO(3) it can be shown that only unitary representations arise [HA]. The unitary operator representing the rotation $R \in SO(3)$ will be denoted by U(R). The group property of the rotation matrices R leads to the group property up to a factor of $\{U(R)\}$. More specifically, for R_1 and $R_2 \in SO(3)$ one has

$$U(R_2)U(R_1) = \pm U(R_2R_1). \tag{11.2}$$

This is an example of a <u>projective representation</u> in contrast to a <u>vector representation</u>, which will satisfy an equation similar to (11.2) with ± replaced by + only. The appearance of signs ± is related to the double connectedness of the group SO(3) and for details on these considerations the reader is referred to [HA]. The vector representations of SO(3) are sometimes called the single-valued representations and representations satisfying (11.2) are referred to as the double-valued representations.

The transformation $\underline{x} \mapsto R\underline{x}$ induces in a natural way a vector representation of SO(3) in the Hilbert space $L^2(\mathbb{R}^3)$. For this one simply sets for $f \in L^2(\mathbb{R}^3)$ and $R \in SO(3)$:

$$(U(R)f)(\underline{x}) = f(R^{-1}\underline{x}) \text{ a.e.} \tag{11.3}$$

Since the Lebesgue measure on $\mathcal{R}^3$ (i.e. the volume element) is rotationally invariant, it follows easily that (11.3) defines a unitary operator $U(R)$ for every $R \in SO(3)$. Furthermore one can verify that the property $(R_2R_1)^{-1} = R_1^{-1}R_2^{-1}$ for R_1, R_2 in $SO(3)$ leads to the relation

$$U(R_2)U(R_1) = U(R_2R_1), \tag{11.4}$$

which shows that (11.3) defines a vector representation in $L^2(\mathcal{R}^3)$ of the rotation group $SO(3)$. The following lemma relates this representation with the Fourier transformation $\mathcal{F}$.

<u>LEMMA 11.1</u> : Let $f \in L^2(\mathcal{R}^3)$. Then $(\mathcal{F}U(R)f)(\underline{k}) = (\mathcal{F}f)(R^{-1}\underline{k})$ for almost all $\underline{k}$ and each $R \in SO(3)$.

<u>Proof</u> : From (2.26) we know that

$$(\mathcal{F}U(R)f)(\underline{k}) = \underset{n\to\infty}{\ell.i.m}\ (2\pi)^{-3/2}\int_{|\underline{x}|\leq n} e^{-i\underline{k}\cdot\underline{x}}(U(R)f)(\underline{x})d^3x$$

$$\text{and } (\mathcal{F}f)(R^{-1}\underline{k}) = \underset{n\to\infty}{\ell.i.m}\ (2\pi)^{-3/2}\int_{|\underline{x}|\leq n} e^{-iR^{-1}\underline{k}\cdot\underline{x}}f(\underline{x})d^3x.$$

The desired result follows by making the change of variable $\underline{x} \to \underline{y} = R\underline{x}$ in the second expression and using the identity $R^{-1}\underline{k}\cdot\underline{x} = \underline{k}\cdot R\underline{x}$. #

We next give sufficient conditions on the potential V so that the scattering system determined by K_0 and $H = K_0 + V$ is invariant under the rotation group $SO(3)$ in the sense of Section 4-2. We shall call a measurable function $\phi : \mathcal{R}^3 \to \mathcal{C}$ <u>spherically symmetric</u> if its values depend only on $r \equiv |\underline{x}|$. With such a function ϕ we can associate a function $\psi : [0,\infty) \to \mathcal{C}$ by setting $\psi(r) = \phi(\underline{x})$. We shall use the same symbol for both these functions. Then we have

PROPOSITION 11.2 : Let K_0 be the free Hamiltonian in $L^2(\mathcal{R}^3)$, the potential V be spherically symmetric and let $V = V_1 + V_2$ with $V_1 \varepsilon L^2(\mathcal{R}^3)$ and $V_2 \varepsilon L^\infty(\mathcal{R}^3)$. Furthermore assume that the self-adjoint operator $H = K_0 + V$ has no singularly continuous spectrum. Then the pair (K_0, H) is invariant under the rotation group, i.e. it verifies (4.21) and (4.22), where $M_\infty(H)$ is defined by (7.3).

Proof : (i) By Lemma 11.1 we see that $\int |\underline{k}^2(FU(R)f)(\underline{k})|^2 d^3k = \int |\underline{k}^2 Ff(R^{-1}\underline{k})|^2 d^3k = \int |\underline{k}^2 (Ff)(\underline{k})|^2 d^3k$, where we have used the invariance of the Lebesgue measure under rotations and also the relation $|R\underline{k}|^2 = |\underline{k}|^2$. Recalling the definition (3.28) of $D(K_0)$ we infer that U(R) leaves $D(K_0)$ invariant, i.e. $U(R)D(K_0) \subseteq D(K_0)$ for every $R \varepsilon SO(3)$. For any $f \varepsilon D(K_0)$, we have $(FK_0U(R)f)(\underline{k}) = \underline{k}^2(FU(R)f)(\underline{k}) = \underline{k}^2(Ff)(R^{-1}\underline{k})$ and $(FU(R)K_0f)(\underline{k}) = (FK_0f)(R^{-1}\underline{k}) = |R^{-1}\underline{k}|^2(Ff)(R^{-1}\underline{k}) = \underline{k}^2(Ff)(R^{-1}\underline{k})$. This shows that $K_0U(R) = U(R)K_0$ on $D(K_0)$ or that K_0 commutes with U(R) for all $R \varepsilon SO(3)$ in the sense of (5.32).

(ii) Under the conditions imposed on V it follows from Proposition 8.7(a,b) that $H = K_0 + V$ is self-adjoint with $D(H) = D(K_0)$. Therefore from part (i) of this proof we obtain that $U(R)D(H) \subseteq D(H)$. Since V is a spherically symmetric function and since $D(H) = D(K_0) \subseteq D(V)$, we have for every $f \varepsilon D(H)$ and $R \varepsilon SO(3)$, $(VU(R)f)(\underline{x}) = V(\underline{x})(U(R)f)(\underline{x}) = V(|\underline{x}|)f(R^{-1}\underline{x})$ and $(U(R)Vf)(\underline{x}) = (Vf)(R^{-1}\underline{x}) = V(R^{-1}\underline{x})f(R^{-1}\underline{x}) = V(|R^{-1}\underline{x}|)f(R^{-1}\underline{x}) = V(|\underline{x}|)f(R^{-1}\underline{x})$. This shows that $U(R)V = VU(R)$ on the set $D(H) = D(K_0)$ and along with part (i) establishes that $U(R)H \subseteq HU(R)$ for every $R \in SO(3)$.

(iii) Thus by the spectral theorem (Proposition 5.9) $U(R)$ commutes with the spectral families $\{E_\lambda^o\}$ and $\{E_\lambda\}$ associated with K_o and H respectively. Similarly by Proposition 5.10(e) we obtain that $U(R)$ commutes with the evolution groups U_t and V_t. Finally from Proposition 7.9 we know that $E_\infty(H) = E_{ac}(H)$ and hence $E_\infty(H)$ commutes with $U(R)$. Since $E_\infty(K_o) = I$, we have verified the relations (4.21) and (4.22). #

If instead of the scattering subspace $M_\infty(H)$ we use the subspace $\bar{M}_\infty(H)$ (see Section 7-6A), then under the same hypotheses on V as in the above proposition we have that $\bar{M}_\infty(H) = H_c(H)$ and thus $\bar{M}_\infty(H)$ reduces $U(R)$ without any restriction on the spectrum of H.

COROLLARY 11.3 : Assume $V \in \mathcal{V}$ and V spherically symmetric. Then the scattering system determined by K_o and $H = K_o + V$ is invariant under SO(3). In particular for all $R \in SO(3)$

$$U(R)\Omega_\pm = \Omega_\pm U(R) \text{ and } U(R)S = SU(R). \tag{11.5}$$

Proof : The existence of $\Omega_\pm$ and the completeness of the theory was established in Proposition 9.14, and $H_{sc}(H) = \{\theta\}$ by Proposition 10.17. Hence the result follows from Propositions 11.2 and 4.9. #

In order to describe in more detail the structure of a spherically symmetric scattering system, we need to introduce the infinitesimal generators of the rotation group SO(3), called the angular momentum operators. For this we notice that the rotation group has three distinguished abelian subgroups, namely those corresponding to rotations about the three axes of the cartesian coordinate system. These are represented by the unitary operators $U_1(\phi), U_2(\phi)$ and $U_3(\phi)$ respectively where we have written just the angle of rotation instead of the matrix R associated with it. For example, the matrix associated with a rotation by the angle ϕ about the 1-axis is given by

$$R_\phi^1 \equiv \begin{pmatrix} 1 & 0 & 0 \\ 0 & \cos\phi & -\sin\phi \\ 0 & \sin\phi & \cos\phi \end{pmatrix}.$$

The strong continuity (Problem 11.1) of the representation U(R) defined by (11.3) implies the strong continuity of each of the one-parameter unitary subgroups $U_1(\phi)$, $U_2(\phi)$ and $U_3(\phi)$. Thus by Stone's theorem (Proposition 3.1) we obtain the three angular momentum operators L_1, L_2 and L_3 as the self-adjoint infinitesimal generators of these three subgroups.

We state some of the properties of the operators L_j ($j = 1,2,3$) the proofs of which are postponed until Section 11-5B. The restriction of L_j to $S(R^3)$ is a partial differential operator given by

$$L_j = -i(x_k\partial/\partial x_\ell - x_\ell\partial/\partial x_k), \tag{11.6}$$

where (j,k,ℓ) is a cyclic permutation of (1,2,3). Each L_j is essentially self-adjoint on $S(R^3)$ and leaves $S(R^3)$ invariant. The commutation relations between these operators on $S(R^3)$ are as follows :

$$[L_1, L_2] = iL_3 \text{ and its cyclic permutations.} \tag{11.7}$$

A relation such as (11.7) is said to define in $L^2(R^3)$ a

representation of the Lie algebra associated with the Lie group SO(3) (see [HA]). For our purposes we need to define on $\mathcal{S}(\mathbb{R}^3)$ three more operators :

$$\underline{L}^2 = L_1^2 + L_2^2 + L_3^2$$
$$L_\pm = L_1 \pm iL_2. \tag{11.8}$$

$\underline{L}^2$ and $L_\pm$ are densely defined and $\underline{L}^2$, called the <u>total angular momentum</u> operator, is essentially self-adjoint on $\mathcal{S}(\mathbb{R}^3)$. Using (11.7) and the definition (11.8) one easily gets the following commutation relations on $\mathcal{S}(\mathbb{R}^3)$:

$$[\underline{L}^2, L_j] = 0 \text{ for all } j \text{ and } [L_\pm, L_3] = \mp L_\pm .$$

We now make a change of variables from cartesian to spherical polar coordinates given as $x_1 = r\sin\theta\cos\phi$, $x_2 = r\sin\theta\sin\phi$, $x_3 = r\cos\theta$. This transformation induces a map V from $L^2(\mathbb{R}^3)$ onto $L^2([0,\infty), r^2dr) \otimes L^2(S^{(2)})$, namely $(Vf)(r,\theta,\phi) = f(r\sin\theta\cos\phi, r\sin\theta\sin\phi, r\cos\theta)$. In the following we shall not mention V explicitly and by an abuse of notation we shall often write f for Vf. An alternative description of this map is to say that V is the unitary mapping from $L^2(\mathbb{R}^3)$ onto $L^2([0,\infty), L^2(S^{(2)}), r^2dr)$ as in Example 5.24. It is evident that the operators U(R) act non-trivially only on $L^2(S^{(2)})$ and we can write $U(R) = I \otimes \mathcal{U}(R)$ in $L^2([0,\infty), r^2dr) \otimes L^2(S^{(2)})$ (see Section 2-4 for notations). Similarly, the operators L_1, L_2, L_3 and $\underline{L}^2$ act in a non-trivial way only in $L^2(S^{(2)})$. In particular L_j is of the form $L_j = I \otimes \mathcal{L}_j$, where $\mathcal{L}_j$ is the infinitesimal generator of the one-parameter subgroup $\{\mathcal{U}_j(\phi)\}$.

The spherical harmonics $\{Y_{\ell m}\}$ ($\ell = 0,1,2,\ldots;$

$-\ell \leq m \leq \ell$), defined in Section 11-5A, form an orthonormal basis of $L^2(S^{(2)})$. For fixed ℓ and m, we denote by $F_{\ell m}$ the projection whose range is the one-dimensional subspace of $L^2(S^{(2)})$ generated by $Y_{\ell m}$ and set

$$E_{\ell m} = I \otimes F_{\ell m}. \tag{11.9}$$

$E_{\ell m}$ is a projection in $H = L^2([0,\infty),r^2dr) \otimes L^2(S^{(2)})$ by Problem 2.32(c) and $E_{\ell m}E_{\ell' m'} = \delta_{\ell\ell'}\delta_{mm'}E_{\ell m}$ and $\sum_{\ell m} E_{\ell m} = I$ in H. For convenience of presentation we shall write H_r for $L^2([0,\infty),r^2dr)$ and H_a for $L^2(S^{(2)})$. Then $E_{\ell m}$ maps every vector f in H onto a vector of product type, i.e. $(E_{\ell m}f)(r,\underline{\omega}) = f_{\ell m}(r)Y_{\ell m}(\underline{\omega})$, where $f_{\ell m}(r) = \int f(r,\underline{\omega})\bar{Y}_{\ell m}(\underline{\omega})d\omega$. Note that by the Schwarz inequality $\|f_{\ell m}\|_r^2 \equiv \int |f_{\ell m}(r)|^2 r^2 dr \leq \int_0^\infty r^2 dr \int |f(r,\underline{\omega})|^2 d\omega \cdot \int |Y_{\ell m}(\underline{\omega})|^2 d\omega = \|f\|^2$. Furthermore, since $\sum_{\ell m} \|E_{\ell m}f\|^2 = \|f\|^2$, we infer that $\|f\|^2 = \sum_{\ell m} \|f_{\ell m}\|^2$. The range of $E_{\ell m}$, denoted $H_{\ell m}$, is then isomorphic to H_r for each ℓ and m and is called a <u>partial wave subspace</u> for angular momentum ℓ. We investigate the structure of operators commuting with $\{E_{\ell m}\}$ in the next lemma, which is then applied in the following two propositions to various operators occuring in scattering theory.

<u>LEMMA 11.4</u> : Let A be a self-adjoint (bounded and everywhere defined, isometric or unitary) operator in $H_r \otimes H_a$ and assume that A commutes with each $E_{\ell m}$. Then there exists a family $\{A_{\ell m}\}$ of self-adjoint (bounded and everywhere defined, isometric or unitary, respectively) operators in H_r such that $AE_{\ell m} = A_{\ell m} \otimes F_{\ell m}$.

<u>Proof</u> : Let $f \in D(A)$ and fix ℓ,m. Then $E_{\ell m}f = f_{\ell m} \otimes Y_{\ell m} \in D(A)$

and $AE_{\ell m}f = A(f_{\ell m} \otimes Y_{\ell m}) = E_{\ell m}Af = g_{\ell m} \otimes Y_{\ell m}$ for some $g_{\ell m} \in H_r$. The map $f_{\ell m} \mapsto g_{\ell m}$ defines a linear operator $A_{\ell m}$ in H_r with $D(A_{\ell m}) = \{f_{\ell m} \in H_r \,|\, f_{\ell m} \otimes Y_{\ell m} \in D(A)\}$. Note that $E_{\ell m}D(A) = D(A_{\ell m}) \otimes Y_{\ell m}$. If $A \in \mathcal{B}(H)$, then clearly $D(A_{\ell m}) = H_r$. If $D(A)$ is only dense in H, then $D(A_{\ell m})$ is dense in H_r. In fact, if $h \in H_r$, then there exists a sequence $\{h^{(n)}\} \in D(A)$ such that $h^{(n)}$ converges strongly to $h \otimes Y_{\ell m}$ as $n \to \infty$. Hence $\|h_{\ell m}^{(n)} - h\|_r = \|E_{\ell m}h^{(n)} - h \otimes Y_{\ell m}\| \to 0$ as $n \to \infty$.

If A is self-adjoint, then $A_{\ell m}$ is easily seen to be symmetric. Also $(A \pm i)D(A)$ is equal to H by Proposition 2.14, implying that $[(A_{\ell m} \pm i)D(A_{\ell m})] \otimes Y_{\ell m} = (A \pm i)E_{\ell m}D(A) = E_{\ell m}(A \pm i)D(A) = H_{\ell m}$. Hence $(A_{\ell m} \pm i)D(A_{\ell m}) = H_r$, i.e. $A_{\ell m}^* = A_{\ell m}$ by Proposition 2.14. If A is bounded, then $\|A_{\ell m}f_{\ell m}\|_r = \|A(f_{\ell m} \otimes Y_{\ell m})\| \leq \|A\| \, \|f_{\ell m}\|_r$, showing that $A_{\ell m}$ is also bounded. If A is isometric, then $A_{\ell m}$ is obviously isometric. If A is unitary, then $AH = H$, and we find as above $A_{\ell m}H_r = H_r$, proving the unitarity of $A_{\ell m}$.

From the first part of the proof we see that $A_{\ell m} \hat{\otimes} F_{\ell m} \subseteq AE_{\ell m}$. If $A \in \mathcal{B}(H_r \otimes H_a)$, then $A_{\ell m} \otimes F_{\ell m} = AE_{\ell m}$ by Proposition 2.6. If A is self-adjoint, then $A_{\ell m}$ is self-adjoint and hence by Problems 5.10(b) and 11.3, $A_{\ell m} \hat{\otimes} F_{\ell m}$ is essentially self-adjoint. That $AE_{\ell m}$ is self-adjoint follows from Lemma 5.4(b), so that we again get $A_{\ell m} \otimes F_{\ell m} = AE_{\ell m}$. #

<u>PROPOSITION 11.5</u> : The free Hamiltonian K_0 commutes with $E_{\ell m}$ for all ℓ,m. For each ℓ, there exists a self-adjoint operator $K_{0,\ell}$ in H_r such that $K_0E_{\ell m} = K_{0,\ell} \otimes F_{\ell m}$. $K_{0,\ell}$ has simple spectrum with spectral representation space $L^2[0,\infty)$ and its

spectral transformation is given by $(U_{o,\ell}f)_\lambda = 2^{-\frac{1}{2}}\lambda^{\frac{1}{4}}(F_\ell f)(k)$ with

$$(F_\ell f)(k) = (2/\pi)^{\frac{1}{2}}(-i)^\ell \ \ell.i.m \int_0^\infty (kr)^{-1}\hat{j}_\ell(kr)f(r)r^2dr, \quad (11.10)$$

where $f\varepsilon H_r$, $k = \lambda^{\frac{1}{2}}$ and $\hat{j}_\ell$ is the Riccati-Bessel function defined in Section 11-5A.

Proof : (i) Let $f\varepsilon C_0^\infty((0,\infty))$. Since $f\otimes Y_{\ell m} \varepsilon L^1(R^3)$, we can compute its Fourier transform pointwise and write

$$(F(f\otimes Y_{\ell m}))(\underline{k}) = (2\pi)^{-3/2} \int_0^\infty f(r)r^2dr \int e^{-i\underline{k}\cdot\underline{x}} Y_{\ell m}(\underline{\omega})d\omega.$$

Inserting the expansion (11.65) of the plane wave and using the orthogonality relation (11.53) for the functions $Y_{\ell m}$, this becomes

$$(F(f\otimes Y_{\ell m}))(\underline{k}) = (2/\pi)^{\frac{1}{2}}(-i)^\ell Y_{\ell m}(\underline{\omega}_k)\int_0^\infty (kr)^{-1}\hat{j}_\ell(kr)f(r)r^2dr. \quad (11.11)$$

Thus $(F(f\otimes Y_{\ell m}))(\underline{k})$ is of the form $g(k)Y_{\ell m}(\underline{\omega}_k)$. Since $F(f\otimes Y_{\ell m}) \varepsilon L^2(R^3)$, we must have $g\varepsilon L^2([0,\infty),k^2dk)$ and in fact, $g = F_\ell f$ with $\int_0^\infty |g(k)|^2k^2dk = \|f\|_r^2$. If we write the space of Fourier transforms of functions in $L^2(R^3)$ as $L^2([0,\infty),k^2dk) \otimes L^2(S^{(2)})$ and define a projection $\tilde{E}_{\ell m}$ in this space as in (11.9), then the above shows that F maps a dense subset of $H_{\ell m}$ into the range of $\tilde{E}_{\ell m}$. Since F is bounded and defined everywhere, this implies that $FH_{\ell m} \subseteq$ range of $\tilde{E}_{\ell m}$ and hence $\tilde{E}_{\ell m}FE_{\ell m} = FE_{\ell m}$. An identical consideration for F^{-1} leads to $E_{\ell m}F^{-1}\tilde{E}_{\ell m} = F^{-1}\tilde{E}_{\ell m}$. The above two relations combined with the fact $F^{-1} = F^*$ imply that $FE_{\ell m} = \tilde{E}_{\ell m}F$.

Let $\tilde{K}_0$ be the multiplication operator by $\underline{k}^2$ on the Fourier transforms of functions in $L^2(R^3)$. Then $\tilde{K}_0F = FK_0$

and $\tilde{K}_o\tilde{E}_{\ell m} = \tilde{E}_{\ell m}\tilde{K}_o$, so that $FK_oE_{\ell m} = \tilde{K}_oFE_{\ell m} = \tilde{E}_{\ell m}\tilde{K}_oF = FE_{\ell m}K_o$. Thus K_o commutes with $E_{\ell m}$, and by Lemma 11.4 there exists for each ℓ,m a self-adjoint operator $K_{o,\ell m}$ in H_r such that $K_oE_{\ell m} = K_{o,\ell m} \otimes F_{\ell m}$.

(ii) We have seen that F_ℓ defines an isometric operator on a dense subset of H_r. This operator can then be extended to all of H_r as in Sections 2-2 and 2-5, the extension being given by (11.10). We leave it as an exercise to check that F_ℓ is unitary and that the extension verifies

$$(F(f \otimes Y_{\ell m}))(\underline{k}) = (F_\ell f)(k)Y_{\ell m}(\underline{\omega}_k) \tag{11.12}$$

for all $f \varepsilon H_r$ (Problem 11.2). Now let $f \varepsilon D(K_{o,\ell m})$. Then by Lemma 11.4, $f \otimes Y_{\ell m} \varepsilon D(K_o)$ and using (11.12) we have $(F_\ell K_{o,\ell m}f)(k)Y_{\ell m}(\underline{\omega}_k) = [F(K_{o,\ell m}f \otimes Y_{\ell m})](\underline{k}) =$ $[FK_o(f \otimes Y_{\ell m})](\underline{k}) = k^2[F(f \otimes Y_{\ell m})](\underline{k}) = k^2(F_\ell f)(k)Y_{\ell m}(\underline{\omega}_k)$. Thus $(U_{o,\ell}K_{o,\ell m}f)_\lambda = \lambda(U_{o,\ell}f)_\lambda$. It is easy to see that each $U_{o,\ell}$ is unitary from H_r to $L^2[0,\infty)$, which shows that $U_{o,\ell}$ is the spectral transformation of $K_{o,\ell m}$. Since the spectral transformation does not depend on m, it follows that $K_{o,\ell m}$ is independent of m. From this spectral representation it is obvious that $K_{o,\ell}$ has simple spectrum. #

<u>PROPOSITION 11.6</u> : Let V be a spherically symmetric potential such that $H = K_o + V$ is self-adjoint with $D(H) = D(K_o)$ and assume (A1) and (A3). Then (a) for each $\ell = 0,1,2,\ldots$ there exists a self-adjoint operator H_ℓ and a unitary operator S_ℓ in H_r such that

$$E_{\ell m}H = HE_{\ell m} = H_\ell \otimes F_{\ell m} \text{ and } SE_{\ell m} = S_\ell \otimes F_{\ell m}. \tag{11.13}$$

(b) S_ℓ is diagonalizable in the spectral representation of

$K_{0,\ell}$ and is given by

$$S_\ell = \{S_\ell(\lambda)\} = \{\exp[2i\delta_\ell(\lambda)]\}, \tag{11.14}$$

where $\delta_\ell(\lambda)$ is a real-valued measurable function defined on $[0,\infty)$. $\delta_\ell(\lambda)$ is determined up to the addition of an integral multiple of π.

Proof : (a) Since $D(H) = D(K_0)$, $E_{\ell m}D(H) \subseteq D(H)$ by Proposition 11.5. Let $f \in D(K_0) \subseteq D(V)$. Then $(VE_{\ell m}f)(\underline{x}) = V(r)(E_{\ell m}f)(\underline{x}) = V(r)f_{\ell m}(r)Y_{\ell m}(\underline{\omega}_x) = (E_{\ell m}Vf)(\underline{x})$. This along with Proposition 11.5 proves that H commutes with each $E_{\ell m}$. Thus by Lemma 11.4 there exist self-adjoint operators $H_{\ell m}$ in H_r such that $HE_{\ell m} = H_{\ell m} \otimes F_{\ell m}$ and $H_{\ell m} = K_{0,\ell} + V$, where V is the multiplication operator in H_r by V(r). Hence $H_{\ell m}$ is independent of m.

The operators K_0 and H commute with $E_{\ell m}$ and by Proposition 5.10(e) we have that the unitary groups U_t and V_t commute with $E_{\ell m}$. Thus $\Omega_\pm$ and S commute with $E_{\ell m}$. By Lemma 11.4 and Propositions 4.8 and 11.5 there exist unitary operators $U_{t,\ell}$ $[\equiv\exp(-iK_{0,\ell}t)]$, $V_{t,\ell}[\equiv\exp(-iH_\ell t)]$, $S_{\ell m}$ and isometries $\Omega_{\pm,\ell m}$ such that $U_tE_{\ell m} = U_{t,\ell} \otimes F_{\ell m}$, $V_tE_{\ell m} = V_{t,\ell} \otimes F_{\ell m}$, $\Omega_\pm E_{\ell m} = \Omega_{\pm,\ell m} \otimes F_{\ell m}$ and $SE_{\ell m} = S_{\ell m} \otimes F_{\ell m}$. On the other hand by (A1), $\Omega_\pm E_{\ell m} = (\text{s-lim}\, V^*_{t,\ell}U_{t,\ell}) \otimes F_{\ell m}$ as $t \to \pm\infty$, so that $\Omega_{\pm,\ell m} = \text{s-lim}\, V^*_{t,\ell}U_{t,\ell}$ as $t \to \pm\infty$. This implies that $\Omega_{\pm,\ell m}$ and hence $S_{\ell m} \equiv \Omega^*_{+,\ell m}\Omega_{-,\ell m}$ are independent of m.

(b) By (4.17), S commutes with U_t and therefore we have $U_{t,\ell}S_\ell \otimes F_{\ell m} = (U_{t,\ell} \otimes F_{\ell m})(S_\ell \otimes F_{\ell m}) = U_tE_{\ell m}SE_{\ell m} = U_tSE_{\ell m} =$

$SU_tE_{\ell m} = S_\ell U_{t,\ell} \otimes F_{\ell m}$, which shows that $[S_\ell, U_{t,\ell}] = 0$ for each ℓ. Hence as in (4.18), S_ℓ commutes with $K_{0,\ell}$, and since the latter has simple spectrum, S_ℓ is diagonalizable in its spectral representation, i.e. $S_\ell = \{S_\ell(\lambda)\}$, where each $S_\ell(\lambda)$ is a complex number. By Corollary 5.28, $|S_\ell(\lambda)| = 1$ a.e., so that one may define $\delta_\ell(\lambda) = (2i)^{-1}\log S_\ell(\lambda) + n\pi$, where log denotes the principal branch of the logarithm. #

The function δ_ℓ is called the partial wave phase shift (or simply phase shift) for the angular momentum ℓ and is widely used in the analysis of scattering processes. It plays a role similar to the phase shifts occuring in the description of scattering of waves from obstacles in classical physics and will be discussed further in Section 11-4. In the following proposition we express the scattering amplitude in terms of the phase shifts.

PROPOSITION 11.7 : Assume the hypotheses of Proposition 11.6. Then (a) the S-matrix is given by the strongly convergent series in $\mathcal{H}_0$

$$R(\lambda) \equiv S(\lambda) - I_0 = \sum_{\ell m}(e^{2i\delta_\ell(\lambda)} - 1)\tilde{F}_{\ell m} \equiv \sum_{\ell m} R_\ell(\lambda)\tilde{F}_{\ell m}, \quad (11.15)$$

where $\tilde{F}_{\ell m}$ is the projection such that $(\tilde{F}_{\ell m} g)(\underline{\omega}_k) = (Y_{\ell m}, g)_0 \, Y_{\ell m}(\underline{\omega}_k)$. $R(\lambda)$ is compact if and only if $\sin\delta_\ell(\lambda) \to 0$ as $\ell \to \infty$, and it is Hilbert-Schmidt if and only if $\sum_\ell (2\ell+1)\sin^2\delta_\ell(\lambda) < \infty$. (b) The scattering amplitude f has the formal expression :

$$f(\lambda; \underline{\omega}' \to \underline{\omega}) = (2i\lambda^{\frac{1}{2}})^{-1} \sum_\ell (2\ell+1) R_\ell(\lambda) P_\ell(\underline{\omega}\cdot\underline{\omega}'), \quad (11.16)$$

where P_ℓ is the Legendre polynomial of degree ℓ. The series converges uniformly in $\underline{\omega}$, $\underline{\omega}'$ if $\sum_\ell (2\ell+1)|\sin\delta_\ell(\lambda)| < \infty$ and in such a case f is a continuous function of $\underline{\omega}$ and $\underline{\omega}'$. The total cross-section is independent of the initial direction $\underline{\omega}'$ and is given by (for $\lambda > 0$)

$$\sigma_{tot}(\lambda) = 4\pi\lambda^{-1}\sum_{\ell}(2\ell+1)\sin^2\delta_{\ell}(\lambda). \tag{11.17}$$

<u>Proof</u> : (a) Since $|\exp(2i\delta_{\ell}(\lambda))-1| \leq 2$, $\|\sum_{\ell=L}^{N}\sum_{m=-\ell}^{\ell}R_{\ell}(\lambda)\tilde{F}_{\ell m}g\|^2 \leq 4\sum_{\ell=L}^{N}\sum_{m=-\ell}^{\ell}\|\tilde{F}_{\ell m}g\|^2 \to 0$ as $L,N \to \infty$ for every $g\varepsilon\mathcal{H}_0$, showing the strong convergence of the series in (11.15). Clearly $\|\sum_{\ell m}R_{\ell}(\lambda)\tilde{F}_{\ell m}\| \leq 2$. Since $(\sum_{\ell' m'}R_{\ell'}(\lambda)\tilde{F}_{\ell' m'})Y_{\ell m} = R_{\ell}(\lambda)Y_{\ell m}$, it suffices by Lemma 2.27 to show that for each ℓ,m

$$R_{\ell}(\lambda)Y_{\ell m} = R(\lambda)Y_{\ell m}. \tag{11.18}$$

For this we use (11.12), (11.13) and Proposition 11.5 to obtain

$(\mathcal{U}_{0,\ell}g)_{\lambda}(R(\lambda)Y_{\ell m})(\underline{\omega}_k) = [R(\lambda)(\mathcal{U}_{0,\ell}g)_{\lambda}Y_{\ell m}](\underline{\omega}_k) =$

$[R(\lambda)(\mathcal{U}_0(g\otimes Y_{\ell m}))_{\lambda}](\underline{\omega}_k) = [\mathcal{U}_0R(g\otimes Y_{\ell m})]_{\lambda}(\underline{\omega}_k) =$

$[\mathcal{U}_0(R_{\ell}g\otimes Y_{\ell m})]_{\lambda}(\underline{\omega}_k) = (\mathcal{U}_{0,\ell}R_{\ell}g)_{\lambda}Y_{\ell m}(\underline{\omega}_k) = R_{\ell}(\lambda)(\mathcal{U}_{0,\ell}g)_{\lambda}Y_{\ell m}(\underline{\omega}_k)$.

By choosing $g\varepsilon\mathcal{H}_r$ such that $(\mathcal{U}_{0,\ell}g)_{\lambda} \neq 0$ for all $\lambda > 0$, we arrive at (11.18) for almost all λ.

From (11.15) it is evident that $R(\lambda)$ has eigenvalues $\{R_{\ell}(\lambda)\}$, each being $(2\ell+1)$-fold degenerate. If $R(\lambda)$ is compact, then by Proposition 2.25 it follows that $|\sin\delta_{\ell}(\lambda)| = \frac{1}{2}|R_{\ell}(\lambda)| \to 0$ as $\ell \to \infty$. Conversely, if $|R_{\ell}(\lambda)| \to 0$ as $\ell \to \infty$, then $\|R(\lambda)g-\sum_{\ell=0}^{L}\sum_{m=-\ell}^{\ell}R_{\ell}(\lambda)\tilde{F}_{\ell m}g\|^2 = \sum_{\ell=L+1}^{\infty}\sum_{m=-\ell}^{\ell}|R_{\ell}(\lambda)|^2\|\tilde{F}_{\ell m}g\|^2 \leq \sup_{\ell>L}|R_{\ell}(\lambda)|^2\|g\|^2 \to 0$, establishing the norm convergence of the series to $R(\lambda)$ and hence the compactness of $R(\lambda)$. By using (2.68) one verifies similarly that $R(\lambda)$ is Hilbert-Schmidt if and only if $\sum_{\ell}(2\ell+1)\sin^2\delta_{\ell}(\lambda) < \infty$.

(b) Since $\tilde{F}_{\ell m}$ is an integral operator in H_0 with kernel $Y_{\ell m}(\underline{\omega})\bar{Y}_{\ell m}(\underline{\omega}')$, we have by the definition (7.48) that formally $f(\lambda;\underline{\omega}'\to\underline{\omega}) = -2\pi i\lambda^{-\frac{1}{2}}\sum_{\ell m} R_\ell(\lambda) Y_{\ell m}(\underline{\omega})\bar{Y}_{\ell m}(\underline{\omega}')$. Using the addition theorem (11.54) for the spherical harmonics, this formal sum is reduced to the one in (11.16). Now P_ℓ is continuous and bounded uniformly with respect to ℓ (see (11.55)). Thus if $\sum(2\ell+1)|\sin\delta_\ell(\lambda)| < \infty$, then it follows that the series (11.16) converges uniformly in $\underline{\omega}$ and $\underline{\omega}'$ and in such a case the scattering amplitude is continuous in both variables.

In general the series (11.16) need not converge pointwise. However if $\sum_\ell (2\ell+1)\sin^2\delta_\ell(\lambda) < \infty$, then, setting $f_L(\lambda;\underline{\omega}'\to\underline{\omega}) = (2i\lambda^{\frac{1}{2}})^{-1}\sum_{\ell=0}^{L}(2\ell+1)R_\ell(\lambda)P_\ell(\underline{\omega}\cdot\underline{\omega}')$ and using (11.56) we see that for $N > L$, $\int|f_N(\lambda;\underline{\omega}'\to\underline{\omega}) - f_L(\lambda;\underline{\omega}'\to\underline{\omega})|^2 d\omega = 4\pi\lambda^{-1}\sum_{\ell=L}^{N}(2\ell+1)\sin^2\delta_\ell(\lambda) \to 0$ as $N,L\to\infty$. This proves that in this case the series (11.16) converges in the $L^2_{\underline{\omega}}(S^{(2)})$ sense and that $\sigma_{tot}(\lambda,\underline{\omega}') \equiv \int|f(\lambda;\underline{\omega}'\to\underline{\omega})|^2 d\omega = 4\pi\lambda^{-1}\sum_\ell(2\ell+1)\sin^2\delta_\ell(\lambda)$, which is independent of $\underline{\omega}'$. #

In Section 10-2 we have shown that if $V\in Y$, then $R(\lambda)$ is Hilbert-Schmidt. If furthermore V is spherically symmetric, then we conclude from Proposition 11.7(a) that the series $\sum_\ell(2\ell+1)\sin^2\delta_\ell(\lambda) < \infty$ for almost all $\lambda\in(0,\infty)$. Therefore, we have that $|\sin\delta_\ell(\lambda)| = o(\ell^{-\frac{1}{2}})$ as $\ell\to\infty$ for fixed λ. For a potential V satisfying $|V(\underline{x})| \le c(1+|\underline{x}|)^{-1-\eta}$ for some $\eta > 0$, we know from Proposition 10.23 that $R(\lambda)$ is compact and thus if V is also spherically symmetric, then $\sin\delta_\ell(\lambda)\to 0$ as $\ell\to\infty$ for fixed λ.

If we put $\underline{\omega}_k = \underline{\omega}' = \underline{e}_3$, i.e. if we choose the initial direction as the third coordinate axis, then the expression (11.16) takes the simpler form $f(\lambda,\theta) = (2i\lambda^{\frac{1}{2}})^{-1}\sum_\ell(2\ell+1)\cdot R_\ell(\lambda)P_\ell(\cos\theta)$. The polar angle θ is then the angle between the initial and final directions and is called the scattering angle. The series (11.16) is often referred to as the partial wave expansion of the scattering amplitude. From (7.50) we see that in general the differential scattering cross section $d\sigma/d\omega$ has all the partial waves interfering. However in the expression (11.17) of $\sigma_{tot}(\lambda)$, this interference disappears because of the orthogonality of the spherical harmonics. The partial wave expansion (11.16) is useful in applications in those cases where it suffices to keep only the first few terms of the series, which is often the case at low energies.

11-2 SPIN-ORBIT INTERACTIONS

To obtain the scattering theory for a physical system which is invariant under the rotation group, it suffices to solve the self-adjointness and the scattering problem in each partial wave subspace. This will be illustrated in the present section for the spin-orbit coupling.

We begin by expressing the properties of the operators $E_{\ell m}$ in the language of direct sums of Hilbert spaces. As was mentioned in Section 2-4, H may be viewed as the direct sum of a subspace M and its orthogonal complement. By a natural extension of this idea we can consider $L^2(R^3)$ as the direct sum of its subspaces $\{\mathcal{H}_{\ell m}\}$, i.e. $L^2(R^3) = \oplus\, \mathcal{H}_{\ell m}$, where each $\mathcal{H}_{\ell m}$ is isomorphic to $\mathcal{H}_{\pi}$. The next Lemma shows that in order to define the total Hamiltonian H, it is sufficient to obtain a self-adjoint sum of the free Hamiltonian and the interaction in each partial wave subspace.

<u>LEMMA 11.8</u> : (a) Let A_i be self-adjoint in $\mathcal{H}_i$ $(i = 1,2,\ldots)$. Then there exists a unique self-adjoint operator $A \equiv \oplus A_i$ in $\mathcal{H} = \oplus \mathcal{H}_i$ such that each $\mathcal{H}_i$ reduces A and the part of A in $\mathcal{H}_i$ is equal to A_i. Its domain consists of vectors $f = \{f_i\}$ with $f_i \varepsilon D(A_i)$ satisfying $\sum_i \|A_i f_i\|^2 < \infty$, and for such f, $Af = \{A_i f_i\}$. (b) Let $B = \oplus B_i$. Then $B\varepsilon\mathcal{B}(\mathcal{H})$ if and only if each $B_i \varepsilon \mathcal{B}(\mathcal{H}_i)$ and $\sup \|B_i\| < \infty$, in which case $\|B\| = \sup \|B_i\|$.

<u>Proof</u> : The self-adjointness of A follows easily from the definitions (Problem 2.32(a)). To prove its uniqueness, let A' be a self-adjoint operator which is reduced by each $\mathcal{H}_i$ to A_i. For $f = \{f_i\}\varepsilon D(A')$ we have that $(A'f)_i = A_i f_i$, so that $f_i \varepsilon D(A_i)$ and $\sum_i \|A_i f_i\|^2 = \|A'f\|^2 < \infty$. Thus $f\varepsilon D(A)$ and we have $A' \subseteq A$. Since A and A' are self-adjoint, it follows that $A = A'$.

Let $B\varepsilon\mathcal{B}(\mathcal{H})$. Then for $f = (\theta,\ldots,f_j,\theta\ldots)\varepsilon\mathcal{H}$ we have that $\|B_j f_j\| = \|Bf\| \leq \|B\|\,\|f\| = \|B\|\,\|f_j\|$ and hence $\sup \|B_j\| \leq \|B\|$. Conversely, $\|Bg\|^2 = \sum_i \|B_i g_i\|^2 \leq (\sup \|B_i\|)^2 \|g\|^2$ for all $g = \{g_i\}\varepsilon\mathcal{H}$ and hence $\|B\| \leq \sup \|B_i\|$. #

We now apply this lemma to the spin-orbit coupling. Before we proceed to the main problem we say a few words about spin. In Example 3.5 we have described the Hilbert space of states of a non-relativistic particle with spin s (s = 0,1/2,1,3/2,...), namely $H = L^2(R^3, C^{2s+1})$. The (2s+1)-dimensional complex space carries an irreducible projective representation of the rotation group, which is single-valued if s is zero or integral and is double-valued if s is half-integral. The action of the rotation group in H is given as follows : $(U(R)f)_m(\underline{x}) = \sum_{m'} D_{mm'}(R) f_{m'}(R^{-1}\underline{x})$, where $D^{(s)}(R)$ is the (2s+1)×(2s+1) matrix (irreducibly) representing the rotation R (see [HA]). Another way of representing the same Hilbert space which is more convenient for our purposes is to write $H = L^2(R^3) \otimes C^{2s+1}$. Transforming from cartesian to spherical polar coordinates and identifying the functions as we have done in the last section, we see that $H = H_r \otimes H_a \otimes \otimes C^{2s+1}$. Then we can write

$$H = \bigoplus_{\ell=0}^{\infty} H_\ell , \tag{11.19}$$

where $H_\ell = H_r \otimes K_\ell \otimes C^{2s+1}$, with K_ℓ the (2ℓ+1)-dimensional Hilbert space generated by the basis $\{Y_{\ell m}\}_{m=-\ell}^{m=\ell}$.

To define the free Hamiltonian, which is formally $\underline{P}^2$, we introduce the operator $H_{0,\ell} = K_{0,\ell} \otimes I \otimes I$ in H_ℓ. By Proposition 11.5 and Problem 2.39, $H_{0,\ell}$ is self-adjoint. Now we define the free Hamiltonian H_0 to be $H_0 = \oplus_\ell H_{0,\ell}$, which is self-adjoint by Lemma 11.8.

We can think of a hydrogen-like atom as made up of an electron moving around a heavy nuclear core and thus giving rise to a current $e\underline{v}$, where e is the charge of an electron and $\underline{v}$ its velocity. By the laws of classical electrodynamics, this leads to a magnetic field $\underline{B}$ proportional to the angular momentum $\underline{L}$ of the electron. Since an electron is experimentally known to have an intrinsic magnetic moment $\underline{\mu}$ proportional to its spin $\underline{S}$ (with $s = \frac{1}{2}$), we have an additional contribution to the energy of the electron by a term $-\underline{B}\cdot\underline{\mu}$, which is proportional to $\underline{L}\cdot\underline{S}$ (see also Example 3.7). This term signifies an interaction between the orbital motion of the electron and its intrinsic spin and has the name spin-orbit coupling. From a theoretical point of view a more satisfactory way of arriving at this term is to consider the relativistic Dirac equation for an electron in an electromagnetic field and its non-relativistic limit. In atomic physics, such a term leads to a small correction in the energy levels. Similar interactions are also of importance in nuclear physics. One can also consider more complicated spin-orbit interactions, like $(\underline{L}\cdot\underline{S})^2$ (see for example [LL, Section 67]). Here we deal with only the simplest one. In such a case, the total Hamiltonian is written formally as

$$H = \underline{P}^2 + V_0(r) + V_1(r)\underline{L}\cdot\underline{S}. \tag{11.20}$$

Our first object would be to define H as a self-adjoint operator. Since we have already considered the term V_0 in Section 8-1, we omit here the term V_0. Let $V_1 : [0,\infty) \to \mathcal{R}$ be a measurable function finite almost everywhere. Then the associated maximal multiplication operator V_1 in $\mathcal{H}_r$ is self-adjoint by Proposition 2.16. Setting

$$H_{I,\ell} = \sum_{j=1}^{3} V_1 \otimes L_j \otimes S_j \text{ in } \mathcal{H}_\ell, \tag{11.21}$$

it can be seen that $H_{I,\ell}$ is self-adjoint in $\mathcal{H}_\ell$ since L_j and S_j are bounded self-adjoint operators in $\mathcal{K}_\ell$ and $\mathcal{C}^{2s+1}$ respectively (see Problems 3.4 and 11.3). Write $H_I = \oplus H_{I,\ell}$ and observe that by Lemma 11.8, H_I is self-adjoint in $\mathcal{H}$ and represents the third term in (11.20). The next proposition allows us to define the sum in (11.20) as a self-adjoint operator.

PROPOSITION 11.9 : Let $\mathcal{H} = \oplus \mathcal{H}_\ell$ as in (11.19) and let $H_0 = \oplus H_{0,\ell}$. Suppose that the interaction term has the form $H_I = \oplus H_{I,\ell}$, where $H_{I,\ell}$ is given by (11.21), and that V_1 defines a K_0-compact operator in $L^2(\mathcal{R}^3)$. Then $H_\ell = H_{0,\ell} + H_{I,\ell}$ is self-adjoint in $\mathcal{H}_\ell$, and the self-adjoint operator $H = \oplus H_\ell$ defines the total Hamiltonian given by (11.20) (with $V_0 = 0$).

Proof : Since $D(K_0) \subseteq D(V_1)$ in $L^2(\mathcal{R}^3)$ and since V_1 is spherically symmetric, it follows from Lemma 11.4 that $D(K_{0,\ell}) \subseteq D(V_1)$ in $\mathcal{H}_r$. The operators L_j, S_j are bounded in $\mathcal{K}_\ell$ and $\mathbb{C}^{2s+1}$ respectively, so that $D(H_{0,\ell}) = D(K_{0,\ell}) \hat{\otimes} \mathcal{K}_\ell \otimes \mathbb{C}^{2s+1} \subseteq D(V_1) \hat{\otimes} \mathcal{K}_\ell \otimes \mathbb{C}^{2s+1} = D(H_{I,\ell})$ in $\mathcal{H}_\ell$ (see Problem 2.39). To prove that $H_\ell = H_{0,\ell} + H_{I,\ell}$ is self-adjoint, it suffices to show that given any $\eta > 0$, we can find an integer n such that $\|H_{I,\ell}(H_{0,\ell}+in)^{-1}\| < \eta$ and then apply Propositions 8.2 and 8.5. Looking upon V_1 alternatively as an operator in $\mathcal{H}_r$ or in $L^2(\mathcal{R}^3)$, we have for $f \in \mathcal{H}_r$, $[V_1(K_{0,\ell}+in)^{-1}f] \otimes Y_{\ell m} = V_1(K_0+in)^{-1}(f \otimes Y_{\ell m})$, so that $\|V_1(K_{0,\ell}+in)^{-1}\|_r \leq \|V_1(K_0+in)^{-1}\|$. We also observe that $H_{I,\ell}(K_{0,\ell}+in)^{-1} = \sum_j (V_1 \otimes L_j \otimes S_j) \cdot [(K_{0,\ell}+in)^{-1} \otimes I \otimes I] = \sum_j V_1(K_{0,\ell}+in)^{-1} \otimes L_j \otimes S_j$ in $\mathcal{H}_\ell$ (the last equality follows easily from the finite-dimensionality of $\mathcal{K}_\ell \otimes \mathbb{C}^{2s+1}$). Now $\|L_j\|_{\mathcal{K}_\ell} = \ell$ and $\|S_j\| = s$, so that by Problem 2.35,

$$\|H_{I,\ell}(H_{0,\ell}+in)^{-1}\|_{\mathcal{H}_\ell} \leq 3\ell s \, \|V_1(K_{0,\ell}+in)^{-1}\|_r$$

$$\leq 3\ell s \, \|V_1(K_0+in)^{-1}\|_{L^2(\mathcal{R}^3)}. \tag{11.22}$$

Since V_1 is K_0-compact in $L^2(R^3)$, by Proposition 8.4 there exists for fixed ℓ and s an integer n such that $\|V_1(K_0+in)^{-1}\| < (3\ell s)^{-1}$. Thus from (11.22) we conclude that $H_{I,\ell}$ is $H_{0,\ell}$-bounded with $H_{0,\ell}$-bound less than 1, and by Proposition 8.5, $H_\ell = H_{0,\ell}+H_{I,\ell}$ is self-adjoint in $\mathcal{H}_\ell$. Finally Lemma 11.8 defines $H = \oplus H_\ell$ as the unique self-adjoint operator in $\mathcal{H}$ associated with the family $\{H_\ell\}$. #

The self-adjoint operator H constructed in the above proposition gives us the total Hamiltonian H for the spin-orbit coupling corresponding to the formal expression (11.20) with $V_0 = 0$. Note that, even when V_1 is K_0-compact in $L^2(R^3)$, it does not necessarily follow that H_I is H_0-compact in $\mathcal{H}$. In order to study the scattering theory for the pair (H,H_0) in $\mathcal{H}$, we set $\Omega(t) = \exp(iHt)\exp(-iH_0 t)$ in $\mathcal{H}$ and $\Omega_\ell(t) = \exp(iH_\ell t)\exp(-iH_{0,\ell}t)$ in $\mathcal{H}_\ell$ and note that $\Omega(t) = \oplus_\ell \Omega_\ell(t)$.

PROPOSITION 11.10 : Let H_0 and H_I be as in Proposition 11.9 with V_1 a spherically symmetric function belonging to $L^2(R^3)$. Then $H = H_0 + H_I$ is self-adjoint and the wave operators exist.

Proof : (i) Since $V_1 \varepsilon L^2(R^3)$, it follows from Proposition 8.7(a) that V_1 is K_0-compact in $L^2(R^3)$ and thus by Proposition 11.9, $H = H_0 + H_I = \oplus H_\ell$ is self-adjoint.

(ii) To prove the existence of $\Omega_\pm$, it suffices to establish the existence of $\Omega_{\pm,\ell} \equiv$ s-lim $\Omega_\ell(t)$ as $t \to \pm\infty$ for each ℓ. In fact, if s-lim $\Omega_\ell(t)$ exists as $t \to \pm\infty$ for each ℓ, the s-lim $\Omega(t)f$ exists as $t \to \pm\infty$ for all vectors f of the form $f = (\theta,\ldots,f_\ell,\theta,\ldots)$. Since these vectors form a fundamental set in $\mathcal{H}$, $\Omega_\pm$ exist by Proposition 2.17.

(iii) From Proposition 8.14 we know that $\Omega_{\pm,\ell}$ exist if $\|H_{I,\ell}\exp(-iH_{0,\ell}t)f_\ell\|$ is integrable at $t = \pm\infty$ for all f_ℓ in some fundamental set in $\mathcal{H}_\ell$. Let $g\varepsilon C_0^\infty((0,\infty))$ and h be any vector in C^{2s+1}. Then $g \otimes Y_{\ell m} \varepsilon \mathcal{S}(\mathcal{R}^3)$ and the vectors of the form $f_\ell = g \otimes Y_{\ell m} \otimes h$ constitute a fundamental set in $\mathcal{H}_\ell$. Proceeding as in the derivation of the inequality (11.22), we get $\|H_{I,\ell}\exp(-iH_{0,\ell}t)f_\ell\| \leq (3\ell s)\,\|h\|\,\|V_1\exp(-iK_0t)(g\otimes Y_{\ell m})\|_{L^2(\mathcal{R}^3)}$. Since $V_1 \varepsilon L^2(\mathcal{R}^3)$, $\|V_1\exp(-iK_0t)(g \otimes Y_{\ell m})\|_{L^2(\mathcal{R}^3)}$ is integrable at $t = \pm\infty$ by Corollary 8.15 and thus $\Omega_{\pm,\ell}$ exist. Then by part (ii) of this proof $\Omega_\pm$ exist and are equal to $\oplus\, \Omega_{\pm,\ell}$. #

It is not difficult to see that the scattering theory is asymptotically complete for the system (H,H_0) if it is so in each partial wave (Problem 11.4). Moreover one can show that if V_1 is spherically symmetric and belongs to the class $\mathcal{V}$, then the theory is asymptotically complete (Problem 11.5). In such a case, one obtains a unitary S-operator in $\mathcal{H}$. The only new feature is that $R(\lambda)$ is a map from $L^2(S^{(2)}) \otimes C^{2s+1}$ into itself for almost all $\lambda\varepsilon[0,\infty)$. Similarly the scattering amplitude $f(\lambda;\underline{\omega}' \to \underline{\omega})$ is a $(2s+1)\times(2s+1)$ matrix function of $\lambda,\underline{\omega}$ and $\underline{\omega}'$. The diagonal elements of this matrix are called the <u>spin-nonflip amplitudes</u> and the off-diagonal ones the <u>spin-flip amplitudes</u>, the reason being that the off-diagonal elements give the amplitude when the incoming and outgoing spin states are different. The differential cross section is then given as follows :

$$\frac{d\sigma}{d\omega}(\lambda;\underline{\omega}',h' \to \underline{\omega},h) = |(h,f(\lambda;\underline{\omega}' \to \underline{\omega})h')_{C^{2s+1}}|^2,$$

where $\underline{\omega}',h'$ and $\underline{\omega},h$ are the incoming and outgoing momentum directions and spin states respectively. In scattering experiments the incoming particle beam is often unpolarized, and in this case one has to average over the incoming spin states to obtain

$$\frac{d\sigma}{d\omega}(\lambda;\underline{\omega}'\to\underline{\omega},h) = (2s+1)^{-1}(h,f(\lambda;\underline{\omega}'\to\underline{\omega})f^*(\lambda;\underline{\omega}'\to\underline{\omega})h)_{C^{2s+1}}.$$

If in addition the detector is not set up to monitor outgoing spin states, then the final spin states have to be summed over and one has

$$\frac{d\sigma}{d\omega}(\lambda;\underline{\omega}'\to\underline{\omega}) = (2s+1)^{-1}\,\|f(\lambda;\underline{\omega}'\to\underline{\omega})\|_{HS}^2,$$

where the Hilbert-Schmidt norm refers to the space C^{2s+1} only. For further discussion of this and details on the case of electrons ($s = \frac{1}{2}$), the reader is referred to [T, Chapters 5 and 7].

11-3 RADIAL SCHRÖDINGER OPERATORS

We have seen in Section 11-1 that for each ℓ, $K_{o,\ell}$ and H_ℓ are self-adjoint operators in $\mathcal{H}_r$ whenever the potential V satisfies suitable assumptions. We easily see that these are actually ordinary differential operators and here we shall study them in more detail. Let C_{oo}^∞ denote the class of infinitely differentiable functions with compact support in $(0,\infty)$ (i.e. functions in $C^\infty(0,\infty)$ each of which vanishes in some neighborhood of the origin and of infinity). Then a simple computation, using the expression for the Laplacian in spherical coordinates (i.e. $\Delta = d^2/dr^2 + 2r^{-1}d/dr - r^{-2}\underline{L}^2$), shows that for $f\in C_{oo}^\infty$

$$(K_{o,\ell}f)(r) = -f''(r) - 2r^{-1}f'(r) + \ell(\ell+1)r^{-2}f(r) \text{ and}$$

$$(H_\ell f)(r) = -f''(r) - 2r^{-1}f'(r) + [\ell(\ell+1)r^{-2} + V(r)]f(r), \quad (11.23)$$

where $f'(r) = df(r)/dr$ and $f''(r) = d^2f(r)/dr^2$. It is convenient to work in $\mathcal{H}_+ \equiv L^2[0,\infty)$ instead of $\mathcal{H}_r \equiv L^2([0,\infty),r^2dr)$, and we make the transformation $\hat{U} : \mathcal{H}_r \to \mathcal{H}_+$ by setting $(\hat{U}f)(r) = rf(r)$. Then $\hat{U}$ is an unitary map from $\mathcal{H}_r$ onto $\mathcal{H}_+$ and leaves C_{00}^∞ invariant.

The above map $\hat{U}$ transforms the operators in (11.23) into the so-called minimal operators in $\mathcal{H}_+$, denoted $\hat{K}_{o,\ell}$ and $\hat{H}_\ell$. We have $D(\hat{K}_{o,\ell}) = D(\hat{H}_\ell) = C_{00}^\infty$, and for $f\varepsilon C_{00}^\infty$

$$(\hat{K}_{o,\ell}f)(r) = -f''(r) + \ell(\ell+1)r^{-2}f(r) \text{ and}$$

$$(\hat{H}_\ell f)(r) = -f''(r) + [\ell(\ell+1)r^{-2} + V(r)]f(r). \quad (11.24)$$

Here $V\varepsilon L^2_{loc}((0,\infty))$, so that $Vf\varepsilon\mathcal{H}_+$. C_{00}^∞ is dense in $\mathcal{H}_+$ (see Problem 2.8) and, integrating by parts, one finds for all $f,g\varepsilon C_{00}^\infty$ that $(g,\hat{K}_{o,\ell}f)_+ = (\hat{K}_{o,\ell}g,f)_+$ and $(g,\hat{H}_\ell f)_+ = (\hat{H}_\ell g,f)_+$, so that $\hat{K}_{o,\ell}$ and $\hat{H}_\ell$ are symmetric operators.

Next we define the maximal operators $K'_{o,\ell}$ and H'_ℓ by setting $D(K'_{o,\ell}) = \{f\varepsilon\mathcal{H}_+ | f,f'$ are absolutely continuous and $-f''+\ell(\ell+1)r^{-2}f\varepsilon\mathcal{H}_+\}$, $D(H'_\ell) = \{f\varepsilon\mathcal{H}_+ | f,f'$ are absolutely continuous and $-f'' + (\ell(\ell+1)r^{-2}+V)f\varepsilon\mathcal{H}_+\}$ and writing for $f\varepsilon D(K'_{o,\ell})$, $g\varepsilon D(H'_\ell)$

$$K'_{o,\ell}f = -f''+\ell(\ell+1)r^{-2}f, \quad H'_\ell g = -g''+(\ell(\ell+1)r^{-2}+V)g. \quad (11.25)$$

Clearly $\hat{K}_{o,\ell} \subseteq K'_{o,\ell}$ and $\hat{H}_\ell \subseteq H'_\ell$. Also note that the maximal operators $K'_{o,\ell}$ and H'_ℓ need not be symmetric. The next proposition, which we state without proof, relates the maximal and the minimal operators.

PROPOSITION 11.11 : Let $\hat{K}_{0,\ell}$, $\hat{H}_\ell$ and $K'_{0,\ell}$, H'_ℓ be the minimal and maximal operators defined above. Then $K'_{0,\ell} = \hat{K}^*_{0,\ell}$ and $H'_\ell = \hat{H}^*_\ell$.

From the definitions (11.24), (11.25), it follows, on integrating by parts, that for $f \varepsilon C^\infty_{00}$ and $g \varepsilon D(K'_{0,\ell})$, $(g, \hat{K}_{0,\ell} f)_+ = (K'_{0,\ell} g, f)_+$ and hence $K'_{0,\ell} \subseteq \hat{K}^*_{0,\ell}$. Similarly one gets $H'_\ell \subseteq \hat{H}^*_\ell$. The converse inclusion depends on properties of ordinary differential equations, and we refer the reader to [NA; paragraph 17.4]. We see from Proposition 11.11 that the maximal operators $K'_{0,\ell}$ and H'_ℓ are closed. Now we investigate whether the minimal operators $\hat{K}_{0,\ell}$ and $\hat{H}_\ell$ are essentially self-adjoint.

PROPOSITION 11.12 : (a) $\hat{K}_{0,0}$ is not essentially self-adjoint whereas $\hat{K}_{0,\ell}$ is essentially self-adjoint for each $\ell \geq 1$.
(b) Let V be spherically symmetric and belong to $L^2(R^3) + L^\infty(R^3)$. Then $\hat{H}_0$ is not essentially self-adjoint and $\hat{H}_\ell$ is essentially self-adjoint for $\ell \geq 1$.

Proof : (a) By Corollary 2.15, it suffices to see if the ranges of $\hat{K}_{0,\ell} \pm i$ are dense. The ranges of $\hat{K}_{0,\ell} \pm i$ are dense if and only if the only vector $g \varepsilon D(\hat{K}^*_{0,\ell})$ verifying $((\hat{K}^*_{0,\ell} \mp i)g, f)_+ = 0$ for all $f \varepsilon C^\infty_{00}$ is $g = \theta$. Since C^∞_{00} is dense in H_+, this involves the study of the solutions of the equations $(K'_{0,\ell} \mp i)g = \theta$ in $D(K'_{0,\ell})$, where we have used the result of Proposition 11.11. We do this for the minus sign only, the other case being similar.

If $K'_{0,\ell} g = ig$, then g satisfies the ordinary differential equation : $-g'' + \ell(\ell+1)r^{-2} g = ig$. This equation is related to Bessel's equation (11.57) and the solution is $g = \alpha_1 r^{\frac{1}{2}} J_{\ell+\frac{1}{2}}(r\sqrt{i}) + \alpha_2 r^{\frac{1}{2}} N_{\ell+\frac{1}{2}}(r\sqrt{i})$, where $J_{\ell+\frac{1}{2}}$ and $N_{\ell+\frac{1}{2}}$ are the Bessel and Neumann functions respectively, and we have chosen the branch of the square root such that $\mathrm{Im}\sqrt{i} > 0$. The asymptotic behaviour of these functions is given in Section 11-5A. The function $r^{\frac{1}{2}} H^{(1)}_{\ell+\frac{1}{2}}(r\sqrt{i}) \equiv r^{\frac{1}{2}}[J_{\ell+\frac{1}{2}}(r\sqrt{i}) + i\, N_{\ell+\frac{1}{2}}(r\sqrt{i})]$ lies in $L^2(1,\infty)$ for all ℓ, whereas $r^{\frac{1}{2}} H^{(2)}_{\ell+\frac{1}{2}}(r\sqrt{i}) \equiv r^{\frac{1}{2}}[J_{\ell+\frac{1}{2}}(r\sqrt{i}) - iN_{\ell+\frac{1}{2}}(r\sqrt{i})]$ does not for any ℓ. Thus g must be a multiple of $r^{\frac{1}{2}} H^{(1)}_{\ell+\frac{1}{2}}(r\sqrt{i})$. Now $r^{\frac{1}{2}} J_{\ell+\frac{1}{2}}(r\sqrt{i}) \in L^2(0,1)$ for all ℓ and $r^{\frac{1}{2}} N_{\ell+\frac{1}{2}}(r\sqrt{i}) \notin L^2(0,1)$ except for $\ell = 0$. Therefore for $\ell \geq 1$, the only solution of $(K'_{0,\ell} - i)g = \theta$ in $D(K'_{0,\ell}) \subseteq H_+$ is $g = \theta$, and for $\ell = 0$, the same equation has a nontrivial solution in $D(K'_{0,0})$, namely $g = \exp[(i-1)r/\sqrt{2}]$. Hence $\hat{K}_{0,\ell}$ is essentially self-adjoint for each $\ell \geq 1$ and $\hat{K}_{0,0}$ is not.

(b) Since $V \in L^2_{loc}(R^3)$, it follows that $V \in L^2_{loc}((0,\infty))$ and we can define the minimal operator $\hat{H}_\ell$ by (11.24). Let $f \in C^\infty_{00}$. Then by Proposition 11.5, $\hat{U}^{-1} f \otimes Y_{\ell m} \in D(K_0)$ and by Proposition 8.7(a,b), $\|Vf\|_+ = \|V(\hat{U}^{-1} f \otimes Y_{\ell m})\| \leq \alpha \|\hat{U}^{-1} f \otimes Y_{\ell m}\| + \beta \|K_0(\hat{U}^{-1} f \otimes Y_{\ell m})\|$, where β can be chosen arbitrarily small. Hence by Proposition 11.5, $\|Vf\|_+ \leq \beta \|\hat{K}_{0,\ell} f\|_+ + \alpha \|f\|_+$. For $\ell \geq 1$, $\hat{K}_{0,\ell}$ is essentially self-adjoint, so that one arrives at the essential self-adjointness of $\hat{H}_\ell$ by a result similar to the Kato-Rellich theorem (see Problem 8.4). To deal with $\ell = 0$ we note from Remark 8.6 that since V has K_0-bound 0, V is $\hat{H}_\ell$-bounded for all ℓ with $\hat{H}_\ell$-bound less than 1. Thus, if $\hat{H}_0$ were essentially self-adjoint, $\hat{K}_{0,0} = \hat{H}_0 - V$ would be essentially self-adjoint by another application of Problem 8.4, which contradicts (a). #

The lack of self-adjointness of $\hat{K}_{0,0}$ is intuitively explained as follows : the absence of the centrifugal barrier (the $\ell(\ell+1)r^{-2}$ term) allows the particle to reach the origin, so that one needs to put a boundary condition at r = 0. These boundary conditions determine the relative change of phase due to reflection of a wave packet at the origin. We consider the following restrictions of $K'_{0,0}$:

$$D(K'_{0,0}(a)) = \{f\varepsilon D(K'_{0,0})|f'(0) + af(0) = 0\}; \ -\infty < a < \infty$$

and
$$D(K'_{0,0}(\infty)) = \{f\varepsilon D(K'_{0,0})|f(0) = 0\}. \qquad (11.26)$$

The action of the operators on their respective domains is the same as that of $K'_{0,0}$. $[K'_{00}(a) \pm i]$ is invertible for each a in $(-\infty,\infty]$, because the only solution in H_+ of $(K'_{0,0} \pm i)g = \theta$ is $g(r) = \exp[(\mp i-1)r/\sqrt{2}]$, which satisfies none of the above boundary conditions. In fact all operators in (11.26) are self-adjoint, and they are the only self-adjoint extensions of $\hat{K}_{0,0}$. For a proof of this in the general case, we refer to [NA], and we shall show below that $\hat{K}'_{0,0}(\infty)$ is self-adjoint and coincides with $\hat{U}K_{0,0}\hat{U}^{-1}$, which is related to the free Hamiltonian K_0 by Proposition 11.5.

By Lemma 11.4, $g\varepsilon D(K_{0,0}) \subseteq H_r$ if and only if $g \otimes Y_{00} \varepsilon D(K_0) \subseteq L^2(R^3)$. This along with Proposition 3.10 implies that g must be bounded. Therefore $f \equiv \hat{U}g \varepsilon D(\hat{U}K_{0,0}\hat{U}^{-1}) \subseteq H_+$ means that f(0) = 0, since f(r) = rg(r). On the other hand since $\hat{U}K_{0,0}\hat{U}^{-1}$ is self-adjoint and since $K'_{0,0} = \hat{K}^*_{0,0}$, we have that $\hat{K}_{0,0} \subseteq \hat{U}K_{0,0}\hat{U}^{-1} \subseteq \hat{K}^*_{0,0} = K'_{0,0}$. Thus $\hat{U}K_{0,0}\hat{U}^{-1}$ is a restriction of $K'_{0,0}$ satisfying the same boundary conditions as $K'_{0,0}(\infty)$, i.e. $\hat{U}K_{0,0}\hat{U}^{-1} \subseteq K'_{0,0}(\infty)$. Now let $h\varepsilon D(K'_{0,0}(\infty))$.

Then by Proposition 2.14 there exists $f \varepsilon D(\hat{U}K_{0,0}\hat{U}^{-1})$ such that $[K'_{0,0}(\infty) + i]h = [\hat{U}K_{0,0}\hat{U}^{-1} + i]f$, i.e. $[K'_{0,0}(\infty) + i](h-f) = \theta$. Since $[K'_{0,0}(\infty) + i]$ is invertible, it follows that $h = f \varepsilon D(\hat{U}K_{0,0}\hat{U}^{-1})$, showing that $K'_{0,0}(\infty) = \hat{U}K_{0,0}\hat{U}^{-1}$.

For $\ell \neq 0$, the boundary condition $f(0) = 0$ is automatically satisfied. Clearly $\hat{K}_{0,\ell} \subseteq \hat{U}K_{0,\ell}\hat{U}^{-1}$ and by Proposition 11.12(a), $\hat{K}_{0,\ell}$ is essentially self-adjoint in this case. Hence $\overline{\hat{K}_{0,\ell}} = \hat{U}K_{0,\ell}\hat{U}^{-1}$ and as in the last paragraph, every vector $f \varepsilon D(\hat{U}K_{0,\ell}\hat{U}^{-1}) = D(\overline{\hat{K}_{0,\ell}}) = D(\hat{K}^*_{0,\ell}) = D(K'_{0,\ell})$ must have the property that $f(0) = 0$.

We should mention that ordinary differential operators have been studied extensively. In particular there exist quite detailed methods for determining the nature of $\hat{H}_\ell$ with more singular potentials than those appearing in Proposition 11.12 (see e.g. [DS II], [NA]). Also the simplicity of the spectrum of any self-adjoint extension of $\hat{H}_\ell$ for each ℓ has been established under various assumptions on V (see Weidmann [1], Amrein and Georgescu [2]). This along with Problem 11.4 and Propositions 11.5 and 9.15 leads to a proof of asymptotic completeness for these potentials.

11-4 PARTIAL WAVE EIGENFUNCTIONS AND PHASE SHIFTS

Having established that the self-adjoint operators $K_{0,\ell}$ and H_ℓ are ordinary differential operators in an appropriate sense, we now want to study their eigenfunctions. In this section, we shall assume that V is spherically symmetric and in $\mathcal{V}$. Since the function A in the definition of the class $\mathcal{V}$ is spherically symmetric, so is the function B, and without loss of generality we may assume B_1 and B_2 to be spherically symmetric. The assumptions made on B_1 and B_2 in (9.27) then imply that

$$b_1^2 \equiv \int_0^\infty B_1(r)^2 r^2 dr < \infty \text{ and } b_2^2 \equiv \int_0^\infty B_2(r)^2 dr < \infty. \tag{11.27}$$

The Green's functions $G^o_{\lambda\pm io}(\underline{x},\underline{y}) = (4\pi)^{-1}|\underline{x}-\underline{y}|^{-1}\cdot \exp(\pm i\lambda^{\frac{1}{2}}|\underline{x}-\underline{y}|)$ depend only on $r = |\underline{x}|$, $r' = |\underline{y}|$ and $\cos\theta = \underline{\omega}_x \cdot \underline{\omega}_y$, and belong to $L^2([0,\pi], \sin\theta d\theta)$ for all $r \neq r'$. Thus for fixed r and r' such that $r \neq r'$, they can be expanded in terms of the Legendre polynomials $\{P_\ell\}$ which form a basis of $L^2([0,\pi],\sin\theta d\theta)$:

$$G^o_{\lambda\pm io}(\underline{x},\underline{y}) = (4\pi rr')^{-1}\sum_\ell (2\ell+1)G^o_{\ell,\lambda\pm io}(r,r')P_\ell(\cos\theta). \tag{11.28}$$

The coefficients $G^o_{\ell,\lambda\pm io}(r,r')$ are called the partial wave (retarded and advanced) Green's functions, and an elementary calculation (Problem 11.6) shows that

$$G^o_{\ell,\lambda\pm io}(r,r') = k^{-1}\cdot \begin{cases} \hat{j}_\ell(kr)\hat{h}^\pm_\ell(kr') & \text{if } r < r' \\ \hat{j}_\ell(kr')\hat{h}^\pm_\ell(kr) & \text{if } r > r' \end{cases} \tag{11.29}$$

where $\hat{j}_\ell$ and $\hat{h}^\pm_\ell$ are the Riccati-Bessel and Riccati-Hankel functions respectively (see Section 11-5A), and $k = \lambda^{\frac{1}{2}}$.

Let $\underline{k}$ be such that $\underline{k}^2 \notin \Gamma_o$ and let $\psi^\pm_{\underline{k}}(\underline{x})$ be the eigenfunctions defined in Section 10-1, where we have used the notation introduced on page 395. Since obviously $\psi^o_{R\underline{k}}(R\underline{x}) = \psi^o_{\underline{k}}(\underline{x})$ and $W^\pm_\lambda$ commute with U(R), we have by (10.30) that $\psi^\pm_{\underline{k}}(\underline{x}) = \psi^\pm_{R\underline{k}}(R\underline{x})$ for all $R\in SO(3)$. By choosing appropriate rotations R in this equality, one finds that, like $\psi_k^o(\underline{x})$, $\psi_k^\pm(\underline{x})$ is a function of k,r and $\underline{\omega}_k\cdot\underline{\omega}_x$ only. Moreover, since $A\psi_k^\pm \in L^2(R^3)$ (Proposition 10.6), $\psi_k^\pm(\underline{x}) \in L^2[-1,+1]$ as a function of $\cos(\underline{\omega}_k\cdot\underline{\omega}_x)$ for fixed k and almost all r. The verification of these statements is given as Problem 11.7. We may now write

$$\psi_k^\pm(\underline{x}) = (2\pi)^{-3/2}(kr)^{-1}\sum_\ell(2\ell+1)i^\ell\psi_{\ell,k}^\pm(r)P_\ell(\underline{\omega}_k\cdot\underline{\omega}_x), \quad (11.30)$$

which is similar to the expansion (11.64) for $\psi_k^o(\underline{x})$. The coefficients $\psi_{\ell,k}^\pm$ are called the partial wave eigenfunctions.

Substituting (11.28), (11.30) and (11.64) in the Lippmann-Schwinger equation (10.36) and using the addition theorem (11.54) and the orthogonality of the spherical harmonics, we have

$$\psi_{\ell,k}^\pm(r) = \hat{j}_\ell(kr) - \int_0^\infty G_{\ell,\lambda\mp io}^o(r,r')V(r')\psi_{\ell,k}^\pm(r')dr'. \quad (11.31)$$

The equation (11.31) is the partial wave version of the Lippmann-Schwinger equation, since $\hat{j}_\ell$ is the eigenfunction of $\hat{U}K_{o,\ell}\hat{U}^{-1}$. It follows from (11.30) and (11.56) that

$$2(2\pi)^{-3}k^{-2}\sum_\ell(2\ell+1)\,||A\psi_{\ell,k}^\pm||_+^2 = ||A\psi_k^\pm||^2 < \infty, \quad (11.32)$$

showing that $A\psi_{\ell,k}^\pm \in H_+$ for each ℓ. The next proposition investigates some properties of the eigenfunctions $\psi_{\ell,k}^\pm$. In its proof we shall use the following bounds on the functions $\hat{j}_\ell$ and $\hat{h}_\ell^\pm$ for making various estimates (see Section 11-5A) :

$$|\hat{j}_\ell(kr)| \le c_1|kr/(1+kr)|^{\ell+1} \text{ and } |\hat{h}_\ell^\pm(kr)| \le c_2|kr/(1+kr)|^{-\ell}, \quad (11.33)$$

where the constants c_1 and c_2 depend only on ℓ.

PROPOSITION 11.13 : Let Δ be a compact subset of $(0,\infty)$ such that $k^2 \notin \Gamma_0$ for all $k \in \Delta$ and let V be spherically symmetric and belong to $\mathcal{V}$. Then for every ℓ, (a) $(k,r) \mapsto \psi^{\pm}_{\ell,k}(r)$ defines two bounded continuous functions on $\Delta \times [0,\infty)$, (b) for every fixed $k \in \Delta$, $\psi^{\pm}_{\ell,k}(r) = O((kr/1+kr)^{\ell+1})$ as $r \to 0$. (c) $\psi^{\pm}_{\ell,k}$ and $\psi^{\pm\prime}_{\ell,k}$ are locally absolutely continuous and $\psi^{\pm}_{\ell,k}$ satisfy the <u>radial Schrödinger equation</u> : $\psi'' - [\ell(\ell+1)r^{-2} + V(r) - k^2]\psi = 0$.

<u>Proof</u> : Since $\psi^{+}_{-k} = \overline{\psi^{-}_{k}}$ (Problem 10.6), one gets from (11.30) that $\psi^{+}_{\ell,k} = \overline{\psi^{-}_{\ell,k}}$, and hence it suffices to prove the proposition for $\psi^{-}_{\ell,k}$.

(a) Considering that $\hat{j}_\ell(kr)$ is bounded and continuous in k and r, we only need to study the integral in (11.31). Using (11.29) we consider the two parts of this integral separately. By the Schwarz inequality, (11.27), (11.32) and (11.33) we have

$$|k^{-1}\hat{h}^{+}_{\ell}(kr)\int_0^r \hat{j}_\ell(ks)V(s)\psi^{-}_{\ell,k}(s)ds| \leq \|A\psi^{-}_{\ell,k}\|_{+} k^{-1}|\hat{h}^{+}_{\ell}(kr)| \cdot$$

$$(\int_0^r |\hat{j}_\ell(ks)|^2 B(s)^2 ds)^{\frac{1}{2}} \leq 2^{\frac{1}{2}} c_1 \|A\psi^{-}_{\ell,k}\|_{+} k^{-1}|\hat{h}^{+}_{\ell}(kr)| \cdot$$

$$\{k^2(kr/1+kr)^{2\ell}\int_0^r s^2 B_1(s)^2 ds + (kr/1+kr)^{2\ell+2}\int_0^r B_2(s)^2 ds\}^{\frac{1}{2}}$$

$$\leq 2^{\frac{1}{2}} c_1 c_2 \|A\psi^{-}_{\ell,k}\|_{+}(b_1^2 + k^{-2}b_2^2)^{\frac{1}{2}} < \infty.$$

Similarly $k^{-1}\hat{j}_\ell(kr)\int_r^\infty \hat{h}^{+}(ks)V(s)\psi^{-}_{\ell,k}(s)ds$ can be shown to have the same bound. The continuity of $\psi^{-}_{\ell,k}(r)$ in k and r follows from the continuity of $A\psi^{-}_{\ell,k}$ in $\mathcal{H}_{+}$-norm (see Lemma 10.7) and is left as an exercise (Problem 11.8).

(b) Using the fact that $\psi^{-}_{\ell,k}$ is bounded and proceeding as in the estimate of part (a), we find

$$|k^{-1}\hat{h}^+_\ell(kr)\int_0^r \hat{j}_\ell(ks)V(s)\psi^-_{\ell,k}(s)ds| \leq \|\psi^-_{\ell,k}\|_\infty k^{-1}|\hat{h}^+_\ell(kr)| \cdot$$

$$\{(\int_0^r |\hat{j}_\ell(ks)|^2 s^{-2}ds)^{\frac{1}{2}}(\int_0^r s^2 B_1(s)^2 ds)^{\frac{1}{2}} +$$

$$(\int_0^r |j_\ell(ks)|^2 ds)^{\frac{1}{2}}(\int_0^r B_2(s)^2 ds)^{\frac{1}{2}}\} \leq c_1 \|\psi^-_{\ell,k}\|_\infty k^{-1}|\hat{h}^+_\ell(kr)|\cdot$$

$$\{b_1 k^{\frac{1}{2}}(\int_0^{kr}(x/1+x)^{2\ell+2}x^{-2}dx)^{\frac{1}{2}} + b_2 k^{-\frac{1}{2}}(\int_0^{kr}(x/1+x)^{2\ell+2}dx)^{\frac{1}{2}}\}$$

$$\leq (2\ell+1)^{-\frac{1}{2}}c_1 c_2 \|\psi^-_{\ell,k}\|_\infty k^{-1}\{b_1 k^{\frac{1}{2}}(kr/1+kr)^{\frac{1}{2}} + b_2 k^{-\frac{1}{2}}kr(kr/1+kr)^{\frac{1}{2}}\},$$

where we have used the simple result that for each integer $n \neq 1$, $\int_0^t (x/1+x)^n x^{-2}dx = (n-1)^{-1}(t/1+t)^{n-1}$. Thus for $0 < r \leq 1$, we have $|k^{-1}\hat{h}^+_\ell(kr)\int_0^r \hat{j}_\ell(ks)V(s)\psi^-_{\ell,k}(s)ds| \leq c_3(\ell,k)(kr/1+kr)^{\frac{1}{2}}$. A similar estimate, using $|\int_1^\infty V(r)dr| \leq (\int_1^\infty A(r)^2 dr)^{\frac{1}{2}} \cdot (\int_1^\infty r^2 B_1(r)^2 dr)^{\frac{1}{2}} + |\int_1^\infty B_2(r)A(r)dr| < \infty$, leads to $|k^{-1}\hat{j}_\ell(kr)\int_r^\infty \hat{h}^+_\ell(ks)V(s)\psi^-_{\ell,k}(s)ds| \leq c_4(\ell,k)|\hat{j}_\ell(kr)|\{\int_1^\infty |V(s)|ds + \int_r^1 (ks/1+ks)^{-\ell}|B(s)|ds\} \leq c_5(\ell,k)\{\hat{j}_\ell(kr) + (kr/1+kr)^{\frac{1}{2}}\}$ for $0 < r \leq 1$. In view of (11.31) and (11.33), it follows that $|\psi^-_{\ell,k}(r)| \leq c_6(\ell,k)\cdot(kr/1+kr)^{\frac{1}{2}}$ for $0 < r \leq 1$. We iterate this procedure another 2ℓ times, each time gaining a factor of $(kr/1+kr)^{\frac{1}{2}}$, so that $|\psi^-_{\ell,k}(r)| \leq c_7(\ell,k)(kr/1+kr)^{\ell+\frac{1}{2}}$ for $0 < r \leq 1$ (in each iteration the previous bound on $\psi^-_{\ell,k}$ is combined with $\hat{j}_\ell$ or $\hat{h}^+_\ell$). A further iteration produces another gain of $(kr/1+kr)^{\frac{1}{2}}$ in the estimation of the first integral. On the other hand the term coming from B_1 in the estimation of the second integral has the form $\int_r^1 ds(ks/1+ks)^{\frac{1}{2}}B_1(s)$, which after an application of the Schwarz inequality leads to a logarithmic behaviour. Thus $|\psi^-_{\ell,k}(r)| \leq c_8(\ell,k)(kr/1+kr)^{\ell+1}\cdot |\log(kr/1+kr)|^{\frac{1}{2}} \leq c_9(\ell,k)(kr/1+kr)^{\ell+1-\delta}$, since $x^{2\delta}\log x$ is bounded for $0 < x \leq 1$, $\delta > 0$. Finally, another iteration on the term coming from B_1 in the second integral gives $\int_r^1 (ks/1+ks)^{1-\delta} B_1(s)ds \le b_1(\int_r^1 (ks/1+ks)^{2-2\delta} s^{-2} ds)^{\frac{1}{2}}$, which is finite for $0 < r \le 1$ if $\delta < \frac{1}{2}$. With this we have proven (b). Notice that because of (a) such a bound is in fact valid for all $r \ge 0$.

(c) By (b) and (11.33), the integrals $\int_0^\infty \hat{h}^{\mp}(ks)V(s) \cdot \psi^{\pm}_{\ell,k}(s)ds$ converge and we can rewrite (11.31) as

$$\psi^{\pm}_{\ell,k}(r) = [1-k^{-1}\int_0^\infty \hat{h}^{\mp}_{\ell}(ks)V(s)\psi^{\pm}_{\ell,k}(s)ds]\hat{j}_\ell(kr) - \int_0^r G^{o'}_{\ell,\lambda}(r,s)V(s)\psi^{\pm}_{\ell,k}(s)ds, \tag{11.34}$$

where $G^{o'}_{\ell,\lambda}(r,s) = k^{-1}[\hat{n}_\ell(kr)\hat{j}_\ell(ks)-\hat{j}_\ell(kr)\hat{n}_\ell(ks)]$. (11.35)

It also follows from (b) that $\int_0^r G^{o'}_{\ell,\lambda}(r,s)V(s)\psi^{\pm}_{\ell,k}(s)ds$ exist for all $0 \le r < \infty$ and we find from [R, page 106] that $\psi^{\pm}_{\ell,k}$ are locally absolutely continuous, since $\hat{j}_\ell$ and $\hat{n}_\ell$ are C^∞-functions on $(0,\infty)$. Differentiating (11.34) once, one has $\psi^{\pm\prime}_{\ell,k}(r) = \hat{j}'_\ell(kr)[1-k^{-1}\int_0^\infty \hat{h}^{\mp}_\ell(ks)V(s)\psi^{\pm}_{\ell,k}(s)ds] - k^{-1}[\hat{n}'_\ell(kr)\int_0^r \hat{j}_\ell(ks)V(s)\psi^{\pm}_{\ell,k}(s)ds - \hat{j}'_\ell(kr)\int_0^r \hat{n}_\ell(ks)V(s)\psi^{\pm}_{\ell,k}(s)ds]$, which shows that $\psi^{\pm\prime}_{\ell,k}$ are also locally absolutely continuous. Since $\hat{n}_\ell(kr)d/dr\,\hat{j}_\ell(kr) - \hat{j}_\ell(kr)d/dr\,\hat{n}_\ell(kr) = k$ (= the Wronskian of $\hat{j}_\ell$ and $\hat{n}_\ell$), and since $d^2\hat{j}_\ell(kr)/dr^2 - [\ell(\ell+1)r^{-2}-k^2]\hat{j}_\ell(kr) = 0$, we obtain by another differentiation that $\psi^{\pm}_{\ell,k}$ satisfy the given differential equation. #

It is traditional to start with the ordinary differential equation of Proposition 11.13(c) and work with its so-called <u>regular solution</u> $\phi_{\ell,k}$ defined by the real boundary condition

$$\lim_{r\to 0} r^{-\ell-1}\,\phi_{\ell,k}(r) = k^{\ell+1}/(2\ell+1)!!^{*)}\ (k > 0). \qquad (11.36)$$

The associated integral equation is of the Volterra type (see [AR], Chapter 3) :

$$\phi_{\ell,k}(r) = \hat{j}_\ell(kr) - \int_0^r G^{o\prime}_{\ell,\lambda}(r,s)V(s)\phi_{\ell,k}(s)ds, \qquad (11.37)$$

which resembles closely the equation (11.34) satisfied by $\psi^{\pm}_{\ell,k}$. A solution of (11.37) can be constructed by an iteration procedure (see [AR] or [N]). The uniqueness of the solution can be established as in the proof of Proposition 11.14(a). Note that all results concerning $\phi_{\ell,k}$ are true for each $k > 0$. We shall next relate $\phi_{\ell,k}$ to $\psi^{-}_{\ell,k}$.

<u>PROPOSITION 11.14</u> : Assume the hypotheses of Proposition 11.13 and let $k\varepsilon\Delta$. Then (a) $1-k^{-1}\int_0^\infty \hat{h}^+_\ell(ks)V(s)\psi^-_{\ell,k}(s)ds \neq 0$. (b) Let $\mathscr{f}_\ell(k) = [1-k^{-1}\int_0^\infty \hat{h}^+_\ell(ks)\,V(s)\,\psi^-_{\ell,k}(s)ds]^{-1}$. Then $\phi_{\ell,k} = \mathscr{f}_\ell(k)\psi^-_{\ell,k}$, and $0 < |\mathscr{f}_\ell(k)| < \infty$.

<u>Proof</u> : (a) We see from Proposition 11.13(b) that $|\int_0^\infty \hat{h}^+_\ell(ks)V(s)\psi^-_{\ell,k}(s)ds| < \infty$. Thus $\mathscr{f}_\ell(k) \neq 0$. Assume that $1-k^{-1}\int_0^\infty \hat{h}^+_\ell(ks)V(s)\psi^-_{\ell,k}(s)ds = 0$. Then it follows from (11.34) that $\psi^-_{\ell,k}$ satisfies the homogeneous equation : $\psi^-_{\ell,k}(r) = -\int_0^r G^{o\prime}_{\ell,\lambda}(r,s)V(s)\psi^-_{\ell,k}(s)ds$. Iterating this equation n times we obtain

$$\psi^-_{\ell,k}(r) = (-1)^n\int_0^r dr_n\int_0^{r_n} dr_{n-1}\cdots\int_0^{r_2} dr_1\,\cdot$$

$$G^{o\prime}_{\ell,\lambda}(r,r_n)V(r_n)G^{o\prime}_{\ell,\lambda}(r_n,r_{n-1})\cdots V(r_1)\psi^-_{\ell,k}(r_1).$$

---

*) $(2\ell+1)!! \equiv (2\ell+1)(2\ell-1)\cdots 5\cdot 3\cdot 1.$

By using the bound $|G^{o'}_{\ell,\lambda}(r,s)| \le \alpha k^{-1}(kr/1+kr)^{\ell+1}(ks/1+ks)^{-\ell}$ for $r \ge s$, Proposition 11.13(b) and the fact that $\int_0^r dr_n \int_0^{r_n} dr_{n-1} \cdots \int_0^{r_2} dr_1 f(r_n)\cdots f(r_1) = (n!)^{-1}[\int_0^r ds f(s)]^n$, we arrive at

$$|\psi^-_{\ell,k}(r)| \le \beta(kr/1+kr)^{\ell+1}(n!)^{-1}M(r)^n. \tag{11.38}$$

Here $M(r) = k^{-1}\int_0^r ds|(ks/1+ks)V(s)| \le (b_1+k^{-1}b_2)[\int_0^\infty A(s)^2 ds]^{\frac{1}{2}} < \infty$. Since n is arbitrary in the right-hand side of (11.38), this implies that $\psi^-_{\ell,k}(r) = 0$ for all r, which contradicts the hypothesis : $1-k^{-1}\int_0^\infty \hat{h}^+_\ell(ks)V(s)\psi^-_{\ell,k}(s)ds = 0$. Hence $|\mathfrak{f}_\ell(k)|<\infty$.

(b) From the definition of $\mathfrak{f}_\ell(k)$ and (11.34) we find that both $\mathfrak{f}_\ell(k)\psi^-_{\ell,k}$ and $\phi_{\ell,k}$ satisfy the same integral equation (11.37). By the uniqueness of the solution of (11.37) we arrive at (b). #

The function $\mathfrak{f}_\ell$ is called the <u>Jost function</u> for the angular momentum ℓ and was introduced by Jost [1]. By combinning the definition of $\mathfrak{f}_\ell$ with the relation $\phi_{\ell,k} = \mathfrak{f}_\ell(k)\psi^-_{\ell,k}$ for $k^2 \notin \Gamma_o$, one obtains

$$\mathfrak{f}_\ell(k) = 1 + k^{-1}\int_0^\infty \hat{h}^+_\ell(ks)V(s)\phi_{\ell,k}(s)ds. \tag{11.39}$$

Note that the integral in (11.39) converges for all $k > 0$ and not only for $k^2 \varepsilon (0,\infty)-\Gamma_o$, since $|\phi_{\ell,k}(r)| \le$ constant$\cdot (kr/1+kr)^{\ell+1}$ for all $k > 0$ (Problem 11.9). From this one can prove the absence of positive energy eigenvalues for the Hamiltonian H_ℓ for each ℓ (Problem 11.10).

Next we study the partial wave scattering matrix or equivalently the phase shifts and the asymptotic behaviour of the eigenfuntion $\psi^-_{\ell,k}(r)$ as $r \to \infty$. In order to derive expressions similar to those in Section 10-2, we first obtain the partial wave components of $M_A(\lambda)$ and $W_\lambda^{\pm}$.

LEMMA 11.15 : Let $V\varepsilon\mathcal{V}$ and be spherically symmetric. Then
(a) $W_\lambda^{\pm} = \sum_{\ell m}\hat{U}^* W_{\ell,\lambda}^{\pm}\hat{U} \otimes F_{\ell m}$, where $W_{\ell,\lambda}^{\pm}$ are Hilbert-Schmidt operators in $\mathcal{H}_+$ with kernels $B(r)G^{o}_{\ell,\lambda\pm io}(r,s)A(s)$. $W_{\ell,\lambda}^{\pm}$ depend continuously on λ in Hilbert-Schmidt norm, uniformly in ℓ.
(b) Let $f\varepsilon\mathcal{H}_r$ and $\lambda > 0$. Then

$$[M_A(\lambda)(f \otimes Y_{\ell m})](\omega_{-k}) = [M_{A,\ell}(\lambda)\hat{U}f]Y_{\ell m}(\omega_{-k}), \qquad (11.40)$$

where each $M_{A,\ell}(\lambda)$ is an operator from $\mathcal{H}_+$ to $\mathcal{C}$ defined by

$$M_{A,\ell}(\lambda)g = (\pi k)^{-\frac{1}{2}}(-i)^{\ell}\int_0^{\infty}\hat{j}_{\ell}(kr)A(r)g(r)dr, (k=\lambda^{\frac{1}{2}}). \qquad (11.41)$$

$M_{A,\ell}(\lambda)$ is a continuous function of λ. A similar statement holds for $M_B(\lambda)$.

Proof : (a) A,B and R_z^o commute with each $E_{\ell m}$, the last one by Proposition 11.5. Therefore W_z and consequently $W_\lambda^{\pm}$ commute with each $E_{\ell m}$, so that by Lemma 11.4 we write $W_\lambda^{\pm} = \sum_{\ell m} W_\lambda^{\pm} E_{\ell m} = \sum_{\ell m} \mathcal{W}_{\lambda,\ell m}^{\pm} \otimes F_{\ell m}$. As in Proposition 11.5, $\mathcal{W}_{\lambda,\ell m}^{\pm}$ are independent of m and we set $W_{\ell,\lambda}^{\pm} \equiv \hat{U}\mathcal{W}_{\lambda,\ell m}^{\pm}\hat{U}^*\varepsilon\mathcal{B}(\mathcal{H}_+)$. Since

$$||W_{\ell,\lambda}^{\pm}||_{HS} = ||\hat{U}^* W_{\ell,\lambda}^{\pm}\hat{U} \otimes F_{\ell m}||_{HS} = ||W_\lambda^{\pm}E_{\ell m}||_{HS} \leq ||W_\lambda^{\pm}||_{HS}, \qquad (11.42)$$

we have $W_{\ell,\lambda}^{\pm}\varepsilon\mathcal{B}_2(\mathcal{H}_+)$ by Lemma 9.13. The continuity of $W_{\ell,\lambda}^{\pm}$ follows similarly. The kernels of $W_{\ell,\lambda}^{\pm}$ can be computed from those of $W_\lambda^{\pm}$ by using (11.28) and (11.54).

(b) Using definition (10.6) of $M_A(\lambda)$ and replacing f by Af in (11.11) we arrive at (11.40). Since $A\varepsilon L^2(0,\infty)$ and $\hat{j}_\ell$ is bounded, the integral in (11.41) exists for each $g\varepsilon\mathcal{H}_+$. From (11.40) it follows that $|[M_{A,\ell}(\lambda)-M_{A,\ell}(\mu)]\hat{U}f| =$

$|([M_A(\lambda)-M_A(\mu)]^*Y_{\ell m}, f \otimes Y_{\ell m})| \leq \|M_A(\lambda)-M_A(\lambda)\| \, \|f\|_{h}$, showing the continuity of $M_{A,\ell}(\lambda)$, uniformly in ℓ, by Lemma 10.1. To verify that the integral in (11.41) exists with B_1 replacing A, note that by (11.33), $\int_0^\infty |\hat{j}_\ell(kr)|^2 B_1(r)^2 dr \leq c_1 k^2 \int_0^\infty r^2 B_1(r)^2 dr < \infty$. #

We mention that the operators $W_\lambda^\pm$ and $M_A(\lambda)$ can also be viewed as direct sums : $W_\lambda^\pm = \oplus_{\ell m} \hat{U}^* W_{\ell,\lambda}^\pm \hat{U}$ in $H = \oplus H_{\ell m}$ and $M_A(\lambda) = \oplus_{\ell m} M_{A,\ell}(\lambda)$ as in (10.70).

PROPOSITION 11.16 : Let $V \varepsilon \mathcal{V}$ and be spherically symmetric. Then (a) the partial wave S-matrix for all $\lambda > 0$ is given by

$$R_\ell(\lambda) \equiv S_\ell(\lambda)-1 = -2\pi i \, M_{B,\ell}(\lambda)(I+W_{\ell,\lambda}^-)^{-1^*} M_{A,\ell}(\lambda)^* \tag{11.43}$$

$$= -2\pi i \, M_{A,\ell}(\lambda)(I+W_{\ell,\lambda}^+)^{-1} M_{B,\ell}(\lambda)^*. \tag{11.44}$$

Furthermore R_ℓ is a continuous complex-valued function on $(0,\infty)$. (b) An alternative expression for $R_\ell(\lambda)$ is

$$R_\ell(\lambda) = -2ik^{-1}\int_0^\infty \hat{j}_\ell(kr)V(r)\psi_{\ell,k}^-(r)dr. \tag{11.45}$$

(c) The eigenfunction $\psi_{\ell,k}^-$ admits the following asymptotic expansion for each ℓ and each $k > 0$ as $r \to \infty$:

$$\psi_{\ell,k}^-(r) = e^{i\delta_\ell(\lambda)}\sin(kr - \tfrac{1}{2}\ell\pi + \delta_\ell(\lambda)) + O(r^{-1}) + O(r^{-\frac{1}{2}-\nu}) + O(r^{\frac{1}{2}-\nu-\delta}), \tag{11.46}$$

where ν and δ are the constants appearing in the definition (9.27) of class $\mathcal{V}$. The regular solution $\phi_{\ell,k}$ also has a similar asymptotic behaviour. (d) For each $k > 0$, $S_\ell(k^2) = \overline{\mathfrak{f}_\ell(k)}/\mathfrak{f}_\ell(k)$, where $\mathfrak{f}_\ell$ is the Jost function.

Proof : (a) One has $I + W_\lambda^\pm = \sum_{\ell m} \hat{U}^*(I+W_{\ell,\lambda}^\pm)\hat{U} \otimes F_{\ell m}$. If $I+W_{\ell,\lambda}^+$ is not invertible for some ℓ and $\lambda > 0$, then neither is $I+W_\lambda^+$, so that by Proposition 10.17, λ is an eigenvalue of H. Then by (11.13) there exists ℓ such that λ is an eigenvalue of H_ℓ. But for any ℓ, H_ℓ has no positive eigenvalue (Problem 11.10). Therefore $(I+W_{\ell,\lambda}^\pm)^{-1}$ exist for all ℓ and $\lambda > 0$, showing that $(I+W_\lambda^\pm)^{-1}$ exist and are equal to $\sum_{\ell m} \hat{U}^*(I+W_{\ell,\lambda}^\pm)^{-1}\hat{U} \otimes F_{\ell m}$. This also shows that $\Gamma_0 \cap (0,\infty) = \emptyset$ and that $(I-W_\lambda^\pm)^{-1} \in \mathcal{B}(\mathcal{H})$, hence $(I+W_{\ell,\lambda}^\pm)^{-1} \in \mathcal{B}(\mathcal{H}_+)$ for all ℓ and $\lambda > 0$. By a calculation similar to (11.42), one finds that

$$\|(I+W_{\ell,\lambda}^\pm)^{-1}\| \leq \|(I+W_\lambda^\pm)^{-1}\| \text{ for all } \ell \text{ and } \lambda > 0. \tag{11.47}$$

From (11.18) and (10.49) we have $R_\ell(\lambda) = (Y_{\ell m}, R(\lambda)Y_{\ell m})_0 = -2\pi i(M_A(\lambda)^*Y_{\ell m}, (I+W_\lambda^+)^{-1}M_B(\lambda)^*Y_{\ell m})$. Now $M_{A,\ell}(\lambda)^*$ maps $\mathbb{C}$ into $\mathcal{H}_+$. By an abuse of notation, we denote by $M_{A,\ell}(\lambda)^*$ the image of the complex number 1 under the action of the operator $M_{A,\ell}(\lambda)^*$. Then one easily gets from (11.40) that

$$M_A(\lambda)^*Y_{\ell m} = \hat{U}^*M_{A,\ell}(\lambda)^* \otimes Y_{\ell m} \in \mathcal{H}_r \otimes \mathcal{H}_a. \tag{11.48}$$

By inserting (11.48) and $(I+W_\lambda^+)^{-1} = \sum_{\ell m} \hat{U}^*(I+W_{\ell,\lambda}^+)^{-1}\hat{U} \otimes F_{\ell m}$ in the above expression for $R_\ell(\lambda)$, we arrive at (11.44). Since $|R_\ell(\lambda)-R_\ell(\mu)| \leq \|R(\lambda)-R(\mu)\|$, the continuity of $R_\ell(\lambda)$ on $(0,\infty)$, uniformly in ℓ, follows from the continuity of $R(\lambda)$ (Proposition 10.12(a)) and the fact that $\Gamma_0 \cap (0,\infty) = \emptyset$. (11.45) results from combining (10.50) with (11.16) and using the expansions (11.30) and (11.64) (Problem 11.13).

(c) Since $\Gamma_0 \cap (0,\infty) = \emptyset$, the integral equation (11.31) is valid for all $k > 0$. By using (11.29) and (11.45), it can be rewritten as

$$\psi^-_{\ell,k}(r) = [\hat{j}_\ell(kr) + (2i)^{-1}R_\ell(\lambda)\hat{h}^+_\ell(kr)] + k^{-1}\hat{h}^+_\ell(kr)\cdot$$

$$\int_r^\infty \hat{j}_\ell(ks)V(s)\psi^-_{\ell,k}(s)ds - k^{-1}\hat{j}_\ell(kr)\int_r^\infty \hat{h}^+_\ell(ks)V(s)\psi^-_{\ell,k}(s)ds.$$

By Proposition 11.13(a), $\psi^-_{\ell,k}$ is bounded for each ℓ and $k > 0$. Using the bounds (11.33) for $\hat{j}_\ell$ and $\hat{h}^+_\ell$, we see that for $r > 1$, $|\hat{h}^+_\ell(kr)\int_r^\infty \hat{j}_\ell(ks)V(s)\psi^-_{\ell,k}(s)ds| \leq \alpha\int_r^\infty |V(s)|ds \leq c_3 r^{-\frac{1}{2}-\nu} + c_4 r^{\frac{1}{2}-\nu-\delta}$. An identical bound is obtained for $\hat{j}_\ell(kr)\int_r^\infty \hat{h}^+_\ell(ks)V(s)\psi^-_{\ell,k}(s)ds$. On the other hand since $R_\ell(\lambda) = \exp(2i\delta_\ell)-1$, $\hat{j}_\ell(kr) + (2i)^{-1}R_\ell(\lambda)\hat{h}^+_\ell(kr) = (2i)^{-1}\exp(i\delta_\ell)\cdot [\exp(i\delta_\ell)\hat{h}^+_\ell(kr) - \exp(-i\delta_\ell)\hat{h}^-_\ell(kr)]$. By using the asymptotic expansion (11.63) of $\hat{h}^\pm_\ell$, we get (11.46). The asymptotic behaviour of $\phi_{\ell,k}(r)$ follows by combining (11.46) with Proposition 11.14(b).

(d) Since $\phi_{\ell,k}$ satisfies a differential equation with real coefficients and a real boundary condition (11.36), it must be real for all r. Another way of arriving at the same result is to consider the integral equation (11.37) and note that both the inhomogeneous term $\hat{j}_\ell$ and the kernel $G^{0\prime}_{\ell,\lambda}$ are real. This means that $0 = \phi_{\ell,k}(r) - \overline{\phi_{\ell,k}(r)} = \mathfrak{f}_\ell(k)\psi^-_{\ell,k}(r) - \overline{\mathfrak{f}_\ell(k)\psi^-_{\ell,k}(r)}$ for all $r\varepsilon[0,\infty)$, in particular as $r\to\infty$. Putting this together with the asymptotic expansion (11.46) of $\psi^-_{\ell,k}(r)$ we obtain $S_\ell(\lambda) \equiv \exp[2i\delta_\ell(\lambda)] = \overline{\mathfrak{f}_\ell(k)}/\mathfrak{f}_\ell(k)$, with $k^2 = \lambda$. #

<u>Remark 11.17</u> : (a) It can be seen from the proofs of this section that most of the results extend to a class of potentials larger than $\mathcal{V}$. In fact, they are valid for all V

satisfying the integrability conditions (see [AR]) :

$$\int_0^1 r|V(r)|dr < \infty \text{ and } \int_1^\infty |V(r)|dr < \infty.$$

The study of Jost functions and phase shifts has been extended to potentials that may be highly singular at the origin (i.e. verifying only $\int_a^\infty |V(r)|dr < \infty$ for each $a > 0$). See Amrein and Georgescu [2].

(b) We have shown in (11.46) that $\psi^-_{\ell,k}(r)$ behaves as $\exp(i\delta_\ell)\sin(kr-\frac{1}{2}\ell\pi+\delta_\ell)$ as $r \to \infty$. On the other hand the free eigenfunction $\hat{j}_\ell(kr)$ has a behaviour like $\sin(kr-\frac{1}{2}\ell\pi)$ at large r. Thus the effect of the interaction is to produce a change by the amount δ_ℓ in the phase of the eigenfunction at large distances, whence the name "phase shift" for δ_ℓ.

(c) By Problem 11.10, the Hamiltonians H_ℓ have no positive eigenvalues. To treat the case of $\lambda = 0$, we have to change the boundary condition (11.36) of the regular solution $\phi_{\ell,k}$ to

$$\lim_{r\to 0} r^{-\ell-1}\phi_{\ell,o}(r) = 1. \tag{11.49}$$

Then one obtains the associated integral equation ([AR]) :

$$\phi_{\ell,o}(r) = r^{\ell+1}-(2\ell+1)^{-1}\int_0^r (r^{-\ell}s^{\ell+1}-r^{\ell+1}s^{-\ell})V(s)\phi_{\ell,o}(s)ds. \tag{11.50}$$

If $\lambda = 0$ is an exceptional point, then there exists at least one ℓ such that $I + W_{\ell,o}$ is not invertible. For potentials behaving at infinity like $r^{-5/2-\eta}$ $(\eta > 0)$, it can be shown that $\lambda = 0$ is an eigenvalue of H_ℓ if $\ell \geq 1$, whereas if $\ell = 0$, it is not. In the latter case, the ordinary differential equation $(d^2/dr^2-V)\phi_{o,o}(r) = 0$ has a bounded solution which tends to a non-zero constant as $r \to \infty$ and hence the solution is not square-integrable. Such a solution is said to represent a quasi-bound state (see Remark 10.18(b)) and has been used to describe the singlet neutron-proton state. For an intuitive description of this phenomenon, the reader is referred to [LL, Section 109]. A rigorous analysis of the preceding statements is given as an exercise (Problem 11.11). There can also be bound states at $\lambda = 0$ in the $\ell = 0$ partial wave if the potential goes to zero at infinity slower than $r^{-5/2}$ (see [N, page 440]).

One often studies the regular solution $\phi_{\ell,k}(r)$ and the Jost functions $\mathcal{f}_\ell(k)$ for complex values of ℓ and k and associates, for instance, the bound states of H_ℓ with the zeros of $\mathcal{f}_\ell(k)$ for $k = i\kappa$, $\kappa > 0$. We do not discuss this here and refer to [AR] and [N] for various analytic properties of $\mathcal{f}_\ell(k)$ and $S_\ell(k)$. It can be shown that if the potential behaves like $\exp(-\alpha r)$ as $r \to \infty$ with $\alpha > 0$, then $\mathcal{f}_\ell(k)$ and $S_\ell(k)$ can be analytically continued into the strip $0 \leq |\mathrm{Im}k| < \alpha/2$.

Another phenomenon which is of great interest is that of <u>resonance scattering</u>. The partial wave total scattering cross section $\sigma_\ell \equiv 4\pi(2\ell+1)\lambda^{-1}\sin^2\delta_\ell(\lambda)$ shows sharp local maxima at certain energies $\lambda = \lambda_r$. This may happen when $\delta_\ell(\lambda_r)$ is close to $\pi/2$. The intuitive picture of a resonance is to associate with it a long-lived metastable state, which implies a large positive time delay. Thus by (7.30) a resonance at energy $\lambda = \lambda_r$ is characterized, in addition to $\delta_\ell(\lambda_r) \approx \pi/2 \pmod{\pi}$, by the property that $d\delta_\ell(\lambda_r)/d\lambda$ is positive and large. The simplest model of a resonance is to assume $\delta_0(\lambda) \approx \arctan[-\frac{1}{2}\Gamma(\lambda-\lambda_r)^{-1}]$ in the neighbourhood of λ_r, where $\Gamma > 0$ and we have taken $\ell = 0$ for simplicity. Then $d\delta_0(\lambda_r)/d\lambda = 2\Gamma^{-1}$, which is positive and can be large if Γ is small. This leads to $S_0(\lambda) \approx (\lambda-\lambda_r-i\Gamma/2)(\lambda-\lambda_r+i\Gamma/2)^{-1}$ and $\sigma_0(\lambda) \approx (4\pi/\lambda)(\Gamma/2)^2[(\lambda-\lambda_r)^2 + (\Gamma/2)^2]^{-1}$. The latter expression is called the Breit-Wigner formula and $\Gamma/2$ is known as the half-width of the resonance. For more realistic models and further discussions, see [N] and Simon [1].

In Proposition 11.16(a) we have seen that $\exp(2i\delta_\ell(\lambda)) \equiv 1 + R_\ell(\lambda)$ is a continuous function of λ for each ℓ and $\lambda > 0$. However $\delta_\ell(\lambda)$ is determined modulo π. We shall see in Section 12-3 that $\exp(2i\delta_\ell(\lambda)) \to 1$ as $\lambda \to +0$ and as $\lambda \to \infty$. We choose that branch of the logarithm which makes δ_ℓ a continuous function on $(0,\infty)$ and $\delta_\ell(\infty) = 0$ for all ℓ. With this choice, one can derive an important result connecting the total number n_ℓ of eigenvalues (including the zero-energy eigenvalues) of the Hamiltonian H_ℓ to the phase shift δ_ℓ at zero energy. This is known as Levinson's theorem and it states that $\delta_\ell(0) = \pi n_\ell$ for $\ell \geq 1$ and $\delta_0(0) = \pi(n_0+\frac{1}{2})$ or $= \pi n_0$ according to whether there is a quasi-bound state at $\lambda = 0$ or not. For a proof of this theorem in the case of spherically symmetric potentials, see [N], and for more general potentials, see Dreyfus [1] and Newton [2].

11-5 NOTES AND SUPPLEMENTARY MATERIAL

A. We give here the definitions of the spherical

harmonics and Bessel functions and summarize some of their properties.

The spherical harmonics $Y_{\ell m}$ ($\ell = 0,1,2,\ldots; -\ell \leq m \leq \ell$) are defined as :

$$Y_{\ell m}(\theta,\phi) = (-1)^m \left(\frac{2\ell+1}{4\pi} \frac{(\ell-m)!}{(\ell+m)!}\right)^{\frac{1}{2}} P_\ell^m(\cos\theta)e^{im\phi} \quad \text{for } m \geq 0$$

$$\text{and} \quad Y_{\ell m}(\theta,\phi) = (-1)^m \overline{Y_{\ell -m}(\theta,\phi)} \quad \text{for } m < 0, \tag{11.51}$$

where $0 \leq \theta \leq \pi$, $0 \leq \phi \leq 2\pi$ and P_ℓ^m are the associated Legendre functions. These functions P_ℓ^m are the regular solutions of the differential equation ($-1 \leq t \leq 1$) :

$$[(1-t^2)d^2/dt^2-2t\ d/dt + \ell(\ell+1)-m^2/(1-t^2)]P_\ell^m = 0. \tag{11.52}$$

The functions $\{Y_{\ell m}\}$ defined above form an orthonormal set :

$$\int_0^{2\pi} d\phi \int_0^\pi \overline{Y_{\ell' m'}(\theta,\phi)} Y_{\ell m}(\theta,\phi)\sin\theta d\theta = \delta_{\ell\ell'}\delta_{mm'} \tag{11.53}$$

and under a reflection about the origin, i.e. $(\theta,\phi) \mapsto (\pi-\theta,\pi+\phi)$, they behave as $Y_{\ell m}(\pi-\theta,\pi+\phi) = (-1)^\ell Y_{\ell m}(\theta,\phi)$. If $\underline{\omega}$ and $\underline{\omega}'$ are two unit vectors with polar angles (θ,ϕ) and (θ',ϕ') respectively with respect to a fixed coordinate system, then the addition theorem states that

$$P_\ell(\underline{\omega}\cdot\underline{\omega}') = 4\pi/(2\ell+1)\sum_{m=-\ell}^{\ell} \overline{Y_{\ell m}(\theta,\phi)} Y_{\ell m}(\theta',\phi'), \tag{11.54}$$

where $P_\ell \equiv P_\ell^0$ is the Legendre polynomial of degree ℓ. Legendre polynomials are continuous and bounded uniformly with respect to their degree, i.e.

$$|P_\ell(\cos\theta)| \leq 1 \text{ for all } \ell \text{ and } \theta\varepsilon[0,\pi]. \tag{11.55}$$

For a proof of these properties and of the fact that the

spherical harmonics form an orthonormal basis of $L^2(S^{(2)})$, the reader is referred to [SA]. In particular, the functions $\{[(2\ell+1)/2]^{\frac{1}{2}}P_\ell(\cdot)\}$ form an orthonormal basis of $L^2[-1,+1]$ so that

$$\int_{-1}^{1} P_\ell(t)P_{\ell'}(t)dt = 2/(2\ell+1)\delta_{\ell\ell'}. \tag{11.56}$$

Bessel's differential equation for $\nu\varepsilon C$ is :

$$d^2y/dx^2 + x^{-1}dy/dx + (1-\nu^2x^{-2})y = 0, \ x\varepsilon[0,\infty). \tag{11.57}$$

Observe that $x = 0$ is a regular point and $x = \infty$ an irregular point, all other points being ordinary points of this equation. Of the two fundamental solutions, the one that is regular (i.e. finite) at the origin is J_ν, called the <u>Bessel function</u> (also known as the Bessel function of the first kind) of order ν. The other solution which is regular at $x = \infty$ but not at $x = 0$ is N_ν, the <u>Neumann function</u> (or Bessel function of the second kind) of order ν. When $\nu = \ell + \frac{1}{2}$, ℓ an integer or zero, one finds that $N_\nu(x) = (-1)^{\ell-1}J_{-\ell-\frac{1}{2}}(x)$. For an elementary description of J_ν and N_ν and their properties, we refer the reader to [WW].

For our purpose it is convenient to set $\nu = \ell + \frac{1}{2}$, with $\ell = 0,1,2,\ldots$ and define the <u>Riccati-Bessel</u> and <u>Riccati-Neumann functions</u> $\hat{j}_\ell(x)$ and $\hat{n}_\ell(x)$ respectively as

$$\begin{aligned} \hat{j}_\ell(x) &\equiv (\pi x/2)^{\frac{1}{2}}J_{\ell+\frac{1}{2}}(x) \\ \hat{n}_\ell(x) &\equiv -(\pi x/2)^{\frac{1}{2}}N_{\ell+\frac{1}{2}}(x) = (-1)^\ell(\pi x/2)^{\frac{1}{2}}J_{-\ell-\frac{1}{2}}(x). \end{aligned} \tag{11.58}$$

It is easy to check that $\hat{j}_\ell$ and $\hat{n}_\ell$ satisfy an equation similar to the radial Schrödinger equation, namely

$$d^2y/dx^2 + (1-\ell(\ell+1)x^{-2})y = 0. \tag{11.59}$$

The fact that $\hat{j}_0(x) = \sin x$ and $\hat{n}_0(x) = \cos x$ and the relations :

$$\hat{j}_\ell(x) = -(-x)^{\ell+1}(x^{-1}d/dx)^\ell(x^{-1}\hat{j}_0(x)) \text{ and}$$
$$\hat{n}_\ell(x) = -(-x)^{\ell+1}(x^{-1}d/dx)^\ell(x^{-1}\hat{n}_0(x)), \tag{11.60}$$

show that $\hat{j}_\ell$ and $\hat{n}_\ell$ are real functions and allow one to compute them for any order ℓ. It can be shown from (11.60) that $\hat{j}_\ell(x)$ behaves like $x^{\ell+1}$ as $x \to 0$ and remains bounded at infinity. Since $\hat{j}_\ell$ is a continuous function, this implies the bound $|\hat{j}_\ell(x)| \leq c(x/1+x)^{\ell+1}$, where c depends on ℓ. Similarly one gets the bound (11.33) for $\hat{h}^\pm_\ell(x)$. The asymptotic behaviours of these functions as $x \to 0$ are given by :

$$\hat{j}_\ell(x) = (2^\ell \ell!/(2\ell+1)!)x^{\ell+1}[1+0(x^2)]$$
$$\hat{n}_\ell(x) = (2\ell)!/(2^\ell \ell!)x^{-\ell}[1+0(x^2)]. \tag{11.61}$$

Next we introduce the <u>Riccati-Hankel functions</u>, denoted $\hat{h}^\pm_\ell$, as

$$\hat{h}^\pm_\ell(x) = \hat{n}_\ell(x) \pm i\hat{j}_\ell(x). \tag{11.62}$$

These can be shown to have the asymptotic behaviour :

$$\hat{h}^\pm_\ell(x) = \exp(\pm i(x - \tfrac{1}{2}\ell\pi))[1+0(x^{-1})] \text{ as } x \to \infty. \tag{11.63}$$

For a more detailed asymptotic expansion as $x \to \infty$, see [WW, Section 17.5].

Next we consider the expansion of a plane wave in spherical harmonics. The function $\exp(i\underline{k}\cdot\underline{x}) = \exp(ikr\cdot\cos\theta)$ is a bounded function of $\cos\theta$ for fixed k and r, where we have set $\cos\theta = \underline{\omega}_k\cdot\underline{\omega}_x$. Thus it admits an expansion in terms of the orthonormal basis $\{[(2\ell+1)/2]^{\frac{1}{2}}P_\ell(\cos\theta)\}$ of $L^2([0,\pi],\sin\theta d\theta)$, i.e. $\exp(ikr\cdot\cos\theta) = \sum_\ell c_\ell(kr)P_\ell(\cos\theta)$, where $c_\ell(kr)$ are the coefficients of expansion. To calculate $c_\ell(kr)$ we note that $c_\ell(kr) = (2\ell+1)/2 \int_{-1}^{1} \exp(ikrt)P_\ell(t)dt$. Then using properties of the Legendre polynomials P_ℓ, one finds that $c_\ell(kr) = (2\ell+1)i^\ell(kr)^{-1}\hat{j}_\ell(kr)$ (for details, see [WY, Section 6.4]). This leads to

$$\exp(ikr\cdot\cos\theta) = (kr)^{-1}\sum_\ell(2\ell+1)i^\ell\hat{j}_\ell(kr)P_\ell(\cos\theta). \qquad (11.64)$$

By the addition theorem (11.54) of the spherical harmonics we can reexpress (11.64) in terms of $\underline{\omega}_k$ and $\underline{\omega}_x$ as

$$\exp(i\underline{k}\cdot\underline{x}) = (4\pi/kr)\sum_{\ell m}i^\ell\hat{j}_\ell(kr)\overline{Y_{\ell m}(\underline{\omega}_k)}Y_{\ell m}(\underline{\omega}_x). \qquad (11.65)$$

Finally, we give a useful expression for a definite integral involving the square of $\hat{j}_\ell$ which will be needed in Problem 12.4 and to discuss the behaviour of the phase shifts for large ℓ. For a proof, see [WA, Section 13.41].

$$\int_0^\infty \frac{[\hat{j}_\ell(x)]^2dx}{x^{1+\beta}} = \frac{\pi\Gamma(\beta)\Gamma(\ell+1-\beta/2)}{2^{\beta+1}[\Gamma((\beta+1)/2)]^2\Gamma(\ell+1+\beta/2)}, \qquad (11.66)$$

where $2\ell+2 > \beta > 0$ and Γ denotes the gamma function.

B. Next we study the self-adjointness properties of the angular momentum operators L_j $(j = 1,2,3)$. By an abuse of notation we shall not distinguish between a function $f\in L^2(R^3)$ and the associated function in $H_r \otimes H_a$. We shall use the notation C_{00}^∞ for $C_0^\infty((0,\infty))$ and denote by $\mathcal{D}$ the set of arbitrary finite linear combinations of the spherical harmonics $\{Y_{\ell m}\}$. Then $C_{00}^\infty \hat{\otimes} \mathcal{D}$ is dense in $H_r \otimes H_a$ and in $L^2(R^3)$ (on identifying the functions as stated above), because both C_{00}^{∞} and $\mathcal{D}$ are dense in H_r and H_a respectively. Also it is clear that

$$C_{00}^{\infty} \hat{\otimes} \mathcal{D} \subseteq C_0^{\infty}(R^3-\{0\}) \subseteq S(R^3).$$

Since $|R\underline{x}| = |\underline{x}|$, it follows that $U(R) = I \otimes \mathcal{U}(R)$ in $H_r \otimes H_a$ for all $R \in SO(3)$, and $\{\mathcal{U}(R)\}$ forms a group of unitary operators in H_a. Consider the one-parameter subgroup $\mathcal{U}_3(\phi') \equiv \mathcal{U}(R_3(\phi'))$, where $R_3(\phi')$ is the matrix associated with a rotation by ϕ' about the 3rd axis. Clearly $(\mathcal{U}_3(\phi')Y_{\ell m})(\theta,\phi) = Y_{\ell m}(\theta,\phi-\phi') = e^{-im\phi'}Y_{\ell m}(\theta,\phi)$. By Stone's theorem and Proposition 5.11, $\mathcal{U}_3(\phi') = \exp(-iL_3\phi')$, L_3 being a self-adjoint operator in H_a, so that each $Y_{\ell m} \in D(L_3)$ and $L_3 Y_{\ell m} = m\, Y_{\ell m}$. It follows that $(L_3 \pm i)\mathcal{D} = \mathcal{D}$ and hence by Corollary 2.15, L_3 is essentially self-adjoint on $\mathcal{D}$. Similarly $(L_3 \pm i)C_{00}^{\infty} \hat{\otimes} \mathcal{D} = C_{00}^{\infty} \hat{\otimes} \mathcal{D}$, implying that the restriction $\hat{L}_3$ of $L_3 \equiv I \otimes L_3$ to $C_{00}^{\infty} \hat{\otimes} \mathcal{D}$ is essentially self-adjoint. Also $(L_3 Y_{\ell m})(\theta,\phi) = -i\partial Y_{\ell m}(\theta,\phi)/\partial\phi$ and transforming back to the Cartesian coordinate system, we find that $\hat{L}_3 = -i(x_1\partial/\partial x_2 - x_2\partial/\partial x_1)$. An integration by parts gives for $f \in S(R^3)$ and $g \in C_{00}^{\infty} \hat{\otimes} \mathcal{D}$, $(f,\hat{L}_3 g) = (-i[x_1\partial/\partial x_2 - x_2\partial/\partial x_1]f,g)$, showing that $f \in D(\hat{L}_3^*) = D(L_3)$, and $(L_3 f)(\underline{x}) = -i[x_1\partial f(\underline{x})/\partial x_2 - x_2\partial f(\underline{x})/\partial x_1]$. Thus L_3 is essentially self-adjoint on $S(R^3)$ and leaves $S(R^3)$ invariant.

Now, choosing a spherical polar coordinate system with the 1-axis as the polar axis and proceeding exactly as above, we arrive at L_1, the self-adjoint generator of the subgroup of rotation about the 1-axis and conclude that $L_1 \equiv I \otimes L_1$ is essentially self-adjoint on $S(R^3)$, leaves $S(R^3)$ invariant and that L_1 equals the differential operator

$-i(x_2\partial/\partial x_3 - x_3\partial/\partial x_2)$ on $\mathcal{S}(\mathbb{R}^3)$. A similar conclusion holds for L_2 which establishes (11.6). Thus each L_j is essentially self-adjoint on $\mathcal{S}(\mathbb{R}^3)$ and $L_j\mathcal{S}(\mathbb{R}^3) \subseteq \mathcal{S}(\mathbb{R}^3)$ for j = 1,2,3. It is now elementary to verify that the operators L_j satisfy the commutation relation (11.7) on $\mathcal{S}(\mathbb{R}^3)$.

Next, we can define $L_\pm$ on $\mathcal{S}(\mathbb{R}^3)$ as in (11.8) and note that by making a transformation to the spherical coordinate system, we can write $L_\pm = I \otimes \mathcal{L}_\pm$ on $C_{00}^\infty \hat{\otimes} \mathcal{D}$, where $\mathcal{L}_\pm$ are the differential operators given by

$$\mathcal{L}_\pm = \pm e^{\pm i\phi}(\partial/\partial\theta \pm i \cot\theta\, \partial/\partial\phi) \text{ on } \mathcal{D}. \tag{11.67}$$

A lengthy computation using (11.51) and (11.52) shows that

$$\mathcal{L}_\pm Y_{\ell m} = [\ell(\ell+1)-m(m \pm 1)]^{\frac{1}{2}} Y_{\ell\, m\pm1}, \tag{11.68}$$

from which it follows that every vector in $\mathcal{H}_{\ell m} \equiv \mathcal{H}_r \otimes Y_{\ell m}$ is in the domain of the closure of $L_\pm$, which we denote also by $L_\pm$. Similarly it follows that $\mathcal{H}_{\ell m} \subseteq D(L_j)$ (j = 1,2,3) and that any bounded operator B that commutes with the family $\{U(R)\}$ will satisfy $L_jBf = BL_jf$ for each j and all $f\in\mathcal{H}_{\ell m}$.

Since $L_j\mathcal{S}(\mathbb{R}^3) \subseteq \mathcal{S}(\mathbb{R}^3)$ for j = 1,2,3, we can define $\underline{L}^2$ on $\mathcal{S}(\mathbb{R}^3)$ by (11.8) and observe that $\underline{L}^2$ leaves $\mathcal{S}(\mathbb{R}^3)$ invariant. On transforming to spherical coordinates, we see that $\underline{L}^2$ coincides with $I \hat{\otimes} \underline{\mathcal{L}}^2$ on $C_{00}^\infty \hat{\otimes} \mathcal{D} \subseteq \mathcal{S}(\mathbb{R}^3)$, where $\underline{\mathcal{L}}^2$ is the differential operator on $\mathcal{D}$ given as

$$\underline{\mathcal{L}}^2 = -[(\sin\theta)^{-1}\partial/\partial\theta(\sin\theta\partial/\partial\theta) + (\sin\theta)^{-2}\partial^2/\partial\phi^2]. \tag{11.69}$$

It is now easy to check that for each ℓ, m,

$$\underline{\mathcal{L}}^2 Y_{\ell m} = \ell(\ell+1)Y_{\ell m}, \tag{11.70}$$

from which it follows that $\underline{L}^2$ is essentially self-adjoint on $\mathcal{D}$. Hence, as before the restriction of $\underline{L}^2 = I \otimes \underline{L}^2$ to $S(R^3)$ is essentially self-adjoint. Thus the self-adjoint operators $L_j (j = 1,2,3)$ and $\underline{L}^2$ have purely discrete spectrum, accumulating at infinity. They however do not form a mutually commuting set. But $[L_j, \underline{L}^2] = 0$ on $S(R^3)$ for all j, and following the usual convention we diagonalize $\underline{L}^2$ and L_3 simultaneously. Let $\{e_k\}$ be an orthonormal basis of H_r. Then the vectors $\{e_k \otimes Y_{\ell m}\}$ are the common eigenvectors of $\underline{L}^2$ and L_3 with eigenvalues $\ell(\ell+1)$ and m respectively and form an orthonormal basis of $H_r \otimes H_a$. Though none of the operators $L_j (j = 1,2,3)$, $\underline{L}^2$ is bounded in H, it is easy to see that they are bounded in the subspace $\oplus_{m=-\ell}^{\ell} H_{\ell m} = H_r \otimes K_\ell$, where K_ℓ is the $(2\ell+1)$-dimensional Hilbert space defined in Section 11-2. In other words L_j and $\underline{L}^2$ may be represented as $(2\ell+1)\times(2\ell+1)$ matrices in each K_ℓ.

PROBLEMS

11.1 : Prove that (11.3) defines a strongly continuous unitary representation of SO(3).

11.2 : (a) Show that the range of F_ℓ given by (11.10) is $L^2([0,\infty), k^2 dk)$, find an expression for the inverse of F_ℓ and verify (11.12). (b) Prove that $\Theta\tilde{E}_{\ell m} = (-1)^{\ell-m}\tilde{E}_{\ell m}\Theta$, where Θ is the time-reversal operator introduced in Problem 8.8.

11.3 : Let A be self-adjoint in H_1 and B self-adjoint in H_2 with $H_p(B) = H_2$. Show that $A \hat{\otimes} B$ is essentially self-adjoint. (For a more general result, see [RS I, Thm. VIII.33].)

11.4 : Let $H = \oplus_{i=1}^{\infty} H_i$. Assume that each H_i reduces $U_t \equiv \exp(-iH_0 t)$, $V_t \equiv \exp(-iHt)$, $E_{ac}(H_0)$ and $E_{ac}(H)$. Denote by $U_{t,i}$, $V_{t,i}$, $E_{ac}(H_{o,i})$ and $E_{ac}(H_i)$ the parts of these operators in $\mathcal{H}_i$. Show that $\Omega_\pm$ exist if and only if $\Omega_{\pm,i}$ exist for each i. Furthermore $\Omega_\pm$ are complete if and only if $\Omega_{\pm,i}$ are complete for each i.

11.5 : Assume the hypotheses of Proposition 11.9 and that $V_1 \in \mathcal{Y}$. Show that the scattering theory for the pair (H,H_o) is asymptotically complete. (Hint : Problem 11.4.)

11.6 : Verify (11.29).

11.7 : Let V be spherically symmetric and in $\mathcal{Y}$. Show that for $k^2 \notin \Gamma_o$, $\psi^\pm_{\underline{k}}(\underline{x})$ is a function of k,r and $\underline{\omega}_k \cdot \underline{\omega}_x$ only and that $\psi^\pm_{\underline{k}} \in L^2[-1,+1]$ as a function of $\cos(\underline{\omega}_k \cdot \underline{\omega}_x)$ for fixed k and almost all r.

11.8 : Prove the continuity part of Proposition 11.13(a).

11.9† : Assume that $\int_0^1 |rV(r)|dr < \infty$ and $\int_1^\infty |V(r)|dr < \infty$. (a) Solve the Volterra equation (11.37) by iteration and show that $|\phi_{\ell,k}(r)| \leq c(\ell)(kr/1+kr)^{\ell+1}$ for all $k > 0$ and all r. (Hint : Use (11.33), (11.35) and verify the bound $|G^{o\,\prime}_{\ell,\lambda}(r,s)| \leq \alpha k^{-1}(kr/1+kr)^{\ell+1}(ks/1+ks)^{-\ell}$ for $r \geq s$. Reference : [AR]) (b) Show that the integral in (11.39) converges for all $k > 0$.

11.10†: Let V be spherically symmetric and in $\mathcal{Y}$. (a) Show that for $k > 0$, $\mathfrak{f}_\ell(k) \neq 0$. (Hint : (i) $\mathfrak{f}_\ell(k) = 0$ if and only if $\phi_{\ell,k}$ verifies the homogeneous equation associated with (11.31). (ii) Assume $\mathfrak{f}_\ell(k) = 0$. Use $B\phi_{\ell,k} \in \mathcal{H}_+$ and Proposition 10.17, deduce that $\phi_{\ell,k}$ is an eigenvector of H_ℓ with eigenvalue k^2. Prove that $\int_r^\infty \hat{j}_\ell(ks)V(s)\phi_{\ell,k}(s)ds$ and $\int_r^\infty \hat{h}^+_\ell(ks)V(s)\phi_{\ell,k}(s)ds$ are square-integrable to conclude from (11.37) that $\int_0^\infty \hat{j}_\ell(ks)V(s)\phi_{\ell,k}(s)ds = 0$. Thus obtain $\phi_{\ell,k}(r) = -k^{-1}\hat{j}_\ell(kr)\cdot \int_r^\infty \hat{h}^+_\ell(ks)V(s)\phi_{\ell,k}(s)ds + k^{-1}\hat{h}^+_\ell(kr)\int_r^\infty \hat{j}_\ell(ks)V(s)\phi_{\ell,k}(s)ds$. (11.71) (iii) With the notation $\|f\|_{[a,b]} = \sup_{r\in[a,b]}|f(r)|$, choose R so large that $\|\hat{h}^+_\ell(k\cdot)\|_{[R,\infty)}\|\hat{j}_\ell(k\cdot)\|_{[R,\infty)}\int_R^\infty |V(s)|ds < 1/4$ and show from (11.71) that $|\phi_{\ell,k}(r)| \leq 2^{-1}\|\phi_{\ell,k}\|_{[R,\infty)}$ for all $r \geq R$, so that $\phi_{\ell,k}(r) = 0$ for all $r \geq R$. By repeating this procedure on a sequence of sufficiently small intervals covering (0,R), prove that $\phi_{\ell,k} = 0$.)

(b) Establish the absence of positive eigenvalues of H. (Hint : Let $H_\ell g = k^2 g$. Use Lemma 9.2 and Problem 10.8(b) to deduce that $|g(r)| \leq \text{const}\cdot(r/1+r)$. Both g and $\phi_{\ell,k}$ verify the radial Schrödinger equation and vanish at $r = 0$, so that $\phi_{\ell,k} = \alpha g$ satisfies the homogeneous equation associated with (11.31).)

11.11[†] : Assume V spherically symmetric, $V \varepsilon \mathcal{Y}$ and that $\lambda = 0$ is in Γ_0. (a) Show that there is at least one ℓ such that $I+W_{\ell,0}$ is not invertible. (b) Assume in addition that $\int_0^\infty r^{3/2+\eta}|V(r)|dr < \infty$ for some $\eta > 0$. Show that, if $\ell \neq 0$, $\lambda = 0$ is an eigenvalue of H_ℓ, whereas if $\ell = 0$, it is not. (Hint : (i) Establish the integral equation (11.50) for $\phi_{\ell,0}$ and solve it by iteration to find $|\phi_{\ell,0}(r)| \leq c(\ell)r^{\ell+1}$. (ii) As in Problem 11.10, deduce that $\mathfrak{f}_\ell(0) \equiv 1+(2\ell+1)^{-1} \cdot \int_0^\infty s^{-\ell}V(s)\phi_{\ell,0}(s)ds = 0$. (iii) $\phi_{\ell,0} \varepsilon L^\infty(0,\infty), \mathfrak{f}_\ell(0) = 0$, (11.50) and the condition on V imply that, if $\ell > \frac{1}{2}$, $\phi_{\ell,0} \varepsilon L^2(0,\infty)$. If $\ell = 0$, $\phi_{0,0} \varepsilon L^2(0,\infty)$ would imply $\int_0^\infty sV(s)\phi_{0,0}(s)ds = 0$, leading to $\phi_{0,0}(r) = -r\int_r^\infty V(s)\phi_{0,0}(s)ds + \int_r^\infty sV(s)\phi_{0,0}(s)ds$. Then proceed as in Problem 11.10.)

11.12 : (a) Under the assumptions of Problem 11.11, show that $g = G^0_{0,0}Af$ satisfies the radial Schrödinger equation with $k = 0$ and is bounded, where f is the exceptional vector. (Hint : See Problem 11.11(b).) Notice that by Problem 11.11(b), g cannot be square-integrable. (b) Let V be a square-well potential and $H = K_0 + \gamma V$. Show that for certain values of γ, H has a quasi-bound state at zero energy.

11.13 : Prove Proposition 11.16(b).

CHAPTER 12 : SCATTERING AT HIGH AND AT LOW ENERGIES

In Proposition 10.12 we have derived some exact expressions for the on-shell S-matrix $S(\lambda)$. However neither of the expressions (10.49) and (10.50) is very useful for the purposes of explicit computation of the scattering amplitude f or the differential scattering cross-section $d\sigma/d\omega$. This is because of the appearance of the operator $(I+W_\lambda^+)^{-1}$ and the total eigenfunction $\psi_{\underline{k}\omega}^-(\underline{x})$ in (10.49) and (10.50) respectively. The explicit knowledge of these two quantities amounts to an explicit solution (in terms of special functions) of the time-independent Schrödinger equation (10.3), which is not possible in general. Thus it is useful to develop approximation schemes which will allow one to compute the scattering amplitude or differential cross-section approximately in terms of known quantities. Various such approximation schemes exist, each suitable for a specific class of applications and each with its associated domain of validity. In this context we mention the Born approximation, eikonal approximation, WKB approximation, effective range and partial wave approximations. While the first two approximations are useful at high energies, the last two are valid at low energies. For a comprehensive account of such schemes and their physical applications, the reader is referred to [JO] and [MK].

The scope of this chapter is to deduce some rigorous results on the Born approximation (Section 12-1), the scattering at low energy (Section 12-2) and the high and low energy behaviours of the phase shifts (Section 12-3).

12-1 THE BORN APPROXIMATION. SCATTERING AT HIGH ENERGIES

For the purposes of this section it is useful to replace the potential V by γV, where γ is a real coupling constant. Then for $V \varepsilon \mathcal{V}$, (10.49) is to be rewritten as

$$R(\lambda,\gamma) = -2\pi i\gamma\, M_A(\lambda)(I+\gamma W_\lambda^+)^{-1} M_B(\lambda)^*; \quad \lambda \varepsilon (0,\infty) - \Gamma_0. \qquad (12.1)$$

If $\|\gamma W_\lambda^+\| < 1$ then $(I+\gamma W_\lambda^+)^{-1}$ can be expanded in a geometric series (Proposition 2.20) and (12.1) can be written as a series, viz.

$$R(\lambda,\gamma) = \sum_{j=1}^{\infty} R^{(j)}(\lambda,\gamma), \qquad (12.2)$$

where $$R^{(j)}(\lambda,\gamma) = -2\pi i\gamma^j M_A(\lambda)(-W_\lambda^+)^{j-1} M_B(\lambda)^*.$$

The series (12.2) is called the Born series for the R-matrix.

One can instead write down the Born series for S-matrix as $S(\lambda,\gamma) \equiv I_0 + R(\lambda,\gamma) = I_0 + \sum_{j=1}^{\infty} R^{(j)}(\lambda,\gamma)$.

We now study the convergence of the Born series as a function of λ and γ. First note that $\|W_\lambda^\pm\|_{HS} = \|W_0^\pm\|_{HS}^2 = (16\pi^2)^{-1}\int\int |B(\underline{x})|^2 |\underline{x}-\underline{y}|^{-2} |A(\underline{y})|^2 d^3x d^3y$, which by Lemma 9.13(c) is finite and is obviously independent of λ. Thus in general we do not expect the series $\sum_1^\infty (-W_\lambda^+)^{j-1}$ to converge in Hilbert-Schmidt norm. However the following proposition, essentially due to Zemach and Klein [1], helps us conclude convergence in operator norm for λ large enough.

PROPOSITION 12.1 : Let $V \in \mathcal{Y}$ and $W_\lambda^\pm$ be the Hilbert-Schmidt operators given in Lemma 9.13 with kernels $W_\lambda^\pm(\underline{x},\underline{y}) = (4\pi)^{-1} B(\underline{x}) |\underline{x}-\underline{y}|^{-1} \exp(\pm i\lambda^{\frac{1}{2}} |\underline{x}-\underline{y}|) A(\underline{y})$. Then $W_\lambda^\pm$ converges to zero in operator norm as $\lambda \to \infty$.

Proof : Since $W_\lambda^\pm$ are Hilbert-Schmidt operators, they are bounded and hence by Lemma 5.30(b) and Proposition 2.32(a), $\|W_\lambda^\pm\|^4 = \|W_\lambda^{\pm *} W_\lambda^\pm\|^2 \leq \|W_\lambda^{+*} W_\lambda^\pm\|_{HS}^2$. The kernel of the Hilbert-Schmidt operator $W_\lambda^{\pm *} W_\lambda^\pm$ is given by Proposition 7.20 as follows :

$$(W_\lambda^{\pm *} W_\lambda^\pm)(\underline{x},\underline{y}) = \int d^3u \, \overline{W_\lambda^\pm(\underline{u},\underline{x})} \, W_\lambda^\pm(\underline{u},\underline{y})$$

$$= (4\pi)^{-2} \int d^3u \, \exp[\pm i\lambda^{\frac{1}{2}} (|\underline{y}-\underline{u}| - |\underline{x}-\underline{u}|)] \cdot$$

$$\{|\underline{x}-\underline{u}|^{-1} |\underline{y}-\underline{u}|^{-1} A(\underline{x}) |B(\underline{u})|^2 A(\underline{y})\}.$$

By (2.67) this leads to the following expression for the Hilbert-Schmidt norm of $W_\lambda^{\pm *} W_\lambda^\pm$:

$$\|W_\lambda^{\pm}{}^*W_\lambda^{\pm}\|_{HS}^2 =$$

$$(4\pi)^{-4}\int d^3x d^3y d^3u d^3v \exp[\pm i\lambda^{\frac{1}{2}}(|\underline{y}-\underline{u}|-|\underline{y}-\underline{v}|-|\underline{x}-\underline{u}|+|\underline{x}-\underline{v}|)]\cdot$$
$$\{|\underline{x}-\underline{u}|^{-1}|\underline{x}-\underline{v}|^{-1}|\underline{y}-\underline{u}|^{-1}|\underline{y}-\underline{v}|^{-1}|A(\underline{x})|^2|B(\underline{u})|^2|B(\underline{v})|^2|A(\underline{y})|^2\}. \tag{12.3}$$

In the above we have interchanged various orders of integration which is allowed by Fubini's theorem [R]. That the expression $\{\cdot\}$ is integrable can be seen first by applying the Schwarz inequality to the two integrals of the type $\int d^3x(A(\underline{x})|\underline{x}-\underline{u}|^{-1})(A(\underline{x})|\underline{x}-\underline{v}|^{-1})$ and then using Lemma 9.13. Choosing a new variable $\rho = |\underline{y}-\underline{u}|-|\underline{y}-\underline{v}|-|\underline{x}-\underline{u}|+|\underline{x}-\underline{v}|$, we can rewrite (12.3) as

$$\|W_\lambda^{\pm}{}^*W_\lambda^{\pm}\|_{HS}^2 = (4\pi)^{-4}\int_{-\infty}^{\infty} f(\rho)\exp(\pm i\lambda^{\frac{1}{2}}\rho)d\rho, \tag{12.4}$$

where $f(\rho)$ is the integral of the expression $\{\cdot\}$ times the appropriate Jacobian factor with respect to all the variables other than ρ. Thus $\int|f(\rho)|d\rho = \int d^3x d^3y d^3u d^3v|\{\cdot\}|$, which we have seen to be finite, and by the Riemann-Lebesgue lemma [R, page 90] the expression in (12.4) converges to zero as $\lambda \to \infty$. This implies that $\|W_\lambda^{\pm}\| \to 0$ as $\lambda \to \infty$. #

The next proposition deals with the behaviour of W_z, defined in Lemma 9.13, as $|z| \to \infty$ along any direction in the complex plane except the positive real axis and is to be compared with the results of Proposition 12.1.

<u>PROPOSITION 12.2</u> : Let $W_z = BR_z^oA$ be defined for all $z\varepsilon\rho(K_o)$ as in Lemma 9.13. Then W_z converges to zero in Hilbert-Schmidt norm as z tends to infinity along any ray in the complex plane other than the positive real axis.

The proof of this Proposition is based on the choice of the branch of the square root of complex z such that $\mathrm{Im}\sqrt{z} > 0$ and an application of the dominated convergence theorem (Problem 12.1). An immediate consequence of Propositions 12.1 and 12.2 is the boundedness of the exceptional set Γ_o introduced in Section 9-1 :

COROLLARY 12.3 : Let $V \varepsilon \mathcal{V}$. Then the set $\Gamma_o = \{\lambda \varepsilon R \mid$ either $(I+W_\lambda^+)$ or $(I+W_\lambda^-)$ is not invertible$\}$, a closed set of Lebesgue measure zero, is bounded.

Proof : Since by Propositions 12.1 and 12.2, $\|W_\lambda^\pm\| \to 0$ as $|\lambda| \to \infty$, there exists a positive number λ_o (depending on A and B only) such that $\|W_\lambda^+\| < 1$ for $|\lambda| > \lambda_o$. By Proposition 2.20 it follows that for $|\lambda| > \lambda_o$, $(I+W_\lambda^\pm)^{-1} = \sum_{j=1}^{\infty}(-W_\lambda^\pm)^{j-1}$ is in $\mathcal{B}(\mathcal{H})$. Thus $\{\lambda \varepsilon R \mid |\lambda| > \lambda_o\} \cap \Gamma_o = \emptyset$. #

By Remark 9.7 and Corollary 12.3 we see that $\sigma_s(H) \subseteq [-\lambda_o', \lambda_o']$, where λ_o' is the least of all positive λ_o such that $\|W_\lambda^\pm\| < 1$ for $|\lambda| > \lambda_o$. This along with Proposition 10.17 implies that the singular spectrum of H consists of only eigenvalues all contained in the interval $[-\lambda_o', \lambda_o']$. Now we come to the question of the convergence of the Born series (12.2).

PROPOSITION 12.4 : Let $V \varepsilon \mathcal{V}$. (a) For each $\lambda > 0$, the Born series is convergent for small coupling constants (i.e. for $|\gamma| < \|W_\lambda^+\|^{-1}$). (b) There exists a $\gamma_o > 0$ such that the Born series is convergent for all $\lambda > 0$ and all $|\gamma| < \gamma_o$. (c) For fixed γ, there exists $0 < \lambda_o < \infty$ such that the Born series converges for each $\lambda > \lambda_0$. In all three cases the convergence is in the Hilbert-Schmidt norm.

Proof : It follows from (9.28) and (9.22) that for each λ, $\|W_\lambda^\pm\| \neq 0$ unless $V = 0$. Thus we can assume $\|W_\lambda^\pm\| \neq 0$, and if $\|\gamma W_\lambda^+\| = |\gamma|\,\|W_\lambda^+\| < 1$ then the series $\sum(-\gamma W_\lambda^+)^{j-1}$ converges in norm. Since by Lemmas 10.1 and 10.21 we know that $M_A(\lambda)$ is Hilbert-Schmidt and $M_B(\lambda)$ is bounded, it follows by (2.63) that the series (12.2) converges in Hilbert-Schmidt norm, whenever $|\gamma| < \|W_\lambda^+\|^{-1}$. By Lemma 9.13(c), $\|W_\lambda^+\| \leq \|W_\lambda^+\|_{HS} = \|W_0^+\|_{HS}$ for all λ. Choose γ_0 such that $\sup_{\lambda\geq 0}\|W_\lambda^+\| = \gamma_0^{-1}$. This proves (a) and (b).

By Proposition 12.1 there exists $0 < \lambda_0 < \infty$ such that $\|W_\lambda^+\| < |\gamma|^{-1}$ for all $\lambda > \lambda_0$. Then as in (a) we conclude that the Born series converges in Hilbert-Schmidt norm. #

Proposition 12.4(a) shows that the Born series (12.2) converges in Hilbert-Schmidt norm also for a complex coupling constant γ satisfying $\|\gamma W_\lambda^+\| < 1$ and therefore defines $R(\lambda,\gamma)$ as a Hilbert-Schmidt operator-valued analytic function in the disc $\{\gamma \,\big|\, |\gamma| < \|W_\lambda^+\|^{-1}\}$ in the complex γ-plane for every fixed energy λ.

The most interesting situation for which the Born series converges is for high energies as shown in Proposition 12.4(c). For further discussion, let us call the partial sum $R_n(\lambda) \equiv \sum_{j=1}^n R^{(j)}(\lambda)$ the <u>n^{th} order Born approximation</u> of $R(\lambda)$. The first order Born approximation is traditionally called the <u>Born approximation</u>. Using the identity $a^2-b^2 = 2a(a-b)-(a-b)^2$ and the triangle inequality (analogue of

(2.89)) for the Hilbert-Schmidt norm, we get that

$$\left| \|R(\lambda)\|_{HS}^2 - \|R_n(\lambda)\|_{HS}^2 \right| \leq 2\|R(\lambda)\|_{HS}\|R(\lambda)-R_n(\lambda)\|_{HS} + \|R(\lambda)-R_n(\lambda)\|_{HS}^2,$$

which by Proposition 12.4(c) converges to zero as $n \to \infty$ for λ fixed but large enough. Using (7.69) we can express this result in terms of the average total scattering cross-section $\bar{\sigma}$. If we define the n-th order Born approximation of $\bar{\sigma}$ as $\bar{\sigma}_n(\lambda) = \pi\lambda^{-1}\|R_n(\lambda)\|_{HS}^2$, then, given $\delta > 0$ however small, $|\bar{\sigma}(\lambda)-\bar{\sigma}_n(\lambda)| < \delta$ whenever $n > n_o(\delta,\lambda)$ for fixed $\lambda > \lambda_o$ (λ_o is given by Proposition 12.4(c)). The next proposition gives the asymptotic behaviour of $\bar{\sigma}(\lambda)$ and $\bar{\sigma}_1(\lambda)$ as $\lambda \to \infty$.

PROPOSITION 12.5 : Let $V \in \mathcal{V}$. Then (a) $\bar{\sigma}(\lambda) = O(\lambda^{-1})$ and $\bar{\sigma}_1(\lambda) = O(\lambda^{-1})$ as $\lambda \to \infty$. (b) If furthermore $B_1 \in L^2(R^3) \cap L^\infty(R^3)$, then $S(\lambda)$ converges to I_o in norm as $\lambda \to \infty$.

Proof : (a) By Lemma 10.22, we have

$$M_B(\lambda)^* M_A(\lambda) = (2\pi i)^{-1}(W_\lambda^+ - W_\lambda^-). \tag{12.5}$$

From (12.2) we see that

$$R^{(1)}(\lambda) = -2\pi i\, M_A(\lambda)M_B(\lambda)^* = -2\pi i\, M_B(\lambda)M_A(\lambda)^*, \tag{12.6}$$

where the last equality follows by considering the two kernels. Using (12.6), the cyclicity of the trace [Problem 2.36], (12.5) and (2.95), one obtains

$$\|R^{(1)}(\lambda)\|_{HS}^2 = \mathrm{Tr}_o R^{(1)}(\lambda)^* R^{(1)}(\lambda) = |2\pi i|^2 \mathrm{Tr}[M_B(\lambda)^* M_A(\lambda)]^2$$

$$= \mathrm{Tr}(W_\lambda^+ - W_\lambda^-)^2 \leq \|W_\lambda^+ - W_\lambda^-\|_{HS}^2, \tag{12.7}$$

which is bounded by Lemma 9.13(c). Similarly for $R(\lambda)$ we have

$$\|R(\lambda)\|_{HS}^2 =$$

$$4\pi^2 \mathrm{Tr}_0\{[M_B(\lambda)(I+W_\lambda^-)^{-1*}M_A(\lambda)^*]^*[M_A(\lambda)(I+W_\lambda^+)^{-1}M_B(\lambda)^*]\}$$

$$= \mathrm{Tr}(W_\lambda^+-W_\lambda^-)(I+W_\lambda^-)^{-1}(W_\lambda^+-W_\lambda^-)(I+W_\lambda^+)^{-1}$$

$$\leq \|W_\lambda^+-W_\lambda^-\|_{HS}^2 \,\|(I+W_\lambda^-)^{-1}\|\,\|(I+W_\lambda^+)^{-1}\|, \tag{12.8}$$

where we have used both expressions (10.48) and (10.49) for $R(\lambda)$, and (2.63) and (2.95). Thus (12.8) along with Lemma 9.13(c) and Proposition 12.1 leads to the result that $\|R(\lambda)\|_{HS}^2 \leq$ constant for $\lambda > \lambda_0$ and we have (a) by virtue of the definitions of $\bar{\sigma}$ and $\bar{\sigma}_1$.

(b) Assume $B_2 = 0$ and $B_1 \varepsilon L^2(R^3) \cap L^\infty(R^3)$. Then we know from Lemma 9.8 that $\|BR^o_{\lambda\pm io}B\|$ is bounded in λ. Since $BR^o_{\lambda+i\eta}B-BR^o_{\lambda-i\eta}B$ is Hilbert-Schmidt and converges in Hilbert-Schmidt norm as $\eta \to +0$ to $BR^o_{\lambda+io}B-BR^o_{\lambda-io}B$, the latter operator is Hilbert-Schmidt (Problem 12.9). A comparison of its kernel with that of $M_B(\lambda)^*M_B(\lambda)$ shows that $M_B(\lambda)^*M_B(\lambda) = (2\pi i)^{-1}[BR^o_{\lambda+io}B-BR^o_{\lambda-io}B]$. By Lemma 5.30(b), we have $\|M_B(\lambda)\|^2 = (2\pi)^{-1}\|BR^o_{\lambda+io}B-BR^o_{\lambda-io}B\| \leq$ constant, independent of λ. Similarly by Lemma 10.22, $\|M_A(\lambda)\|^2 = (2\pi)^{-1}\|AR^o_{\lambda+io}A-AR^o_{\lambda-io}A\|$. Furthermore by setting $B = A$ in Proposition 12.1, we find that $\|AR^o_{\lambda\pm io}A\| \to 0$ as $\lambda \to \infty$. Therefore

$$\|R(\lambda)\|^2 \leq 4\pi^2\|M_A(\lambda)\|^2\|(I+W_\lambda^+)^{-1}\|^2\|M_B(\lambda)\|^2$$

$$\leq 2\pi\,\|AR^o_{\lambda+io}A-AR^o_{\lambda-io}A\|\,\|(I+W_\lambda^+)^{-1}\|^2\|M_B(\lambda)\|^2 \to 0, \tag{12.9}$$

as $\lambda \to \infty$. Since $R(\lambda) = S(\lambda)-I_o$, we have the desired result.

The proof of (b) with $B_2 \neq 0$ is left as an exercise (Problem 12.5). The hypothesis of (b) can be considerably relaxed. It suffices, for instance, that $B_1 \varepsilon L^2(R^3) \cap L^{2+\delta}(R^3)$

for some $\delta > 0$ (see Remark following Problem 12.9). #

From (12.6) it is obvious that $R^{(1)}(\lambda)^* = -R^{(1)}(\lambda)$ and hence $R^{(1)}(\lambda)+R^{(1)}(\lambda)^* = 0$, while $\|R^{(1)}(\lambda)\|_{HS}^2 \neq 0$ in general. This, when compared with (7.59), is often expressed by saying that the Born approximation violates unitarity.

We have seen in Proposition 12.4(c) that the Born series (12.2) is convergent for each sufficiently high but fixed energy. This does not mean that (12.2) is a high energy asymptotic expansion for $R(\lambda)$. For example the Born approximation need not satisfy $\|R(\lambda)-R^{(1)}(\lambda)\|_{HS} = o(\|R^{(1)}(\lambda)\|_{HS})$ as $\lambda \to \infty$. For the scattering cross section, Proposition 12.5 shows trivially that $|\bar{\sigma}(\lambda)-\bar{\sigma}_1(\lambda)| \to 0$ as $\lambda \to \infty$, which again does not imply that $\bar{\sigma}_1(\lambda)$ is a good approximation to $\bar{\sigma}(\lambda)$ at high energies. In order that some function σ_{appr} be a good approximation to $\bar{\sigma}(\lambda)$, one needs to know for instance that $|\bar{\sigma}(\lambda)-\sigma_{appr}(\lambda)| = o(\sigma_{appr}(\lambda))$ as $\lambda \to \infty$. A more satisfactory scheme is to look for an asymptotic expansion for $\bar{\sigma}(\lambda)$ of the form :

$$\bar{\sigma}(\lambda) = c_1\lambda^{-1}+c_2\lambda^{-2}+o(\lambda^{-2}), \tag{12.10}$$

where c_1 and c_2 are constants, independent of λ. Calculations for the Yukawa potential and their comparison with results based on numerical integration of the Schrödinger equation make it very plausible that the requirement $|\bar{\sigma}(\lambda)-\bar{\sigma}_1(\lambda)| = o(\bar{\sigma}_1(\lambda))$ as $\lambda \to \infty$ need not be verified (see [JO, Section 8.6.1]). An asymptotic expansion of the form (12.10) for $\bar{\sigma}(\lambda)$ (with $c_1 \neq 0$ in general) is given by the eikonal approximation (see Hunziker [1]).

It is interesting to note that for separable interactions, introduced in Section 8-3, the Born series also provides an asymptotic expansion as $\lambda \to \infty$. In fact from (8.27) we see that the R-matrix in this case is given by

$$R(\lambda) = R^{(1)}(\lambda)[1+\gamma(h,R^{o}_{\lambda+io}h)]^{-1},$$

where the Born approximation $R^{(1)}(\lambda)$ is expressed as $R^{(1)}(\lambda)f = -2\pi i\gamma h_\lambda(h_\lambda,f)$ for every $f\varepsilon L^2(S^{(2)})$ and we have used the notations of Section 8-3. Since by Lemma 8.19 the function $t\mapsto(h,U_t h)$ is integrable, we know from (8.22) and the Riemann-Lebesgue lemma that $(h,R^{o}_{\lambda+io}h) \to 0$ as $\lambda \to \infty$. Thus there exists λ_o with $0 < \lambda_o < \infty$ such that $|\gamma(h,R^{o}_{\lambda+io}h)| < 1$ for all $\lambda > \lambda_o$. $|\bar{\sigma}(\lambda)-\bar{\sigma}_1(\lambda)| = \bar{\sigma}(\lambda)\Big|\,|1+\gamma(h,R^{o}_{\lambda+io}h)|^2 - 1\Big|$, and if $\bar{\sigma}(\lambda) \neq 0$ then this implies that

$$[\bar{\sigma}(\lambda)]^{-1}|\bar{\sigma}(\lambda)-\bar{\sigma}_1(\lambda)| \leq \text{constant}\cdot|(h,R^{o}_{\lambda+io}h)|. \qquad (12.11)$$

The inequality (12.11) shows that in this model the error term $\bar{\sigma}(\lambda)-\bar{\sigma}_1(\lambda)$ is small compared to $\bar{\sigma}(\lambda)$ as $\lambda \to \infty$.

Born series similar to (12.2) can be written down for other quantities relevant to a scattering problem, e.g. for R_z, $G_{\lambda\pm io}(\underline{x},\underline{y})$, $\psi^{\pm}_{k}(\underline{x})$, $f(\lambda;\underline{\omega}'\to\underline{\omega})$. (We remind the reader that for the rest of this section and also in the next section, we shall employ the notation introduced in page 395.) In all these examples, there are Born series for small coupling constant and high energies as in Proposition 12.4. We consider here some of the properties of the Born approximation for $\psi^{\pm}_{k}$, $f(\lambda;\underline{\omega}'\to\underline{\omega})$ and the differential cross-section $d\sigma/d\omega$, referring to Problem 12.3 for some of the others. For the rest of this section we assume that $V\varepsilon\mathcal{Y}$ and $B_2 = 0$. We set $k = \lambda^{\frac{1}{2}}$.

PROPOSITION 12.6 : Let $V \varepsilon \mathcal{V}$ and $B_2 = 0$. (a) For every $\underline{\omega} \varepsilon S^{(2)}$ and fixed $\lambda > \lambda_0$ (λ_0 as defined in the proof of Corollary 12.3) the series $\sum_{j=1}^{\infty} (-W_\lambda^{\pm *})^{j-1} A\psi_k^o$ converges strongly in $L^2(R^3)$ to $A\psi_k^{\pm}$. (b) The Born approximation ψ_k^o to $\psi_k^{\pm}$ is a high energy approximation in the sense that

$$||A\psi_k^{\pm} - A\psi_k^o|| = o(||A\psi_k^o||), \quad \text{as } \lambda \to \infty. \tag{12.12}$$

(c) The Born series for the scattering amplitude : $f(\lambda;\underline{\omega}' \to \underline{\omega}) = \sum_{j=1}^{\infty} f^{(j)}(\lambda;\underline{\omega}' \to \underline{\omega})$ converges uniformly in $\underline{\omega},\underline{\omega}'$ and λ for $\lambda > \lambda_0$, where $f^{(j)}$ is given by

$$f^{(j)}(\lambda;\underline{\omega}' \to \underline{\omega}) = -2\pi^2 (B\psi_{k\underline{\omega}}^o, (-W_\lambda^{-*})^{j-1} A\psi_{k\underline{\omega}'}^o). \tag{12.13}$$

Proof : The functions $A\psi_k^o$ and $A\psi_k^{\pm}$ are in $L^2(R^3)$ for every $\lambda \varepsilon (0,\infty) - \Gamma_0$ and $\underline{\omega} \varepsilon S^{(2)}$, the latter by Proposition 10.6. Since by Corollary 12.3, no $\lambda > \lambda_0$ belongs to Γ_0 and since by Proposition 12.1, $||W_\lambda^{\pm}|| \leq \delta < 1$ for all $\lambda > \lambda_0$, we infer that the series $\sum_{j=1}^{\infty} (-W_\lambda^{\pm *})^{j-1} A\psi_k^o$ converges strongly to $(I+W_\lambda^{\pm})^{-1*} A\psi_k^o$ which equals $A\psi_k^+$ by (10.30).

From the series it is easy to compute the difference $A\psi_k^{\pm} - A\psi_k^o$ to obtain

$$||A\psi_k^{\pm} - A\psi_k^o|| \leq ||A\psi_k^o|| \sum_{j=1}^{\infty} ||W_\lambda^{\pm}||^j,$$

which is majorized by $||W_\lambda^{\pm}|| [I - ||W_\lambda^{\pm}||]^{-1} ||A\psi_k^o||$, since $||W_\lambda^{\pm}|| \leq \delta < 1$ for $\lambda > \lambda_0$. Now $||A\psi_k^o|| = (2\pi)^{-3/2} ||A||_2 \neq 0$, and we have that $||A\psi_k^o||^{-1} ||A\psi_k^{\pm} - A\psi_k^o|| \leq \text{constant} \cdot ||W_\lambda^{\pm}|| \to 0$ as $\lambda \to \infty$ by Proposition 12.1 and we arrive at (12.12).

By (10.50) the scattering amplitude can be written as

$f(\lambda;\underline{\omega}' \to \underline{\omega}) = -2\pi^2(B\psi^{o}_{k\underline{\omega}}, A\psi^{-}_{k\underline{\omega}'})$, where we have noted that $B\psi^{o}_{k}$ and $A\psi^{-}_{k'}$ are in $L^2(R^3)$ for fixed $\underline{\omega},\underline{\omega}'$. Since by part (a) of this proof the series for $A\psi^{-}_{k'}$ converges strongly for $\lambda > \lambda_o$, we can substitute this series in the above expression for f and obtain the series given by (12.13). Now $\|B\psi^{o}_{k}\|$ and $\|A\psi^{o}_{k'}\|$ are independent of $\lambda,\underline{\omega},\underline{\omega}'$ and for all $\lambda > \lambda_o$, $\|W^{-}_{\lambda}\| \leq \delta < 1$. From this the uniform convergence of the Born series for $f(\lambda;\underline{\omega}' \to \underline{\omega})$ follows easily by applying the Schwarz inequality to (12.13). #

The results of the above proposition allow us to write down the n-th order Born approximation for the differential scattering cross-section $d\sigma/d\omega\,(\lambda;\underline{\omega}' \to \underline{\omega})$ given by (7.50). Setting,

$$f_n(\lambda;\underline{\omega}' \to \underline{\omega}) = \sum_{j=1}^{n} f^{(j)}(\lambda;\underline{\omega}' \to \underline{\omega}), \tag{12.14}$$

where $f^{(j)}$ are given by (12.13), we define the n-th order Born approximation $d\sigma_n/d\omega$ as $d\sigma_n/d\omega\,(\lambda;\underline{\omega}' \to \underline{\omega}) = |f_n(\lambda;\underline{\omega}' \to \underline{\omega})|^2$. The first Born approximation is $d\sigma_1/d\omega = |f^{(1)}(\lambda;\underline{\omega}' \to \underline{\omega})|^2$, where

$$f^{(1)}(\lambda;\underline{\omega}' \to \underline{\omega}) = -2\pi^2 \int \overline{\psi^{o}_{k\underline{\omega}}(\underline{x})}\; V(\underline{x})\; \psi^{o}_{k\underline{\omega}'}(\underline{x})\, d^3x$$

$$= -(4\pi)^{-1} \int e^{-ik(\underline{\omega} - \underline{\omega}')\cdot\underline{x}}\; V(\underline{x})\, d^3x = -(\pi/2)^{\frac{1}{2}}\, \tilde{V}\,(\lambda^{\frac{1}{2}}(\underline{\omega} - \underline{\omega}')). \tag{12.15}$$

By hypothesis $V \in L^1(R^3)$, so that by the Riemann-Lebesgue lemma $\tilde{V}(\lambda^{\frac{1}{2}}(\underline{\omega} - \underline{\omega}'))$ tends to zero as $\lambda \to \infty$ whenever $\underline{\omega} \neq \underline{\omega}'$, implying that $d\sigma_1/d\omega(\lambda;\underline{\omega}' \to \underline{\omega}) \to 0$ as $\lambda \to \infty$ for $\underline{\omega} \neq \underline{\omega}'$.

However, $f(\lambda;\underline{\omega}' \to \underline{\omega})$ and hence also $d\sigma/d\omega(\lambda)$ converges to zero as $\lambda \to \infty$ if $\underline{\omega} \neq \underline{\omega}'$ (Problem 12.2). In fact every term in the Born series for f converges to zero as $\lambda \to \infty$ except in the forward direction. Thus the statement that the difference $|d\sigma/d\omega - d\sigma_1/d\omega|$ tends to zero as $\lambda \to \infty$ is not very interesting, and one has to aim at establishing estimates of the type (12.10). We do not know of any such results.

From Proposition 12.4(c) we know that the Born series (12.2) for $R(\lambda;\gamma)$ converges when λ is large enough. But it may well happen that for the convergence of the Born series, the positive number λ_0 is much larger than m^2, where m is the reduced mass. In such a case, when the kinetic energy greatly exceeds m^2, we are in the domain of relativistic kinematics and our theory of non-relativistic scattering is no more applicable. For a discussion of the high energy behaviour in relativistic scattering, the reader is referred to Martin [2].

12-2 SCATTERING AT LOW ENERGIES

Here we consider the scattering amplitude $f(\lambda;\underline{\omega}' \to \underline{\omega})$ and the differential cross-section $d\sigma/d\omega(\lambda;\underline{\omega}' \to \underline{\omega})$ as $\lambda \to +0$. The low energy behaviour of f, particularly its insensitivity to the details of the shape of the potential, was used extensively in the early days of nuclear physics. For a discussion of these applications, namely the <u>effective range</u> theories, the reader is referred to [JO, Section 11.6].

We start with the expressions (10.49) and (10.50) and find that they do not make sense for $\lambda = 0$ if the point 0 is in the exceptional set, i.e. if $I+W_0$ is not invertible (note that $W_0^+ = W_0^- \equiv W_0$). For simplicity, we shall make the assumption here that the point 0 does not belong to Γ_0. By Lemma 9.13, W_0 is Hilbert-Schmidt with kernel $W_0(\underline{x},\underline{y}) = (4\pi)^{-1}B(\underline{x})|\underline{x}-\underline{y}|^{-1}A(\underline{y})$. We set

$$u(\underline{x}) \equiv \psi_0^0(\underline{x}) = (2\pi)^{-3/2},\psi_0(\underline{x}) = A^{-1}(I+W_0)^{-1^*}Au, \qquad (12.16)$$

and study some of the properties of ψ_0.

PROPOSITION 12.7 : Let $V\varepsilon\mathcal{Y}$ and $0\notin\Gamma_0$. Then $\psi_0\varepsilon E'$ and $\psi_k^{\pm}$ converges in E'-norm to ψ_0 as $\lambda = k^2 \to +0$, uniformly in $\underline{\omega}$. If, furthermore, $B_2 = 0$ and $|\underline{x}|A(\underline{x})\varepsilon L^2(R^3)$, then $\|\psi_k^{\pm} - \psi_0\|_{E'} = 0(\lambda^{\frac{1}{2}})$, uniformly in $\underline{\omega}$.

Proof : We use the notations of Section 10-1 and recall from (10.21) that $\|\psi\|_{E'} = \|A\psi\|$. Since $0\notin\Gamma_0$ and $Au\varepsilon L^2(R^3)$, it follows from (12.16) that $\psi_0\varepsilon E'$. Since Γ_0 is closed, its complement must be a countable union of disjoint open intervals [R, Proposition 2.8] and thus the point 0 must belong to one of these intervals, say Δ'. Next we consider a compact interval $\Delta \equiv [0,\kappa] \subseteq \Delta'$ and let $\underline{k}\varepsilon\Delta^{(3)}$. The uniform continuity in E'-norm of $\psi_k^{\pm}$ for $\underline{k}$ in $\Delta^{(3)}$ follows as in Lemma 10.7, since $\Delta^{(3)}$ is a compact set. In particular this shows that $\|\psi_k^{\pm}-\psi_0\|_{E'} \to 0$ as $\lambda \to +0$, uniformly in $\underline{\omega}$.

Let $\lambda\varepsilon\Delta$ and $B_2 = 0$. Then $\|W_\lambda^{\pm}-W_0\|_{HS}^2 = (16\pi^2)^{-1}\int|B_1(\underline{x})|^2|\underline{x}-\underline{y}|^{-2}|A(\underline{y})|^2|\exp(\pm i\sqrt{\lambda}|\underline{x}-\underline{y}|)-1|^2d^3xd^3y \le (16\pi^2)^{-1}\lambda\|B_1\|_2^2\,\|A\|_2^2$, where we have used the inequality $|\exp(i\alpha\beta)-1| \le |\alpha\beta|$. Similarly we can write

$$\|A\psi_k^0 - Au\|^2 = (2\pi)^{-3}\int|A(\underline{x})|^2|\exp(i\underline{k}\cdot\underline{x})-1|^2d^3x$$
$$\le (2\pi)^{-3}\lambda\int|\underline{x}|^2|A(\underline{x})|^2d^3x. \qquad (12.17)$$

Since $|\underline{x}|A(\underline{x})\varepsilon L^2(R^3)$, we obtain $\|A\psi_k^0 - Au\| = 0(\lambda^{\frac{1}{2}})$. By (10.30) and (12.16)

$$\|A\psi_k^{\pm} - A\psi_o\| \le \|(I+W_\lambda^{\pm})^{-1}\| \, \|W_\lambda^{\pm}-W_o\| \, \|(I+W_o)^{-1}\| \, \|A\psi_k^o\|$$
$$+ \|(I+W_o)^{-1}\| \, \|A\psi_k^o - Au\|. \qquad (12.18)$$

Noting that $\|A\psi_k^o\| = (2\pi)^{-3/2}\|A\|_2$ and that $\|(I+W_\lambda^{\pm})^{-1}\|$ is bounded for $\lambda\varepsilon\Delta$ by Lemma 9.5(a), the estimate $\|A\psi_k^{\pm} - A\psi_o\| = 0(\lambda^{\frac{1}{2}})$ follows from (12.18). #

Remark 12.8 : We can set up the Lippmann-Schwinger equation for ψ_o as in (10.36). Under the additional hypothesis that $B_1\varepsilon L^\infty(R^3)$ one can conclude as in Proposition 10.11 that $\psi_k^{\pm}(\underline{x})$ is bounded for $(\underline{x},\underline{k}) \varepsilon R^3 \times \Delta^{(3)}$ and $\psi_k^{\pm} - \psi_k^o$ is uniformly continuous for $(\underline{x},\underline{k})$ in the same domain, where $\Delta = [0,\kappa]$. It then follows that ψ_o is bounded and uniformly continuous in R^3, since u is.

The next proposition deals with the limit of the scattering amplitude $f(\lambda;\underline{\omega}' \to \underline{\omega})$ and also its asymptotic behaviour as $\lambda \to +0$.

PROPOSITION 12.9 : Let $V\varepsilon\mathcal{Y}$ and $B_2 = 0$. (a) The scattering amplitude $f(\lambda;\underline{\omega}' \to \underline{\omega})$ converges to a constant $\alpha \equiv -(\pi/2)^{\frac{1}{2}}\cdot \int V(\underline{x})\psi_o(\underline{x})d^3x$ as $\lambda\to +0$, uniformly in $\underline{\omega}$ and $\underline{\omega}'$. (b) The S-matrix $S(\lambda)$ converges to I_o in Hilbert-Schmidt norm and $\|S(\lambda)-I_o\|_{HS} = 0(\lambda^{\frac{1}{2}})$ as $\lambda \to +0$. (c) If furthermore $B = CB'$ with $B'\varepsilon L^2(R^3)$ and C the multiplication operator by $(1+|\underline{x}|)^{-1}$, then $|f(\lambda;\underline{\omega}' \to \underline{\omega})-\alpha| = 0(\lambda^{\frac{1}{2}})$ as $\lambda\to +0$, uniformly in $\underline{\omega}$ and $\underline{\omega}'$.

Proof : By (10.50) we have $f(\lambda;\underline{\omega}' \to \underline{\omega}) = -2\pi^2(B_1\psi_k^o, A\psi_{k'}^-)$, where both $B_1\psi_k^o$ and $A\psi_{k'}^-$ belong to $L^2(R^3)$ for fixed $\underline{k}$ and $\underline{k}'$.

By Proposition 12.7, $A\psi_{\bar{k}'}$ converges strongly to $A\psi_0$ as $\lambda \to +0$, uniformly in $\underline{\omega}'$. As in (10.35), one finds that $B_1\psi_k^0$ is strongly continuous in $\underline{k} = \lambda^{\frac{1}{2}}\underline{\omega}$ for $\lambda\varepsilon\Delta$, where Δ is as defined in the proof of Proposition 12.7. This shows that $\|B_1\psi_k^0 - B_1u\|$ converges to 0 as $\lambda \to 0$, uniformly in $\underline{\omega}$. Therefore $f(\lambda;\underline{\omega}' \to \underline{\omega})$ converges to $-2\pi^2(B_1u,A\psi_0) = -(\pi/2)^{\frac{1}{2}}\int V(\underline{x})\psi_0(\underline{x})d^3x = \alpha$ as $\lambda \to +0$, uniformly in $\underline{\omega}$ and $\underline{\omega}'$. Thus $|f(\lambda;\underline{\omega}' \to \underline{\omega})|$ is bounded in all variables in a neighbourhood of $\lambda = 0$, so that we have $\|R(\lambda)\|_{HS}^2 = (2\pi)^{-2}\lambda\int|f(\lambda;\underline{\omega}' \to \underline{\omega})|^2 d\omega d\omega' = O(\lambda)$. This proves (a) and (b).

To estimate $f(\lambda;\underline{\omega}' \to \underline{\omega})-\alpha$, we notice that for $B = CB'$ $|f(\lambda;\underline{\omega}' \to \underline{\omega})-\alpha| = 2\pi^2|(CB'\psi_k^0 - CB'u,A\psi_0) + (B'\psi_k^0,CA\psi_{\bar{k}'} - CA\psi_0)| \leq 2\pi^2\|A\psi_0\|\,\|CB'\psi_k^0 - CB'u\| + (\pi/2)^{\frac{1}{2}}\|B'\|_2\|CA\psi_{\bar{k}'} - CA\psi_0\|$. The first term in the above inequality is seen to be $O(\sqrt{\lambda})$ by an estimate as in (12.17), whereas the second one is $O(\lambda^{\frac{1}{2}})$ by Proposition 12.7. Since these estimates are uniform in $\underline{\omega}$ and $\underline{\omega}'$ respectively, we have (c). #

Proposition 12.9 leads to a very convenient expression for the scattering at low energies. The differential cross section has the behaviour : $d\sigma/d\omega(\lambda;\underline{\omega}' \to \underline{\omega}) = \alpha^2 + O(\lambda^{\frac{1}{2}})$ as $\lambda \to +0$. The constant $-\alpha$ has the dimension of length and is called the <u>scattering length</u>. The uniformity of the convergence of $f(\lambda;\underline{\omega}' \to \underline{\omega})$ to α has the physical consequence that the scattering at low energy is isotropic even if the potential is not spherically symmetric. Intuitively this is due to the fact that at very low energies the wavelength of the quantum mechanical particle is so large that it cannot feel the detailed structure of the potential. Also note that

$\sigma_{tot}(\lambda;\underline{\omega}')$ converges to $4\pi\alpha^2$ as $\lambda\to+0$, uniformly in $\underline{\omega}'$ (i.e. uniformly with respect to the initial direction).

12-3 HIGH AND LOW ENERGY BEHAVIOUR OF THE PHASE SHIFTS

In this section we consider spherically symmetric potentials for which the results of the preceding sections can be somewhat strengthened. It must be borne in mind, however, that all of the bounds derived below are valid for the scattering amplitude in each partial wave and do not necessarily imply identical bounds for the total scattering amplitude. We shall assume that $V\varepsilon\mathcal{V}$ and use the notations and results of Section 11-4.

PROPOSITION 12.10 : Assume that $\lambda = 0$ is not an exceptional point. (a) $R_\ell(\lambda) \equiv \exp[2i\delta_\ell(\lambda)-1] = o(1)$ as $\lambda\to+0$ for each ℓ. (b) If $B_2 = 0$, then as $\lambda\to+0$, $R_\ell(\lambda) = 0(\lambda^{\frac{1}{2}})$ for $\ell = 0$ and $R_\ell(\lambda) = o(\lambda^{\frac{1}{2}})$ for $\ell \geq 1$. (c) If $B_2 = 0$ and $\nu > 2\ell_0 + 3/2$ for some positive integer ℓ_0, then $R_\ell(\lambda) = 0(\lambda^{\ell+\frac{1}{2}})$ for all $\ell \leq \ell_0$ and $R_\ell(\lambda) = 0(\lambda^{\ell_o+\frac{1}{2}})$ for each $\ell > \ell_0$ as $\lambda\to+0$.

Proof : Since 0 is assumed not to be an exceptional point, $\|(I+W^{\pm}_{\ell,\lambda})^{-1}\|$ are bounded uniformly in ℓ and $\lambda\varepsilon\Delta$ by (11.47), where Δ is the interval $[0,\kappa]$ as defined in the proof of Proposition 12.7. From (11.44) it follows that we need to estimate only $\|M_{A,\ell}(\lambda)\|$ and $\|M_{B,\ell}(\lambda)\|$ for fixed ℓ as $\lambda\to+0$.

Recalling the expression (11.41) for the kernels of $M_{A,\ell}(\lambda)$ and using the bound (11.33) for $\hat{j}_\ell$, we have that

$\|M_{A,\ell}(\lambda)\|^2 = (\pi\lambda^{\frac{1}{2}})^{-1}\int_0^\infty A(r)^2|\hat{j}_\ell(kr)|^2 dr \leq c_1\pi^{-1}\lambda^{\frac{1}{2}}\int_0^\infty r^2 A(r)^2 dr,$ for all ℓ. If $\ell \geq 1$ we can improve this estimate slightly by writing

$$\|M_{A,\ell}(\lambda)\|^2 = \pi^{-1}\lambda^{\frac{1}{2}}\int_0^\infty r^2 A(r)^2|\hat{j}_\ell(kr)/kr|^2 dr. \tag{12.19}$$

By (11.33), $|\hat{j}_\ell(kr)/kr| \leq c_1|kr/(1+kr)|^\ell$, which converges to zero as $\lambda \to +0$ for all r. Since $\int_0^\infty r^2 A(r)^2 dr < \infty$, it follows from (12.19) and the Lebesgue dominated convergence theorem that $\|M_{A,\ell}(\lambda)\|^2 = o(\lambda^{\frac{1}{2}})$. A similar set of estimates is also true for $M_{B_1,\ell}(\lambda)$ because $\int_0^\infty B_1(r)^2 r^2 dr < \infty$, and this implies (b).

For the case when $B_2 \neq 0$, we need to estimate also $\|M_{B_2,\ell}(\lambda)\|$. Since $\int_0^\infty B_2(r)^2 dr < \infty$, $|\hat{j}_\ell(kr)| \leq c_1$ and $\hat{j}_\ell(kr) \to 0$ as $k \to +0$, a similar consideration as above shows that for all ℓ, $\lambda^{\frac{1}{2}}\|M_{B_2,\ell}(\lambda)\|^2 = o(1)$ as $\lambda \to +0$. On the other hand we have seen that $\|M_{A,\ell}(\lambda)\| = 0(\lambda^{\frac{1}{4}})$ and these two together lead to (a).

The hypothesis of (c) implies that one may write $V = A'B'$ where $\int_0^\infty r^{2\ell+2}(|A'(r)|^2 + |B'(r)|^2)dr < \infty$ for all $\ell \leq \ell_0$. To see this, set $B'(r) = B(r)(1+r)^{-\ell_0}$. Now as in (b), we have for $\ell \leq \ell_0$, $\|M_{A',\ell}(\lambda)\|^2 \leq c_1(\pi\sqrt{\lambda})^{-1}\int_0^\infty A'(r)^2(kr)^{2\ell+2}dr = 0(\lambda^{\ell+\frac{1}{2}})$ as $\lambda \to +0$. The same estimate holds for $M_{B',\ell}(\lambda)$, which proves the first part of (c). The second follows from the inequality : $|\hat{j}_\ell(kr)| \leq c_1(\ell)\cdot(kr)^{\ell_0+1}$ for each $\ell \geq \ell_0$. #

Since the constant in the bound (11.33) for the Riccati-Bessel function $\hat{j}_\ell$ depends on ℓ, none of the estimates of the above proposition is uniform in ℓ. Considering the identity $R_\ell(\lambda) = 2ie^{i\delta_\ell}\sin\delta_\ell$ one concludes from Proposition 12.10 that, as $\lambda \to +0$, $\sin\delta_\ell(\lambda)$ goes to zero, which shows in particular that $\delta_\ell(\lambda) \to n_\ell\pi$ as $\lambda \to +0$, where n_ℓ is an integer.

Now $\delta \leq 2\cdot\sin\delta$ for all δ with $0 \leq \delta < \pi/4$ and hence the results of Proposition 12.10 can also be given in terms of $|\delta_\ell(\lambda)-n_\ell\pi|$ instead of $|R_\ell(\lambda)|$. For a heuristic derivation of these estimates, see [LL, Section 108].

Next we derive some results on the high energy behaviour of $R_\ell(\lambda)$.

PROPOSITION 12.11 : (a) If B_1 satisfies $\int_0^1 B_1(r)^2 r\,dr < \infty$, then $R_\ell(\lambda) = O(\lambda^{-\frac{1}{4}})$ as $\lambda \to \infty$. (b) If instead $B_1 \varepsilon L^\infty[0,\infty)$, then $R_\ell(\lambda) = O(\lambda^{-\frac{1}{2}})$ as $\lambda \to \infty$. (c) Under the hypothesis of (b), $R_\ell(\lambda)$ has the asymptotic expansion :

$$R_\ell(\lambda) = -i\lambda^{-\frac{1}{2}}\int_0^\infty V(r)dr + o(\lambda^{-\frac{1}{2}}). \tag{12.20}$$

Proof : By Proposition 12.1, we have that $\|W_\lambda^\pm\| = o(1)$ as $\lambda \to \infty$ and $\|(I+W_\lambda^\pm)^{-1}\|$ is uniformly bounded for $\lambda > \lambda_0$. Therefore, $\|W_{\ell,\lambda}^\pm\| = o(1)$, uniformly in ℓ as $\lambda \to \infty$ and $\|(I+W_{\ell,\lambda}^\pm)^{-1}\|$ are uniformly bounded in ℓ and λ for $\lambda > \lambda_0$. This and (11.44) imply that, to obtain bounds for $R_\ell(\lambda)$ as $\lambda \to \infty$, it suffices to estimate $M_{B,\ell}(\lambda)$ and $M_{A,\ell}(\lambda)$.

Since $\hat{j}_\ell(kr)$ is uniformly bounded and $\int_0^\infty A(r)^2 dr$ and $\int_0^\infty B_2(r)^2 dr$ are finite, it follows from (11.41) that $\|M_{A,\ell}(\lambda)\|^2 = O(\lambda^{-\frac{1}{2}})$ and $\|M_{B_2,\ell}(\lambda)\|^2 = O(\lambda^{-\frac{1}{2}})$. For B_1, we find that

$$\|M_{B_1,\ell}(\lambda)\|^2 = (\pi\lambda^{\frac{1}{2}})^{-1}\int_0^\infty B_1(r)^2|\hat{j}_\ell(kr)|^2 dr$$

$$\leq (\pi\lambda^{\frac{1}{2}})^{-1}\int_0^1 B_1(r)^2|\hat{j}_\ell(kr)|^2 dr + (\pi\lambda^{\frac{1}{2}})^{-1}\int_1^\infty B_1(r)^2 dr. \tag{12.21}$$

We note that $(kr)^{-\frac{1}{2}}\hat{j}_\ell(kr)$ is bounded in k and r and hence $\lambda^{-\frac{1}{2}}\int_0^1 B_1(r)^2|\hat{j}_\ell(kr)|^2 dr \leq c\int_0^1 B_1(r)^2 r\,dr < \infty$. Also since

$\int_1^\infty B_1(r)^2 dr \leq \int_1^\infty B_1(r)^2 r^2 dr < \infty$, one finds from (12.21) that $\|M_{B,\ell}(\lambda)\| = 0(1)$ as $\lambda \to \infty$ and we have proven (a). If B_1 is bounded, then $\int_0^1 B_1(r)^2 dr < \infty$ and we have instead $\|M_{B,\ell}(\lambda)\| = 0(\lambda^{-\frac{1}{4}})$ which leads to the result in (b).

We write (11.44) as

$$R_\ell(\lambda) = -2\pi i\, M_{A,\ell}(\lambda) M_{B,\ell}(\lambda)^* +$$

$$+ 2\pi i\, M_{A,\ell}(\lambda) W^+_{\ell,\lambda}(I+W^+_{\ell,\lambda})^{-1} M_{B,\ell}(\lambda)^*. \tag{12.22}$$

Since $\|W^+_{\ell,\lambda}\| \to 0$ as $\lambda \to \infty$ and since by (b), $\|M_{A,\ell}(\lambda)\| = 0(\lambda^{-\frac{1}{4}})$ and $\|M_{B,\ell}(\lambda)\| = 0(\lambda^{-\frac{1}{4}})$ as $\lambda \to \infty$, we obtain that

$$\|M_{A,\ell}(\lambda) W^+_{\ell,\lambda}(I+W^+_{\ell,\lambda})^{-1} M_{B,\ell}(\lambda)^*\| = o(\lambda^{-\frac{1}{2}}). \tag{12.23}$$

From the expression (11.41) for the kernel of $M_{A,\ell}(\lambda)$ and a similar one for that of $M_{B,\ell}(\lambda)$ we get the so-called <u>partial wave Born approximation</u> for $R_\ell(\lambda)$:

$$-2\pi i\, M_{A,\ell}(\lambda) M_{B,\ell}(\lambda)^* = -2i\lambda^{-\frac{1}{2}} \int_0^\infty dr\, V(r) |\hat{j}_\ell(kr)|^2. \tag{12.24}$$

From the asymptotic behaviour of the Riccati-Bessel function (see Section 11-5A) we know that $\hat{j}_\ell(kr)^2 - \sin^2(kr - \frac{1}{2}\ell\pi)$ converges to zero for all $r \neq 0$ as $k = \lambda^{\frac{1}{2}} \to \infty$. Since under the hypothesis of (b), $\int_0^\infty |V(r)| dr < \infty$, it follows by the Lebesgue dominated convergence theorem that $\int_0^\infty dr\, V(r)\{|\hat{j}_\ell(kr)|^2 - \sin^2(kr - \frac{1}{2}\ell\pi)\}$ converges to zero as $\lambda \to \infty$. On the other hand $\int_0^\infty dr\, V(r) \sin^2(kr - \frac{1}{2}\ell\pi) = \frac{1}{2}\int_0^\infty V(r) dr - \frac{1}{2}\int_0^\infty dr\, V(r) \cos(2kr - \ell\pi) = \frac{1}{2}\int_0^\infty V(r) dr + o(1)$, where we have used the Riemann-Lebesgue lemma in the last step. This result combined with (12.24) implies that $-2\pi i M_{A,\ell}(\lambda) M_{B,\ell}(\lambda)^* = -i\lambda^{-\frac{1}{2}} \int_0^\infty V(r) dr + o(\lambda^{-\frac{1}{2}})$, which along with (12.22) and (12.23) leads to (12.20). #

Remark 12.12 : (a) As with the Proposition 12.10, none of the 0- and o-estimates are uniform with respect to ℓ. It is possible to get bounds uniform in ℓ for $M_{A,\ell}(\lambda)$ or $M_{B,\ell}(\lambda)$. More generally, if $|\phi(r)| \leq \alpha(1+r)^{-\nu}$ $(\nu > \frac{1}{2})$, then $\|M_{\phi,\ell}(\lambda)\| \leq c\lambda^{\frac{1}{2}(\nu-1)}\ell^{(\frac{1}{2}-\nu)}$ for $\ell \neq 0$ and $2\ell + 3 > 2\nu$, where c is a constant independent of λ and ℓ (see Misra, Speiser and Targonski [1] and Problem 12.4). From this it follows that for every λ with $0 < \lambda < \infty$, $\|M_{A,\ell}(\lambda)\|$ behaves like $\ell^{-1-\eta}$ for some $\eta > 0$ and hence $\sum_\ell (2\ell+1)|R_\ell(\lambda)|^2 < \infty$, since the series $\sum_\ell \ell^{-1-2\eta}$ converges. This gives another verification of the fact that $R(\lambda)$ is Hilbert-Schmidt for spherically symmetric potentials in $\mathcal{V}$.

(b) The asymptotic expansion (12.20) is expected to be valid for all potentials V satisfying $\int_0^\infty |V(r)|dr < \infty$. See [N, Section 12.1.3] for such results. Since $R_\ell(\lambda) \to 0$ as $\lambda \to \infty$, we deduce that $\delta_\ell(\lambda) \to m_\ell\pi$ as $\lambda \to \infty$, m_ℓ being an integer. All the results of Proposition 12.11 can as well be expressed with $\delta_\ell - m_\ell\pi$ replacing $R_\ell(\lambda)$. It is usual to set $m_\ell = 0$ for all ℓ. The relation (12.20) then implies that for large λ, $\delta_\ell(\lambda)$ is negative (respectively positive) if the potential is everywhere positive (respectively negative), in particular if the potential is purely repulsive (respectively attractive). It can be shown that this result is in fact true for all energies.

The Yukawa potential : $V(r) = \gamma r^{-1} \cdot \exp(-\alpha r)$ with $\alpha > 0$ does not satisfy any of the conditions of Proposition 12.11. It is clear intuitively as well as from (12.21) that the difficulty stems from the behaviour of V at r = 0. In this example, we can take $B_1(r) = r^{-1}$ for $0 \leq r \leq 1$ and

$B_1(r) = 0$ elsewhere, to obtain $\lambda^{-\frac{1}{2}}\int_0^1 dr\, B_1(r)^2 |\hat{j}_\ell(kr)|^2 \leq \int_0^\infty d\rho |\hat{j}_\ell(\rho)/\rho|^2 < \infty$ for all ℓ. Thus for Yukawa potential also we get an estimate similar to that in Proposition 12.11(a), namely $R_\ell(\lambda) = O(\lambda^{-\frac{1}{4}})$ as $\lambda \to \infty$ for all ℓ.

PROBLEMS

12.1 : Prove Proposition 12.2.

12.2 : (a) Show that, under the assumptions of Proposition 12.6, $f(\lambda;\underline{\omega}'\to\underline{\omega})$ and $f^{(j)}(\lambda;\underline{\omega}'\to\underline{\omega})$ converge to zero as $\lambda\to\infty$, uniformly in $\underline{\omega}$ and $\underline{\omega}'$ away from the forward direction, for all j. (b) Calculate the rate of decay of $d\sigma_1/d\omega(\lambda;\underline{\omega}'\to\underline{\omega})$ as $\lambda\to\infty$ for a Yukawa potential.

12.3 : Write down the Born series for R_z and $G_{\lambda\pm io}(\underline{x},\underline{y})$ and discuss their convergence. (Hint : Section 10-4$\underline{B}$).

12.4†: (a) Let $|\phi(r)| \leq \alpha(1+r)^{-\nu}$ with $\nu > \frac{1}{2}$. Show that $\|M_{\phi,\ell}(\lambda)\| \leq c\,\lambda^{\frac{1}{2}(\nu-1)}\ell^{\frac{1}{2}-\nu}$ for $\ell \neq 0$ and $2\ell+3 > 2\nu$. (Hint : Use (11.66) and Binet's expression for the gamma function [WW, page 249] : $\log\Gamma(z) = (z-\frac{1}{2})\log z - z + \frac{1}{2}\log 2\pi + \int_0^\infty [\frac{1}{2}-t^{-1}+(e^t-1)^{-1}]t^{-1}e^{-tz}dt$.) (b) Verify directly that $M_C(\lambda)$, defined in Lemma 10.21, is compact for each $\lambda\varepsilon(0,\infty)$.

12.5 : Show that $\|M_{B_2}(\lambda)\| \leq c(\eta)\cdot\lambda^{-\frac{1}{4}+\eta}$ for each $\eta\varepsilon(0,\delta]$ and use it to prove Proposition 12.5(b) with $B_2 \neq 0$. (Hint : Use Problem 12.4, Lemma 11.8(b) and the fact that, if $\phi(\underline{x}) = \psi(\underline{x})\rho(\underline{x})$ with $\rho\varepsilon L^\infty(\mathbb{R}^3)$, then $M_\phi(\lambda) = M_\psi(\lambda)\rho(\underline{Q})$.)

12.6 : Let $V\varepsilon\mathcal{V}$, $B_2 = 0$ and $0\notin\Gamma_0$. Show that $\|\psi^\pm_{k\underline{\omega}} - \psi_0\|_{\mathcal{E}'} = o(1)$, uniformly in $\underline{\omega}$, as $\lambda\to 0$. (Hint : Proposition 12.7 and (3.63).)

12.7 (Distorted wave Born approximation) : Assume all the hypotheses of Problem 10.12. (a) Show that $\|B(H_1-\lambda\pm io)^{-1}A\|\to 0$ as $\lambda\to\infty$. (b) Write down the Born series as in (12.2) for $R^+(H,H_1)(\lambda)$ and discuss its convergence. (c) Insert the Born approximation for $R^+(H,H_1)(\lambda)$ into (10.75) and deduce the so-called distorted wave Born approximation (DWBA) for the scattering amplitude : $f(H,K_0)(\lambda;\underline{\omega}'\to\underline{\omega}) = f(H_1,K_0)(\lambda;\underline{\omega}'\to\underline{\omega}) - 2\pi^2\gamma\int\overline{\psi^+_{k\underline{\omega}}(\underline{x})}V(\underline{x})\psi^-_{k\underline{\omega}'}(\underline{x})d^3x$, where $\psi^{\pm}$ are the eigenfunctions of $H_{1,ac}$. (Hint : $M^+_\phi(\lambda)^*S(H_1,K_0)(\lambda) = M^-_\phi(\lambda)^*$.)

Remark : The DWBA is useful if the scattering problem for the pair (H_1,K_0) can be solved explicitly and γV can be considered to be a "small" perturbation. It is often applied to the case where W is a Coulomb potential.

12.8†: Find sufficient conditions on V_1 in Proposition 11.10 for $R(\lambda)$ to be Hilbert-Schmidt in $L^2(S^{(2)})\otimes\mathbb{C}^{2s+1}$. (Hint : Use the estimate in Remark 12.12(a).)

12.9 : Let $V\varepsilon\mathcal{Y}$. Show that $[B_1R^o_{\lambda+i\eta}B_1 - B_1R^o_{\lambda-i\eta}B_1]$ converges in Hilbert-Schmidt norm as $\eta\to+0$ to $[B_1R^o_{\lambda+io}B_1 - B_1R^o_{\lambda-io}B_1]$, and that the latter operator is equal to $2\pi i\, M_{B_1}(\lambda)^* M_{B_1}(\lambda)$.

Remark : In Proposition 12.5(b) we required in addition that $B_1\varepsilon L^\infty(\mathbb{R}^3)$, which led to $\|M_{B_1}(\lambda)\| \leq c$ for all $\lambda\varepsilon[0,\infty)$ by Lemma 9.8. This can be relaxed to the hypothesis that $B_1\varepsilon L^2(\mathbb{R}^3)\cap L^{2+\delta}(\mathbb{R}^3)$ for some $\delta\varepsilon(0,1]$ (this class includes for example the Yukawa potential). Then by the Hölder inequality [RS II, page 32], $\|B_1f\|_p \leq \|B_1\|_{2+\delta}\|f\|_2$ with $p^{-1} = (2+\delta)^{-1}+2^{-1}$. By Lemma 5.30, $\|M_{B_1}(\lambda)\|^2 = \sup_{\|f\|=1}(f,M_{B_1}(\lambda)^*M_{B_1}(\lambda)f)$. Using (10.8) and (3.63), one obtains for $0\leq\nu\leq 1$, $\|M_{B_1}(\lambda)f\|^2 \leq c_1(\nu)\lambda^{\nu/2}\int|(B_1f)(\underline{x})||(B_1f)(\underline{y})||\underline{x}-\underline{y}|^{\nu-1}d^3xd^3y$. Now choosing $\nu = 1-3\delta/(2+\delta)$ and applying the Sobolev inequality [RS II, page 31], one gets $\|M_{B_1}(\lambda)\|^2 \leq c_2(\nu)\lambda^{\nu/2}$. By Problem 12.5, $\|M_A(\lambda)\| \leq c_3(\eta)\lambda^{-\frac{1}{4}+\eta}$ for each $\eta\varepsilon(0,1]$, so that by choosing η sufficiently small, one finds from (12.9) that $\|R(\lambda)\|\to 0$ as $\lambda\to\infty$.

CHAPTER 13 : SCATTERING THEORY FOR LONG RANGE POTENTIALS

It was mentioned in Sections 4-3 and 8-2 that the family $\{\exp(iHt)\exp(-iK_0t)\}$ does not have strong limits as $t \to \pm\infty$ if $H = K_0 + V$ and V is a Coulomb or other long range potential. Section 4-3 also contained a description of various possibilities of formulating the asymptotic condition other than requiring the existence of strong limits of $\{V_t^* U_t E_\infty(H_0)\}$ as $t \to \pm\infty$. In the present chapter we elaborate on these points. Section 13-1 is devoted to proving the existence of the so-called generalized wave operators for long range potential scattering as well as the non-existence of the ordinary ones. In Section 13-2 we reconsider the formulation of the asymptotic condition in terms of observables, in particular its relation to the generalized wave operators of Section 13-1.

13-1 WAVE OPERATORS FOR LONG RANGE POTENTIALS

The problem of scattering by a Coulomb potential $V_c(\underline{x}) = \gamma|\underline{x}|^{-1}$ is of great interest in atomic and nuclear physics. The associated time-independent Schrödinger equation has been known from the early days of quantum mechanics to be exactly solvable in terms of special functions. The Coulomb eigenfunctions $\psi_{k}^{c}(\underline{x})$, which are the solutions of equation (10.3) for $\lambda = \underline{k}^2 > 0$, have the following asymptotic behaviour (see e.g. [LL, Section 112]) :

$$\psi_{k}^{c-}(\underline{x}) \simeq (2\pi)^{-3/2} e^{i[\underline{k}\cdot\underline{x}+\gamma k^{-1}\log(kr-\underline{k}\cdot\underline{x})]}[1-i\gamma^2 k^{-2}(kr-\underline{k}\cdot\underline{x})^{-1}]$$

$$+ (2\pi)^{-3/2} f_c(k,\theta) r^{-1} e^{i(kr-\gamma k^{-1}\log 2kr)} . \qquad (13.1)$$

Here θ is the angle between $\underline{k}$ and $\underline{x}$ and is held non-zero while $r\to\infty$. A comparison between (10.4) and (13.1) shows clearly that the dominant part of (13.1) as well as the scattered spherical wave have a logarithmic "distortion" in their phases. The distorted plane wave $\exp[i\underline{k}\cdot\underline{x}+i\gamma k^{-1}\log(kr-\underline{k}\cdot\underline{x})]$ is not an eigenfunction of K_0 corresponding to the spectral point $\lambda = \underline{k}^2$. One therefore expects that also in the time-dependent formalism the total evolution $\exp(-iH_c t)$ with $H_c = K_0 + V_c$ does not quite look like the free evolution $\exp(-iK_0 t)$ as $t\to\pm\infty$, but rather like some distortion of the free evolution.

For a satisfactory theory one formulates the asymptotic condition in terms of observables and finds that under suitable assumptions this implies the existence of a family of unitary operators $\{Y_t\}$, which are functions of K_0 or of $\underline{P}$ such that

$$\text{s-lim } V_t^* Y_t \equiv \Omega_\pm \text{ exist as } t \to \pm\infty. \tag{13.2}$$

This will be explained in the next section. Conversely, if there exists some such family $\{Y_t\}$ satisfying (13.2), it is easy to verify that $V_t^* A V_t f$ converges strongly to $\Omega_\pm A \Omega_\pm^*$ as $t \to \pm\infty$ for each A in $\mathcal{B}(\mathcal{H})$ that commutes with K_0 or $\underline{P}$ respectively and for all f in the range of $\Omega_\pm$ (see Proposition 4.11). $\{Y_t\}$ represents the above-mentioned distortion of the free evolution and will be called the <u>modified free evolution</u>, and $\Omega_\pm$ are the generalized wave operators in the sense of Section 4-3. We next give a formal method for arriving at a candidate for $\{Y_t\}$.

As in Proposition 8.14 the convergence of $V_t^* Y_t$ is shown by proving that $\int_{\pm a}^{\pm\infty} \| d/dt\, V_t^* Y_t f \| dt < \infty$ for all vectors f in some fundamental set. By writing $Y_t = U_t \exp(-iX_t)$ and $X_t \equiv X_t(\underline{P})$, one has

$$\| d/dt\, V_t^* Y_t f \| = \| (V - d/dt\, X_t) Y_t f \| = \| (U_t^* V U_t - d/dt\, X_t) \exp(-iX_t) f \|, \tag{13.3}$$

since U_t commutes with $d/dt\, X_t(\underline{P})$. From the definitions (3.46) of the unitary operators C_t and Q_t and from the relation (3.47), i.e. $U_t = C_t Q_t$, it follows easily that if $\phi \varepsilon L^\infty(\mathcal{R}^3)$ or $\phi \varepsilon L^2(\mathcal{R}^3)$, then

$$C_t^* \phi(\underline{Q}) C_t = \phi(2t\underline{P}), \qquad U_t^* \phi(\underline{Q}) U_t = Q_t^* \phi(2t\underline{P}) Q_t. \tag{13.4}$$

We also know from Proposition 3.17 that Q_t and Q_t^* converge strongly to I as $t \to \pm\infty$. Therefore by (13.4), $U_t^* V(\underline{Q}) U_t - V(2t\underline{P})$ will converge strongly to zero under suitable hypotheses on V. Thus if

$$d/dt\, X_t(\underline{P}) = V(2t\underline{P}), \tag{13.5}$$

then we find from (13.3) that $\|d/dt\ V_t^*U_t f\|$ converges to zero as $t \to \pm\infty$ and may even be integrable at infinity. As we shall see below, X_t satisfying (13.5) gives a good candidate for Y_t for potentials behaving like $|\underline{x}|^{-\beta}$ at infinity, with $\frac{1}{2} < \beta \leq 1$. In particular for a Coulomb potential, it gives $X_t = \frac{1}{2}\gamma(\text{sign } t)\ K_0^{-\frac{1}{2}} \log|t|$. This operator was introduced by Dollard [1] to prove the existence of the generalized Coulomb wave operators.

It is clear that for spherically symmetric potentials, $\{X_t\}$ and $\{Y_t\}$ are functions of K_0 and hence commute with all the constants of the free motion, whereas in general $\{X_t\}$ are functions of $\underline{P}$, and $\{Y_t\}$ commute with all bounded operators that commute with $\underline{P}$. For short range potentials, i.e. $\beta > 1$, we see from (13.5) that X_t converges to zero as $t \to \pm\infty$, showing that the asymptotic behaviour of V_t in such a case is given by U_t as expected.

From now on we always assume that the potential V can be written as a sum of two real parts V_S and V_L, i.e.

$$V = V_S + V_L, \text{ with } (1+|\underline{x}|^2)V_S \in L^2(R^3) \qquad (13.6)$$

and that V_L is four times continuously differentiable and satisfies *)

$$|D^m V_L(\underline{x})| \leq c(1+|\underline{x}|)^{-|m|-\alpha} \text{ for some } \alpha \text{ with } 0 < \alpha < 1 \text{ and } 0 \leq |m| \leq 4. \qquad (13.7)$$

---

*) We write $m = (m_1, m_2, m_3)$, $|m| = m_1+m_2+m_3$ and $D^m f(\underline{x}) = (\partial^{m_1}/\partial x_1^{m_1})(\partial^{m_2}/\partial x_2^{m_2})(\partial^{m_3}/\partial x_3^{m_3}) f(\underline{x})$, where m_1, m_2, m_3 are non-negative integers.

Clearly V_L defined above is a bounded operator and since (13.6) implies $V_S \varepsilon L^2(R^3)$, V_S is K_0-compact, so that by Proposition 8.7, $H = K_0 + V \equiv K_0 + V_S + V_L$ is self-adjoint with $D(H) = D(K_0)$.

We can consider (13.5) as an equation for the function $X_t(\underline{k})$ and integrate it to obtain $X_t(\underline{k}) - X_o(\underline{k}) = \int_0^t V_L(2s\underline{k})ds$, where $X_o(\underline{k})$ is the constant of integration and we have used only the long range part of the interaction, namely V_L, in (13.5). In the expression $Y_t = U_t \exp(-iX_t)$, the constant of integration X_o just multiplies by a time-independent operator and has no influence on the convergence problem of (13.2). So we drop X_o and write

$$X_t(\underline{k}) = \int_0^t V_L(2s\underline{k})ds. \tag{13.8}$$

Such an ambiguity is a general feature of generalized wave operators (see Problem 4.5).

Example 13.1 : Let $V(\underline{x}) = \gamma|\underline{x}|^{-\beta}$, $\gamma \varepsilon R$ and $0 < \beta \leq 1$ and let $\psi \varepsilon C^\infty(R^3)$ be real-valued, spherically symmetric and such that $\psi(|\underline{x}|) = 0$ for $|\underline{x}| \leq 1$ and $\psi(|\underline{x}|) = 1$ for $|\underline{x}| \geq 2$. Set $V_L = \psi V$ and $V_S = (1-\psi)V$. Then notice that if $\beta < 1$, (13.7) is satisfied with arbitrary $|m|$ and $\alpha = \beta$. If on the other hand $\beta = 1$, we may take in (13.7) any $\alpha \varepsilon (\frac{1}{2},1)$ and any $|m|$.

The following lemma studies the operator family $\{X_t\}$ and some of its properties.

LEMMA 13.2 : (a) The function X_t given by (13.8) is four times continuously differentiable, and $D^m X_t$ is bounded for each t and $0 \leq |m| \leq 4$. Furthermore, for all $k > 0$

$$|(D^m X_t)(\underline{k})| \leq c2^{-\alpha}(1-\alpha)^{-1} k^{-|m|-\alpha}|t|^{1-\alpha}, \quad 0 \leq |m| \leq 4. \quad (13.9)$$

(b) For each t and $0 \leq |m| \leq 4$, the operator $D^m X_t$ defined by

$$[F(D^m X_t)f](\underline{k}) = (D^m X_t)(\underline{k})\tilde{f}(\underline{k}) \qquad (13.10)$$

is a self-adjoint operator in $\mathcal{B}(\mathcal{H})$. Moreover the family $\{D^m X_t\}$ is norm-continuous in t. (c) $\{\exp(-iX_t)\}$ is feebly oscillating, i.e. for every fixed τ, $[\exp(-iX_{t+\tau}) - \exp(-iX_t)]$ converges strongly to 0 as $t \to \pm\infty$. (d) The unitary operator family $\{\exp(-iX_t)\}$ is strongly differentiable and $d/dt \exp(-iX_t) = -iV_L(2t\underline{P})\exp(-iX_t)$.

Proof : (a) Since V_L is four times continuously differentiable, so is $X_t(\underline{k})$ and it follows easily that $(D^m X_t)(\underline{k}) = \int_0^t (2s)^{|m|}(D^m V_L)(2s\underline{k})ds$ for $0 \leq |m| \leq 4$. By (13.7), $D^m V_L$ is bounded, and hence $D^m X_t$ is also bounded for each fixed t. We have also from (13.7) that $|(D^m X_t)(\underline{k})| \leq c \int_0^{|t|}(2s)^{|m|}(2sk)^{-|m|-\alpha}ds$, which gives (13.9).

(b) For each m and t, the function $D^m X_t$ is measurable, real-valued and bounded. Therefore the first part of (b) is obtained by using Proposition 2.16. The norm-continuity of $\{D^m X_t\}$ in t follows from the continuity of $(D^m X_t)(\underline{k})$ in t, uniformly in $\underline{k}$.

(c) Note that for $k > 0$ and $|t| > \tau > 0$, $|\exp(-iX_{t+\tau}(\underline{k})) - \exp(-iX_t(\underline{k}))| \leq |X_{t+\tau}(\underline{k}) - X_t(\underline{k})| = |\int_0^\tau V_L(2(s+t)\underline{k})ds| \leq c(2k)^{-\alpha}\int_0^\tau |t+s|^{-\alpha}ds \leq c(2k)^{-\alpha}\tau(|t|-\tau)^{-\alpha} \to 0$ as $|t| \to \infty$. Thus by the Lebesgue dominated convergence theorem $\|[\exp(-iX_{t+\tau}) - \exp(-iX_t)]f\|^2 = \int d^3k|\exp(-iX_{t+\tau}(\underline{k})) - \exp(-iX_t(\underline{k}))|^2|\tilde{f}(\underline{k})|^2$ converges to 0 as

$|t| \to \infty$, since $|\exp(-iX_{t+\tau}(\underline{k})) - \exp(-iX_t(\underline{k}))| \leq 2$ for all $\underline{k}$. The proof for $\tau < 0$ is similar.

(d) We have for $\tau > 0$

$$\|\tau^{-1}[\exp(-iX_{t+\tau}) - \exp(-iX_t)]f + iV_L(2t\underline{P})\exp(-iX_t)f\|^2$$

$$= \int d^3k|\tau^{-1}[\exp(-iX_{t+\tau}(\underline{k})) - \exp(-iX_t(\underline{k}))] + \\ + iV_L(2t\underline{k})\exp(-iX_t(\underline{k}))|^2|\tilde{f}(\underline{k})|^2 \quad .$$

Since by (13.8), $d/dt \exp(-iX_t(\underline{k})) = -iV_L(2t\underline{k})\exp(-iX_t(\underline{k}))$ for each $\underline{k}$, it follows that the integrand in the above integral converges pointwise to 0 as $\tau \to 0$. Also we find as in (c) that $|\tau^{-1}[\exp(-iX_{t+\tau}(\underline{k})) - \exp(-iX_t(\underline{k}))] \leq |\tau^{-1}\int_0^\tau V_L(2(s+t)\underline{k})ds| \leq c$. Since V_L is bounded and $|\tilde{f}(\underline{k})|^2$ is integrable, the result follows by the Lebesgue dominated convergence theorem. A similar conclusion holds for $\tau < 0$ and $\tau \to 0$. #

We next obtain some preliminary estimates necessary to establish the existence of the generalized wave operators. For this we shall work with functions whose Fourier transforms vanish in some neighbourhood of the origin, or in physical terms with states in which the momentum is bounded away from zero. Thus we define, for $n = 0,1,2,\ldots$,
$\mathcal{D}_n = \{f\varepsilon L^2(R^3)|\tilde{f}\varepsilon C_0^n(R^3-\{0\})\}$, where $C_0^n(R^3-\{0\})$ is the set of all n times continuously differentiable functions of compact support on R^3 each of which vanishes in some neighbourhood of the origin. Since $\mathcal{D}_{n+1} \subseteq \mathcal{D}_n$ and $C_0^\infty(R^3-\{0\})$ is dense in $L^2(R^3)$ (which can be seen as in Problem 2.8), and since the Fourier transformation is unitary, it follows that each $\mathcal{D}_n$ is dense in $L^2(R^3)$. We also set $f_t \equiv \exp(-iX_t)f$ for any $f\varepsilon L^2(R^3)$.

LEMMA 13.3 : Let $\frac{1}{2} < \alpha < 1$ in (13.7), $\rho\varepsilon(0,1]$ and $|t| \geq 1$.
(a) If $f\varepsilon\mathcal{D}_n$, $0 \leq n \leq 3$, then

$$\|(I+\rho|\underline{Q}|)^{-n} U_{t/\rho} f_t\| \leq c_1 |t|^{-n} \sum_{|m|=0}^{n} \|D^m \tilde{f}\|.$$

(b) If $f\varepsilon\mathcal{D}_{n+1}$, $0 \leq n \leq 2$, then

$$\|(I+\rho|\underline{Q}|)^{-n-\alpha} U_{t/\rho} f_t\| \leq c_2 |t|^{-n-\frac{1}{2}} \sum_{|m|=0}^{n+1} \|D^m \tilde{f}\|.$$

(c) For $f\varepsilon\mathcal{D}_2$, $\|V_S U_t f_t\| \leq c_3 |t|^{-2\alpha} \|(1+|\underline{x}|^2)V_S\|_2 \sum_{|m|=0}^{2} \|D^m \tilde{f}\|_1$.

Here c_1, c_2, c_3 are finite constants depending only on the support of $\tilde{f}$.

Proof : We first note some properties of $\mathcal{D}_n$ which will be used in this and the following proofs. If $h\varepsilon\mathcal{D}_n$ and $\kappa\varepsilon R$, then $P_i|\underline{P}|^\kappa h$ is again in $\mathcal{D}_n$, for each $i = 1,2,3$. If $h\varepsilon\mathcal{D}_0$, then $\|F|\underline{P}|^\kappa h\|_p \leq c(\kappa)\ \|\tilde{h}\|_p$ for $p = 1,2$ where for each κ, $c(\kappa)$ depends only on the support of h.

(a) We use the following identity on $\mathcal{D}_1$: $\rho Q_i U_{t/\rho} = U_{t/\rho}(2tP_i+\rho Q_i)$ for $i = 1,2,3$, which we get by remembering that $(FQ_i h)(\underline{k}) = i\partial\tilde{h}(\underline{k})/\partial k_i$ for all $\tilde{h}\varepsilon C_0^1(R^3-\{0\})$. Multiplying the above identity on the right by P_i and summing over $i = 1,2,3$ we have $\rho\sum_i Q_i U_{t/\rho} P_i = U_{t/\rho}(2t\underline{P}^2 + \rho\sum_i Q_i P_i)$. Thus for any $f\varepsilon\mathcal{D}_1$,

$$U_{t/\rho} f_t = \frac{\rho}{2t}\sum_i Q_i U_{t/\rho} P_i |\underline{P}|^{-2} f_t - \frac{\rho}{2t} U_{t/\rho} \sum_i Q_i P_i |\underline{P}|^{-2} f_t =$$

$$\frac{\rho}{2t}\sum_i Q_i U_{t/\rho} P_i |\underline{P}|^{-2} f_t - \frac{\rho}{2t} U_{t/\rho} g_t - \frac{\rho}{2t} U_{t/\rho} \sum_i (\partial_i X_t) P_i |\underline{P}|^{-2} f_t, \tag{13.11}$$

where $\tilde{g}(\underline{k}) = i \operatorname{div}(\underline{k}k^{-2}\tilde{f}(\underline{k}))$ and $\partial_i \equiv \partial/\partial k_i$.

For n = 0, the inequality (a) is obviously satisfied.

For general n we prove it by induction. So assume that it is satisfied for some fixed integer n, $0 \le n \le 2$. Then using $\|(I+\rho|\underline{Q}|)^{-1}\rho Q_i\| \le 1$ for each i, we obtain from (13.11) that

$$\|(I+\rho|\underline{Q}|)^{-n-1}U_{t/\rho}f_t\| \le |2t|^{-1}\sum_i \|(I+\rho|\underline{Q}|)^{-n}U_{t/\rho}P_i|\underline{P}|^{-2}f_t\| + |2t|^{-1}\|(I+\rho|\underline{Q}|)^{-n}U_{t/\rho}g_t\| + |2t|^{-1}\sum_i \|(I+\rho|\underline{Q}|)^{-n-1}U_{t/\rho}(\partial_i X_t)\cdot P_i|\underline{P}|^{-2}f_t\|. \tag{13.12}$$

If $f\varepsilon\mathcal{D}_{n+1}$, then $P_i|\underline{P}|^{-2}f$ and g are in $\mathcal{D}_n$ and so is $(\partial_i X_t)P_i|\underline{P}|^{-2}f$ by Lemma 13.2(a). Thus we deduce from (13.12) by using the inequality (a) for n that

$$\|(I+\rho|\underline{Q}|)^{-n-1}U_{t/\rho}f_t\| \le c_4|t|^{-n-1}[\sum_i\sum_{|m|=0}^{n} \|D^m(k_i k^{-2}\tilde{f})\| + \sum_{|m|=0}^{n} \|D^m\tilde{g}\|] + c_5|t|^{-n-1}\sum_i\sum_{|m|=0}^{n}\|D^m(\partial_i X_t)(\underline{k})k_i k^{-2}\tilde{f}\|. \tag{13.13}$$

One next uses Leibniz' rule for the derivatives of the various products, the fact that $|k_i| \le k$ and the inequality (13.9) to obtain

$$\|(I+\rho|\underline{Q}|)^{-n-1}U_{t/\rho}f_t\| \le c_6|t|^{-n-1}\sum_{|m|=0}^{n+1} \|D^m\tilde{f}\| + c_7|t|^{-n-\alpha}\sum_{|m|=0}^{n}\|D^m\tilde{f}\|, \tag{13.14}$$

where c_6 and c_7 depend only on the support of f. By Lemma 13.2(a), $\partial_i X_t \,\varepsilon\, C^3$, hence $(\partial_i X_t)P_i|\underline{P}|^{-2}f\varepsilon\mathcal{D}_{n+1}$, so that we can apply (13.14) to $(\partial_i X_t)P_i|\underline{P}|^{-2}f$ instead of f and find

$$\begin{aligned}&\|(I+\rho|\underline{Q}|)^{-n-1}U_{t/\rho}(\partial_i X_t)P_i|\underline{P}|^{-2}f_t\| \\ &\le c_6|t|^{-n-1}\sum_{|m|=0}^{n+1} \|D^m(\partial_i X_t)(\underline{k})k_i k^{-2}\tilde{f}\| + c_7|t|^{-n-\alpha}\sum_{|m|=0}^{n} \|D^m(\partial_i X_t)(\underline{k})k_i k^{-2}\tilde{f}\| \\ &\le c_8|t|^{-n-\alpha}\sum_{|m|=0}^{n+1} \|D^m\tilde{f}\| + c_9|t|^{-n+1-2\alpha}\sum_{|m|=0}^{n} \|D^m\tilde{f}\|.\end{aligned} \tag{13.15}$$

Finally, substituting (13.15) in the last term of (13.12) and using (a) for n to estimate the first two terms on the right hand side of (13.12) as in (13.13), we conclude that $\|(1+\rho|\underline{Q}|)^{-n-1}U_{t/\rho}f_t\| \leq d_1(|t|^{-n-1}+|t|^{-n-1-\alpha})\sum_{|m|=0}^{n+1}\|D^m\tilde{f}\| + c_9|t|^{-n-2\alpha}\sum_{|m|=0}^{n}\|D^m\tilde{f}\|$. Since $2\alpha > 1$, this completes the proof of (a) by induction.

(b) Let $0 \leq n \leq 2$. Since $(1+\rho|\underline{Q}|)^{-\kappa}$ is a bounded self-adjoint operator for all $\kappa > 0$, it follows by Schwarz' inequality that

$$\|(I+\rho|\underline{Q}|)^{-n-\alpha}U_{t/\rho}f_t\|^2 = ((I+\rho|\underline{Q}|)^{-n}U_{t/\rho}f_t, (I+\rho|\underline{Q}|)^{-n-2\alpha}U_{t/\rho}f_t)$$

$$\leq \|(I+\rho|\underline{Q}|)^{-n}U_{t/\rho}f_t\|\,\|(I+\rho|\underline{Q}|)^{-n-1}U_{t/\rho}f_t\|, \qquad (13.16)$$

where we have noticed that for $2\alpha > 1$, $\|(I+\rho|\underline{Q}|)^{1-2\alpha}\| \leq 1$. Then (b) results from inserting the estimates (a) in (13.16).

(c) First let $f\epsilon\mathcal{D}_1$. Multiplying (13.11) by $V_S(\underline{Q})$ on the left and setting $\rho = 1$, one has as in (13.12)

$$\|V_SU_tf_t\| \leq |2t|^{-1}\sum_i \|\,|\underline{Q}|V_SU_tP_i|\underline{P}|^{-2}f_t\| + |2t|^{-1}\|V_SU_tg_t\|$$

$$+ |2t|^{-1}\sum_i \|V_SU_t(\partial_iX_t)P_i|\underline{P}|^{-2}f_t\|. \qquad (13.17)$$

Note next that for any $\phi\epsilon L^2(R^3)$ and a vector $h\epsilon L^2(R^3)$ with $\tilde{h}\epsilon L^1(R^3)$, $\|\phi(\underline{Q})h\| \leq \|\phi\|_2\|h\|_\infty \leq (2\pi)^{-3/2}\|\phi\|_2\|\tilde{h}\|_1$. Also (13.6) implies that both V_S and $|\underline{x}|V_S$ are in $L^2(R^3)$. Using these observations, the inequality (13.9) and the support properties of f, we can estimate the right hand side of (13.17) as

$$\|V_SU_tf_t\| \leq d_2|t|^{-1}(\|\,|\underline{x}|V_S\|_2\|\tilde{f}\|_1 + \|V_S\|_2\|\tilde{g}\|_1) +$$

$$d_3|t|^{-\alpha}\|V_S\|_2\|\tilde{f}\|_1. \qquad (13.18)$$

Now we replace f in (13.18) by $(\partial_i X_t)P_i|\underline{P}|^{-2}f$, use the definition of g, Leibniz' rule as in (a) and (13.9) to obtain

$$\|V_S U_t(\partial_i X_t)P_i|\underline{P}|^{-2}f_t\| \le d_4|t|^{-\alpha}(\||\underline{x}|V_S\|_2\|\tilde{f}\|_1 + \|V_S\|_2\|\tilde{f}\|_1 + \sum_i\|V_S\|_2\|\partial_i\tilde{f}\|_1) + d_5|t|^{1-2\alpha}\|V_S\|_2\|\tilde{f}\|_1. \qquad (13.19)$$

Substituting (13.19) into (13.17) and estimating the first two terms of (13.17) as in (13.18), we arrive at

$$\|V_S U_t f_t\| \le d_6|t|^{-1}\|(1+|\underline{x}|)V_S\|_2(\|\tilde{f}\|_1 + \sum_i\|\partial_i\tilde{f}\|_1). \qquad (13.20)$$

Now let $f\varepsilon\mathcal{D}_2$. From (13.19) the last term in (13.17) can be seen to be bounded by $d_7|t|^{-2\alpha}$. Since $g\varepsilon\mathcal{D}_1$, we may apply (13.20) to g and find that the second term on the right hand side of (13.17) is bounded by $d_8|t|^{-2}$. An identical estimate for the first term on the right hand side of (13.17) is obtained by replacing, in (13.20), V_S by $|\underline{Q}|V_S$ and f by $P_i|P|^{-2}f$ and noticing that $\||\underline{x}|(1+|\underline{x}|)V_S\|_2 \le \sqrt{2}\|(1+|\underline{x}|^2)V_S\|_2$. This proves (c) except for the explicit form of the constant. The latter can easily be derived and will not be used in what follows. #

We can now prove the main theorem of this section, namely the existence of the generalized wave operators as strong limits (13.2). For this we set $Y_t \equiv U_t\exp(-iX_t)$ with X_t defined by (13.8) and Lemma 13.2(b).

PROPOSITION 13.4 : (a) Assume (13.6) and (13.7) with $\frac{1}{2} < \alpha < 1$. The strong limits (13.2) exist and define two isometries $\Omega_\pm$.
(b) $\Omega_\pm$ intertwine H and K_o, i.e. $H\Omega_\pm = \Omega_\pm K_o$.

Proof : (a) Since $\|V_t^* Y_t\| = 1$ for all t, it suffices by

Proposition 2.17 to prove convergence in (13.2) on the dense set $\mathcal{D}_3$. By Proposition 4.15, this follows if we show that $V_t^*Y_tf$ is strongly differentiable and $\int_{\pm 1}^{\pm\infty} \|d/dt\ V_t^*Y_tf\|\ dt < \infty$ for each $f \in \mathcal{D}_3$.

Since Y_t maps $\mathcal{D}_3$ into itself and $\mathcal{D}_3 \subseteq D(H_0) = D(H)$, we have from Stone's theorem (Proposition 3.1) and Lemma 13.2(d) that for each $f \in \mathcal{D}_3$, $V_t^*Y_tf$ is strongly differentiable and $d/dt\ V_t^*Y_tf = V_t^*(iH - iH_0 - idX_t/dt)Y_tf = iV_t^*(V_S + V_L - V_L(2t\underline{P}))\cdot Y_tf$. Thus $\|d/dt\ V_t^*Y_tf\| \leq Z_S(t) + Z_L(t)$, where $Z_S(t) =$

$$\|V_SY_tf\| \equiv \|V_SU_tf_t\| \text{ and } Z_L(t) = \|(V_L - V_L(2t\underline{P}))Y_tf\|. \tag{13.21}$$

$$\text{By (13.4), } Z_L(t) = \|[U_t^*V_L(\underline{Q})U_t - V_L(2t\underline{P})]f_t\|$$
$$= \|[Q_t^*V_L(2t\underline{P})Q_t - V_L(2t\underline{P})]f_t\|. \tag{13.22}$$

Now $f \in \mathcal{D}_3$ implies that f_t, $Q_\tau f_t$ and $V_L(2t\underline{P})Q_\tau f_t$ are in $D(|\underline{Q}|^2)$, for all t,τ in $(-\infty,\infty)$ *). Also $Q_jf_t \equiv Q_j\ \exp(-iX_t)f = \exp(-iX_t)\ [(\partial_jX_t) + Q_j]f, (j = 1,2,3)$ and $[V_L(2t\underline{P}), |\underline{Q}|^2]f = 4t^2(\Delta V_L)(2t\underline{P})f - 4it\ \sum_j(\partial_jV_L)(2t\underline{P})Q_jf$ for each $f \in \mathcal{D}_3$. Remembering that $Q_{t/\rho} = \exp(i\rho|\underline{Q}|^2/4t)$, this yields for any $f \in \mathcal{D}_3$ that

$$[Q_t^*V_L(2t\underline{P})Q_t - V_L(2t\underline{P})]f_t = \int_0^1 d\rho\ d/d\rho(Q_{t/\rho}^*V_L(2t\underline{P})Q_{t/\rho}f_t) =$$

$$= i/4t \int_0^1 d\rho\ Q_{t/\rho}^*[V_L(2t\underline{P}), |\underline{Q}|^2]Q_{t/\rho}f_t$$

$$= \int_0^1 d\rho\ \sum_i Q_{t/\rho}^*(\partial_iV_L)(2t\underline{P})Q_{t/\rho}\exp(-iX_t)(Q_i + \partial_iX_t)f$$

$$+ it\int_0^1 d\rho\ Q_{t/\rho}^*(\Delta V_L)(2t\underline{P})Q_{t/\rho}f_t. \tag{13.23}$$

---

*) Note that Q_t $(-\infty < t < \infty)$ is the operator defined in (3.46), while Q_i $(i = 1,2,3)$ are the three components of the position operator.

Putting together (13.22) and (13.23), using (13.4) and (13.7) we obtain

$$Z_L(t) \leq \int_0^1 d\rho \sum_i \|(\partial_i V_L)(\rho \underline{Q}) U_{t/\rho} \exp(-iX_t)(Q_i + \partial_i X_t) f\| + |t| \int_0^1 d\rho \|\Delta V_L(\rho \underline{Q}) U_{t/\rho} f_t\| \leq c \int_0^1 d\rho \sum_i \|(I+\rho|\underline{Q}|)^{-1-\alpha} U_{t/\rho} \exp(-iX_t) \cdot (Q_i + \partial_i X_t) f\| + c|t| \int_0^1 d\rho \|(I+\rho|\underline{Q}|)^{-2-\alpha} U_{t/\rho} f_t\|. \tag{13.24}$$

From Lemma 13.3 we know that for $|t| \geq 1$, $|t| \|(I+\rho|\underline{Q}|)^{-2-\alpha} U_{t/\rho} f_t\|$ and $\|(I+\rho|\underline{Q}|)^{-1-\alpha} U_{t/\rho}(Q_i f)_t\|$ are bounded by $c_2|t|^{-3/2}$ while $\|V_S U_t f_t\| \leq c_3|t|^{-2\alpha}$. On the other hand $\|(I+\rho|\underline{Q}|)^{-1-\alpha} U_{t/\rho} \exp(-iX_t)(\partial_i X_t) f\| \leq c_2|t|^{-3/2} \cdot \sum_{|m|=0}^2 \|D^m \mathcal{F}[(\partial_i X_t) f]\| \leq d|t|^{-\frac{1}{2}-\alpha}$ by (13.9). For $\alpha > \frac{1}{2}$, these estimates together with (13.21) and (13.24) imply the integrability of $\|d/dt\, V_t^* Y_t f\|$ at $t = \pm\infty$.

(b) Since X_t commutes with U_t, we have for fixed $\tau \in \mathcal{R}$ and any $f \in \mathcal{H}$ that $V_\tau^* \Omega_\pm f = \text{s-lim}\, V_\tau^* V_t^* Y_t f$

$= \text{s-lim}\, V_{t+\tau}^* U_t [\exp(-iX_t) - \exp(-iX_{t+\tau})] f + \text{s-lim}\, V_{t+\tau}^* Y_{t+\tau} U_\tau^*$

$= \text{s-lim}\, V_{t+\tau}^* U_t [\exp(-iX_t) - \exp(-iX_{t+\tau})] f + \Omega_\pm U_\tau^*$, as $t \to \pm\infty$.

By virtue of Lemma 13.2(c) this means that $V_\tau^* \Omega_\pm = \Omega_\pm U_\tau^*$, from which one deduces $H\Omega_\pm = \Omega_\pm K_o$ as in the proof of Proposition 4.4. #

<u>Remark 13.5</u> : (a) As we have mentioned in Example 13.1, for the Coulomb potential we can choose any α with $\frac{1}{2} < \alpha < 1$ in the bounds of (13.7) and hence the results of Lemma 13.3 and Proposition 13.4 are valid in this case.

(b) The hypothesis (13.6) on V_S can be relaxed as far as the behaviour at infinity is concerned. In fact if $V_S = V_{S1} + V_{S2}$, with V_{S1} satisfying (13.6) and $|V_{S2}(\underline{x})| \leq d_1(1+|\underline{x}|)^{-1-\eta}$ for some $\eta > 0$, then Lemma 13.3(c) is valid for V_{S1}, and for V_{S2} one has to interpolate the estimates of Lemma 13.3(a) between n = 1 and n = 2. It can be shown (Problem 13.1) that for all $f \varepsilon \mathcal{D}_2$

$$\|(I + |\underline{Q}|)^{-1-\eta} U_t f_t\| \leq d_2 |t|^{-1-2^{-\ell}} \sum_{|m|=0}^{2} \|D^m \tilde{f}\|, \quad (13.25)$$

where ℓ is the smallest integer satisfying $2^{-\ell} < \eta$. Since the bound in (13.25) is integrable in t, one has the existence of the generalized wave operators in this case also.

From Example 13.1, Remark 13.5(a) and Proposition 13.4 we conclude that for $V(\underline{x}) = \gamma|\underline{x}|^{-\beta}$ $(\frac{1}{2} < \beta \leq 1)$, the generalized wave operators exist and have the usual intertwining property. Now for a subset of these potentials we show that the ordinary wave operators do not exist, i.e. $\{V_t^* U_t\}$ do not have strong limits as $t \to \pm \infty$.

<u>PROPOSITION 13.6</u> : Let $V(\underline{x}) = \gamma|\underline{x}|^{-\beta}$. Then (a) if $0 < \beta \leq 1$, $\{\exp(iX_t)\}$ converges weakly to zero as $t \to \pm \infty$, (b) if $3/4 < \beta \leq 1$, $\{V_t^* U_t\}$ converges weakly to zero but does not have strong limits as $t \to \pm \infty$.

<u>Proof</u> (for $t \to +\infty$) : (a) By Example 13.1 we have for k > 0 and t large enough, $X_t(\underline{k}) = (2k)^{-1}\int_0^{2kt} \psi(r)V(r)dr = \gamma(2k)^{-1} \cdot \int_1^2 \psi(r) r^{-\beta} dr + \gamma(2k)^{-1}\int_2^{2kt} r^{-\beta} dr$. If $0 < \beta < 1$, then $X_t(k)$ is of the form $a(k) + c_1 k^{-\beta} t^{1-\beta}$ while if $\beta = 1$, then $X_t(k) = b(k) + c_2 k^{-1} \log t$, where a(k) and b(k) are continuous

real-valued functions on $(0,\infty)$.

Let $f \varepsilon \mathcal{D}_\infty$ and $g \varepsilon \mathcal{H}$. Then $(g, \exp(iX_t)f) =$
$= \int \exp(ic_1 k^{-\beta} t^{1-\beta}) \bar{\tilde{g}}(\underline{k}) \exp(ia(k)) \tilde{f}(\underline{k}) d^3k$ if $0 < \beta < 1$ and
$(g, \exp(iX_t)f) = \int \exp(ic_2 k^{-1} \log t) \bar{\tilde{g}}(k) \exp(ib(k)) \tilde{f}(\underline{k}) d^3k$ if
$\beta = 1$. In either case we can write $(g, \exp(iX_t)f) =$
$\int \exp[ik^{-\beta}\phi(t,\beta)] h(\underline{k}) d^3k$, where $\phi(t,\beta) = c_1 t^{1-\beta}$ if $0 < \beta < 1$
and $\phi(t,\beta) = c_2 \log t$ if $\beta = 1$ and $h \varepsilon L^1(\mathcal{R}^3)$. Making
the change of variable $y = k^{-\beta}$, we note that
$w(y) \equiv -\beta^{-1} \int d\omega\, h(y^{-1/\beta}\underline{\omega}) y^{-3/\beta - 1} \varepsilon L^1([0,\infty), dy)$ and
$(g, \exp(iX_t)f) = \int \exp(iy\phi(t,\beta)) w(y) dy$. This converges to zero by the Riemann-Lebesgue lemma as $\phi(t,\beta) \to \infty$, which happens when $t \to +\infty$. Since $\|\exp(iX_t)\| = 1$ and $\mathcal{D}_\infty$ is dense, this implies weak convergence of $\{\exp(iX_t)\}$ to 0 as $t \to +\infty$.

(b) Let $f \varepsilon \mathcal{D}_\infty$. Then $\|\Omega_+ \exp(iX_s)f - V_s^* U_s f\| =$
$\|(\Omega_+ - V_s^* Y_s)\exp(iX_s)f\| \leq \int_s^\infty dt\, \|d/dt(V_t^* Y_t)\exp(iX_s)f\|$. Notice that $\exp(iX_s)f \varepsilon \mathcal{D}_3$. Proceeding as in the proof of Proposition 13.4(a), one can show (Problem 13.2) that the above integral converges to zero as $s \to \infty$ if $3/4 < \alpha < 1$. Next, for any $g \varepsilon \mathcal{H}$, we write $(g, V_s^* U_s f) = (\Omega_+^* g, \exp(iX_s)f) - (g, [\Omega_+ \exp(iX_s) - V_s^* U_s]f)$, which converges to zero as $s \to \infty$ by part (a) and the preceding observation. Thus $V_s^* U_s f$ converges weakly to zero as $s \to \infty$ for all $f \varepsilon \mathcal{D}_\infty$. Since $\|V_s^* U_s\| = 1$, one can extend the result to all vectors $f \varepsilon \mathcal{H}$. Now suppose $V_s^* U_s$ converges strongly to the isometry Ω_+' as $s \to \infty$. Then it certainly converges weakly to Ω_+', which must be zero by the above result. This is impossible since $\|\Omega_+'\| = 1$. #

To end this section, we briefly describe some other work on scattering theory for long range potentials. The existence of generalized wave operators was first proven for the Coulomb potential by Dollard [1]. The method for constructing X_t presented in this section breaks down for $0 < \alpha \leq 1/2$. In such a case, one has to replace (13.5) by the more general equation :

$$dX_t/dt = V(2t\underline{P} + \underline{\nabla} X_t(\underline{P})). \qquad (13.26)$$

To obtain X_t from (13.26) one can either try to solve this non-linear equation, which was done by Hörmander [1], or set up an iteration scheme following Buslaev and Matveev [1] and Alsholm [1]. The iteration scheme is as follows : $X_t^{(o)} = 0$, $X_t^{(m)}(\underline{k}) = \int_0^t V(2s\underline{k} + \underline{\nabla} X_s^{(m-1)}(\underline{k}))ds$. Then one uses the solution X_t of (13.26) or some $X_t^{(m)}$ to prove the existence of the generalized wave operators. Our estimates in Lemma 13.3 and Proposition 13.4 are similar to those of Alsholm [1], which can also be used to extend Proposition 13.6(b) to values of β less than 3/4. Generalized wave operators for interactions given by pseudo-differential operators (which include momentum dependent potentials) have been obtained by Berthier and Collet [1].

The operator $\exp(-iX_t)$ is sometimes called a <u>time-dependent "dressing" transformation</u>. There exist constructions of generalized wave operators using <u>time-independent "dressing" transformations</u>. This was first introduced by Mulherin and Zinnes [1] for the Coulomb potential for which they give two densely defined operators $K_\pm$ as

$$(K_\pm f)(\underline{x}) = (2\pi)^{-3/2} \int d^3k \exp[i\underline{k}\cdot\underline{x} \mp i\gamma k^{-1}\log(kr \pm \underline{k}\cdot\underline{x})]\tilde{f}(\underline{k})$$

with f in a suitable dense subset of $L^2(\mathbb{R}^3)$. Note that the exponential in the above expression is nothing but the first term in the asymptotic expansion (13.1). Then they show that the s-lim $V_t^* K_\pm U_t$ exist as $t \to \pm\infty$ and coincide with the generalized wave operators obtained by the method given here. Their approach has been extended by Georgescu [1] to a large class of spherically symmetric long range potentials.

The asymptotic completeness of the theory for the Coulomb potential was proven by Dollard [1] using an eigenfunction expansion. More generally, for any spherically symmetric long range potential V, one can study the parts H_ℓ and $K_{0,\ell}$ of H and K_0 respectively in the partial wave subspaces $\mathcal{H}_{\ell m}$ and try to establish the asymptotic completeness of the theory in each $\mathcal{H}_{\ell m}$ (see Sections 11-1 and 11-3). For this one finds from Weidmann [1] or Georgescu [1] that under very moderate assumptions on V, H_ℓ has simple spectrum, and thus one can apply Proposition 9.15 to prove asymptotic completeness of the theory for each ℓ. Spectral properties of Hamiltonians with non-spherically symmetric long range potentials were studied by Lavine [3], and Ikebe [2] has given a construction of the spectral representation of such Hamiltonians.

For long range potentials one can apply the methods of Chapter 6 and derive time-independent equations. However, in contrast to the case of short range potentials, the Lippmann-Schwinger equation has no inhomogeneous term (see e.g. Prugovečki and Zorbas [1]). For $V(\underline{x}) = \gamma|\underline{x}|^{-\beta}$ $(\frac{1}{2} < \beta \leq 1)$, this can be seen as follows. One writes on $D(H) = D(K_0)$,

$\Omega_+ = R^o_z(H-V-z)\Omega_+$ and thus $\Omega_+ = \int \Omega_+ dE^0_\lambda = -\int R^o_{\lambda-i\eta} V\Omega_+ dE^0_\lambda +$ $\int R^o_{\lambda-i\eta}\Omega_+ K_0 dE^0_\lambda - \int R^o_{\lambda-i\eta}\Omega_+ \lambda dE^0_\lambda + i\eta \int R^o_{\lambda-i\eta} dE_\lambda \Omega_+$, where we have used the relation $H\Omega_+ = \Omega_+ K_0$. The second and third terms on the right hand side of the above expression cancel because of Lemma 6.1 and we have by (6.12) that for $\eta > 0$

$$\Omega_+ = -\int R^o_{\lambda-i\eta} V\Omega_+ dE^0_\lambda + \eta\int_0^\infty dt\ \exp(-\eta t)U_t^* V_t \Omega_+ = -\int R^o_{\lambda-i\eta} V\Omega_+ dE^0_\lambda$$

$$+ \eta\int_0^\infty dt\ \exp(-\eta t)U_t^* V_t(\Omega_+ - V_t^* Y_t) + \eta\int_0^\infty dt\ \exp(-\eta t - iX_t).$$

By Proposition 13.4, $U_t^* V_t(\Omega_+ - V_t^* Y_t)$ converges strongly to 0 as $t\to\infty$ and hence $\eta\int_0^\infty dt\ \exp(-\eta t)U_t^* V_t(\Omega_+ - V_t^* Y_t)$ converges strongly to zero as $\eta \to +0$. For short range potentials as we have observed before, $X_t \to 0$ as $t \to \infty$ and the last term in the expression for Ω_+ gives I as $\eta \to +0$, leading to (6.23). But for long range potentials, we know from Proposition 13.6(a) that $\exp(-iX_t)$ converges weakly to zero as $t \to \infty$ and hence $\eta\int_0^\infty \exp(-\eta t - iX_t)dt$ does the same as $\eta \to +0$, giving the following analogue of the Lippmann-Schwinger equation,

$$\Omega_\pm = -\underset{\eta\to+0}{\text{w-lim}} \int R^o_{\lambda\mp i\eta} V\Omega_+ dE^0_\lambda \text{ on } D(K_0). \qquad (13.27)$$

It is also interesting to note that for a long range potential, the weak limit of the right hand side of (6.39) is zero and hence T_z does not lead to the scattering amplitude (Chandler and Gibson [3]). On the other hand, the scattering into cones formula (7.33) remains unchanged (Martin [1] and Problem 13.5). If the long range part of the potential is spherically symmetric, the ambiguity in $\Omega_\pm$ (see Problem 4.5) is a unitary operator-valued function of K_0. Let $S = \Omega_+^*\Omega_-$ and $S' = \Omega_+'^*\Omega_-'$ be any two possible scattering operators. Since they both commute with K_0, they are related by $S' = US$, where U is a unitary function of H_0. It follows that in this case the ambiguity in $\Omega_\pm$ does not show up in the scattering into cones formula (7.33). Finally we should say that no general expression for the scattering cross section is known for long range potentials.

13-2 FURTHER DISCUSSION OF THE ASYMPTOTIC CONDITION

In the preceding section we described how generalized wave operators for long range potentials can be obtained as strong limits of $\{V_t^* Y_t\}$ as $t \to \pm\infty$, where the choice of Y_t was motivated by the convergence criterion of Proposition 4.15. In the present section we show on an abstract level that, whenever there exist generalized wave operators $\Omega_\pm$ verifying (4.35) for a suitable set of operators A, then these wave operators are obtainable as strong limits in the above way by introducing one or possibly two suitably modified unperturbed evolutions $\{Y_t^\pm\}$. In the proofs we shall make use of the hypothesis that the Hilbert space $\mathcal{H}$ is separable.

The definition of the generalized wave operators in Section 4-3 involved a set (denoted here by $\mathcal{A}^1$) of bounded observables in $\mathcal{M}_\infty(H_0)$ commuting with the unperturbed evolution group $U_t = \exp(-iH_0 t)$ such that the following limits exist for all A in this set $\mathcal{A}^1$:

$$\omega_\pm(A) \equiv \underset{t\to\pm\infty}{\text{s-lim}} \; V_t^* A V_t E_\infty(H) . \tag{13.28}$$

Clearly these limits also exist on the set $\mathcal{A}^2$ of all finite linear combinations of operators in $\mathcal{A}^1$, and $\omega_\pm$ are linear maps on A^2. Notice that, as opposed to A^1, A^2 contains non-self-adjoint operators. By using (4.3) one finds as in Proposition 4.3 that, for each $A \in \mathcal{B}(\mathcal{H})$ for which the limits in (13.28) exist,

$$V_\tau \omega_\pm(A) = \omega_\pm(A) V_\tau \quad \text{for all } \tau \in \mathbb{R}. \tag{13.29}$$

By Problem 3.11 this implies that $\omega_\pm(A)H \subseteq H\omega_\pm(A)$, hence by Proposition 5.9, $E_\lambda \omega_\pm(A) = \omega_\pm(A)E_\lambda$ for all $\lambda \in \mathbb{R}$. Thus, if $E_\infty(H)$ is a spectral projection of H (which is often the case, cf. Section 7-1), then one has for all $A \in A^2$

$$E_\infty(H)\omega_\pm(A) = \omega_\pm(A)E_\infty(H) = \omega_\pm(A). \tag{13.30}$$

In what follows we always assume (13.30) to hold. Together with Corollary 2.19 it implies that

$$\omega_\pm(A) = \operatorname*{s-lim}_{t\to\pm\infty} E_\infty(H)V_t^* A V_t E_\infty(H). \tag{13.31}$$

Now let A and B be such that the limits $\omega_\pm(A)$ and $\omega_\pm(B)$ exist. By using (13.30) one gets for each $f \in \mathcal{M}_\infty(H)$ that $\|V_t^* ABV_t f - \omega_\pm(A)\omega_\pm(B)f\| \le \|V_t^* AV_t\| \, \|V_t^* BV_t f - \omega_\pm(B)f\| + \|[V_t^* AV_t E_\infty(H) - \omega_\pm(A)]\omega_\pm(B)f\| \to 0$ as $t \to \pm\infty$, so that

$$\omega_\pm(AB) = \omega_\pm(A)\omega_\pm(B). \tag{13.32}$$

Thus it is natural to require convergence in (13.28) for all operators in the algebra A^3 generated by A^2. Furthermore, taking the adjoint of (13.31), one sees that
$\text{w-lim}\, E_\infty(H)V_t^* A^* V_t E_\infty(H) = [\omega_\pm(A)]^*$ as $t \to \pm\infty$ for each $A \in A^3$. If A^* is also in A^3, then the preceding sequence converges strongly and hence also weakly to $\omega_\pm(A^*)$, showing that

$$\omega_\pm(A^*) = [\omega_\pm(A)]^*. \tag{13.33}$$

In view of this we shall henceforth assume that the limits in (13.28) exist for all A in some $*$-algebra*) $\mathcal{A}$ in $\mathcal{B}(\mathcal{H})$. If in addition $\mathcal{A}$ is complete in the operator norm (i.e. if each uniform Cauchy sequence $\{A_n\}\varepsilon\mathcal{A}$ has a uniform limit $A\varepsilon\mathcal{A}$), then $\mathcal{A}$ is a C*-algebra**). We shall return to the algebraic properties of $\mathcal{A}$ further on and add here the following definitions. The commutant $\mathcal{B}'$ of a subset $\mathcal{B}$ of $\mathcal{B}(\mathcal{H})$ is the set of all operators $A\varepsilon\mathcal{B}(\mathcal{H})$ verifying $AB = BA$ for all $B\varepsilon\mathcal{B}$. The commutant $(\mathcal{B}')'$ of $\mathcal{B}'$ will be denoted by $\mathcal{B}''$, etc. It is easy to see that the commutant of a $*$-algebra is a C*-algebra (Problem 13.7). If E is a projection in $\mathcal{B}'$, then $\mathcal{B}_E$ denotes the part of $\mathcal{B}$ in $E\mathcal{H}$ (i.e. all operators of the form $BE = EBE$ with $B\varepsilon\mathcal{B}$, viewed as operators in $E\mathcal{H}$) and $\mathcal{B}_E'$ the commutant of $\mathcal{B}_E$ in $E\mathcal{H}$. We also denote by $\mathcal{A}_0$ the commutant in $\mathcal{M}_\infty(H_0)$ of $\{U_tE_\infty(H_0)\}$, i.e. the set of all operators in $\mathcal{B}(\mathcal{M}_\infty(H_0))$ that commute with U_t for all $t\varepsilon\mathbb{R}$. With this we may summarize as follows the considerations made so far.

PROPOSITION 13.7 : Let $\mathcal{A}$ be a $*$-algebra and suppose that $\omega_\pm$ are defined on $\mathcal{A}$ by (13.28). If (13.30) is verified for all $A\varepsilon\mathcal{A}$, then $\omega_\pm$ are $*$-homomorphisms (i.e. linear maps verifying (13.32) and (13.33)) from $\mathcal{A}$ into $\{V_t\}'_{E_\infty(H)}$ and

$$\|\omega_\pm(A)\| \leq \|A\| . \qquad (13.34)$$

---

*) $\mathcal{A}$ is a $*$-algebra if, whenever $A,B\varepsilon\mathcal{A}$ and $\alpha\varepsilon\mathbb{C}$, then $\alpha A+B\varepsilon\mathcal{A}$, $A^*\varepsilon\mathcal{A}$ and $AB\varepsilon\mathcal{A}$.

**) The requirements $\|A^*\| = \|A\|$ and $\|A^*A\| = \|A\|^2$ in the abstract definition of a C*-algebra are verified by Proposition 2.8 and Lemma 5.30(b).

If one assumes in addition that

$$\omega_{\pm}(A) \neq 0 \text{ unless } A = 0, \qquad (13.35)$$

then $\omega_{\pm}$ are $*$-isomorphisms.

(13.34) follows from the definition (13.28) and Proposition 2.1. The condition (13.35) means essentially that, if the asymptotic values of an observable are zero in all states, then the observable is the zero operator. In order for the asymptotic condition to be sufficiently interesting, one must also assume that the algebra $\mathcal{A}$ is not too small. Now it is clear that increasing $\mathcal{A}$ decreases the commutant $\mathcal{A}'$. We shall impose a restriction on $\mathcal{A}'$, namely that $\mathcal{A}'$ be <u>abelian</u> (i.e. $BC = CB$ for all $B, C \in \mathcal{A}'$). We shall discuss the physical meaning of this requirement further on. Here we give one of its mathematical consequences. We say that a vector e in $\mathcal{H}$ is <u>cyclic</u> for a linear subset $\mathcal{B}$ of $\mathcal{B}(\mathcal{H})$ if the set $\mathcal{B}e \equiv \{Be \mid B \in \mathcal{B}\}$ is dense in $\mathcal{H}$, and that e is <u>separating</u> for $\mathcal{B}$ if $Be = \theta$ and $B \in \mathcal{B}$ together imply $B = 0$.

<u>LEMMA 13.8</u> : Let $\mathcal{B}$ be a $*$-algebra in $\mathcal{H}$ such that $I \in \mathcal{B}$.
(a) e is cyclic for $\mathcal{B}$ if and only if e is separating for $\mathcal{B}'$.
(b) If $\mathcal{B}'$ is abelian, there exists a cyclic vector for $\mathcal{B}$.

<u>Proof</u> : (i) Fix $f \in \mathcal{H}$. Then $\mathcal{B}f$ is a linear manifold, and we denote by F the projection whose range is the closure of $\mathcal{B}f$. If $A, B \in \mathcal{B}$, then $ABf \in \mathcal{B}f$, i.e. A maps $\mathcal{B}f$ and hence its closure $F\mathcal{H}$ into $F\mathcal{H}$. If in addition A is self-adjoint, it follows from Lemma 5.4(a) that A commutes with F. Since by (5.107) every operator in $\mathcal{B}$ is a linear combination of two self-adjoint ones, we have $F \in \mathcal{B}'$.

(ii) Suppose e is separating for $\mathcal{B}'$. Let E be the projection whose range is the closure of $\mathcal{B}e$. Since $I\epsilon\mathcal{B}$, we have $Ee = e$, i.e. $(I-E)e = \theta$. By (i), $(I-E)\epsilon\mathcal{B}'$, so that $I - E = 0$ by the definition of a separating vector. Hence e is cyclic for $\mathcal{B}$. Conversely, if e is cyclic for $\mathcal{B}$ and $A\epsilon\mathcal{B}'$ is such that $Ae = \theta$, then $ABe = BAe = \theta$ for all $B\epsilon\mathcal{B}$. Since $\mathcal{B}e$ is dense, we have $A = 0$ by Lemma 2.27. This establishes (a).

(iii) To prove (b), we choose an orthonormal basis $\{g_k\}$ of $\mathcal{H}$ and set $e_1 = g_1$ and E_1 the orthogonal projection whose range is the closure of $\mathcal{B}e_1$. If $E_1 \neq I$, we denote by ℓ the smallest integer such that $g_\ell \notin E_1\mathcal{H}$ and set $e_2 = \frac{1}{2}(I-E_1)g_\ell$ and E_2 the orthogonal projection with range $\overline{\mathcal{B}e_2}$. We have $E_2E_1 = 0$, since $(Be_2,E_1f) = (e_2,BE_1f) = (E_1e_2,Bf) = 0$ for all $B\epsilon\mathcal{B}$ and $f\epsilon\mathcal{H}$. Continuing in this way by taking $e_k = k^{-1}(I-\sum_{i=1}^{k-1}E_i)g_s$, where s is the smallest number such that $g_s \notin \sum_{i=1}^{k-1}E_i\mathcal{H}$, one arrives at a finite or countably infinite set $\{E_k\}$ of projections verifying $E_jE_k = \delta_{jk}E_k$ and $\sum_k E_k = I$, since each basis vector g_i belongs to $\sum_k E_k\mathcal{H}$.

Define $e = \sum_k e_k$. By (2.10) $\|e\|^2 = \sum_k\|e_k\|^2 \leq \sum_{k=1}^{\infty}k^{-2} < \infty$. It remains to show that e is separating for $\mathcal{B}'$. So assume $Ae = \theta$ for some $A\epsilon\mathcal{B}'$. Since $\mathcal{B}'$ is abelian by hypothesis and $E_k\epsilon\mathcal{B}'$ by (i), we find for any $B\epsilon\mathcal{B}$ that $\theta = BAe = ABe = \sum_k ABE_ke_k = \sum_k E_kABe_k$. Hence $\|ABe\|^2 = \sum_k \|ABe_k\|^2 = 0$, i.e. $ABe_k = \theta$ for each k and each $B\epsilon\mathcal{B}$. Since $\mathcal{B}e_k$ is dense in $E_k\mathcal{H}$, we have $AE_k = 0$ for each k, i.e. $A = \sum_k AE_k = 0$. #

Before continuing, we introduce another type of $*$-algebra. For this we notice that it follows immediately from the definition of the commutant that $\mathcal{B} \subseteq \mathcal{B}''$, $\mathcal{B}' = \mathcal{B}'''$, $\mathcal{B}'' = \mathcal{B}''''$, etc. A special type of $*$-algebra is therefore that where the first inclusion is also an equality, i.e. where $\mathcal{B} = \mathcal{B}''$. A $*$-algebra verifying this condition is called a von Neumann algebra. By Problem 13.7(c) every von Neumann algebra is a C*-algebra. The converse is however not true, as will be seen from the examples given later. Clearly a von Neumann algebra contains I, and the commutant of any $*$-algebra is a von Neumann algebra. With this we come to the principal theorem of this section.

PROPOSITION 13.9 : Let $\mathcal{A}$ be a von Neumann algebra on $M_\infty(H_0)$ such that $\mathcal{A}'_{E_\infty(H_0)}$ is abelian. Suppose that the limits $\omega_\pm(A)$ exist for each $A \in \mathcal{A}$, and that there are two partial isometries $\Omega_\pm$ with initial set $M_\infty(H_0)$ and range $M_\infty(H)$ such that $\omega_\pm(A) = \Omega_\pm A \Omega_\pm^*$ for all $A \in \mathcal{A}$. Then there exist two families $\{Y_t^\pm\}$ of closed operators in $M_\infty(H_0)$ and a dense set $\mathcal{D}$ in $M_\infty(H_0)$ verifying

(α) $\mathcal{D} \subseteq D(Y_t^\pm) \subseteq M_\infty(H_0)$ for all $t \in R$,

(β) $Y_t^\pm A f = A Y_t^\pm f$ for all $f \in D(Y_t^\pm)$ and all $A \in \mathcal{A}$,

(γ) $\lim \|Y_t^\pm f\| = \|f\|$ as $t \to \pm\infty$ for each $f \in \mathcal{D}$,

(δ) the functions $t \mapsto Y_t^\pm f$ are strongly continuous for each $f \in \mathcal{D}$, uniformly in $t \in R$,

(ε) $\Omega_\pm f = \text{s-lim}\, V_t^* Y_t^\pm f$ as $t \to \pm\infty$ for each $f \in \mathcal{D}$.

This proposition states that the generalized wave operators can be obtained as $\text{s-lim}\, V_t^* Y_t^\pm$ on a dense set. The operators $Y_t^\pm$ need not be bounded, but (γ) means that they are asymptotically isometric on a dense set. Note that in general this latter property is sufficient to ensure that, if s-lim $V_t^* Y_t^{\pm}$ as $t \to \pm\infty$ exist, then their closures are isometric. We shall comment further on in this section on the possibility of choosing $Y_t^{\pm}$ bounded or unitary.

<u>Proof</u> : We prove the proposition for $t \to +\infty$ and drop the + sign. We shall write A'_E for $A'_{E_\infty(H_0)}$.

(i) By Lemma 13.8 there exists a vector $e \in M_\infty(H_0)$ such that Ae is dense in $M_\infty(H_0)$. We assume $\|e\| = 1$ and set $\mathcal{D} = Ae$. Let F be the projection whose range is the closure of $A'_E e$. By (i) of the proof of Lemma 13.8, $F \in A''_E = A$, and clearly $Fe = e$. Furthermore, if $Ae = \theta$ for some $A \in A$, then $AF = 0$. Indeed, $Ae = \theta$ implies $ABe = BAe = \theta$ for all $B \in A'_E$, i.e. $Af = \theta$ for all f in a dense set in $F\mathcal{H}$.

(ii) We define for each t an operator Z_t with $D(Z_t) = \mathcal{D}$ by

$$Z_t Ae = AFV_t \Omega e, \text{ with } A \in A. \tag{13.36}$$

In order for this definition to make sense, it must be verified that, whenever $A_1 e = A_2 e$, then $(A_1 - A_2)FV_t\Omega e = \theta$. This is indeed the case, since $(A_1 - A_2)e = \theta$ implies $(A_1 - A_2)F = 0$ by (i). Clearly the function $t \mapsto Z_t Ae$ is strongly continuous, uniformly in t by Problem 3.1. Also Z_t is densely defined in $M_\infty(H_0)$ and its range lies in $M_\infty(H_0)$ since $A = E_\infty(H_0)A$ by (4.33). Z_t commutes with A on $\mathcal{D}$: if $B \in A$ and $f = Ae \in \mathcal{D}$, then $Z_t Bf = Z_t BAe = BAFV_t\Omega e = BZ_t Ae = BZ_t f$. We shall show below that Z_t is closable and set $Y_t \equiv \overline{Z}_t$ in $M_\infty(H_0)$.

(iii) If $f = Ae$, then $\|(\Omega - V_t^* Y_t)f\| = \|(\Omega AF - V_t^* AFV_t \Omega)e\| = \|[\Omega AF\Omega^* - V_t^* AFV_t E_\infty(H)]\Omega e\|$, since $E_\infty(H)\Omega = \Omega$. This converges to zero as $t \to \infty$ since $AF \varepsilon A$, which gives (ε). (γ) follows from (ε) and Proposition 2.1 : $\lim \|Y_t F\| = \lim \|V_t^* Y_t f\| = \|\Omega f\| = \|f\|$ as $t \to +\infty$.

(iv) We now show that Z_t is closable. Since A_E' is abelian, we get that for $B \varepsilon A_E'$, $\|B^*e\|^2 = (B^*e, B^*e) = (e, BB^*e) = (Be, Be) = \|Be\|^2$. By this, the fact that $\{Be | B\varepsilon A_E'\}$ is dense in $F\mathcal{H}$, Problem 5.13(a) and the Schwarz inequality we have for any $A \varepsilon A$

$$\|FA^*e\| = \sup_{\substack{Be \neq \theta \\ B\varepsilon A_E'}} \frac{|(Be, FA^*e)|}{\|Be\|} = \sup_{\substack{Be \neq \theta \\ B\varepsilon A_E'}} \frac{|(Ae, B^*e)|}{\|B^*e\|} \leq \|Ae\| . \quad (13.37)$$

Now suppose that the sequence $\{A_n\} \varepsilon A$ is such that s-lim $A_n e = \theta$ and that s-lim $Z_t A_n e \equiv g$ exists as $n \to \infty$. Then for any $C \varepsilon A$, $(Ce, g) = \lim (Ce, A_n FV_t \Omega e)$ as $n \to \infty$. Now by (13.37) $|(Ce, A_n FV_t \Omega e)| \leq \|FA_n^* Ce\| \, \|V_t \Omega e\| \leq \|C^* A_n e\| \leq \|C^*\| \, \|A_n e\| \to 0$ as $n \to \infty$, so that $(Ce, g) = 0$. Hence g is orthogonal to $\mathcal{D} = Ae$, and since $g \varepsilon M_\infty(H_0)$, we have $g = \theta$ by Proposition 2.2, which proves that Z_t is closable.

(v) It remains to prove (β). Let $f \varepsilon D(Y_t) = D(\overline{Z}_t)$. By the definition of the closure, there exists $\{A_n\} \varepsilon A$ such that s-lim $A_n e = f$ and s-lim $Z_t A_n e = Y_t f$ as $n \to \infty$. Let $C \varepsilon A$. Then $Cf =$ s-lim $CA_n e$ and $CY_t f =$ s-lim $CZ_t A_n e =$ s-lim $Z_t CA_n e$ as $n \to \infty$, since $Z_t C = CZ_t$ on $\mathcal{D}$. Thus, by Lemma 2.5, $Cf \varepsilon D(Y_t)$ and $Y_t Cf =$ s-lim $Z_t CA_n e = CY_t f$. #

We now introduce as in Problem 4.4 the two one-parameter unitary groups in $M_\infty(H_0)$:

$$W_t^{\pm} = \Omega_\pm^* V_t \Omega_\pm , \tag{13.38}$$

whose infinitesimal generators $K_\pm$ are the renormalized unperturbed Hamiltonians acting in $M_\infty(H_0)$ (see Section 4-3). The sequences $\{V_t^* W_t^{\pm} E_\infty(H_0)\}$ may be strongly convergent or not as $t \to \pm\infty$ (for instance in potential scattering $W_t^{\pm} = U_t$, and the strong limits of $V_t^* U_t$ as $t \to \pm\infty$ exist for short range but not for long range potentials, see Proposition 13.6). One can show as in Problem 13.6 that s-lim $V_t^* W_t^{+} E_\infty(H_0)$ as $t \to +\infty$ exists if and only if s-lim $V_t^* W_\tau^{+} V_t E_\infty(H) = V_\tau E_\infty(H)$, uniformly in $\tau \geq 0$, as $t \to +\infty$, and similarly for $t \to -\infty$. If $\{V_t^* W_t^{\pm} E_\infty(H_0)\}$ are not strongly convergent, then the $Y_t^{\pm}$ differ from $W_t^{\pm}$ in an essential way. To exhibit this difference, we define $G_t^{\pm} = (W_t^{\pm})^* Y_t^{\pm}$, so that $Y_t^{\pm} = W_t^{\pm} G_t^{\pm}$. As in Lemma 13.2(c), the families $\{G_t^{\pm}\}$ have the property of being feebly oscillating. More precisely :

<u>COROLLARY 13.10</u> : For any $\tau \varepsilon R$ and any $f \varepsilon D$, one has

$$\lim_{t \to \pm\infty} \| G_{t+\tau}^{\pm} f - G_t^{\pm} f \| = 0. \tag{13.39}$$

<u>Proof</u> (for the + sign) : By Problem 4.4, $AW_\tau = W_\tau A$ for all $A \varepsilon A$, hence $W_\tau \varepsilon A_E'$. Since $A_E' \subseteq A$ (see Problem 13.7), we have s-lim $V_t^* W_\tau V_t E_\infty(H) = \Omega W_\tau \Omega^* = V_\tau E_\infty(H)$ as $t \to +\infty$.

Now let $f = Ae \varepsilon D$. From (13.36) one gets $\| G_{t+\tau} f - G_t f \| = \| W_{t+\tau}^* AFV_{t+\tau} \Omega e - W_t^* AFV_t \Omega e \| = \| AF(V_{t+\tau} - W_\tau V_t) \Omega e \| \leq \| AF \| \, \| (V_\tau - V_t^* W_\tau V_t) \Omega e \| \to 0$ as $t \to +\infty$. #

(13.38) leads to the intertwining relation $V_t \Omega_\pm = \Omega_\pm W_t^{\pm}$, which implies that $W_t^+ S = SW_t^-$, i.e. the scattering operator intertwines W_t^+ and W_t^-. In particular, if $W_t^+ = W_t^- = \exp(-iKt)$, then S commutes with the renormalized unperturbed Hamiltonian K. This is so for example in potential scattering if the scattering system is time reversal invariant, in which case S is also unitary (see Proposition 6 of Amrein, Martin and Misra [1]). Another result along these lines which applies in particular to potential scattering with long range potentials is the following :

PROPOSITION 13.11 : Assume the hypotheses of Proposition 13.9 with $A \subseteq A_0$ and $M_\infty(H) \subseteq \mathcal{H}_{ac}(H)$. If $(H-z)^{-1} - (H_0-z)^{-1}$ is compact for some z with $\operatorname{Im} z \neq 0$, then $W_t^+ = W_t^- = U_t E_\infty(H_0) = \exp(-iH_0 t) E_\infty(H_0)$, i.e. there is no energy renormalization.

Proof : We write again A_E' for $A'_{E_\infty(H_0)}$.

(i) Since $A \subseteq A_0 = \{U_t\}_E'$, we have $(A_0)_E' \subseteq A_E'$. Now $A_E' \subseteq A$, since A_E' is abelian and $A = A_E''$. Because $R_z^0 \varepsilon (A_0)_E'$, we have $R_z^0 \varepsilon A$.

(ii) The operator $H/M_\infty(H)$ is absolutely continuous and $(R_z - R_z^0) \varepsilon \mathcal{B}_0$, so that by Lemma 7.3, s-lim $V_t^*(R_z - R_z^0) V_t E_\infty(H) = 0$ as $t \to \pm\infty$. By (i) we have $R_z^0 \varepsilon A$, hence $R_z E_\infty(H) = \Omega_\pm R_z^0 \Omega_\pm^*$ or $\Omega_\pm^* R_z \Omega_\pm = R_z^0 E_\infty(H_0)$.

(iii) From the definition (13.38) and (3.58) one obtains $(K_\pm - z)^{-1} E_\infty(H_0) = \Omega_\pm^* R_z \Omega_\pm$. Together with the result of (ii) and Proposition 5.10(f), this leads to $(K_\pm / M_\infty(H_0) - z)^{-1} = (H_0 / M_\infty(H_0) - z)^{-1}$, whence $K_\pm E_\infty(H_0) = H_0 E_\infty(H_0)$ and consequently $W_t^\pm = U_t E_\infty(H_0)$. #

We return for a moment to operator algebras. As an example, we consider potential scattering and introduce the following two $*$-subalgebras of $\mathcal{B}(\mathcal{H})$: A_1 is the set of all bounded continuous functions of the momentum operators P_1, P_2, P_3, and A_2 the set of all essentially bounded measurable functions of P_1, P_2, P_3 (these operators are simply defined as multiplication operators in $\tilde{L}^2(\mathbb{R}^3)$ *)). To determine the commutants of these algebras, we introduce the spectral representation of P_1, i.e. we write $\tilde{L}^2(\mathbb{R}^3) = L^2(\mathbb{R},\tilde{L}^2(\mathbb{R}^2))$, where the variable in $\mathbb{R}$ is k_1 and those in $\mathbb{R}^2$ are k_2 and k_3. If $B\varepsilon\mathcal{B}(\mathcal{H})$ commutes with A_1 or with A_2, then it commutes with all operators of the form $\exp(i\underline{a}\cdot\underline{P})$, $\underline{a}\varepsilon\mathbb{R}^3$. It follows as in Corollary 3.2 that $BP_k \subseteq P_kB$. Hence by Proposition 5.27, B is decomposable in $L^2(\mathbb{R},\tilde{L}^2(\mathbb{R}^2))$, i.e. $B = \{B(k_1)\}$ with $\text{ess sup}_{k_1} \|B(k_1)\| = \|B\|$. $B(k_1)$ commutes with P_2 and P_3 in $\tilde{L}^2(\mathbb{R}^2)$ for almost all k_1. By introducing successively the spectral representations of $\tilde{L}^2(\mathbb{R}^2)$ relative to P_2 and of $\tilde{L}^2(\mathbb{R}^1)$ relative to P_3, one finds by repeating the above argument that $B = \{B(k_1,k_2,k_3)\}$ in $\tilde{L}^2(\mathbb{R}^3) \equiv L^2(\mathbb{R}^3,\mathbb{C})$, i.e. B is diagonalisable in $\tilde{L}^2(\mathbb{R}^3)$, each $B(k_1,k_2,k_3)$ is just a complex number and $\text{ess sup}|B(k_1,k_2,k_3)| = \|B\|$. Hence B is an essentially bounded function of P_1, P_2, P_3. On the other hand each essentially bounded function of P_1, P_2, P_3 commutes with A_1 and A_2, so that $A_1' = A_2' = A_2$.

---

*) We use the notation $\tilde{L}^2(\mathbb{R}^n)$ to denote the representation of the Hilbert space $L^2(\mathbb{R}^n)$ in which the momentum operators are diagonal, i.e. if $f(\underline{x}) \varepsilon L^2(\mathbb{R}^n)$, then $\tilde{f}(\underline{k})$ is considered to be a function in $\tilde{L}^2(\mathbb{R}^n)$.

It follows from the above that $A_2'' = A_2$, i.e. A_2 is a von Neumann algebra which is in addition identical with its commutant. The latter property is expressed by saying that A_2 is <u>maximal abelian</u> (i.e. the elements of A_2 commute pairwise, and A_2 is a maximal set in $B(H)$ having this property). In physical terms the operators P_1, P_2 and P_3 form a complete set of commuting observables, i.e. each other observable that can be diagonalized simultaneously with P_1, P_2 and P_3 must be a function of P_1, P_2 and P_3. The condition that $A'_{E_\infty(H_0)}$ be abelian which we impose in this section means that A contains at least a complete set of commuting observables in $M_\infty(H_0)$ (see [DI, Chapter I.1.7]).

The algebra A_1 is a C*-algebra contained in A_2 (Problem 13.8). Since $A_1' = A_2$, $A_1'' = A_2' = A_2$. Since A_1 is a proper subalgebra of A_2, A_1 is strictly smaller than A_1'', i.e. A_1 is not a von Neumann algebra. It can be shown that A_2 is the weak closure as well as the strong closure of A_1 [NM, §34.2]. The algebra A_0, i.e. the commutant of $\{\exp(-iK_0t)\}_{t\in R}$, is a von Neumann algebra which is larger than A_1 and A_2, since it contains in particular all essentially bounded functions of the angular momentum operators L_1, L_2 and L_3.

The operators $Y_t^{\pm}$ constructed in the proof of Proposition 13.9 are in general unbounded. In applications one can usually find $Y_t^{\pm}$ that are unitary (see Section 13-1). It is possible to prove abstractly that $Y_t^{\pm}$ can be chosen unitary on $M_\infty(H_0)$ if one requires in addition that the weak limits of $E_\infty(H)V_t^*AV_tE_\infty(H)$ exist uniformly in $\{A\in A \mid \|A\| = 1\}$ or uniformly on all projections of A and that $A = A'_{E_\infty(H_0)}$, see

Mourre [1]. Without any assumption of this type one can show that $Y_t^{\pm}$ may be taken to be bounded (but not necessarily uniformly in t). We shall now indicate how this strengthening of Proposition 13.9 is possible and at the same time discuss further the relation between $Y_t^{\pm}$ and the algebra A.

PROPOSITION 13.12 : Assume all the hypotheses of Proposition 13.9. Then there exist two strongly continuous families $\{\mathcal{Y}_t^{\pm}\}$ of bounded operators in $M_\infty(H_0)$ and a dense set $\mathcal{D}$ in $M_\infty(H_0)$ such that (γ) and (ε) of Proposition 13.9 are verified on $\mathcal{D}$ and such that $\mathcal{Y}_t^{\pm} \varepsilon A'_{E_\infty(H_0)}$ for each t.

Sketch of the proof (for the + sign) : As before we set $E_\infty(H_0) = E$, and we use the polar decomposition of a closed operator which is a generalization of Proposition 5.6 : If C is a densely defined closed operator, then C*C is positive self-adjoint, $D(|C|) = D(C)$, where $|C| = (C^*C)^{\frac{1}{2}}$ is defined as in (5.113), and $C = \mathcal{U}|C|$, where $\mathcal{U}$ is a partial isometry with initial set $\overline{|C|\mathcal{H}}$ and final set $\overline{C\mathcal{H}}$. (For a proof using only concepts given in this book, we refer to [NM, §5.9.VI and §21.1.II]).

(i) Let $Y_t = \mathcal{U}_t|Y_t|$ be the polar decomposition of Y_t and $\{E_\lambda^t\}$ the spectral family of $|Y_t|$ in $M_\infty(H_0)$, where $\{Y_t\}$ is the family of closed operators given in Proposition 13.9. By (β) of that Proposition we have $BY_t \subseteq Y_tB$ for each $B\varepsilon A$, which implies that $B^*Y_t^* \subseteq Y_t^*B^*$. Since A is a $*$-algebra, this means that $AY_t^* \subseteq Y_t^*A$ for each $A\varepsilon A$. From the above one may deduce that $AY_t^*Y_t \subseteq Y_t^*Y_tA$ for each $A\varepsilon A$, so that by Proposition 5.9 each $A\varepsilon A$ commutes with the spectral family of $Y_t^*Y_t = |Y_t|^2$, hence also with that of $|Y_t|$ and with $|Y_t|$ itself. Thus $AE_\lambda^t = E_\lambda^t A$ for each $A \in \mathcal{A}$ and all real t and λ.

It follows that, for $f \in D(Y_t) = D(|Y_t|)$ and $A \in \mathcal{A}$, $AY_t f = A\mathcal{U}_t |Y_t| f = Y_t Af = \mathcal{U}_t |Y_t| Af = \mathcal{U}_t A |Y_t| f$, hence $A\mathcal{U}_t = \mathcal{U}_t A$ on $\overline{|Y_t|\mathcal{H}}$. Now $(|Y_t|\mathcal{H})^\perp = N(|Y_t|)$ (see Problem 5.5), and A leaves $N(|Y_t|)$ invariant, since $N(|Y_t|) = E_{\{0\}}^t \mathcal{H}$. Since $\mathcal{U}_t g = \theta$ for all $g \in N(|Y_t|)$, we have have $A\mathcal{U}_t g = \mathcal{U}_t Ag = \theta$ for all such g, which implies that $A\mathcal{U}_t = \mathcal{U}_t A$ for each $A \in \mathcal{A}$.

(ii) Let $\{f_k\}$ be a countable dense set in $\mathcal{M}_\infty(H_0)$, and for each k choose a sequence $\{B_n^k\}_{n=1}^\infty$ in $\mathcal{A}$ such that $f_k = \text{s-lim } B_n^k e$ as $n \to \infty$, where e is the cyclic vector for $\mathcal{A}$ appearing in the proof of Proposition 13.9. The set of operators $\mathcal{B} = \{B_n^k\}_{k,n=1}^\infty$ is a countable subset of $\mathcal{A}$ such that $\mathcal{B}e$ is dense in $\mathcal{M}_\infty(H_0)$. We set $\mathcal{D} = \mathcal{B}e$ and let $\{A_m\}$, m = 1,2,..., be an enumeration of the elements of $\mathcal{B}$.

(iii) For each t, choose a number $\lambda_t \in (0,\infty)$ such that $\|(I-E_{\lambda_t}^t)Y_t A_m e\| < 1/t$ for all $m \leq t$, which is possible since $A_m e \in D(Y_t)$ and s-lim $E_\mu^t = I$ as $\mu \to \infty$. Define $\mathcal{Y}_t = Y_t E_{\lambda_t}^t$. Clearly $\mathcal{Y}_t \in \mathcal{B}(\mathcal{H})$, and $\mathcal{Y}_t A = A\mathcal{Y}_t$ for each $A \in \mathcal{A}$ by (i).

We have seen in (i) that $\mathcal{U}_t$ and E_λ^t are in $\mathcal{A}_E'$. Since $\mathcal{A}_E'$ is abelian, $\mathcal{U}_t$ and E_λ^t commute, hence $\mathcal{Y}_t = E_{\lambda_t}^t Y_t$ on $D(Y_t)$. Now let $f = A_n e \in \mathcal{D}$. Then $\mathcal{Y}_t f = Y_t A_n e - (I-E_{\lambda_t}^t) Y_t A_n e$, so that by (2.88)

$$\Big| \|\mathcal{Y}_t f\| - \|Y_t f\| \Big| \leq \|(I-E_{\lambda_t}^t) Y_t A_n e\| < 1/t \text{ for all } t \geq n.$$

Hence $\lim \|\mathcal{Y}_t f\| = \lim \|Y_t f\| = \|f\|$ as $t \to +\infty$, which shows that

(γ) holds for $\{\mathcal{Y}_t\}$. Similarly $\|V_t^*\mathcal{Y}_t f - \Omega f\| \leq \|V_t^* Y_t f - \Omega f\| + \|V_t^*(I-E_{\lambda_t}^t)Y_t A_n e\| \to 0$ as $t \to +\infty$, showing that s-lim $V_t^*\mathcal{Y}_t f = \Omega f$ for all $f \in \mathcal{D}$.

Thus the family $\{\mathcal{Y}_t\}$ verifies all properties required in the proposition except possibly the strong continuity. The latter can be obtained by a slight modification of the above construction (Problem 13.10). #

The preceding proof shows that the spectral family $\{E_\lambda^t\}$ of $|Y_t|$ and the partial isometry $\mathcal{U}_t$ appearing in the polar decomposition of Y_t commute with the algebra A. If Y_t is bounded, this means that $Y_t \in A'_{E_\infty(H_0)}$. If Y_t is unbounded, one says that Y_t is <u>affiliated</u> with $A'_{E_\infty(H_0)}$. In the example where $A = A_0$, we may introduce the spectral representation of $H_0/M_\infty(H_0)$ in which A_0 consists of all bounded decomposable operators and $(A_0)'_{E_\infty(H_0)}$ of all bounded diagonalizable operators. It follows then that Y_t is a function of $H_0/M_\infty(H_0)$ in the sense of functional calculus.

As a last topic we briefly consider the following fundamental question : What additional conditions are sufficient to deduce that the $*$-isomorphisms $\omega_\pm$ in Proposition 13.7 are <u>spatial</u>, i.e. of the form $\omega_\pm(A) = \Omega_\pm A \Omega_\pm^*$, where $\Omega_\pm$ are partial isometries with initial set $M_\infty(H_0)$ and range $M_\infty(H)$? We assume again that A is a C*-algebra or a von Neumann algebra on $M_\infty(H_0)$ and that $A'_{E_\infty(H_0)}$ is abelian and consider the + sign. If Ω exists, then $\omega(A) \equiv \{\Omega A \Omega^* | A \in A\}$ is a C*-algebra or a von Neumann algebra respectively in $M_\infty(H)$, and $[\omega(A)]'_{E_\infty(H)}$ is abelian. If e is a cyclic vector for $\mathcal{A}$ in $M_\infty(H_0)$, then $e' \equiv \Omega e$ is cyclic for $\omega(\mathcal{A})$ in $M_\infty(H)$, and $\omega(A)e' = \Omega Ae$ for all $A\varepsilon\mathcal{A}$. Conversely, if ω is a $*$-isomorphism from $\mathcal{A}$ into $\mathcal{B}(M_\infty(H))$ and there is a cyclic vector e for $\mathcal{A}$ in $M_\infty(H_0)$ and a cyclic vector e' for $\omega(\mathcal{A})$ in $M_\infty(H)$ such that

$$(e,Ae) = (e',\omega(A)e') \quad \text{for all } A\varepsilon\mathcal{A}, \tag{13.40}$$

one may define a linear operator Ω by setting $\Omega Ae = \omega(A)e'$ for $A\varepsilon\mathcal{A}$ and $\Omega f = \theta$ for $f\varepsilon M_\infty(H_0)^\perp$. It follows that for each $B\varepsilon\mathcal{A}$, $\omega(B)\omega(A)e' = \omega(BA)e' = \Omega BAe = \Omega B\Omega^*\Omega Ae = \Omega B\Omega^*\omega(A)e'$ and $\|\Omega Ae\|^2 = \|\omega(A)e'\|^2 = (e',\omega(A^*A)e') = (e,A^*Ae) = \|Ae\|^2$. These two facts together with Problem 2.15 and Proposition 2.10 imply that the closure of Ω is a generalized wave operator.

We thus see that the following two additional conditions in Proposition 13.7 are necessary and sufficient for the existence of Ω : (a) $\omega(\mathcal{A})$ must have a cyclic vector relative to $M_\infty(H)$, (b) there exist cyclic vectors e and e' for $\mathcal{A}$ and $\omega(\mathcal{A})$ respectively such that (13.40) holds. In view of Lemma 13.8, (a) is implied by the requirement that $[\omega(\mathcal{A})]'_{E_\infty(H)}$ be abelian (physically this means that the asymptotic algebra $\omega(\mathcal{A})$ contains a complete set of commuting observables in $M_\infty(H)$). If $\omega(\mathcal{A})$ is a von Neumann algebra, this last requirement together with the hypothesis that $\mathcal{A}'_{E_\infty(H_0)}$ be abelian also implies (b) [DI, Chapter III.3.2, Corollary 1], hence the existence of the generalized wave operator. This is not so if $\omega(\mathcal{A})$ is only a C*-algebra (Problem 13.11).

Finally we mention that if $\mathcal{A}$ is a C*-algebra then so is $\omega(\mathcal{A})$, and if $\mathcal{A}$ is a von Neumann algebra then so is $\omega(\mathcal{A})$.

For a proof we refer to [NM, Corollary of Theorem 24.6] for a C*-algebra and to Lemma 1 of the article of Amrein, Georgescu and Martin in [EN] for a von Neumann algebra. The proof in the latter case involves the ultraweak continuity of the homomorphism ω. It should be added that von Neumann algebras are distinguished from C*-algebras by certain topological properties. A C*-algebra is closed in operator norm, whereas a von Neumann algebra is also closed in various other topologies such as the strong, the weak and the ultraweak ones [DI]. These topological properties allow one to deduce stronger results if A is assumed to be a von Neumann algebra.

In potential scattering, the existence of the homomorphisms $\omega_{\pm}$ on the C*-algebra A_1 of all continuous bounded functions of P_1, P_2 and P_3 was obtained by Lavine by means of trace methods [1] and relative smoothness [3]. For potentials with spherically symmetric long range part, the existence and asymptotic completeness of $\Omega_{\pm}$ was then established by Thomas [2] by using properties of the representation in $L^2(R^3)$ of the Euclidean group in R^3. A somewhat different algebraic approach from ours to potential scattering has been proposed by Corbett [1]. An algebraic determination of the Coulomb phase shifts was given by Grosse, Grümm, Narnhofer and Thirring [1].

PROBLEMS

13.1 : Prove the estimate (13.25). (Hint : Use the inequality $\|(I+|\underline{Q}|)^{-1-\eta}U_t f_t\|^2 \leq ((I+|\underline{Q}|)^{-1}U_t f_t, (I+|\underline{Q}|)^{-1-2^{-\ell+1}}U_t f_t)$, apply Lemma 13.3(a) and iterate.)

13.2 : Show that, under the hypotheses of Proposition 13.6(b), one has $\|\Omega_{\pm}\exp(iX_s)f - V_s^* U_s f\| \to 0$ as $s\to\pm\infty$ for all $f\varepsilon\mathcal{D}_\infty$.

13.3 : (a) Let $V = V_1+V_2$ with $(1+|\underline{x}|)^3 V_1(\underline{x}) \varepsilon L^2(\mathcal{R}^3)$ and $|V_2(\underline{x})| \leq c(1+|\underline{x}|)^{-2-\eta}$, $\eta > 0$. Show that for all $f\varepsilon\mathcal{D}_3$, $\|VU_t f\| \leq d|t|^{-2-2^{-\ell}}\sum_{|m|=0}^{3} \|D^m\tilde{f}\|_1$, where $2^{-\ell} < \eta$. (Hint : Adapt Lemma 13.3(a,c) and interpolate as in Problem 13.1.)
(b) Assume that f and Sf are in $\mathcal{D}_3$. Verify the conditions (α), (β) of Proposition 7.11 for $H_o = K_o$ under the above assumptions on V.
(c)† Assume in addition that V is spherically symmetric and $\int_1^\infty r^3|V(r)|dr < \infty$. Show that there is a dense set $\mathcal{E}$ in $\mathcal{D}_3$ such that $Sf\varepsilon\mathcal{D}_3$ for all $f\varepsilon\mathcal{E}$. (Hint : Equations (11.37), (11.39), Proposition 11.16(d) and Problem 11.9(a).)

13.4 : (a) Under the hypotheses of Proposition 13.9, show that the operators $W_t^{\pm}$ defined in Problem 4.4 are independent of the ambiguity of $\Omega_{\pm}$ discussed in Problem 4.5. (b) Assume furthermore that $S\varepsilon\mathcal{A}$ (for example that $\mathcal{A} = \mathcal{A}_o$ and $W_t^{\pm} = U_t$). Then the map $A\mapsto S^*AS$ on $\mathcal{A}$ is independent of the above ambiguity.

13.5 : Assume the hypotheses of Proposition 13.4 and (A3). Prove the scattering into cones formula (7.33). (Hint : Show that $\|U_t f_t - C_t f_t\| \to 0$ as $t\to+\infty$ by using the relation $(\mathcal{Q}_t - I)g_t = \int_0^1 d\rho\; d/d\rho\; \mathcal{Q}_{t/\rho} g_t$ for $g\varepsilon\mathcal{D}_2$.)

13.6 : Show that $\Omega_+ = \text{s-lim}\, V_t^* U_t E_\infty(H_o)$ as $t\to+\infty$ exists if and only if $\|U_t^* V_\tau U_t f - U_\tau f\| \to 0$ as $t\to+\infty$ uniformly in $\tau \geq 0$ for each $f\varepsilon\mathcal{M}_\infty(H_o)$.

13.7 : (a) Prove that $\mathcal{B}(\mathcal{H})$ is a C*-algebra. (Hint : A uniform Cauchy sequence in $\mathcal{B}(\mathcal{H})$ has a strong limit defining a linear operator A with $D(A) = \mathcal{H}$. Verify that A is bounded without using the uniform boundedness principle). (b) Convince yourself that $\mathcal{B}(\mathcal{H})$ is a von Neumann algebra. (c) Show that the commutant of a $\star$-algebra is a C*-algebra. [Hint : See (a)]. (d) If $\mathcal{A}$ is a von Neumann algebra and $\mathcal{A}'$ is abelian, then $\mathcal{A}' \subseteq \mathcal{A}$.

13.8 : (a) Let $\{A_i\}$ be a sequence in $\mathcal{A}_1$ (the set of all continuous bounded functions of P_1, P_2, P_3) which is Cauchy in operator norm. Show that the limit operator is also in $\mathcal{A}_1$. (Hint : Use Problem 13.7(a) and Proposition 2.16). (b) Show that every operator $\phi(\underline{P})$ in $\mathcal{A}_2$ is the strong limit of a sequence $\{\phi_n(\underline{P})\}\varepsilon\mathcal{A}_1$. (Hint : The function ϕ can be approximated in $L^2(\Delta)$ by a uniformly bounded sequence of C_0^∞-functions for each compact Δ. Use Proposition 2.17).

13.9 : Assume the hypotheses of Proposition 13.9. Let K_+ be the infinitesimal generator of the unitary group $\{W_t^+\}$ and $\{E_\lambda^+\}$ its spectral family. (a) Prove that there exists a cyclic vector e for $\mathcal{A}$ relative to $\mathcal{M}_\infty(H_0)$ such that $e\,\varepsilon\, D(K_+)$. (b) Let $G_t^+ = (W_t^+)^* Y_t^+$, where Y_t^+ is the closure of Z_t given in (13.36) with $e\,\varepsilon\, D(K_+)$. Prove that s-lim $G_t^+ f = \theta$ as $t\to+\infty$ for each f in the dense set $\mathcal{D}_0 = \{E_\Delta^+ Ae \mid A\varepsilon\mathcal{A},\ \Delta \text{ compact}\}$.
(c) Use this to prove again that (13.39) holds for all $f\,\varepsilon\,\mathcal{D}_0$.

13.10 : Show that the families $\{\mathcal{Y}_t^\pm\}$ in Proposition 13.12 can be chosen such as to be strongly continuous. (Hint : Let $\{t_n\}$ be an increasing sequence of real numbers such that $t_n\to\infty$ and $(t_{n+1}-t_n)\to 0$ as $n\to\infty$. For $t\varepsilon[t_n,t_{n+1}]$, define a modified family $\{\hat{\mathcal{Y}}_t\}$ as a linear combination of $\mathcal{Y}_{t_n}$ and $\mathcal{Y}_{t_{n+1}}$, where $\mathcal{Y}_t$ are the operators constructed in the proof of Proposition 13.12).

13.11 : (a) Show that there exists a subset Δ of [0,1] with all the properties mentioned on page 377. (An example is given in the lectures of Amrein in [LM]). (b) Let $\mathcal{A}$ and $\mathcal{B}$ be the C*-algebras of all continuous functions of Q in $L^2(\Delta)$ and $L^2(0,1)$ respectively, where Δ is as in (a). Let ω be the natural isomorphism of $\mathcal{A}$ onto $\mathcal{B}$ [i.e. $\omega^{-1}(B) = BE_\Delta$ for $B\,\varepsilon\,\mathcal{B}$, where $(E_\Delta f)(x) = \chi_\Delta(x) f(x)$]. Show that ω is a $*$-isomorphism, that the condition (a) on page 560 is verified, while the condition (b) is not. (Hint : If ϕ is continuous, then $\sup_{x\varepsilon[0,1]}|\phi(x)| = \sup_{x\varepsilon\Delta}|\phi(x)|$, and use a reasoning similar to that in Section 2-5$\underline{A}$.)

PART IV

MULTICHANNEL SCATTERING SYSTEMS

CHAPTER 14 : GENERAL FORMULATION OF MULTICHANNEL SCATTERING

This chapter contains a description of the general structure and the typical properties of multichannel scattering systems. Applications of this theory to N-body potential scattering will be postponed until the next chapter. Section 14-1 is intended as an introduction giving the physical background and acquainting the reader with the notion of clustering of particles in many-particle systems. In Section 14-2 we formulate the asymptotic condition in an abstract setting and deduce the properties of the wave and scattering operators. In Section 14-3 we introduce the concept of scattering channels, and in the final section we list some of the time-independent equations for multichannel scattering systems.

In this chapter we shall encounter operators of the form $A = A_1 \otimes I \otimes \dots \otimes I + \dots + I \otimes \dots \otimes I \otimes A_n$ in a tensor product space $H = H_1 \otimes H_2 \otimes \ldots \otimes H_n$, where each A_k is self-adjoint in H_k. It can be shown that A is essentially self-adjoint on $D(A_1) \hat{\otimes} \ldots \hat{\otimes} D(A_n)$, hence also on the larger domain $\cap_k D(I \otimes \ldots \otimes A_k \otimes \ldots \otimes I)$. Moreover, if each A_k is bounded below, then the operator sum A is self-adjoint. (A proof of these facts is indicated in Problem 14.8; for a more general theorem see [RS, Chapter VIII.10]). In our applications the operators A_k are Hamiltonians of certain physical systems. We shall tacitly assume that all occuring Hamiltonians are bounded below (which is usually the case in applications), so that sums of the above form will always be self-adjoint. Our statements are easily adapted to the case where this hypothesis is dropped if one replaces such operator sums by their closures.

14-1 CLUSTERING OF PARTICLES

In Section 4-1 we defined a simple scattering system by the property that the time evolution of each scattering state of H can be approximated at large negative and positive times by states whose time evolution is governed by a single unperturbed evolution group. We then saw in Chapters 7, 8 and 9 that the concept of a simple scattering system is suitable for the description of physical systems composed of only two elementary*) subsystems, for instance two elementary particles. In that case it is usual to distinguish two kinds of states of the system, the bound states, in which the two subsystems stay bound to each other at all times, and the scattering

*) The term "elementary" means here that such a system will never be viewed as being itself composed of subsystems.

states, in which the two subsystems move farther and farther apart from each other as $t \to \pm\infty$.

If a physical system consists of more than two elementary constituents, this simple picture is no longer sufficient. As an example, let us consider a system composed of three elementary subsystems labelled a, b and c. Bound states will again be characterized by the property that all three subsystems stay close together at all times, and a scattering state will be such that at large negative and positive times the system splits into at least two parts whose relative distance becomes larger and larger as $t \to \pm\infty$. It is seen that there are various possibilities for this splitting up into parts, and accordingly one will distinguish various types of scattering states.

It may happen that each of the three constituents will move separately from the other two, with the relative distance between any two constituents becoming larger and larger as $t \to +\infty$. The total time evolution of such a scattering state will be asymptotically approximated by that of the three freely moving subsystems. We shall formally characterize such scattering states by $\{[a],[b],[c]\}$ (the square brackets indicate the individual parts into which the system splits up). Another possibility is that where a and b stay bound together at large times and the composite system $[a,b]$ moves farther and farther away from c as $t \to +\infty$, i.e. states which may be designated by $\{[a,b],[c]\}$. The total time evolution of such a state will be asymptotically approximated by the free evolution of a certain state of a two-component system consisting of the composite system $[a,b]$ and of the elementary subsystem c. This free evolution group will be different from that indicated above for states of the type $\{[a],[b],[c]\}$, since it will necessarily involve the forces that keep a and b bound together, whereas the free evolution of the three independent subsystems contains no forces between them. It must be pointed out that in this description a composite system like $[a,b]$ is considered as a single physical object, which may have an internal structure in the sense that it may be in various bound states. Its free motion is determined (up to the binding energy) by the kinetic energy operator of its center of mass. It is also seen that the preceding description of the splitting of the three-particle system into $[a,b]$ and $[c]$ involves three different subspaces : the sets of scattering states of H of the type $\{[a,b],[c]\}$ for $t\to\pm\infty$ respectively (these two sets may be different), and the set of possible states of the two-component system consisting of $[a,b]$ and $[c]$ (the latter set contains the states by which the states of the first two are asymptotically approximated).

Thus in the preceding example one has to distinguish (for each sign of the time) four types of scattering states of H, namely $\{[a],[b],[c]\}$, $\{[a,b],[c]\}$, $\{[a,c],[b]\}$ and $\{[a],[b,c]\}$, and the asymptotic description of the total time evolution of each type will be in terms of a different unperturbed evolution group. A similar situation will be encountered when one considers a system consisting of $N \geq 3$ elementary subsystems; the number of possibilities of splitting the system into at least two (elementary or composite) subsystems will again determine the number of possible types of scattering states. As a simple physical example (apart from those indicated in Chapter 1) we may consider the scattering of a proton by a deuteron. This involves three elementary particles, namely two protons and a neutron, and the deuteron is a composite system given as a bound state of a proton and the neutron. Because of the indistinguishability of the two protons and because no proton-proton bound states exist, there are only two types of scattering states, those being composed asymptotically of two free protons and a free neutron and those consisting asymptotically of a proton and a deuteron. It should be added that the total number of particles (i.e. elementary subsystems) in the type of multichannel theory considered here is fixed, so that such a theory will not describe the creation of additional particles such as mesons at high energies.

It is seen from this example that the various types of scattering states may lead asymptotically to quite different physical situations. The first step towards a scattering theory is therefore to introduce the states which appropriately describe these situations, i.e. which, when evolving under an appropriate unperturbed evolution group, may give the correct asymptotic description of the different types of scattering states of the total Hamiltonian H. In the case of a simple scattering system these states were determined as those in $\mathcal{M}_\infty(H_0)$, and their time evolution was given by $U_t = \exp(-iH_0t)$. In the present context there will be several such subspaces (called cluster subspaces in the sequel), and with each of them there will be associated a one-parameter unitary group leaving the cluster subspace invariant. The infinitesimal generator of this group will be called a cluster Hamiltonian. For the time being we shall content ourselves with a description of the characteristic properties of these cluster subspaces and cluster Hamiltonians. The reader who would like to see an example may already consult Section 15-1 where potential scattering for N-particle systems is treated explicitly.

We consider a system consisting of N distinguishable*) particles, each described by the states and observables associated with a Hilbert space H_k (k = 1, ..., N). The Hilbert space of the entire system is then $H = H_1 \otimes \ldots \otimes H_N$. There will be a cluster subspace of H for each possible partition of the set of numbers {1,...,N} into n subsets with $2 \leq n \leq N$. These subsets will be called <u>clusters</u> and denoted by (1), (2), ...,(n). The corresponding partition will also be designated by D = {(1),...,(n)} or occasionally by D_n to indicate that the number of clusters in the partition is n. Note that for a given n there may be several different partitions. Since each particle is labelled by a number, we may associate with each cluster (k) a subset of the N particles, namely those whose number appears in (k), as well as a Hilbert space $H^{(k)}$, the tensor product of the spaces H_j with $j \in (k)$. Thus each partition leads to a division of the N particles into at least two subsets. These subsets may consist of 1, 2, ..., N-1 particles and will occasionally be called clusters of

---

*) The considerations that follow can be adapted to the case of identical particles obeying Bose-Einstein or Fermi-Dirac statistics by considering only subspaces formed of suitably symmetrized or antisymmetrized states, cf. [T, Chapter 22].

particles. $\mathcal{H}^{(k)}$ is the Hilbert space corresponding to the set of particles associated with the cluster (k), and the N-particle Hilbert space $\mathcal{H}$ may be written as

$$\mathcal{H} = \mathcal{H}^{(1)} \otimes \mathcal{H}^{(2)} \otimes \ldots \otimes \mathcal{H}^{(n)}. \tag{14.1}$$

A <u>cluster Hamiltonian</u>*) is a self-adjoint operator of the form $H^D = H^{(1)} \otimes I \otimes \cdots \otimes I + \ldots + I \otimes \cdots \otimes I \otimes H^{(n)}$ in the representation (14.1) of $\mathcal{H}$ as a tensor product, where for each k, $H^{(k)}$ is a Hamiltonian for the set of particles forming the cluster (k), and $H^{(k)}$ acts in $\mathcal{H}^{(k)}$. The associated one-parameter unitary group (the <u>cluster evolution group</u>) then has the form

$$U_t^D \equiv \exp(-iH^D t) = \exp(-iH^{(1)}t) \otimes \cdots \otimes \exp(-iH^{(n)}t), \tag{14.2}$$

which means that each cluster of particles moves independently of the other ones. ((14.2) may be checked by noticing that the infinitesimal generator of the right-hand side coincides with H^D on $D(H^{(1)}) \hat{\otimes} \cdots \hat{\otimes} D(H^{(n)})$ and remembering that H^D is essentially self-adjoint on this subset of $\mathcal{H}$. See Problem 14.9).

Consider now one of the self-adjoint Hamiltonians $H^{(k)}$ acting in $\mathcal{H}^{(k)}$. We define $M_0(H^{(k)})$ to be the subspace of $\mathcal{H}^{(k)}$ consisting of all the bound states associated with $H^{(k)}$ (i.e. the set of states in which all the particles belonging to (k) stay close together at all times relative to the evolution group $\{\exp(-iH^{(k)}t)\}$. These states may be defined as in (7.4) by using the relevant position operators; see Section 15-1).

*) A more precise term would be "clustering Hamiltonian", since H^D is associated with a definite clustering of the N particles rather than with an individual cluster.

The cluster subspace associated with the cluster Hamiltonian H^D is determined as the following subspace of H in the representation (14.1) *) :

$$M^D = M_0(H^{(1)}) \otimes M_0(H^{(2)}) \otimes \ldots \otimes M_0(H^{(n)}). \qquad (14.3)$$

A word must be added about clusters consisting of only one particle. In this context a particle is considered to be a bound state of itself, i.e. if (k) contains only one element, then $M_0(H^{(k)})$ is taken to be the entire one-particle space $H^{(k)}$ (which is identical with one of the original spaces H_i).

Since $M_0(H^{(k)})$ reduces $H^{(k)}$ for each k, M^D reduces H^D, or in other words the cluster evolution group $\{U_t^D\}$ leaves M^D invariant (Proposition 5.10(e)). The states in M^D describe n subsystems (composite or elementary) each of which forms a "scattering fragment" and which move independently of one another under the cluster evolution group $\{U_t^D\}$. In order for this interpretation to be consistent, we must require that the cluster evolution group does not lead to binding between clusters (if it did, the corresponding bound states would simply be included in a different cluster subspace).

The set of cluster subspaces $\{M^D\}$ assumes the role that was played in Section 4-1 by the subset $M_\infty(H_0)$ of scattering states of the unperturbed Hamiltonian. The total number of these cluster subspaces is finite and does not exceed the number M_N of possible ways of partitioning the set {1, ..., N} into at least two clusters. It may however be considerably

---

*) If M_k are linear manifolds in H_k (k = 1,2), then $M_1 \otimes M_2$ is the subspace of $H_1 \otimes H_2$ spanned by $M_1 \hat{\otimes} M_2$.

smaller than M_N, since certain clusters of particles may have no bound states at all. Thus $\mathcal{M}^D \neq \{\theta\}$ if and only if for each cluster (k) of more than one particle appearing in the partition D, the Hamiltonian $H^{(k)}$ has at least one bound state. If all interparticle forces are repulsive, there will be no cluster bound states at all, so that the only nontrivial cluster subspace will be the one corresponding to N free particles, i.e. $\mathcal{M}^{D_N} = \mathcal{H}$, and the scattering theory will be that of a simple scattering system*).

Since the state vectors in the cluster subspaces evolving under the cluster evolution group will be used to approximate the different types of scattering states of H at large times, it is clear that the cluster Hamiltonians must be somehow related to the total Hamiltonian H. A simple prescription to obtain a cluster Hamiltonian H^D from H would be to simply drop in H all parts of the interaction involving particles belonging to different clusters in the partition D. This is possible for instance if the total interaction is a sum of terms each of which involves only two particles, e.g. a sum of pair potentials depending on the relative distance of the respective pair of particles. One will then retain in H^D only those potentials that correspond to a pair of particles belonging to the same cluster (cf. Section 15-1). In more general situations one may have to introduce some kind of effective interaction between the particles in each cluster to define the operators H^D.

---

*) For N = 2, this does not quite lead to the situation considered in Section 4-1, since there it was admitted that the unperturbed Hamiltonian H_0 might itself have bound states, whereas in the present context this possibility is excluded.

In general the relative position in H of two different cluster subspaces is quite arbitrary, although in certain cases an inclusion relation may exist. For example we have already seen that the cluster subspace corresponding to the partition D_N into N single particles is the entire Hilbert space, so that $M^D \subseteq M^{D_N} = H$ for any partition D.

For the mathematical description of multichannel scattering systems it is convenient to introduce in addition to the physical Hilbert space H an auxiliary Hilbert space called the asymptotic Hilbert space H_{as}. It is constructed by taking the direct sum of all cluster subspaces, i.e.

$$H_{as} = \oplus M^D, \tag{14.4}$$

where the sum extends over all partitions D of $\{1,\ldots,N\}$ into at least two clusters. Henceforth we shall write M^D when M^D is viewed as a subspace of H and M^D_{as} when it is viewed as a subspace of H_{as}. Similarly a vector $f^D \varepsilon M^D$ will be denoted by f^D_{as} when viewed as a vector in H_{as}.

Since each summand in (14.4) is identified with a subspace of H, one may define a linear operator J, the injection operator, from H_{as} to H, as follows : the restriction of J to a summand H_{as} is the identification map, and this is extended by linearity to linear combinations of vectors in different summands. Thus, if $f^D_{as} \varepsilon M^D_{as}$, $Jf^D_{as} = f^D \varepsilon H$, and

$$J(\textstyle\sum_D f^D_{as}) = \sum_D f^D. \tag{14.5}$$

Since the restriction of J to each M^D_{as} has norm 1 and the sum in (14.4) is finite, J is a bounded operator from H_{as} to H. By Remark 10.2, its adjoint J* is well defined, bounded and maps H into H_{as} (see Problem 14.3).

It is usual to further subdivide the cluster subspaces into the so-called channel subspaces. This subdivision is based physically on the specification of the internal structure or the binding energy of the different clusters of particles, which are sometimes also determined in experiments. This subdivision will be discussed in Section 14-3, after the formulation of the asymptotic condition for which it suffices to have at one's disposal the cluster subspaces.

14-2 THE ASYMPTOTIC CONDITION

In this section we shall formulate the asymptotic condition for multichannel scattering systems and describe some of its consequences. Most of these are proved in exactly the same way as the corresponding results for simple scattering systems given in Chapter 4. In order to avoid repetition, we shall indicate only those proofs that differ from preceding ones and content ourselves with simply stating the other results.

The mathematical objects needed for a multichannel scattering theory are

(i) a separable Hilbert space H,

(ii) a strongly continuous unitary one-parameter group $\{V_t\}$, called the total evolution group, and a subspace $M_\infty(H)$ of scattering states associated with it,

(iii) for each D belonging to some index set Ξ, a subspace M^D of H and a strongly continuous unitary one-parameter group $\{U_t^D\}$ leaving M^D invariant.

The subspaces $\{M^D\}$ will be called cluster subspaces, the groups $\{U_t^D\}$ cluster evolution groups. They may for instance be the objects defined in the preceding section, in which case Ξ is a finite set. However, instead of these objects one could just as well take the channel subspaces and the associated channel evolution groups that will be introduced in Section 14-3 (in which case Ξ may be countably infinite) or some other set of subspaces and evolution groups. We denote by $E_\infty(H)$ the projection with range $M_\infty(H)$ and by E^D that with range M^D. By assumption we have $V_t E_\infty(H) = E_\infty(H)V_t$ and $U_t^D E^D = E^D U_t^D$ for each $D \varepsilon \Xi$.

The picture of multichannel scattering is similar to that explained in Section 4-1. The basic observation made in a multichannel scattering experiment is that of the arrangement of the particles into clusters (i.e. the type of scattering fragments) at large times together with the measurement of the momenta of these fragments. Correspondingly there should exist for each $D \varepsilon \Xi$ an observable on the set of scattering states $M_\infty(H)$ determining whether the time evolution of a state will asymptotically lead to the type of clustering determined by D or not. In fact there should be two such observables for each D, represented by two self-adjoint operators $B_\pm^D$, corresponding to $t \to +\infty$ and $t \to -\infty$ respectively. In addition there will be suitable momentum operators (or other observables) defined on each M^D.

Let us discuss here the first type of observables. We say that a scattering state g of H leads to clustering D as $t \to +\infty$ if there exists $f_+^D \varepsilon M^D$ such that $\|V_t g - U_t^D f_+^D\| \to 0$ as

$t \to +\infty$. The set of all states $g \in M_\infty(H)$ verifying this condition is clearly a subspace N_+^D of $M_\infty(H)$. The observable B_+^D is required to be such that the subspace N_+^D is the eigenspace of B_+^D corresponding to a definite eigenvalue, say to 1. Thus $N_+^D = \{g \in M_\infty(H) \mid B_+^D g = g\}$. Similarly one defines $N_-^D = \{g \in M_\infty(H) \mid \|V_t g - U_t^D f_-^D\| \to 0 \text{ as } t \to -\infty \text{ for some } f_-^D \in M^D\}$ and requires that $N_-^D = \{g \in M_\infty(H) \mid B_-^D g = g\}$. We denote by $F_\pm^D$ the orthogonal projections with range $N_\pm^D$ and notice that $F_\pm^D$ are the spectral projections of $B_\pm^D$ associated with the Borel set $\{1\}$. It will not be necessary to specify the observables $B_\pm^D$ any further.

It is usually assumed in the interpretation of experiments that the observations of the different arrangements of particles into clusters are compatible measurements. Thus one will require that the observables $\{B_+^D\}_{D \in \Xi}$ form a set of mutually commuting operators, and similarly for $\{B_-^D\}_{D \in \Xi}$. The commutativity of two self-adjoint operators implies that of their spectral projections (Proposition 5.9). Hence this requirement leads in particular to

$$F_+^C F_+^D = F_+^D F_+^C, \quad F_-^C F_-^D = F_-^D F_-^C \quad \text{for all } C, D \in \Xi. \tag{14.6}$$

If the asymptotic description of the scattering states by means of the cluster evolution groups is to be meaningful, no scattering state of H should lead simultaneously to two different clusterings. This means that one has to require in addition that

$$N_+^C \cap N_+^D = \{\theta\}, \quad N_-^C \cap N_-^D = \{\theta\} \text{ whenever } C \neq D. \tag{14.7}$$

This together with (14.6) has the following consequence. Let

$C \neq D$ and let $g^D \varepsilon N_+^D$, i.e. $F_+^D g^D = g^D$. Consider the vector $F_+^C g^D \varepsilon N_+^C$. We have by (14.6) $F_+^D F_+^C g^D = F_+^C F_+^D g^D = F_+^C g^D$, so that the vector $F_+^C g^D$ is an eigenvector of F_+^D with eigenvalue 1, i.e. $F_+^C g^D \varepsilon N_+^D$. Thus $F_+^C g^D \varepsilon N_+^D \cap N_+^C$. Hence by (14.7) $F_+^C g^D = \theta$ for each $g^D \varepsilon N_+^D$. This means that N_+^D is orthogonal to N_+^C whenever $C \neq D$. Similarly the subspaces $\{N_-^D\}$ are mutually orthogonal. In terms of the projections $\{F_\pm^D\}$ this means that

$$F_+^C F_+^D = 0, \quad F_-^C F_-^D = 0 \quad \text{whenever } C \neq D. \tag{14.8}$$

Since (14.8) clearly implies (14.6) and (14.7), we shall henceforth replace the two conditions (14.6) and (14.7) by (14.8).

A last hypothesis will be the asymptotic completeness of the theory, i.e. that the set of cluster subspaces with associated cluster evolution groups is sufficiently large to furnish an asymptotic description of all scattering states of H*). This may be simply restated in terms of the subspaces $\{N_\pm^D\}$ by requiring that the direct sum of all N_+^D as well as that of all N_-^D is the whole of $M_\infty(H)$.

The preceding considerations lead to the following formulation of the asymptotic condition : For each $D\varepsilon\Xi$ there exists a pair of orthogonal projections $F_\pm^D$ verifying (14.8) such that $W_\pm^D \equiv$ s-lim $U_t^{D*} V_t F_\pm^D$ as $t \to \pm\infty$ exist, with (range $W_\pm^D$) $\subseteq M^D$, and such that $\sum_D F_+^D = \sum_D F_-^D = E_\infty(H)$.

The operators $W_\pm^D$ are partial isometries with initial

---

*) In a more general theory one may have to consider two families $\{M_\pm^D\}$ of cluster subspaces and two families $\{U_{t,\pm}^D\}$ of cluster evolution groups.

sets $N_{\pm}^{D}$ and final sets $M'^{D}_{\pm} \subseteq M^{D}$. Since in principle every state in M^{D} may be prepared as a possible initial state (provided that M^{D} is suitably defined, e.g. as in Section 14-1), one must have $M'^{D}_{-} = M^{D}$. As in the case of simple scattering systems, it is convenient to require that $M'^{D}_{+} = M^{D}$. If this is the case, one may rewrite the asymptotic condition by using the cluster wave operators $\Omega_{\pm}^{D} = \text{s-lim}\ V_t^* U_t^D E^D$ rather than the operators $W_{\pm}^{D}$. This leads to the most usual form of the asymptotic condition consisting of the postulates (B1) - (B3) below.

(i) The following limits, called <u>cluster wave operators</u>, exist for each $D \varepsilon \Xi$:

$$\Omega_{\pm}^{D} = \underset{t \to \pm\infty}{\text{s-lim}}\ V_t^* U_t^D E^D. \tag{B1}$$

It follows that $\Omega_{\pm}^{D}$ are partial isometries with initial set M^{D} and final set $F_{\pm}^{D} H$, where

$$F_{\pm}^{D} = \Omega_{\pm}^{D} \Omega_{\pm}^{D*}. \tag{14.9}$$

(ii) $$F_{+}^{C} F_{+}^{D} = 0, \quad F_{-}^{C} F_{-}^{D} = 0 \text{ whenever } C \neq D. \tag{B2}$$

(iii) $$\sum_{D \varepsilon \Xi} F_{+}^{D} = \sum_{D \varepsilon \Xi} F_{-}^{D} = E_{\infty}(H). \tag{B3}$$

Notice that $W_{\pm}^{D} = \Omega_{\pm}^{D*}$.

The description of the scattering process may now be given as follows. Suppose that the initial state is $f_{-}^{C} \varepsilon M^{C}$. Let $g = \Omega_{-}^{C} f_{-}^{C}$, i.e. $\|V_t g - U_t^C f_{-}^C\| \to 0$ as $t \to -\infty$. One has to decompose g into a sum of states each of which will lead to a definite clustering as $t \to +\infty$. Such a decomposition is possible, namely $g = \sum_D F_{+}^{D} g$, and it is unique. For each $F_{+}^{D} g$ there exists a vector $f_{+}^{D} \varepsilon M^{D}$, namely $f_{+}^{D} = \Omega_{+}^{D*} g$, such that $\|V_t F_{+}^{D} g - U_t^D f_{+}^{D}\| \to 0$

as $t \to +\infty$. We see from this that the effect of the scattering may be described by associating with each vector $f^C \equiv f_-^C$ in $\mathcal{M}^C$ a collection of vectors $\{f_+^D\}_{D\varepsilon\Xi}$ with $f_+^D \varepsilon \mathcal{M}^D$. This correspondence is obviously linear, and f_+^D is interpreted as the part of the final state (at time t = 0) with clustering D. From the definition of f_+^D given above we see that

$$f_+^D = \Omega_+^D{}^*\Omega_-^C f_-^C \equiv S_{DC} f_-^C. \tag{14.10}$$

Thus the analogue of the scattering operator of Section 4-1 is a collection $\{S_{DC}\}$ $(C,D\varepsilon\Xi)$ of linear operators, where S_{DC} maps $\mathcal{M}^C$ into $\mathcal{M}^D$ and is defined as

$$S_{DC} = \Omega_+^D{}^*\Omega_-^C. \tag{14.11}$$

Since $\Omega_\pm^D$ are partial isometries, we have

$$\|S_{DC}\| \leq \|\Omega_+^D\| \, \|\Omega_-^C\| = 1. \tag{14.12}$$

The operators S_{CC} describe the part of the scattering in which the clustering of the particles remains unchanged, whereas the operators S_{DC} with $D \neq C$ describe the various possibilities of rearrangement scattering. In order to have rearrangement scattering, it is necessary and sufficient that $F_+^D \neq F_-^D$ for some $D\varepsilon\Xi$.

It should be repeated that the ranges of the different cluster wave operators (with fixed sign + or -) are mutually orthogonal, whereas their initial sets are in general in quite arbitrary relative positions in $\mathcal{H}$. We now mention some simple properties of the wave and scattering operators (Problem 14.4).

PROPOSITION 14.1 : Let $\Omega_\pm^D$ be defined by (B1). Then

(a) $\Omega_\pm^D$ are partial isometries with initial set M^D, i.e.

$$\Omega_\pm^{D*}\Omega_\pm^D = E^D \tag{14.13}$$

and with range $F_\pm^D H$, where $F_\pm^D = \Omega_\pm^D \Omega_\pm^{D*}$.

(b) The adjoints of $\Omega_\pm^D$ are given by

$$\Omega_\pm^{D*} = \underset{t\to\pm\infty}{\text{s-lim}}\, U_t^{D*} V_t F_\pm^D. \tag{14.14}$$

(c) $\Omega_\pm^D$ intertwine H and H^D, i.e. one has for all $t\varepsilon R$

$$V_t \Omega_\pm^D = \Omega_\pm^D U_t^D, \quad \Omega_\pm^{D*} V_t = U_t^D \Omega_\pm^{D*}. \tag{14.15}$$

(d) If $f^C \varepsilon D(H^C)$, then $\Omega_\pm^C f^C \varepsilon D(H)$ and

$$H\Omega_\pm^C f^C = \Omega_\pm^C H^C f^C. \tag{14.16}$$

(e) The projections $F_\pm^D$ reduce $\{V_t\}$, i.e.

$$F_\pm^D V_t = V_t F_\pm^D \quad \text{for all } t\varepsilon R. \tag{14.17}$$

(f) If (B2) is also verified, then

$$\Omega_+^{C*}\Omega_+^D = 0, \quad \Omega_-^{C*}\Omega_-^D = 0 \quad \text{whenever } C \neq D. \tag{14.18}$$

PROPOSITION 14.2 : Assume that (B1), (B2) and (B3) hold, and let S_{DC} be defined by (14.11). Then

(a) S_{DC} intertwines H^D and H^C, i.e.

$$U_t^D S_{DC} = S_{DC} U_t^C \quad \text{for all } t\varepsilon R, \tag{14.19}$$

and for each $f^C \varepsilon D(H^C)$ one has $S_{DC} f^C \varepsilon D(H^D)$ and

$$H^D S_{DC} f^C = S_{DC} H^C f^C. \tag{14.20}$$

(b)
$$\sum_{D\varepsilon\Xi} S^*_{DC} S_{DC'} = \delta_{CC'} E^C \tag{14.21}$$

$$\sum_{D\varepsilon\Xi} S_{CD} S^*_{C'D} = \delta_{CC'} E^C, \tag{14.22}$$

where $\delta_{CC'} = 1$ if $C = C'$ and $\delta_{CC'} = 0$ if $C \neq C'$.

Proof of part (b) : We use successively (14.11), (14.9), (B3), (14.13) and (14.18) to get

$$\sum_D S^*_{DC} S_{DC'} = \sum_D \Omega_-^{C*} \Omega_+^D \Omega_+^{D*} \Omega_-^{C'} = \Omega_-^{C*}(\sum_D F_+^D)\Omega_-^{C'}$$
$$= \Omega_-^{C*} E_\infty(H)\Omega_-^{C'} = \Omega_-^{C*}\Omega_-^{C'} = \delta_{CC'} E^C.$$

This proves (14.21). The proof of (14.22) is similar. #

Proposition 14.2(a) expresses the conservation of energy in the scattering process. If $\{E_\lambda^C\}$ denotes the spectral family of H^C, then one has for any Borel set $\Delta \subseteq R$

$$E_\Delta^D S_{DC} f^C = S_{DC} E_\Delta^C f^C. \tag{14.23}$$

If f^C has spectral support in a small interval Δ (with respect to H^C), i.e. if $f^C = E_\Delta^C f^C$, then $S_{DC} f^C$ has spectral support in the same interval Δ (with respect to H^D). It should be pointed out that H^D may contain a part describing the kinetic energy of the different clusters as well as a part describing their internal motion and that Proposition 14.2(a) signifies only the conservation of their sum. Hence this theory admits the possibility of conversion of kinetic energy into binding energy and vice versa (see also Section 14-3).

Proposition 14.2(b) expresses the unitarity relations for multichannel scattering, which are the multichannel analogue of Proposition 4.8. (14.21) implies that for each $f^C \varepsilon M^C$

$$\sum_{D\varepsilon\Xi}\|S_{DC}f^C\|^2 = \sum_{D\varepsilon\Xi}(f^C, S_{DC}^* S_{DC}f^C) = \|E^C f^C\|^2 = \|f^C\|^2. \qquad (14.24)$$

Thus, given any initial state $f_-^C \varepsilon M^C$, the squares of the norms of the final states $\{f_+^D\}$ ($f_+^D = S_{DC}f_-^C$) associated with f_-^C sum up to $\|f_-^C\|^2$. A consequence of (14.24) is that in general $\|S_{DC}f^C\|^2 < \|f^C\|^2$, so that the individual operators S_{DC} are not partially isometric.

Some of the preceding results can be expressed more concisely in a <u>two-Hilbert space formulation</u> of the theory involving the Hilbert space H and the asymptotic Hilbert space H_{as} introduced in (14.4). To motivate the use of H_{as}, we consider two vectors $f^C \varepsilon M^C$ and $f^D \varepsilon M^D$ with $C \neq D$. Since by (B1), Problem 2.4 and (B2)

$$(U_t^D f^D, U_t^C f^C) = (V_t^* U_t^D f^D, V_t^* U_t^C f^C) \to (\Omega_\pm^D f^D, \Omega_\pm^C f^C) = 0 \qquad (14.25)$$

as $t \to \pm\infty$, we see that $U_t^C f^C$ and $U_t^D f^D$ become orthogonal to each other as $t \to \pm\infty$. Thus, if one is only interested in the behaviour of the cluster states at large times, it is natural to introduce the direct sum of the subspaces $\{M^D\}$, which is nothing but the space H_{as} defined in (14.4).

We now define an <u>asymptotic Hamiltonian</u> H_{as} acting in H_{as} by

$$H_{as} = \oplus_{D\varepsilon\Xi} H^D E^D, \qquad (14.26)$$

which means that on each M_{as}^D, H_{as} coincides with the corresponding cluster Hamiltonian. Similarly one introduces the <u>asymptotic evolution group</u> as

$$U_t^{as} = \exp(-iH_{as}t) = \oplus_{D\varepsilon\Xi} U_t^D E^D. \qquad (14.27)$$

It is now easy to see (Problem 14.2) that (B1) is equivalent to the requirement of the existence of

$$\underset{t\to\pm\infty}{\text{s-lim}}\; V_t^* J U_t^{as} \equiv \Omega_\pm, \tag{14.28}$$

provided that $||J|| < \infty$ (which is the case for the operator J defined in (14.5)). Notice that the operators $V_t^* J U_t^{as}$ and $\Omega_\pm$ map $\mathcal{H}_{as}$ into $\mathcal{H}$ and that the restrictions of $\Omega_\pm$ to the subspace $\mathcal{M}_{as}^D$ of $\mathcal{H}_{as}$ coincide, up to the identification map between $\mathcal{M}_{as}^D$ and $\mathcal{M}^D$, with the cluster wave operators $\Omega_\pm^D$: If $f_{as}^D \varepsilon \mathcal{M}_{as}^D$, then

$$\Omega_\pm f_{as}^D = \Omega_\pm^D f^D \varepsilon F_\pm^D \mathcal{H}, \text{ where } f^D = Jf_{as}^D . \tag{14.29}$$

<u>PROPOSITION 14.3</u> : If (B1)-(B3) are verified, then $\Omega_\pm$ are isometric operators from $\mathcal{H}_{as}$ to $\mathcal{H}$ with range $\mathcal{M}_\infty(\mathrm{H})$, i.e.

$$\Omega_\pm^* \Omega_\pm = I_{as}, \quad \Omega_\pm \Omega_\pm^* = E_\infty(\mathrm{H}). \tag{14.30}$$

The restriction of $\Omega_\pm$ to $\mathcal{M}_{as}^D$ maps $\mathcal{M}_{as}^D$ onto $F_\pm^D\mathcal{H}$ and is given by (14.29), and the restriction of $\Omega_\pm^*$ to $F_\pm^D$ maps $F_\pm^D\mathcal{H}$ onto $\mathcal{M}_{as}^D$ and is given by

$$\Omega_\pm^* F_\pm^D g = (\Omega_\pm^{D*} g)_{as} \varepsilon \mathcal{M}_{as}^D. \tag{14.31}$$

<u>Proof</u> (for Ω_+) : Let $f = \oplus_D f_{as}^D \varepsilon \mathcal{H}_{as}$. Then by (14.29) $\Omega_+ f = \sum_D \Omega_+ f_{as}^D = \sum_D \Omega_+^D f^D$. By (B2), $\Omega_+^D f^D$ is orthogonal to $\Omega_+^C f^C$ if $C \neq D$. Hence by (2.10) and (14.4)

$$||\Omega_+ f||^2 = \sum_D ||\Omega_+^D f^D||^2 = \sum_D ||f^D||^2 = \sum_D ||f_{as}^D||^2 = ||f||^2_{\mathcal{H}_{as}}$$

which proves the isometry of Ω_+.

(14.29) and (B3) imply that the range of Ω_+ is $\mathcal{M}_\infty(\mathrm{H})$, so that $\Omega_+\Omega_+^* = E_\infty(H)$ by the two-Hilbert space form of Proposition 2.11. To prove (14.31), let $g \varepsilon F_+^D \mathcal{H}$. Then $g = \Omega_+^D f^D = \Omega_+ f_{as}^D$ for some $f^D \varepsilon M^D$. Thus by (14.13) $\Omega_+^* g = \Omega_+^*\Omega_+ f_{as}^D = f_{as}^D = (\Omega_+^{D*}g)_{as}$. #

We may now define a scattering operator in $\mathcal{H}_{as}$ by

$$S = \Omega_+^*\Omega_- . \qquad (14.32)$$

S is unitary (provided that (B1)-(B3) are verified). Since $V_t\Omega_\pm = \Omega_\pm U_t^{as}$, S commutes with H_{as}. The action of S on M_{as} can be computed from (14.29) and (14.31) to be

$$Sf_{as}^C = \underset{D\varepsilon\Xi}{\oplus} S_{DC}f_{as}^C , \qquad (14.33)$$

i.e. the component of the state Sf_{as}^C in M_{as}^D is nothing but $(S_{DC}f^C)_{as}$ (notice that in (14.33) S_{DC} is interpreted as an operator mapping M_{as}^C into M_{as}^D). Thus the correspondence $f^C \mapsto \{S_{DC}f^C\}$ characterizing the effect of the scattering has been incorporated into a single operator S acting in the auxiliary Hilbert space $\mathcal{H}_{as}$.

The reader will have noticed that in the two-Hilbert space form, multichannel scattering theory looks very much like a simple scattering system, the only apparent difference being the use of two Hilbert spaces instead of a single one. That this is only a formal analogy and not a complete equivalence is shown in Problem 14.3.

We end this section with short comments on symmetries and cluster observables. A multichannel scattering system is said to be invariant under a group G if there exists a representation of G by unitary or anti-unitary operators U(γ) in

$\mathcal{H}$ such that each $U(\gamma)$ commutes with $E_\infty(H)$, H, E^D and H^D, in the sense of (4.21) and (4.22), for all $D\varepsilon\Xi$.

PROPOSITION 14.4 : Suppose that a multichannel scattering system is invariant under a group G. Let $\gamma\varepsilon G$. Then

$$S_{DC}U(\gamma) = U(\gamma)S_{DC} \text{ if } U(\gamma) \text{ is unitary,} \tag{14.34}$$

$$S^*_{CD}U(\gamma) = U(\gamma)S_{DC} \text{ if } U(\gamma) \text{ is anti-unitary.} \tag{14.35}$$

In particular, if a scattering system is invariant under the time-reversal operator Θ, then $|(f^D,S_{DC}f^C)| = |(f^D,\Theta^*S^*_{CD}\Theta f^C)| = |(\Theta f^C,S_{CD}\Theta f^D)|$, so that the probability for scattering from $f^C\varepsilon\mathcal{M}^C$ to $f^D\varepsilon\mathcal{M}^D$ is the same as that from Θf^D to Θf^C.

A cluster observable is a bounded self-adjoint operator A^D in $\mathcal{M}^D$ (i.e. with domain $\mathcal{M}^D$ and range in $\mathcal{M}^D$) which commutes with U_t^D. It is clear that the asymptotic condition implies that for each cluster observable

$$\underset{t\to\pm\infty}{\text{s-lim}}\ V_t^*A^DV_tF_\pm^D = \Omega_\pm^DA^D\Omega_\pm^{D*}. \tag{14.36}$$

A formulation of the asymptotic condition based on the requirement of the convergence of cluster observables as in (14.36) rather than on (B1) is given in the article of Amrein, Georgescu and Martin in [EN]. Such a generalization is necessary to incorporate long range potentials in multichannel scattering theory (Dollard [1]).

14-3 SCATTERING CHANNELS

As we pointed out at the end of Section 14-1, one frequently specifies not only the type of clustering of the scattering states of H at large times but also the internal structure (binding energy, spin etc.) of the various clusters of particles. Mathematically this amounts to further subdividing the cluster subspaces M^D into the so-called channel subspaces. To do this, we resume our discussion of clustering in Section 14-1; in particular we assume that H^D is given by (14.2) and M^D by (14.3) in the Hilbert space $\mathcal{H}$ defined by (14.1).

Let us consider the Hilbert space $\mathcal{H}^{(k)}$ associated with a cluster (k) consisting of two or more particles and the corresponding Hamiltonian $H^{(k)}$. One first has to separate the center-of-mass motion of the cluster (k) from its internal motion. For this one introduces instead of the momentum operators of the individual particles forming the cluster a new set of momentum operators in $\mathcal{H}^{(k)}$ such that one of them, the total momentum operator $\underline{P}^{(k)}_{tot}$, is the sum of the individual momentum operators and the remaining ones are a set of linearly independent operators representing relative momenta between these particles (or equivalently one may choose a set of coordinates such that one of them is the center-of-mass coordinate of the particles forming the cluster and the remaining ones are a set of linearly independent relative coordinates). As was done in Section 7-5 for the case of two particles, one will then write

$$\mathcal{H}^{(k)} = \mathcal{H}^{(k)}_{CM} \otimes \mathcal{H}^{(k)}_{rel}, \qquad (14.37)$$

where the states in $\mathcal{H}_{CM}^{(k)}$ describe the center of mass of the cluster and the states in $\mathcal{H}_{rel}^{(k)}$ its internal structure.

We now assume that the time evolution of the states of the center of mass is independent of that of the internal states, i.e. that

$$\exp(-iH^{(k)}t) = \exp(-iH_{CM}^{(k)}t) \otimes \exp(-iH_{rel}^{(k)}t), \tag{14.38}$$

or equivalently that $H^{(k)}$ has the following form in the representation (14.37) of $\mathcal{H}^{(k)}$:

$$H^{(k)} = H_{CM}^{(k)} \otimes I + I \otimes H_{rel}^{(k)}. \tag{14.39}$$

$\{\exp(-iH_{CM}^{(k)}t)\}$ will describe the free movement of the cluster as a whole and $\{\exp(-iH_{rel}^{(k)}t)\}$ the relative movement of the particles forming the cluster. In particular $H_{rel}^{(k)}$ will contain the interaction between these particles. The subspace $\mathcal{M}_o(H^{(k)})$ of $\mathcal{H}^{(k)}$ will be of the form

$$\mathcal{M}_o(H^{(k)}) = \mathcal{H}_{CM}^{(k)} \otimes \mathcal{M}_o(H_{rel}^{(k)}) \tag{14.40}$$

in the representation (14.37), where $\mathcal{M}_o(H_{rel}^{(k)})$ is the subspace of $\mathcal{H}_{rel}^{(k)}$ formed by the bound states of $H_{rel}^{(k)}$. The bound states of the particles in (k) have an internal structure whenever $\mathcal{M}_o(H_{rel}^{(k)})$ is more than one-dimensional. To specify this internal structure, one selects in $\mathcal{M}_o(H_{rel}^{(k)})$ an orthonormal basis $\{e_j^{(k)}\}$, $j = 1,\ldots, \dim \mathcal{M}_o(H_{rel}^{(k)})$.

In a channel subspace each cluster consisting of two or more particles is in one of the states of the respective orthonormal basis. We use the index α (or β,γ) to label the scattering channels. A channel is determined as follows : One first chooses a partition D of $\{1,\ldots,N\}$ into $n \geq 2$ subsets.

Then, if (k_1), ..., (k_m) denote the clusters in D consisting of two or more particles, one selects for each of them a basis vector $e_{j_s}^{(k_s)}$ $(s = 1,\ldots,m)$. Thus one may write

$$\alpha = \{D;\ e_{j_1}^{(k_1)},\ \ldots,\ e_{j_m}^{(k_m)}\},\ \text{with } m = m(D). \tag{14.41}$$

The associated <u>channel subspace</u> M^{α} is of the form

$$M^{\alpha} = M^{(1),\alpha} \otimes M^{(2),\alpha} \otimes \ldots \otimes M^{(n),\alpha} \tag{14.42}$$

in the representation (14.1), where $M^{(k),\alpha} = H^{(k)}$ if (k) consists of only one particle and, if $k = k_s$ for some $s = 1, \ldots, m$, $M^{(k_s),\alpha} = H_{CM}^{(k_s)} \otimes e_{j_s}^{(k_s)}$ in the space (14.37).

The set of all channels is obtained by taking in (14.41) all possible choices of a partition and all possible combinations of basis vectors. We shall write $\alpha \div D$ if α is a channel associated with the partition D as in (14.41). Thus $\alpha \div D$ means that the clusters occuring in α are the same as those of the partition D. We also write $\alpha \div \beta$ if there exists a partition D such that $\alpha \div D$ and $\beta \div D$. It is clear from the definition of the scattering channels that the cluster subspace M^D is the direct sum of the channel subspaces determined by the partition D, i.e.

$$M^D = \bigoplus_{\alpha \div D} M^{\alpha}. \tag{14.43}$$

The basis $\{e_j^{(k)}\}$ of $M_0(H_{rel}^{(k)})$ may for instance be chosen as the common eigenvectors of a complete set of commuting observables in $M_0(H_{rel}^{(k)})$ [J]. It is useful to take as one of these observables the Hamiltonian $H_{rel}^{(k)}$, since the bound states associated with $H_{rel}^{(k)}$ usually coincide with the subspace spanned by its eigenvectors and the eigenvalues correspond to the binding energies. If none of these eigenvalues is degenerate as (k) varies over all possible clusters (i.e. if the spectral multiplicity of each eigenvalue of all the $H_{rel}^{(k)}$ is 1), then the set of channel subspaces is uniquely determined by this choice of $\{e_j^{(k)}\}$. On the other hand, if some of the eigenvalues are degenerate, the channel subspaces depend on the choice of further observables such as spins. The number of scattering channels is determined by the number of eigenvalues (including multiplicities) of the various operators $H_{rel}^{(k)}$. If each of these operators has only a finite number of eigenvalues of finite spectral multiplicity, the total number of channels is finite. If $\mathcal{H}_p(H_{rel}^{(k)})$ is infinite-dimensional for at least one cluster (k), then the number of channels is countably infinite (since $\mathcal{H}$ is separable).

From now on we shall always assume that the basis vectors $\{e_j^{(k)}\}$ are eigenvectors of $H_{rel}^{(k)}$. This allows us to introduce for each channel α a channel Hamiltonian H^α. For this we notice that under the preceding assumption $\mathcal{M}^\alpha$ reduces the corresponding cluster Hamiltonian H^D, and it is easy to see that the part of H^D in $\mathcal{M}^\alpha$ is self-adjoint and given by (see Problem 14.10; $H_{CM}^{(k)}$ is now viewed as an operator in $\mathcal{H}$)

$$H^D/\mathcal{M}^\alpha = \sum_{k=1}^{n} H_{CM}^{(k)}/\mathcal{M}^\alpha + \sum_{s=1}^{m} \lambda_s^\alpha, \text{ with } \alpha \div D. \qquad (14.44)$$

Here n is the number of clusters in the partition D appearing in the definition of α, m is the number of clusters in D consisting of two or more particles and $\{\lambda_s^\alpha\}$ are the eigenvalues associated with the eigenvectors in (14.41), i.e.

$$H_{rel}^{(k_s)} e_{j_s}^{(k_s)} = \lambda_s^\alpha e_{j_s}^{(k_s)} . \tag{14.45}$$

On M^α the channel Hamiltonian H^α is simply given by (14.44). It is convenient to define the following operator H^α on the entire Hilbert space :

$$H^\alpha = \sum_{k=1}^n H_{CM}^{(k)} + \sum_{s=1}^m \lambda_s^\alpha \equiv H_{kin}^\alpha + \lambda^\alpha. \tag{14.46}$$

H^α and H^D $(\alpha \div D)$ coincide on M^α but may be quite different on the orthogonal complement of M^D. H_{kin}^α represents the total kinetic energy of the fragments of channel α and λ^α their total binding energy.

The self-adjoint operator H^α defines a unitary channel evolution group $U_t^\alpha \equiv \exp(-iH^\alpha t)$. The part of U_t^α in M^α describes the free movement of the clusters (1), ..., (n) while the internal state of each cluster consisting of more than one particle is stationary.

Let E^α be the projection with range M^α. Then $U_t^\alpha E^\alpha = E^\alpha U_t^\alpha$, and we may define the channel wave operators by

$$\Omega_\pm^\alpha = \underset{t\to\pm\infty}{\text{s-lim}}\, V_t^* U_t^\alpha E^\alpha. \tag{14.47}$$

Their ranges are $F_\pm^\alpha H$, where

$$F_\pm^\alpha = \Omega_\pm^\alpha \Omega_\pm^{\alpha *}. \tag{14.48}$$

Since $H^\alpha E^\alpha = H^D E^\alpha$ by (14.44), we get

$$\Omega_\pm^\alpha = \Omega_\pm^D E^\alpha \ (\alpha \div D) \text{ and } \Omega_\pm^D = \sum_{\alpha \div D} \Omega_\pm^\alpha . \tag{14.49}$$

From the relation (14.43) one has

$$E^\alpha E^\beta = 0 \quad \text{whenever } \alpha \neq \beta \text{ and } \alpha \div \beta. \tag{14.50}$$

(14.49), (14.13) and (14.50) imply that, if $f^\alpha \varepsilon M^\alpha$, $f^\beta \varepsilon M^\beta$,

$\alpha \neq \beta$ and $\alpha \div \beta$, then $(\Omega_+^\alpha f^\alpha, \Omega_+^\beta f^\beta) = (\Omega_+^D f^\alpha, \Omega_+^D f^\beta) = (f^\alpha, f^\beta) = 0$. Hence the ranges of Ω_+^α and Ω_+^β are orthogonal if $\alpha \neq \beta$ and $\alpha \div \beta$, in other words $F_+^\alpha F_+^\beta = 0$. On the other hand, if α and β belong to different partitions, then $F_+^\alpha F_+^\beta = 0$ by (14.48) and (B2). Thus, if (B1)-(B3) are verified, we have

$$F_+^\alpha F_+^\beta = 0 \;,\; F_-^\alpha F_-^\beta = 0 \quad \text{whenever } \alpha \neq \beta, \tag{14.51}$$

$$\sum_\alpha F_+^\alpha = \sum_\alpha F_-^\alpha = E_\infty(H). \tag{14.52}$$

The preceding considerations show that, if (B1)-(B3) hold for the cluster subspaces and cluster evolution groups, then they are also verified for the channel subspaces and channel evolution groups. Thus all statements in Propositions 14.1 and 14.2 remain true if one replaces everywhere the cluster indices C and D by channel indices α and β. The channel scattering operators are defined as

$$S_{\beta\alpha} = \Omega_+^{\beta *}\Omega_-^\alpha \;, \tag{14.53}$$

and in terms of the channels the scattering is described by associating with each vector $f_-^\alpha \varepsilon M^\alpha$ a family of vectors $\{f_+^\beta\}$ with $f_+^\beta = S_{\beta\alpha} f_-^\alpha \varepsilon M^\beta$. A subscript α will mean that we use an equation of Section 14-2 or 14-4 for channels instead of clusters. Thus for example $(14.9)_\alpha$ is the same as equation (14.48).

The operator $S_{\alpha\alpha}$ describes the elastic scattering in channel α, i.e. involving no change of clustering and no change of the internal structure of the scattering fragments. In particular their binding energies are the same before and after the scattering, so that by (14.46) and $(14.20)_\alpha$ their total kinetic energy H_{kin}^α is conserved. The operators

$S_{\beta\alpha}$ with $\alpha \div \beta$ and $\alpha \neq \beta$ describe <u>inelastic scattering</u> between the fragments of channel α. The fragments remain the same, but at least one of them undergoes a change of its internal state. In particular the binding energies and hence the total kinetic energy of the fragments may change. Finally, if the partitions defining α and β are different, then $S_{\beta\alpha}$ describes <u>rearrangement scattering</u>.

Given any pair of channels α,β, scattering from α into β may not be possible for small values of the total kinetic energy of the fragments in the initial channel α. To illustrate this, we take a sharp value $\lambda^{\alpha}_{\text{kin}}$ of this total kinetic energy. Then by $(14.20)_{\alpha}$ and (14.46), $\lambda^{\alpha}_{\text{kin}} + \lambda^{\alpha} = \lambda^{\beta}_{\text{kin}} + \lambda^{\beta}$. Since the kinetic energy is positive, this relation cannot be verified if $\lambda^{\alpha}_{\text{kin}} < \lambda^{\beta} - \lambda^{\alpha}$. Thus if $\lambda^{\beta} - \lambda^{\alpha} > 0$, there is no scattering from channel α into channel β whenever $\lambda^{\alpha}_{\text{kin}} < \lambda^{\beta} - \lambda^{\alpha}$, i.e.

$$S_{\beta\alpha}\hat{E}^{\alpha}_{\lambda} = 0 \quad \text{for all } \lambda < \lambda^{\beta} - \lambda^{\alpha}, \tag{14.54}$$

where $\{\hat{F}^{\alpha}_{\lambda}\}$ denotes the spectral family of H^{α}_{kin}. The minimal total (i.e. kinetic plus binding) energy in channel α at which scattering into channel β becomes possible in the above sense is called the <u>threshold</u> energy $\lambda(\alpha \to \beta)$ for scattering from channel α to channel β. Thus under the above circumstances

$$\lambda(\alpha \to \beta) = \max\{\lambda^{\alpha},\lambda^{\beta}\}. \tag{14.55}$$

Finally we should say that the two-Hilbert space form of multichannel scattering defined in terms of channels is exactly the same as that given in Section 14-2 in terms of clusters. This follows easily from the preceding remarks and from (14.43) which implies that $\mathcal{H}_{as} = \oplus_\alpha \mathcal{M}^\alpha$.

14-4 TIME-INDEPENDENT MULTICHANNEL SCATTERING THEORY

In this section we assemble the time-independent equations for cluster wave and scattering operators. These equations are quite similar to and have the same meaning as those of Chapter 6. The proofs are essentially identical with those of the corresponding equations in Chapter 6 and will not be repeated. The basic equations for the wave operators are the following :

PROPOSITION 14.5 : Suppose that (B1) is verified. Then

$$\Omega_\pm^D = \underset{\eta\to+0}{\text{s-lim}}\ (\pm i\eta) \int_{\mathbb{R}} R_{\lambda\mp i\eta}\, dE_\lambda^D E^D \tag{14.56}$$

and

$$\Omega_\pm^{D*} = \underset{\eta\to+0}{\text{w-lim}}\ (\pm i\eta) \int_{\mathbb{R}} E^D R^D_{\lambda\mp i\eta}\, dE_\lambda, \tag{14.57}$$

where $R_z^D = (H^D - z)^{-1}$.

Remark 14.6 : In contrast to (6.19), the limit as $\eta\to+0$ in (14.57) is only a weak limit. This is due to the fact that in a simple scattering system verifying (A1) and (A3) $\Omega_\pm^* =$ s-lim $U_t^* V_t E_\infty(H)$, whereas for a multichannel system one has only $\Omega_\pm^{D*} =$ w-lim $E^D U_t^{D*} V_t$ as $t\to\pm\infty$, except in the case where there is only one non-zero cluster wave operator (Problem 14.3). (14.57) is true with a strong limit if one multiplies the equation from the right by $F_\pm^D$ (Proposition 4.2). This is however inconvenient, since the stationary expression for

S_{DC} will then contain the projection F_+^D which is difficult to handle mathematically. Also the weak limits are quite sufficient for deducing the scattering amplitude (cf. Sections 8-3, 10-2 and 16-2). The spectral integrals in (14.57) exist of course as strong limits by Proposition 6.3.

Further equations for $\Omega_\pm^D$ can be obtained from Proposition 14.5 by inserting an expression for R_z in terms of the cluster resolvents R_z^D. There are various possibilities for writing such expressions (e.g. see Chapter 16). A simple one is the second resolvent equation for the pair R_z and R_z^D, leading to the Lippmann-Schwinger equations as in Proposition 6.5. We assume that $D(H) = D(H^D)$ and write

$$V^D = H - H^D. \tag{14.58}$$

PROPOSITION 14.7 : Assume (B1) and $D(H) = D(H^D)$. Then

$$\Omega_\pm^D = E^D - \underset{\eta\to+0}{\text{s-lim}} \int_{\mathcal{R}} R_{\lambda\mp i\eta} V^D dE_\lambda^D E^D, \tag{14.59}$$

$$\Omega_\pm^D = E^D - \underset{\eta\to+0}{\text{s-lim}} \int_{\mathcal{R}} R^D_{\lambda\mp i\eta} V^D \Omega_\pm^D dE_\lambda^D, \tag{14.60}$$

and

$$\Omega_\pm^{D*} = E^D - \underset{\eta\to+0}{\text{w-lim}} \int_{\mathcal{R}} E^D R^D_{\lambda\mp i\eta} V^D dE_\lambda. \tag{14.61}$$

PROPOSITION 14.8 : Assume (B1)-(B3) and $D(H) = D(H^C) = D(H^D)$. Then

$$S_{DC} = \delta_{DC} E^C + \underset{\eta\to+0}{\text{w-lim}} \int_{\mathcal{R}} E^D (R^D_{\lambda-i\eta} - R^D_{\lambda+i\eta}) V^D \Omega_-^C dE_\lambda^C \tag{14.62}$$

$$= \delta_{DC} E^C + \underset{\eta\to+0}{\text{w-lim}} \int_{\mathcal{R}} (R^D_{\lambda-i\eta} - R^D_{\lambda+i\eta}) \Omega_+^{D*} V^C dE_\lambda^C E^C. \tag{14.63}$$

The proof is analogous to that of Proposition 6.7 once one notices that $S_{DC} - \delta_{DC} E^C = (\Omega_+^{D*} - \Omega_-^{D*}) \Omega_-^C$ by virtue of

(14.11) and (14.18).

PROPOSITION 14.9 : Under the assumptions of Proposition 14.8, one has

$$S_{DC} = \delta_{DC}E^C + \text{w-}\lim_{\eta\to+0}\ \text{s-}\lim_{\delta\to+0} \int_{\mathbb{R}} E^D(R^D_{\lambda-i\eta} - R^D_{\lambda+i\eta})(V^D - V^D R_{\lambda+i\delta} V^C)\, dE^C_\lambda E^C$$

$$= \delta_{DC}E^C + \text{w-}\lim_{\eta\to+0}\ \text{s-}\lim_{\delta\to+0} \int_{\mathbb{R}} E^D(R^D_{\lambda-i\eta} - R^D_{\lambda+i\eta})(V^C - V^D R_{\lambda+i\delta} V^C)\, dE^C_\lambda E^C .$$

The proof of the first identity is the same as that of Proposition 6.11 except that instead of (6.43) one obtains $i\int R^D_{\lambda\mp i\eta}\, dE^C_\lambda = \int_0^{\pm\infty} dt\ e^{\mp\eta t} U^{D*}_t U^C_t \in \mathcal{B}(\mathcal{H})$ by Proposition 6.3. The second identity follows from the first one and the fact that (Problem 14.6)

$$\text{w-}\lim_{\eta\to+0} \int_{\mathbb{R}} E^D(R^D_{\lambda-i\eta} - R^D_{\lambda+i\eta})(V^D - V^C)\, dE^C_\lambda E^C = 0. \qquad (14.64)$$

It is seen that the operator $T_z = V - VR_zV$ given in (6.44) for a simple scattering system is here replaced by either one of the two families of operators $\{T^{\underline{D}C}_z\}$ and $\{T^{D\underline{C}}_z\}$ defined as

$$T^{\underline{D}C}_z = V^D - V^D R_z V^C, \quad T^{D\underline{C}}_z = V^C - V^D R_z V^C. \qquad (14.65)$$

Of course $T^{\underline{D}C}_z \neq T^{D\underline{C}}_z$ if $C \neq D$, but either of them may be used in Proposition 14.9 to calculate S_{DC}. The last two propositions indicate that the scattering amplitude for scattering from $\mathcal{M}^C$ to $\mathcal{M}^D$ should be proportional to the on-shell part of $V^D\Omega^C_-$, $T^{\underline{D}C}_z$ and $T^{D\underline{C}}_z$ (see Section 15-3 for the precise meaning of on-shell in this more complex situation).

Time-independent equations for the channel wave and scattering operators are simply obtained from the preceding ones by replacing the projections E^D by projections whose ranges are the corresponding channel subspaces. Thus one obtains for example from (14.60)

$$\Omega_{\pm}^{\alpha} = E^{\alpha} - \underset{\eta\to+0}{\text{s-lim}} \int_{\mathbb{R}} R^{D}_{\lambda\mp i\eta} V^{D}\Omega_{\pm}^{\alpha} dE_{\lambda}^{\alpha}, \quad \alpha \div D, \tag{14.66}$$

where $\{E_{\lambda}^{\alpha}\}$ is the spectral family of H^{α}. Similarly

$$S_{\alpha\beta} - \delta_{\alpha\beta}E^{\beta} = \underset{\eta\to+0}{\text{w-lim}} \int_{\mathbb{R}} E^{\alpha}(R^{\alpha}_{\lambda-i\eta} - R^{\alpha}_{\lambda+i\eta}) V^{D}\Omega_{-}^{\beta} dE_{\lambda}^{\beta}$$

$$= \underset{\eta\to+0}{\text{w-lim}}\ \underset{\delta\to+0}{\text{s-lim}} \int_{\mathbb{R}} E^{\alpha}(R^{\alpha}_{\lambda-i\eta} - R^{\alpha}_{\lambda+i\eta}) T_{z}^{DC} dE_{\lambda}^{\beta} E^{\beta}, \tag{14.67}$$

with $\alpha \div D$ and $\beta \div C$. For other time-independent equations the reader may consult Chandler and Gibson [1].

PROBLEMS

14.1 : Consider a system consisting of four distinct particles interacting via spherically symmetric square-well pair potentials. (a) Enumerate all possible partitions and write down the associated cluster Hamiltonians and cluster subspaces. (b) Enumerate all channels and give the channel Hamiltonians. Discuss the dependence on the coupling constants. (Notice that the number of eigenvalues for a two-body Hamiltonian with a square-well potential is finite.)

14.2 : Assume $\|J\| < \infty$. Verify that then (B1) is equivalent to the existence of the limits in (14.28).

14.3 : (a) Show that the injection map J defined in (14.5) is bounded and find an explicit expression for its adjoint J*. (b) (B1)-(B3) and $\|J\| < \infty$ imply that w-lim $U_t^{as*}J^*V_tE_\infty(H) = \Omega_{\pm}^*$ as $t \to \pm\infty$. Show that, if one assumes in addition that $\mathcal{H}^D = \mathcal{H}$ for some $D\in\Xi$, then the strong convergence of $\{U_t^{as*}J^*V_tE_\infty(H)\}$ implies that there is only one cluster subspace. Compare this with Proposition 4.2.

14.4 : Prove Propositions 14.1 and 14.2(a).

14.5 : Assume (B1), (B2), $\sum_D F_-^D \le \sum_D F_+^D$ and that there exists an anti-unitary operator U verifying the analogues of (4.21) and (4.22) for all cluster Hamiltonians. Prove that the unitarity relations (14.21) and (14.22) hold. (Hint : Proposition 4.10.)

14.6 : Prove (14.64). (Hint : Write $V^D - V^C = H^C - H^D$, use Lemma 6.1, Proposition 6.3 and (14.25).)

14.7 : Prove some of the theorems in Section 14-4.

14.8† : Let A_i be self-adjoint operators in $\mathcal{H}_i$ $(i=1,2)$ and let $\{E_{\lambda,i}\}$ be the spectral family of A_i. (a) Show that $\{I \otimes E_{\lambda,2}\}$ is the spectral family of the self-adjoint operator $I \otimes A_2$ (see Problem 2.39).
(b) Show that $C_\pm \equiv \int (A_1+\lambda\pm i)^{-1} \otimes dE_{\lambda,2}$ exists and is in $\mathcal{B}(\mathcal{H}_1 \otimes \mathcal{H}_2)$. (Hint : Proposition 6.3).

(c) Let $f\varepsilon\mathcal{H}_1$, $g\varepsilon\mathcal{H}_2$ and denote by A the closure of the symmetric operator $A_1 \hat{\otimes} I + I \hat{\otimes} A_2$. Show that $C_\pm(f \otimes g) \,\varepsilon\, D(A)$ and $(A \pm i)C_\pm(f \otimes g) = f \otimes g$. (Hint : $C_\pm f \otimes g = \lim_{n\to\infty} \lim_{|\Pi|\to 0} \sum_k (A+\lambda_k \pm i)^{-1} f \otimes E_{(-n,n),2}\, g$. Use also Lemma 6.2).
(d) Deduce from (c) that $A_1 \hat{\otimes} I + I \hat{\otimes} A_2$ is essentially self-adjoint. (e) If moreover A_1 and A_2 are bounded below, then $A_1 \otimes I + I \otimes A_2$ is self-adjoint. (Hint : With no loss of generality, assume $A_1, A_2 \ge 0$. Then $((A_1 \otimes I)g,(I \otimes A_2)g) = ((A_1^{\frac{1}{2}} \otimes I)g,(I \otimes A_2)(A_1^{\frac{1}{2}} \otimes I)g) \ge 0$ for all $g\varepsilon D(A_1) \hat{\otimes} D(A_2)$.)

(f) Let A_k be self-adjoint in $\mathcal{H}_k$ $(k = 1,\dots,n)$. Show by induction that $A = A_1 \otimes I \otimes \cdots \otimes I + \dots + I \otimes \cdots \otimes I \otimes A_n$ is essentially self-adjoint on $D(A_1) \hat{\otimes} \cdots \hat{\otimes} D(A_n)$ and is self-adjoint as an operator sum if each A_k is bounded below.

14.9 : Verify that the right-hand side of (14.2) defines a strongly continuous one-parameter unitary group and prove that its infinitesimal generator is H^D.

14.10 : Let $\mathcal{M}^\alpha$ be defined as in the paragraph containing equation (14.44). Show that $\mathcal{M}^\alpha$ reduces each $\mathcal{H}_{CM}^{(k)}$ and each $\mathcal{H}_{rel}^{(k)}$, hence also H^D. Deduce from this that $H^D/\mathcal{M}^\alpha$ is self-adjoint and given by equation (14.44).

CHAPTER 15 : MULTICHANNEL POTENTIAL SCATTERING

In this chapter we elaborate certain aspects of multichannel scattering theory for a system consisting of N distinguishable non-relativistic spinless particles. The interaction is assumed to be given by a sum of pair potentials each of which is a function of the relative position of the two particles forming the respective pair. The results that we give in this chapter are analogous to those of Sections 8-1, 8-2, 7-1, 7-3 and 4-3 and can be derived by adapting the earlier proofs to the more complex situation of N-particle systems. As an additional technical step one often has to express the N-particle Hilbert space and the relevant operators as tensor products in a suitable way. Two such representations with which the reader should be familiar are those given in the proofs of Propositions 15.1 and 15.7.

In Section 15-1 we prove the self-adjointness of the total and the cluster Hamiltonians before and after removing the center-of-mass motion. We then show that the subspace of bound states of each subsystem is spanned by the eigenvectors of the Hamiltonian describing the internal motion of this subsystem, which allows us to introduce the scattering channels in the usual way. We conclude the first section with some remarks on Hunziker's theorem characterizing the essential spectrum of multiparticle Hamiltonians. In Section 15-2 we verify (B1) and (B2), i.e. the existence of the cluster wave operators and the orthogonality of their ranges. In Section 15-3 we introduce as new observables those determining whether at large positive times the particles or clusters emerge in given cones and study the properties of these observables. This section also contains a discussion of multiparticle scattering cross sections. In Section 15-4 we describe among other things results about spectral properties and asymptotic completeness of multiparticle systems.

15-1 MULTIPARTICLE HAMILTONIANS

We consider N distinguishable spinless particles of masses $m_1, \ldots, m_N$. The Hilbert space $\mathcal{H}_j$, formed of the state vectors of the j-th particle, is isomorphic to $L^2(\mathcal{R}^3)$. The variable in this space will be denoted by $\underline{x}_j$, that occuring in the Fourier transforms of the functions in $L^2(\mathcal{R}^3)$ by $\underline{k}_j$ $(j=1,\ldots,N)$. We shall use the notation $\tilde{L}^2(\mathcal{R}^n)$ to denote the representation of the Hilbert space $L^2(\mathcal{R}^n)$ in which the momentum operators are diagonal (see page 555).

The Hilbert space for the system of N particles is

$$\mathcal{H} = \mathcal{H}_1 \otimes \mathcal{H}_2 \otimes \ldots \otimes \mathcal{H}_N. \qquad (15.1)$$

It may be identified with $L^2(\mathbb{R}^{3N})$ (cf. Section 2-4), the independent variables being $\underline{x}_1, \ldots, \underline{x}_N$. The multiplication operator in $L^2(\mathbb{R}^{3N})$ by $\underline{x}_j$ is the (three-component) position operator $\underline{Q}_j$ of the j-th particle, and the multiplication operator by $\underline{k}_j$ in $\tilde{L}^2(\mathbb{R}^{3N})$ is its momentum operator $\underline{P}_j$. In addition we shall need the kinetic energy operator of the j-th particle which is given as $K_{o,j} = (2m_j)^{-1}\underline{P}_j^2$.

The <u>free Hamiltonian</u> H_o of the N-particle system is the sum of the kinetic energy operators of the individual particles, i.e.

$$H_o = \sum_{j=1}^N K_{o,j}. \qquad (15.2)$$

H_o is simply the maximal multiplication operator in $\tilde{L}^2(\mathbb{R}^{3N})$ by $\sum_{j=1}^N (2m_j)^{-1}\underline{k}_j^2$, hence by Proposition 2.16 it is self-adjoint and

$$D(H_o) = \{f \varepsilon L^2(\mathbb{R}^{3N}) \,\big|\, \sum_{j=1}^N (2m_j)^{-1}\underline{k}_j^2 \tilde{f}(\underline{k}_1,\ldots,\underline{k}_N) \varepsilon \tilde{L}^2(\mathbb{R}^{3N})\}. \qquad (15.3)$$

H_o is a positive operator and is spectrally absolutely continuous with $\sigma(H_o) = \sigma_{ac}(H_o) = [0,\infty)$ by Problem 5.7.

To define the interaction, we consider for each pair $\{k,\ell\}$, $1 \leq k < \ell \leq N$, a real-valued measurable function $V_{k\ell}$ defined on $\mathbb{R}^3$. These functions will be called <u>pair potentials</u>. Each pair potential defines a self-adjoint operator $V_{k\ell}$ in $L^2(\mathbb{R}^{3N})$, the maximal multiplication operator by $V_{k\ell}(\underline{x}_k - \underline{x}_\ell)$. The <u>total interaction</u> is the sum of these pair interactions, i.e. the maximal multiplication operator by $\sum_{k<\ell} V_{k\ell}(\underline{x}_k - \underline{x}_\ell)$:

$$V = \sum_{1 \leq k < \ell \leq N} V_{k\ell}. \tag{15.4}$$

Our first task is to give a meaning to the sum $H = H_0 + V$ as a self-adjoint operator.

PROPOSITION 15.1 : Let $V^o_{k\ell}$ denote the maximal multiplication operator in $L^2(\mathbb{R}^3)$ by the pair potential $V_{k\ell}(\underline{x})$, and let K_0 be the free Hamiltonian in $L^2(\mathbb{R}^3)$ (i.e. the operator (3.29) with $h/2\pi = 2m = 1$). Assume that each $V^o_{k\ell}$ is K_o-compact in $L^2(\mathbb{R}^3)$. Then

(a) V is H_0-bounded in $L^2(\mathbb{R}^{3N})$ with H_0-bound 0,

(b) $H = H_0 + V$ is self-adjoint in $L^2(\mathbb{R}^{3N})$ with $D(H) = D(H_0)$,

(c) H is bounded below.

Proof : (b) and (c) follow from (a) by the theorem of Rellich-Kato (Proposition 8.5(a)). To prove (a), it suffices to show that for each $\delta > 0$ there exists a number $\gamma > 0$ such that $\|V(H_0+\gamma)^{-1}\| < \delta$, cf. (8.3). Now $\|V(H_0+\gamma)^{-1}\| \leq \sum_{k<\ell}\|V_{k\ell}(H_0+\gamma)^{-1}\|$. Since the sum if finite, it is sufficient to prove that for each pair $\{k,\ell\}$

$$\lim_{\gamma\to\infty} \|V_{k\ell}(H_0+\gamma)^{-1}\| = 0. \tag{15.5}$$

To establish (15.5), we set for simplicity $k = 1$, $\ell = 2$ and introduce the total and the relative momentum operators $\underline{P}^{[1,2]}_{tot} = \underline{P}_1 + \underline{P}_2$ and $\underline{P}^{[1,2]}_{rel} = (m_1 \underline{P}_2 - m_2 \underline{P}_1)/M_{12}$ of the pair of particles $\{1,2\}$ as in (7.80), where $M_{12} = m_1 + m_2$. By (7.81) we may then write

$$H_0 = \sum_{j=3}^N (2m_j)^{-1}\underline{P}_j^2 + \frac{1}{2M_{12}} [\underline{P}^{[1,2]}_{tot}]^2 + \frac{1}{2m_{12}} [\underline{P}^{[1,2]}_{rel}]^2 \tag{15.6}$$

with $m_{12} = m_1 m_2/M_{12}$. We define $A_{12}(z)$ to be the multiplication operator in $\tilde{L}^2(R^{3N})$ by $[(2m_{12}M_{12}^2)^{-1}(m_1\underline{k}_2 - m_2\underline{k}_1)^2 - z]^{-1}$. Then

$$V_{12}(H_0+\gamma)^{-1} = V_{12}A_{12}(i\gamma)\left(\frac{1}{2m_{12}}[\underline{P}_{rel}^{[1,2]}]^2(H_0+\gamma)^{-1} - i\gamma(H_0+\gamma)^{-1}\right).$$

By (15.6) and (5.52), each of the two operators in the square bracket is a multiplication operator in $\tilde{L}^2(R^{3N})$ by a function the absolute value of which is majorized by 1 (for each $\gamma > 0$). Hence by Proposition 2.16(b) the norm of the operator in the square bracket is less than 2 for each $\gamma > 0$.

As in Section 7-5, we now write $H_1 \otimes H_2$ as $H_{rel}^{[1,2]} \otimes H_{CM}^{[1,2]}$, so that

$$H = H_{rel}^{[1,2]} \otimes H_{CM}^{[1,2]} \otimes H_3 \otimes \ldots \otimes H_N. \tag{15.7}$$

In this decomposition of H we have

$$V_{12}A_{12}(i\gamma) = V_{12}^o[(2m_{12})^{-1}K_0 - i\gamma]^{-1} \otimes I \otimes \ldots \otimes I. \tag{15.8}$$

Hence by Problem 2.35, $\|V_{12}A_{12}(i\gamma)\| = \|V_{12}^o[(2m_{12})^{-1}K_0 - i\gamma]^{-1}\|$, and the latter converges to zero as $\gamma \to \infty$ by our assumption on V_{12} and the proof of Proposition 8.4. Consequently $\|V_{12}(H_0+\gamma)^{-1}\| \leq 2\|V_{12}A_{12}(i\gamma)\| \to 0$ as $\gamma \to \infty$, which proves (15.5) for the pair {1,2}. The proof for the other pairs $\{k,\ell\}$ is the same. #

<u>COROLLARY 15.2</u> : Suppose that each pair potential $V_{k\ell}$ may be written as $V_{k\ell} = V_{k\ell}^{(1)} + V_{k\ell}^{(2)}$ such that $V_{k\ell}^{(1)} \varepsilon L^2(R^3)$ and $V_{k\ell}^{(2)} \varepsilon L^\infty(R^3)$. Then (a) $H = H_0+V$ is self-adjoint in $L^2(R^{3N})$ with $D(H) = D(H_0)$ and bounded below, (b) V is H_0-bounded with H_0-bound 0.

<u>Proof</u> : Let $V^{(1)} = \sum_{k<\ell}V_{k\ell}^{(1)}$, $V^{(2)} = \sum_{k<\ell}V_{k\ell}^{(2)}$. Since $V_{k\ell}^{(1)}$ is

K_0-compact in $L^2(R^3)$ by Proposition 8.7(a), $H_0+V^{(1)}$ is self-adjoint and bounded below by Proposition 15.1. Since $V^{(2)} \varepsilon B(H)$ by Proposition 2.16, $H_0 + V^{(1)} + V^{(2)}$ is self-adjoint and bounded below. Part (b) follows from Proposition 15.1(a) and the fact that $V^{(2)} \varepsilon B(H)$. #

It is sometimes useful to consider only the relative movement of the N particles. We now outline the method of removing the center-of-mass motion but leave it to the reader to verify the details (Problem 15.1). One takes as new variables the <u>center-of-mass coordinate</u> $\underline{x}_{CM} = M^{-1}\sum_{k=1}^{N} m_k \underline{x}_k$, where $M = \sum_{k=1}^{N} m_k$, and a set of N-1 linearly independent relative coordinates $\underline{y}_1,\ldots,\underline{y}_{N-1}$ between the N particles. The coordinates $\underline{y}_k$ are linear combinations of the relative distances $\{\underline{x}_j-\underline{x}_\ell\}$, chosen such that the Jacobian determinant of the change of variables is 1. There are many ways of selecting the set $\{\underline{y}_k\}$, since the number of relative distances $\{\underline{x}_j-\underline{x}_\ell\}$ between pairs of particles exceeds N-1. (Notice that, if we write $\underline{y}_k = \sum_{\ell=1}^{N} \alpha_{k\ell}\underline{x}_\ell$, then $\sum_{\ell=1}^{N} \alpha_{k\ell} = 0$.) By expressing the functions of $L^2(R^{3N})$ in the new set of variables $\{\underline{x}_{CM},\underline{y}_1,\ldots,\underline{y}_{N-1}\}$, one may write $L^2(R^{3N}) = L^2(R^3) \otimes L^2(R^{3N-3})$ where the first factor is the Hilbert space of states of the center of mass and the second one that of the relative states of the N particles. The momentum operator of the center of mass is $\underline{P}_{tot} = \sum_{k=1}^{N} \underline{P}_k$, and with each relative coordinate $\underline{y}_k$ one may associate a three-component relative momentum operator $\underline{P}_k$, the self-adjoint operator determined by $-i\nabla_{\underline{y}_k}$. $\underline{P}_k$ will be viewed as an operator in $L^2(R^{3N-3})$ *). If one denotes by $\underline{Q}_k$

*) For the sake of clarity we shall distinguish in this section

the multiplication operator in $L^2(\mathcal{R}^{3N-3})$ by $\underline{y}_k$, then

$$[P_{kj}, Q_{\ell r}] \subseteq -i\delta_{k\ell}\delta_{jr} \quad (k,\ell = 1,\ldots,N-1;\ j,r = 1,2,3). \tag{15.9}$$

As in (7.81), one finds that the free Hamiltonian may be written as (see also Problem 2.39)

$$H_o = H_{o,CM} + I \otimes H_{o,rel} \tag{15.10}$$

with

$$H_{o,CM} = \frac{1}{2M} \underline{P}^2_{tot} = \frac{1}{2M} K_o \otimes I, \quad H_{o,rel} = \sum_{j,k=1}^{N-1} a_{jk} \underline{P}_j \cdot \underline{P}_k. \tag{15.11}$$

Here the matrix $\{a_{jk}\}$ is real, symmetric and positive definite, and we have explicitly used the fact that the free Hamiltonian for the relative motion acts non-trivially only in $L^2(\mathcal{R}^{3N-3})$. The matrix $\{a_{jk}\}$ is in general not diagonal, but it is possible to choose the set $\{\underline{y}_k\}$ such that it becomes diagonal. One way to achieve this is to take for $\underline{y}_k$ the relative distance between the particle k+1 and the center of mass of the first k particles, i.e.

$$\underline{y}_k = \underline{x}_{k+1} - M_k^{-1} \sum_{j=1}^{k} m_j \underline{x}_j, \quad M_k = \sum_{j=1}^{k} m_j, \tag{15.12}$$

in which case (Problem 15.1)

$$H_{o,rel} = \sum_{k=1}^{N-1} \frac{1}{2} \left(\frac{1}{m_{k+1}} + \frac{1}{M_k}\right) \underline{P}_k^2 \equiv \sum_{k=1}^{N-1} \gamma_k \underline{P}_k^2 . \tag{15.13}$$

Notice that there exist numbers $0 < a < b < \infty$ such that

$$a \sum_{k=1}^{N-1} \underline{P}_k^2 \leq H_{o,rel} \leq b \sum_{k=1}^{N-1} \underline{P}_k^2. \tag{15.14}$$

---

*) Ctd.

operators acting in $\mathcal{H}_{rel} = L^2(\mathcal{R}^{3N-3})$ by the use of a script letter or by adding the subscript "rel". This convention will be dropped in the remainder of the book, since it will always be clear whether we work in $\mathcal{H}$ or in $\mathcal{H}_{rel}$.

The interaction V clearly acts nontrivially only in the relative variables, i.e. we have $V = I \otimes \mathcal{V}$ in the above-mentioned decomposition : $L^2(\mathcal{R}^{3N}) = L^2(\mathcal{R}^3) \otimes L^2(\mathcal{R}^{3N-3})$. The total Hamiltonian will be of the form

$$H = H_{o,CM} + I \otimes H_{rel} \quad \text{with} \quad H_{rel} = H_{o,rel} + \mathcal{V}. \tag{15.15}$$

PROPOSITION 15.3 : Under the assumptions of Corollary 15.2, $\mathcal{V}$ is $H_{o,rel}$-bounded with $H_{o,rel}$-bound 0 and the relative Hamiltonian H_{rel} is a self-adjoint operator in $L^2(\mathcal{R}^{3N-3})$ with $D(H_{rel}) = D(H_{o,rel})$. Also H_{rel} is bounded below.

Proof : Let $V^{(1)} = I \otimes \mathcal{V}^{(1)}$ and $V^{(2)} = I \otimes \mathcal{V}^{(2)}$ be as in the proof of Corollary 15.2. Since $\mathcal{V}^{(2)} \varepsilon \mathcal{B}(\mathcal{H}_{rel})$, it suffices to show that $\mathcal{V}^{(1)}$ is $H_{o,rel}$-bounded with $H_{o,rel}$-bound 0. For this, we fix $g \varepsilon L^2(\mathcal{R}^3) \cap D(K_o)$ with $\|g\| = 1$. If $\delta > 0$, there exists $\gamma > 0$ such that $\|V(H_o + \gamma)^{-1}\| < \delta$ by (15.5). If $f \varepsilon D(H_{o,rel}) \subseteq L^2(\mathcal{R}^{3N-3})$, then $g \otimes f \varepsilon D(H_o)$ by (15.10) and (15.11). Hence

$$\begin{aligned} \|\mathcal{V}^{(1)} f\| &= \|(I \otimes \mathcal{V}^{(1)})(g \otimes f)\| < \delta \, \|(H_o + \gamma)(g \otimes f)\| \\ &\leq \delta \|g \otimes H_{o,rel} f\| + \frac{\delta}{2M} \|K_o g \otimes f\| + \gamma\delta \, \|g \otimes f\| \\ &= \delta \|H_{o,rel} f\| + \frac{\delta}{2M} \{\|K_o g\| + 2M\gamma\} \|f\|. \quad \# \end{aligned}$$

The next question that we consider is that of the characterization of the bound states of H_{rel}. For this purpose we introduce for each $r > 0$ a projection F_r in $L^2(\mathcal{R}^{3N-3})$ by setting

$$(F_r f)(\underline{y}_1, \ldots, \underline{y}_{N-1}) = \chi_r(\underline{y}_1, \ldots, \underline{y}_{N-1}) f(\underline{y}_1, \ldots, \underline{y}_{N-1}), \tag{15.16}$$

where χ_r is the characteristic function of the set

$\{(y_1,\ldots,y_{N-1}) \mid \sum_{k=1}^{N-1} y_k^2 < r^2\}$. The subspaces $M_\infty(H_{rel})$ of the scattering states and $M_o(H_{rel})$ of the bound states of H_{rel} are defined as in (7.3) and (7.4) respectively, by using the projections (15.16). The following theorem will relate these subspaces to spectral subspaces of H_{rel}. Roughly speaking, the states in $M_o(H_{rel})$ are those in which the N particles stay together at all times, whereas the states in $M_\infty(H_{rel})$ are characterized by the property that at least one relative distance between particles becomes large at large times.

PROPOSITION 15.4 : Assume that H_{rel} has no singularly continuous spectrum and that each pair potential belongs to $L^2(R^3) + L^\infty(R^3)$ *). Then one has $M_o(H_{rel}) = H_p(H_{rel})$ and $M_\infty(H_{rel}) = H_{ac}(H_{rel})$.

Proof : It follows from Proposition 15.3 and Remark 8.6 that V is H_{rel}-bounded and that there exists $\alpha > 0$ such that $-\alpha \in \rho(H_{rel})$. We then get from (6.37) that

$$F_r(H_{rel} + \alpha)^{-1} = F_r(H_{o,rel} + \alpha)^{-1}[I - V(H_{rel} + \alpha)^{-1}]. \quad (15.17)$$

The operator in the square bracket is in $B(H_{rel})$ by Remark 8.3. For $0 < R < \infty$, let G_R be the multiplication operator by χ_R in $\tilde{L}^2(R^{3N-3})$. By Lemma 7.6 applied to $L^2(R^{3N-3})$, $F_r G_R$ is compact, hence $F_r G_R (H_{o,rel} + \alpha)^{-1} \in B_o$. Now

$$\|F_r(H_{o,rel} + \alpha)^{-1} - F_r G_R (H_{o,rel} + \alpha)^{-1}\| \leq \|F_r\| \, \|(I - G_R)(H_{o,rel} + \alpha)^{-1}\|$$
$$\leq (aR^2 + \alpha)^{-1} \to 0 \text{ as } R \to \infty,$$

*) A function $V : R^n \to R$ is said to belong to $L^p(R^n) + L^q(R^n)$ if it can be written as $V = V_1 + V_2$ with $V_1 \in L^p(R^n)$ and $V_2 \in L^q(R^n)$.

where the last estimate is obtained by using the first inequality in (15.14) and applying Proposition 2.16(b). Thus in view of Proposition 2.22(d), we have $F_r(H_{o,rel} + \alpha)^{-1} \varepsilon \mathcal{B}_0$. Hence $F_r(H_{rel} + \alpha)^{-1} \varepsilon \mathcal{B}_0$ by (15.17) and Proposition 2.22(b), and the desired result now follows from Corollary 7.5. #

Remark 15.5 : The essential point of Proposition 15.4 for our present purposes is the fact that the spectrum of the part of H_{rel} in $\mathcal{M}_o(H_{rel})$ consists only of eigenvalues, since this will permit us to introduce the channel subspaces and channel Hamiltonians by the method indicated in Section 14-3 (see below). In this context we should point out that the identity $\mathcal{M}_o(H_{rel}) = \mathcal{H}_p(H_{rel})$ can easily be established without the assumption that $\mathcal{H}_{sc}(H_{rel}) = \{\theta\}$. This is important because the absence of the singularly continuous spectrum of H_{rel} has been proved only for some particular classes of N-body Hamiltonians (see Section 15-4B). To prove the above identity, one introduces a subspace $\bar{\mathcal{M}}_\infty(H_{rel})$ as in Section 7-6A, i.e. the set of states for which at least one relative coordinate becomes large in the time average when the interval over which the time average is taken tends to infinity. One can then prove as in Proposition 15.4 that $\bar{\mathcal{M}}_\infty(H_{rel}) = \mathcal{H}_c(H_{rel})$ and $\mathcal{M}_o(H_{rel}) = \mathcal{H}_p(H_{rel})$; cf. Amrein and Georgescu [1] for the details.

We now proceed to define the cluster and channel Hamiltonians and subspaces. Let D be a partition of $\{1,\ldots,N\}$ into $n \geq 2$ clusters (1), ..., (n). To define the associated cluster Hamiltonian H^D, one drops in V all interactions between particles belonging to different clusters. In other words one defines

$$H^D = H_0 + \sum_{k=1}^{n} \sum_{\substack{i,\ell\varepsilon(k)\\ i<\ell}} V_{i\ell} = H_0 + V - V^D, \tag{15.18}$$

where V^D is the operator used in (14.58). In the sum over k in (15.18) only the clusters consisting of at least two particles contribute. If we assume that all pair potentials $V_{k\ell}$ verify the hypotheses of Corollary 15.2, then each H^D is self-adjoint with $D(H^D) = D(H_0)$ by Corollary 15.2 (since the pair potentials appearing in (15.18) form a subset of all pair potentials and thus necessarily satisfy the hypotheses of that Corollary).

One may also consider the Hilbert space $\mathcal{H}^{(k)}$ corresponding to the set of particles forming the cluster (k) which was used in (14.1). The Hamiltonian $H^{(k)}$ in $\mathcal{H}^{(k)}$ describing these particles is of the form

$$H^{(k)} = H_0^{(k)} + \sum_{\substack{i,\ell\varepsilon(k)\\ i<\ell}} V_{i\ell}, \tag{15.19}$$

hence it is self-adjoint on $D(H_0^{(k)})$ by Corollary 15.2. From (15.18) and (15.19) one obtains that in the decomposition (14.1)

$$H^D = H^{(1)} \otimes I \otimes \cdots \otimes I + \ldots + I \otimes \cdots \otimes I \otimes H^{(n)}. \tag{15.20}$$

We leave it as an exercice to prove that this identity is indeed true on $D(H_0) = D(H^D)$ (Problem 15.2).

The <u>cluster subspace</u> $\mathcal{M}^D$ associated with the partition D is given by (14.3). For each cluster (k) consisting of more than one particle, we may write $\mathcal{H}^{(k)} = \mathcal{H}_{CM}^{(k)} \otimes \mathcal{H}_{rel}^{(k)}$ and define $H_{rel}^{(k)}$ in $\mathcal{H}_{rel}^{(k)}$ as in (15.15). By applying Proposition 15.4 to the subsystem (k), we obtain $\mathcal{M}_0(H^{(k)}) = \mathcal{H}_{CM}^{(k)} \otimes \mathcal{M}_0(H_{rel}^{(k)}) = \mathcal{H}_{CM}^{(k)} \otimes \mathcal{H}_p(H_{rel}^{(k)})$, so that

$$\mathcal{M}^D = \mathcal{H}_{CM}^{(1)} \otimes \mathcal{H}_p(H_{rel}^{(1)}) \otimes \ldots \otimes \mathcal{H}_{CM}^{(n)} \otimes \mathcal{H}_p(H_{rel}^{(n)}). \tag{15.21}$$

Here $\mathcal{H}_{CM}^{(k)}$ is the Hilbert space describing the center of mass of the k-th cluster and $\mathcal{H}_p(H_{rel}^{(k)})$ is the subspace of $\mathcal{H}_{rel}^{(k)}$ spanned by all the eigenvectors of $H_{rel}^{(k)}$. If (k) consists of only one particle, the corresponding factor $\mathcal{H}_p(H_{rel}^{(k)})$ is absent.

The channel subspaces and channel Hamiltonians can now be defined exactly as in Section 14-3 and do not require any further discussion.

We end this section with some remarks regarding the essential spectrum of the relative Hamiltonian H_{rel}. In Proposition 15.1 we assumed that each pair potential $V_{k\ell}$ is K_0-compact in $L^2(\mathcal{R}^3)$. Now it follows from Lemma 15.13 (Section 15-4) that, even though the first factor in (15.8) is a compact operator in $\mathcal{H}_{rel}^{[1,2]}$, $V_{12}A_{12}(i\gamma)$ is <u>not</u> compact. Thus, even if, as in Proposition 15.1, each pair potential is K_0-compact in $L^2(\mathcal{R}^3)$, the total interaction is not expected to be H_0-compact. The same situation is encountered after removing the center-of-mass motion of the N-particle system, since there will be at least one factor I in (15.8) unless $N = 2$. Consequently Proposition 8.5(b) cannot be applied, i.e. if $N > 2$ one cannot infer the invariance of the essential spectrum of the relative Hamiltonian. Thus $\sigma_e(H_{rel})$ may be different from $\sigma_e(H_{o,rel}) = [0,\infty)$.

To understand why this is so, consider a system of three particles such that only V_{12} is non-zero. The Hamiltonian describing the relative motion of the particles 1 and 2 may have negative eigenvalues. Consider a scattering channel α for the partition $D = \{[1,2],[3]\}$ defined by choosing a bound state of [1,2] corresponding to a negative eigenvalue $\mu_\alpha < 0$. Since the subsystem [1,2] moves freely relative to the third particle, the relative kinetic energy may assume an arbitrary positive value. On the channel subspace M^α the total Hamiltonian H_{rel} in the center-of-mass system is the sum of the relative kinetic energy and the bound state energy μ_α of [1,2] by (14.44), so that its continuous spectrum will extend from μ_α to $+\infty$. So in this case the continuous spectrum of H_{rel} will contain the interval $[\lambda_{min}^{[1,2]}, +\infty)$, where $\lambda_{min}^{[1,2]}$ is the lowest negative eigenvalue of the relative Hamiltonian for [1,2], and $\lambda_{min}^{[1,2]} = 0$ if this relative Hamiltonian has no negative eigenvalues. The only other type of scattering state of H_{rel} is that corresponding asymptotically to three free independently moving particles. Since no binding energies are involved here, these states will have positive total energy. Hence we expect that in this example $\sigma_{ac}(H_{rel}) = \sigma_e(H_{rel}) = [\lambda_{min}^{[1,2]}, \infty)$.

Similarly for a general N-particle system one may express the essential spectrum of H_{rel} in terms of the lower bounds of the spectra of all cluster Hamiltonians. The spectrum of each cluster Hamiltonian H^D is absolutely continuous and equal to some interval $[\lambda^D, \infty)$ with $\lambda^D \leq 0$ (see Problem 15.3). If s-lim $\exp(iH_{rel}t)\exp(-iH_{rel}^D t)$ as $t \to \infty$ exist on the entire space $\mathcal{H}_{rel}$, then by Proposition 5.21 the absolutely continuous spectrum of H_{rel} will contain all the intervals $[\lambda^D,\infty)$, i.e. $[\lambda_0,\infty) \subseteq \sigma_{ac}(H_{rel}) \subseteq \sigma_e(H_{rel})$, where λ_0 is the smallest of the numbers λ^D, $D \neq D_1$. λ_0 may also be characterized as $\lambda_0 = \inf \lambda^\alpha$, where λ^α is the total binding energy associated with the channel α and the infimum is taken over all scattering channels α. One expects that under suitable assumptions on the pair potentials, the spectrum of H_{rel} below λ_0 will be discrete, hence that $\sigma_e(H_{rel}) = [\lambda_0,\infty)$. This is known as <u>Hunziker's theorem</u>. We shall prove it in this chapter for a particular class of pair potentials. References and further results on spectral properties can be found in Section 15-4.

<u>PROPOSITION 15.6</u> : Assume that each $V_{k\ell}$ belongs to $L^2(\mathbb{R}^3) + L^p(\mathbb{R}^3)$ for some $2 < p < 3$. Then

$$\sigma_e(H_{rel}) = [\lambda_0,\infty) \quad \text{with } \lambda_0 = \min_{D\neq D_1} \lambda^D, \tag{15.22}$$

where $[\lambda^D,\infty) = \sigma(H^D_{rel}) = \sigma(H^D)$.

<u>Proof</u> : The existence of the cluster wave operators is proved in Section 15-2, so that $[\lambda_0,\infty) \subseteq \sigma_e(H_{rel})$ by the preceding discussion. The opposite inclusion follows from Proposition 15.14 (Section 15-4), since each $V_{k\ell}$ is K_0-compact by Proposition 8.7 and the discussion preceding Proposition 8.30. The spectrum of H^D is studied in Problem 15.3. #

15-2 THE CLUSTER WAVE OPERATORS

In this section we assume that each pair potential is in $L^2(R^3)$. The operators H, H^D and H_{rel} then have all the properties derived in Section 15-1. Our goal here is to prove the existence of the cluster and channel wave operators and the orthogonality of their ranges. Some factorization properties of these wave operators when the relative distances between the clusters in the initial state become large will be established in Section 15-4D.

The first proposition asserts the existence of the strong limits of $\{\exp(iHt)\exp(-iH^Dt)\}$ as $t \to \pm\infty$ on the entire Hilbert space. This obviously implies the existence of the cluster wave operators $\Omega_\pm^D$ and of the channel wave operators $\Omega_\pm^\alpha$, since they are just the restrictions of the above limits to M^D and M^α with $D \div \alpha$ respectively.

<u>PROPOSITION 15.7</u> : Assume that each pair potential belongs to $L^2(R^3)$, and let D be an arbitrary partition of $\{1,\ldots,N\}$ into at least two subsets. Then s-lim $V_t^* U_t^D f$ exist as $t \to \pm\infty$ for all $f \in H$; hence (B1) is verified.

<u>Proof</u> (for $t \to +\infty$) : By Proposition 8.14 it suffices to show that $\int_1^\infty \|(H-H^D)U_t^D f\|\, dt < \infty$ for all f in some fundamental set $\mathcal{D} \subseteq D(H^D) = D(H_0)$. Since $H-H^D$ is a finite sum of pair potentials, it suffices to show that $\|V_{k\ell} U_t^D f\| \in L^1(1,\infty)$ for each such potential $V_{k\ell}$ and all $f \in \mathcal{D}$. By (15.18), $V_{k\ell}$ depends on the relative distance $\underline{x}_k - \underline{x}_\ell$ of two particles belonging to different clusters in D. To be specific, we assume for the

moment that the cluster containing the k-th particle is (1) in $D = \{(1),\ldots,(n)\}$.

We factorize $H^{(1)}$ as in (14.37) into $H^{(1)}_{CM} \otimes H^{(1)}_{rel}$ and write

$$H = H^{(1)}_{CM} \otimes H^{(1)}_{rel} \otimes G \text{ with } G = H^{(2)} \otimes \ldots \otimes H^{(n)}. \tag{15.23}$$

The factor $H^{(1)}_{rel}$ is present only if (1) consists of more than one particle. The variable in $H^{(1)}_{CM}$ is $\underline{X}^{(1)}_{CM}$, those in $H^{(1)}_{rel}$ are collectively denoted by x_{rel} and those in G by y. $\underline{x}_\ell$ coincides with one of the variables in the set $\{y\}$, whereas $\underline{x}_k$ is a linear combination of $\underline{X}^{(1)}_{CM}$ and of the variables in x_{rel}, the coefficient of $\underline{X}^{(1)}_{CM}$ in this linear combination being positive. Hence $\underline{x}_{k\ell} = \underline{x}_k - \underline{x}_\ell$ is a linear combination of $\underline{X}^{(1)}_{CM}$, $\underline{x}_\ell$ and the variables in x_{rel} such that the Jacobian determinant J for the change of variables $\{\underline{X}^{(1)}_{CM}, x_{rel}, y\} \to \{\underline{x}_{k\ell}, x_{rel}, y\}$ is a positive constant.

Consider a vector of the form $f = e \otimes g \otimes h$ in the representation (15.23) of H, with $e \in S(R^3)$, $g \in D(H^{(1)}_{o,rel})$ and $h \in D(H^G_o)$, where H^G_o is the kinetic energy operator in G of the set of particles belonging to the clusters (2),...,(n). The cluster evolution group factorizes in the representation (15.23), i.e.

$$U^D_t = U^{(1)}_{t,CM} \otimes U^{(1)}_{t,rel} \otimes U^G_t. \tag{15.24}$$

Here $U^{(1)}_{t,CM} = \exp[-i(2M_1)^{-1}K_o t]$ as an operator in $L^2(R^3)$, where K_o is the operator (3.29) with $h/2\pi = 2m = 1$ and M_1 the total mass of the cluster (1). This property of $U^{(1)}_{t,CM}$ implies together with Corollary 3.13 that $|[U^{(1)}_{t,CM}e](\underline{X}^{(1)}_{CM})| \leq c|t|^{-3/2}\cdot \|e\|_1$ for all values of $\underline{X}^{(1)}_{CM}$. This, together with the above change of variables, leads to

$$\|V_{k\ell}U_t^D f\|^2 = \int d^3X_{CM}^{(1)} dx_{rel} dy |V_{k\ell}(\underline{x}_{k\ell})|^2 |(U_t^D f)(\underline{X}_{CM}^{(1)}, x_{rel}, y)|^2$$

$$\leq J \int d^3x_{k\ell} dx_{rel} dy |V_{k\ell}(\underline{x}_{k\ell})|^2 c^2 |t|^{-3} \|e\|_1^2 |[U_{t,rel}^{(1)} g](x_{rel})|^2 \cdot$$

$$|(U_t^G h)(y)|^2 = c^2 J |t|^{-3} \|e\|_1^2 \|V\|_2^2 \|g\|^2 \|h\|^2.$$

Since all quantities appearing in the last expression are finite, we have $\|V_{k\ell}U_t^D f\| \in L^1(1,\infty)$.

A similar argument can be applied for each $V_{k\ell}$, and it is easily seen that one may take for $\mathcal{D}$ the set of vectors of the form $f = e^{(1)} \otimes g^{(1)} \otimes \ldots \otimes e^{(n)} \otimes g^{(n)}$ in the representation

$$\mathcal{H} = \mathcal{H}_{CM}^{(1)} \otimes \mathcal{H}_{rel}^{(1)} \otimes \ldots \otimes \mathcal{H}_{CM}^{(n)} \otimes \mathcal{H}_{rel}^{(n)}, \qquad (15.25)$$

with $e^{(k)} \in \mathcal{S}(\mathbb{R}^3)$ and $g^{(k)} \in D(H_{o,rel}^{(k)})$. #

By combining the method of the preceding proof with that of Proposition 8.30, one arrives at the existence of s-lim $V_t^* U_t^D$ as $t \to \pm\infty$ under the hypothesis that each pair potential belongs to $L^2(\mathbb{R}^3) + L^p(\mathbb{R}^3)$ for some $2 < p < 3$.

We next give a preliminary result for proving the orthogonality of the ranges of the cluster wave operators.

<u>LEMMA 15.8</u> : Let $\alpha \div C$, $\beta \div D$ with $C \neq D$. Then w-lim $U_t^{\alpha *} U_t^{\beta} = 0$ as $t \to \pm\infty$.

<u>Proof</u> : We have $U_t^{\alpha *} U_t^{\beta} = \exp[i(H^{\alpha} - H^{\beta})t]$. Thus, by Lemma 5.20, it suffices to show that $H^{\alpha} - H^{\beta}$ is spectrally absolutely continuous. Now by (14.46)

$H^\alpha - H^\beta = H^\alpha_{kin} - H^\beta_{kin} + \mu = \sum_{(j)\varepsilon C}[\underline{P}^{(j)}_{tot}]^2 - \sum_{(\ell)\varepsilon D}[\underline{P}^{(\ell)}_{tot}]^2 + \mu,$

where $\mu = \lambda^\alpha - \lambda^\beta$ and $\underline{P}^{(j)}_{tot} = \sum_{i\varepsilon(j)}\underline{P}_i$ is the momentum operator of the center of mass of the cluster (j). Thus $H^\alpha - H^\beta - \mu$ is a quadratic form in $\underline{P}_1,\ldots,\underline{P}_n$ which is easily seen to be of the form $H^\alpha - H^\beta - \mu = 2\sum_{j<\ell} a_{j\ell}\underline{P}_j\cdot\underline{P}_\ell$, where $a_{j\ell}$ assumes the values $0,\pm 1$ depending on the partitions C and D. Since $C \neq D$, we may assume without loss of generality that the clusters $(j_1)\varepsilon C$ and $(\ell_1)\varepsilon D$ containing particle 1 are different, and that (j_1) contains also particle 2 whereas (ℓ_1) does not. Then the term $2\underline{P}_1\cdot\underline{P}_2$ is present in H^α but not in H^β, so that $a_{12} = +1$.

Thus $H^\alpha - H^\beta$ is the multiplication operator in $\tilde{L}^2(\mathcal{R}^{3N})$ by the function $\psi(\underline{k}_1,\ldots,\underline{k}_n) = 2\sum_{j<\ell} a_{j\ell}\underline{k}_j\cdot\underline{k}_\ell + \mu$ with $a_{12} = +1$. It follows that $\partial\psi/\partial k_{11} = 2k_{21} + 2\sum_{\ell=3}^{N} a_{1\ell}k_{\ell 1} \neq 0$ almost everywhere in $\mathcal{R}^{3N}$, hence that $|\mathrm{grad}\psi| \neq 0$ a.e. in $\mathcal{R}^{3N}$, so that $H^\alpha - H^\beta$ is absolutely continuous by Problem 5.7. #

<u>PROPOSITION 15.9</u> : If each pair potential is in $L^2(\mathcal{R}^3)$ (or in $L^2(\mathcal{R}^3) + L^p(\mathcal{R}^3)$ with $2 < p < 3$), then (B2) is verified, i.e. $F^C_\pm F^D_\pm = \delta_{CD}F^C_\pm$.

<u>Proof</u> (for the + sign) : We must show that $(f, F^C_+ F^D_+ g) \equiv (F^C_+ f, F^D_+ g) = 0$ for all $f,g\varepsilon\mathcal{H}$ and $C \neq D$. By (14.49), $F^C_+ f$ and $F^D_+ g$ are of the form $F^C_+ f = \sum_{\alpha\div C}\Omega^\alpha_+ f^\alpha$, $F^D_+ g = \sum_{\beta\div D}\Omega^\beta_+ g^\beta$ with $f^\alpha\varepsilon M^\alpha$, $g^\beta\varepsilon M^\beta$. Hence it is enough to prove that $(\Omega^\alpha_+ f^\alpha, \Omega^\beta_+ g^\beta) = 0$ for all $f^\alpha\varepsilon M^\alpha$, $f^\beta\varepsilon M^\beta$ and all α,β belonging to different partitions (i.e. $\alpha\div C$, $\beta\div D$ with $C \neq D$). Now by (14.47) and Problem 2.4(a)

$$(\Omega_+^\alpha f^\alpha, \Omega_+^\beta f^\beta) = \lim_{t\to\infty} (V_t^* U_t^\alpha f^\alpha, V_t^* U_t^\beta f^\beta) = \lim_{t\to\infty} (f^\alpha, U_t^{\alpha *} U_t^\beta f^\beta),$$

which is zero by Lemma 15.8 (see also Problem 15.10). #

The preceding two propositions establish the first two parts of the asymptotic condition, viz. (B1) and (B2), for $V_{k\ell} \varepsilon L^2(R^3) + L^p(R^3)$, $2 < p < 3$. As in the case of simple potential scattering (Chapters 8 and 9), the proof of asymptotic completeness is considerably more difficult. We shall comment on this point in Section 15-4.

So far we have worked in the Hilbert space $H = L^2(R^{3N})$ given by (15.1) which describes the states of the N particles. After removal of the center-of-mass motion, the scattering theory may also be done in $H_{rel} = L^2(R^{3N-3})$. It is clear from (15.10) and (15.15) that $\Omega_\pm^D = I \otimes \Omega_{\pm,rel}^D$ in the representation $H = H_{CM} \otimes H_{rel}$, where $\Omega_{\pm,rel}^D = \text{s-lim}\, \exp(iH_{rel}t)\exp(-iH_{o,rel}t)E_{rel}^D$ as $t \to \pm\infty$ and $E^D = I \otimes E_{rel}^D$. This shows that it is sufficient to solve the scattering problem in H_{rel} and that the asymptotic condition may be formulated in H or in H_{rel} without changing the physical content of the theory.

15-3 SCATTERING INTO CONES. CROSS SECTIONS

The basic observations in multichannel experiments are the arrangement of the particles into clusters and the direction of motion and kinetic energy of these clusters at large times. In Section 14-2 we discussed in some detail the first type of observable and were led to introduce, for each partition $D = \{(1),\dots,(n)\}$, $n \geq 2$, a pair of projections $F_{\pm}^{D}$ verifying (14.6) and (14.8). The vectors in the range of $F_{\pm}^{D}$ are scattering states which, as $t \to \pm\infty$, will describe n freely moving clusters $(1),\dots,(n)$.

We shall now consider additional observables $B_{C_1 \dots C_n}^{D}$ which allow one to determine whether or not at large positive times the particles in cluster (1) lie in cone C_1, those in cluster (2) in cone C_2, etc. In the first part of this section we show that these observables are all compatible with each other and with the projections F_{+}^{D}, and then that the measurement of $B_{C_1 \dots C_n}^{D}$ in a state leading to clustering D amounts to a measurement of the total momenta of the individual clusters. In the second part we shall discuss cross sections, for which we shall work with momentum observables in the center-of-mass frame as we did in Chapter 7 for the case of two particles.

Let $D = \{(1),\dots(n)\}$, $n \geq 2$ be a partition and let $C_1,\dots,C_n$ be cones in $\mathbb{R}^3$ with apex at the origin. We denote by $F_{C_1 \dots C_n}^{D}$ the projection in $L^2(\mathbb{R}^{3N})$ whose range is the states in which each particle belonging to cluster (k) is localized in C_k $(k = 1,\dots,n)$. An explicit representation of this projection will be given in the proof of Lemma 15.10 in Section 15-4. In the same way as in Section 4-3, the observable $B_{C_1 \dots C_n}^{D}$ is the limit of $V_t^* F_{C_1 \dots C_n}^{D} V_t E_\infty(H)$ as $t \to +\infty$. To prove the existence of this limit, it is convenient to first replace $\{V_t\}$ by a simpler evolution group, namely a cluster evolution group $\{U_t^L\}$, and then to use the existence of the cluster wave operators Ω_{+}^{L} to pass from $\{U_t^L\}$ to $\{V_t\}$.

In the following lemma we state the properties of the limits of $U_t^{L*}F^D_{C_1\ldots C_n}U_t^L E^L$, the physical meaning of which is quite transparent. We shall postpone the proof of this lemma till Section 15-4.

Suppose that L is of the form $L = \{(1)',\ldots,(m)'\}$, $m \geq 2$. Let $f^L \varepsilon M^L$. Since U_t^L leaves M^L invariant, the particles of each cluster (k)' in the state $U_t^L f^L$ will stay close together at all times. In order for the state $U_t^L f^L$ to belong to the range of the projection $F^D_{C_1\ldots C_n}$, each particle in (k)' must lie in one of the cones (determined by D), which is possible if and only if the cluster (k)' is localized in the intersection of all cones C_ℓ for which $(\ell)\cap(k)' \neq \emptyset$. We call this new cone Δ_k, i.e. we set $\Delta_k = \cap C_\ell$ with $(\ell)\cap(k)' \neq \emptyset$. It is also to be expected that asymptotically the position of a particle may be replaced by the position of the center of mass of the cluster (k)' containing that particle. Hence a vector $f^L \varepsilon M^L$ will be in the range of the limit of $U_t^{L*}F^D_{C_1\ldots C_n}U_t^L$ if and only if the total momentum of each cluster (k)' of L lies in the corresponding cone Δ_k. This is the principal result of Lemma 15.10. For its statement we introduce the projection

$$G^{DL}_{\Delta_1\ldots\Delta_m} = G_{\Delta_1} \otimes I \otimes G_{\Delta_2} \otimes I \otimes \ldots \otimes G_{\Delta_m} \otimes I \tag{15.26}$$

in the representation (15.25) of $H = L^2(R^{3N})$ with respect to L, where G_{Δ_k} is the multiplication operator by the characteristic function of Δ_k in $\tilde{L}^2(R^3)$ as in (7.31).

<u>LEMMA 15.10</u> : Let $D,L,C_1,\ldots,C_n$ be as above, and assume that $M_o(H_{rel}^{(k)'}) = H_p(H_{rel}^{(k)'})$ for each cluster (k)' in the partition L. Then

$$\underset{t\to+\infty}{\text{s-lim}}\ U_t^{L*} F^D_{C_1\ldots C_n} U_t^L E^L = G^{DL}_{\Delta_1\ldots\Delta_m} E^L = E^L G^{DL}_{\Delta_1\ldots\Delta_m} . \quad (15.27)$$

We may now determine the observables $B^D_{C_1\ldots C_n}$ mentioned at the beginning as well as their properties.

<u>PROPOSITION 15.11</u> : Let $D = \{(1),\ldots,(n)\}$, $n \geq 2$, be a partition and $C_1,\ldots,C_n$ cones in $\mathcal{R}^3$. Assume that (B1) - (B3) hold and that $\mathcal{M}_0(H^{(k)'}_{rel}) = \mathcal{H}_p(H^{(k)'}_{rel})$ for each cluster (k)' consisting of less than N elements. Then

$$B^D_{C_1\ldots C_n} \equiv \underset{t\to+\infty}{\text{s-lim}}\ V_t^* F^D_{C_1\ldots C_n} V_t E_\infty(H) = \sum_L \Omega_+^L G^{DL}_{\Delta_1\ldots\Delta_m} \Omega_+^{L*}. \quad (15.28)$$

The self-adjoint operators $B^D_{C_1\ldots C_n}$, as D and $C_1,\ldots,C_n$ vary, commute with each other and with each F_+^L.

<u>Proof</u> : (i) We set for convenience $F^D_{C_1\ldots C_n} = F_{\{C\}}$ and $G^{DL}_{\Delta_1\ldots\Delta_m} = G^L_{\{\Delta\}}$. Let $f\varepsilon\mathcal{M}_\infty(H)$. Then by (B3)

$$\|V_t^* F_{\{C\}} V_t f - \sum_L \Omega_+^L G^L_{\{\Delta\}} \Omega_+^{L*} f\| = \|\sum_L (F_{\{C\}} V_t F_+^L - V_t \Omega_+^L G^L_{\{\Delta\}} \Omega_+^{L*}) f\|$$

$$\leq \sum_L \|U_t^{L*} F_{\{C\}} U_t^L U_t^{L*} V_t F_+^L f - U_t^{L*} V_t \Omega_+^L G^L_{\{\Delta\}} \Omega_+^{L*} f\| . \quad (15.29)$$

By (14.14) and Lemma 15.10, $\text{s-lim}\ U_t^{L*} V_t \Omega_+^L = \Omega_+^{L*}\Omega_+^L = E^L$ and

$$\text{s-lim}\ U_t^{L*} F_{\{C\}} U_t^L U_t^{L*} V_t F_+^L = G^L_{\{\Delta\}} E^L \Omega_+^{L*} +$$

$$+ \text{s-lim}\ U_t^{L*} F_{\{C\}} U_t^L (I-E^L) U_t^{L*} V_t F_+^L \quad \text{as } t\to+\infty .$$

The last limit is zero, since by (14.14)

$$\|U_t^{L*} F_{\{C\}} U_t^L (I - E^L) U_t^{L*} V_t F_+^L f\|$$

$$\leq \|(I-E^L) U_t^{L*} V_t F_+^L f\| \to \|(I-E^L)\Omega_+^{L*} f\| = 0.$$

By using these results, Proposition 2.1 and the fact that the sum in (15.29) is finite, one sees that the last term of (15.29) converges as $t \to +\infty$ to $\sum_L \| G^L_{\{\Delta\}} E^L \Omega^{L*}_+ f - E^L G^L_{\{\Delta\}} \Omega^{L*}_+ f \|$, which is zero since E^L and $G^L_{\{\Delta\}}$ commute.

(ii) Let D and L be two partitions. Then

$$B^D_{C_1 \ldots C_n} F^L_+ - F^L_+ B^D_{C_1 \ldots C_n} = \sum_J \Omega^J_+ G^J_{\{\Delta\}} \Omega^{J*}_+ F^L_+ - \sum_J F^L_+ \Omega^J_+ G^J_{\{\Delta\}} \Omega^{J*}_+ ,$$

which is zero since $\Omega^{J*}_+ F^L_+ = \delta_{JL} \Omega^{L*}_+$ and $F^L_+ \Omega^J_+ = \delta_{JL} \Omega^L_+$ by (B2). Similarly one shows that $B^D_{C_1 \ldots C_n}$ commutes with $B^L_{C'_1 \ldots C'_m}$ (Problem 15.5). #

PROPOSITION 15.12 : Assume all the hypotheses of Proposition 15.11. Let $P_D(f^L; C_1, \ldots, C_n)$ be the probability that, if the scattering is initiated in the state $f^L \in M^L$, the final state will have clustering $D = \{(1), \ldots, (n)\}$ and each particle of the cluster (k) will be localized in the cone C_k as $t \to +\infty$ (k=1,...,n), i.e.

$$P_D(f^L; C_1, \ldots, C_n) \equiv \lim_{t \to +\infty} \| F^D_{C_1 \ldots C_n} V_t F^D_+ \Omega^L_- f^L \|^2 . \tag{15.30}$$

Then $P_D(f^L; C_1, \ldots, C_n)$ is just the probability that, in the part $S_{DL} f^L$ of the final state corresponding to the clustering D, the total momentum of each cluster (k) lies in the corresponding cone C_k, i.e.

$$P_D(f^L; C_1, \ldots, C_n) = \| G^{DD}_{C_1 \ldots C_n} S_{DL} f^L \|^2 . \tag{15.31}$$

Proof : (15.30), (15.28) and $F^D_+ E_\infty(H) = F^D_+$ imply that

$$P_D(f^L; C_1, \ldots, C_n) = \| \sum_J \Omega^J_+ G^{DJ}_{\Delta_1 \ldots \Delta_m} \Omega^{J*}_+ F^D_+ \Omega^L_- f^L \|^2 =$$

$$= \| \sum_J \delta_{JD} \Omega_+^J G^{DJ}_{\Delta_1 \ldots \Delta_m} S_{JL} f^L \|^2 = \| \Omega_+^D G^{DD}_{\Delta_1 \ldots \Delta_n} S_{DL} f^L \|^2 .$$

(15.31) follows from this since $G^{DD}_{\Delta_1 \ldots \Delta_n} = G^{DD}_{C_1 \ldots C_n}$ and since $g \equiv G^{DD}_{C_1 \ldots C_n} S_{DL} f^L \varepsilon M^D$, so that $\| \Omega_+^D g \| = \| g \|$. #

Except for (B3), the hypotheses of Propositions 15.11 and 15.12 have all been verified in the preceding two sections for pair potentials belonging to $L^2(R^3)$ (see Propositions 15.7, 15.9, 15.4 and Remark 15.5).

We now turn to the second topic of this section, namely cross sections. As in Chapter 7, we shall work in the center-of-mass coordinate system, i.e. the relevant Hilbert space is $H_{rel} = L^2(R^{3N-3})$ as introduced in Section 15-1. We shall use channels rather than clusterings to describe the cross sections, which somewhat simplifies the discussion. Since the mathematical steps are essentially the same as those for single-channel systems given in Chapter 7, we shall not repeat them but simply specify the necessary generalizations.

We restrict ourselves to the most common situation, namely that of scattering with two-body initial states. A remark about the general case can be found in Section 15-4D. Let L be a partition of $\{1,\ldots,N\}$ into two subsets and α a channel such that $\alpha \div L$. The states in $M^\alpha \subseteq H_{rel}$ are such that the particles of each cluster of L are in a fixed bound state. We denote by K^α the part of H^α in M^α. K^α represents, apart from the binding energy λ^α, the kinetic energy operator for the relative motion of the two clusters of L. More precisely, M^α may be identified with $L^2(R^3)$ in such a way that the momentum operator in $L^2(R^3)$ represents the relative momentum $\underline{P}^\alpha$ of the centers of mass of the two clusters of L. We then have for the operator K^α in $L^2(R^3)$

$$K^\alpha = \frac{1}{2\mu_\alpha} K_0 + \lambda^\alpha, \tag{15.32}$$

where μ_α is the reduced mass of the two clusters. The spectral representation of K^α is $G^\alpha = L^2([\lambda^\alpha,\infty), L^2(S^{(2)}))$, which is similar to that of K_0 given in (5.94). If $f = \{f_\lambda\} \in G^\alpha$, we shall write as before $f_\lambda(\underline{\omega})$ for the value of the function f_λ at the point $\underline{\omega} \in S^{(2)}$. In this representation the relative momentum operator is given by

$$(P_k^\alpha f)_\lambda(\underline{\omega}) = [2\mu_\alpha(\lambda-\lambda^\alpha)]^{\frac{1}{2}}\omega_k f_\lambda(\underline{\omega}), \quad k = 1,2,3. \tag{15.33}$$

Thus $\underline{\omega}$ denotes the direction and $[2\mu_\alpha(\lambda-\lambda_\alpha)]^{\frac{1}{2}}$ the absolute value of the relative momentum associated with the point $(\lambda,\underline{\omega})$, whereas $\lambda-\lambda^\alpha$ is the relative kinetic energy.

Similarly, to describe the final states, let D be an arbitrary partition of $\{1,\ldots,N\}$ into $n \geq 2$ clusters and let $\beta \div D$. M^β may be identified with $L^2(R^{3n-3})$, and the multiplication operators by the independent variables in $\tilde{L}^2(R^{3n-3})$ represent relative momenta between the n clusters. The part K^β of H^β in M^β is of the form $K^\beta = \sum_{k\ell} a_{k\ell} \underline{P}_k \cdot \underline{P}_\ell + \lambda^\beta$. The quadratic form $\sum a_{k\ell} \underline{p}_k \cdot \underline{p}_\ell$ is positive definite, and the equation $\sum_{k\ell} a_{k\ell} \underline{p}_k \cdot \underline{p}_\ell = c \; (c > 0)$ defines a (3n-4)-dimensional ellipsoid $E^{(3n-4)}$ embedded in R^{3n-3}. K^β is absolutely continuous (see the proof of Lemma 15.8), and its spectral representation is of the form $G^\beta = L^2([\lambda^\beta,\infty), L^2(E^{(3n-4)}))$, where $L^2(E^{(3n-4)})$ is defined with respect to a suitably chosen measure on $E^{(3n-4)}$, which we shall denote by $d\xi$. If

$g = \{g_\lambda\} \varepsilon G^\beta$, we write $g_\lambda(\underline{\xi})$ for the value of g_λ at the point $\underline{\xi} \varepsilon E^{(3n-4)}$. It is clear that the relative momenta can be expressed in terms of λ and $\underline{\xi}$, although for the present discussion this relation is not needed explicitly. An example will be given in Section 16-2.

In the present context we write H_0^α for $L^2(S^{(2)})$ and H_0^β for $L^2(E^{(3n-4)})$. We define in analogy with (7.38)

$$R_{\beta\alpha} = S_{\beta\alpha} - \delta_{\beta\alpha} E^\alpha, \tag{15.34}$$

i.e. $R_{\beta\alpha} = S_{\beta\alpha}$ if $\alpha \neq \beta$ and $R_{\alpha\alpha} = S_{\alpha\alpha} - E^\alpha$. The motivation for the subtraction of E^α in the elastic part is the same as that in Section 7-3. $R_{\beta\alpha}$ maps M^α into M^β and is zero on the orthogonal complement of M^α. It is convenient to view $R_{\beta\alpha}$ simply as an operator from the Hilbert space M^α to the Hilbert space M^β, which we shall do for the remainder of this section.

It follows from the intertwining relation $(14.19)_\alpha$ that for all $t \varepsilon R$

$$\exp(-iK^\beta t) R_{\beta\alpha} = R_{\beta\alpha} \exp(-iK^\alpha t), \tag{15.35}$$

i.e. $R_{\beta\alpha}$ intertwines K^α and K^β. This should be compared to the case of a single-channel system (Section 7-3) where we had only one free Hamiltonian K_0 and instead of (15.35) the relation $U_t R = RU_t$. In that case R was a decomposable operator in the spectral representation of K_0, i.e. $R = \{R(\lambda)\}$, where $R(\lambda)$ was an operator in $L^2(S^{(2)})$. The generalization of this result to the present case is the following : There exists a family $\{R_{\beta\alpha}(\lambda)\}$ of operators from H_0^α to H_0^β, defined for $\lambda\varepsilon\sigma(K^{\alpha})\cap\sigma(K^{\beta})$, such that for all $f = \{f_{\lambda}\}\varepsilon G^{\alpha}$

$$(R_{\beta\alpha}f)_{\lambda} = R_{\beta\alpha}(\lambda)f_{\lambda}. \tag{15.36}$$

The proof of this statement is a simple adaptation of that of Proposition 5.27 to the present situation (Problem 15.6).

The fact that $R_{\beta\alpha}(\lambda)$ is defined only for $\lambda\,\varepsilon\,\sigma(K^{\alpha})\cap\sigma(K^{\beta}) = [\lambda^{\alpha},\infty)\cap[\lambda^{\beta},\infty)$ corresponds to the following physical circumstances : If $\lambda^{\alpha} < \lambda^{\beta}$, then there is no scattering from channel α to channel β at energies λ lying between λ^{α} and λ^{β}; at such energies channel β is not open. If $\lambda^{\alpha} > \lambda^{\beta}$ and $\lambda\varepsilon(\lambda^{\beta},\lambda^{\alpha})$, then there are no initial states in channel α having energy λ.

As in (7.42) we make the hypothesis that for almost all λ and each channel β, $R_{\beta\alpha}(\lambda)$ is a Hilbert-Schmidt operator. If this is the case, there exist functions $R_{\beta\alpha}(\lambda;\underline{\xi},\underline{\omega})$ such that for almost all λ and for any $f\varepsilon G^{\alpha}$

$$\|R_{\beta\alpha}(\lambda)\|^2_{HS} = \int d\xi d\omega\,|R_{\beta\alpha}(\lambda;\underline{\xi},\underline{\omega})|^2 < \infty, \tag{15.37}$$

and

$$(R_{\beta\alpha}f)_{\lambda}(\underline{\xi}) = \int d\omega R_{\beta\alpha}(\lambda;\underline{\xi},\underline{\omega})f_{\lambda}(\underline{\omega}). \tag{15.38}$$

Let Δ be a subset of the relative momentum space R^{3n-3} of the clusters of β, and $f^{\alpha}\varepsilon M^{\alpha}$ an initial state such that $\tilde{f}^{\alpha}(\underline{k})$ has support in a small cone C_0. In analogy with (7.39) and in view of (15.31), we take the following expression for the probability that the associated final state will be in channel β with the relative momentum of the clusters lying in Δ :

$$P_{\beta}(f^{\alpha},\Delta) = \int_{\Delta} d\lambda\; d\xi\,|(R_{\beta\alpha}f^{\alpha})_{\lambda}(\underline{\xi})|^2. \tag{15.39}$$

Here, if $\beta = \alpha$, Δ has to be such that it does not intersect C_0. As in Section 7-3, one translates f^α in the impact parameter plane and defines

$$\sigma_\beta(f^\alpha,\Delta) = \int d^2a\, P_\beta(f^\alpha_{\underline{a}},\Delta) = \int d^2a \int_\Delta d\lambda\, d\xi |(R_{\beta\alpha} f^\alpha_{\underline{a}})_\lambda(\underline{\xi})|^2,$$

which, under suitable assumptions on $\tilde{f}^\alpha$, may be transformed into (cf. (7.47))

$$\sigma_\beta(f^\alpha,\Delta) = (2\pi)^2 \int_\Delta d\xi\, d\lambda \int d\omega |R_{\beta\alpha}(\lambda;\underline{\xi},\underline{\omega})|^2 |f^\alpha_\lambda(\underline{\omega})|^2 (k^2 \cos\theta')^{-1}$$

with $k^2 = 2\mu_\alpha(\lambda-\lambda^\alpha)$. This leads to the following expression for the <u>differential scattering cross section</u> from channel α to channel β :

$$d\sigma_{\alpha\to\beta}/d\xi(\lambda;\underline{\omega}\to\underline{\xi}) = |f_{\alpha\to\beta}(\lambda;\underline{\omega}\to\underline{\xi})|^2, \tag{15.40}$$

with $$f_{\alpha\to\beta}(\lambda;\underline{\omega}\to\underline{\xi}) = -2\pi i[2\mu_\alpha(\lambda-\lambda^\alpha)]^{-\frac{1}{2}} R_{\beta\alpha}(\lambda;\underline{\xi},\underline{\omega}). \tag{15.41}$$

The interpretation of these equations is as in Proposition 7.18 and will not be repeated. The function $f_{\alpha\to\beta}$ is called the <u>scattering amplitude</u> for the ordered pair of channels $\{\alpha,\beta\}$.

We can now define the elastic cross section σ^α_{el}, the inelastic cross section σ^α_{inel}, the rearrangement cross section σ^α_{rearr} and the total cross section σ^α_{tot} for scattering initiated in channel α ($\alpha \div L$) at relative energy λ and relative initial direction $\underline{\omega}$ by

$$\sigma^\alpha_{el}(\lambda;\underline{\omega}) = \int d\omega'\, d\sigma_{\alpha\to\alpha}/d\omega(\lambda;\underline{\omega}\to\underline{\omega}') \tag{15.42}$$

$$\sigma^\alpha_{inel}(\lambda;\underline{\omega}) = \sum_{\alpha\neq\beta\div L} \int d\omega'\, d\sigma_{\alpha\to\beta}/d\omega(\lambda;\underline{\omega}\to\underline{\omega}') \tag{15.43}$$

$$\sigma^\alpha_{rearr}(\lambda;\underline{\omega}) = \sum_{D\neq L,\, \beta\div D} \int d\xi\, d\sigma_{\alpha\to\beta}/d\xi(\lambda;\underline{\omega}\to\underline{\xi}) \tag{15.44}$$

$$\sigma^\alpha_{tot}(\lambda;\underline{\omega}) = \sum_\gamma \int d\xi\, d\sigma_{\alpha\to\gamma}/d\xi(\lambda;\underline{\omega}\to\underline{\xi}), \tag{15.45}$$

where the last sum is over all channels. Clearly $\sigma^{\alpha}_{tot} = \sigma^{\alpha}_{el} + \sigma^{\alpha}_{inel} + \sigma^{\alpha}_{rearr}$ *).

By averaging these cross sections over all initial directions $\underline{\omega}$ as in (7.68), one may express them in terms of the Hilbert-Schmidt norm of $R_{\beta\alpha}$ as in (7.69). For example

$$\bar{\sigma}_{\alpha}(\lambda) \equiv \frac{1}{4\pi}\int d\omega\, \sigma^{\alpha}_{tot}(\lambda;\underline{\omega}) = \pi\, [2\mu_{\alpha}(\lambda-\lambda^{\alpha})]^{-1} \sum_{\gamma} \| R_{\gamma\alpha}(\lambda)\|^{2}_{HS}. \quad (15.46)$$

Thus if, for a given $\lambda > \lambda^{\alpha}$, each $R_{\gamma\alpha}(\lambda)$ is a Hilbert-Schmidt operator, and if the total number of channels is finite, then $\bar{\sigma}_{\alpha}(\lambda) < \infty$. More about the properties of $R_{\beta\alpha}(\lambda)$ and the number of channels will be said in Chapter 16.

The branching ratio for scattering from channel α to channel β is defined as

$$b_{\alpha\to\beta}(\lambda;\underline{\omega}) = \sigma_{\alpha\to\beta}(\lambda;\underline{\omega})\,[\sigma^{\alpha}_{tot}(\lambda,\underline{\omega})]^{-1}$$

and gives the probability for scattering into channel β at energy λ and initial direction $\underline{\omega}$. [$\sigma_{\alpha\to\beta}(\lambda;\underline{\omega})$ is the contribution from channel β to the sum in (15.45)]. The optical theorem for multichannel scattering follows easily from (14.21) (Problem 15.7) :

$$\sigma^{\alpha}_{tot}(\lambda;\underline{\omega}) = 4\pi[2\mu_{\alpha}(\lambda-\lambda^{\alpha})]^{-\frac{1}{2}} \mathrm{Im}\, f_{\alpha\to\alpha}(\lambda;\underline{\omega}\to\underline{\omega}). \quad (15.47)$$

It relates the imaginary part of the elastic scattering amplitude in the forward direction to the total cross section for all possible final channels and is valid if all scattering amplitudes are continuous functions of their arguments.

---

*) In the literature one often defines the inelastic cross section as $\sigma^{\alpha}_{tot} - \sigma^{\alpha}_{el}$.

15-4 NOTES AND SUPPLEMENTARY MATERIAL

A. We first give a lemma cited in Section 15-1.

LEMMA 15.13 : Let $\mathcal{H}_1$, $\mathcal{H}_2$ be Hilbert spaces with dim $\mathcal{H}_2 = \infty$. Let B be a compact operator in $\mathcal{H}_1$ and let I be the identity operator in $\mathcal{H}_2$. Then $B \otimes I$ is not compact in $\mathcal{H}_1 \otimes \mathcal{H}_2$ unless $B = 0$.

Proof : Let $\{e_k\}_{k=1}^{\infty}$ be an orthonormal basis of $\mathcal{H}_2$ and $f \in \mathcal{H}_1$. Define $g_k = f \otimes e_k$. Since $(g_k, g_\ell) = \delta_{k\ell} \|f\|^2$, the infinite sequence $\{g_k\}$ converges weakly to zero in $\mathcal{H}_1 \otimes \mathcal{H}_2$ (cf. the remark following eq. (2.11)). Let $h_k = (B \otimes I)g_k = Bf \otimes e_k$. We have $\|h_k\| = \|Bf\| \, \|e_k\| = \|Bf\|$. Assume $B \otimes I$ to be compact. Then $\|h_k\| \to 0$ as $k \to \infty$ by Proposition 2.23. Hence $\|Bf\| = 0$. Since this holds for each $f \in \mathcal{H}_1$, we have $B = 0$. #

B. We now discuss some spectral properties of multiparticle Hamiltonians. We begin with a theorem about their essential spectrum which was used to prove Proposition 15.6.

PROPOSITION 15.14 : Let H_{rel} be the relative Hamiltonian for an N-particle system for which each pair interaction $V^o_{k\ell}$ is K_o-compact in $L^2(\mathbb{R}^3)$. Then

$$\sigma_e(H_{rel}) \subseteq \bigcup_{D \neq D_1} \sigma_e(H^D_{rel}). \tag{15.48}$$

Proof : Throughout this proof we work in $\mathcal{H}_{rel} = L^2(\mathbb{R}^{3N-3})$. To simplify the notation, we shall write H for H_{rel} and H^D for H^D_{rel}.

(i) Let $\lambda\varepsilon\sigma_e(H)$. By Lemma 5.19 there exists a sequence $\{f_n\}_{n=1}^{\infty}\varepsilon H_{rel}$ such that $\|f_n\| = 1$, w-lim $f_n = \theta$ and lim $\|(H-\lambda)f_n\| = 0$ as $n\to\infty$. We have to show that there exists a partition $D \neq D_1$ such that $\lambda\varepsilon\sigma_e(H^D)$, i.e. by Lemma 5.19 to construct a sequence $\{g_s\}\varepsilon H_{rel}$ verifying $\|g_s\| = 1$, w-lim $g_s = \theta$ and $\|(H^D-\lambda)g_s\| \to 0$ as $s\to\infty$ for some $D \neq D_1$.

Let Ξ be the set of all partitions of $\{1,\ldots,N\}$ into at least two subsets and Z^+ the set of positive integers. If ϕ is a map from R^{3N-3} to R and $r\varepsilon Z^+$, we denote by ϕ_r the multiplication operator in $L^2(R^{3N-3})$ by the function $\phi_r(\underline{y}_1,\ldots,\underline{y}_{N-1}) = \phi(r^{-1}\underline{y}_1,\ldots,r^{-1}\underline{y}_{N-1})$, $\{\underline{y}_k\}$ being some relative coordinates. Also, let $F_r^{k\ell}$ be the projection whose range is the subspace of states in which the relative distance between the pair of particles $\{k,\ell\}$ is less than r, i.e. the multiplication operator by the characteristic function of the set $\{(\underline{y}_1,\ldots,\underline{y}_{N-1})\,\big|\,|\underline{x}_k - \underline{x}_\ell| \leq r\}$ in H_{rel}. We shall give in (vi) below for each $C\,\varepsilon\,\Xi$ a function $\phi^C\varepsilon C^\infty(R^{3N-3})$ such that (a) $|\phi^C(\underline{y}_1,\ldots,\underline{y}_{N-1})| \leq 1$, (b) the function $\phi^0 \equiv 1-\sum_{C\varepsilon\Xi}\phi^C$ belongs to $C_0^\infty(R^{3N-3})$ and (c) $\phi_r^C = (I-F_r^{k\ell})\phi_r^C$ for all pairs $\{k,\ell\}$ not belonging to the same cluster in C and all r. For the moment we assume that such functions exist and proceed with the construction of the sequence $\{g_s\}$.

(ii) If A is an H-bounded operator, we have

$$\|Af_n\| = \|A(H+i)^{-1}[(H-\lambda)f_n + (\lambda+i)f_n]\| \tag{15.49}$$

$$\leq \|A(H+i)^{-1}\|\,\|(H-\lambda)f_n\| + |\lambda+i|\,\|A(H+i)^{-1}f_n\|. \tag{15.50}$$

The first term on the right-hand side converges to zero as

$n \to \infty$ by the definition of $\{f_n\}$. If moreover A is H-compact, the second term also converges to zero by Proposition 2.23. In particular, since $\phi^o \varepsilon C_o^\infty(R^{3N-3})$, ϕ_r^o is H-compact (see the proof of Proposition 15.4), so that s-lim $\phi_r^o f_n = \theta$ as $n \to \infty$, for each r.

(iii) It follows that for each r, there exists a number N(r) such that $\|\phi_r^o f_n\| < \frac{1}{2}$ for all $n > N(r)$. We choose a sequence $\{n_r\}_{r=1}^\infty$ such that $n_r > \max\{N(r), n_{r-1}\}$. Clearly $n_r \to \infty$ as $r \to \infty$. Next, let δ be such that $0 < \delta^{-1} < (2c_N)^{-1}$, where c_N denotes the number of elements of Ξ. Since

$$1 = \|f_{n_r}\| = \|(\phi_r^o + \sum_{C\varepsilon\Xi}\phi_r^C)f_{n_r}\| \le \|\phi_r^o f_{n_r}\| + \sum_{C\varepsilon\Xi}\|\phi_r^C f_{n_r}\|$$

and $\|\phi_r^o f_{n_r}\| < \frac{1}{2}$, there exists for each r at least one partition C(r) such that $\|\phi_r^{C(r)} f_{n_r}\| > \delta^{-1}$. We fix one such partition C(r) for each r.

The correspondence $r \mapsto C(r)$ defines a map Ψ with domain Z^+ and range in Ξ. Since Ξ is a finite set, there exists a partition $D \varepsilon \Xi$ such that $\Psi^{-1}(D)$ is an infinite subset of Z^+. In what follows, s varies over $\Psi^{-1}(D)$. We define $\{g_s\}$, $s\varepsilon\Psi^{-1}(D)$, by $g_s = \phi_s^D f_{n_s}/\|\phi_s^D f_{n_s}\|$. Clearly $\|g_s\| = 1$ and $\|\phi_s^D f_{n_s}\|^{-1} < \delta$ for all s. Also, if $\{k,\ell\}$ is a pair of particles not belonging to the same cluster in D, then

$$|(h,g_s)| \le \delta|(\phi_s^{D*}h, f_{n_s})| \le \delta\|\phi_s^{D*}h\| \le \delta\,\|\phi_s^D\|\,\|(I-F_s^{k\ell})h\|,$$

which converges to zero as $s \to \infty$ since $\|\phi_s^D\| \le 1$ and s-lim $F_s^{k\ell} = I$ as $s \to \infty$. Hence w-lim $g_s = \theta$ as $s \to \infty$.

(iv) We next show that $\|(H-\lambda)g_s\| \to 0$ as $s\to\infty$. For this, we use (15.13) and (7.86) to get

$$\|(H-\lambda)g_s\| \leq \delta\,\|(H-\lambda)\phi_s^D f_{n_s}\| \leq \delta\,\|\phi_s^D(H-\lambda)f_{n_s}\| +$$

$$\sum_k \gamma_k\,\delta\|(\Delta_k\phi_s^D)f_{n_s}\| + 2\delta\sum_{k=1}^{N-1}\gamma_k\,\|(\underline{\nabla}_k\phi_s^D)\cdot\,\underline{P}_k f_{n_s}\|.$$

It suffices to verify that each member of the last term tends to zero as $s\to\infty$. For the first one this follows from the properties of $\{f_{n_s}\}$. For the second one we use the fact that $(\Delta_k\phi_s^D)(\underline{y}_1,\ldots,\underline{y}_{N-1}) = s^{-2}(\Delta_k\phi^D)(s^{-1}\underline{y}_1,\ldots,s^{-1}\underline{y}_{N-1})$, hence that $\|(\Delta_k\phi_s^D)f_{n_s}\| \leq s^{-2}\,\|\Delta_k\phi^D\|_\infty\|f_{n_s}\|$ by Proposition 2.16. Similarly the third term is majorized by $2\delta s^{-1}\sum_{k=1}^{N-1}\sum_{\ell=1}^{3}\,\|\partial\phi^D/\partial y_{k,\ell}\|_\infty\|P_{k,\ell}f_{n_s}\|$, which is seen to converge to zero as $s\to\infty$ since $P_{k,\ell}$ is H-bounded, so that $\|P_{k,\ell}f_{n_s}\|$ is uniformly bounded in s by (15.49).

(v) It remains to show that $\|(H^D-\lambda)g_s\| \to 0$ as $s\to\infty$. For this we write

$$\|(H^D-\lambda)g_s\| \leq \|(H-\lambda)g_s\| + \|V^D g_s\|$$

$$\leq \|(H-\lambda)g_s\| + \delta\|V^D\phi_s^D f_{n_s}\|. \tag{15.51}$$

The first term in (15.51) converges to zero by (iv). The second term is a finite sum of numbers of the form $\|V_{k\ell}\phi_s^D f_{n_s}\|$, with k and ℓ belonging to different clusters in D. For each of them we have

$$\|V_{k\ell}\phi_s^D f_{n_s}\| \leq \|\phi^D\|_\infty\|(I-F_s^{k\ell})V_{k\ell}(H+i)^{-1}\|\,\|(H+i)f_{n_s}\|.$$

Now $\|(H+i)f_{n_s}\| \leq \|(H-\lambda)f_{n_s}\| + |\lambda+i|\;\|f_{n_s}\|$, which is bounded uniformly in s. Also, by setting $\mathcal{H}_{rel} = \mathcal{H}_{rel}^{[k,\ell]} \otimes L^2(\mathbb{R}^{3N-6})$

similarly to (15.7), one may write

$$(I-F_s^{k\ell})V_{k\ell}(H+i)^{-1} = (I-F_s^{k\ell})V_{k\ell}(H_o+\gamma)^{-1}(H_o+\gamma)(H+i)^{-1}$$
$$= (I-F_s^{k\ell})V_{k\ell}A_{k\ell}(i\gamma)B,$$

where $B\varepsilon\mathcal{B}(\mathcal{H})$ and $A_{k\ell}$ is defined as in the proof of Proposition 15.1. One has as in (15.8) that

$$(I-F_s^{k\ell})V_{k\ell}A_{k\ell}(i\gamma) = \{(I-F_s^o)V_{k\ell}^o[(2m_{k\ell})^{-1}K_o - i\gamma]^{-1}\} \otimes I,$$

where F_s^o is the multiplication operator by the characteristic function of the set $\{\underline{x} \,|\, |\underline{x}| \leq s\}$ in $\mathcal{H}_{rel}^{[k,\ell]} = L^2(\mathcal{R}^3)$. Since s-lim $(I-F_s^o) = 0$ as $s\to\infty$ and $V_{k\ell}^o[(2m_{k\ell})^{-1}K_o - i\gamma]^{-1}\varepsilon\mathcal{B}_o$ by hypothesis, $\|(I-F_s^{k\ell})V_{k\ell}A_{k\ell}(i\gamma)\| \to 0$ as $s\to\infty$ by Lemma 8.23, showing that the second term in (15.50) converges to zero as $s\to\infty$.

(vi) Finally we construct the functions ϕ^C as announced in (i). Let $\psi\varepsilon C^\infty(0,\infty)$ be such that $|\psi(t)| \leq 1$, $\psi(t) = 0$ for $0 \leq t \leq 1$ and $\psi(t) = 1$ for $t \geq 2$. For $C \varepsilon \Xi$, define $\rho^C : \mathcal{R}^{3N-3} \to \mathcal{R}$ by $\rho^C(\underline{y}_1,\ldots,\underline{y}_{N-1}) = \Pi_{k,\ell}\psi(|\underline{x}_k - x_\ell|)$, the product being over all pairs $\{k,\ell\}$ not belonging to the same cluster in C. (States in the range of ρ_r^C are such that the relative distance between any pair $\{k,\ell\}$ not belonging to the same cluster is larger than r). Let $\{L_k\}_{k=1}^{c_N}$ be an enumeration of the elements of Ξ. Define $\phi^{L_1} = \rho^{L_1}$, $\phi^{L_k} = \rho^{L_k}\Pi_{i=1}^{k-1}(1-\rho^{L_i})$ for $2 \leq k \leq c_N$. It is a matter of simple algebra to verify that $\phi^o \equiv 1 - \sum_k \phi^{L_k} = \Pi_{i=1}^{c_N}(1-\rho^{L_i}) \varepsilon C_o(\mathcal{R}^{3N-3})$, and the other conditions imposed on ϕ^C are also easily checked. #

Hunziker's theorem (Proposition 15.6) can be proved under weaker assumptions on the interaction and without using the existence of the wave operators. In particular, under the hypotheses of Proposition 15.14, the inclusion in (15.48) is indeed an equality. The proof of the reverse inclusion is usually based on a suitable resolvent equation (such equations are discussed in Chapter 16). Some relevant references are Hunziker [3], Combes [1] , Balslev [1], [SM], [JW] and Enss [1].

If in addition the cluster wave operators exist, then $\sigma_{ac}(H_{rel}) = [\lambda_0,\infty)$. The absence of the singularly continuous spectrum has been established for so-called dilatation-analytic potentials and arbitrary N by Balslev and Combes [1] and Balslev [1], as well as by Faddeev [F] for certain 3-body systems. The spectrum below λ_0 is discrete. The situation here is similar to the 2-body case : If each pair potential tends to zero at infinity faster then $|x|^{-2-\eta}$ for some $\eta > 0$, the number of eigenvalues of H_{rel} below λ_0 is finite, otherwise it may be infinite. We refer to the papers by Sigal [1] for results and further references. If the two-body potentials do not oscillate too rapidly, H_{rel} has no positive eigenvalues. On the other hand eigenvalues of H_{rel} may occur in $[\lambda_0,0]$, often for reasons of symmetry. For the question of eigenvalues embedded in $\sigma_e(H_{rel})$, the reader is referred to Albeverio [1]. See also Problem 15.4.

The restriction to 2-body potentials has been made in this chapter only for convenience of presentation and because these are the most commonly used interactions. The various aspects of scattering theory can also be developped for k-body interactions with $k \geq 2$; see e.g. Nelson [1], Combes [1].

C. The problem of asymptotic completeness for multiparticle systems (either in the form (B3) or in the form $\sum_{D\in\Xi} F_{\pm}^{D} = I \otimes E_{ac}(H_{rel})$ in $\mathcal{H}_{CM} \otimes \mathcal{H}_{rel}$; cf. the beginning of Section 9-1) has been approached on three different levels. (a) If all pair potentials are repulsive, no subsystem will have bound states. Hence there will be only one scattering channel, corresponding to the partition D_N, in which each particle forms a cluster by itself. The same is true for non-repulsive potentials, provided that they tend to zero sufficiently rapidly at infinity, if the coupling constant is very small. In both cases the wave operators $\Omega_{\pm}$ for the only scattering channel will be unitary (see also Proposition 9.4), implying also that H_{rel} is absolutely continuous. Such a scattering system is formally a simple scattering system. Precise statements and additional references can be found in the paper by Ferrero, de Pazzis and Robinson [1].

(b) For the 3-body problem, strong asymptotic completeness was first obtained by Faddeev [F]. His proof involves an analysis of the relevant operators in a Banach space of Hölder continuous functions of the momenta and accordingly the conditions imposed on the pair potentials are sufficiently rapid decrease at infinity and Hölder continuity of their Fourier transforms. In recent years the three-body problem was also treated by stationary methods in configuration space under certain integrability or decrease assumptions on the pair potential functions $V_{k\ell}(\underline{x})$ (see Ginibre and Moulin [1], Thomas [3], Newton [1], Howland [1] and Mourre [2]). This method is more in the line of our Chapters 9 and 10. We shall indicate in Chapter 16 how the necessary estimates on the resolvent of H_{rel} near the real axis may be obtained.

(c) The method of Faddeev has been extended to the N-body problem ($N > 3$) by Hepp [1] and Sigal [2]. These proofs involve hypotheses on spectral properties of the relative Hamiltonians of subsystems consisting of less than N particles, similar to the hypotheses (H1)-(H4) made in Section 16-1. A verification of these hypotheses has recently been announced by Sigal [3] for dilatation-analytic potentials decreasing at infinity faster than $|x|^{-2}$, for all values of the coupling constant except possibly a discrete set.

Finally we should add that the division of the scattering states of H into subsets leading asymptotically to the different clusterings (see Section 14-2) may be defined without resorting to the cluster wave operators by using the relative position operators. Such a definition in terms of the particle positions is more general and very close to the physical picture described in Section 14-1 and can be obtained by adapting the definitions of Sections 7-1 and 15-1 to express mathematically the notions of "bound together" and "far separated at large times"; see Grawert and Petzold [1], Deift and Simon [2].

For questions of multiparticle scattering systems with long range potentials, the reader may consult

Dollard [1,4], Chandler and Gibson [2,3] and the article by Amrein, Georgescu and Martin in [EN].

D. We now make some comments about cluster properties of the wave operators which are the multichannel analogue of Corollary 8.17. Let $D = \{(1),\ldots,(n)\}$ be a partition and let $U_D(\underline{a}_1,\ldots,\underline{a}_n)$ be the unitary operator describing the translation of the clusters $(1),\ldots,(n)$ by $\underline{a}_1,\ldots,\underline{a}_n$ respectively (i.e. each particle in cluster (k) is translated by $\underline{a}_k$). U_D is given by

$$U_D(\underset{\sim}{a}_D) \equiv U_D(\underline{a}_1,\ldots,\underline{a}_n) = \exp(i\textstyle\sum_{k=1}^{n}\underline{a}_k\cdot\underline{P}_{tot}^{(k)}), \qquad (15.52)$$

where we have written $\underset{\sim}{a}_D$ for $\{\underline{a}_1,\ldots,\underline{a}_n\}$. We also set $d_D = \min_{j\neq k}|\underline{a}_j - \underline{a}_k|$.

A partition C is said to be finer than D, written $C \underset{\sim}{<} D$, if it can be obtained from D by further subdividing the clusters in D. Let $C \underset{\sim}{<} D$ and fix a state vector $f^C \in M^C$. Consider the scattering for the initial state $U_D(\underset{\sim}{a}_D)f^C$ when d_D is very large. The potentials between different clusters in D (hence between certain groups of clusters in C) will then be ineffective, and in the limit where $d_D \to \infty$, one should obtain the same scattering theory as that in which all intercluster potentials in the partition D have been set equal to zero (i.e. $V^D = 0$). Thus the particles in each cluster of D will scatter among themselves independently of those in the other clusters, so that the scattering theory will be entirely determined by that of the individual clusters of D. In the next proposition we indicate a proof of this fact and refer to Hunziker [2] for further results. We use the notation $\Omega_\pm(A,H^C) = \text{s-lim}\exp(iAt)U_t^C E^C$ as $t \to \pm\infty$, with $A = A^*$.

PROPOSITION 15.15 : Let $D \neq D_1$ be a partition and $C \lesssim D$. Assume the hypotheses of Proposition 15.9. Then

(a) $$\underset{d_D\to\infty}{\text{s-lim}}\ U_D(\underset{\sim}{a}_D)^* V_t U_D(\underset{\sim}{a}_D) = U_t^D, \tag{15.53}$$

(b) $$\underset{d_D\to\infty}{\text{s-lim}}\ U_D(\underset{\sim}{a}_D)^* \Omega_\pm(H,H^C) U_D(\underset{\sim}{a}_D) = \Omega_\pm(H^D,H^C). \tag{15.54}$$

(a) means that in the limit as $d_D \to \infty$, the dynamics of the system are entirely determined by H^D. A relation like (15.53) could be used in certain situations to define the cluster evolution groups U_t^D from H and the translation groups of the individual particles. It can be shown that (15.53) holds uniformly in t, $-\infty < t < +\infty$ (see Hunziker [2] or Problem 8.10). The decoupling of the different clusters of D in (15.54) becomes more transparent if we write $\Omega_\pm(H^D,H^C)$ in the representation (14.1) of $\mathcal{H}$ with respect to D :

$$\Omega_\pm(H^D,H^C) = \Omega_\pm^{(1)} \otimes \cdots \otimes \Omega_\pm^{(n)}$$

where $\Omega_\pm^{(k)}$ is the wave operator for the subsystem (k) in $\mathcal{H}^{(k)}$ between $H^{(k)}$ and $H_C^{(k)} = H_o^{(k)} + \sum_{i\ell} V_{i\ell}$, where both i and ℓ are in (k) and belong to the same cluster in C.

Proof : Represent $\mathcal{H}$ with respect to $C = \{(1),\ldots,(m)\}$ in the form (15.25). E^C is of the form $E^C = I \otimes E_o^{(1)} \otimes \cdots \otimes I \otimes E_o^{(m)}$, whereas U_D is a function of the total mementa of the clusters in C since $C \lesssim D$, i.e. $U_D = U_D^{(1)} \otimes I \otimes \cdots \otimes U_D^{(m)} \otimes I$. Hence $E^C U_D = U_D E^C$. (15.54) can be deduced from this equation and Proposition 8.16, and (15.53) is then contained in part (iii) of the proof of that proposition.

To apply Proposition 8.16, one makes the following substitutions : $H_0 \mathrel{\hat{=}} H^C$, $H \mathrel{\hat{=}} H^D$, $V_0 \mathrel{\hat{=}} V^C - V^D$, $E_\infty(H_0) \mathrel{\hat{=}} E^C$,

$V_n \mathrel{\hat{=}} U_D(\underset{\sim}{a}_D)^*V^C U_D(\underset{\sim}{a}_D) \equiv V(\underset{\sim}{a}_D)$. Notice that $V(\underset{\sim}{a}_D) = V_0 + U_D(\underset{\sim}{a}_D)^*V^D U(\underset{\sim}{a}_D)$. The hypotheses (a) and (b) of Proposition 8.16 are verified as a consequence of Proposition 15.1, and (d) holds by virtue of the proof of Proposition 15.7. (c) is established by combining the arguments of the proof of Proposition 15.1 with those of Corollary 8.17 (Problem 15.8). #

These cluster properties disclose some of the difficulties that one encounters when considering cross sections for n-body initial channels with $n > 2$. One can still define a cross section in the usual way by displacing the initial state in a suitably chosen $(3n-4)$-dimensional hyperplane (a generalized impact parameter plane) and integrating over all such displacements. However, if the displacement is very large, it may happen that some of the n clusters, say $(1),\ldots,(\ell)$, get scattered among themselves independently of the remaining ones, and similarly for $(\ell+1),\ldots,(n)$. These disconnected processes will conserve the total momentum $\underset{\sim}{p}_{tot}^{(1)\ldots(\ell)}$ of $\{(1),\ldots,(\ell)\}$ as well as $\underset{\sim}{p}_{tot}^{(\ell+1)\ldots(n)}$, i.e. the scattering amplitude will include a term that contains corresponding δ-functions. To arrive at a finite cross section, one will therefore have to exclude in the final channels those values of the relative momenta for which such disconnected processes are permitted by the energy and momentum conservation laws. (This is the analogue of excluding the forward direction in the case of two particles, see Section 7-3.) More about these questions may be found in the lecture notes of Hunziker in [BB] and in Newton and Shtokhamer [1].

E. Proof of Lemma 15.10 : $F^D_{C_1\ldots C_n}$ is the multiplication operator by $\chi^D_{C_1\ldots C_n}$ in $L^2(R^{3N})$, where $\chi^D_{C_1\ldots C_n}(\underline{x}_1,\ldots,\underline{x}_N) = 1$ if $\underline{x}_i$ belongs to the cone C_k such that $i\varepsilon(k)$, for each $i = 1,\ldots,N$, and $\chi^D_{C_1\ldots C_n} = 0$ otherwise. Thus each operator appearing in equation (15.27) factorizes in the representation (14.1) of H with respect to the partition L (i.e. $H = H^{(1)'} \otimes \ldots \otimes H^{(m)'}$). Hence by Problem 2.35, (15.27) holds provided that we can prove convergence for each factor.

If (k)' consists of only one particle, say particle ℓ, then $\Delta_k = C_j$ with $\ell\varepsilon(j)$ and $U_t^{(k)'} = \exp(-iK_0 t/2m_\ell)$. Convergence of the factor acting in $H^{(k)'}$ then follows from (3.51). If (k)' consists of $\nu \geq 2$ particles, the problem reduces to the following one. Let $C_1,\ldots,C_\nu$ be closed cones in R^3, $\Delta = \cap_{i=1}^{\nu} C_i$, $H^o = L^2(R^{3\nu}) = H^o_{CM} \otimes H^o_{rel}$, $U_t = U_{t,CM} \otimes U_{t,rel}$ with $U_{t,CM} = \exp(-iK_0 t/2m)$, $m > 0$, $G^o_\Delta = G_\Delta \otimes I$ with G_Δ given by (7.31), $F_{C_1\ldots C_\nu}$ as above and $E_o = I \otimes E_o(H_{rel})$.*) Show that s-lim $U_t^* F_{C_1\ldots C_\nu} U_t E_o = G^o_\Delta E_o$ as $t \to +\infty$.

Since all appearing operators have norm 1, it suffices (as in Proposition 2.17) to prove convergence on the fundamental set of vectors in $E_o H$ of the form $g = f \otimes e_k$, where $f\varepsilon L^2(R^3)$ and $\{e_k\}$ forms an orthonormal basis of $M_o(H_{rel})$. Since by hypothesis $M_o(H_{rel}) = H_p(H_{rel})$, we may assume that each e_k is an eigenvector of H_{rel}. We then have, with the notation $F_{\{C\}} = F_{C_1\ldots C_\nu}$,

---

*) $E_o(H)$ is the projection whose range is $M_o(H)$.

$$\|U_t^* F_{\{C\}} U_t g - G_\Delta^o g\| = \|(F_{\{C\}} - G_\Delta^o)(U_{t,CM} f \otimes e_k)\|$$

$$\leq \|(F_{\{C\}} - F_\Delta^o)(U_{t,CM} f \otimes e_k)\| + \|(F_\Delta^o - G_\Delta^o)(U_{t,CM} f \otimes e_k)\|, \quad (15.55)$$

where $F_\Delta^o = F_\Delta \otimes I$, F_Δ being the multiplication operator in $L^2(R^3)$ by χ_Δ. The second member of the last term of (15.55) is equal to $\|(F_\Delta U_{t,CM} - U_{t,CM} G_\Delta)f\|$, which by Proposition 3.17 has the same limit as $\|(F_\Delta C_{t/2m} - C_{t/2m} G_\Delta)f\|$ for $t \to +\infty$. Now it follows from (3.46) that $F_\Delta C_\tau = C_\tau G_\Delta$ if $\tau > 0$, so that the second term in (15.55) converges to zero as $t \to +\infty$.

It remains to prove the convergence to zero of the first term in (15.55). For this, we replace as above $U_{t,CM}$ by $C_{t/2m}$. Using the variables $\{\underline{x}_{CM}, \underline{y}_1, \ldots, \underline{y}_{\nu-1}\}$ and setting $\underline{z} = m\underline{x}_{CM}/t$, one obtains

$$\|(F_{\{C\}} - F_\Delta^o)(C_{t/2m} f \otimes e_k)\|^2 =$$

$$\int d^3z\, d^{3\nu-3}y\, |\tilde{f}(\underline{z})|^2 |e_k(y)|^2 [\chi_{C_1 \ldots C_\nu}(\underline{x}_1, \ldots, \underline{x}_\nu) - \chi_\Delta(\underline{x}_{CM})].$$

Clearly $\underline{x}_{CM} \in \Delta$ iff $\underline{z} \in \Delta$, so that $\chi_\Delta(\underline{x}_{CM}) = \chi_\Delta(\underline{z})$. Also $\underline{x}_i = \alpha_i \underline{x}_{CM} + \sum_{k=1}^{\nu-1} \beta_{ik} \underline{y}_k = m^{-1} \alpha_i t\, \underline{z} + \sum_{k=1}^{\nu-1} \beta_{ik} \underline{y}_k$ with $\alpha_i > 0$ (the t-dependence of the integrand lies in $\chi_{C_1 \ldots C_\nu}$ in this implicit way). Fixing $\underline{y}_k \in R^3$ and $\underline{z} \neq 0$, we have : (i) if $\underline{z}$ is in the interior of Δ, then $\underline{x}_i \in \Delta \subseteq C_i$ for all i if t is large enough, (ii) if $\underline{z} \notin \Delta$, then $\underline{z} \notin C_j$ for some j, hence $\underline{x}_j \notin C_j$ for that j and all $t > t_0$. It follows that, as $t \to +\infty$, the integrand above converges pointwise to zero for almost all $\{\underline{z}, y\}$. Since it is bounded uniformly in t by $2|\tilde{f}(\underline{z})|^2 |e_k(y)|^2$, the integral converges to zero as $t \to +\infty$ by the Lebesgue dominated convergence theorem. #

From the preceding proof one sees that s-lim $U_t^{\alpha *} F^D_{C_1 \ldots C_n} U_t^\alpha$ as $t \to +\infty$ exists on the entire Hilbert space and equals $G^{DL}_{\Delta_1 \ldots \Delta_m}$, for any channel $\alpha \div L$. Further results about scattering into cones are given in Dollard [4].

PROBLEMS

15.1 : (a) In $L^2(\mathbb{R}^{3N})$, define $\underline{y}_N = \underline{x}_{CM} = \sum_{k=1}^N m_k \underline{x}_k / M$ and $\underline{y}_\ell = \sum_{k=1}^N \alpha_{\ell k} \underline{x}_k$ ($\ell = 1,\ldots,N-1$) such that $\sum_{k=1}^N \alpha_{\ell k} = 0$. Define the $N \times N$ matrix $B = (b_{ik})$ by $b_{ik} = \alpha_{ik}$ for $i = 1,\ldots,N-1$, $b_{Nk} = m_k/M$, and assume that $\det B = 1$. Let $\underline{X}_k$ and $\underline{Y}_k$ be the the multiplication operators by $\underline{x}_k$ and $\underline{y}_k$ respectively, $\underline{P}_k$ and $\underline{Z}_k$ the differential operators $-i\underline{\nabla}_{x_k}$ and $-i\underline{\nabla}_{y_k}$ respectively. Show that $[Z_{jr}, Y_{ks}] \subseteq \delta_{jk}\delta_{rs} I$ ($j,k = 1,\ldots,N; r,s = 1,2,3$), $\underline{Z}_N = \sum_{k=1}^N \underline{P}_k$, and that $\sum_{k=1}^N (2m_k)^{-1} \underline{P}_k^2 = (2M)^{-1} \underline{Z}_N^2 + \sum_{j,k=1}^{N-1} a_{jk} \underline{Z}_j \cdot \underline{Z}_k$, where (a_{jk}) is a positive definite symmetric matrix. (Hint : Writing in vector notation $y = Bx$, one finds $P = B^T Z$. Use the fact that $\sum_k \alpha_{\ell k} = 0$ if $\ell \neq N$ and $\sum_k \alpha_{\ell k} = 1$ if $\ell = N$, and that, if C is the matrix obtained from (α_{ik}) by crossing out one column, then $\det C = \pm 1$).

(b) Verify (15.13).

15.2 : Show that the right-hand side of (15.20) is equal to H^D on $D(H_0)$.

15.3 : Show that for $D \neq D_1$, $\sigma(H^D) = \sigma(H^D_{rel}) = [\lambda^D, \infty)$ for some $\lambda^D \leq 0$. (Hint : Use Lemma 5.19).

15.4 : Construct an example of a three-particle Hamiltonian such that $\sigma_e(H_{rel}) = [\mu, \infty)$ for some $\mu < 0$ and such that H_{rel} has a negative eigenvalue λ embedded in $\sigma_e(H_{rel})$, i.e. $\mu < \lambda < 0$. (Hint : Take particle 1 to be infinitely heavy and $V_{23} = 0$).

15.5 : Let $D = \{(1),\ldots,(n)\}$, $J = \{(1)',\ldots,(m)'\}$ be partitions and $C_1,\ldots,C_n,C_1',\ldots,C_m'$ cones in $\mathcal{R}^3$. Show that, under the hypothesis of Proposition 15.11, $B^D_{C_1,\ldots C_n}$ commutes with $B^J_{C_1'\ldots C_m'}$.

15.6†: (a) Let $\{W_t\}$ and $\{Z_t\}$ be two continuous unitary groups with absolutely continuous infinitesimal generators of uniform spectral multiplicity. Let $A \in \mathcal{B}(\mathcal{H})$ be such that $W_t A = A Z_t$ for all $t \in \mathcal{R}$. Adapt the statement and the proof of Proposition 5.27 to this situation. (Hint : Use (9.42).) (b) Prove (15.36).

15.7 : Prove the optical theorem (15.47).

15.8 : Fill in the missing details in the proof of Proposition 15.15.

15.9 : Prove the result of Proposition 15.7 under the assumption (a) that each pair potential belongs to $L^p(\mathcal{R}^3)$ with $2 < p < 3$, (b)† that each pair potential verifies the hypothesis of Proposition 8.31. (Hint : Replace ϕ in the proof of Proposition 8.31 by a product of suitable C^∞-functions).

15.10 : Establish Proposition 15.4 and Lemma 15.8 under the hypothesis that each pair potential belongs to $L^2(\mathcal{R}^3) + L^p(\mathcal{R}^3)$ with $2 < p < \infty$. (Hint : If $V \in L^p(\mathcal{R}^n)$ with $p > 2$, then $V \in L^2(\mathcal{R}^n) + L^\infty(\mathcal{R}^n)$.)

CHAPTER 16 : THE THREE-BODY PROBLEM

The results of Chapter 15 are the multiparticle analogues of those derived in Chapters 7 and 8 for simple potential scattering, covering the self-adjointness of the Hamiltonian, the determination of its essential spectrum, the study of bound states and scattering states, the definition of cross sections and the existence of the wave operators. It remains to elaborate the multiparticle counterparts of the topics treated in Chapters 9, 10 and 12, namely asymptotic completeness, eigenfunction expansions, finiteness of the total cross section, properties of the scattering amplitudes, spectral properties of the Hamiltonian H_{rel} and approximation methods. For simple potential scattering the basic prerequisites for dealing with these questions were (i) a control over the behaviour of the resolvent R_z of the total Hamiltonian as z approached the real axis, and (ii) the existence of suitable operators M_ϕ mapping $H = L^2(R^3)$ into $H_0 = L^2(S^{(2)})$, the Hilbert space appearing in the spectral representation of the free Hamiltonian K_0. Similarly, for multiparticle systems, the main task is to analyze R_z near the real axis and to obtain the counterparts of the operators M_ϕ. Once this has been accomplished, many of the questions mentioned above can be handled by proceeding as in the earlier chapters.

In Section 16-1 we describe a method for studying the three-particle resolvent near the real axis. This method requires a more detailed knowledge of the spectral properties of the two-body subsystems than that obtained in Chapter 10, and we shall simply refer to the literature for proofs of these properties. In Section 16-2 we define the operators M_ϕ and give as an application the proof of the finiteness of the total cross section and some properties of the channel S-matrices $S_{\alpha\beta}(\lambda)$. Lack of space prevents us from treating the remaining questions. Some relevant references will be mentioned in Section 16-3.

16-1 THE THREE-PARTICLE RESOLVENT

In this chapter we use the same notations as in Chapter 15, with $N = 3$, and we shall always work in the Hilbert space $H_{rel} = L^2(R^6)$ which we shall simply denote by H. We use the letters b,c and d to denote pairs of particles (i.e. b, c and d assume the three values $\{1,2\}$, $\{1,3\}$ and $\{2,3\}$). For fixed $d = \{k,\ell\}$, we use relative coordinates

$\underline{x}_d$ and $\underline{y}_d$ such that $\underline{x}_d = \underline{x}_k - \underline{x}_\ell$ and $\underline{y}_d$ is the relative distance between the center of mass of the pair $\{k,\ell\}$ and the third particle as in (15.12). We denote by $\underline{P}_d$ and $\underline{K}_d$ the associated momentum operators (i.e. the self-adjoint operators determined by $-i\nabla_{\underline{x}_d}$ and $-i\nabla_{\underline{y}_d}$ respectively). By (15.13) the free Hamiltonian has the form $H_o = \gamma_d \underline{P}_d^2 + \eta_d \underline{K}_d^2$, where γ_d and η_d are positive constants depending on the masses of the particles.

In the first two sections we assume that each pair interaction V_d may be factorized as $V_d = A_d B_d$, where A_d and B_d are multiplication operators in $L^2(\mathcal{R}^6)$ by functions $A(\underline{x}_d)$ and $B_d(\underline{x}_d)$ verifying $A(\underline{x}_d) = (1+|\underline{x}_d|)^{-\nu}$ for some fixed $\nu > 3/2$ and $B_d(\cdot) \in L^2(\mathcal{R}^3) \cap L^\infty(\mathcal{R}^3)$. We denote by A_d^o and B_d^o the multiplication operators associated with A_d and B_d respectively in $L^2(\mathcal{R}^3)$ and set $V_d^o = A_d^o B_d^o$. The preceding conditions imply that $V_d^o(\cdot) \in L^1(\mathcal{R}^3) \cap L^\infty(\mathcal{R}^3)$, and the set of all such potentials will be denoted by $\mathcal{V}_{1\nu}^\infty$. We shall also need the operators C_d which are defined to be the multiplication operators in $L^2(\mathcal{R}^6)$ by $C(\underline{y}_d) = (1+|\underline{y}_d|)^{-\nu}$.

To control the two-particle resolvent near the real axis, we had to require that (C1) $W_z \equiv BR_z^o A \in \mathcal{B}_o$, (C2) $I+W_z$ is invertible for $\operatorname{Im} z \neq 0$ and (C3) u-lim $W_{\lambda \pm i\eta}$ as $\eta \to +0$ exists, which allowed us to conclude the existence of $(I+W_{\lambda\pm io})^{-1}$ for almost all $\lambda \in \mathcal{R}$ (see Lemma 9.5). The three-particle resolvent will be handled in a similar way by proving the existence of a certain operator inverse for almost all $\lambda \in \mathcal{R}$. However, due to the more complex structure of three-particle systems, Lemma 9.5 cannot be applied so directly, and it is necessary to suitably split up the resolvent into other operators whose boundary values on the real axis can be shown to exist. To isolate these difficulties we shall derive the relevant expression for the three-particle resolvent in three consecutive steps.

The first possibility that suggests itself is to use the second resolvent equation (6.37) for the three-particle resolvent, which gives

$$B_c R_z A_d = B_c R_z^o A_d - \sum_b B_c R_z^o A_b B_b R_z A_d. \tag{16.1}$$

By defining $W_z^{cd} = B_c R_z^o A_d$ and $Y_z'^{cd} = B_c R_z A_d$, this may be rewritten as

$$Y_z'^{cd} = W_z^{cd} - \sum_b W_z^{cb} Y_z'^{bd}. \tag{16.2}$$

It is convenient to write this as an equation for a single operator in a larger Hilbert space K. Let K be the direct sum of three copies of H : $K = H \oplus H \oplus H$. We label the first, second and third summand by {1,2}, {1,3} and {2,3} respectively. Then, if $\{A^{cd}\}$ is a family of nine operators in H, one may define an operator A in K as the operator-valued matrix with elements A^{cd}, where A^{cd} is viewed as an operator from H_d to H_c. We shall denote by $\underline{\underline{A}}$ the diagonal part of A, i.e. $\underline{\underline{A}}^{cd} = A^{cd}\delta_{cd}$.

Thus, if Y_z' and W_z are the operators in K obtained from $\{Y_z'^{cd}\}$ and $\{W_z^{cd}\}$ respectively, (16.2) may be reexpressed as the following operator equation in K :

$$Y_z' = W_z - W_z Y_z'. \tag{16.3}$$

Its solution is $Y'_z = (I+W_z)^{-1}W_z$, provided that the inverse exists. In order to study the limit $z\to\lambda\pm i0$, we first examine to what extent W_z verifies the hypotheses of Lemma 9.5. The next lemma shows that W_z is compact if and only if each W_z^{cd} is. Its proof is simple and left as an exercise (Problem 16.1).

LEMMA 16.1 : Let $\{A^{cd}\}$ be a family of linear operators in $\mathcal{H}$ and A the associated operator in $\mathcal{K}$. Then A is in $\mathcal{B}(\mathcal{K})$ if and only if each A^{cd} is in $\mathcal{B}(\mathcal{H})$ and A is compact if and only if each A^{cd} is compact in $\mathcal{H}$.

PROPOSITION 16.2 : Assume that each pair potential is in $\mathcal{Y}_{1\nu}^{\infty}$ $(\nu > 3/2)$ and let Im $z \neq 0$. Then

(a) if $c \neq d$, W_z^{cd} is compact,

(b) $W_z^{cc} \in \mathcal{B}(\mathcal{H})$ but W_z^{cc} is not compact unless $V_c = 0$,

(c) for all c and d, $\{W_{\lambda\pm i\eta}^{cd}\}$ converges in operator norm as $\eta\to+0$, the convergence being uniform in $\lambda\in\mathbb{R}$,

(d) $I+W_z$ is invertible in $\mathcal{K}$, and $(I+W_z)^{-1} = (I-Y'_z) \in \mathcal{B}(\mathcal{K})$.

Proof : It follows from Proposition 2.16 that $A_d, B_d \in \mathcal{B}(\mathcal{H})$, so that $W_z^{cd} = B_c R_z^o A_d \in \mathcal{B}(\mathcal{H})$ for all c,d and $z\in\rho(H_o)$. Thus $W_z \in \mathcal{B}(\mathcal{K})$ by Lemma 16.1. Similarly $Y'_z \in \mathcal{B}(\mathcal{K})$ if $z\in\rho(H)$. (16.3) implies that $(I+W_z)(I-Y'_z) = I$. Similarly one deduces from (6.36) that $(I-Y'_z)(I+W_z) = I$, which proves (d).

Fix c and assume that $W_z^{cc}\in\mathcal{B}_o$. Define $Z_t = \exp(-i\underline{K}_c^2 t)$. $\underline{K}_c^2$ is the self-adjoint multiplication operator in $\tilde{L}^2(\mathbb{R}^6)$ by the square of the second variable, i.e. $\underline{K}_c^2\tilde{f}(\underline{p}_c,\underline{k}_c) = \underline{k}_c^2\tilde{f}(\underline{p}_c,\underline{k}_c)$.

By Problem 5.7, $\underline{K}_c^2$ is absolutely continuous, so that w-lim $Z_t = 0$ as $t \to \infty$ by Lemma 5.20. Hence $\lim \|W_z^{cc} Z_t f\| = 0$ as $t \to \infty$ for all $f \varepsilon H$ by Proposition 2.23. Clearly Z_t commutes with R_z^o. It also commutes with the multiplication operator by $\underline{x}_c$ (see (15.9)) and hence with A_c and B_c, so that $\|W_z^{cc} Z_t f\| = \|Z_t W_z^{cc} f\| = \|W_z^{cc} f\|$. It follows that $W_z^{cc} f = \theta$ for all $f \varepsilon H$, i.e. $W_z^{cc} = 0$. Hence W_z^{cc} cannot be compact unless it is zero. It is not difficult to verify that $W_z^{cc} = 0$ implies $V_c = 0$.

With this we have established (b). (a) and (c) are special cases of the following lemma the proof of which can be found in Section 16-3. #

<u>LEMMA 16.3</u> : Let $\Phi, \Psi : R^3 \to C$ be in $L^2(R^3) \cap L^\infty(R^3)$ and let A_d, C_d and B_c be the multiplication operators by $\Phi(\underline{x}_d)$, $\Phi(\underline{y}_d)$ and $\Psi(\underline{x}_c)$ respectively. Let $z \varepsilon \rho(K_o)$.

(a) $B_c R_z^o A_d \varepsilon \mathcal{B}_o$ if $c \neq d$, and $B_c R_z^o C_d \varepsilon \mathcal{B}_o$ for any c,d.

(b) If $c \neq d$, then $\{B_c R^o_{\lambda \pm i\eta} C_d\}$ converges in operator norm as $\eta \to +0$, uniformly in $\lambda \varepsilon R$.

(c) For any c and d, $\{B_c R^o_{\lambda \pm i\eta} A_d\}$ converges in operator norm as $\eta \to +0$, uniformly in $\lambda \varepsilon R$.

It follows from Lemma 16.1 and Proposition 16.2 that $\{W_z\}$ verifies (C2) and (C3) but not (C1) unless all pair potentials are zero. One may show that one may still get a result similar to Lemma 9.5 if some power of W_z is compact [KA, Ch. XIII.5]. It is also easy to check that this is not the case (Problem 16.2), so that the equation

(16.3) is not suitable for studying the operators $Y'_{\lambda\pm io}$. We shall now deduce a different equation in which the operator that has to be inverted does not contain the diagonal elements W_z^{cc} of the matrix $\{W_z^{cd}\}$. Such an equation can be obtained at the price of introducing the resolvents of the cluster Hamiltonians H^D. In the three-body problem there are only four partitions involved, namely {[1],[2],[3]}, {[1,2],[3]}, {[1,3],[2]} and {[1],[2,3]}, and it is convenient to label the last three by indicating which pair of particles is combined into a cluster, i.e. by using the label d introduced above. Thus for instance for d = {1,2}, we have $R_z^d = (H_o+V_d-z)^{-1} = (H_o+V_{12}-z)^{-1}$.

We now introduce the following families of operators in $\mathcal{H}$:

$$X_z^{cd} = B_c R_z^c A_d \qquad (16.4)$$

and the associated operator $X_z = \{X_z^{cd}\}$ in $\mathcal{K}$. X_z^{cd} is similar to W_z^{cd}, except that the free resolvent R_z^o has been replaced by the cluster resolvent R_z^c. By (6.37) we have

$$R_z = R_z^c - R_z^c(V - V_c)R_z, \qquad (16.5)$$

leading to

$$Y'_z = X_z - (X_z - \underline{\underline{X}}_z)Y'_z. \qquad (16.6)$$

Instead of Y'_z, it is more convenient to study the operator $Y_z = \{Y_z^{cd}\}$, where $Y_z^{cd} = B_c R_z C_d$. If we also set $O_z = \{O_z^{cd}\}$ with $O_z^{cd} = B_c R_z^c C_d$, then we get the following equation for Y_z from (16.5) :

$$Y_z = O_z - (X_z - \underline{\underline{X}}_z)Y_z. \qquad (16.7)$$

To proceed, we have to study the operator $X_z - \underline{\underline{X}}_z$.

PROPOSITION 16.4 : If Im $z \neq 0$, then $(I+X_z-\underline{\underline{X}}_z)$ is invertible.

Proof : From Proposition 6.9 we have

$$R_z^c = R_z^o - R_z^o V_c R_z^c = R_z^o - R_z^c V_c R_z^o. \qquad (16.8)$$

The first equation in (16.8) leads to $X_z = W_z - \underline{\underline{W}}_z X_z$, which itself implies that $\underline{\underline{X}}_z = \underline{\underline{W}}_z - \underline{\underline{W}}_z \underline{\underline{X}}_z$. From these last two identities it follows that

$$(I+\underline{\underline{W}}_z)(I+X_z-\underline{\underline{X}}_z) = I + X_z - \underline{\underline{X}}_z + \underline{\underline{W}}_z + \underline{\underline{W}}_z X_z - \underline{\underline{W}}_z \underline{\underline{X}}_z$$

$$= I + X_z + \underline{\underline{W}}_z X_z = I + W_z.$$

Thus, if $f \in \mathcal{K}$ and $(I+X_z-\underline{\underline{X}}_z)f = \theta$, we have $(I+W_z)f = \theta$, which implies $f = \theta$ since $I + W_z$ is invertible by Proposition 16.2. This shows that $(I + X_z - \underline{\underline{X}}_z)$ is invertible. #

Though $(X_z - \underline{\underline{X}}_z) \in \mathcal{B}_0$ (see Problem 16.3), we face a new difficulty when trying to apply Lemma 9.5 to these operators. To see this, let c be one of the partitions of {1,2,3} into two subsets, $\mathcal{M}^c$ the cluster subspace associated with the cluster Hamiltonian $H^c = H_0 + V_c$ and E^c the projection whose range is $\mathcal{M}^c$. The elements of the operator-valued matrix $X_z - \underline{\underline{X}}_z$ have the form

$$B_c R_z^c A_d = B_c(I-E^c)R_z^c A_d + B_c E^c R_z^c A_d \quad (c \neq d). \qquad (16.9)$$

The first term, which corresponds to the continuous spectrum of the two-particle Hamiltonian $\gamma_c \underline{P}_c^2 + V_c^o$, has properties similar to $B_c R_z^o A_d$. In particular it converges in the operator norm as $z \to \lambda \pm i0$, as we shall see below. To study the second term, we notice that $E^c = \sum_{\alpha \div c} E^\alpha$ and that by (14.44)

$$E^\alpha R_z^c = R_z^c E^\alpha = (\eta_c \underline{K}_c^2 + \lambda^\alpha - z)^{-1} E^\alpha \quad (\alpha \div c). \qquad (16.10)$$

Since B_c commutes with $\underline{K}_c$ by (15.9), we obtain $B_c R_z^c E^\alpha A_d = (\eta_c \underline{K}_c^2 + \lambda^\alpha - z)^{-1} B_c E^\alpha A_d$. The last operator clearly becomes unbounded when $z \to \mu \pm i0$ for each $\mu > \lambda^\alpha$. Thus $X_z - \underline{\underline{X}}_z$ will not converge to a bounded operator as z tends to the real axis whenever one of the two-body subsystems has bound states, i.e. whenever there are scattering channels other than the one corresponding to the partition $D_3 = \{(1),(2),(3)\}$.

This shows that $X_z - \underline{\underline{X}}_z$ will not verify (C3) and that this difficulty arises from the part of the resolvent R_z^c in the cluster subspace M^c. To deal with this problem, one has to modify the equation (16.7) in such a way that this part of the cluster resolvent R_z^c can be handled differently. For this, it is convenient to double the Hilbert space K, i.e. to introduce the space $\hat{K} = K \oplus K$, and to use its first and second summand to treat the operators $R_z^c(I-E^c)$ and $R_z^c E^c$ respectively. We shall denote by $\hat{E}_0$ and $\hat{E}_1$ the projections whose ranges are the first and second summands in $\hat{K}$ respectively. An operator $\hat{A}$ in $\hat{K}$ may be written as a matrix $\hat{A} = \{A_{ik}\}$, i,k = 0,1. Here each A_{ik} is an operator in K (i.e. A_{ik} is itself an operator-valued 3×3 matrix), and if $f = f_0 \oplus f_1 \in \hat{K}$, then $\hat{A}f = (A_{00}f_0 + A_{01}f_1) \oplus (A_{10}f_0 + A_{11}f_1)$.

For each z with Im z ≠ 0 we define an operator K_z from K to $\hat{K}$ and an operator J_z from $\hat{K}$ to K as follows :

$$K_z g = K_z^0 g \oplus K_z^1 g, \quad J_z(f_0 \oplus f_1) = f_0 + \underline{\underline{G}}_z f_1. \tag{16.11}$$

Here $f_0, f_1, g \in K$ and K_z^0, K_z^1 and $\underline{\underline{G}}_z$ are operators in K given by

$$K_z^0 = \{B_c(I-E^c)R_z^c C_d\}, \quad K_z^1 = \{C_c^{-1} E^c C_d\}, \tag{16.12}$$

$$\underline{\underline{G}}_z = \{B_c E^c R_z^c C_c \delta_{cd}\}. \tag{16.13}$$

We also introduce the operator $\hat{D}_z = \{D_{z,ik}\}$ acting in $\hat{K}$ by

$$D_{z,00} = \{B_c(I-E^c)R_z^c A_d(1-\delta_{cd})\},\quad D_{z,01} = D_{z,00}\underline{\underline{G}}_z, \tag{16.14}$$

$$D_{z,10} = \{C_c^{-1}E^c A_d(1-\delta_{cd})\},\quad D_{z,11} = D_{z,10}\underline{\underline{G}}_z. \tag{16.15}$$

K_z^1 and $D_{z,10}$ contain the unbounded operators C_c^{-1}. However a part of the unboundedness of C_c^{-1} is compensated for by C_d or A_d and the rest by the decrease at large distances of the eigenvectors of $\gamma_c \underline{P}_c^2 + V_c^0$ appearing in E^c, so that the closures of K_z^1 and $D_{z,10}$ belong to $\mathcal{B}(K)$. This result will be assumed for the moment and proved further on (Lemma 16.13). $D_{z,00}$ does not contain the part of R_z^c in M^c and is thus expected to have boundary values on the real axis. The operators $D_{z,01}$ and $D_{z,11}$ have the same property, as we shall see in Proposition 16.6. We now relate these operators to those appearing in (16.7).

PROPOSITION 16.5 :

(a)
$$J_z K_z = 0_z. \tag{16.16}$$

(b)
$$J_z \hat{D}_z = (X_z - \underline{\underline{X}}_z)J_z. \tag{16.17}$$

(c) $I + \hat{D}_z$ is invertible.

Proof : (16.16) is a direct consequence of the definitions.

(b) We have $J_z \hat{D}_z \hat{E}_0 = (D_{z,00} + \underline{\underline{G}}_z D_{z,10})\hat{E}_0$, the operator in the bracket being viewed as an operator in K. Now

$$D_{z,00} + \underline{\underline{G}}_z D_{z,10} = \{(1-\delta_{cd})[B_c(I-E^c)R_z^c A_d + B_c R_z^c E^c A_d]\}$$

$$= \{(1-\delta_{cd})B_c R_z^c A_d\} = X_z - \underline{\underline{X}}_z,$$

so that $J_z \hat{D}_z \hat{E}_0 = (X_z - \underline{\underline{X}}_z) J_z \hat{E}_0$. Similarly one gets

$$D_{z,01} + \underline{\underline{G}}_z D_{z,11} = (D_{z,00} + \underline{\underline{G}}_z D_{z,10})\underline{\underline{G}}_z = (X_z - \underline{\underline{X}}_z)\underline{\underline{G}}_z,$$

implying that $J_z \hat{D}_z \hat{E}_1 = (X_z - \underline{\underline{X}}_z) J_z \hat{E}_1$.

(c) Suppose $f = f_0 \oplus f_1 \in \hat{K}$ and $(I+\hat{D}_z)f = \theta$, i.e.

$$(I + D_{z,00})f_0 + D_{z,01}f_1 = \theta = D_{z,10}f_0 + (I + D_{z,11})f_1. \qquad (16.18)$$

It follows from (16.17) that $[I+(X_z-\underline{\underline{X}}_z)]J_z f = J_z(I+\hat{D}_z)f = \theta$, which implies together with Proposition 16.4 that $J_z f = \theta$, i.e.

$$f_0 + \underline{\underline{G}}_z f_1 = \theta. \qquad (16.19)$$

By inserting (16.19) into (16.18) and using (16.15), we obtain $\theta = (-D_{z,10}\underline{\underline{G}}_z + I + D_{z,11})f_1 = f_1$. By combining this with (16.19), we arrive at $f_0 = -\underline{\underline{G}}_z f_1 = \theta$, i.e. $f = \theta$. Thus $(I+\hat{D}_z)$ is invertible. #

The preceding proposition allows us to express Y_z in terms of $\hat{D}_z$. For this we define $\hat{L}_z = (I+\hat{D}_z)^{-1}$ for $\operatorname{Im} z \neq 0$. We shall show below that $\hat{L}_z \varepsilon \mathcal{B}(\hat{K})$. Clearly $\hat{L}_z$ verifies the equation $\hat{L}_z = I - \hat{D}_z \hat{L}_z$. Combined with (16.16) and (16.17) it leads to

$$J_z \hat{L}_z K_z = J_z K_z - J_z \hat{D}_z \hat{L}_z K_z = O_z - (X_z - \underline{\underline{X}}_z) J_z \hat{L}_z K_z,$$

which implies together with (16.7) that

$$[I + (X_z - \underline{\underline{X}}_z)](Y_z - J_z \hat{L}_z K_z) = 0.$$

Since $I + X_z - \underline{\underline{X}}_z$ is invertible (Proposition 16.4), one arrives at

$$Y_z = J_z \hat{L}_z K_z = J_z (I + \hat{D}_z)^{-1} K_z . \tag{16.20}$$

In order to apply Lemma 9.5 to $\{\hat{D}_z\}$, we have to establish the compactness of the operators $D_{z,ik}$ and their uniform convergence as $z \to \lambda \pm i0$. We collect these properties in the next proposition but postpone their proof till Section 16-3. This proof will involve the knowledge of certain spectral properties of the two-particle Hamiltonians $\gamma_c \underline{P}_c^2 + V_c^o$ in $L^2(\mathcal{R}^3)$, namely that for each c

(H1) $\gamma_c \underline{P}_c^2 + V_c^o$ has no singularly continuous spectrum,

(H2) the point spectrum of $\gamma_c \underline{P}_c^2 + V_c^o$ consists of a finite number of strictly negative eigenvalues of finite multiplicity, or it is empty,

(H3) the corresponding eigenfunctions e_k^c of $\gamma_c \underline{P}_c^2 + V_c^o$ in $L^2(\mathcal{R}^3)$ verify $(1+|\underline{x}|)^{2\nu} e_k^c(\underline{x}) \varepsilon L^2(\mathcal{R}^3)$.

(H4) $I + B_c^o(\gamma_c \underline{P}_c^2)^{-1} A_c^o$ is invertible in $L^2(\mathcal{R}^3)$, i.e. $\lambda = 0$ is not in the exceptional set Γ_o^c associated with the family of operators $\{B_c^o(\gamma_c \underline{P}_c^2 - z)^{-1} A_c^o\}$ in $L^2(\mathcal{R}^3)$.

(H1), (H2) and (H4) are equivalent to the condition that Γ_o^c does not intersect $[0,\infty)$. For potentials in $\mathcal{V}_{1\nu}^{\infty}$, (H1) was proved in Proposition 10.17 and the absence of strictly positive eigenvalues by Kato [1]. Also the number of negative eigenvalues, multiplicities counted, is known to be finite (see the article of Simon in [LI].) The rapid decrease of the eigenfunctions as $|\underline{x}| \to \infty$ is also known to hold, see e.g. O'Connor [1]. From this we see that for our potentials (H1)-(H3) are verified, with the possible exception of (H2) at $\lambda = 0$, i.e. $\lambda = 0$ may be an eigenvalue of $\gamma_c \underline{P}_c^2 + V_c^0$. If this is the case or if (H4) does not hold, (H1)-(H4) will be fulfilled after an arbitrarily small change of the coupling constant, by Remark 10.18(b). (H2) and (H3) are needed to compensate for the unbounded operator C_c^{-1} appearing in K_z^1 and D_z (Lemma 16.13). The most extensive use of these spectral properties is made in Lemma 16.15 to show that $B_c(I-E^c)R^c_{\lambda\pm i\eta}A_c$ converges as $\eta \to +0$.

PROPOSITION 16.6 : Let $V^o_{k\ell} \in \mathcal{V}^\infty_{1\nu}$ and assume (H1) - (H4). Let $\operatorname{Im} z \neq 0$. Then (a) $D_{z,10} \in \mathcal{B}(\mathcal{K})$. (b) The operators $D_{z,00}$, $D_{z,01}$ and $D_{z,11}$ are compact operators in $\mathcal{K}$. (c) For all i,k, $\{D_{\lambda\pm i\eta,ik}\}$ converges in operator norm, uniformly in $\lambda \in \mathcal{R}$, as $\eta \to +0$.

Lemma 9.5 cannot be applied directly to $\hat{D}_z$, since $D_{z,10}$ is not compact. This is however only a minor complication, as can be seen from the following proposition. For its statement we define $\hat{D}_{\lambda\pm io} = \text{u-lim}\ \hat{D}_{\lambda\pm i\eta}$ as $\eta \to +0$, $\Gamma_o^\pm = \{\lambda \in \mathcal{R} \mid (I+\hat{D}_{\lambda\pm io}) \text{ is not invertible}\}$ and $\Gamma_o = \Gamma_o^+ \cup \Gamma_o^-$.

PROPOSITION 16.7 : Under the assumptions of Proposition 16.6, we have (a) $(I+\hat{D}_z)^{-1} \in \mathcal{B}(\hat{\mathcal{K}})$ for all $z \notin \Gamma_o$, (b) $(I+\hat{D}_z)^{-1}$ is uniformly bounded and uniformly continuous in operator norm on any compact subset of the closed upper or lower half plane not intersecting Γ_o (see Lemma 9.5(a)), (c) Γ_o is a closed set of Lebesgue measure zero.

Proof : We write

$$\hat{D}_z = \hat{N} + \hat{M}_z \equiv \begin{pmatrix} 0 & 0 \\ D_{z,10} & 0 \end{pmatrix} + \begin{pmatrix} D_{z,00} & D_{z,01} \\ 0 & D_{z,11} \end{pmatrix}.$$

One has $\hat{N}^2 = 0$, hence $(I-\hat{N})(I+\hat{N}) = (I+\hat{N})(I-\hat{N}) = I$, i.e. $(I+\hat{N})^{-1} = (I-\hat{N})\varepsilon\mathcal{B}(\hat{\mathcal{K}})$. Thus

$$I + \hat{D}_z = (I+\hat{N})[I+(I-\hat{N})\hat{M}_z] \equiv (I+\hat{N})(I+\hat{\omega}_z). \tag{16.21}$$

This implies that $I + \hat{D}_z$ is invertible if and only if $I + \hat{\omega}_z$ is, and that

$$(I+\hat{D}_z)^{-1} = (I+\hat{\omega}_z)^{-1}(I-\hat{N}),\quad (I+\hat{\omega}_z)^{-1} = (I+\hat{D}_z)^{-1}(I+\hat{N}). \tag{16.22}$$

It is seen from this and Proposition 16.5(c) that $I + \hat{\omega}_z$ is invertible whenever $\operatorname{Im} z \neq 0$.

As in Lemma 16.1, we see by using Proposition 16.6(b) that $\hat{M}_z$ is compact. Hence $\hat{\omega}_z \equiv (I-\hat{N})\hat{M}_z \varepsilon \mathcal{B}_o$. By combining the preceding statements with Proposition 16.6(c), we see that the family $\{\hat{\omega}_z\}$ verifies all the hypotheses of Lemma 9.5 and that the exceptional set associated with it is precisely Γ_o. It follows that $(I+\hat{\omega}_z)^{-1}\varepsilon\mathcal{B}(\hat{\mathcal{K}})$ if $z\notin\Gamma_o$, which together with (16.22) implies (a). (b) and (c) follow from Lemma 9.5 and (16.22). #

Proposition 16.7 shows that $\hat{L}_z = (I+\hat{D}_z)^{-1}$ has boundary values on the real axis away from a closed set of measure zero Γ_o. On the other hand $\underline{\underline{G}}_z$ and the diagonal part of K_z^o do not converge as $z\to\lambda\pm io$, so that $Y_z = J_z\hat{L}_zK_z$ will not have boundary values unless all the projections E^c are zero. Nevertheless the existence of the boundary values of $\hat{L}_z$ suffices to obtain the scattering amplitude (see Section 16-2) and to show that $(f,R_z g)$ converges as $z\to\lambda\pm i0$ for $\lambda\notin\Gamma_0$ and f,g in some dense set $\mathcal{D}$. We end this Section by indicating a proof of this latter result some applications of which will be cited in Section 16-3A. We define Ψ to be the multiplication operator in $L^2(\mathbb{R}^6)$ by $(1+|\underline{x}_{12}|^2 + |\underline{y}_{12}|^2)^{-\nu/2}$. $\mathcal{D}$ may be for instance the range of Ψ.

PROPOSITION 16.8 : Let $V^0_{k\ell}\epsilon\mathcal{Y}^\infty_{1\nu}$ and assume (H1)-(H4). Then $\{\Psi R_{\lambda\pm i\eta}\Psi\}$ converges in operator norm as $\eta\to+0$ whenever $\lambda\notin\Gamma_0$, the convergence being uniform in λ on compact subsets of $\mathbb{R}-\Gamma_0$.

Sketch of the proof : It is easy to see that $A_c^{-1}\Psi$, $C_c^{-1}\Psi$ and the closures of ΨA_c^{-1} and ΨC_c^{-1} are in $\mathcal{B}(\mathcal{H})$ for any c. By using (6.37), $Y_z^{cd} = B_c R_z C_d$ and (16.20), one finds

$$\Psi R_z\Psi = (\Psi A_c^{-1})(A_c R_z^o A_c)(A_c^{-1}\Psi) - \textstyle\sum_d(\Psi R_z^o A_d)(J_z\hat{L}_z K_z)^{dd}(C_d^{-1}\Psi).$$

One obtains the boundary values of the first term from Lemma 16.3(c), those of $(K_z^o)^{cd}C_d^{-1}\Psi = B_c(I-E^c)R_z^c A_c(A_c^{-1}\Psi)$ from Lemma 16.15 and those of $\Psi R_z^o A_d \underline{G}_z^{dd} = \Psi R_z^o V_d R_z^d E^d C_d =$ $\Psi R_z^o C_d E^d - (\Psi C_d^{-1})C_d R_z^d E^d C_d$ from Lemma 16.3(b) and Lemma 16.14. #

Remark : In the above and in various other proofs given in this chapter we use the same notation for an operator like ΨA_c^{-1} and its closure, and for the closure of a product of operators we make use of Problem 2.17(d) without mentioning it.

16-2 THREE-PARTICLE CROSS SECTIONS

As an application of the results of Section 16-1, we prove here the finiteness and continuity of the averaged total cross section $\bar{\sigma}_\alpha(\lambda)$ for each two-body initial channel α and all $\lambda > \lambda^\alpha$ not belonging to Γ_0. We again assume that the pair potentials are in $\mathcal{V}^\infty_{1\nu}$ and verify (H1)-(H4).

We begin by introducing for each channel α certain operators mapping $L^2(\mathbb{R}^6)$ into $\mathcal{H}^\alpha_0$, where $\mathcal{H}^\alpha_0 = L^2(S^{(2)})$ or $L^2(E^{(5)})$. They are the analogues of the operators $M_\phi(\lambda)$ defined in Section 10-1 and will be denoted $M^\alpha(\Phi,\lambda)$ for typographical reasons.

We first consider two-body channels. Let $\alpha \div c$, and let e_α be the eigenvector in $L^2(\mathbb{R}^3)$ of $\gamma_c \underline{P}_c^2 + V_c^0$ defining the channel α (see (14.41)). Furthermore let $\mathcal{U}^\alpha_0 : \mathcal{M}^\alpha \to \mathcal{G}^\alpha = L^2([\lambda^\alpha,\infty), L^2(S^{(2)}))$ be the spectral transformation of K^α (see Section 15-3). By (14.42) one has

$$(E^\alpha f)(\underline{x}_c,\underline{y}_c) = e_\alpha(\underline{x}_c) \int d^3x_c' \; \bar{e}_\alpha(\underline{x}_c')f(\underline{x}_c',\underline{y}_c). \qquad (16.23)$$

We define $\mathcal{U}^\alpha : L^2(\mathbb{R}^6) \to \mathcal{G}$ by $\mathcal{U}^\alpha = \mathcal{U}^\alpha_0 E^\alpha$, i.e. (see 5.95)

$$(\mathcal{U}^\alpha f)_\lambda(\underline{\omega}) = \mu_\alpha^{\frac{1}{2}} k_\alpha^{\frac{1}{2}} \, [\mathcal{F}_y \int \bar{e}_\alpha(\underline{x}_c')f(\underline{x}_c',\underline{y}_c)d^3x_c'](k_\alpha\underline{\omega}), \qquad (16.24)$$

where $k_\alpha = [2\mu_\alpha(\lambda-\lambda^\alpha)]^{\frac{1}{2}}$ and $\mathcal{F}_y$ denotes the Fourier transformation with respect to the variable $\underline{y}_c$. Now let ϕ be a measurable function from $\mathbb{R}^6$ to $\mathbb{R}$ and denote by Φ the maximal multiplication operator in $L^2(\mathbb{R}^6)$ by $\phi(\underline{x}_c,\underline{y}_c)$. The operator $M^\alpha(\Phi,\lambda) : L^2(\mathbb{R}^6) \to \mathcal{H}^\alpha_0 \equiv L^2(S^{(2)})$ will be defined similarly to (10.6) by setting for $f \in D(\Phi)$

$$[M^{\alpha}(\Phi,\lambda)f](\underline{\omega}) = (U^{\alpha}\Phi f)_{\lambda}(\underline{\omega}). \tag{16.25}$$

Clearly the closure of such an operator, also denoted by $M^{\alpha}(\Phi,\lambda)$, is an integral operator with kernel

$$(2\pi)^{-3/2}\,\mu_{\alpha}^{\frac{1}{2}}k_{\alpha}^{\frac{1}{2}}\,\bar{e}_{\alpha}(\underline{x}_c)\phi(\underline{x}_c,\underline{y}_c)\exp[-ik_{\alpha}\underline{\omega}\cdot\underline{y}_c]. \tag{16.26}$$

We shall only need functions ϕ for which this kernel is square-integrable, so that $M^{\alpha}(\Phi,\lambda)\in\mathcal{B}_2$ and

$$\|M^{\alpha}(\Phi,\lambda)\|_{HS}^{2} = (2\pi^{2})^{-1}\mu_{\alpha}k_{\alpha}\int d^3x_c d^3y_c\,|e_{\alpha}(\underline{x}_c)|^{2}|\phi(\underline{x}_c,\underline{y}_c)|^{2}. \tag{16.27}$$

Some such functions are considered in the following lemma.

LEMMA 16.9 : Let $\alpha \div c$, $d \neq c$ and b arbitrary. Then

(a) $M^{\alpha}(A_d,\lambda)$, $M^{\alpha}(B_d,\lambda)$, $M^{\alpha}(A_cA_d,\lambda)$, $M^{\alpha}(A_c^{-1}B_d,\lambda)$, $M^{\alpha}(C_d^{-1}A_dB_d,\lambda)$ and $M^{\alpha}(A_b^{-1}A_dB_d,\lambda)$ are Hilbert-Schmidt operators from $L^2(R^6)$ to $L^2(S^{(2)})$ which depend continuously on λ in Hilbert-Schmidt norm*).

(b) Each of the preceding six operators $M^{\alpha}(\Phi,\lambda)$ has all the properties given in Lemma 10.3 if K_0 is replaced by K^{α}, U_0 by U_0^{α} and g by $E^{\alpha}g$ in that lemma. For example one has $B_d^*E^{\alpha}g = \int M^{\alpha}(B_d,\lambda)^*(U_0^{\alpha}E^{\alpha}g)_{\lambda}d\lambda$ and w-lim $B_d^*E^{\alpha}(R_{\lambda+i\eta}^{\alpha}-R_{\lambda-i\eta}^{\alpha})g = 2\pi i\, M^{\alpha}(B_d,\lambda)^*(U_0^{\alpha}E^{\alpha}g)_{\lambda}$ as $\eta\to+0$ for all $g\in S(R^6)$.

Proof : Since $c \neq d$, we have $\underline{x}_d = \gamma\underline{x}_c + \delta\underline{y}_c$ with $\gamma,\delta \neq 0$. Hence by (16.27), $\|M^{\alpha}(A_d,\lambda)\|_{HS}^{2} = ck_{\alpha}\|e_{\alpha}\|^{2}\,\|A_d^{0}\|_2^{2} < \infty$. Similarly $\|M^{\alpha}(A_c^{-1}B_d,\lambda)\|_{HS}^{2} = c_{\alpha}k_{\alpha}\|B_d^{0}\|_2^{2}\,\|(1+|\underline{x}|)^{\nu}e_{\alpha}(\underline{x})\|^{2}$, which is

*) It is understood that all occuring operators A_d, B_d etc. are expressed as multiplication operators in the variables $\underline{x}_c$ and $\underline{y}_c$.

finite by (H3). Finally $\|M^{\alpha}(C_d^{-1}A_dB_d,\lambda)\|_{HS}^2 \leq c_\alpha k_\alpha \kappa^{2\nu} \|B_d^o\|_2^2 \cdot \|(1+|\underline{x}|)^{\nu} e_\alpha(\underline{x})\|^2$ by using the inequality (16.32) as in the proof of Lemma 16.13, and similarly for $M^{\alpha}(A_b^{-1}A_dB_d,\lambda)$. This shows that the operators in (a) are Hilbert-Schmidt. Their continuity in Hilbert-Schmidt norm follows from the preceding estimates and the Lebesgue dominated convergence theorem, and the proof of (b) is the same as that of Lemma 10.3 if one uses the identity $B_d^* U_t^{\alpha} E^{\alpha} = B_d^* \exp(-i\lambda^{\alpha} t - i\eta_c \underline{K}_c^2 t) E^{\alpha}$ and an estimate similar to (16.31). #

We now consider the channel corresponding to three free particles which we distinguish by setting $\alpha = 0$. It is convenient to introduce three spectral transformations $\mathcal{U}_d^o$ for H_o mapping $L^2(\mathbb{R}^6)$ onto $\mathcal{G}_d^o = L^2([0,\infty), L^2(\mathcal{E}^{(5)}))$, where $d = \{1,2\}$, $\{1,3\}$ and $\{2,3\}$. To obtain $\mathcal{U}_d^o$, we replace the six variables $\underline{p}_d$, $\underline{k}_d$ by $\lambda = \gamma_d \underline{p}_d^2 + \eta_d \underline{k}_d^2$, $\underline{\omega} = \underline{p}_d/|\underline{p}_d|$ and $\underline{q} = (\eta_d/\lambda)^{\frac{1}{2}} \underline{k}_d$ (for instance for $d = \{1,2\}$, these variables represent the total kinetic energy λ, the direction $\underline{\omega}$ of the relative momentum of particles 1 and 2, the direction $\underline{\omega}_k$ of the relative momentum of particle 3 and the center of mass of the pair $\{1,2\}$, and the fraction $q^2 = \eta_d \underline{k}_d^2/\lambda$ of the kinetic energy carried by the relative motion of particle 3 with respect to the center of mass of $\{1,2\}$). The Jacobian determinant for this change of variables is easily calculated, leading to $d^3p\, d^3k = h_d(\lambda,q)^2 d\lambda\, d\omega\, d^3q$ with $h_d(\lambda,q)^2 = \frac{1}{2}(\gamma_d\eta_d)^{-3/2}(1-q^2)^{\frac{1}{2}}\lambda^2 q^2$. We notice that $0 \leq q \leq 1$ and that the variables $\underline{\omega}$, $\underline{q}$ parametrize the ellipsoid $\mathcal{E}^{(5)} = \{\underline{p}_d, \underline{k}_d | \gamma_d \underline{p}_d^2 + \eta_d \underline{k}_d^2 = 1\}$. We now define

$$(\mathcal{U}_d^o f)_\lambda(\underline{\omega},\underline{q}) = h_d(\lambda,q)\tilde{f}([\gamma_d^{-1}\lambda(1-q^2)]^{\frac{1}{2}}\underline{\omega}, (\lambda/\eta_d)^{\frac{1}{2}}\underline{q}), \tag{16.28}$$

where the integral in the definition of the Fourier transform is over the variables $\underline{x}_d$ and $\underline{y}_d$. The unitarity of $\mathcal{U}_d^o$ follows from that of $\mathcal{F}$. In particular $\|\mathcal{U}_d^o f\|^2 = \int d\lambda \, \|(\mathcal{U}_d^o f)_\lambda\|_o^2 = \int d\lambda \int d\omega \, d^3q \, |(\mathcal{U}_d^o f)_\lambda(\underline{\omega},\underline{q})|^2 = \|f\|^2$.

We shall need the operators $M^o(A_d,\lambda)$ and $M^o(B_d,\lambda)$ mapping $L^2(\mathcal{R}^6)$ into $L^2(\mathcal{E}^{(5)})$. They are first defined on $\mathcal{S}(\mathcal{R}^6)$ by

$$[M^o(A_d,\lambda)f](\underline{\omega},\underline{q}) = (\mathcal{U}_d^o A_d f)_\lambda(\underline{\omega},\underline{q}) \tag{16.29}$$

and similarly for $M^o(B_d,\lambda)$. We shall see that the operators defined in (16.29) are closable (in fact bounded) and denote their closures also by $M^o(A_d,\lambda)$ and $M^o(B_d,\lambda)$ respectively. Their properties are as follows.

<u>LEMMA 16.10</u> : (a) $M^o(A_d,\lambda)$ and $M^o(B_d,\lambda)$ are bounded and everywhere defined operators from $L^2(\mathcal{R}^6)$ to $L^2(\mathcal{E}^{(5)})$, and they are strongly continuous functions of λ. (b) They have all the properties given in Lemma 10.3 if K_o is replaced by H_o and $\mathcal{U}_o$ by $\mathcal{U}_d^o$ in that lemma. In particular $A_d^* g = \int M^o(A_d,\lambda)^* (\mathcal{U}_d^o g)_\lambda \, d\lambda$ and $\text{w-lim}\, A_d^*(R^o_{\lambda+i\eta} - R^o_{\lambda-i\eta})g = 2\pi i \, M^o(A_d,\lambda)^*(\mathcal{U}_d^o g)_\lambda$ as $\eta \to +0$ for all $g \in \mathcal{S}(\mathcal{R}^6)$.

<u>Proof</u> : By the change of variables $\underline{q} \to \underline{k} = (\lambda/\eta_d)^{\frac{1}{2}}\underline{q}$ we find for $f \in \mathcal{S}(\mathcal{R}^6)$ that

$$\|M^o(A_d,\lambda)f\|_o^2 = \int d^3q \, d\omega \, |(\mathcal{U}_d^o A_d f)_\lambda(\underline{\omega},\underline{q})|^2 = (\eta_d/\lambda)^{3/2} \cdot$$

$$\int d^3k \, d\omega \, \chi_\lambda(\eta_d \underline{k}^2) h_d(\lambda, (\eta_d/\lambda)^{\frac{1}{2}} k)^2 \, |(\mathcal{F} A_d f)(\gamma_d^{-\frac{1}{2}}[\lambda - \eta_d \underline{k}^2]^{\frac{1}{2}} \underline{\omega}, \underline{k})|^2,$$

where $\chi_\lambda(u) = 1$ if $u \leq \lambda$ and $\chi_\lambda(u) = 0$ if $u > \lambda$. For each fixed $\underline{k}$ with $\eta_d \underline{k}^2 < \lambda$ we get as in the proof of Lemma 10.1 that

$$|(\mathcal{F}A_d f)(\gamma_d^{-\frac{1}{2}}[\lambda-\eta_d \underline{k}^2]^{\frac{1}{2}}\underline{\omega},\underline{k})|^2 \leq (2\pi)^{-3} \|A_d^o\|_2^2 \int d^3p|\tilde{f}(\underline{p},\underline{k})|^2.$$

Since $h_d(\lambda,q)^2 \leq c_d\lambda^2$ for all $0 \leq q \leq 1$, we arrive at $\|M^o(A_d,\lambda)f\|_o^2 \leq \alpha_d\lambda^2 \|A_d^o\|_2^2 \|f\|^2$, proving that $M^o(A_d,\lambda)$ is bounded. Thus by Proposition 2.6 it admits a closure in $\mathcal{B}(L^2(\mathcal{R}^6), L^2(\mathcal{E}^{(5)}))$ with the same norm. Its strong continuity can be established with a similar estimate by using the Lebesgue dominated convergence theorem and Proposition 2.17. Therefore $M^o(A_d,\lambda)^*$ is weakly continuous, hence for each $g \in \mathcal{S}(\mathcal{R}^6)$ the numerical-valued Riemann integral $\int d\lambda (M^o(A_d,\lambda)^* (U_d^o g)_\lambda, f)$ exists for all $f \in L^2(\mathcal{R}^6)$ and defines a bounded linear functional. By Proposition 2.3, $\int d\lambda\, M^o(A_d,\lambda)^* (U_d^o g)_\lambda$ is a vector in $L^2(\mathcal{R}^6)$. The rest of the proof of (b) follows the lines of that of Lemma 10.3 (Problem 16.5). #

<u>PROPOSITION 16.11</u> : Let $V_{ik}^o \in \mathcal{Y}_{1\nu}^\infty$ and assume (H1)-(H4). Let β be a two-body channel. Then $R_{\alpha\beta}(\lambda)$ is a Hilbert-Schmidt operator for each final channel α and each λ with $\lambda^\beta < \lambda \notin \Gamma_o$, and $\{R_{\alpha\beta}(\lambda)\}$ is continuous in Hilbert-Schmidt norm on $(\lambda^\beta,\infty)-\Gamma_o$. The averaged total cross-section for the initial channel β is finite (i.e. $\bar{\sigma}_\beta(\lambda) < \infty$) and a continuous function of λ away from Γ_o.

<u>Proof</u> : The proof is similar to that of Proposition 10.12(a). We assume $\beta \div c$ and denote by $\lambda_+ < 0$ the eigenvalue nearest to 0 of all two-body Hamiltonians of the form $\gamma_d \underline{P}_d^2 + V_d^o$ in $L^2(\mathcal{R}^3)$. Let $\Delta = [\sigma,\mu] \subseteq (\lambda^\beta,\infty)$ be an interval of length $0 < |\Delta| < |\lambda_+|$ such that $\Delta \cap \Gamma_o = \emptyset$, and let $f,g \in \mathcal{S}(\mathcal{R}^6)$. From (14.19) and the time-independent equation (14.67) for $R_{\alpha\beta}$

one obtains

$$(f, R_{\alpha\beta} E_\Delta^\beta g) = (E_\Delta^\alpha f, R_{\alpha\beta} g)$$

$$= \lim_{\eta\to+0} \lim_{\delta\to+0} \int_\Delta ([R_{\lambda+i\eta}^\alpha - R_{\lambda-i\eta}^\alpha] E^\alpha f, T_{\lambda+i\delta}^{D\underline{c}} dE_\lambda^\beta E^\beta g) \qquad (16.30)$$

if $\alpha \div D$. The operator $T_z^{D\underline{c}} = V^c - V^D R_z V^c$ is a sum of terms of the form $V_b = A_b B_b$ or $V_b R_z V_d = A_b Y_z^{bd} C_d^{-1} A_d B_d$. To each of these summands one applies the argument of the proof of Proposition 10.12(a). We shall not repeat this argument but simply remark that its prerequisites are (i) the statements of Lemma 10.3 (which are verified in the present context by Lemmas 16.9 and 16.10) and (ii) a control over Y_z^{bd} near the real axis (which will be obtained from the results of Sections 16-1 and 16-3).

We shall show that each summand in $T_z^{D\underline{c}}$ gives a Hilbert-Schmidt contribution to $R_{\alpha\beta}(\lambda)$. For the three-particle channel (α = o) we shall first choose that spectral transformation $\mathcal{U}_b^o$ which bears the same label b as the potential appearing on the left of a summand. Since we must express $R_{o\beta}$ as an operator from $\mathcal{G}^\beta$ to $\mathcal{G}_d^o$ for a fixed d, each such term will have to be multiplied from the left by $\mathcal{U}_d^o \mathcal{U}_b^{o*}$. Now $\mathcal{U}_d^o \mathcal{U}_b^{o*}$ is decomposable as a map from $\mathcal{G}_b^o$ to $\mathcal{G}_d^o$ (Problem 15.6), i.e. $\mathcal{U}_d^o \mathcal{U}_b^{o*} = \{U_{db}(\lambda)\}$, and it turns out that the unitary operators $U_{db}(\lambda)$ are independent of λ (see Problem 16.6(a); in fact U_{db} is induced by the corresponding coordinate transformation on $\mathcal{E}^{(5)}$). We shall not display the operators U_{db} in what follows. It is clear that they have no influence on the Hilbert-Schmidt nature and the continuity of $R_{o\beta}(\lambda)$.

We first study the z-independent terms in T_z^{Dc}. We set c = {1,2}, which implies $V^c = V_{13} + V_{23}$. First let D be a two-body clustering, say D = d. If d = c, we write $V_{13} = A_{13}B_{13}$ and see that the contribution of V_{13} to $R_{\alpha\beta}(\lambda)$ ($\alpha \div d$, $\lambda \in \Delta$) is $-2\pi i\, M^\alpha(A_d,\lambda)\, M^\beta(B_d,\lambda)^*$, which is trace class since each factor is Hilbert-Schmidt by Lemma 16.9. V_{23} is handled similarly. If d ≠ c, say d = {2,3}, then V_{13} is again handled as above, whereas for V_{23} one writes $V_{23} = (A_{23}A_{12})(A_{12}^{-1}B_{23})$, leading to the following contribution to $R_{\alpha\beta}(\lambda)$: $-2\pi i\, M^\alpha(A_{23}A_{12},\lambda)M^\beta(A_{12}^{-1}B_{23},\lambda)^*$. This is again in $\mathcal{B}_1$ by Lemma 16.9. If α = 0, then for instance V_{13} contributes a term $-2\pi i\, M^o(A_{13},\lambda)M^\beta(B_{13},\lambda)^*$ to $R_{o\beta}(\lambda)$. This is a Hilbert-Schmidt operator by Lemmas 16.9 and 16.10.

We now consider the z-dependent terms in T_z^{Dc} which are of the form $A_a(J_z\hat{L}_zK_z)^{ab}C_b^{-1}A_bB_b$ with b = {1,3} or {2,3}. By Proposition 16.7, $\hat{L}_{\lambda+i\delta}$ has norm-continuous boundary values for $\lambda\in\Delta$. We next study the operators on the right of $\hat{L}_z$. K_z^1 is bounded and z-independent, so that contributions to $R_{\alpha\beta}(\lambda)$ involving K_z^1 will have a factor $M^\beta(C_b^{-1}A_bB_b,\lambda)^* \in \mathcal{B}_2$ on the right (see Lemma 16.9). Terms coming from K_z^o have the form $[B_d(I-E^d)R_z^dA_d][A_d^{-1}A_bB_b]$ with b as above. The first factor has norm-continuous boundary values by Lemma 16.15, and the second one introduces a factor $M^\beta(A_d^{-1}A_bB_b,\lambda)^* \in \mathcal{B}_2$ in the corresponding term in $R_{\alpha\beta}(\lambda)$. Thus the right-most factor in any contribution to $R_{\alpha\beta}(\lambda)$ is always a Hilbert-Schmidt operator.

It remains to show that the operators on the left of $\hat{L}_z$ produce bounded factors in $R_{\alpha\beta}(\lambda)$. Since J_z^o is the identity operator, contributions involving J_z^o will have an operator $M^\alpha(A_a,\lambda)$ on the left, which is in $\mathcal{B}_2$ if $\alpha \neq o$ and bounded if $\alpha = o$. For terms involving $J_z^1 = \underline{\underline{G}}_z$, we first take $\alpha \neq o$, say $\alpha \div d$. Then $A_a \underline{\underline{G}}_z^{aa} = (A_a B_a C_a^{-1})(C_a E^a R_z^a C_a)$, where $a \neq d$ by the definition of V^d. Here the second operator has norm-continuous boundary values by Lemma 16.14, and the first one leads to a factor $M^\alpha(A_a B_a C_a^{-1},\lambda) \in \mathcal{B}_2$ in the corresponding term in $R_{\alpha\beta}(\lambda)$. Finally, for $\alpha = o$, we denote by F_μ^a the spectral projection of $\eta_a \underline{K}_a^2$ associated with the interval $[0,\mu]$ and notice that $E_\Delta^o = F_\mu^a E_\Delta^o$, since $\lambda\varepsilon\Delta = [\sigma,\mu]$ implies $\eta_a \underline{k}_a^2 = \lambda - \gamma_a \underline{p}_a^2 \leq \mu$. Now F_μ^a commutes with R_z^o and V_a, so that in each term in (16.30) containing $V_a \underline{\underline{G}}_z^{aa}$ one may replace $A_a \underline{\underline{G}}_z^{aa}$ by $A_a F_\mu^a \underline{\underline{G}}_z^{aa} = \sum_{\gamma \div a} A_a B_a F_\mu^a R_z^\gamma E^\gamma C_a$. But the operator

$$F_\mu^a R_{\lambda+i\delta}^\gamma E^\gamma = F_\mu^a(\eta_a \underline{K}_a^2 + \lambda^\gamma - \lambda - i\delta)^{-1} E^\gamma = F_\mu^a(\eta_a \underline{K}_a^2 F_\mu^a + \lambda^\gamma - \lambda - i\delta)^{-1} E^\gamma$$

has norm-continuous boundary values for $\lambda\varepsilon\Delta$, since $\lambda-\lambda^\gamma \geq \lambda-\lambda_+ > \mu$ for $\lambda\varepsilon\Delta$, so that $\lambda-\lambda^\gamma$ belongs to the resolvent set of $\eta_a \underline{K}_a^2 F_\mu^a$. Thus terms in $R_{o\beta}(\lambda)$ involving $\underline{\underline{G}}_z$ begin on the left by $\sum_{\gamma \div a} M^o(B_a,\lambda) A_a [F_\mu^a R_{\lambda+io}^\gamma] E^\gamma C_a$, which is bounded.

The above implies that $R_{\alpha\beta}(\lambda) \in \mathcal{B}_2$ for each $\lambda\varepsilon\Delta$, hence, by varying Δ, for each $\lambda > \lambda^\beta$ with $\lambda \notin \Gamma_o$. Now the operators $M^\beta(\Phi,\lambda)$ are continuous in Hilbert-Schmidt norm by Lemma 16.9, $M^o(\Phi,\lambda)$ is strongly continuous by Lemma 16.10 and the boundary values of the remaining factors in $R_{\alpha\beta}(\lambda)$ are continuous in operator norm by Lemmas 16.14, 16.15 and Proposition 16.7. Hence each term in $R_{\alpha\beta}(\lambda)$ is of the form $D(\lambda) M^\beta(\Phi,\lambda)^*$, where $D(\lambda)$ is strongly continuous. From this one deduces the continuity of $R_{\alpha\beta}(\lambda)$ in Hilbert-Schmidt norm by applying Lemma 8.23 to $R_{\alpha\beta}(\lambda)^*$. The finiteness of $\bar{\sigma}_\beta(\lambda)$ is obtained from the finiteness of the number of channels and (15.46). #

Finally we should mention that, if $\lambda^\alpha < \lambda^\beta$, the total cross section for scattering from channel β to channel α may be infinite at the threshold $\lambda = \lambda^\beta$, due to the factor $(\lambda-\lambda^\beta)^{-1}$ appearing in (15.46).

16-3 NOTES AND SUPPLEMENTARY MATERIAL

A. The first rigorous study of the three-particle resolvent is due to Faddeev [F]. This, together with additional references, was already mentioned in Section 15-4C. The analysis given in Section 16-1 is a Hilbert space adaptation of Faddeev's method developped by Ginibre and Moulin [1]. Various generalizations of the so-called Faddeev equations (a set of coupled equations similar to (16.7) for the operators $M_z^{cd} = V_d\delta_{cd} - V_cR_zV_d$, see Problem 16.7) have been proposed for N > 3 (e.g. Yakubovskii [1], Sigal [2], Grassberger and Sandhas [1]).

Proposition 16.8 has been deduced by Ginibre and Moulin [1] for a larger class of potentials (which includes $\mathcal{V}$) by a more sophisticated choice of the operators C_d and more elaborate estimates. The result of Proposition 16.8 can be used to derive spectral properties of the three-particle Hamiltonian (see [F], Ginibre and Moulin [1], Combescure and Ginibre [1] and references given therein) and to prove asymptotic completeness. For the completeness proof one shows that $\|\sum_D F_\pm^D E_\Delta \Psi f\|^2 = \|\Omega_\pm^{0*} E_\Delta \Psi f\|^2 + \sum_C \|\Omega_\pm^{C*} E_\Delta \Psi f\|^2$ for all $f\epsilon\mathcal{H}$ and all closed intervals Δ not intersecting Γ_0, where $\Omega_\pm^0$ = s-lim $\exp(iHt)\exp(-iH_0 t)$ as $t \to \pm\infty$. This may be done by a stationary method similar to that of Proposition 9.17, using the results of Sections 16-1 and 16-3. One then obtains as in Proposition 9.6 that $\sum_D F_\pm^D = E_{ac}(H) = E_{\mathcal{R}-\Gamma_0}$. For details we refer to Thomas [3] and Ginibre and Moulin [1]. For properties of eigenfunctions of three-particle systems the reader may consult [F], [T], Merkuriev, Gignoux and Laverne [1] and the references quoted there.

An interesting feature of multi-particle Hamiltonians is the fact that they may have an infinite number of negative eigenvalues even when the pair potentials are of very short range. This phenomenon is known as the Efimov effect and occurs for example in three-particle systems when each two-body Hamiltonian $\gamma_C \underline{P}_C^2 + V_C^0$ in $L^2(\mathcal{R}^3)$ has absolutely continuous spectrum but at least two of these Hamiltonians have a quasi-bound state at zero energy (see Remark 10.18(b)). One then has $\sigma_e(H) = [0,\infty)$ and the negative spectrum of H consists of an infinite number of eigenvalues (for details see Yafaev [1]).

A different equation for the N-particle resolvent was introduced by Weinberg [1] and van Winter [1]. It can be obtained by using the second resolvent equation for the various pairs of subsystems, or else by resumming the Born series $R_z = R_z^0[I + \sum_{n=1}^\infty (-VR_z^0)^n]$ for R_z in particular way, leading to an equation of the form $R_z = I_z + F_z R_z$, where F_z is compact. This equation has been useful for proving spectral properties of N-particle Hamiltonians (e.g. Hunziker [3], van Winter and Brascamp [1], Balslev and Combes [1], van Winter [2]). For $N = 3$ and in the notation of Section 16-1, the Weinberg-van Winter equation for Y_z' is obtained by inserting (16.6) into the right-hand side of (16.3), i.e. $Y_z' = (W_z - W_z X_z) + W_z(X_z - \underline{\underline{X}}_z)Y_z'$. It is seen that $W_z(X_z - \underline{\underline{X}}_z)$ is compact but need not converge as $z \to \lambda \pm i0$. Also it was pointed out by Federbush [1] that $I - W_z(X_z - \underline{\underline{X}}_z)$ is not necessarily invertible for $\mathrm{Im}\, z \neq 0$.

B. We now give the proofs that were omitted in Section 16-1 and also include some auxiliary lemmas.

Proof of Lemma 16.3 : (a) We use the variables $\{\underline{x}_c, \underline{y}_c\}$ to parametrize $\mathcal{R}^6$. If $d \neq c$, one has $\underline{x}_d = \alpha\underline{x}_c + \beta\underline{y}_c$ for some non-zero real constants α and β. We denote by E_N^o the spectral projection of H_0 associated with the interval $[0,N]$. $R_z^o E_N^o$ is the multiplication operator in $\tilde{L}^2(\mathcal{R}^6)$ by

$$h(\underline{p}_c, \underline{k}_c) \equiv (\gamma_c \underline{p}_c^2 + \eta_c \underline{k}_c^2 - z)^{-1} \chi_{[0,N]}(\gamma_c \underline{p}_c^2 + \eta_c \underline{k}_c^2) .$$

Clearly $h \varepsilon L^2(\mathcal{R}^6)$. Hence $B_c R_z^o E_N^o A_d$ is an integral operator in $L^2(\mathcal{R}^6)$ with kernel $(2\pi)^{-3}\Psi(\underline{x}_c)\tilde{h}(\underline{x}_c' - \underline{x}_c, \underline{y}_c' - \underline{y}_c)\Phi(\alpha\underline{x}_c' + \beta\underline{y}_c')$. By writing $\|B_c R_z^o E_N^o A_d\|_{HS}^2$ in terms of this kernel and integrating successively over the variables $\underline{y}_c$, $\underline{y}_c'$, $\underline{x}_c'$ and $\underline{x}_c$ one obtains

$$\|B_c R_z^o E_N^o A_d\|_{HS} = (2\pi\beta)^{-3} \|\Psi\|_{L^2(\mathcal{R}^3)} \|\Phi\|_{L^2(\mathcal{R}^3)} \|h\|_{L^2(\mathcal{R}^6)} ,$$

so that $B_c R_z^o E_N^o A_d \varepsilon \mathcal{B}_o$. On the other hand, by Proposition 2.16

and (5.52), $\|B_c R_z^o A_d - B_c R_z^o E_N^o A_d\| \leq \|B_c\| \|A_d\| \|R_z^o(I-E_N^o)\| = \|\Psi\|_\infty \|\Phi\|_\infty \sup_{\lambda \geq N} |\lambda - z|^{-1} \to 0$ as $N \to \infty$. Hence $B_c R_z^o A_d \in \mathcal{B}_o$ by Proposition 2.22. Similarly $B_c R_z^o C_d \in \mathcal{B}_o$ for any c,d.

(c) We have $B_c U_t A_d = B_c \exp(-i\eta_c \underline{K}_c^2 t)\exp(-i\gamma_c \underline{P}_c^2 t) A_d$. Since by (15.9), B_c commutes with $\underline{K}_c$, we get $\|B_c U_t A_d\| = \|B_c \exp(-i\gamma_c \underline{P}_c^2 t) A_d\| = \|A_d \exp(i\gamma_c \underline{P}_c^2 t) B_c\|$. Now one obtains as in the proof of Lemma 9.8 that

$$|[\exp(i\gamma_c \underline{P}_c^2 t) B_c f](\underline{x}_c, \underline{y}_c)|^2 \leq C|t|^{-3} \|\Psi\|_2^2 \int d^3x_c' |f(\underline{x}_c', \underline{y}_c)|^2,$$

$$\|A_d \exp(i\gamma_c \underline{P}_c^2 t) B_c f\|^2 \leq C|t|^{-3} \|\Psi\|_2^2 \int d^3x_c d^3y_c |\Phi(\alpha \underline{x}_c + \beta \underline{y}_c)|^2 \cdot$$

$$\int d^3x_c' |f(\underline{x}_c', \underline{y}_c)|^2 = C\alpha^{-3} |t|^{-3} \|\Psi\|_2^2 \|\Phi\|_2^2 \|f\|^2, \qquad (16.31)$$

where $\alpha = 1$ and $\beta = 0$ if $c = d$. Since also $\|B_c U_t A_d\| \leq \|\Psi\|_\infty \|\Phi\|_\infty$, we have $\|B_c U_t A_d\| \in L^1(\mathbb{R})$ as in Lemma 9.8. The remainder of the proof of (c) is identical with that given in Lemma 9.8. (b) is obtained similarly. #

<u>LEMMA 16.12</u> : Let $\underline{u}, \underline{v} \in \mathbb{R}^3$ and $\alpha, \beta \in \mathbb{R}$. Set $u = |\underline{u}|$, $v = |\underline{v}|$ and $\kappa/\sqrt{2} = \max\{1, |\alpha|, |\beta|\}$. Then

$$(1 + |\alpha\underline{u} + \beta\underline{v}|) \leq \kappa(1+u)(1+v). \qquad (16.32)$$

<u>Proof</u> : $(1+|\alpha\underline{u}+\beta\underline{v}|)^2 \leq 2[1 + |\alpha\underline{u}+\beta\underline{v}|^2]$

$\leq 2[1+\alpha^2u^2+\beta^2v^2+2|\alpha||\beta|uv] \leq \kappa^2(1+u^2+2uv+v^2)$

$\leq \kappa^2[(1+u)(1+v)]^2$. #

LEMMA 16.13 : Assume (H2) and (H3). Then (a) The closures of $A_c^{-1}E^c$, $E_cA_c^{-1}$, $E_cA_c^{-2}$ and $A_cC_c^{-1}C_d$ are in $\mathcal{B}(H)$ for any c,d. If $c \neq d$, the closure of $A_cC_c^{-1}A_d$ is in $\mathcal{B}(H)$. (b) K_z^1 and $D_{z,10}$ are in $\mathcal{B}(K)$.

Proof : (a) Remember that C_c is the multiplication operator by $(1+|\underline{y}_c|)^{-\nu}$. Also $E^c = \sum_{\alpha \div c} E^\alpha$, the sum being finite by (H2). If $\alpha \div c$, we deduce from this and (16.23) that

$$C_cE^\alpha = E^\alpha C_c, \quad C_cE^c = E^cC_c, \quad C_c^{-1}E^c = E^cC_c^{-1}. \tag{16.33}$$

Let $f \varepsilon D(A_c^{-2})$. By using the Schwarz inequality, we obtain from (16.23) that

$$\|E^\alpha A_c^{-2}f\|^2 \leq \|e_\alpha\|^2 \int d^3x_c'(1+|\underline{x}_c'|)^{4\nu}|e_\alpha(\underline{x}_c')|^2 \|f\|^2.$$

The integral is finite by (H3), so that $\|E^\alpha A_c^{-2}f\| \leq M_\alpha\|f\|$ with $M_\alpha < \infty$. It follows that $\|E^cA_c^{-2}f\| \leq \sum_\alpha \|E^\alpha A_c^{-2}f\| \leq \sum_\alpha M_\alpha \|f\| = M_c \|f\|$, with $M_c < \infty$ since the sum if finite. Thus $E^cA_c^{-2}$ is densely defined and bounded, so that by Proposition 2.6 its closure (which we also denote by $E_cA_c^{-2}$) is in $\mathcal{B}(H)$. It follows that $E_cA_c^{-1} = E_cA_c^{-2}A_c \varepsilon \mathcal{B}(H)$. Similarly one shows that $A_c^{-1}E^c \varepsilon \mathcal{B}(H)$.

$A_cC_c^{-1}C_d$ is the multiplication operator by $(1+|\underline{x}_c|)^{-\nu}(1+|\underline{y}_c|)^{\nu}(1+|\underline{y}_d|)^{-\nu}$. One has $\underline{y}_c = \alpha\underline{x}_c + \beta\underline{y}_d$ with $\alpha, \beta \varepsilon R$. It follows from (16.32) that the above function is bounded by a constant. Hence $A_cC_c^{-1}C_d$ is bounded by Proposition 2.16 and its closure (also denoted by $A_cC_c^{-1}A_d$) is in $\mathcal{B}(H)$. Similarly $A_cC_c^{-1}A_d \varepsilon \mathcal{B}(H)$ if $c \neq d$.

(b) One has $K^1_{z,cd} = C_c^{-1}E^cC_d = E^cC_c^{-1}C_d = E^cA_c^{-1}A_cC_c^{-1}C_d$, so that $\|K^1_{z,cd}\| \le \|E^cA_c^{-1}\|\,\|A_cC_c^{-1}C_d\| < \infty$. Hence $K^1_z \varepsilon \mathcal{B}(\mathcal{K})$. Similarly one shows that $D_{z,10}\varepsilon\mathcal{B}(\mathcal{K})$. #

We next study an operator related to $\underline{\underline{G}}_z$.

<u>LEMMA 16.14</u> : Assume (H2). Then $C_dE^dR^d_zC_d$ is compact and u-lim $C_dE^dR^d_{\lambda\pm i\eta}C_d$ exists, uniformly in $\lambda\varepsilon\mathcal{R}$, as $\eta\to+0$.

<u>Proof</u> : We have $C_dE^dR^d_zC_d = \sum_{\alpha\div d}C_dE^\alpha R^d_zC_d$. By (16.10) and (16.33) each term of this sum is of the form

$$C_dE^\alpha R^d_zC_d = P_{e_\alpha}\otimes C^o_d(\eta_d\underline{K}^2_d + \lambda^\alpha - z)^{-1}C^o_d \equiv \eta_d^{-1}P_{e_\alpha}\otimes W^d_{(z-\lambda^\alpha)/\eta_d} \tag{16.34}$$

in the representation $L^2(\mathcal{R}^6) = L^2(\mathcal{R}^3)\otimes L^2(\mathcal{R}^3)$ with respect to the variables $\underline{x}_d$ and $\underline{y}_d$. P_{e_α} is the projection in $L^2(\mathcal{R}^3)$ whose range is the one-dimensional subspace spanned by the eigenvector e_α associated with the channel α, and W^d_ξ is a special case of the operator $A_1R^o_zA_2$ considered in Lemma 9.8. W^d_ξ is Hilbert-Schmidt in $L^2(\mathcal{R}^3)$ by Proposition 7.7, hence compact. It follows that $P_{e_\alpha}\otimes W^d_\xi$ is compact in $L^2(\mathcal{R}^6)$ (Problem 16.8), so that the operator in (16.34) is compact. Since the number of channels is finite by (H2), $C_dE^dR^d_zC_d$ is compact. The preceding considerations also imply the existence of the boundary values of $C_dE^dR^d_zC_d$, since by Lemma 9.8(c) u-lim $W^d_{\lambda\pm i\eta}$ exists uniformly in $\lambda\varepsilon\mathcal{R}$ as $\eta\to+0$. #

<u>LEMMA 16.15</u> : Assume (H1), (H2) and (H4). Then $\{B_d(I-E^d)R^d_{\lambda\pm i\eta}A_d\}$ converges in operator norm as $\eta\to+0$, the convergence being uniform in λ.

Proof (for $\lambda+i\eta$) : (i) Consider the Hamiltonian $Z_d = \gamma_d \underline{P}_d^2 + V_d^o$ in $L^2(\mathcal{R}^3)$. Let $-2\tau < 0$ be its eigenvalue nearest to 0 and $E_{[-\tau,\infty)}$ its spectral projection for the interval $[-\tau,\infty)$. (If the point spectrum of Z_d is empty, one may take $\tau = 1$.) By (H1) and (H2), $E_{[-\tau,\infty)} = E_{ac}(Z_d)$. We claim that $\{B_d^o E_{ac}(Z_d)(Z_d-\mu-i\eta)^{-1}A_d^o\}$ converges in operator norm as $\eta \to +0$, uniformly in $\mu\varepsilon\mathcal{R}$.

To see this, we notice that $E_{ac}(Z_d)(Z_d-\mu-i\eta)^{-1} = E_{ac}(Z_d)(Z_{d,ac}-\mu-i\eta)^{-1}$. In the region $\mu < -\tau$, this is a holomorphic function of $z = \mu+i\eta$ which is bounded in operator norm by τ and converges to zero as $\mu \to -\infty$ by (5.52). Since $A_d^o, B_d^o \varepsilon \mathcal{B}(L^2(\mathcal{R}^3))$, it follows that our assertion holds for $\mu < -\tau$. For $\mu \geq -\tau$ we write $B_d^o E_{ac}(Z_d)(Z_d-\mu-i\eta)^{-1}A_d^o = B_d^o(Z_d-\mu-i\eta)^{-1}A_d^o - \sum_{k=1}^n B_d^o(\lambda_k-\mu-i\eta)^{-1}E_{\{\lambda_k\}}A_d^o$, where $\lambda_1,\ldots,\lambda_n$ are the eigenvalues of Z_d and $E_{\{\lambda_k\}}$ the associated eigenprojections. The second term on the right-hand side of the last equation has the required boundary values for $\mu \geq -\tau$ because $\mu-\lambda_k \geq \tau > 0$. (This term is absent if Z_d has no point spectrum.) By (9.20) the first term is

$$B_d^o(Z_d-\mu-i\eta)^{-1}A_d^o = I-[I+B_d^o(\gamma_d \underline{P}_d^2 - \mu - i\eta)^{-1}A_d^o]^{-1}. \qquad (16.35)$$

By Lemma 9.5 the operator on the right-hand side has boundary values for all $\mu \geq -\tau$, uniformly on any compact set, because $\Gamma_o^d \cap [-\tau,\infty) = \emptyset$ by (H1), (H2) and (H4). In addition $\|B_d^o(\gamma_d \underline{P}_d^2-\mu-i\eta)^{-1}A_d^o\| \to 0$ as $\mu \to \infty$, uniformly in $0 \leq \eta \leq 1$, by Propositions 12.1 and 12.2. Hence $\|B_d^o(Z_d-\mu-i\eta)^{-1}A_d^o\| \to 0$ as $\mu \to \infty$, uniformly in $0 \leq \eta \leq 1$, so that the boundary values in operator norm of the operator in (16.35) exist uniformly in μ on $[-\tau,\infty)$. This completes the verification of our claim.

(ii) We identify $\tilde{L}^2(R^6)$ with the spectral representation of $\underline{K}_d$, i.e. $\tilde{L}^2(R^6) = L^2(\tilde{L}^2(R^3), d^3k_d)$. Thus any $\tilde{f}(\underline{p}_d, \underline{k}_d)$ is, for almost every $\underline{k}_d$, viewed as belonging to $\tilde{L}^2(R^3)$ as a function of $\underline{p}_d$. Since $U^d_{\lambda+i\eta} \equiv B_d(I-E^d)(\gamma_d \underline{P}^2_d + V_d + \eta_d \underline{K}^2_d - \lambda - i\eta)^{-1} A_d$ commutes with $\underline{K}_d$ by (15.9) and (16.23), it is decomposable in the above representation of $\tilde{L}^2(R^6)$ and given by the following $\underline{k}_d$-dependent family of operators in $L^2(R^3)$:

$$U^d_{\lambda+i\eta} = \{B^o_d E_{ac}(Z_d)(Z_d + \eta_d \underline{k}^2_d - \lambda - i\eta)^{-1} A^o_d\} \equiv \{U_{k_d}(\lambda+i\eta)\}.$$

From part (i) of the proof we see that u-lim $U_k(\lambda+i\eta)$ exists as $\eta \to +0$, uniformly in λ and k, and that $\|U_k(\lambda+i0)\| \to 0$ as $\underline{k}^2 \to \infty$. Hence the decomposable operator $U^d_{\lambda+i0}$ defined by $\{U_{k_d}(\lambda+i0)\}$ is in $B(L^2(R^6))$ by Proposition 5.25 and

$$\|U_{\lambda+i\eta} - U_{\lambda+i0}\|_{L^2(R^6)} = \sup_{\underline{k}\in R^3} \|U_k(\lambda+i\eta) - U_k(\lambda+i0)\|_{L^2(R^3)}$$

converges to zero as $\eta \to +0$, uniformly in $\lambda \in R$. #

Proof of Proposition 16.6 : (a) was proved in Lemma 16.13. For (b), we first consider an element of the matrix $D_{z,00}$ (without the factor $1-\delta_{cd}$) which we rewrite by using the second equation in (16.8) and by defining $M^{cd}_z = A_c R^o_z A_d$:

$$B_c(I-E^c)R^c_z A_d = B_c R^o_z A_d - B_c E^c R^o_z A_d - B_c(I-E^c)R^c_z A_c B_c R^o_z A_d =$$

$$[I - B_c(I-E^c)R^c_z A_c]W^{cd}_z - B_c E^c A^{-1}_c M^{cd}_z. \qquad (16.36)$$

Since $c \neq d$, W^{cd}_z and M^{cd}_z are compact by Lemma 16.3(a). Since $B_c E_c A_c^{-1} \in B(H)$ by Lemma 16.13, $D_{z,00}$ is compact if $\text{Im}\, z \neq 0$.

Since $\underline{\underline{G}}_z \varepsilon \mathcal{B}(K)$, we also have $D_{z,10} = D_{z,00} \underline{\underline{G}}_z \varepsilon \mathcal{B}_0$.

We now consider a term of $D_{z,11}$. These have the form

$$C_c^{-1}E^c A_d B_d E^d R_z^d C_d = (C_c^{-1}E^c A_d B_d C_d^{-1}E^d)(C_d E^d R_z^d C_d). \qquad (16.37)$$

The second factor is compact by Lemma 16.14, so that is suffices to show that the first one is in $\mathcal{B}(H)$. This follows easily by writing it as $(E_c A_c^{-2})(A_c C_c^{-1} A_d) B_d (A_c C_d^{-1} A_d)(A_d^{-1} E^d)$, which is in $\mathcal{B}(H)$ by Lemma 16.13 and since $B_d \varepsilon \mathcal{B}(H)$.

We now prove part (c) of the proposition, namely the existence of the boundary values of D_z as $z \to \lambda + i0$, uniformly in λ on the real line R. For $D_{z,11}$ this follows from (16.37) and Lemma 16.14. For $D_{z,00}$ the desired result is obtained from (16.36) by using Lemma 16.15 and Lemma 16.3(c). The elements of $D_{z,01}$ are obtained by multiplying (16.36) on the right by $B_d E^d R_z^d C_d$. In view of Lemma 16.15, it suffices to show that $Q_z^{cd} \equiv W_z^{cd} B_d E^d R_z^d C_d$ and $\mathcal{Q}_z^{cd} \equiv M_z^{cd} B_d E^d R_z^d C_d$ with $c \neq d$ have boundary values. We treat Q_z^{cd} which by (16.8) and (16.34) equals

$$Q_z^{cd} = B_c R_z^o V_d R_z^d E^d C_d = B_c R_z^o C_d E^d - B_c R_z^d C_d E^d .$$

For the first term on the right-hand side the required limit exists by Lemma 16.3(b). The proof of the existence of the second limit is similar to that of Lemma 16.3(b). One writes $B_c R_z^d C_d E^d = \sum_{\alpha \div d} B_c R_z^d E^\alpha C_d$, uses (16.10) and (3.58) to obtain for instance

$$B_c R_{\lambda+i\eta}^d C_d E^d = \sum_{\alpha \div d} i \int_0^\infty dt e^{i(\lambda+i\eta-\lambda^\alpha)t} B_c \exp(-i\eta_d \underline{K}_d^2 t) C_d E^\alpha$$

and notices that $\| B_c \exp(-i\eta_d \underline{K}_d^2 t) C_d \| \varepsilon L^1(R)$. #

PROBLEMS

16.1 : Prove Lemma 16.1.

16.2 : Let W_z be as in Proposition 16.2. Show that for any positive integer n, $(W_z)^n$ is not compact unless all pair potentials are zero.

16.3 : Assume the hypotheses of Proposition 16.2 and that $\Gamma_0^c = \emptyset$ for each c. Show that $X_z - \underline{\underline{X}}_z \varepsilon \mathcal{B}_0$ and write out the proof of the existence of u-lim$(X_{\lambda\pm i\eta} - \underline{\underline{X}}_{\lambda\pm i\eta})$ as $\eta \to +0$, uniformly in λ. Use (16.6) and show that $Y'_{\lambda\pm i\eta}$ has boundary values in operator norm. (Hint : Lemma 9.5.)

16.4 : Fill in the missing details in the proof of Proposition 16.8.

16.5 : Complete the proof of Lemma 16.10.

16.6 : (a) Let $\mathcal{U}_d^o$ be as in (16.28). Show that $\mathcal{U}_d^o \mathcal{U}_b^{o*} = \{U_{db}\}$ as a decomposable operator from $\mathcal{G}_b^o$ to $\mathcal{G}_d^o$.

(b) Obtain an expression for $R_{\alpha\beta}(\lambda)$ in terms of the operators used in the proof of Proposition 16.11.

16.7 : Let $M_z^{cd} = V_d \delta_{cd} - V_c R_z V_d$. Use (16.3) to establish the Faddeev equations

$$M_z^{cd} = V_d \delta_{cd} - V_c R_z^o \sum_b M_z^{bd}. \tag{16.38}$$

16.8 : Let $A \varepsilon \mathcal{B}_o(\mathcal{H}_1)$ and $B \varepsilon \mathcal{B}_o(\mathcal{H}_2)$. Then $A \otimes B$ is compact in $\mathcal{H}_1 \otimes \mathcal{H}_2$.

BIBLIOGRAPHY

BOOKS AND MONOGRAPHS

[AG] AKHIEZER, N.I. and GLAZMAN, I.M. : Theory of linear operators in Hilbert space (English translation), Vols. I and II, Frederick Ungar, New York (1963).

[AR] de ALFARO, V. and REGGE, T. : Potential scattering, North-Holland, Amsterdam (1965).

[B] BEREZANSKII, J.M. : Expansions in eigenfunctions of self-adjoint operators, Am. Math. Soc. translations of Math. Monographs, Vol. 17, Providence, R.I. (1968).

[BB] BARUT, A.O. and BRITTIN, W.E., eds. : Lectures in theoretical physics, Vol. X-A, Gordon and Breach, New York (1968).

[CS] CHADAN, K. and SABATIER, P.C. : Inverse problems in quantum scattering theory, Springer, New York (1977).

[D] DINCULEANU, N. : Vector measures, Pergamon Press, Oxford (1962).

[DA] DAVIES, E.B. : Quantum theory of open systems, Academic Press, London (1976).

[DI] DIXMIER, J. : Les algèbres d'opérateurs dans l'espace Hilbertien, 2nd edition, Gauthier-Villars, Paris (1969).

[DS] DUNFORD, N. and SCHWARTZ, J.T. : Linear operators, Parts I and II, Interscience, New York (1958 and 1963).

[E] EDEN, R.J., LANDSHOFF, P.V., OLIVE, D.I. and POLKINGHORNE, J.C. : The analytic S-matrix, Cambridge Univ. Press, Cambridge (1966).

[EM] EMCH, G.G. : Algebraic methods in statistical physics and quantum field theory, Wiley-Interscience, New York (1972).

[EN] ENZ, C.P. and MEHRA, J., eds. : Physical reality and mathematical description, Reidel, Dordrecht-Holland (1974).

[F] FADDEEV, L.D. : Mathematical aspects of the three-body problem in quantum scattering theory, Israel program of scientific transl.,Jerusalem (1965).

[GS] GELFAND, I.M. and SHILOV, G.E. : Generalized functions, Vols. II and III (English translation), Academic Press, New York (1968).

[GV] GELFAND, I.M. and VILENKIN, N.Y. : Generalized functions, Vol. IV (English translation), Academic Press, New York (1968).

[H] HALMOS, P.R. : Finite-dimensional vector spaces, Springer, New York (1970).

[HA] HAMERMESH, M. : Group theory and its application to physical problems, Addison-Wesley, Reading, Massachusetts (1962).

[HP] HILLE, E. and PHILLIPS, R.S. : Functional analysis and semi-groups, Am. Math. Soc. Colloq. Publ., Vol. 31, Providence, R.I. (1957).

[J] JAUCH, J.M. : Foundations of quantum mechanics, Addison-Wesley, Reading, Massachusetts (1968).

[JO] JOACHAIN, C.J. : Quantum collision theory, North-Holland, Amsterdam (1975).

[JW] JÖRGENS, K. and WEIDMANN, J. : Spectral properties of Hamiltonian operators, Lecture Notes in Mathematics, Vol. 313, Springer, Berlin (1973).

[K] KATO, T. : Perturbation theory for linear operators, 2nd edition, Springer, New York (1976).

[KA] KANTOROVICH, L.V. and AKILOV, G.P. : Functional analysis in normed spaces, Pergamon, Oxford (1964).

[L] LANG, S. : Analysis I and II, Addison-Wesley, Reading, Massachusetts (1968 and 1969).

[LI] LIEB, E.H., SIMON, B. and WIGHTMAN, A.S., eds. : Studies in mathematical physics, Princeton Univ. Press, Princeton (1976).

[LL] LANDAU, L.D. and LIFSHITZ, E.M. : Quantum mechanics, non-relativistic theory (English translation), Pergamon, Oxford (1958).

[LM] La VITA, J.A. and MARCHAND, J.P., eds. : Scattering theory in mathematical physics, Reidel, Dordrecht-Holland (1974).

[M] MACKEY, G.W. : Mathematical foundations of quantum mechanics, W.A. Benjamin, Advanced Book Program, Reading, Massachusetts (1969).

[MK] MIGDAL, A.B. and KRAINOV, V.P. : Approximation methods in quantum physics, W.A. Benjamin, New York (1969).

[MS] McSHANE, E.J. : Integration, Princeton Univ. Press, Princeton (1947).

[MU] MUSKHELISHVILI, N.I. : Singular integral equations, Noordhoff, Groningen (1953).

[N] NEWTON, R.G. : Scattering theory of waves and particles, McGraw-Hill, New York (1966).

[NA] NAIMARK, M.A. : Linear differential operators, Parts I and II (English translation), G.G. Harrap, London (1968).

[NM] NAIMARK, M.A. : Normed Rings, Noordhoff, Groningen (1959).

[P] PIRON, C. : Foundations of quantum physics, W.A. Benjamin, Advanced Book Program, Reading, Massachusetts (1975).

[PE] PEARCY, C., ed. : Topics in operator theory, Am. Math. Soc., Providence, R.I. (1974).

[PU] PUTNAM, C.R. : Commutator properties of Hilbert space operators and related topics, Springer, Berlin (1967).

[R] ROYDEN, H.L. : Real analysis, Macmillan, New York (1963).

[RN] RIESZ, F. and SZ.-NAGY, B. : Functional analysis (English translation), Frederick Ungar, New York (1965).

[RO] ROBINSON, D.W. : The thermodynamic pressure in quantum statistical mechanics, Lecture Notes in Physics, Vol. 9, Springer, Berlin (1971).

[RS] REED, M. and SIMON, B. : Methods of modern mathematical physics, Vols. I and II, Academic Press, New York (1972 and 1975).

[RU] RUDIN, W. : Real and complex analysis, McGraw-Hill, New York (1966).

[SA] SANSONE, G. : Orthogonal functions (English translation), Interscience, New York (1959).

[SI] SIMMS, D.J. : Lie groups and quantum mechanics, Lecture Notes in Mathematics, Vol. 52, Springer, New York (1968).

[SM] SIMON, B. : Quantum mechanics for Hamiltonians defined as quadratic forms, Princeton Univ. Press, Princeton (1971).

[SO] SOBOLEV, S.L. : Applications of functional analysis in mathematical physics, Am. Math. Soc. translations of Math. Monographs, Vol. 7, Providence, R.I. (1963).

[T] TAYLOR, J.R. : Scattering theory, Wiley, New York (1972).

[TI] TITCHMARSH, E.C. : Eigenfunction expansion associated with second-order differential equations, Parts I and II, Oxford Univ. Press, Oxford (1962 and 1958).

[TU] THIRRING, W. and URBAN, P., eds. : The Schrödinger equation, Springer, Wien (1977).

[WA] WATSON, G.N. : Theory of Bessel functions, Cambridge Univ. Press, Cambridge (1944).

[WW] WHITTAKER, E.T. and WATSON, G.N. : A course of modern analysis, Cambridge Univ. Press, Cambridge (1963).

[WY] WYLD, H.W. : Mathematical methods for physicists, W.A. Benjamin, Advanced Book Program, Reading, Massachusetts (1976).

ARTICLES

AGMON, S. : [1] Ann. Scuola Normale Sup. Pisa, Serie IV, 2, 151 (1975).
AGUILAR, J. and COMBES, J.M. : [1] Comm. Math. Phys. 22, 269 (1971).
ALBEVERIO, S. : [1] Ann. Phys. 71, 167 (1972).
ALSHOLM, P. : [1] J. Math. Anal. Appl. 59, 550 (1977).
ALSHOLM, P. and SCHMIDT, G. : [1] Arch. Rational Mech. Anal. 40, 281 (1971).
AMREIN, W.O. and GEORGESCU, V. : [1] Helv. Phys. Acta 46, 635 (1973). [2] ibid. 47, 517 (1974).
AMREIN, W.O., GEORGESCU, V. and JAUCH, J.M. : [1] Helv. Phys. Acta 44, 407 (1971).
AMREIN, W.O., MARTIN, Ph. and MISRA, B. : [1] Helv. Phys. Acta 43, 313 (1970).
ARONSZAJN, N. : [1] Am. J. Math 79, 597 (1957).
BALSLEV, E. : [1] Ann. Phys. 73, 49 (1972).
BALSLEV, E. and COMBES, J.M. : [1] Comm. Math. Phys. 22, 280 (1971).
BAUMGÄRTEL, H. : [1] Math. Nachr. 58, 279 (1973). [2] ibid. 62, 103 (1974).
BERTHIER, A.M. and COLLET, P. : [1] Ann. Inst. Henri Poincaré, Sec. A, 26, 279 (1977).
BIRMAN, M.S. : [1] Math. USSR Izvestija (English translation) 2, 879 (1968).
BIRMAN, M.S. and ENTINA, S.B. : [1] Math. USSR Izvestija (English translation) 1, 391 (1967).
BIRMAN, M.S. and SOLOMJAK, M.Z. : [1] Topics in Math. Phys. (English translation) 1, 15 and 2, 19 (1969).
BOLLÉ, D. and OSBORN, T.A. : [1] Phys. Rev. D13, 299 (1976).
BUSLAEV, V.S. and MATVEEV, V.B. : [1] Theor. Math. Phys. (English translation) 1, 367 (1970).
CHANDLER, C. and GIBSON, A.G. : [1] J. Math. Phys. 14, 1328 (1973). [2] ibid. 15, 291 (1974). [3] ibid. 15, 1366 (1974).
COMBES, J.M. : [1] Comm. Math. Phys. 12, 283 (1969).
COMBES, J.M., NEWTON, R.G. and STOKHAMER, R. : [1] Phys. Rev. D11, 366 (1975).
COMBESCURE, M. and GINIBRE, J. : [1] Ann. Phys. 101, 355 (1976). [2] Scattering and local absorption for the Schrödinger operator, to appear in J. Functional Anal.
COOK, J.M. : [1] J. Math. and Phys. 36, 82 (1957).
CORBETT, J.V. : [1] Phys. Rev. D1, 3331 (1970).
DAVIES, E.B. : [1] Arch. Rational Mech. Anal. 63, 261 (1977).
DEIFT, P. and SIMON, B. : [1] J. Functional Anal. 23, 218 (1976). [2] A time-dependent approach to the completeness of multiparticle quantum systems, preprint, New York Univ. and Yeshiva Univ. (1977).
DOLLARD, J.D. : [1] J. Math. Phys. 5, 729 (1964). [2] ibid. 9, 620 (1968). [3] Comm. Math. Phys. 12, 193 (1969). [4] J. Math. Phys. 14, 708 (1973).
DREYFUS, T. : [1] The determinant of the scattering matrix and its relation to the number of eigenvalues, to appear in J. Math. Anal. Appl.
ENSS, V. : [1] Comm. Math. Phys. 52, 233 (1977).
FARIS, W.G. : [1] Helv. Phys. Acta 45, 1074 (1972).
FEDERBUSH, P.G. : [1] Phys. Rev. 148, 1551 (1966).
FERRERO, P., de PAZZIS, O. and ROBINSON, D.W. : [1] Ann. Inst. Henri Poincaré, Sec. A, 21, 217 (1974).
FRIEDRICHS, K.O. : [1] Comm. Pure Appl. Math. 1, 361 (1948).
GALINDO-TIXAIRE, A. : [1] Helv. Phys. Acta 32, 412 (1959).
GEORGESCU, V. : [1] Méthodes stationnaires pour des potentiels à longue portée à symétrie sphérique, thèse, Univ. Genève (1974).
GINIBRE, J. and MOULIN, M. : [1] Ann. Inst. Henri Poincaré, Sec. A, 21, 97 (1974).
GRAWERT, G. and PETZOLD, J. : [1] Z. Naturforsch. 15a, 311 (1960).

GRASSBERGER, P. and SANDHAS, W. : [1] Nucl. Phys. B2, 181 (1967).
GREEN, T.A. and LANFORD III, O.E. : [1] J. Math. Phys. 1, 139 (1960).
GROSSE, H., GRÜMM, H.R., NARNHOFER, H. and THIRRING, W. : [1] Acta Phys. Austriaca 40, 97 (1940).
GUSTAFSON, K. and JOHNSON, G. : [1] Helv. Phys. Acta 47, 163 (1974).
HACK, M.N. : [1] Nuovo Cimento 9, 731 (1958).
HEPP, K. : [1] Helv. Phys. Acta 42, 425 (1969).
HÖRMANDER, L. : [1] Math. Z. 146, 69 (1976).
HORWITZ, L.P., La VITA, J. and MARCHAND, J.P. : [1] Rocky Mountain J. Math. 1, 225 (1971).
HOWLAND, J.S. : [1] J. Functional Anal. 22, 250 (1976).
HUNZIKER, W. : [1] Helv. Phys. Acta 36, 838 (1963). [2] J. Math. Phys. 6, 6 (1965). [3] Helv. Phys. Acta 39, 451 (1966). [4] ibid. 40, 1052 (1967).
IKEBE, T. : [1] Arch. Rational Mech. Anal. 5, 1 (1960). [2] J. Functional Anal. 20, 158 (1975).
JANSEN, K.H. and KALF, H. : [1] Comm. Pure Appl. Math. 28, 747 (1975).
JAUCH, J.M. : [1] Helv. Phys. Acta 31, 127 and 661 (1958).
JAUCH, J.M., LAVINE, R.B. and NEWTON, R.G. : [1] Helv. Phys. Acta 45, 220 (1972).
JAUCH, J.M., MISRA, B. and GIBSON, A.G. : [1] Helv. Phys. Acta 41, 513 (1968).
JAUCH, J.M., MISRA, B. and SINHA, K.B. : [1] Helv. Phys. Acta 45, 398 (1972).
JAUCH, J.M. and SINHA, K.B. : [1] Helv. Phys. Acta 45, 580 (1972).
JAUCH, J.M. and ZINNES, I.I. : [1] Nuovo Cimento 11, 553 (1959).
JOST, R. : [1] Helv. Phys. Acta 20, 256 (1947).
KATO, T. : [1] Comm. Pure Appl. Math. 12, 403 (1959). [2] Math. Ann. 162, 258 (1966). [3] Studia Math. 31, 535 (1968).
KATO, T. and KURODA, S.T. : [1] Rocky Mountain J. Math. 1, 127 (1971).
KUPSCH, J. and SANDHAS, W. : [1] Comm. Math. Phys. 2, 147 (1966).
KURODA, S.T. : [1] Nuovo Cimento 12, 431 (1959). [2] Construction of eigenfunction expansions by the perturbation method and its application to n-dimensional Schrödinger operators, MRC Technical report no. 744, Univ. Wisconsin (1967). [3] Oxford Quart. J. 21, 81 (1970). [4] J. Fac. Sci. Univ. Tokyo, Sec. I, 17, 315 (1970). [5] J. Math. Soc. Japan 25, 222 (1973).
LAVINE, R.B. : [1] J. Functional Anal. 5, 368 (1970). [2] Indiana Univ. Math. J. 21, 643 (1972). [3] J. Functional Anal. 12, 30 (1973).
LIPPMANN, B.A. and SCHWINGER, J. : [1] Phys. Rev. 79, 469 (1950).
MARTIN, Ph.A. : [1] Helv. Phys. Acta 45, 794 (1972). [2] Nuovo Cimento 17A, 73 (1973). [3] Comm. Math. Phys. 47, 221 (1976).
MARTIN, Ph.A. and MISRA, B. : [1] J. Math. Phys. 14, 997 (1973).
MATVEEV, V.B. and SKRIGANOV, M.M. : [1] Soviet Math. Dokl. (English translation) 13, 185 (1972).
MERKURIEV, S.P., GIGNOUX, C. and LAVERNE, A. : [1] Ann. Phys. 99, 30 (1976).
MISRA, B., SPEISER, D. and TARGONSKI, G. : [1] Helv. Phys. Acta 36, 963 (1963).
MØLLER, C. : [1] Kgl. Danske Videnskab. Selskab., Mat.-Fys. Medd. 23, 1 (1945) and 22, 19 (1946).
MOURRE, E. : [1] Ann. Inst. Henri Poincaré, Sec. A, 18, 215 (1973). [2] ibid. 26, 219 (1977).
MULHERIN, D. and ZINNES, I.I. : [1] J. Math. Phys. 11, 1402 (1970).
NELSON, E. : [1] J. Math. Phys. 5, 332 (1964).
NEUMANN, J. von, and WIGNER, E.P. : [1] Z. Phys. 30, 465 (1929).
NEWTON, R.G. : [1] J. Math. Phys. 12, 1552 (1971). [2] ibid. 18, 1348 (1977).
NEWTON, R.G. and STOKHAMER, R. : [1] Phys. Rev. A14, 642 (1976).

OBERMANN, P. and WOLLENBERG, M. : [1] Abel wave operators II. Wave operators for functions of operators, to appear in J. Functional Anal.

O'CONNOR, A.J. : [1] Comm. Math. Phys. 32, 319 (1973).

PEARSON, D.B. : [1] Nuovo Cimento 2A, 853 (1971). [2] Helv. Phys. Acta 47, 249 (1974). [3] Comm. Math. Phys. 40, 125 (1975). [4] Helv. Phys. Acta 48, 639 (1975). [5] A generalisation of the Birman trace theorem, to appear in J. Functional Anal.

POVZNER, A.Y. : [1] Am. Math. Soc. translations, Ser. 2, 60, 1 (1966).

PRUGOVEČKI, E. and ZORBAS, J. : [1] J. Math. Phys. 14, 1398 (1973).

REJTO, P.A. : [1] J. Math. Anal. Appl. 17, 453 and 20, 145 (1967). [2] Helv. Phys. Acta 44, 708 (1971).

SCHECHTER, M. : [1] Ann. Mat. Pura Appl. 105, 313 (1975).

SIGAL, I.M. : [1] Comm. Math. Phys. 48, 137 and 155 (1976). [2] Mathematical foundations of quantum scattering theory for multiparticle systems, to appear in Mem. Am. Math. Soc. [3] On quantum mechanics of many-body systems with dilation-analytic potentials, preprint, ETH-Zurich (1977).

SIMON, B. : [1] Ann. Math. 97, 247 (1973). [2] Math. Z. 131, 361 (1973).

SIMANDER, C.G. : [1] Math. Z. 138, 53 (1974).

SINHA, K.B. : [1] Ann. Inst. Henri Poincaré, Sec. A, 26, 263 (1977).

THOMAS, L.E. : [1] Comm. Math. Phys. 33, 335 (1973). [2] J. Functional Anal. 15, 364 (1974). [3] Ann. Phys. 90, 127 (1975).

VILLARROEL, D. : [1] Nuovo Cimento 70A, 175 (1970).

WEIDMANN, J. : [1] Math. Z. 98, 268 (1967).

WEINBERG, S. : [1] Phys. Rev. 113B, 232 (1964).

WIGNER, E.P. : [1] Ann. Math. 40, 149 (1939). [2] Phys. Rev. 98, 145 (1955).

WILCOX, C.R. : [1] J. Functional Anal. 12, 257 (1973).

WINTER, C. van : [1] Kgl. Danske Selskab., Mat.-Fys. Skrift. 2, no. 8 (1964) and 2, no. 10 (1965). [2] J. Math. Anal. Appl. 47, 633 (1974), 48, 368 (1974) and 49, 88 (1975).

WINTER, C. van, and BRASCAMP, H.J. : [1] Comm. Math. Phys. 11, 19 (1968).

WOLLENBERG, M. : [1] Pacific J. Math. 59, 303 (1975).

WÜST, R. : [1] Math. Z. 119, 276 (1971).

YAFAEV, D.R. : [1] Math. USSR Sbornik (English translation) 23, 535 (1974).

YAKUBOVSKII, O.A. : [1] Soviet J. Nucl. Phys. (English translation) 5, 937 (1967).

ZEMACH, C. and KLEIN, A. : [1] Nuovo Cimento 10, 1078 (1958).

NOTATION INDEX

SUBJECT INDEX

Of Related Interest

Foundations of Quantum Mechanics

J. M. JAUCH

This advanced text on elementary quantum mechanics is designed to acquaint the reader with the modern approach to the subject, and with the mathematical tools used in this approach. A notable feature of the book is its use of modern mathematical language to a greater degree than is customary in other texts. In addition to standard reference material, this book provides the student with the results of recent research on the foundations of quantum mechanics, which has been conducted in Geneva over the past few years. The text is chiefly concerned with conceptual foundations rather than applications or approximations, and the discussion is restricted to the general aspects of nonrelativistic theory. Every chapter contains supplementary information in the form of problems, approximately three hundred and fifty appearing throughout the book. These are designed to reinforce and supplement ideas and challenge the student to participate actively and critically.

1968 (third printing, 1977), xii, 299 pp., illus., hardbound
ISBN 0-201-03298-8

Classical Charged Particles

Foundations of Their Theory

F. ROHRLICH

Syracuse University

Contents: Philosopy and Logic of Physical Theory. A Short History of the Classical Theory of Charged Particles. Foundations of Classical Mechanics. The Maxwell-Lorentz Field. Electromagnetic Radiation. The Charged Particle. Generalizations. The Relation of the Classical Lorentz-Invariant Charged-Particle Theory to Other Levels of Theory. The Theory's Structure and Place in Physics. The Space-Time of Special Relativity. The Space-Time of General Relativity. Author Index. Subject Index.

"Professor Rohrlich is the co-author of the widely used graduate textbook, *The Theory of Photons and Electrons,* and is eminently qualified, by virtue of his intensive and continuing research on the theory of the electron, to write this important and unique book. A scholarly text for the study of the classical or non-quantum theory of charged particles, it has the avowed three-fold objective of (*a*) presenting a consistent and coherent classical theory of point charges, while emphasizing the many difficulties encountered in the past that have now been overcome, (*b*) teaching the foundations of this theory, whose numerous applications have many excellent complementary expositions, and (*c*) giving a better understanding of the nature of physical theory, its logic and its structure. As the aim of the book is to set the theory of charged particles into its proper place in the web of physical theory, it is unlike any other book on electromagnetic theory, of which there are legion. It is a brilliant tour de force and deserves the serious attention of all graduate students of physics and teachers of electromagnetic theory."

–*American Scientist*

1965, xiv, 305 pp., illus., hardbound ISBN 0-201-06492-8

Addison-Wesley/W.A. Benjamin, Inc.
ADVANCED BOOK PROGRAM
Reading, Massachusetts 01867

QC20 179188
.7
.S3 Amrein
.A47 Scattering theory in
quantum mechanics

Randall Library – UNCW
QC20.7.S3 A47
NXWW
Amrein / Scattering theory in quantum mechanics :
304900241332V